Contents

Design of Machine Elements

C.S. Sharma
Formerly Professor
Department of Mechanical Engineering
Jai Narain Vyas University
Jodhpur

Kamlesh Purohit
Associate Professor
Department of Mechanical Engineering
Jai Narain Vyas University
Jodhpur

Prentice-Hall of India Private Limited
New Delhi - 110001
2005

Rs. 295.00

DESIGN OF MACHINE ELEMENTS
C.S. Sharma and Kamlesh Purohit

ISBN-81-203-1955-9

The export rights of this book are vested solely with the publisher.

Third Printing ... ... **October, 2005**

Published by Asoke K. Ghosh, Prentice-Hall of India Private Limited, M-97, Connaught Circus, New Delhi-110001 and Printed by Jay Print Pack Private Limited, New Delhi-110015.

Preface

Design of machine elements is one of the principal subjects being taught to undergraduate students in mechanical, production, and industrial engineering disciplines. It deals with the application of scientific principles from various fields of engineering and requires imagination of the designer to create new technical feats, which can perform specific functions with maximum economy. In the context of today's global competitiveness, the designer's task has become more difficult than ever before. Today, the designer is required to deliver the design from the conceptual stage to the final finished form in the shortest possible time to remain competitive. Although various analytical tools, namely CAD, Optimization, FEM are available but their application is still limited due to non-availability of enough resources and text material.

This book is intended to bridge the gap between conventional design methodology and new tools like CAD. It contains twenty chapters. The first chapter gives introduction to engineering design where design philosophy and design morphology are discussed. The second chapter deals with the fundamentals of computer graphics along with commercial CAD package, AutoCAD and its programming language AutoLISP. The third chapter deals with engineering materials, including the BIS method of designation of various steels. The fourth and fifth chapters deal with strength of materials and fundamental principles of machine design. The manufacturing aspects of design are covered in the sixth chapter. The designs of various types of joints, namely riveted, welded, pin, cotter, and knuckle have been discussed in the seventh chapter. The design aspects of screw fasteners and power screws are discussed in the eighth chapter. The ninth chapter deals with the design of various types of springs. The mechanical power transmission elements, namely belts, drives, gears, shafts, keys and splines, couplings, clutches and brakes are discussed in chapters ten to fourteen. The design procedures of various types of bearings are explained in the fifteenth chapter. The design aspects of pressure vessels are covered in the sixteenth chapter and those of I.C. engine components are discussed in the seventeenth chapter. The eighteenth chapter deals with the design of flywheels and rotating discs. The nineteenth chapter explains the optimization methods in design. The use of an optimization tool is also explained in this chapter. The final chapter introduces the finite element method (FEM)—an engineering analysis tool used by the designers.

Computer programs have been included in the relevant chapters to illustrate the computer-aided design procedures. An abundance of thoroughly worked examples, interspersed throughout the text, enhance students' understanding of the topics. End-of-chapter exercises and multiple choice questions are given for thorough practice by the students. The salient features of the book are:

- Indian standards are used extensively throughout the book.
- Efforts have been made to give balanced presentation of theoretical discussions coupled with illustrations of design procedures using numerous examples.

- A comprehensive treatment of manufacturing aspects in design has been presented which includes: (i) limits, fits, and tolerances, (ii) reliability, (iii) failure mode and effect analysis, and (iv) ergonomic considerations.
- Exhaustive coverage of computer-aided design and drafting is presented. Many chapters have at least one topic on CAD and include computer programs written in C++ language.
- A separate chapter on design optimization includes optimization of machine elements using Microsoft Excel 2000.
- The current engineering analysis tool—Finite Element Method—is introduced.

It is but natural that some errors might have crept into a work of such volume. We would appreciate if such errors and shortcomings are brought to our notice. Suggestions for improvements to the book would also be welcome.

We wish to acknowledge indebtedness to Bureau of Indian Standards for several BIS standards used in the book. Many thanks go to other authors and publishers whose books have been consulted during the preparation of the manuscript.

Finally, we thank the editorial and production staff of Prentice-Hall of India, New Delhi for their continuous cooperation and help in publishing this book.

C.S. SHARMA
KAMLESH PUROHIT

Chapter 1 Introduction to Engineering Design

1.1 INTRODUCTION

Engineering design is the culmination of an engineering education; without the design industry we would not have new or improved products; basic information resulting from research would not be put to the use of human beings and the progress of the society would be halted. Engineering design is found in all professions. It is a dynamic field with various types of design specialists. In industry we find the product designer, system designer, apparatus designer, tool designer, ergonomic designer, and many others. A natural question arises: what is design? Over the years the art of design has been described in different ways by various designers. Some of the important definitions enumerated by them are given below:

> Engineering design is a purposeful activity directed towards the goal of fulfilling human needs, particularly those which can be met by the technological factors of our culture.
>
> Design is an iterative-creative decision making process directed towards the fulfilment of human needs.
>
> Design is an intellectual activity which leads to organization of human abilities and physical resources to create things to satisfy man's needs.
>
> The process of applying various techniques for the purpose of defining a device, a process or a system in sufficient detail to permit its physical realization.
>
> The task of creating the plan which enables a product to be made in such a way that it not only meets the stipulated conditions but also permits its manufacture in the most economical way.

These definitions of design provide a general description to highlight various aspects of engineering design. Essentially, engineering design is an intellectual activity of an extremely complex kind which involves almost all the spheres of knowledge; thus, a single definition of design cannot throw light on every aspect of it.

Engineering design is an activity which fullfils the human need but in order to fulfil the need, a designer has to conceive and implement a plan with innovative approach, utilizing the available physical resources. All these plans must be defined in sufficient detail so that they can be physically realized in the most economical way. Engineering design is an involved process when the appropriate technology is complex, its application is not obvious and when the prediction and optimization of the outcome require analytical procedures. Engineering always requires synthesis of technical, human and economic factors in addition to the consideration of environmental, social and political factors, whenever they are relevant.

1.2 PHILOSOPHY OF ENGINEERING DESIGN

Philosophy is a body of principles and general concepts which underlies a given branch of learning. It includes the application of these principles in the domain of their relevance. In other words, philosophy is a consistent and integrated personal attitude towards learning.

To develop a philosophy of engineering design, we must find out those principles and concepts which are of the greatest generality, consistent with usefulness and can lead to a discipline of design. We should formulate a methodology whereby the discipline of design can be applied in the most general sense. The principles and concepts on which we must rely are formed on the collective experiences of mankind. Hence it is inevitable that choice of principles and their formulation will be coloured or biased. The origin of philosophy is empirical but its test is pragmatic. The philosophy must include an evaluating scheme which guides and enables the formulation of a specific criterion. This evaluative element is essentially a feedback mechanism which serves to indicate how well the principles have been applied in a particular instance and to reveal the shortcomings so that an improved application of the principles can be made.

Thus, the philosophy of engineering design comprises three major parts, namely:

1. A set of consistent principles and their logical derivation.
2. An operational discipline which results in action.
3. A critical feedback apparatus which measures the advantages, reveals the shortcomings, and illuminates the direction of improvement.

Figure 1.1 shows the elements of engineering design philosophy. Based upon the collective experience of various designers over a number of years, the following general principles have been derived for engineering design. The list is not intended to be a rigid set of formal self-evident propositions.

1. Need
2. Physical realizability
3. Economic viability
4. Design criteria
5. Morphology and design process
6. Reduction of uncertainty
7. Minimum commitment to future design decisions
8. Communication

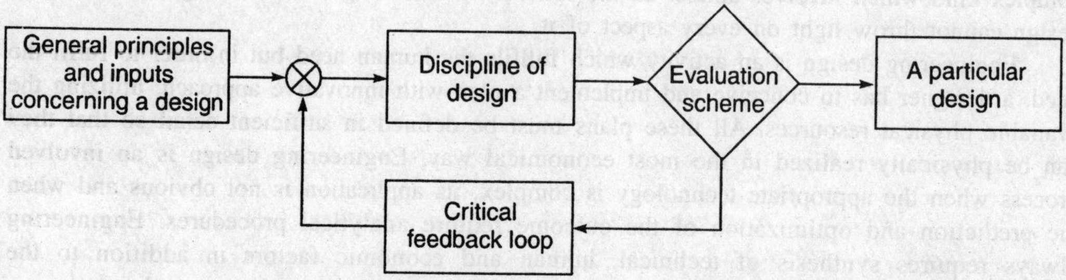

Figure 1.1 Elements of engineering design philosophy.

1.3 KIND OF DESIGN WORK

The technological innovations and scientific discoveries are not only taking place at a rapid pace but also are quickly available for technological exploration. Based upon the rate of technological changes, the design work can be broadly classified into two categories, namely (i) design by evolution and (ii) design by innovation.

1.3.1 Design by Evolution

In the past the design tended to evolve over a long span of time. The rate of development was slow. Products remained static for quite some time. If changes or some improvements were made, these were small and gradual. Each change only made a small improvement in a product which already existed. Most of the new designs were copies of existing designs with minor changes to suit the needs of the local requirements. This leisurely pace of technology change reduced the risk of making major errors. The circumstances rarely demanded the utmost skill and analytical capabilities of the designer. Such a situation is referred to as *design by evolution* in which the technical risk and stakes are proportionately low.

In design by evolution the designer has to weigh a number of conflicting requirements such as the following so that the ultimate product is reliable and satisfies the needs of the customers.

1. Improvement in performance, quality, and appearance
2. Reduction in overall cost
3. Changed market requirements and trend of competition
4. Use of new or improved materials and manufacturing technologies
5. Improved functionality of the product

1.3.2 Design by Innovation

In today's scenario, design by evolution conditions is not valid in many spheres of design activity. The competition has presented a stern and relentless challenge to the designer. The gradual improvement of the product is now less likely to meet the demands of competition especially in the fields of computers, electronics, and instrumentation. Present circumstances require bolder and faster improvements; consequently the present-day designer faces a higher technical risk and has greater stakes. Nowadays designs are by innovation, i.e. following any scientific discovery when a new product is developed, its proper use may dictate an almost complete break with past practice. A new product is developed on ideas which were never practised or tried. As the risk of technical error is high, the skills and knowledge required have to be much greater. To help designers face this challenging environment, various analytical tools such as computer-aided design, drafting, finite element method, and simulation techniques have been developed to reduce the risk of technical error. However, the job of the designers is still not fully simplified by these developments. They encounter a host of problems which are peculiar to the process of design. These are:

1. Need of developing, organizing, and evaluating information almost always in the face of uncertainty.
2. Necessity of taking into account the complicated interaction of components.
3. Constant requirement to make predictions in terms of design criteria.
4. Need to work within the constraints of an economic framework.

1.4 DESIGN PROCESS

The term design process is also sometimes referred to as morphology of design. It is the methodology of design by which ideas about need are projected creatively into ideas about things and which in turn are translated into engineering prescriptions for transforming suitable resources into useful physical objects. Any project in its implementation goes through a series of major phases and generally the next phase is not initiated till the previous one is complete. The methodology of design refers to the study of chronological structure of the design project.

A typical design project progresses through a series of phases in order to achieve the end product. Figure 1.2 shows the various phases of the design process. These phases are:

1. Conceptual design
2. Embodiment design
3. Detailed design

A brief discussion on these phases is presented in the subsections below.

1.4.1 Conceptual Design

Conceptual design is the first phase of the design process in which the problem is identified through abstraction. It involves establishment of structure, search for suitable solutions, principles and their combination into concept variants. The concept variants that have been elaborated must now be evaluated. The variants that do not satisfy the demands of the specification have to be eliminated; the rest must be judged by the application of specific criteria. During this phase, the chief criteria are of technical nature. On the basis of the evaluation, the best solution concept can be selected. The various steps of conceptual design are:

1. To determine whether the need is original, whether it is valid, has current existence, or has strong evidence of latent existence.
2. To explore the design problem engendered by the need and to identify its elements such as working parameters, constraints, and major design criteria.
3. To conceive a number of plausible solutions to the problem.
4. To sort out the potentially useful solution out of plausible ones on the basis of (a) technical suitability, (b) physical realizability, and (c) economic feasibility.

In conclusion, the conceptual design indicates whether a current or potential need exists; what the design problem is; what are plausible solutions of it? In other words, the conceptual design explores the feasibility of the design project.

1.4.2 Embodiment Design

The second phase of the design process is embodiment design. In this phase preliminary design of the system starts with the set of useful solutions which were evolved in the conceptual design phase. The purpose of the embodiment design phase is to establish which of the proposed alternatives is the best design concept. Each of the alternative solutions is subjected to test of analyses until the evidence suggests that either the particular solution is inferior to others or is superior to all. The surviving solution is accepted for further examination.

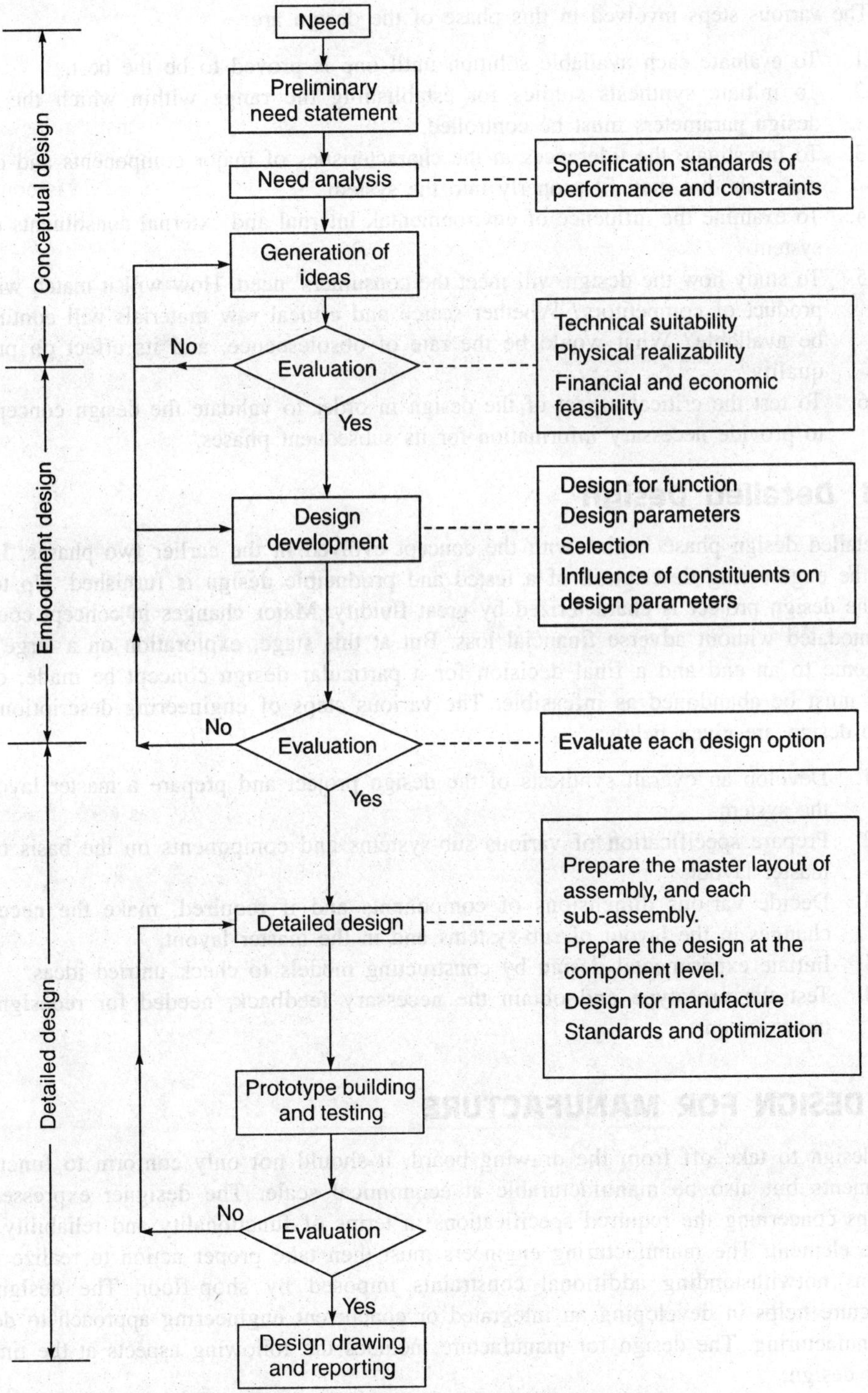

Figure 1.2 Design process diagram.

The various steps involved in this phase of the design are:

1. To evaluate each available solution until one is proved to be the best.
2. To initiate synthesis studies for establishing the range within which the major design parameters must be controlled.
3. To investigate the tolerances in the characteristics of major components and critical materials that may fit properly into the system.
4. To examine the influence of environmental, internal and external constituents on the system.
5. To study how the design will meet the consumers' need. How will it match with the product of competitors? Whether scarce and critical raw materials will continue to be available? What would be the rate of obsolescence, and its effect on product quality?
6. To test the critical aspect of the design in order to validate the design concept and to provide necessary information for its subsequent phases.

1.4.3 Detailed Design

The detailed design phase begins with the concept evolved in the earlier two phases. In this phase the engineering description of a tested and producible design is furnished. Up to this point the design project is characterized by great fluidity. Major changes in concept could be accommodated without adverse financial loss. But at this stage, exploration on a large scale must come to an end and a final decision for a particular design concept be made, or the project must be abandoned as infeasible. The various steps of engineering description of a feasible design are given below:

1. Develop an overall synthesis of the design project and prepare a master layout of the system.
2. Prepare specification of various sub-systems and components on the basis of the master layout.
3. Decide various dimensions of components and if required, make the necessary changes in the layout of sub-systems and in the master layout.
4. Initiate experimental design by constructing models to check untried ideas.
5. Test the prototype and obtain the necessary feedback, needed for redesign and improvement.

1.5 DESIGN FOR MANUFACTURE

For a design to take off from the drawing board, it should not only conform to functional requirements but also be manufacturable at economical scale. The designer expresses his intentions concerning the required specifications in terms of functionality and reliability of a machine element. The manufacturing engineers must then take proper action to realize these intentions notwithstanding additional constraints imposed by shop-floor. The design for manufacture helps in developing an integrated or concurrent engineering approach to design and manufacturing. The design for manufacture includes the following aspects at the time of detailed design:

1. Preparing the manufacturing process plan for every part, sub-assembly, and final assembly. It includes selection of machine tools, cutting tools, jigs, fixtures, cutting conditions, and determination of tool path.

2. Designing for quality control and information flow system.
3. Designing for packaging to take care of outer shape of the product, economy in transportation, and its efficient handling.
4. Providing facility for modular addition to the system to enhance its capacity.
5. Designing to reduce the rate of obsolescence.
6. Designing to use the component at several levels of application.

1.6 TRAITS OF A GOOD DESIGNER

As the designer develops and converts his idea into finished plans for a machine, he must bring into play his expertise in areas roughly classified as:

1. Technical knowledge
2. Industrial experience
3. Human factors

The technical knowledge necessary to design a machine varies with the type and field of application and no designer can become expert in all the fields. The knowledge of mechanisms, mechanics of machine elements, thermodynamics, fluid mechanics and electrical circuits is essential to every type of design. In many situations the designer has to rely upon experimental data. The ability to analyze and use experimental data is an important characteristic of the designer.

Industrial experience of existing designs is essential for gaining a thorough understanding of existing and new designs. The previous design should not hinder the creative ability of the designer. He must also have knowledge of industry standards. A designer should acquire the habit of discussing problems with technicians, salespersons, and operators, etc. because different viewpoints put across by them will always help in developing a well-balanced design.

The consideration of human factors at the product design stage is also essential to produce an acceptable product. The aspects like human safety, comfort and ease of operation, human performance, reliability, quality control methods must all be examined and incorporated in the design. If human aspects are considered properly at the design stage, then the majority of human errors can be reduced. Human factors at the design stage must aim at reducing losses from accident and misuse, ensuring user acceptance, improving human performance, manpower utilization, and economy of operation, etc.

EXERCISES

1. What do you mean by engineering design? Explain with a suitable example.
2. Explain the concept of philosophy of engineering design.
3. What do you understand by the following terms: design by evolution and design by innovation? Explain with suitable examples.
4. What human needs require consideration in a design problem?
5. How can you analyze the need for a given design?

6. Formulate the 'need' statements for each of the following products:
   (a) Bicycle (b) Voltage stabilizer
   (c) Personal computer (d) Pen
7. The design of a bus body is required to be such that it can carry seventy passengers with maximum comfort possible. Give at least three conceptual designs.
8. What do you mean by feasibility of design?
9. What do you understand by morphology of design?
10. What are the basic traits of a good designer?

MULTIPLE CHOICE QUESTIONS

1. In the need analysis stage of design, all information is
   (a) reliable in nature (b) statistical
   (c) rational (d) dispensable
2. Looking for an acceptable solution is the purpose of
   (a) need analysis (b) preliminary design
   (c) feasibility study (d) detailed design
3. Information refers to
   (a) facts (b) data
   (c) disorganized knowledge (d) all of these
4. Which of the following information is verifiable, unambiguous, documentable, and checked by several sources?
   (a) Hard (b) Soft
   (c) Rational (d) Dispensable
5. The need is stated and validated during the following stage of design:
   (a) Feasibility study (b) Embodiment design
   (c) Detailed design (d) Implementation
6. The steps like preparation, concentration, incubation, illumination, and verification are encountered in the following method of design:
   (a) Evolution design (b) Industrial design
   (c) Creative design (d) Scientific design
7. The statement, "Design is a progression from the abstract to the concrete giving a vertical structure to a design project," means:
   (a) Design criteria (b) Design process
   (c) Morphology (d) Need
8. Custom, habit, and tradition are enemies of
   (a) creativity (b) innovation
   (c) synthesis (d) scientific solution
9. The technology of work design based on human biological sciences is called
   (a) anatomy (b) psychology
   (c) ergonomics (d) aesthetics

Chapter 2 Computer-Aided Design and Drafting

2.1 INTRODUCTION

Computer-aided design (CAD) uses the mathematical and graphic processing power of the computer to assist the engineer in the creation, modification, analysis and display of the design. Many factors have contributed to CAD technology becoming a necessary tool in the engineering world. The combination of human creativity with the computer speed at processing complex equations and managing technical databases provides the design efficiency that has made CAD such a popular design tool. The CAD is often thought of simply as computer-aided drafting and its use as an electronic drawing-board is a powerful tool in itself. The function of a CAD system extends far beyond this capability to represent and manipulate graphics. Geometric modelling, engineering analysis, simulation and the communication of the design information can all be performed using CAD.

In every branch of engineering, prior to the implementation of CAD, design has traditionally been accomplished manually. Hence it was time consuming and labour intensive. The CAD has now made the design process much more smooth and efficient. The computer is well suited to design work especially in four areas, namely (i) geometric modelling, (ii) engineering analysis, (iii) automated testing, and (iv) automated drafting. Figure 2.1 illustrates the relationship between the CAD technology and the design process.

Geometric modelling is one of the keystones of the CAD system. It uses mathematical description of geometric elements to facilitate the representation and manipulation of graphical images on the computer display screen.

Engineering analysis can be performed using the various analytical methods available to the designer in a CAD system. The Finite Element Analysis (FEA) is one of the most powerful numerical analysis programs used to solve complex problems in many engineering and scientific fields such as structural analysis, thermal analysis, fluid mechanics, and so on.

Automated drafting capabilities in the CAD systems facilitate presentation, which is the final stage of the design process. The CAD data stored in the computer memory can be sent to a plotter or printer for a hard copy of the detailed and assembly drawing, bill of material and cross-sectioned views of the designed parts.

Although CAD has made the design process less tedious and more efficient than the traditional methods, the fundamental design process in general remains unchanged. Nevertheless, the CAD is a powerful time-saving tool.

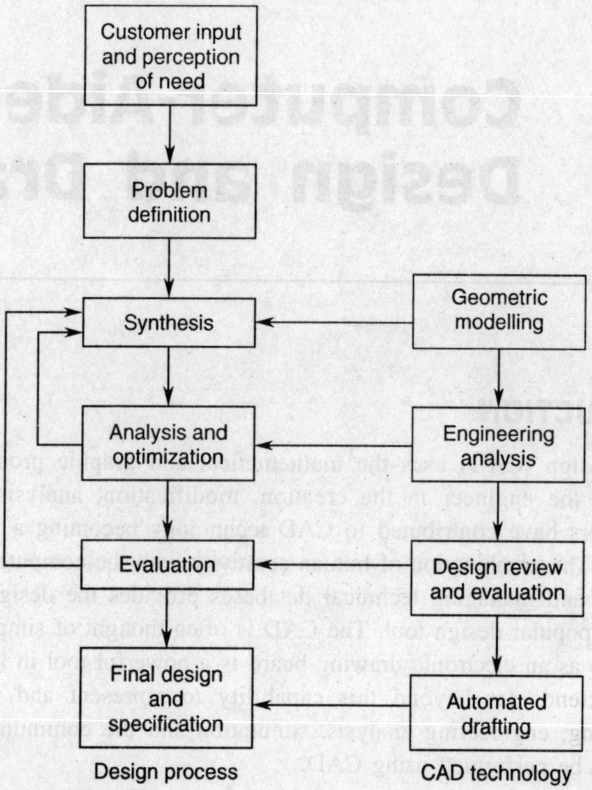

Figure 2.1 Application of the computer to the design process.

2.2 THE COMPUTER

The computer is an electronic data processing machine capable of receiving input, storing sets of instructions for solving problems, and generating output with high speed and accuracy. Without going into the details of the computer, as most readers already know about it, we describe below the use of CAD software to design a machine for a specific application.

2.2.1 CAD Software

The CAD software is general-purpose software that features all of the programs needed for CAD applications. The CAD software is generally classified into three categories: (i) graphic software, (ii) solid modelling, and (iii) analysis software. The graphic and solid modelling software is used to generate a design and represent it on the screen, whereas the analysis software makes use of the stored data relating to the design and applies it to various analytical studies.

Graphic software

The graphic software of a CAD system is nothing but an electronic drawing-board. It facilitates graphical representation of a design on-screen by converting input data into Cartesian coordinates. All CAD systems offer defined geometric elements that can be called into the drawing by the execution of a software command. For example, the circle is mathematically defined by

$$(x - a)^2 + (y - b)^2 = r^2 \tag{2.1}$$

If the user specifies the coordinates of the centre (*a*, *b*) and radius *r*, a circle of a specified size will be represented on the screen. A similar process can be applied to other geometrical shapes.

The combination of individually defined geometric elements enables the designer to create a unique design as per the requirement of the designer's plan. In graphic software, individual or combinations of geometric shapes are converted to cells called *regions*. These regions can then be added and/or subtracted in a number of ways using boolean operations such as union, intersection, subtraction to form a complex geometric shape. Figure 2.2 shows the typical example of boolean operations in which two circles are subtracted from the rectangle.

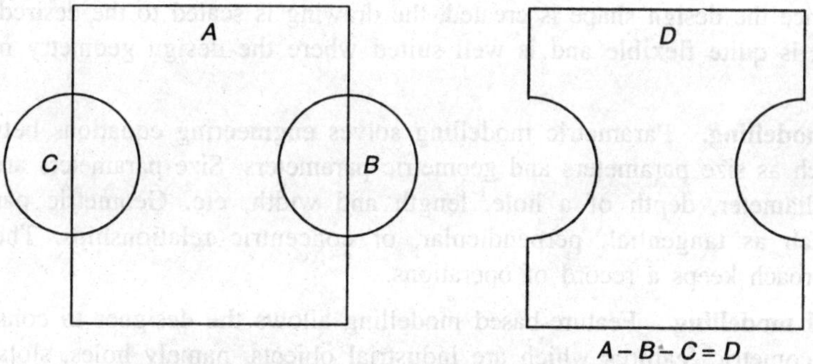

Figure 2.2 An example of boolean operations.

Graphic software supports three-dimensional representation of objects using (i) wireframe and (ii) surface modelling. A wireframe model is a skeletal description of a 3D object. It consists of points, lines and curves that describe the behaviour of the object. There are no surfaces in a wireframe model. The surface modelling defines not only the edges of the 3D object but also its surfaces. In surface modelling, the Non-uniform Rotational B-splines (NURBS) curve surface is most commonly used. Surface modelling is very advantageous due to point-to-point data collections usually required for numerical control (NC) programs in computer-aided manufacturing (CAM) applications.

Most graphic software packages are featured with dimensioning, tolerancing, cross-hatching, scaling, enlargement of view of the part, use of custom- and tailored-made symbols. They also facilitate presentation of drawings, both on-screen and in a hard copy format. Drafting standards specified by a company can also be programmed so that the final draft will comply with those standards.

Some graphic software packages support product data management which allows to make CAD data available interdepartmentally on a computer network. This approach offers faster retrieval of CAD files and automated distribution of design to management, manufacturing engineers, and shop-floor workers for design review.

Solid modelling software

Solid modelling software defines the surfaces of an object, with the added attributes of volume and mass. This allows data to be used in calculating the physical properties of the final product. Solid modelling software uses two methods for creating 3D solid models: (i) constructive solid geometry (CSG) and (ii) boundary representation (B-rep).

In the CSG method, the user can combine solid geometry shapes such as cube, sphere, cylinder, prism, etc. by employing boolean logic operators (union, subtraction and intersection) to generate a more complex part. The boundary representation method utilizes the 2D profile of the part and then using a linear, rotational or compound sweep, the designer extends the profile along a line to define a 3D object with volume.

Software manufacturers generally employ the following methods for solid modelling:

Variational modelling. The variational modelling involves creating the two-dimensional profiles of the design that represent the end view or the cross-section. Using this approach, the designer typically focuses on creating the desired shape with little regard for dimensional parameters. Once the design shape is created, the drawing is scaled to the desired dimensions. This approach is quite flexible and is well suited where the design geometry might change dramatically.

Parametric modelling. Parametric modelling solves engineering equations between sets of parameters such as size parameters and geometric parameters. Size parameters are dimensions such as the diameter, depth of a hole, length and width, etc. Geometric parameters are constraints such as tangential, perpendicular, or concentric relationships. The parametric modelling approach keeps a record of operations.

Feature-based modelling. Feature-based modelling allows the designer to construct a solid model from geometric features which are industrial objects, namely holes, slots, shells, and grooves. For example, a hole can be defined using a 'through hole' feature (see Figure 2.3). Whenever this feature is used, the hole will always be open at both ends. In variational modelling if a hole is created in a plane of specified thickness and if the thickness is increased, the hole would be a blind hole until the designer revises the dimension of the hole depth. A knowledge-based feature modelling is more intelligent than other methods.

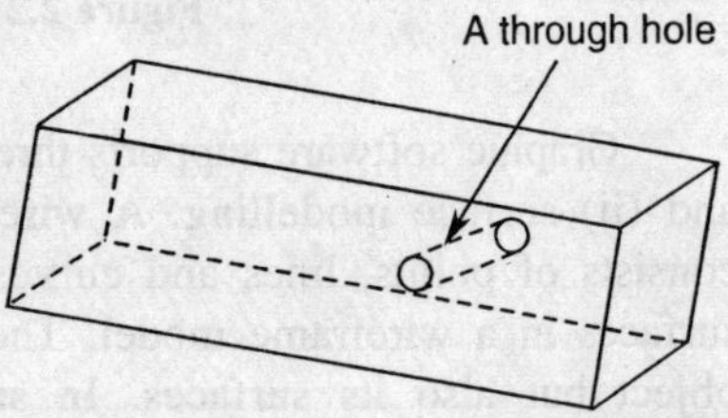

Figure 2.3 Feature-based modelling.

Analysis software

The simulation of the performance of a designed product is an important part of the design process. This performance simulation can be done by solving complex equations using computer-oriented numerical methods. For example, stress, deflection and vibrational analysis of a mechanical structure, steady state and transient thermal analysis of a heat exchanger, and fluid dynamic analysis to study flow over a turbine blade, and so on. The finite element analysis is one such numerical analysis tool which can perform computer-aided engineering (CAE) analysis.

The finite element method divides a given physical or mathematical model into small and simpler elements, performs analysis on each individual element using the required mathematics and forms individual characteristics matrices. It then assembles these characteristics matrices to form a global assembled characteristics matrix of the physical model. Finally, the boundary conditions and the loads are applied to elements and their nodes. After verifying the element connectivity and checking for duplication of nodes, the analysis is performed and the results are obtained.

Commercially, these finite element analysis programs are available as general-purpose FEA packages. The most popular analysis packages are ANSYS, MSC-NASTRAN, ADINA and SAP. These FEA software packages usually consist of three parts: (i) preprocessor, (ii) solver, and (iii) postprocessor. Some FEA packages, namely ANSYS offer a design optimization module for effective synthesizing of the design into an optimized product.

2.2.2 CAD Standards and Translators

In order for CAD applications to run across systems from various vendors and to achieve at least a reasonable high level of integration between CAD, CAE and CAM, the following three main formats are most widely used.

Initial Graphic Exchange Standard (IGES). The IGES is an ANSI standard for digital representation and exchange of information between CAD/CAE/CAM systems. The 2D geometry, the 3D constructive solid geometry (CSG) and the 3D boundary representation (B-rep) solid modelling can all be translated into the IGES format for export/import. Common translating commands IGESIN and IGESOUT are available in the IGES library.

Data Exchange Format (DXF). The DXF format developed by Autodesk Inc. for AutoCAD, is the de facto standard for exchange of data on PC-based systems. The import and export of data can be done through the DXFIN and DXFOUT commands, respectively.

Standards for the Exchange of Product Model Data (STEP). The STEP is an inter-national standard for representing and exchanging a computerized model of a product in neutral form without loss of completeness and integrity throughout the product life cycle. This is documented in number of parts under the ISO : 10303 code, which covers complete STEP technology. For example, part 42 covers the geometry model including wireframe, surface modelling with CSG and B-rep forms. For details, the reader may refer to the original ISO code.

2.3 AutoCAD

The AutoCAD design package is a general-purpose computer-aided design/drafting software available on the PC platform, which can be used to prepare a wide variety of two- and three-dimensional models. The speed and ease with which a drawing model can be prepared and modified using a computer offer a phenomenal advantage over manual preparation. There is virtually no limit to the kind of drawings that we can prepare using AutoCAD. Some of the applications where AutoCAD is widely used today are architectural drawings, interior design, facility planning, line work for fine arts, and engineering drawings in areas of electronics, chemical, civil, mechanical, electrical, and other engineering disciplines.

To get into the software, the user has to click the left mouse button twice at the icon of AutoCAD 2002 in Windows 98 or 2000 based software. After a few seconds, you are taken directly into the drawing editor of the software. At the beginning it displays a menu where the user can select the system of units (i.e. the English or the metric system). The first screen of the

AutoCAD would look like as that shown in Figure 2.4 (a user may find some changes from this due to different settings of the toolbars). The top strip of the screen where few words like file, edit, view, format, draw, modify, etc. are visible is known as the *drop-down menu bar*, where each word represents the name of a group menu. The toolbar represents various commands.

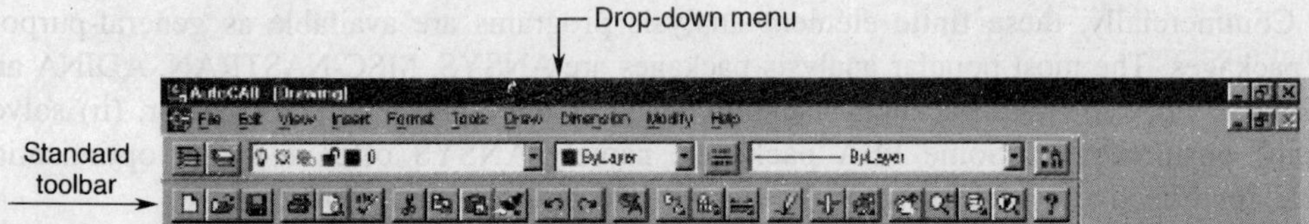

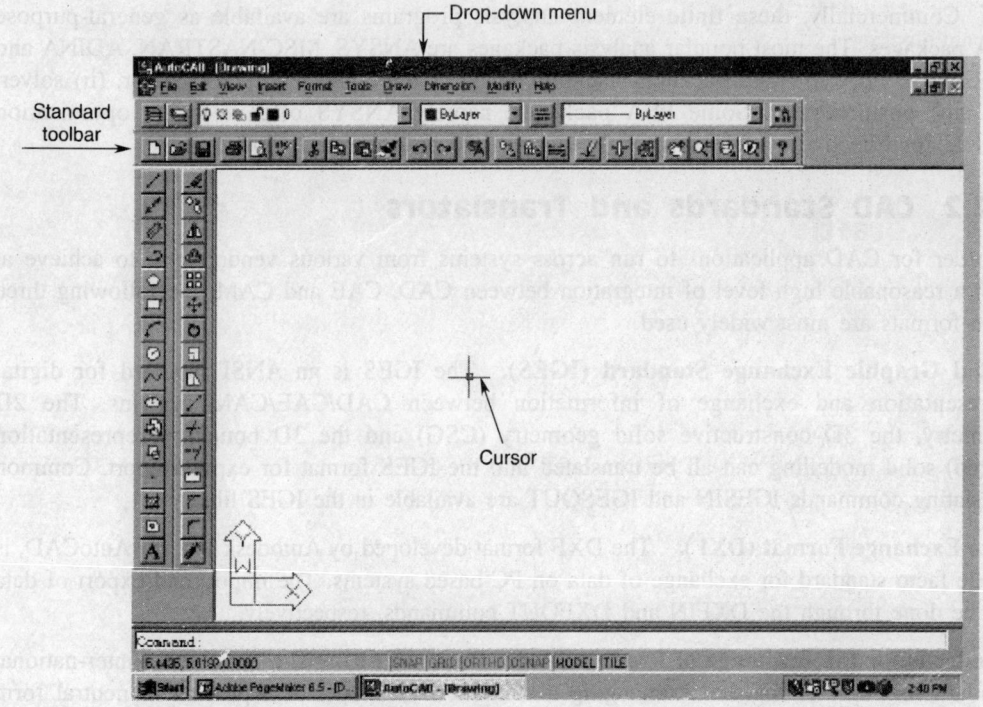

Figure 2.4 First screen of AutoCAD

The largest area of the monitor is drawing area where a user can create a drawing. In the drawing area, the world coordinate system (WCS) icon displays the direction of positive *X*- and *Y*-axes. The positive *Z*-axis is a ray perpendicular to the screen and directed towards the direction of the user. The area at the bottom of the drawing area is known as the *command prompt area.* This area reflects commands that we select either from the toolbar or drop-down menu and the user can interact and supply the data through the keyboard.

In AutoCAD, data entry can be usually accomplished either through picking a point by mouse, or by supplying the coordinates of the point through the keyboard. The keyboard approach is most common; this approach is strongly recommended for production of accurate, dimensioned drawings. The AutoCAD makes use of the following three coordinate systems.

Absolute Cartesian coordinate system. In this coordinate system a point is specified by x, y, and z values which are always measured with reference to global zero point, i.e. lower left corner of the drawing. For example, a 2D point with x value 100.0 and y value 50.0 is supplied as, 100.0, 50.0.

Relative Cartesian coordinate system. In this system, the coordinate of the previously selected point is taken as the reference coordinate and it is assumed to be zero for deciding the coordinate of the next point. For example if the last point is "100.0, 50.0", and from here if we have to draw a horizontal line of 125.0 unit length then the coordinate of the next point is 125.0, 0.0. It is supplied as @ 125.0, 0.0, where @ is a symbol of relative coordinate system.

Polar coordinate system. In a polar coordinate system, to specify a point we require the distance between the previous point and the new proposed point and the angle is measured in the anticlockwise direction from the positive *X*-axis. Its general syntax is

@ length < angle

For example, @ 100 < 45 will draw a line of length 100 units inclined at 45° in anticlockwise direction from the positive *X*-axis taken at the last point.

2.3.1 Drafting Environment

The drafting environment of AutoCAD requires the selection of units, paper size, drawing scale and methods of data entry. AutoCAD UNITS command has no relationship with the working unit. It simply sets the format of data entry and precision of data. The LIMITS command allows the user to specify the working area by supplying the coordinates of the lower-left corner and upper-right corner of the drawing sheet. The user can put on the limit check option which will not allow the work to go outside the area defined by the LIMITS command.

In AutoCAD, data entry through mouse is very easy provided two commands, SNAP and ORTHO, are set properly. In SNAP command, the points entered by means of a mouse can be locked into alignment with an imaginary rectangular grid by the snap mechanism. The snap mode can be turned ON/OFF by the Flip Key F9. The snap resolution can be set by the snap spacing option of the SNAP command. The ORTHO command allows to select a new point to be displaced from the base point along an orthogonal line.

2.3.2 Drawing in CAD Environment

In this section we will take a closer look at some of the commands used for drawing basic entities.

LINE **command.** The LINE command creates a basic entity line between two points. The LINE command asks for the 'first point', against which the user can supply the coordinates of a point; from this point, a line will start and go up to a new point whose coordinates are supplied against the 'next point' option. The coordinates can be supplied either in absolute, relative or polar coordinate system depending upon the availability of data. The LINE command also has 'undo' and 'close' options. The 'undo' option negates the effect of the last coordinates supplied and the 'close' option closes the polygon. For example, if we have to draw an equilateral triangle of 100.0 unit side, the command line sequence is as follows (Figure 2.5):

Command: LINE ↵
First point: 50, 50 ↵
Next point: @ 100, 0 ↵
Next point: @ 100 < 120 ↵
Next point: C ↵

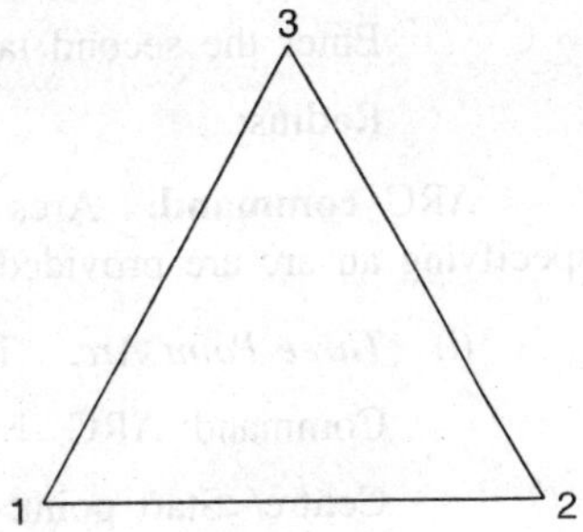

Figure 2.5

CIRCLE **command.** The CIRCLE command can be used in five different ways depending upon the availability of the data as explained below:

(i) *Centre-Radius.* It is a default method in which a circle can be drawn by using the centre of circle and radius (Figure 2.6).

Command: CIRCLE ↵

3P/2P/TTR/<Centre point>: 100, 100 ↵

Diameter/<Radius>: 25 ↵

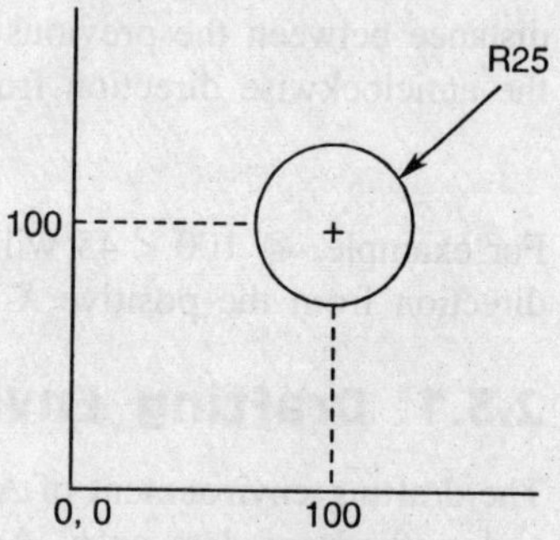

Figure 2.6

(ii) *Centre-Diameter.* In this method, the coordinates of the centre of the circle and its diameter are supplied as input to generate the required circle.

(iii) *Three Point (3P).* When the coordinates of three different points, which do not lie on a straight line, are known we can construct the circle by the 3P option (Figure 2.7).

Command: CIRCLE ↵

3P/2P/TTR/<Centre point>: 3P ↵

First point:

Second point:

Third point:

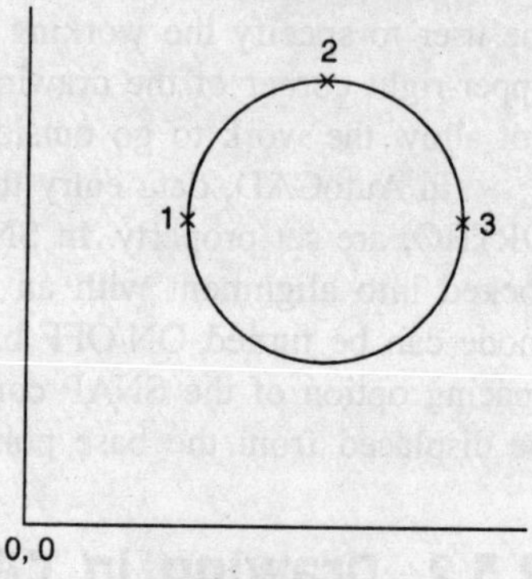

Figure 2.7

(iv) *Two Point (2P).* In the two point (2P) method the coordinates of two diametrically opposite points are supplied as input to draw a circle.

(v) *Tangent, Tangent and Radius (TTR).* When a circle of a specified radius is tangent to two already drawn entities, the TTR option can be selected (Figure 2.8).

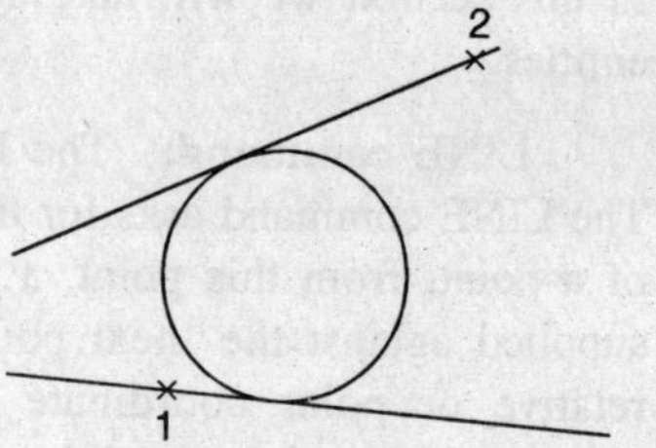

Figure 2.8

Command: CIRCLE ↵

3P/2P/TTR/<Centre point>: TTR ↵

Enter the first tangent spec:

Enter the second tangent spec:

Radius:

ARC **command.** Arcs are partial circles, In AutoCAD, eight different methods of specifying an arc are provided. These are discussed below:

(i) *Three Point Arc.* This method is similar to the 3P method of circle.

Command: ARC ↵

Centre/<Start point>: 40, 40 ↵

Centre/End/<Second point>: 20, 60 ↵

End point: 20, 20

The three point arc may be specified from either direction (Figure 2.9).

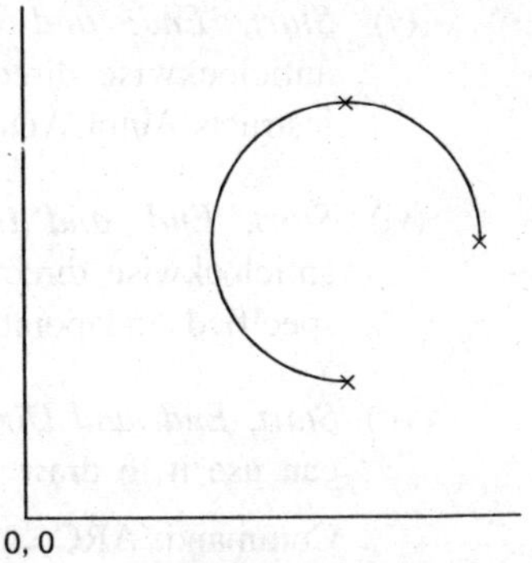

Figure 2.9

(ii) *Start, Centre, and End.* This method specifies an arc in the anticlockwise direction from the start point to the end point. The end point is used to determine the included angle; the arc does not necessarily pass through the end point (Figure 2.10).

Command: ARC ↵

Centre/<Start point>: 40, 30 ↵

Centre/End/<Second point>: C ↵

Centre: 20, 30 ↵

Angle/Length/<End point>: 20, 50 ↵

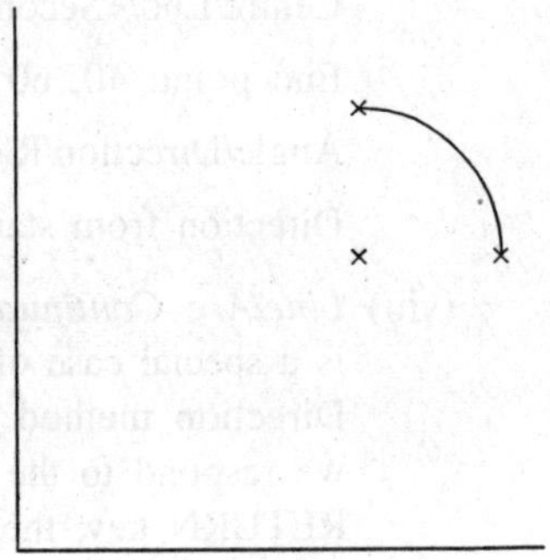

Figure 2.10

(iii) *Start, Centre, and Included Angle.* This method draws an arc with the specified centre and start point, spanning the included angle. Ordinarily, the arc is drawn anticlockwise from the start point. However, if the specified angle is negative, the arc is drawn clockwise (Figure 2.11).

Command: ARC ↵

Centre/<Start point>: 40, 30 ↵

Centre/End/<Second point>: C ↵

Centre: 20, 30 ↵

Angle/Length/<End point>: A ↵

Included Angle: 90

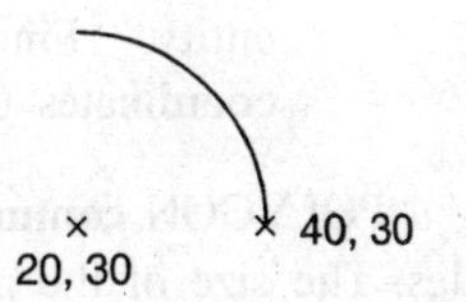

Figure 2.11

(iv) *Start, Centre, and Length of Chord.* For this type of specification, the chord length is used to compute the angle. By default, it draws an arc in the anticlockwise direction. If the length of the chord is negative, then it will draw a major arc, i.e. the included angle is more than 180° (Figure 2.12).

Command: ARC ↵

Centre/<Start point>: 50, 50 ↵

Centre/End/<Second point>: C ↵

Centre: 40, 50 ↵

Angle/Length/<End point>: L ↵

Length of chord: 14.14 or −14.14

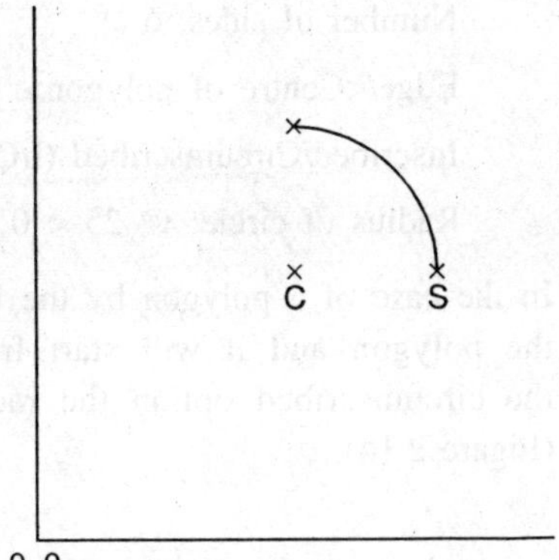

Length of chord = 14.14

Figure 2.12

(v) *Start, End, and Radius.* AutoCAD always draws this type of arc in the anticlockwise direction from the start point. A negative value of the arc radius instructs AutoCAD to draw the major arc.

(vi) *Start, End, and Included Angle.* This type of arc is normally drawn in the anticlockwise direction if the angle is positive or vice versa. Such arcs end at the specified end point.

(vii) *Start, End, and Direction.* This method begins the arc in a specified direction. We can use it to draw an arc tangent to another entity (Figure 2.13).

Command: ARC ↵

Centre/<Start point>: 50, 50 ↵

Centre/End/<Second point>: E ↵

End point: 40, 60

Angle/Direction/Radius/<Centre>: D ↵

Direction from start point: 45 ↵

(viii) *Line/Arc Continuation.* This method is a special case of the Start, End, and Direction method discussed above. If we respond to the first prompt with a RETURN key, the arc's starting point and direction are taken from the end point and ending direction of the last entity. Finally it asks for the coordinates of the end point.

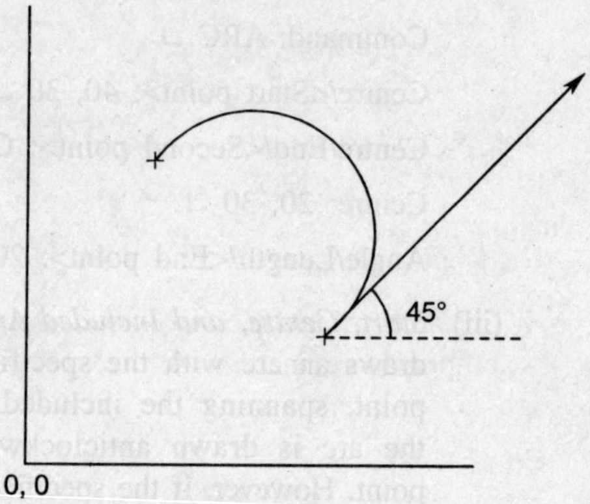

Figure 2.13

POLYGON **command.** The POLYGON command draws a regular polygon of 3 to 1024 sides. The size of the polygon is specified by the radius of the circle or by the length of the edge.

Command: POLYGON ↵

Number of sides: 6 ↵

Edge/<Centre of polygon>: 50, 50 ↵

Inscribed/Circumscribed (I/C): I ↵

Radius of circle: @ 25 < 0 ↵

In the case of a polygon by the inscribed option, the radius point r is taken as the vertex of the polygon and it will start from that point in the anticlockwise direction, whereas in the circumscribed option the radius point will be the mid-point of the edge of polygon (Figure 2.14).

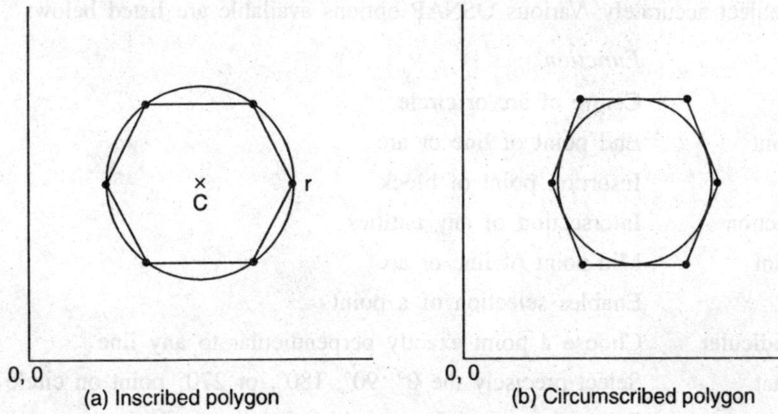

Figure 2.14

RECTANG **command.** If we know the coordinates of two diagonally opposite corners of a rectangle, we can quickly draw a rectangle by the RECTANG command. It also allows us to draw a filleted or chamfered rectangle.

ELLIPSE **command.** The ELLIPSE command lets us draw ellipses. It also allows us to draw an elliptical arc in which the additional inputs of start ângle and ending angle/included angle are required to be supplied.

Command: ELLIPSE ↵

Specify the axis end point or [Arc/Centre]: (pick point P_1)

Specify the other end point of the axis: (pick point P_2)

Specify the distance to the other axis [or Rotation]: (pick point P_3)

The rotation option asks for the rotation angle. The major axis is treated as the diameter line of a circle which will be rotated by a specified angle around the axis. Rotation angle can be between 0 and 89.4° (Figure 2.15).

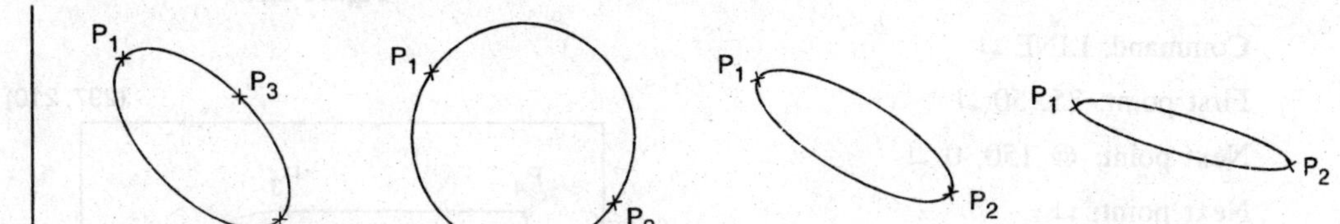

Figure 2.15

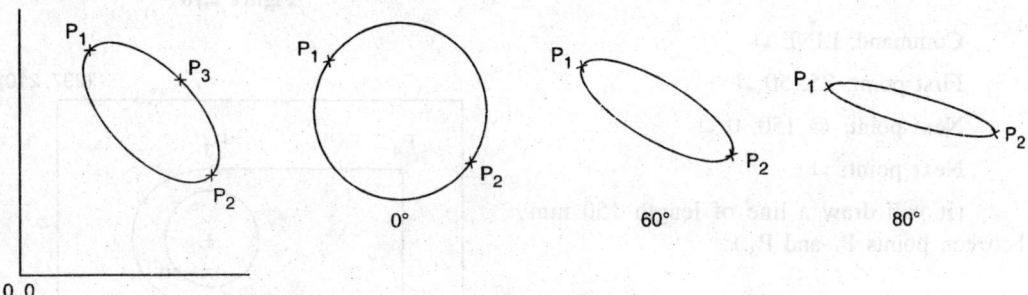

OSNAP **command.** To select an accurate point from a previously drawn entity, AutoCAD provides a tool called the **object snap** or OSNAP. It facilitates the user to pick a point on an object accurately. Various OSNAP options available are listed below:

Option	*Function*
CENter	Centre of arc or circle
ENDpoint	End point of line or arc
INSert	Insertion point of block
INTersection	Intersection of any entities
MIDpoint	Mid-point of line or arc
NODe	Enables selection of a point
PERpendicular	Choose a point exactly perpendicular to any line
QUAdrant	Select precisely the 0°, 90°, 180°, or 270° point on circle
TANgent	Select a tangential point

We can invoke any mode of OSNAP by typing first the three characters shown in uppercase or by selecting the appropriate button on the OSNAP toolbar.

Exercises

Suppose we have to draw the plan view of a simple plate as shown in Figure 2.16. The following sequence of commands needs to be run to create this drawing.

Command: MVSETUP ↵

Enable paper space (Yes/No) <Y>: N ↵

Unit (Scientific/Decimal/Engineering/Architect/Fractional): D ↵

Scale factor: 1 ↵

Paper width: 297 ↵

Paper height: 210 ↵

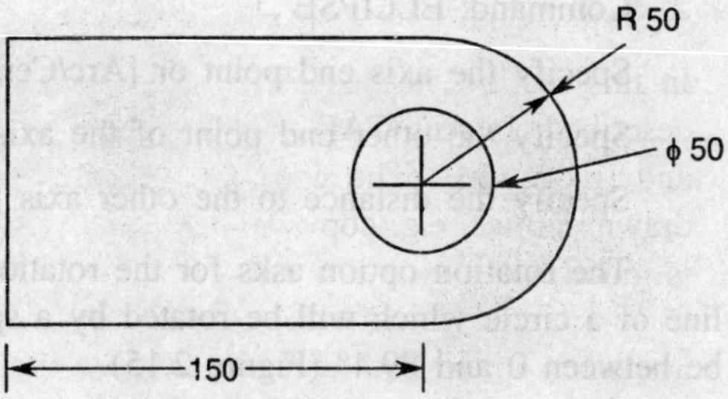

Figure 2.16

Command: LINE ↵

First point: 25, 50 ↵

Next point: @ 150, 0 ↵

Next point: ↵

(It will draw a line of length 150 mm between points P_1 and P_2.)

Command: ARC ↵ (choose Start, Centre, Angle option).

Centre/<Start point>: (choose OSNAP ENDpoint option and pick the line near the end point P_2.)

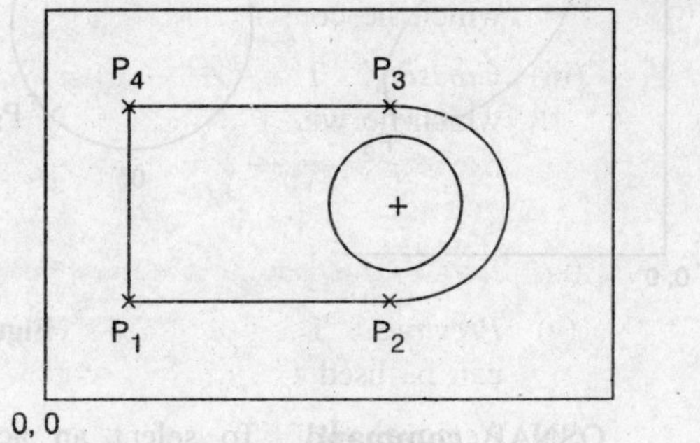

Figure 2.17

Centre/End/<Second point>: C ↵

Centre: @ 50 < 90 ↵

Angle/Length/<End point>: A ↵

Included angle: 180 ↵
(It will create an arc between the points P_2 and P_3.)

Command: LINE ↵

First point: (choose OSNAP ENDpoint option and pick the arc near the point P_3.)

Next point: @ 150 < 180 ↵

Next point: @ 100 < – 90 ↵

Next point: ↵
(It will draw two lines between the points P_3 and P_4 and between the points P_4 and P_1 as shown in Figure 2.17.)

Command: CIRCLE ↵

2P/3P/TTR/<Centre>: (choose OSNAP CENter option and pick the arc.)

Diameter/<Radius>: 25.0 ↵

(It will draw a circle at the centre of the arc with 25 mm radius.)

2.3.3 Editing

In this section various commands which are used to edit or modify the existing drawing are described. In AutoCAD, two types of editing commands are available: (i) constructive editing and (ii) modify. The constructive editing allows us to make new entities using the already drawn entities, e.g. copy, mirror, etc. On the other hand, in modify editing the existing drawing is modified to give it a new look, e.g. trim, break, extend, etc.

In order to perform editing/modifying operations, users have to select objects needed for a specific operation. The following methods of object selection are most commonly used.

(i) *Single entity selection.* In this method a single object is selected by pointing to the entity which we want to select.

(ii) *Window.* In this method, AutoCAD prompts for two corner points. The objects which lie completely within this window are selected.

(iii) *Crossing.* This option is similar to 'window' option but it selects all the objects which lie within or are crossing the boundary of the rectangular window. AutoCAD executes window or crossing option automatically if a point is picked away from the object.

(iv) *Last.* This option selects the most recently created object.

(v) *Previous.* The AutoCAD always remembers the most recent selection and the same can be used for the next command by the previous option.

(vi) *Remove.* It helps to deselect all those objects which we did not want to select.

(vii) *W Polygon/C Polygon.* It allows to create a window polygon or crossing polyon for selection of the object.

COPY **command.** When we want to produce an exact replica of a selected object, we can use the COPY command. The syntax of this command is given below (see Figure 2.18).

Command: COPY ↵
Select object: 'pick object A'
Select object: ↵
<Base point>/Multiple: _CEN of
second point of displacement: @ 100 < 30

More than one copy of the object can be pasted using 'multiple' option of the COPY command.

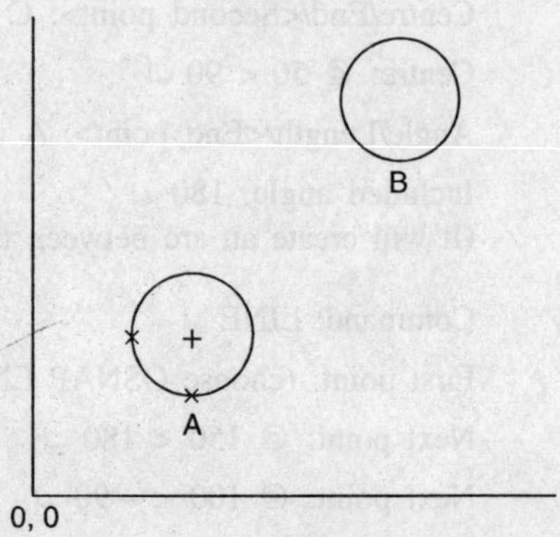

Figure 2.18

ARRAY **command.** When we require multiple copies of an object either in a rectangular matrix of rows and columns or in a polar grid, then the most suitable command is ARRAY (Figure 2.19).

Command: ARRAY ↵
Select object: 'pick object A'
Select object: ↵
Rectangular or Polar (R/P): R ↵
Number of rows: 3 ↵
Number of columns: 4 ↵
Unit cell distance between rows: 25 ↵
Unit cell distance between columns: 25 ↵

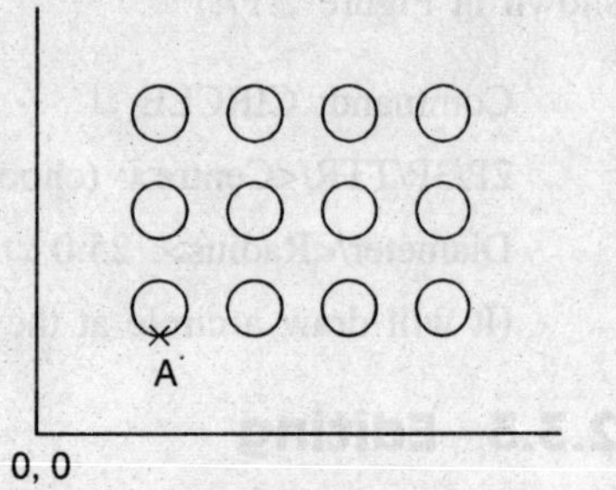

Figure 2.19

The distances between the rows and that between the columns can be negative to achieve other effect.

The command syntax for polar array is given below (Figure 2.20).

Command: ARRAY ↵
Select object: 'pick object A'
Select object: ↵
Rectangular/Polar array (R/P): P ↵
Centre of polar: 'choose centre of circle B'
Number of items: 6 ↵
Angle to fill <360>: ↵
Rotate objects as they are copied <N>: ↵

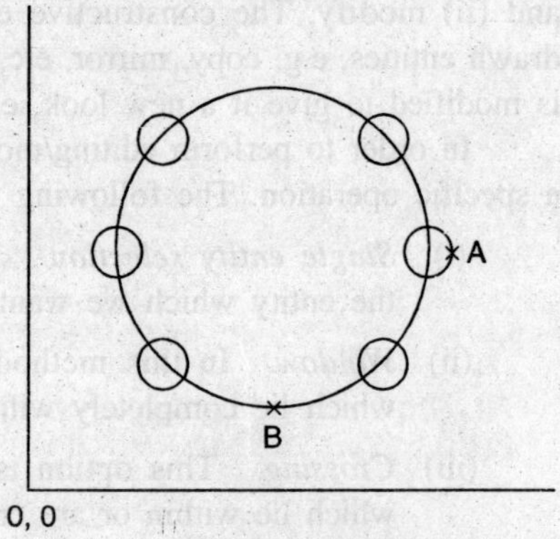

Figure 2.20

MIRROR **command.** The MIRROR command makes the mirror image of the existing object by either deleting or retaining the original object. This command is useful when the object is symmetrical about either axis or both axes (Figure 2.21).

Command: MIRROR ↵

Select object: 'select object by window/crossing option'

Select object: ↵

First point on the mirror line: mid of

Second point: mid of

Delete old object < N >: ↵

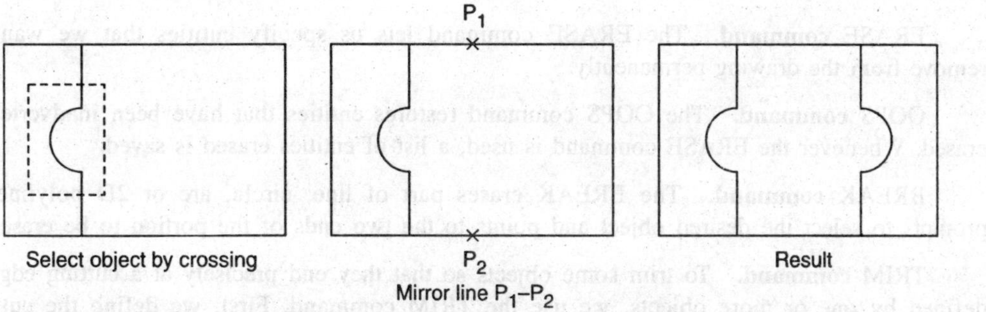

Figure 2.21

Usually when a set of entities are mirrored, the text and attributes are also mirrored. In order not to mirror text and attributes, a variable MIRRTEXT is set to 0 (zero) using the SETVAR command.

FILLET **command.** The FILLET command produces an arc of a specified radius between two entities. The arc radius may be defined by 'Radius' option of the FILLET command. It usually trims or extends objects if they are longer or shorter than what is required. If the entity is a polyline, all sharp corners are filleted with the same radius. A fillet with zero radius produces sharp corners.

CHAMFER **command.** The CHAMFER command trims two intersecting lines by a specified distance and angle and connects these two ends with a new line segment.

OFFSET **command.** The OFFSET command constructs an entity parallel to another entity at either a specified distance or through a specified point. The command syntax is given below:

Command: OFFSET ↵

Offset distance or through < >: 50 ↵

Select object: 'pick an object'

Side of offset: 'select side' (Refer Figure 2.22)

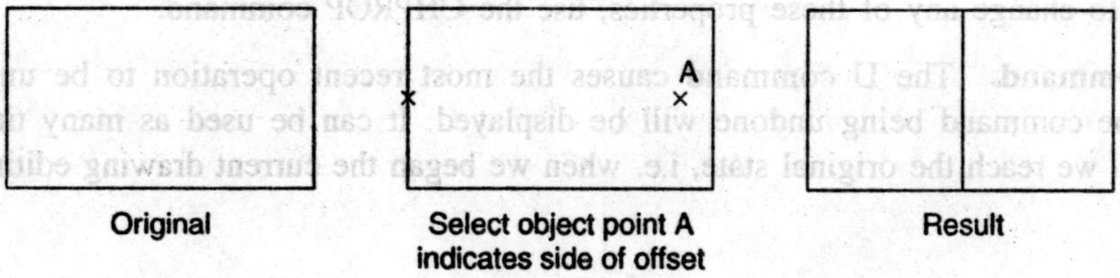

Figure 2.22

MOVE command. The MOVE command allows us to move one or more entities from their present location to a new one without changing their orientation or size. The command syntax is similar to COPY command except for the multiple option.

ROTATE command. It is used to change the orientation of existing entities by rotating them about a specified base point.

SCALE command. The SCALE command allows us to change the size of the existing entities. It is very similar to the ROTATE command.

ERASE command. The ERASE command lets us specify entities that we want to remove from the drawing permanently.

OOPS command. The OOPS command restores entities that have been inadvertently erased. Whenever the ERASE command is used, a list of entities erased is saved.

BREAK command. The BREAK erases part of line, circle, arc or 2D polyline. It prompts to select the desired object and points to the two ends of the portion to be erased.

TRIM command. To trim some objects so that they end precisely at a cutting edge(s) defined by one or more objects, we use the TRIM command. First, we define the cutting edge(s) at which the entities are to be trimmed (Figure 2.23). The syntax of this command is given below:

Command: TRIM ↵

Select cutting edge(s) ...

Select objects:

Select object to trim:

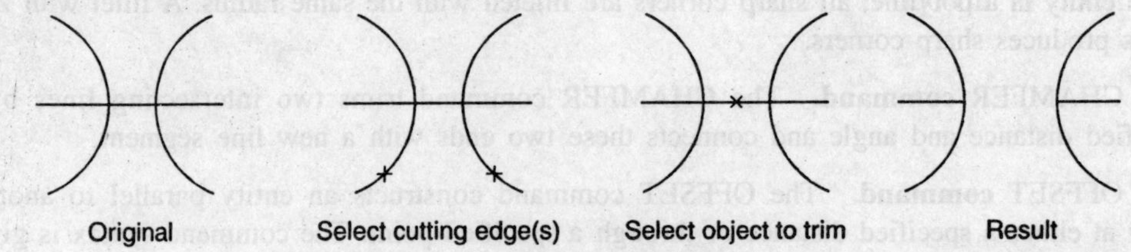

Figure 2.23

EXTEND command. The EXTEND command is complement to the TRIM command. It lengthens the existing object so that it ends precisely to the boundary edge defined by the other object.

CHPROP command. Every entity has an associated layer, colour, linetype, and thickness. To change any of these properties, use the CHPROP command.

U command. The U command causes the most recent operation to be undone. The name of the command being undone will be displayed. It can be used as many times as we want to, till we reach the original state, i.e. when we began the current drawing editing session.

2.3.4 Additional Drawing Commands

In this section we will discuss some of the additional commands used for drawing.

TEXT **Annotation.** The text is an integral part of any drawing. Design notes, tables, information in title block and legends are just a few examples. In AutoCAD, the text can be written by two methods: (i) Single line text using DTEXT command and (ii) Multiline paragraph text through MTEXT command. In DTEXT command whatever we type gets echoed not only at command prompt but also at the location picked on the screen. The command format is given below:

Command: DTEXT ↵

Justification/Style/<Start point>: J ↵

Align/Fit/Centre/Middle/Right/TL/TC/TR/ML/MC/MR/BL/BC/BR:

The various options are described below:

Start point. It is a left justified text.

Command: DTEXT ↵

Justification/Style/<Start point>:

Text height: 5 ↵

Rotation angle <0>: ↵

Text: JODHPUR ↵ × JODHPUR

Text: ↵

Right. It is a right justified text. The text ends at the specified point.

Align. In this option, the text fits between a specified length defined by two end points. The size of characters depends upon the distance between the specified points and the number of characters required to be fitted.

Fit. Fit option is similar to align option but it prompts for text height. It uses the specified text height and adjusts the text width.

Centre. It centres the text base line at a specified point.

Middle. To centre the text both horizontally and vertically at a specified point, the middle option is used. Other options are shown in Figure 2.24.

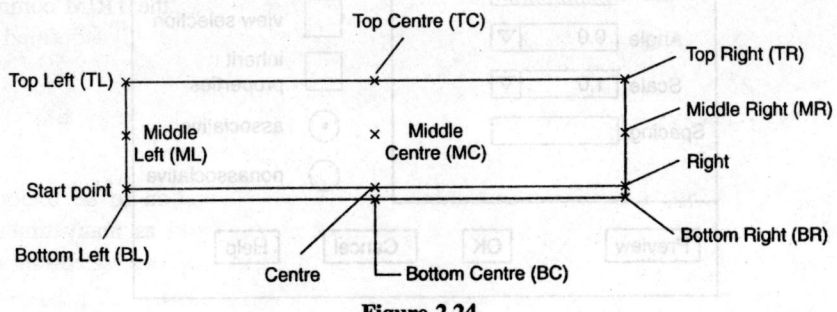

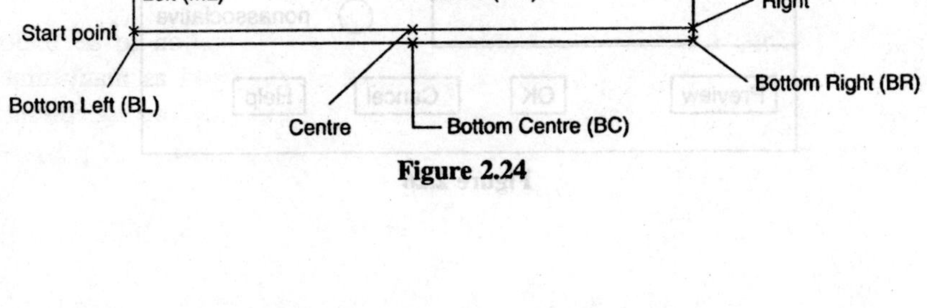

Figure 2.24

A text style is how we want our text font to be seen. This requires additional properties such as oblique angle, height, width factor, backward, upside down, etc. This information put together is called *text style*. Text style can be created by the STYLE command. Already created styles can be changed by style options available in the DTEXT command.

HATCHING. It is a process of drawing section lines in a drawing. The patterns for hatching are defined in ACAD.PAT file. The AutoCAD has vastly improved hatching capability in Release 2002 as a result of the improved BHATCH command. It has three styles of hatching, namely (i) normal, (ii) outermost, and (iii) ignore—the default one being normal (Figure 2.25).

Figure 2.25 Hatching style.

The command format is:

Command : BHATCH ↵

The BHATCH command displays a dialog box as shown in Figure 2.26. The user can select *hatch type* and *pattern*. Different hatch patterns are available. Depending upon the requirement of drawing and hatch type, the user can opt for *hatch angle*, *hatch scale*, and *hatch spacing*. In the hatch command, two methods for the selection of the object are available: (i) pick point and (ii) select object. In the 'pick point' option a point near to the boundary is selected whereas in the 'select object' option the individual objects are selected. Before applying hatch, we can see the preview of hatching. It also allows us to seek the properties from other hatch patterns using the 'inherit properties' option. Hatching can be inserted as a block by using associative composition.

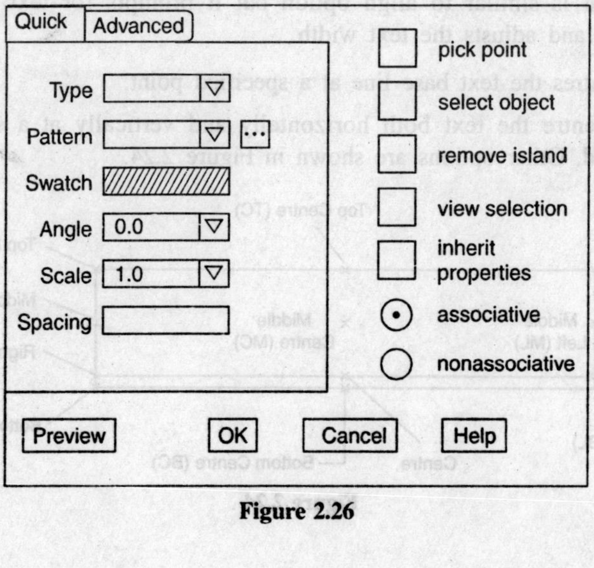

Figure 2.26

2.3.5 Dimensioning

In many applications, a precise drawing plotted to the scale is not sufficient to convey the desired information; annotations must be added showing the length, distance or angle between objects. Dimensioning is the process of adding these annotations to a drawing. In AutoCAD, dimensioning is grouped into: (i) Linear (ii) Aligned (iii) Ordinate (iv) Radius (v) Diameter, and (vi) Angular.

Linear dimensioning. It is done with horizontal, vertical and rotated options. The only difference among these is the angle at which the dimension line is drawn. The command format for linear dimensioning is given below:

Command: LINEAR ↵

Specify first Extension line origin or <select object>: ↵

Select object to dimension:

Specify dimension line location or [MText/Text/Horizontal/Vertical/Angle/Rotated]:

The various options are described below:

MText	It activates the MTEXT command to write the text.
Text	If the user wants to supply dimension text instead of using default or measured text, the text option is used.
Horizontal	It is for horizontal dimensioning.
Vertical	It is for vertical dimensioning, however, depending upon the type of the entity selected and location of the dimension line, it automatically selects either the horizontal or the vertical option.
Rotated	When dimension line is to be located at a specified angle, this option is useful.
Angle	Dimension text can be oriented at some angle.

Aligned dimensioning. When the dimension line is located in line with the drawing line, this type of dimensioning is called the *aligned dimension.* The command format is the same as for the linear dimensioning.

In linear or aligned dimensioning, the following points should be noted.

1. The dimension text is normally centred between the extension lines.
2. If the text does not fit in between the extension lines, the text is placed outside the second extension line. If the extension line origin is selected by selecting object, the second extension line is farthest from the point at which the entity is selected.
3. Normally the text is placed between the dimension line, however, we can write the text above the dimension line as well.
4. By default, all dimension texts follow undirectional approach, i.e. they are placed horizontally. However, they can be aligned to dimension line.

The manner in which dimensions are drawn is controlled by a set of dimension variables. Some of them are simply 'on/off' switches, while others require numeric/string values. The

AutoCAD 2002 provides a simple way to modify dimension variables using the 'dimension style' command available in 'format' drop-down menu.

Ordinate dimensioning. There are times when some drawings need to be created using ordinates. The ordinates are distances specified with respect to a reference point. This method is followed by a number of industries since it leaves the drawing relatively uncluttered and brings out the message perfectly.

Radius dimensioning. Radius dimensions act on arcs and circles (Figure 2.27). The command format is given below:

Command: Radius ↵

Select arc or circle:

Specify dimension line location or [Mtext/Text/Angle]:

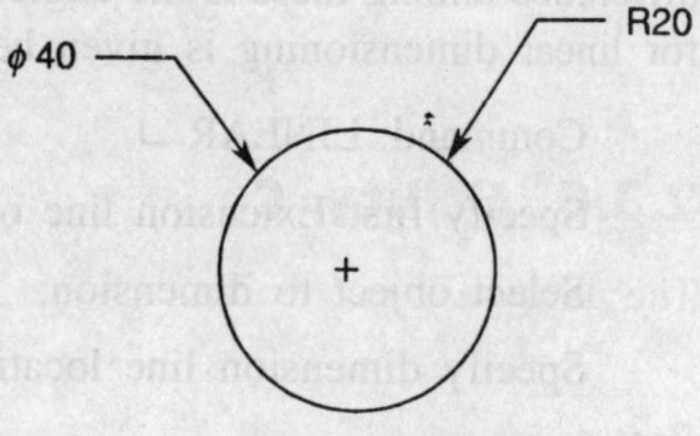

Figure 2.27

Diameter dimensioning. It is similar to radius dimensioning except that the diameter of the circle is dimensioned.

Angular dimensioning. It is used to dimension angles. The dimension line is an arc spanning the angle between two non-parallel straight lines (Figure 2.28). The command format is:

Command: ANGULAR ↵

Select arc, circle, line or <specify vertex>: 'select line'

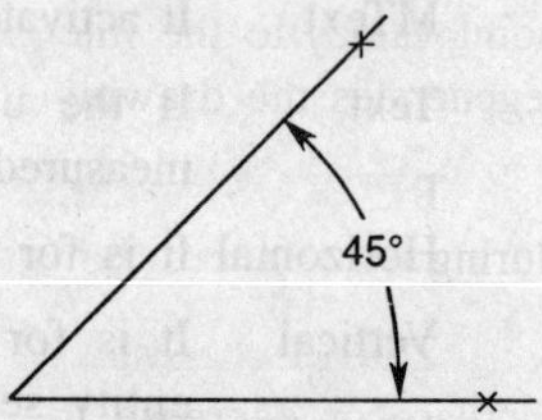

Figure 2.28

Select second line: 'select second line'

Specify dimension arc line location or [Mtext/Text/Angle]:

Angular dimensioning allows to dimension a single line inclined at some angle using the vertex option.

Continuing linear dimensions. Often, a series of related dimensions must be drawn. Sometimes several dimensions are measured from the common base line; at other times one long dimension is broken into shorter segments that add up to the total measurement. The BASELINE and CONTINUE commands are provided to simplify these operations. In these cases, the first dimension is drawn using the linear dimension option. Then, we enter either the BASELINE or the CONTINUE command. The AutoCAD proceeds directly to the "second extension line origin" prompt and then asks for dimension text. When the BASELINE command is used, new dimension line is drawn at a distance of DIMDLI variable from the previous dimension (Figure 2.29).

Leader dimensioning. For some dimensioning, the text may not fit comfortably next to the object it describes. In such a case, it is customary to place the text nearby and draw a leader line from the text to the object (Figure 2.30).

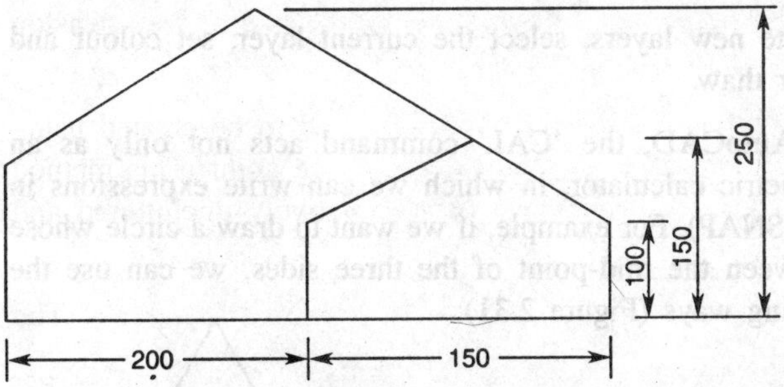

Figure 2.29

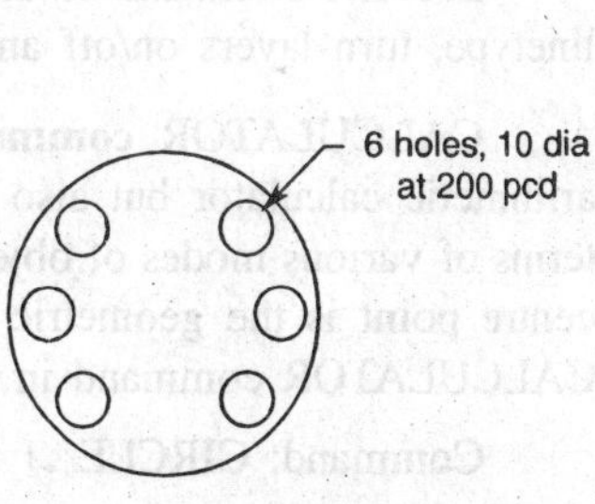

Figure 2.30

2.3.6 Utility Commands

The utility commands in AutoCAD are as follows:

REGEN **command.** The coordinates, angles, radii, etc. are stored in a database as floating point values for greater precision. However, the drawing is placed on the screen using pixels which are integers, therefore, every time data have to be transformed from the floating point values to the integer values. This process is called *regeneration.* The REGEN command regenerates the drawing in the current viewport.

REDRAW **command.** The AutoCAD saves the integer screen coordinates calculated during the regeneration process and can replay them very quickly by the REDRAW command.

ZOOM **command.** The ZOOM command acts like a zoom lens on a camera; it lets us increase or decrease the apparent size of the item we are viewing in the current viewport, although its actual size remains constant. The format of this command is as follows:

Command: ZOOM
All/Centre/Dynamic/Extents/Left/Previous/Window/<Scale>:

Zoom scale. It allows to enter the scale factor; the magnification is computed relative to full view. If a number followed by 'X' is entered, the scale is computed relative to the current view.

Zoom-all. It allows us to see the entire drawing in a current viewport.

Zoom extent. It magnifies the view to the maximum possible extent.

Zoom window. It allows us to specify the area, by selecting the coordinates of two points which we wish to enlarge.

Zoom centre. It allows us to specify a new display window by entering the desired point. Later on, we can specify magnification by a numeric number followed by X.

Zoom previous. For going back to the previous view screen, we can choose the zoom-previous option. Up to ten views can be stored.

LAYER **command.** We can place entities of a drawing on one or more transparent sheets called the layer. A set of layers can hold entities related to a particular aspect of the drawing. Their visibility, colour and linetype can be controlled globally. The same drawing limits, the coordinate system and the zoom factor apply to all layers in a drawing. There is no limit to the number of layers in a drawing nor is there any limit to the number of entities per layer.

LAYER command is used to create new layers, select the current layer, set colour and linetype, turn layers on/off and freeze or thaw.

CALCULATOR **command.** In AutoCAD, the 'CAL' command acts not only as an arithmetic calculator but also as a geometric calculator, in which we can write expressions in terms of various modes of object snap (OSNAP). For example, if we want to draw a circle whose centre point is the geometric mean between the mid-point of the three sides, we can use the CALCULATOR command in the following ways (Figure 2.31):

Command: CIRCLE ↵

3P/2P/TTR/<centre point>: ' CAL ↵

>> expression: (mid + mid + mid)/3

>> select entity for MID point: 'pick p_1'

>> select entity for MID point: 'pick p_2'

>> select entity for MID point: 'pick p_3'

Diameter/<Radius>: 2 ↵

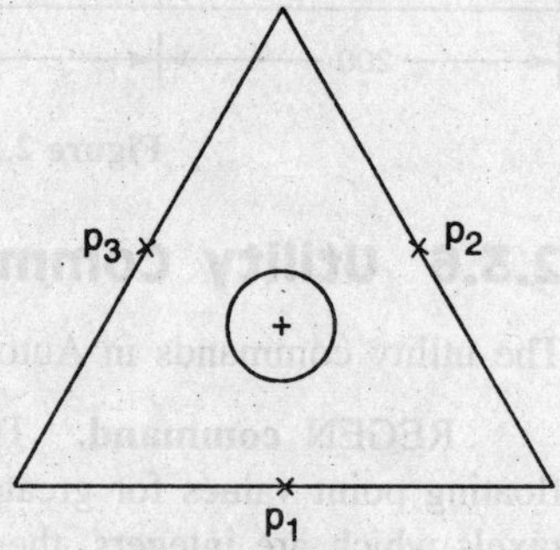

Figure 2.31

2.3.7 Block and Attributes

A block is a set of entities grouped together into a compound object. Once so grouped, these entities are given a block name. A block provides the following advantages: (i) work reduction, (ii) customization, (iii) building a custom library of symbols, and (iv) attaching textual information as attributes. A block created by BLOCK or WBLOCK command can be inserted in the same or any other drawing file by INSERT command. While inserting such blocks the user can resize them by an appropriate scale factor.

Attributes are textual information that vary with each insertion of a block and can be displayed as ordinary text or remain invisible. To us an attribute, we must first create an attribute definition using the ATTDEF command. The attribute definition is a drawing entity that describes the characteristics of the attribute. It is displayed on the screen as a text string. and is called the attribute tag. Once we create the attribute definition, we can specify it as one of the entities to be included in the block definition. Thereafter, whenever we insert the block through the INSERT command, it prompts for the value of the attribute, which may be strings. The attribute information from the drawing can be extracted by the ATTEXT command. The extracted information is written on a disk file in a form suitable to database which can be, later on, processed for generating a bill of material and computing the cost of design.

2.4 AutoLISP

AutoLISP is a special programming language supported by AutoCAD which is used to write instructions carried out by AutoCAD. It is a subset of the common LISP programming language which has been modified to execute the instructions/commands of AutoCAD. It is not a true independent programming language like PASCAL, FORTRAN, or C/C++, because many of the functions of AutoLISP can only be used with AutoCAD. However, AutoLISP does share its syntax structures and functions with its parent programming language LISP.

A LISP routine called the lisp file can have any name that is valid DOS file name and it must have file extension.lsp, e.g. test.lsp. Although an AutoCAD user can work comfortably in designing and drafting with AutoCAD without having any knowledge of AutoLISP, the knowledge of AutoLISP can nonetheless open up new avenues for improving the productivity of the design office. The main advantages, and the situation in which AutoLISP is useful, are the following:

1. To automate the working of AutoCAD and thereby improve its performance.
2. To create new and unique AutoCAD commands.
3. To create special menu macros to perform special operations.
4. To parameterize the drawings used frequently with different dimensions such as keys, keyways, nuts, bolts, steel sections, etc.

Before discussing the syntax of various functions and features of AutoLISP, let us look at some AutoLISP programmes to understand how these work.

Example 1 Getting a message

Consider a simple programme given below. This program when executed will produce the following message at the command prompt of AutoCAD.

"Supply coordinate of a point:"

Program 2.1

```
(defun C:Ex1 ( )
;print a message at command prompt
(prompt "supply coordinate of a point:")
)
```

Let us have a close look at this program. The first opening paranthesis is a special function used by the AutoLISP to tell the computer where the program starts, the closing parenthesis indicates the end of the program. The next line informs the name of the system definition function. It can be without C: as written below.

(defun Ex1 ( )

When the definition function is written with C: it is treated at par with the AutoCAD command and the same can be executed at command prompt, whereas without C: it is a user's defined function. The empty pair of parentheses immediately after the (defun) name indicates that the function has no local arguments or parameters. The semi-colon sign (;) on next line is a comment line. Comments are used in a program to enhance its readability. The comment lines are non-executable statements and they are ignored by the AutoLISP interpreter. The next line is the only executable statement in the program. The (prompt) is an AutoLISP function which displays the message on the screen. The message must be placed between quotes. The output of the statement on command prompt will look like as shown below:

Command: Supply coordinate of a point: nil

Since the (prompt) function is not an input function, it will not accept any input data; it simply gives the message.

Example 2 Draw a line segment

Consider another program which draws a single line segment between two points. The complete program is given below:

Program 2.2

```
; draw a line segment
(defun C: ELINE ( )
(setq p1 (getpoint "First point:")) (terpri)
(setq p2 (getpoint "Second point:"))
(command "LINE" p1 p2 " ")
)
```

The first line is a comment line. The next statement is (defun) which assigns a name to the function. This function is equivalent to the standard AutoCAD command named ELINE. AutoLISP does not differentiate between the lowercase and uppercase letters. The third statement is a little bit complex. It consists of two important functions—(setq) and (getpoint). The (getpoint) function is the standard input function of AutoLISP. Wherever this function appears in the program, it terminates its execution and waits for input information. The words written within the quotes "First point:" appear as the prompt message. The function (setq) is an assignment statement which assigns input information to a variable p_1. The fourth statement is similar to the third statement which receives the coordinates of the second point and assigns them to variable p_2. The function (terpri), which is inserted between the third and the fourth statements, terminates printing and sets the cursor on the next line, so that the next prompt will appear on a new line, otherwise both the inputs will be asked on the same line. The fifth statement is a command statement. This statement tells to use the LINE command and draws a segment of line between two points whose coordinates are assigned to variables p_1 and p_2. Finally, the LINE command is terminated by the quotes " ". The execution of the program looks as shown in Figure 2.32.

```
Command: ELINE
first point: 2, 2 ↵
second point: 4, 4 ↵
command: nil
```

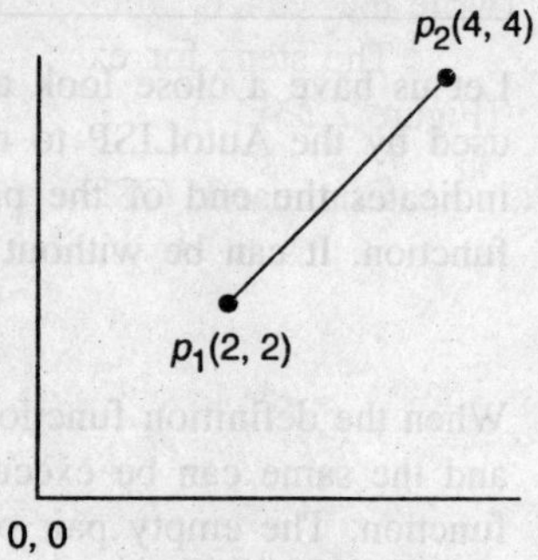

Figure 2.32

Example 3 Draw two concentric circles

This example illustrates how to perform calculations and then draw two concentric circles. It requires the centre point, the inner radius and the wall thickness as the inputs. The program is self-explanatory as it is well commented.

Program 2.3

```
; drawing of two concentric circles
(defun C: CCIRCLE ( )
; input centre of circle
(setq c1 (getpoint "Centre of circle:")) (terpri)
; input radius of inner circle
(setq r1 (getreal "Inner radius:")) (terpri)
; input wall thickness
(setq t (getreal "Wall thickness:"))
; calculate the radius of outer circle
(setq r2 (+r1 t))
; Draw the circles
(command "CIRCLE" c1 r1)
(command "CIRCLE" c1 r2)
(princ)
)
```

The structure of the above program is similar to that of Examples 1 and 2, except that a new input function (getreal) is introduced. It receives a real number through the keyboard. A mathematical executable statement (setq r_2 (+ r_1t)) computes the radius of the outer circle. Mathematically, it is equal to $r_2 = r_1 + t$. Another new function (princ) is introduced. This function ends the program logically and does not allow to write the 'cancel' message at the command prompt.

The steps for execution of this program are given below (Figure 2.33):

```
Command: CCIRCLE ↵
Centre of circle: 4, 4 ↵
Inner radius: 1.0 ↵
Wall thickness: 0.5

Command:
```

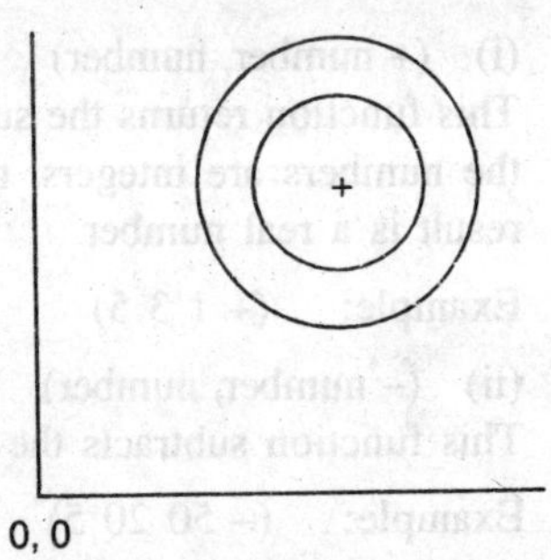

Figure 2.33

The three examples discussed above illustrate that an AutoLISP program is a set of instructions supplied to AutoCAD to perform specific tasks. These instructions can be viewed as a group of building blocks called the *function*. A function may include one or more statements designed to perform a specific task. Therefore, in AutoLISP, the concept of modular programming can be easily used.

2.4.1 Data Types

The AutoLISP supports the following data types:

Integer constant. Integer constants are whole numbers entered without a decimal point. For example, 123, 1960, 24563. AutoLISP integers are 32-bit signed numbers. Although it uses 32-bit values internally, those transferred between AutoLISP and AutoCAD are restricted to 16-bit values.

Real constant. A real constant is a number containing a decimal point having at least 14 digits of precision even though the AutoCAD command line area shows 8 significant digits. For example, 2.567, 0.123 E + 02.

Variables. In AutoLISP, variables are also called *symbols.* AutoLISP uses symbols to store a varying value. A variable can be made up of any characters except ( ), ', ", ; . For example, name, centre, X, p_1, p_2, etc.

List. A group of two or more items is called a *list.* A commonly used list in AutoLISP is coordinates of a point. For example, (15.26, 125.2), is a list containing X and Y coordinates of a 2D point.

Strings. Strings are a group of alphanumeric characters which are not to be evaluated. Strings can be of any length.

File descriptor. File descriptors are alphanumeric labels assigned to files opened by AutoLISP. When an AutoLISP function needs to access a file, its label must be referenced.

2.4.2 Mathematical Functions

Doing mathematical operations in AutoLISP is extremely easy, though the syntax is a little cryptic than what we normally use. The following mathematical functions are most widely used in AutoLISP.

(i) (+ number, number)
This function returns the sum of all numbers; one can use it with real or integer numbers. If all the numbers are integers, the result is an integer. If any of the numbers is a real number, the result is a real number.

Example: (+ 1 3 5) returns 9

(ii) (– number, number)
This function subtracts the second and onward numbers from the first number.

Example: (– 50 20 5) returns 25

(iii) (* number, number)
This function returns the product of all numbers.

Example: (* 2 4 5.0) returns 40.0

(iv) (/ number, number)
This function divides the first number by the second through the last and returns the final quotient.

Example: (/ 200 2 25) returns 4

(v) (abs number)
This function returns the absolute value of the number. The number argument can be real or integer.

Example: (abs – 500) returns 500

(vi) (sin angle)
This function computes the sine of an angle as real. The angle must be expressed in radians.

Example: (sin 1.0) 0.84147

(vii) (cos angle)
This function computes the cosine of an angle, which is expressed in radians.

Example: (cos 0.0) returns 1.0

(viii) (atan num)
This function returns the arctangent of the number given as argument.

Example: (atan 0.5) returns 0.463648

(ix) (exp number)
This function returns the value of e^{number}

Example: (exp 1.0) returns 2.718

(x) (expt base power)
This function returns the base raised to the power.

Example: (expt 5 2) returns $5^2 = 25$

(xi) (log number)
This function calculates the natural log of the number.

(xii) (sqrt number)
This function returns the square root of the number as a real.

(xiii) (float number)
This function converts the integer value into real.

(xiv) (fix number)
This function returns the conversion of the real number into an integer.

(xv) (rem num1 num2)
This function divides num1 by num2 and returns the remainder.

2.4.3 Input Functions

AutoLISP has a rich set of input functions for getting various types of input data. The input functions cause the execution of the program to pause and wait for the user to supply the needed information before continuing with the execution. The following input functions are most common.

(getpoint). This function pauses for the user input of a point. The user can specify a point either by pointing or by entering the coordinate. If the optional base point argument is present, AutoCAD draws a rubber-band line from that point to the current cursor position.

Examples: (setq p_1 (getpoint "specify first point:"))
(setq p_2 (getpoint p_1 "second point:"))

(getcorner). The (getcorner) function gets the coordinate of the input point, similar to the (getpoint). However, it requires a base point argument and draw a rubber-band rectangle from that point to the present cursor position.

Example: (setq p_1 (getpoint "specify first corner:"))
(setq p_2 (getcorner p_1 "other corner:"))

(getdist). This function pauses for the user input of a distance or two points and returns a real number which is the distance between those points.

Example: (setq d (getdist "specify distance:"))

(getreal). This function pauses for the user input of a real number.

Example: (setq t (getreal "wall thickness:"))

(getint). This function pauses for the user input of a 16-bit integer number.

Example: (setq no (getint "specify number of holes:"))

(getstring). This function pauses for the user input of string. If the string is longer than 132 characters, it returns only the first 132 characters.

Example: (setq f1 (getstring "Enter file name:"))

(getangle). This function pauses for the user input of angle and returns that angle in radians. The user can enter the angle in the current angle unit format but the function always returns the angle in radians.

2.4.4 Utility Functions

Besides input and arithmetic functions, AutoLISP has also utility functions which are helpful in efficient programming. Some of the utility functions are described here.

(getvar). This function retrieves the value of an AutoCAD system variable. The variable name must be written within double quotes.

(setvar). This function sets an AutoCAD system variable to the given value.

(distance pt1 pt2). This function calculates the distance between two points.

(angle pt1 pt2). This function calculates the angle of a straight line running between points pt1 and pt2. The returned angle value is in radians.

(polar pt angle distance). This function computes the coordinate of a point whose distance and angle in radians from the *x*-axis are known with reference to the base point pt.

Example: (setq p_1 (getpoint "specify first point:"))
(setq d (getdist "enter distance:"))
(setq a (getangle "enter angle:"))
(setq p_2 (polar p_1 a d))

(\n). This is a new line character function, which displays message on the next new line.

(princ). This function prints the expression on screen without returning any value.

(redraw). This function redraws the drawing on the current viewport.

(graphscr). This function switches the display from text screen to graphic screen.

(textscr). It switches the display from graphic screen to text screen.

(progn). This function usually clubs various statements and forms a block of statements which are executed sequentially where only one statement is expected.

(list). This function takes individual elements and forms a list of elements. It is very useful in forming the coordinate of a point when data are to be taken from different variables.

Example: (setq x (getreal "enter x coordinate:"))
(setq y (getreal "enter y coordinate:"))
(setq p_1 (list x y))
or (setq p_1 ' (x y))

(car list). This function returns the first element of the list.

Example: (car p_1) returns x

(cdr list). This function returns the list containing all elements except the first one.

Example: (setq p_2 ' (x y z a b))
(cdr p_2) returns (y z a b)

(cadr list). This function returns the second element of the list.

Example: (cadr p_2) returns y

(caddr list). It returns the third element of the list.

Example: (caddr p_2) returns z

(command). This function executes the AutoCAD commands from within AutoLISP and always returns nil. The arguments represent AutoCAD commands and their options.

Example: (setq p_1 (getpoint "specify first point:"))
(setq p_2 (getpoint "second point:"))
(command "LINE" p_1 p_2 " ")

2.4.5 Equality/Conditional Functions

We often compare two quantities, and depending on their relation, take certain decisions. These comparisons can be done with the help of a relational or logical operator. AutoLISP has conditional functions to do the job of comparison. The following conditional functions are frequently used. These functions return true (T) if the specified items satisfy the given conditions.

=	equals to
/ =	not equal to
<	less than
< =	less than equal to
>	greater than
> =	greater than equal to

Examples: (= 5 5.0) returns T
(/= "Jodhpur" "Delhi") returns T
(< 10 30) returns T
(< = 10 20) returns T
(> 10 20) returns nil

AutoLISP supports logical functions, which performs tests for 'and', 'equal' and 'or' conditions. For example:

(and). This function returns the logical AND of a list of expressions. It ceases further evaluation and returns nil if any of the expressions evaluates to nil.

(equal). This function determines whether expr1 and expr2 are equal; if they are equal then it returns T, otherwise nil.

(or expr1 expr2). This function returns the logical OR of a list of expressions. The OR function evaluates the expressions from left to right looking for a non-nil expression. If one is found, OR ceases further evaluation and returns T.

We have seen that an AutoLISP program is a set of statements which are normally executed sequentially in the order in which they appear. This happens when no options or no repetitions of certain calculations are necessary. However, in practice, we have a number of situations where we may have to change the order of execution of statements based on certain conditions. This involves a kind of decision making to see whether particular conditions are met.

AutoLISP possesses such decision making capabilities and supports the following functions:

(i) if
(ii) repeat
(iii) while
(iv) cond

A brief discussion on these functions is given below:

(if testexpr thenexpr elseexpr). The (if) function evaluates the 'textexpr' and considers it as the condition, if this condition is satisfied it returns T. The next expression 'thenexpr' is then evaluated, otherwise 'elseexpr' is evaluated. The commonly used conditions are relational or logical conditions.

Example:
```
(setq a 2)
(setq b 3)
(if (= a b) (prompt "a = b") (prompt "a/= b"))
(if ( < a b)
    (progn
    (prompt "a is less than b"
    ); end of then condition
    (prompt "a is greater than b"); end else
); end if
```

(repeat). This function evaluates each expression for a fixed number of times and returns the value of the last expression.

Example:
```
(setq a 5)
(repeat 5
    (setq a (+ a 5))
    (print1 a)
)
```

(while). This function evaluates the test expression and if the test expression is not nil, it evaluates the other expression and then evaluates the test expression again. This continues until the test expression is nil.

Example:
```
(setq a 115)
(setq loop t)
```

```
(while loop
    (setq b (getreal "enter number greater than a":))
    (if (> b a) (setq loop nil))
); end while
```

(cond (test1 result1) ...). This function accepts any number of lists as arguments. It evaluates the first item in each list. It then evaluates those expressions which follow the test that succeeded and returns the value of the last expresssion in the sublist.

```
Example: (cond  ((= a "Y") 1)
                ((= a "y") 1)
                ((= a "N") 0)
                ((= a "n") 0)
         ); end cond
```

2.4.6 Other Useful Functions

In AutoLISP, manipulation of strings, files and entities is an important act. To deal with these, we have special functions described below:

String handling. Strings are text characters which are required in a program. Strings can be manipulated by the following functions:

(strcat). In AutoLISP, string constants are limited to a maximum of 132 characters. However, we can create unlimited length strings by concatenating different strings. The concatenation is done using the (strcat) function.

Example:

```
(setq a "Hello,")
(setq b "My dear")
(setq c (strcat a b)); the variable c holds the string value "Hellow, My dear".
```

(substr string start length). A long string can be broken into smaller strings by the (substr) function. It requires 'start' which is the starting point from which the substring will start. The 'length' is the number of characters we want to separate to form a substring.

```
Example: (setq a "Jodhpur is a Suncity")
         (substr a 1 7)    returns "Jodhpur"
         (substr a 14 7)   returns "Suncity"
```

(strcase). This function returns the case changed string.

(rtos). This function coverts a real number into a string.

(atoi). It is used to convert string value into an integer constant.

(atof). This function converts a string into a floating or real constant.

File handling. For a number of applications, we can create a parametric program which takes its input from a file rather than from the user. For example, if we have to draw a flange coupling, the program can be written in such a way that it accepts shaft diameter as an input from the user. Based upon this diameter, other dimensions of the flange coupling may be taken from the data file. In a parametric program, the data file can be manipulated by the following functions.

(open). This function opens a file for access by the AutoLISP I/O function. It returns a file descriptor to be used by the other I/O functions; therefore, it must be assigned to a symbol using the (setq) function.

Example: (setq f1 (open "input.dat" "r")

AutoLISP files (data) can be opened in three modes: (i) read only mode "r" (ii) write mode "w", and (iii) append mode "a".

(close). This function closes a file and returns nil. After a close of file, the file descriptor is unchanged but is no longer valid.

Example: (close f1)

(read-line). This function reads a string from the data file. If read-line encounters the end of file, it returns nil.

Example: (setq line1 (read-line f1))

(write-line). This function writes a string to the file described by the descriptor. However, the file should be opened in write mode.

Example: (Write-line "Test" f1)
The application of these functions is demonstrated through the following example programs.

Program 2.4 To draw a rectangle when the coordinates of its diagonal corners are known.

```
; program to draw a rectangle

(defun c: RECT (\ p1 p2 p3 p4)
(setq p1 (getpoint "\n specify first corner:"))
(setq p3 (getcorner p1 "\n other corner:"))
(setq p2 (list (car p3) (cadr p1))
(setq p4 (list (car p1) (cadr p3))
(command "PLINE" p1 p2 p3 p4 "c")
(princ)
); end defun
```

Program 2.5 Write a program to generate the drawing shown below.

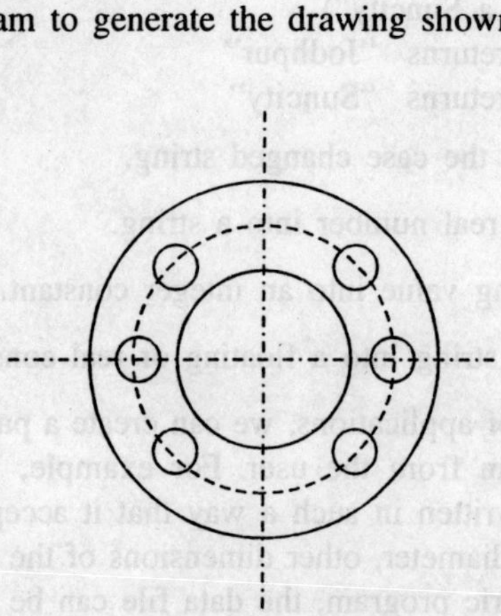

```
; program to generate end view of a flange coupling

(defun c: EFLANGE ( )
(setvar "cmdecho" 0); prevents command echo on screen
; ID = Inside dia, OD = outside dia, pcd = pitch circle dia
; HD = hole dia, N = No. of holes
(initget 1);
(setq p1 (getpoint "\n specify centre:"))
(initget 7);
(setq OD (getreal "\n specify outside dia:"))
(setq ID (getreal "\n specify inside dia:"))
(while > = ID OD
   (setq ID (getreal "\n specify inside dia:"))
); end while
; draw ID and OD circle
(command "CIRCLE" p1 "d" OD)
(command "CIRCLE" p1 "d" ID)
; input of pcd
(setq cycle t
   (while cycle
   (initget 7)
   (setq pcd (getreal "\n enter PCD:"))
      (if (> pcd OD)
         (prompt "PCD greater than OD"); then condition
         (setq x t); else condition
      ); end if
   (if (< pcd ID)
      (prompt "PCD less than ID"); then condition
      (setq y t); else condition
   ); end if
   (if (and x y)
   (setq cycle nil); then
   ); end if
); end while
; draw pcd circle
(command "CIRCLE" p1 "d" pcd)
(command "CHANGE" "L" " " "p" "LT" "centre" " ")
; hole dia input
(setq loop t)
(while loop
   (initget 7)
   (setq hd (getreal "\n hole diameter :"))
      (if (> = hd (- pcd ID))
         (prompt "\n hole cuts edge"); then
         (setq  a  t); else condition
      ); end if
```

```
        (if (< = hd (– OD pcd))
            (prompt "\n hole cuts edge")
            (setq b t); else
        ); end if
        (if (and a b)
            (setq loop nil)
        ); end if
    ); end while
    (initget 7)
    (setq n (getint "\n Number of holes:"))
    (setq hc (polar p1 (dtr 0.0) ( / pcd 2)))
    (command "CIRCLE" hc "d" hd)
    (command "ARRAY" "L" "" "p" p1 n "" "")
    (setvar "cmdecho" 1)
    (princ)
)   ; end defun
```

Program 2.6 Write a program to draw the angle section as shown below:

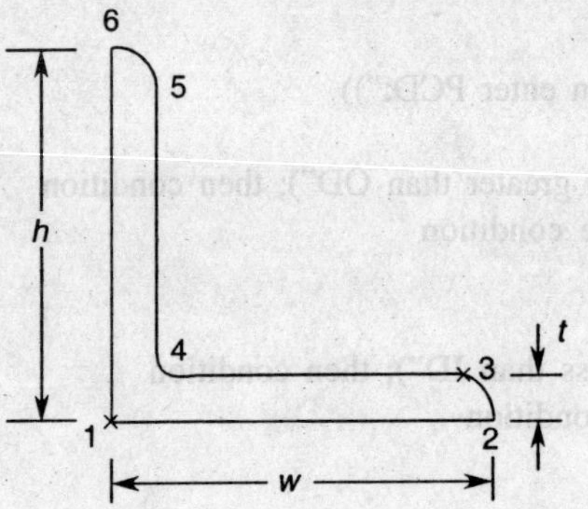


```
(defun C: ANGLE ( )
; program to draw angle section
; h = height of angle, w = width of angle
; t = thickness, r = radius
; input data through keyboard/screen
(setq h (getreal "\n height of angle:"))
(setq w (getreal "\n width of angle:"))
(setq t (getreal "\n thickness of angle:"))
(setq r (getreal "\n radius:"))
(setq p1 (getpoint "\n start point:")
; calculate the coordinates of various points
(setq p2 (polar p1 (dtr 0.0) w))
(setq p3 (polar p2 (dtr 90) t))
(setq p4 (polar p3 (dtr 180) (– w t)))
```

```
     (setq p5 (polar p4 (dtr 90) (– h t)))
     (setq p6 (polar p5 (dtr 180) t))
     ; draw angle section
     (command "LINE" p6 p1 p2 " ")
     (command "PLINE" p2 p3 p4 p5 p6 " ")
     (setq ent1 (entlast))
     (command "FILLET" "R" r)
     (command "FILLET" "P" ent1)
);   end defun.
```

Chapter 3

Engineering Materials

3.1 INTRODUCTION

After formulating the design problem, and selecting the layout of the machine and its necessary mechanisms, it becomes necessary to select the suitable material for each machine element. The selection of the material for a machine member or a structural member has always been a difficult problem for the designer. It requires a lot of experience and engineering judgement. The choice of proper materials requires the consideration of various factors, namely strength, cost, weight, size, and manufacturability. The selection of the material and the manufacturing process for the production of the component must both be considered together because of their interdependence. After chosing the material and the process, the designer decides proportions of components so that the internal stresses, strains or any other parameters such as temperature effect, corrosion, wear resistance, damping capacity, and so forth, depending upon the design criteria, remain within reasonable and satisfactory values compared with the properties of the material. Therefore, the designer should have a thorough knowledge of various properties of materials, their treatments and manufacturing processes so that an optimal design can be created.

3.2 PROPERTIES OF MATERIALS

The properties of materials used in engineering design may be classified as:

Physical and chemical properties. Chemical composition, structure, homogeneity, specific weight, melting point temperature, thermal conductivity, and coefficient of thermal expansion.

Mechanical properties. Elastic limit, modulus of elasticity, strength in tension, compression, bending and torsion, endurance limit, hardness, and wear resistance.

Technological properties. Forgeability, castability, malleability, bending, machinability, weldability, and the like.

For optimum use of engineering materials, it is very important to have a thorough knowledge of their mechanical and technological properties because ultimately these properties decide the physical dimensions of the part and the manufacturing processes to be used for the production of the machine components.

A brief review of the mechanical and technological properties is given in the following subsections.

3.2.1 Mechanical Properties

The mechanical properties are those which indicate how the material would behave when subjected to various types of loads. These properties are established by various tests. Standardized test methods are described in various standards. A good knowledge of mechanical properties permits the designer to determine the size and shape of the machine components. The typical values of properties may vary significantly depending upon the batch of production, so these should be taken as average values.

The most commonly evaluated mechanical properties are discussed below:

Isotropy. A material that displays the same elastic property in all direction is termed *isotropic*.

Elasticity. It is a property of the material by virtue of which a machine component can regain its original shape and size when the external load acting upon it is removed. The modulus of elasticity is an index for evaluation of elasticity.

Strength. The strength of a member is defined as the ability of its material to withstand yielding and/or fracture against applied forces so that it can continue to perform its designed function in the machine. The stress magnitude which corresponds to permanent deformation of a definite amount, usually up to 0.2 per cent of original gauge length, is termed *yield strength* (σ_y).

Ductility. It is a property of the material by virtue of which it can undergo a large permanent deformation in tension before being fractured. It is usually expressed as per cent elongation in 50 mm gauge length.

Malleability. It is a property of the material which permits large plastic deformation in compression without fracture. In other words, a malleable material is the one that can easily be flattened or rolled.

Brittleness. It is a characteristic opposite to ductility. A material may be considered brittle if its elongation prior to rupture in tension is less than 5 per cent in a 50 mm long specimen.

Hardness. It is the ability of the material to resist penetration, plastic indentation, abrasion, or scratching. It is also an indicator of wear resistance under certain conditions. It is usually expressed by numbers which are dependent upon the methods of testing. Mostly, two popular methods, called the Brinell and Rockwell hardness numbers, are in use.

The Rockwell (R_C) and Brinell hardness number (BHN) are related by the following relation:

$$R_C = 88(\text{BHN})^{0.162} - 192 \quad (3.1)$$

Experimental results have shown that the ultimate tensile strength and the surface endurance strength of plain carbon steel are related by the following:

$$\sigma_{ut} \approx 3.45\text{BHN N/mm}^2 \quad (3.2)$$

$$\sigma_{ens} \approx (2.75\text{BHN} - 70) \text{ N/mm}^2 \quad (3.3)$$

The hardness of synthetic materials like rubber, foam, plastic, etc. can be measured by the durometer. The durometer hardness number is based on any arbitrary scale of 0–100.

Resilience. The ability of a material to absorb energy within its elastic limit without any permanent deformation is called *resilience.*

Toughness. The ability of a material to absorb energy before it fractures is called *toughness*. The Bureau of Indian Standards (BIS) has suggested two test procedures to measure the toughness. They are (i) Charpy impact test and (ii) Izod impact test.

Creep. A material under a heavy steady load at a high temperature for a long period begins to deform plastically. This deformation, called *creep*, is time dependent and increases with time until fracture takes place.

Wear. The material on the surface keeps getting dislodged in the form of small particles because of relative motion between mating surfaces. This type of depletion of the material is called *wear*. Wear is attributed to several phenomena, namely: (a) **Scuffing**—This type of wear is caused owing to failure of the lubricant film which creates microwelds between surface asperities. These microwelds are sheared by relative motion between sliding parts, thereby dislodging small particles from surfaces. This type of wear is also called *galling* or *seizing*. (b) **Abrasion**—When hard foreign particles such as dust, metal grit or metal oxides find their way in between two rubbing surfaces, they cause abrasive wear. (c) **Pitting**—The wear caused by cyclic contact stresses between mating surfaces is called *pitting*. (d) **Frettage**—A surface damage caused by small movements between mating surfaces is known as *frettage*.

3.2.2 Technological Properties

Technological properties are those which relate to manufacturing of a machine component, namely machinability, formability, castability, etc. A designer must also have a thorough knowledge of these properties because the material selected for a component should be such that it can easily be manufactured by using the available facilities.

Machinability. The term machinability is used to specify the relative ease with which a given metal can be machined. The machinability rating of a material is usually decided on the basis of the following parameters: (i) power required, (ii) tool-life, (iii) machining time, and (iv) surface quality. Machinability of a metal depends upon its (a) hardness, (b) strength, and (c) chemical composition. The presence of sulphur, lead, and manganese in steel improves its machinability.

Formability. It is an indication of suitability of a metal for a machine part that requires forming. Ductility and tensile strength are important properties which determine the formability of a metal.

Castability. The term castability is an indication of how easy it is to cast. The important factors which affect castability are: (i) **solidification temperature**, (ii) amount of metal **shrinkage** after solidifying, and (iii) the **strength** at the temperature just below which the solidification starts.

Weldability. The term weldability is used to indicate the ease with which a successful weld can be made.

Forgeability. It is an indication of ease involved in forging a machine component. In forging, a larger plastic range is desirable because it allows a longer time to work with the metal before reheating becomes necessary.

3.3 FERROUS METALS

Ferrous metals are alloys of iron, carbon and a certain number of other alloying elements which are present either as natural impurities or as those intentionally added to improve some of the mechanical properties. In ferrous metals, the presence of carbon up to 0.008 per cent at room temperature by weight is taken as pure iron. Steel is an alloy of iron, iron-carbide and other alloying elements. The quantity of carbon in the steel is up to two per cent. Steel, like other materials, has a crystalline structure when viewed under the microscope. The alloy of iron with carbon more than two per cent is called cast iron.

3.3.1 Iron-carbon Diagram

The structure of ferrous metal is described by a number of metallurgical allotropies and these allotropies vary with the percentage of carbon in steel and cast iron and the heating temperature. These allotropies can be easily represented on an iron-carbon diagram, shown in Figure 3.1. It is an important tool which provides a complete picture of phase relation, micro-structure, temperature, and percentage of carbon.

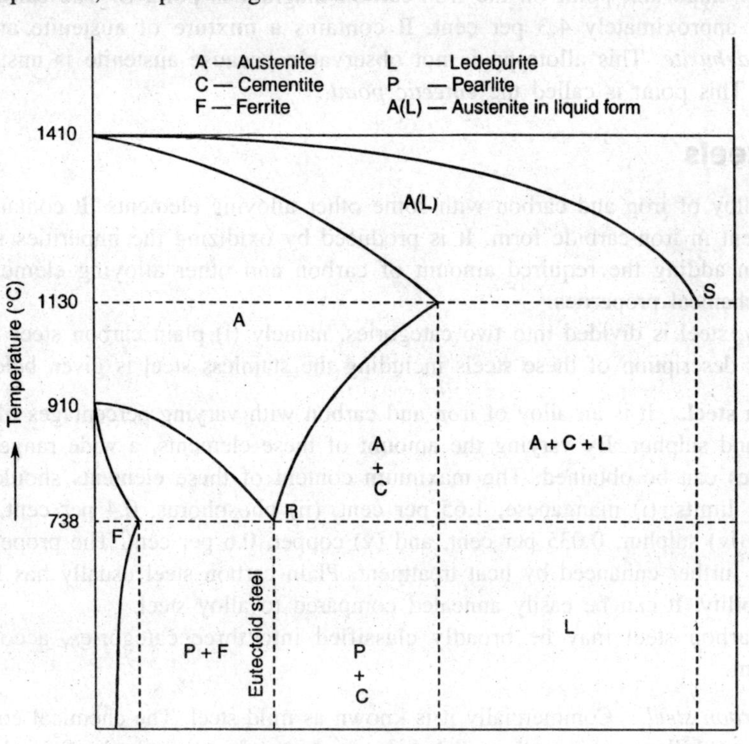

Figure 3.1 Iron-carbon diagram.

At room temperature the iron has body-centred cubic (BCC) structure and has magnetic properties. This form is called alpha (α) iron. As the temperature is increased to 738°C, the crystal structure remains the same but it loses its magnetic properties. Carbon is soluble in alpha iron to a maximum of 0.025 per cent at this temperature and only up to 0.008 per cent at room temperature. The interstitial solid solution with dissolved carbon in alpha iron is commonly called *ferrite.* It is soft, ductile, and has high tensile strength. A further increase in temperature to 910°C or higher up to 1410°C changes iron to face-centred cubic (FCC) structure. Carbon is soluble in gamma (γ) iron to a maximum of about 2 per cent at a temperature of 1130°C. The name given to this interstitial solid solution is *austenite.* At 1410°C iron again takes on the BCC structure and is called the delta (δ) iron.

Under the equilibrium conditions, the carbon is in the form of iron carbide and is called *cementite.* It is hard, brittle, weak in tension, strong in compression, and is the hardest material in the iron-carbon equilibrium diagram.

In the iron-carbon diagram, point R is of special significance, at which iron contains 0.83 per cent carbon at 738°C. This point is called *eutectoid.* It is the lowest point on the diagram at which austenite disappears when cooled slowly. The resulting structure at this point is called *pearlite.* It is a mechanical mixture of ferrite and cementite. When viewed under a microscope, the mixture appears as a laminar layer of cementite and ferrite. Iron with carbon up to 0.83 per cent is called *hypoeutectoid steel* and carbon content more than 0.83 per cent but less than 2 per cent is called *hypereutectoid steel.*

Another important point on the iron-carbon diagram is point S. The carbon content at this point is approximately 4.3 per cent. It contains a mixture of austenite and cementite, known as *ledeburite.* This allotropy is not observable because austenite is unstable at room temperature. This point is called the *eutectic point.*

3.3.2 Steels

Steel is an alloy of iron and carbon with some other alloying elements. It contains carbon up to two per cent in iron-carbide form. It is produced by oxidizing the impurities in molten pig iron and then adding the required amount of carbon and other alloying elements to obtain different mechanical properties.

Broadly, steel is divided into two categories, namely (i) plain carbon steel and (ii) alloy steel. A brief description of these steels including the stainless steel is given below.

Plain carbon steel. It is an alloy of iron and carbon with varying percentages of phosphorus, manganese, and sulphur. By varying the amount of these elements, a wide range of strengths and hardnesses can be obtained. The maximum content of these elements should not exceed the following limits: (i) manganese, 1.65 per cent, (ii) phosphorus, 0.4 per cent, (iii) silicon, 0.6 per cent, (iv) sulphur, 0.035 per cent, and (v) copper, 0.6 per cent. The properties of these steels can be further enhanced by heat treatment. Plain carbon steel usually has high strength and machinability. It can be easily annealed compared to alloy steel.

Plain carbon steel may be broadly classified into three categories, according to the carbon content.

Low carbon steel. Commercially it is known as mild steel. The chemical composition of this steel is as follows: (a) carbon 0.1–0.3 per cent, (b) manganese 0.25–1.0 per cent, (c) phosphorus and sulphur $<$ 0.05 per cent. This steel has better strength, good ductility, and machinability. It is used for a variety of small forgings, welded, cold formed and machined parts. Shallow penetrating hardening can be obtained by carburizing.

Medium carbon steel. Steel with carbon content between 0.3–0.6 per cent is usually called medium carbon steel. This steel is used for higher strengths and cold- or hot-formed parts. It can be welded. However, the welded joint should not be allowed to cool down rapidly to avoid local hardening. Heat treatment with water quenching produces a reasonable depth of hardness.

High carbon steel. The carbon content in these steels is more than 0.7 per cent. They exhibit high yield strength, hardness and brittleness. These properties make them valuable for cutting tools, hand tools, and so forth.

Alloy steel. The strength and hardness of plain carbon steel can be improved by heat treatment but at the cost of ductility and toughness. Further, this process is limited to components of small size. In order to achieve high tensile strength and hardness of large-section parts even at elevated temperatures, alloy steels are used. A plain carbon steel with varying quantities of alloying elements such as carbon, manganese, silicon, sulphur, nickel, chromium, tungsten, vanadium, etc. is termed alloy steel. These alloying elements are added not only to modify the properties of steel as follows but also to permit a greater latitude in heat treatment.

(i) To improve hardness at elevated temperatures (W, V, Cr, Mo)
(ii) To delay the rate at which austenite is transformed into pearlite upon quenching to allow sufficient time for thick section to be hardened throughout (Mn, Cr, W)
(iii) To obtain high strength at elevated temperatures (Mn, Si, Cr, Mo, Ni)
(iv) To check on grain growth in austenite (V, Al)
(v) To improve machining properties (S, P)
(vi) To obtain greater corrosion resistance (Cr, Ni)

The effects of some of the individual alloying elements are listed below:

Chromium. The addition of chromium to steel results in the formation of chromium carbide which is very hard. It alters the grain structure so that toughness of steel is increased. The addition of chromium increases the critical range of temperatures and moves the eutectoid point to the left. It also improves the wear and cutting ability of steel.

Nickel. It also shifts the eutectoid point towards the left on the iron-carbon diagram, thus lowering the carbon content in eutectoid steel. It also increases the critical range of temperatures. It is soluble in both alpha and gamma forms of iron. It increases the strength of steel without loss of ductility. Case hardening results in a better surface hardness. Fine grains are produced. The coefficient of expansion is reduced and resistance to corrosion is increased. Nickel along with chromium improves the toughness of steel.

Molybdenum. When molybdenum is used with other alloying elements, namely nickel, chromium, it forms complex carbides. It is soluble in ferrite, i.e. alpha form of iron. It increases the hardness and toughness of steel to some extent. It decreases the tendency towards temper brittleness.

Tungsten. It is widely used in tool steel as it provides hardening capability even at red hot state. It produces a fine and dense structure and imparts both toughness and hardness.

Vanadium. The addition of small percentage of vanadium forms a complex carbide with steel which improves the abrasion resistance. It is a strong deoxidizer which results in fine-grained structure.

Silicon. When silicon is added to low carbon steel, it produces a brittle material with a low hysteresis loss and high magnetic permeability. It is soluble in ferrite and also acts as a deoxidizing agent.

The mechanical properties and uses of different plain carbon and alloy steels are listed in Table 3.1.

Table 31 Properties of plain carbon and alloy steels

Designation (BIS grade)	*Tensile strength* (N/mm^2)	*Yield strength* (N/mm^2)	*BHN*	*Suggested uses*
C15	360–480	235	140	Cold worked rivets, low stressed components
C20	430–510	250	160	Same as C15
C30	490–590	300	180	Levers, rods, sprocket hubs and fasteners
C40	570–670	325	220	Shafts, crank shafts, bolts, gears
C45	620–700	350	230	Bolts, connecting rods, spindles, gears
C50	650–770	375	250	Cylinders, worm push rods
C60	750	400	260	Hardened screws, nuts, valve springs
37Mn2	600–800	400–600	170–250	Welded structures, shafts, axles, levers, bolts
40Cr1	700–1000	500–700	200–300	Connecting rods, gears, wear resisting parts
40Cr1Mo28	650–1100	500–800	200–350	Shafts, gears, bolts, studs
25Cr3Mo55	800–1200	700–900	250–350	Parts requiring high surface hardness and wear resistance
40Ni3	800–1000	600–700	220–300	Heavy forgings, turbine blades, screws, bolts, nuts
35Mn2Mo28	650–1100	500–800	200–350	Levers, bolts, crank shafts, connecting rods
35Ni1Cr60	650–1000	500–700	200–300	Components of heavy vehicles
30Ni4Cr1	1500	1200	440	Highly stressed gears, axles, aero-engine parts
40Ni2Cr1Mo28	800–1300	600–1200	300–400	High strength machine tool parts, inlet and exhaust valves

Stainless steel. An alloy steel containing at least 12 per cent chromium is called stainless steel. These steels are resistant to many corrosive conditions. Such alloys are also known as high alloy steels. Metallurgically, these steels can be classified into three types:

Ferritic chrome steels. These are alloys of iron, carbon and chromium. Their structure is ferritic and they are non-hardenable by heat treatment. Chromium makes these steels corrosion- and heat-resistant. Being ductile these steels can readily be drawn, formed or bent. Cold working operations, namely forging and rolling improve the yield strength. However, these steels exhibit poor machining properties. The chromium content in these steels is 12–18 per cent.

Martensitic chrome steels. These contain 12–24 per cent chromium, less than 3 per cent nickel and up to 1.1 per cent carbon by weight. This type of steel is also called chrome steel. By rapid cooling, it can be hardened up to 40–60 Rockwell C. Martensitic chrome steel is most commonly used in ball bearings, nozzles, needle valves, etc. requiring corrosion and wear-resistance property.

Austenitic nickel–chrome steels. Austenitic steels retain their austenite allotropy when cooled rapidly from above the transformation range temperature. These steels have higher strength than that of the ferritic chrome steels and higher ductility than that of the martensitic chrome steels. These steels are non-magnetic and have low thermal conductivity. Austenitic nickel–chrome steel can provide improved machinability, formability, weldability, and chemical resistance.

3.3.3 Casting Materials

When carbon content in iron is in the range of 2–6.67 per cent, the resulting alloy is called cast iron (CI). In commercial casting, the carbon content varies between 2–4 per cent. Cast iron is a brittle material and has low ductility; it cannot be cold worked. Different properties of cast iron can be obtained by varying the amount, type, size, and distribution of the various forms of carbon. The carbon in cast iron is generally present in two forms: (i) iron carbide and (ii) graphite as mechanical admixture. In CI, graphite is available in the form of dispersed flakes which occupy approximately 6–10 per cent of the volume. These flakes damage the continuity of the mixture to such an extent that they exert pronounced effect upon the mechanical properties.

Besides carbon, another important alloying element in cast iron is silicon. The absence of silicon keeps the carbon in the form of carbide. The presence of silicon causes the softening effect and reduces the ability of the cast iron to retain carbon in the form of carbide.

There are four primary types of cast iron: (i) grey cast iron, (ii) white cast iron, (iii) malleable cast iron, and (iv) nodular or ductile cast iron.

Grey cast iron

Grey cast iron (CI) is the most widely used of all cast irons. The reasons are its strength in compression; the ease with which it can be cast into any desired shape; the ease with which its strength and hardness can be varied; its machinability; its valuable characteristic of damping out vibrations and its relative inexpensiveness. Grey cast iron is produced by melting low-quality foundary pig, scrapped castings and coke in a furnace and then allowed to cool at a very slow rate. Upon solidification, the cementite, being unstable breaks up into austenite and graphite. The silicon acts as a graphitizer. The graphite is available in the form of irregular-shaped flakes which give the grey cast iron its grey appearance when fractured. The typical composition of grey cast iron is as follows:

carbon 2–4 per cent; silicon 1–3 per cent; manganese 0.2–1 per cent; phosphorus up to 0.8 per cent, and sulphur 0.05–0.15 per cent.

In a grey cast iron casting, approximately 0.8 per cent carbon is available in the form of carbide and the rest in the form of graphite. On account of availability of carbon in both forms, it is possible to have a wide range of properties. However, it does not have a well-defined yield point.

The Bureau of Indian Standards (BIS) in IS : 210–1978 have classified grey cast iron into seven classes and designated them by letters 'FG' followed by their tensile strength in N/mm^2, e.g. FG 150, FG 200, FG 220, FG 260, FG 300, FG 350, and FG 400. Where the chemical constituent silicon is important, then in the designation nomenclature the percentage of silicon is also specified, e.g. FG 350 Si12 means grey cast iron having ultimate tensile strength of 350 N/mm^2 and 12 per cent silicon.

Grey CI is most widely used in the casting of cylinder blocks, heads, housings, flywheels, machine tool structures such as base, bed, columns, and table. Grey CI is difficult to weld due to its tendency to crack. However, it can be welded by preheating it.

White cast iron

In white cast iron the entire carbon content is in the form of cementite and pearlite and no free carbon in the form of graphite is present. Thus, the resulting structure when viewed after fracture, is white in colour. It is, therefore, called white cast iron. The white cast iron can be produced by either of two methods:

(i) Molten grey cast iron is cooled at a very high cooling rate after pouring in the mould. The high cooling rate converts entire carbon to the form of carbide and no free carbon is left.

(ii) By adjusting the composition of grey CI in such a way that carbon and silicon contents are kept low. When the silicon, which acts as graphitizer, percentage is low, it reduces the chances of forming graphite and thus the resulting structure has carbon in the form of cementite and/or pearlite.

When the silicon content in the white cast iron is below 1 per cent, it results into very hard, brittle, and wear resistant cast iron. When chromium is added above 3 per cent, it prevents the formation of graphite and the resulting cast iron has better high temperature resistance and corrosion resistance properties.

White cast iron being hard with poor machinability, it is used in components requiring high abrasion resistance such as brake shoes, plowshaves, and rolls of rolling mills.

Malleable cast iron

When white cast iron with certain composition is heated up to 870°C for an extended period of time, approximately 40 h, depending upon the type and size of the structure and then cooled slowly, the resulting product is called malleable cast iron. The micro-structure of malleable cast iron consists of free carbon nodules and ferrite. Such iron has superior mechanical properties than those of grey cast iron, except wear resistance. A good grade malleable cast iron may have ultimate tensile strength as high as 420 N/mm^2 with elongation as much as 17 per cent. It is readily machinable and can be used for thin-section castings, gear housings, brake pedals, automobile parts, agriculture machine parts, etc.

A malleable cast iron with a small percentage of manganese can retain more carbon in the form of pearlite and the resulting structure is called pearlitic malleable cast iron. It has high ultimate tensile strength and hardness. However, the per cent elongation is relatively low.

According to IS: 210–1978, malleable cast iron may be designated by two letters followed by ultimate tensile strength, e.g.

BM 300, PM 400, WM 200

where BM, PM, and WM stand for black hearth, pearlitic, and white hearth, respectively.

Ductile cast iron

Ductile cast iron or nodular cast iron, as it is sometimes called, contains graphite in the form of spheroids. It exhibits properties very close to those of malleable cast iron. It is made by adding magnesium to the molten melt. It desulphurizes iron and upon solidification, nodular cast iron is formed. The typical composition of ductile cast iron consists of carbon 3–4 per cent, silicon 1–4 per cent, Mn 0.1–0.8 per cent, phosphorus 0.1 per cent, and Mg 0.01 per cent. The matrix of ductile cast iron is predominantly ferritic or mixture of ferrite and pearlite.

Ductile cast iron is stronger, more ductile, tougher, and less porous than grey cast iron. It has good machinability, fluidity, weldability, and wear resistance properties. Its elastic modulus is greater than that of grey cast iron. It is used for fabrication of crankshafts, pistons, cylinder heads, pulleys, etc.

According to IS: 210–1978, ductile or spheroidal nodular cast iron is designated as SG 420/12, which means spheroidal cast iron having ultimate tensile strength 420 N/mm^2 and percentage elongation 12 per cent.

3.4 BIS METHOD OF DESIGNATION OF STEEL

In practice, a large varieties of steel are used. The BIS designate various grades of steel by a system of codification, which has direct relationship with the important properties of steel, chemical composition, and manufacturability.

According to IS: 1762–1978, the code for designation of steel is based on its chemical composition supported by manufacturing characteristics. The manufacturing characteristics are represented by various symbols as described below:

(1) **Method of Deoxidation**

R — for rimming steel
K — for killed steel

If no symbol is used, it means that the steel is of semi-killed type.

(2) **Steel Quality**

Q1 — Non-ageing　　Q2 — Free flakes
Q3 — Grain size controlled　　Q4 — Inclusion controlled

(3) **Degree of Purity**

The maximum percentage of phosphorus and sulphur in steel denotes its purity. P25 means 0.025 per cent phosphorus and sulphur while no symbol means 0.055 per cent phosphorus and sulphur.

(4) **Surface Condition**

S1 — Scarfed　　S2 — Descaled
S3 — Pickled　　S4 — Shot blasted by sand
S5 — Peeled　　S6 — Bright drawn/cold rolled

If no symbol is used, it is assumed that the surface is either rolled or forged.

(5) **Surface Finish**

F1	— General purpose finish	F2	— Full finish
F3	— Exposed	F4	— Unexposed
F5	— Matt surface	F6	— Bright
F7	— Plating	F8	— Unpolished
F9	— Polished	F10	— Polished with blue colour
F11	— Polished with yellow colour	F12	— Mirror finish
F13	— Enamel finish	F14	— Direct annealed finish.

(6) **Weldability Guarantee**

W — Fusion weldability
Wp — Pressure weldability
Wr — Weldable by resistance welding
Ws — Weldable by spot welding

(7) **Formability of Sheet**

D1 — Drawing quality
D2 — Deep drawing quality
D3 — Extra deep drawing

(8) **Various Treatments**

T1	— Shot peened	T2	— Hard drawn
T3	— Normalized	T4	— Controlled rolled
T5	— Annealed	T6	— Patented
T7	— Solution treated	T8	— Solution treated and aged
T9	— Controlled cooled	T10	— Bright annealed
T11	— Spheroidized	T12	— Stress relieved
T13	— Case hardened	T14	— Hardened and tempered

No symbol means that the steel is hot rolled.

(9) When a steel is guaranteed for elevated temperatures, the letter 'H' is used.

For the purpose of designation of steels based on chemical composition, they are grouped into four categories.

(i) Plain carbon steel
(ii) Tool steel
(iii) Free cutting steel
(iv) Alloy steel

The detailed procedure for the designation of (i) plain carbon, and (ii) alloy steels is given below. However for other categories of steels, the reader may refer to the original code.

Plain carbon steel

The designation of plain carbon steel shall consist of the following in the order:

(a) A number indicating 100 times the average percentage of carbon content.
(b) Letter C.
(c) A number indicating 10 times the average percentage of manganese.
(d) Symbols indicating special characteristics.

Examples

1. 20 C 2 K T3 — A killed, normalized steel with average 0.2 per cent carbon, 0.2 per cent manganese and phosphorus, and sulphur up to 0.055 per cent.
2. 37 C 12 G Q1 — A semi-killed, non-aging quality steel having 0.37 per cent carbon, 1.2 per cent manganese, and 0.055 per cent phosphorus and sulphur. This steel is guaranteed for hardenability.
3. 20 C 5 F1 — A general-purpose finished sheet steel having 0.2 per cent carbon and 0.5 per cent manganese. This steel is of semi-killed type having commercial quality of formability.

Alloy steel

According to IS: 1762–1978, alloy steel has been further classified into four categories, namely (i) low and medium alloy steel, (ii) high alloy steel, (iii) free cutting alloy steel, and (iv) alloy tool steel. The BIS system of coding for low and medium alloy steels is given below. However, for other categories of alloy steels the reader may consult the original code.

Low and medium alloy steel. An alloy steel in which the total of the alloying elements does not exceed 10 per cent is called the low and medium alloy steel. The BIS system of designation for these steels consists of:

(a) A number indicating 100 times the average percentage of carbon.
(b) Chemical symbols of the alloying elements, each symbol followed by its percentage content multiplied by a factor. For Cr, Co, Ni, Mn, Si, and W the multiplying factor is 4 whereas for Al, Ba, V, Pb, Cu, Ti, Mo, Zr it is 10. For phosphorus, sulphur and nitrogen, the multiplying factor is 100. The numbers after multiplying are rounded to the nearest integer. The chemical symbols and their percentages are listed in order of decreasing content.
(c) Additional symbols indicating special characteristics, namely purity, hardenability, etc.

Examples

1. 35 Cr 6 Mo 2 G — Alloy steel with guaranteed hardenability, having carbon content 0.35 per cent, 1.5 per cent chromium, and 0.2 per cent molybdenum.
2. 40 Cr 8 Al 1 Mo 2 W — A hot-rolled steel with guaranteed fusion weldability having 0.4 per cent carbon, 2 per cent chromium, 0.1 per cent aluminium, and 0.2 per cent molybdenum.

3.5 HEAT TREATMENT OF STEEL

Heat treatment is defined as heating and cooling of a metal alloy to alter or induce certain desirable mechanical properties into it. It is generally employed for the following purposes:

(i) To improve mechanical properties, e.g. strength, ductility, hardness, etc.
(ii) To produce a hard surface on a ductile core
(iii) To increase resistance to wear, heat and corrosion

(iv) To change or refine grain size
(v) To improve machinability

The most commonly used operations of heat treatment are (a) annealing, (b) normalizing, (c) hardening, and (d) tempering.

A brief description of these operations is given below:

Annealing

According to the American Society of Material Testing (ASMT), annealing is defined as the softening process in which the hypoeutectoid steel is heated 20°C above the upper critical temperature and the hypereutectoid steel is heated 20°C above the lower critical temperature (Figure 3.2). This heated steel is then allowed to cool slowly at the rate of 30°C to 150°C per hour, depending upon the size of the component, below the transformation range in the furnace itself.

The steel upon cooling changes to ferrite and pearlite for hypoeutectoid steel, pearlite for eutectoid steel, and pearlite and cementite for hypereutectoid steel. The basic objectives of annealing are the following:

(i) To soften the metals and increase their ductility
(ii) To refine the grain size due to phase recrystallization
(iii) To improve machinability
(iv) To relieve internal stress

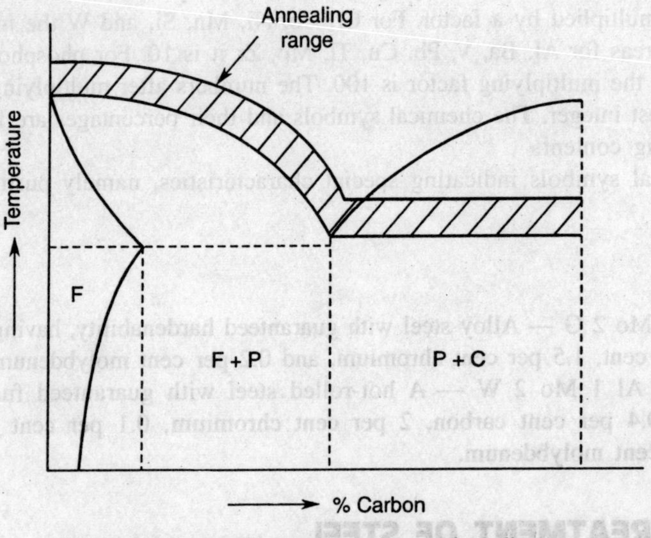

Figure 3.2 Annealing range on iron-carbon diagram.

Normalizing

It is frequently applied as the final heat-treatment process on items which are subjected to relatively high stresses. The parts subjected to normalized treatment have higher yield strength but the ductility is somewhat reduced. According to ASMT, it is defined as the process in which steels are heated to 40–50°C above the upper critical temperature range and held there for a specified time period and this is then followed by cooling in still air at room temperature. In normalizing both hypoeutectoid and hypereutectoid steels, they are heated above the critical temperature as shown in Figure 3.3. The normalized steel consists of ferrite and pearlite for hypoeutectoid steel and pearlite and cementite for hypereutectoid steel. The main objectives of normalizing are:

(i) To increase strength and reduce internal stresses
(ii) To improve machinability
(iii) To eliminate coarse grain structure obtained during forging, rolling, and stamping

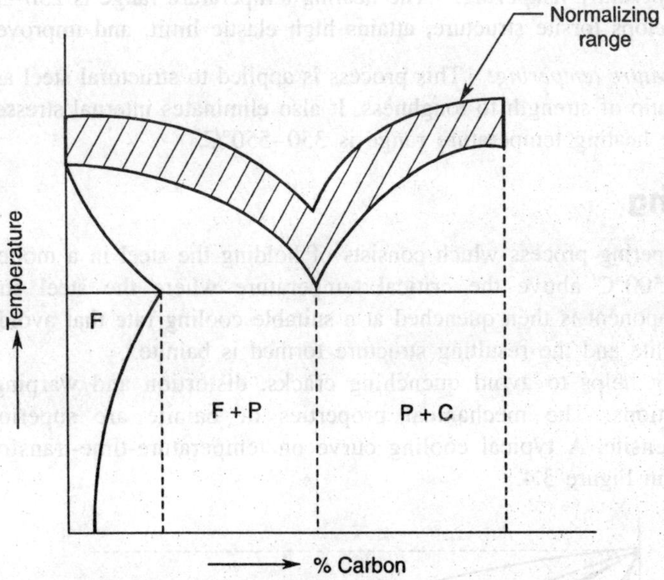

Figure 3.3 Normalizing range on iron-carbon diagram.

Hardening

According to ASMT, hardening is defined as the heating and cooling process in which steel is heated to 20°C above the upper critical temperature in the case of hypoeutectoid steel and to 20°C above the lower critical point in the case of hypereutectoid steel. The steel is soaked at this temperature for a considerable period of time to ensure thorough penetration of heat inside the component, followed by continuous cooling at room temperature through quenching in water, oil, or brine solution.

On heating the steel, the ferrite and pearlite structure of hypoeutectoid steel and pearlite and cementite structure of hypereutectoid steel is transformed into austenite and cementite. Upon cooling at critical rate (about 200°C/min) the austenite is changed to martensite, which is the super-saturated solution of carbon in α-iron and is very hard. The hardness in the steel is produced due to this microstructure.

Tempering

It is an operation used to modify the properties of steel hardened by quenching. According to ASMT, it is a reheating process in which the component is reheated below the critical temperature and then allowed to cool so that the trapped martensite is transformed and the internal stresses relieved.

As per the requirement of applications, the tempering temperature is varied between 200–550°C and accordingly classified into three categories:

Low temperature tempering. Here the heating temperature range is up to 200°C. The reheating microstructure remains martensite but its brittleness is reduced.

Medium temperature tempering. The heating temperature range is 250–350°C. The steel on tempering develops torsite structure, attains high elastic limit, and improved toughness.

High temperature tempering. This process is applied to structural steel as it provides the most favourable ratio of strength to toughness. It also eliminates internal stresses produced due to quenching. The heating temperature range is 350–550°C.

Austempering

It is a special tempering process which consists of holding the steel in a molten salt bath at a temperature 250–500°C above the critical temperature where the steel structure is pure austenite. The component is then quenched at a suitable cooling rate that avoids the formation of ferrite and pearlite and the resulting structure formed is bainite.

Austempering helps to avoid quenching cracks, distortion and warping of small and delicate cross-sections. The mechanical properties of bainite are superior to those of conventional martensite. A typical cooling curve on temperature-time-transformation (TTT) diagram is shown in Figure 3.4.

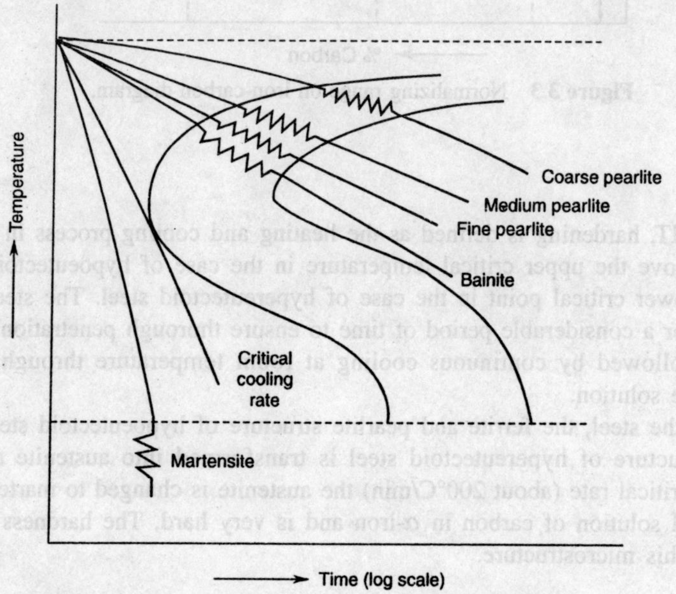

Figure 3.4 Temperature-Time-Transformation (TTT) diagram.

Martempering

In this process, steel is heated above the critical temperature and then suddenly quenched in a molten salt bath at the temperature of 200–300°C. The component is held in the bath until the core and skin temperatures are equalized. Martempering produces martensite in the steel but with minimum residual stresses and distortion.

3.6 NON-FERROUS METALS

In engineering applications, besides steel and cast iron, non-ferrous metals are also used extensively because of their specific properties such as light weight, low coefficient of friction, good thermal and electrical properties, and ease of fabrication. In engineering design, aluminium and copper are the most widely used metals. A brief description of these metals is given below.

Aluminium

Aluminium and its alloys are the most important materials after iron-base alloys. These materials are well-known for their lightness, high strength to weight ratio, corrosion resistance, ductility, and electrical and thermal conductivities.

Pure aluminium has low tensile strength, which is approximately 90 N/mm^2. However, it can be improved by cold working methods. The stiffness of aluminium is about one-third that of steel. Aluminium can be worked by sand casting, die-casting, extrusion, rolling and spinning. Aluminium in pure form is difficult to machine, however, its alloys can be easily machined, soldered, brazed, and welded. The melting point of aluminium is low, about 650°C, compared to that of steel which makes it suitable for casting. Commercially, aluminium is available in the form of plates, rods, bars, tubes, sheets, foils, wires, and other structural shapes.

Aluminium alloys are classified as casting alloys and wrought alloys depending upon the type of alloying elements used and their percentage. In aluminium alloys, the major constituents are copper, silicon, magnesium, iron, etc. The effect of copper as an alloying element is to raise the ultimate strength and endurance limit and to improve the casting and machinability characteristics. Aluminium–silicon alloys have better corrosion resistance and mechanical properites but poorer machinability than Al–Cu alloys. Chemical composition, properties and typical uses of aluminium alloys are given in Table 3.2.

Table 3.2 Characteristics of some aluminium alloys

Composition in per cent	*Characteristics*	*Typical uses*
5.0 Si	Good castability, weldability and corrosion resistance	Carburettor body, brackets, fittings, cooking utensils, etc.
7.0 Cu, 1.7 Zn	Good castability and machinability	General-purpose casting of housings, brackets, etc.
7.0 Cu, 2.0 Si, 1.7 Zn	Provides better strength	Hydraulic brake, pistons, cylinder heads, etc.

(contd.)

Table 3.2 (*contd.*)

Composition in per cent	*Characteristics*	*Typical uses*
10.0 Cu, 0.2 Mg	Retains strength at elevated temperatures, good hardness and wear resistance	Meter parts, bushings, bearing caps, etc.
0.8 Cu, 0.8 Fe 12 Si, 1.0 Mg, 2.5 Ni	Low coefficient of expansion, good machinability and weldability	Automotive pistons
4.0 Cu, 1.5 Mg, 2.0 Ni	Good strength at high temperatures	Aircraft generator housing, motorcycle, diesel engine and aircraft parts
4.5 Cu, 2.5 Si	Good casting and machinability	Aircraft fittings, gun control systems, connecting rods, etc.
3.5 Cu, 9.0 Si	Low coefficient of thermal expansion	Gas meters, regulator parts, I.C. engine cylinder heads, etc.

Copper

Copper is a soft and ductile metal which has very high electrical and thermal conductivities. Being ductile, it can be stamped, rolled into sheets and drawn into wires and tubes. The hardness and strength of copper can be increased by cold working but at the cost of ductility. Copper is resistant to certain types of corrosions and finds extensive use in electrical, refrigeration, air-conditioning, and other heat transfer equipments.

When copper is alloyed with zinc, it is called brass. If it is alloyed with tin, the resulting alloy is termed bronze. Commercially, hundreds of varieties of brass and bronze are available in the market. The properties and typical uses of some of the brasses and bronzes are listed in Table 3.3.

Table 3.3 Characteristics of some copper alloys

Composition in per cent	*Characteristics*	*Typical uses*
Gliding brass—5.0 Zn, 4–6 Pb, 0.75 Ni	Easy cold workability ductile, difficult to machine	Jewellery purposes
Commercial brass—1.0 Zn, 4 Sn, 5.0 Pb	Good cold workability, high ductility	Jewellery and utensils
Red brass—15 Zn	Forged, stamped, high strength at elevated temperatures, corrosion resistance, cheap and good machinability	Water pump impellers, bushings, electrical components, etc.
Low brass—Zn 20	Less costly, poor corrosion resistance, possibility of skin cracking	Deep-drawn parts
Cartridge—Zn 30	Corrosive, maybe season-cracked	Hardware and gears

(*contd.*)

Table 3.3 (*contd.*)

Composition in per cent	*Characteristics*	*Typical uses*
Yellow brass—Zn 35	Ductile, good thermal conductivity	Radiator parts, tubes fabrication
Naval brass—Zn > 35 + Fe, Pb	Corrosion resistance, high strength and toughness	Propeller shafts and valves of marine installations
Phosphor bronze—P 0.3–1, Sn 2.8 + Cu	Very hard and tough, high tensile strength	Worm gears, nuts, springs, bushings, etc.
Aluminium bronze—Al 12, Sn 0.5, Cu + impurities	High strength, corrosion resistance, resistance against fatigue, good cold working, high endurance limit	Worm gears, forged items, valves, guides, etc.
Manganese bronze—Zn 34, Sn 1.5, Mn 3.5, Al 1, Fe 2 + Cu	High strength and toughness	Automobile parts
Gun metals—Sn 10, Pb 2.5, Zn 1 + Cu	Good machinability, resistance to corrosion by water and atmosphere	Bearings, bushes, etc.

3.7 PLASTICS

Plastics are synthetic organic compounds which show diverse properties. There are hardly any areas of modern civilization where they are not used. A designer's interest in plastics concerns their use as structural material as well as their use for special applications. Basically, plastics are compounds of carbon, hydrogen, and oxygen. They are linked together in saturated or unsaturated compounds (Figure 3.5). Ethylene and acetylene are unsaturated compounds whereas ethane and methane belong to saturated organic compounds. When heat, pressure and suitable catalysts are used, the unsaturated bonds of a molecule allow additional similar molecules to join and form a long chain. This process is called *polymerization.* The resulting molecule is called the *polymer.* The small chain molecules are called monomers. A large number of other polymers can be produced by replacing the hydrogen or linking two or more different monomers. When two or more polymers are cross-linked, the process is called *copolymerization.* The resulting plastics are called *thermosetting plastics.* These plastics

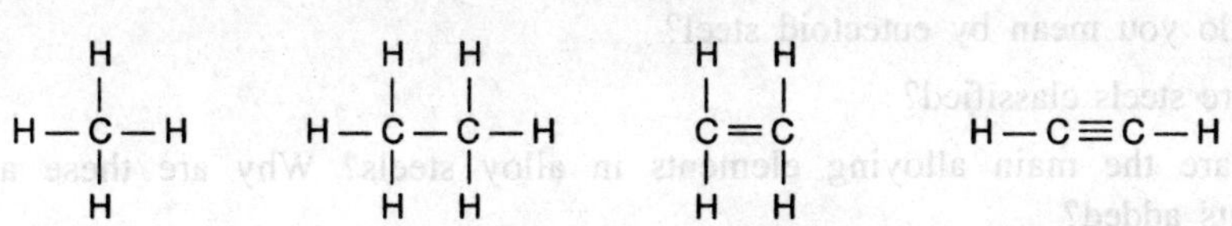

methane   ethane   ethylene   acetylene

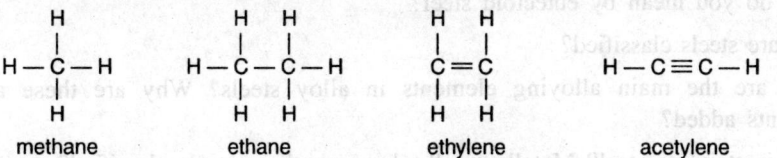

Figure 3.5 Saturated and unsaturated molecules.

experience a chemical change upon the application of heat and pressure, and once formed these plastics cannot be resoftened. Polymers that are not cross-linked are called *thermoplastics.* These plastics become soft on heating and hard on cooling. This cycle of heating and cooling can be repeated without destroying the properties of the material.

Thermosetting and thermoplastic resins when mixed with different fillers, namely paper, cotton, asbestos, fibre glass, result in two groups of plastic materials known as *reinforced plastics* and *laminates.* These plastics possess special characteristics, i.e. high toughness, wear resistance, low dielectric losses, etc. The typical advantages offered by plastics are the following:

1. Ease of shaping and hence low cost of manufacturing
2. Unlimited range of properties
3. High thermal and electrical insulation
4. Corrosion resistant
5. Light weight
6. Greater scope of colour designs through pigmentation
7. Can be electroplated to obtain metal finish.

The general uses of some of the plastic materials are given in Table 3.4.

Table 3.4 Typical uses of plastics

Name of the plas	*Uses*
Textolite laminated fabric	Gear wheels, machine tool slideways, pulleys, bearings, liners
Wood laminate	Pulleys, bearings, large-sized bearings
Farlite	Pipes to convey chemical fluid
Polythylene	Pipes
Compressed wood plastics	Bearing materials, pipes, hand rails
Fibre glass	Hulls of ships, boats, automobile bodies
Ethylene polymers	Electric and radio parts, packings, pipes, and valves

EXERCISES

1. List the factors required to be taken into account for the selection of materials.
2. Define the following terms:

   (a) Ductility (b) Malleability (c) Brittleness
   (d) Resilience (e) Toughness (f) Machinability
   (g) Formability (h) Castability (i) Weldability
   (j) Forgeability
3. Discuss the iron-carbon diagram and various allotropies of steel.
4. What do you mean by eutectoid steel?
5. How are steels classified?
6. What are the main alloying elements in alloy steels? Why are these alloying elements added?
7. What is stainless steel? Metallurgically, how are these steels classified?

8. What is the difference between grey cast iron and white cast iron?
9. What do you mean by white hearth and black hearth cast iron?
10. Discuss the BIS method of designation of steels.
11. Define the following terms:

(a) Annealing (b) Normalizing
(c) Hardening (d) Tempering

12. What do you understand by the term plastic?
13. What is the difference between the terms thermosetting and thermoplastic?
14. What are the advantages of plastics over metals?

MULTIPLE CHOICE QUESTIONS

1. In compression, a piece of brittle material will break

(a) by forming a bulge (b) by shearing along the oblique plane
(c) by being crushed into pieces (d) in the direction of the applied load

2. Malleability of a material is

(a) the ability to undergo a large permanent deformation
(b) the ability to recover its original form
(c) the ability to absorb energy
(d) none of the above

3. Killed steels are those steels

(a) which are destroyed by burning
(b) which can be recycled
(c) in which carbon content is burnt
(d) which are deoxidized with silicon

4. The hardness of steel depends upon

(a) its carbon content
(b) its shape and distribution of carbide
(c) its contents of the alloying elements
(d) all of above factors

5. Stainless steel is resistant to corrosion due to

(a) chromium and nickel (b) sulphur and phosphorus
(c) vanadium and aluminium (d) tungsten

6. The depth of hardness of steel is increased by the addition of

(a) nickel (b) chromium
(c) tungsten (d) vanadium

7. The machining properties of steel are improved by adding

(a) silicon and aluminium (b) sulphur and lead
(c) chromium and nickel (d) tungsten

8. Large forgings, crankshafts, and axles are normally made of steel having carbon content
   (a) 0.05–0.2 per cent (b) 0.2–0.4 per cent
   (c) 0.4–0.55 per cent (d) 0.55–1.0 per cent
9. Taps, dies, and drills contain carbon
   (a) below 0.5 per cent (b) below 1 per cent
   (c) above 1 per cent (d) above 2.2 per cent
10. The tensile strength of wrought iron is
   (a) maximum along the lines of grain distributions
   (b) maximum perpendicular to grain distributions
   (c) uniform in all directions
   (d) unpredictable
11. Brass contains:
   (a) 70 per cent Cu and 30 per cent Zn
   (b) 90 per cent Cu and 10 per cent Al
   (c) 80 per cent Cu and 20 per cent Sn
   (d) 70 per cent Zn and 30 per cent Cu
12. Bronze contains:
   (a) 70 per cent copper and 30 per cent zinc
   (b) 90 per cent copper and 10 per cent tin
   (c) 90 per cent copper and 10 per cent aluminium
   (d) 70 per cent copper and rest tin
13. Heat treatment involving heating of steel above the upper critical temperature and then cooling it in air is known as
   (a) annealing (b) tempering
   (c) normalizing (d) austempering
14. Plastic is
   (a) an organic compound (b) an inorganic compound
   (c) a mixture (d) none of the above
15. Polymer is a
   (a) short chain molecule (b) long chain molecule
   (c) broken chain molecule (d) molecule without chains
16. The weight to strength ratio of plastics compared to steel is
   (a) low (b) high
   (c) equal (d) unpredictable

Chapter 4

Mechanics of Machine Elements

4.1 INTRODUCTION

The main objective of any machine design process is that the machine should function properly to satisfy the needs of the customer and it should be safe against the predicted modes of failure. One of the major difficulties that arises in designing a machine or its components is the selection of an appropriate analytical model which represents the actual behaviour of the proposed design and the predicted system of loading. This requires considerable engineering judgement because if the model selected is not representative of actual conditions, any further analysis would be misleading and a misendeavour. In real life situations, the analytical model is always a compromised attempt to idealize the system by assuming suitable assumptions so that a reasonably simple solution is achieved.

Here for the purpose of review and ready reference some basic topics of strength of materials, which are quite frequently used by the designer, are presented in this chapter.

4.2 LOAD AND STRESS

An engineering design is built up of a number of elements which are in equilibrium under the action of external forces and their reactions at the supports. This pair of forces (the external force and its reaction) is shared by individual elements of the structure and constitutes the load on the element. In other words, the external forces grouped together constitute what is called the *load*, when acting upon a body. In mechanical members these pair of forces try to displace the molecules from their natural equilibrium position. These molecules are bonded by the internal forces of cohesion which offer resistance to balance the external forces. The resisting force per unit area of the element is known as *stress*. When the machine parts are subjected to various types of forces, the different types of stresses are produced. The total influence of the external forces may give rise to three states of stress in the element which are:

(i) Uniaxial stress
(ii) Biaxial stress
(iii) Triaxial stress

4.2.1 Uniaxial Stress

When a bar of given cross-section is subjected to unidirectional load F at the ends, as shown in Figure 4.1, the intermolecular resistance offered by it is called the *uniaxial stress*. The

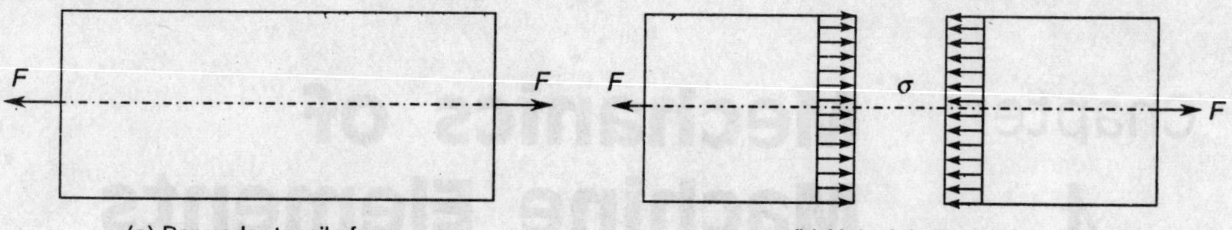

Figure 4.1 A bar subjected to uniaxial stress.

uniaxial stress means that if we cut the bar at any section, we find that the cross-section of the bar is subjected to uniformly distributed stresses of magnitude σ along the line of action of the external force. This stress is called pure tension, compression, or shear stress depending upon the nature of load.

The unidirectional tensile or compressive stress is calculated from the following equation:

$$\sigma = \frac{F}{A} \tag{4.1}$$

where

σ is the unidirectional stress (N/mm^2)

F is the applied force (N)

A is the area of cross-section (mm^2)

When a member is subjected to two equal and opposite forces acting tangentially across the resisting section of the member in such a way that its one layer tends to slide over the other, the member is said to be in the state of *shear*. The resulting stress on this layer is called the *direct shear stress*. For example, two plates held together by means of a rivet, as shown in Figure 4.2, are in the state of shear. The average shear stress produced in the rivet is given by

$$\tau = \frac{F}{A} \tag{4.2}$$

where τ is the shear stress (N/mm^2).

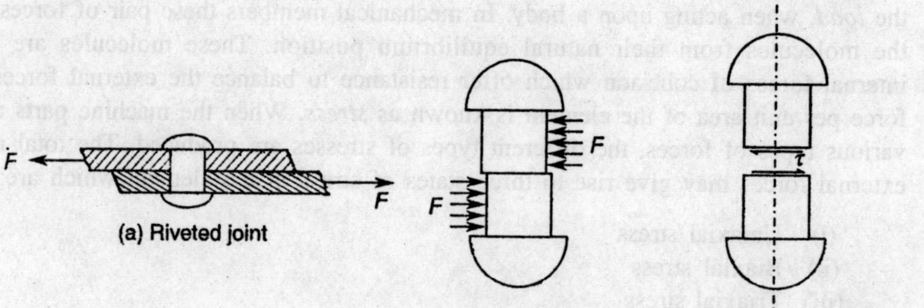

Figure 4.2 A rivet under the state of shear force.

4.2.2 Biaxial Stress

When a component is subjected to different forces in such a way that stresses produced act on two planes perpendicular to each other and there is no stress on the third plane which is perpendicular to these two planes, the above state of stress is then called *biaxial stress.* Figure 4.3 shows a general two-dimensional or biaxial stress element having two normal stresses σ_x, σ_y, all positive, and two shear stresses, τ_{xy} and τ_{yx}. The element is in static equilibrium and hence τ_{xy} is equal to τ_{yx}. The sign convention followed is that outwardly directed normal stresses are tensile or positive. The shear stress is positive if its direction is clockwise. The shear stress acts parallel to the axis of the second subscript, while the first subscript indicates the coordinate normal to the element face.

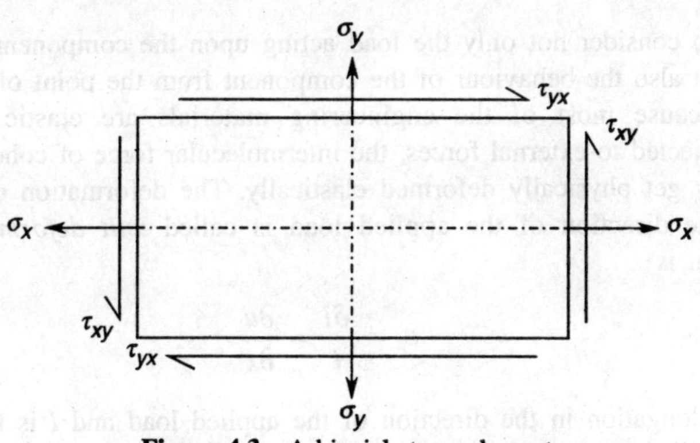

Figure 4.3 A biaxial stress element.

4.2.3 Triaxial Stress

When a cubical element of a deformable body is under the action of external forces, a stress would act on each of its six faces. If these stresses are resolved into normal and tangential components to each of the faces, in all nine stresses will act on the element and it is said to be a triaxial stress element. A triaxial stress element, as shown in Figure 4.4, consists of three

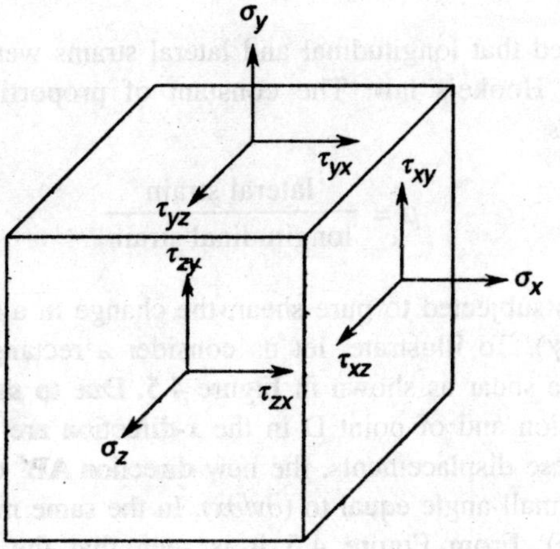

Figure 4.4 A triaxial stress element.

normal stresses σ_x, σ_y, and σ_z and six shear stresses τ_{xy}, τ_{yx}, τ_{yz}, τ_{zy}, τ_{zx}, τ_{xz}, all positive. If the element is in equilibrium, then $\tau_{xy} = \tau_{yx}$, $\tau_{xz} = \tau_{zx}$, and $\tau_{yz} = \tau_{zy}$. The nine stress components at a point on the element can be represented by a second-order tensor T, i.e.

$$T = \begin{bmatrix} \sigma_x & \tau_{xy} & \tau_{xz} \\ \tau_{yx} & \sigma_y & \tau_{yz} \\ \tau_{zx} & \tau_{zy} & \sigma_z \end{bmatrix} \tag{4.3}$$

4.3 ELASTIC STRAIN

A designer has to consider not only the load acting upon the component and thereby stresses produced in it but also the behaviour of the component from the point of view of deformation or deflection because most of the engineering materials are elastic in nature. When a component is subjected to external forces, the intermolecular force of cohesion tries to resist its failure but it may get physically deformed elastically. The deformation of the component per unit length in the direction of the applied load is called *unit deformation* or *strain*. The equation for strain is

$$\varepsilon_x = \frac{\delta l}{l} = \frac{\partial u}{\partial x} \tag{4.4}$$

where δl is the elongation in the direction of the applied load and l is the original length of the member in the same direction.

Further, it is also observed from the experimental evidence that when a member is strained in one direction, say in the x-direction, its lateral dimensions are also changed. Thus, there will be straining in other direction too, i.e. y and z directions. The strains in these directions would be the same for isotropic materials. These lateral strains are:

$$\varepsilon_y = \frac{\partial v}{\partial y} \quad \text{and} \quad \varepsilon_z = \frac{\partial w}{\partial z} \tag{4.5}$$

Poisson demonstrated that longitudinal and lateral strains were proportional to each other within the range of the Hooke's law. The constant of proportionality is expressed as the Poisson's ratio, defined as

$$\mu = \frac{\text{lateral strain}}{\text{longitudinal strain}} \tag{4.6}$$

When an element is subjected to pure shear, the change in a right angle of the element is called the *shear strain* (γ). To illustrate, let us consider a rectangular stress element ABCD, which is subjected to pure shear as shown in Figure 4.5. Due to shear force, the displacements of point B in the y-direction and of point D in the x-direction are $(\partial v/\partial x) \cdot dx$ and $(\partial u/\partial y) \cdot dy$, respectively. Owing to these displacements, the new direction AB′ of the face AB is inclined to the initial direction by a small angle equal to $(\partial v/\partial x)$. In the same manner AD′ is inclined to AD by a small angle $(\partial u/\partial y)$. From Figure 4.5 it is seen that the initial right angle BAD is

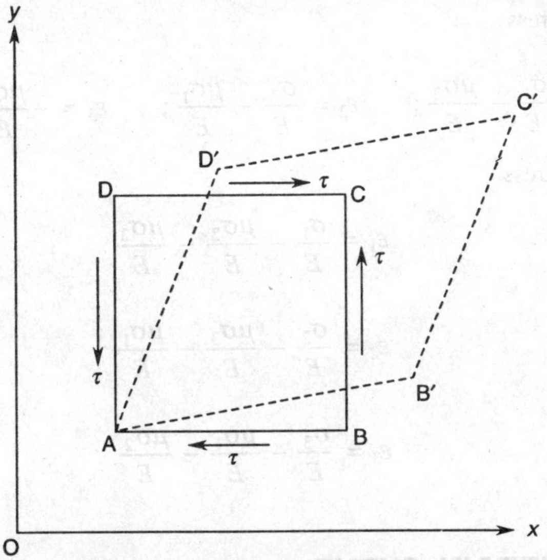

Figure 4.5 Element under pure shear.

diminished by the angle $(\partial u/\partial y) + (\partial v/\partial x)$. This is called the *shear strain* between the planes xz and yz and is written as

$$\gamma_{xy} = \frac{\partial u}{\partial y} + \frac{\partial v}{\partial x} \tag{4.7}$$

Similarly for the three-dimensional element, the shearing strain between the other planes can be obtained as

$$\gamma_{xz} = \frac{\partial u}{\partial z} + \frac{\partial w}{\partial x}; \qquad \gamma_{yz} = \frac{\partial v}{\partial z} + \frac{\partial w}{\partial y} \tag{4.8}$$

In an elastic material the stress and strain are governed by the Hooke's law, which states that within certain limits the stress in a material is proportional to strain. Thus, we can write

$$\sigma = E\varepsilon \tag{4.9}$$

and

$$\tau = G\gamma \tag{4.10}$$

where E and G are constants of proportionality and are called *modulus of elasticity* and *modulus of rigidity*, respectively. These constants are indicative of stiffness or rigidity of the material. The modulus of elasticity E, modulus of rigidity G, and Poisson's ratio μ are related by the following equation:

$$E = 2G(1 + \mu) \tag{4.11}$$

The relations between stress and strain for all three states of stresses are given as:

(i) Unidirectional stress

$$\varepsilon_1 = \frac{\sigma_1}{\varepsilon}; \qquad \varepsilon_2 = -\mu\,\varepsilon_1; \qquad \varepsilon_3 = -\mu\,\varepsilon_1 \tag{4.12}$$

(ii) Biaxial stress

$$\varepsilon_1 = \frac{\sigma_1}{E} - \frac{\mu\sigma_2}{E}; \qquad \varepsilon_2 = \frac{\sigma_2}{E} - \frac{\mu\sigma_1}{E}; \qquad \varepsilon_3 = -\frac{\mu\sigma_1}{E} - \frac{\mu\sigma_2}{E} \tag{4.13}$$

(iii) Triaxial stress

$$\varepsilon_1 = \frac{\sigma_1}{E} - \frac{\mu\sigma_2}{E} - \frac{\mu\sigma_3}{E}$$

$$\varepsilon_2 = \frac{\sigma_2}{E} - \frac{\mu\sigma_3}{E} - \frac{\mu\sigma_1}{E}$$

$$\varepsilon_3 = \frac{\sigma_3}{E} - \frac{\mu\sigma_1}{E} - \frac{\mu\sigma_2}{E} \tag{4.14}$$

4.4 STRESS–STRAIN CURVE

The relation between stress and strain, within the elastic limit, is defined by the Hooke's law which states that stress is proportional to strain. However, beyond the elastic limit the relation is not linear.

Figure 4.6 shows the stress–strain relationship for a ductile material, namely plain carbon steel. In this diagram, the stress calculated on the basis of the original area of cross-section of

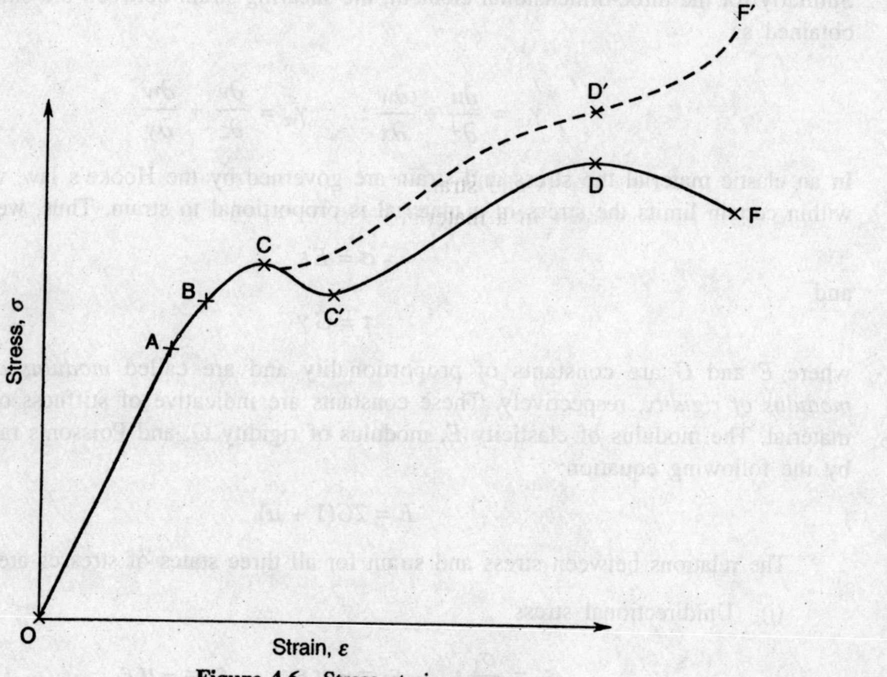

Figure 4.6 Stress–strain curve.

the test specimen is plotted against axial or longitudinal strain. The first part of the curve OA represents *limits of proportionality*. Further increase in load/stress may increase strain more rapidly and is represented by the curve between A and C. The point B is known as the *elastic limit* of the material. The elastic limit represents the maximum value of the stress developed during the tensile test without permanent deformation. For many materials, the elastic and the proportional limits are the same.

The point C on the curve is called the *yield point*. It is a point where yielding of the material begins. The phenomenon of yielding for the plain carbon steel is very peculiar and it exhibits two yield points C and C′. The point C is known as the higher yield point and C′ the lower yield point. The lowest stress associated with the yield point is called the *yield strength*. For a brittle material where there is no well-defined yield point, the yield strength is determined by the offset method. In this method, a line is drawn parallel to the straight portion of the stress–strain curve, offset by 0.2 per cent of strain. The stress corresponding to the intersection of the offset line with the stress–strain curve is known as the yield strength and is denoted by σ_y (see Figure 4.7).

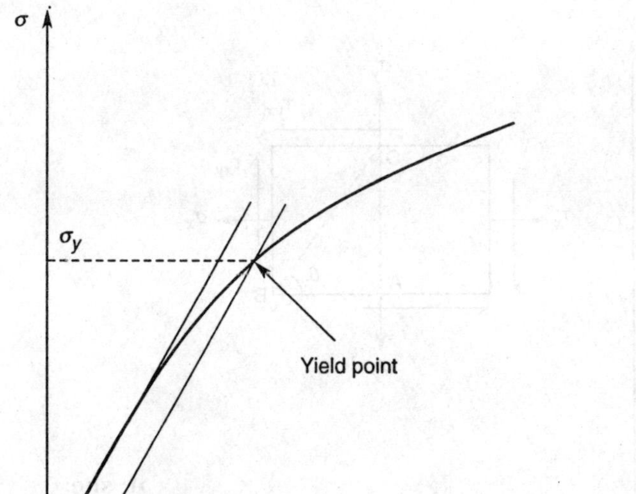

Figure 4.7 Offset method to find yield strength.

The nominal stress corresponding to point D on the stress–strain curve, which is the maximum stress a material can withstand without rupture, is called the *ultimate strength* and is denoted by σ_u. Up to the ultimate strength, the test specimen extends uniformly over its length but if straining is continued, a local reduction in cross-section may occur which may lead to ruputre failure, called the *rupture strength.*

The dashed portion of the curve, shown in Figure 4.6, is called the actual stress–strain curve. The actual stress is computed as load divided by the reduced or actual area of cross-section. The actual stress intensity at ultimate and rupture points is greater than the nominal strength.

4.5 PRINCIPAL STRESSES

In the design of machine components, the uniaxial state of stress is rarely used because most of the components are subjected to a combination of loads such as axial load, bending moment, and twisting moment. The biaxial or triaxial state of stress is the most common condition on the surface of the component. Therefore, the most important problem for the designer is to relate the biaxial or triaxial state of stress and strength of material to achieve sufficient margin of safety.

Let us consider a biaxial stress element as shown in Figure 4.8, in which shear stress $\tau_{xy} = \tau_{yx}$. This stress element is cut by an oblique plane inclined at an angle θ to the x-axis; the revised coordinate axes system taken as x'–y' is rotated by angle θ from the original x–y axes. By passing a plane BC normal to the x'-axis, an element ABC is isolated as shown in Figure 4.8(b). Let the area of plane BC be dA. The areas of planes AC and AB are $dA \sin\theta$ and $dA \cos\theta$, respectively. For static equilibrium of the stress element ABC, the sum of the horizontal and vertical forces along the x' and y' axes should be zero, i.e.

$$\Sigma F_{x'} = 0$$

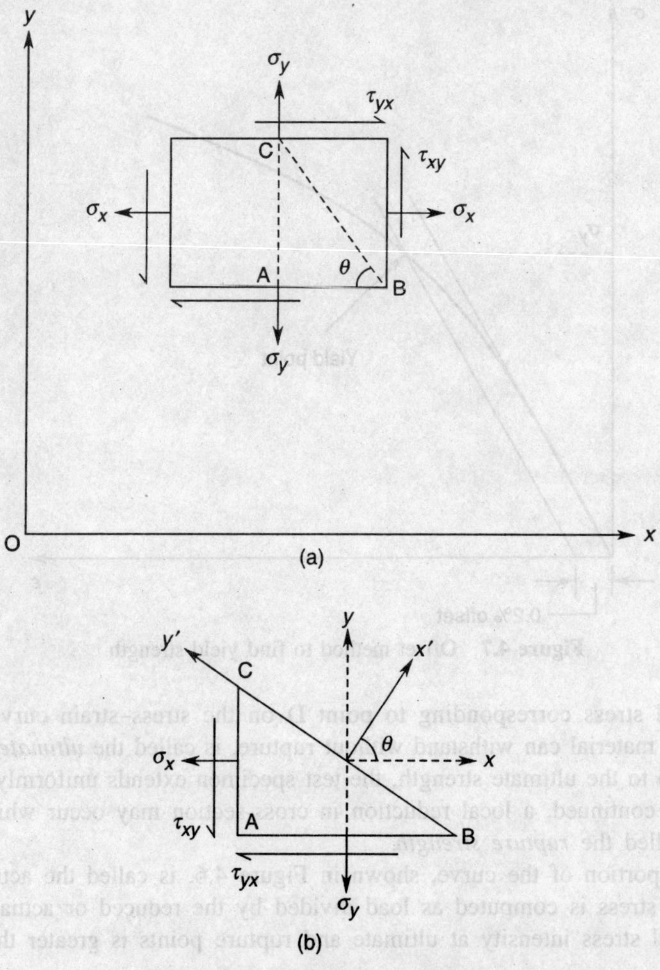

Figure 4.8 Biaxial stress element.

Therefore,

$\sigma dA = \sigma_x\, dA \cos\theta \cdot \cos\theta + \sigma_y\, dA \sin\theta \cdot \sin\theta + \tau_{xy}\, dA \cos\theta \cdot \sin\theta + \tau_{xy}\, dA \sin\theta \cdot \cos\theta$

or

$$\sigma = \frac{\sigma_x + \sigma_y}{2} + \frac{\sigma_x - \sigma_y}{2} \cos 2\theta + \tau_{xy} \sin 2\theta \qquad (4.15)$$

Similarly, for $\Sigma F_{y'} = 0$, we get

$$\tau = -\left(\frac{\sigma_x - \sigma_y}{2}\right) \sin 2\theta + \tau_{xy} \cos 2\theta \qquad (4.16)$$

Equations (4.15) and (4.16) are general relations for normal and shear stresses acting on any plane located at an angle θ.

To locate the principal planes, which are planes of maximum or minimum normal stresses, Eq. (4.15) can be differentiated and equated to zero, which gives

$$\tan 2\theta = \frac{2\tau_{xy}}{\sigma_x - \sigma_y} \qquad (4.17)$$

where θ is the angle of the plane of maximum or minimum normal stress. These two stresses are called the *principal stresses* and the corresponding planes are called the *principal planes*. The shear stresses at the principal planes are zero. Substituting the value of sin 2θ and cos 2θ in Eq. (4.15) from Eq. (4.17), we get the equation of maximum and minimum principal stresses as

$$\sigma_1, \sigma_2 = \frac{\sigma_x + \sigma_y}{2} \pm \sqrt{\left(\frac{\sigma_x - \sigma_y}{2}\right)^2 + \tau_{xy}^2} \qquad (4.18)$$

Similarly, for maximum value of shear stress we differentiate Eq. (4.16) and equate to zero, which gives

$$\tan 2\theta = \tan 2\theta_1 = \frac{-(\sigma_x - \sigma_y)}{2\tau_{xy}} \qquad (4.19)$$

Upon comparing Eq. (4.17) and Eq. (4.19), we observe that

$$\theta_1 = \theta + 45° \qquad (4.20)$$

This means that the angle which locates the plane of maximum shear stress makes an angle of 45° with the plane of the principal stresses. Substituting the values of sin 2θ and cos 2θ in Eq. (4.16) from Eq. (4.19), we get the equation of maximum shear stress as

$$\tau_{max} = \sqrt{\left(\frac{\sigma_x - \sigma_y}{2}\right)^2 + \tau_{xy}^2} \qquad (4.21)$$

The graphical representation of Eqs. (4.18) and (4.21) is called the Mohr's circle (Figure 4.9). It is an effective tool to visualize the state of stresses at a point.

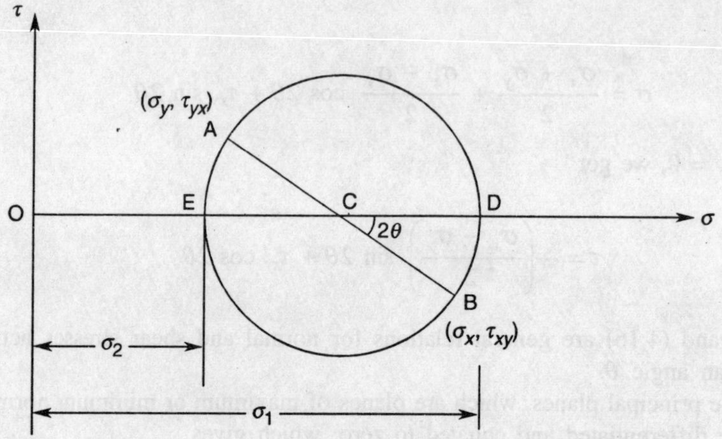

Figure 4.9 Mohr's circle for the biaxial stress element.

The following major observations are made from the Mohr's circle.

1. The centre of the Mohr's circle is located at $\left(\dfrac{\sigma_x + \sigma_y}{2}, 0\right)$.
2. The radius of the circle is given by $\sqrt{\left(\dfrac{\sigma_x - \sigma_y}{2}\right)^2 + \tau_{xy}^2}$ which represents the maximum shear stress. The normal stress for this condition is $(\sigma_1 + \sigma_2)/2$.
3. Summation of stresses $\sigma_x + \sigma_y = \sigma_1 + \sigma_2 =$ constant.
4. If $\sigma_x + \sigma_y = 0$, the centre of the circle coincides with the origin of σ–τ coordinates and the state of pure shear occurs.
5. At maximum and minimum principal stresses, the shear stress τ is zero.

4.6 SHEAR FORCE AND BENDING MOMENT

A beam is a structural member which is subjected to external forces in the transverse direction, i.e. at right angle to its longitudinal axis. Depending upon the type of end supports, beams are classified into the following four types:

(i) *Simply supported beam.* A beam is said to be simply supported when its ends are not allowed to have translation displacement but rotation is allowed [Figure 4.10(a)].

(ii) *Cantilever beam.* A beam whose one end is clamped or fixed and the other end is free as shown in Figure 4.10(b), is called the cantilever beam.

(iii) *Fixed beam.* A fixed beam is one whose both ends are clamped so that they are not free to have translation or rotary motion [Figure 4.10(c)].

(iv) *Continuous beam.* A beam which is supported at more than two supports is called the continuous beam [(Figure 4.10(d)].

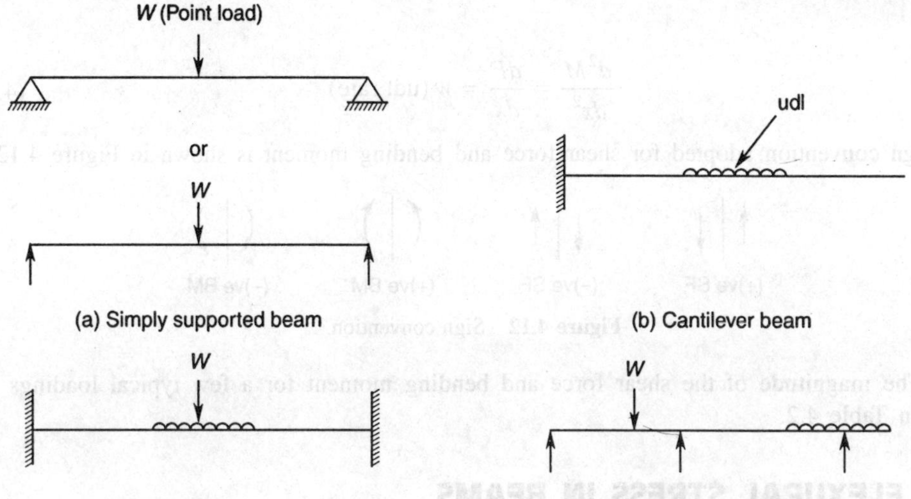

Figure 4.10 Types of beams and loads.

Beams are subjected to point or concentrated loads and/or distributed loads. The distributed load acts over the length of the beam. If the rate of distribution is constant, it is called the uniformly distributed load (udl), for example, self-weight of the beam.

Consider a beam as shown in Figure 4.11(a) subjected to point loads W_1, W_2, and W_3, the action of these loads needs reactions, R_A and R_B at supports A and B, respectively, for equilibrium. If this beam is imagined to be cut into two parts at section X-X from the left support and the right-hand portion is removed, the left-hand portion needs a force F and bending moment M to assure equilibrium as shown in Figure 4.11(b). The shear force F is obtained by algebraic sum of the forces to the left of the cut-section. The bending moment is defined as the algebraic sum of the moments of the forces to the left of the cut-section. The shear force and bending moment are related by the following equations

$$\frac{dM}{dx} = F \tag{4.22}$$

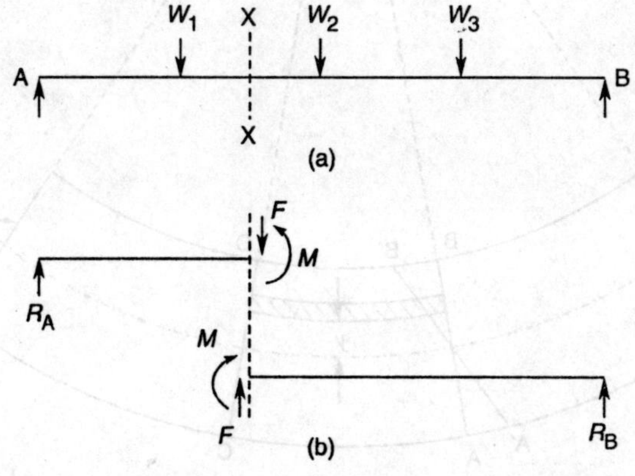

Figure 4.11 Beam subjected to shear force and bending moment.

and

$$\frac{d^2M}{dx^2} = \frac{dF}{dx} = w \text{ (udl rate)} \tag{4.23}$$

The sign convention adopted for shear force and bending moment is shown in Figure 4.12.

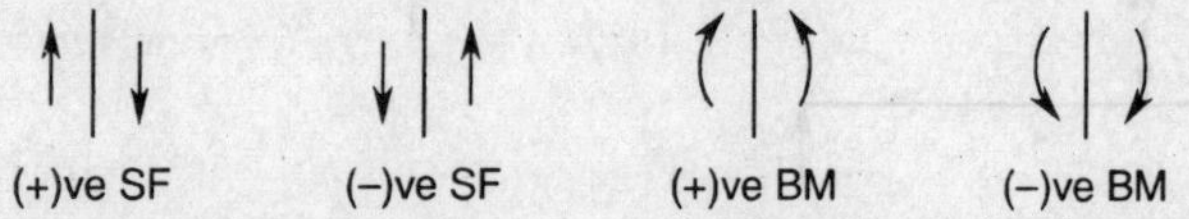

Figure 4.12 Sign convention.

The magnitude of the shear force and bending moment for a few typical loadings are listed in Table 4.2.

4.7 FLEXURAL STRESS IN BEAMS

When a segment of a beam is in equilibrium under the action of a moment alone, this condition is called *pure bending* or *flexure.* For determining the flexural stress in a beam, we assume that the plane of loading is coincident with one of the principal centroidal axes of the cross-section and the plane section remains plane even after bending.

Consider a horizontal prismatic beam having a cross-section which is symmetrical about a vertical axis. The *x*-axis is coincident with the neutral axis of the section. When such a beam is subjected to positive bending moment in a centroidal plane, the upper surface of the beam bends in compression and the neutral axis is then curved as shown in Figure 4.13. Because of

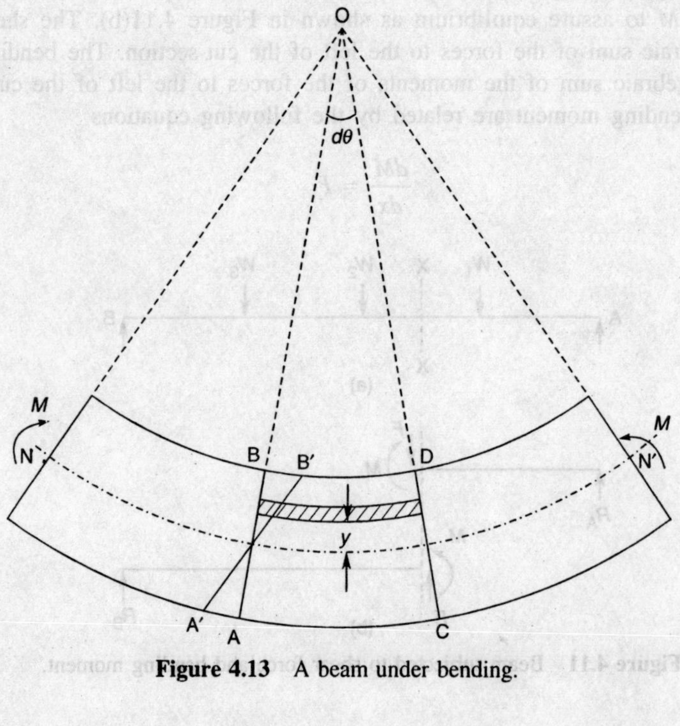

Figure 4.13 A beam under bending.

the curvature, the section AB originally parallel to CD will rotate through a small angle $d\theta$ to A′B′. The A′B′ remains a straight line as we have assumed that a plane section remains plane even after bending. If R is the radius of curvature of the neutral axis, I the moment of inertia and σ the bending stress, the relation between these can be given as

$$\frac{M}{I} = \frac{\sigma}{y} = \frac{E}{R} \tag{4.24}$$

where y is the distance from the neutral axis to the fibre where the stress is calculated and the bending stress

$$\sigma = \frac{M \cdot y}{I} = \frac{M}{Z} \tag{4.25}$$

with $Z = I/y$ = section modulus.

Equation (4.25) states that bending stress is directly proportional to the distance y from the neutral axis to the fibre where the stress is calculated. The moment of inertia and section modulus of various cross-sections are given in Table 4.1.

Table 4.1 Properties of various cross-sections

Section	*Moment of inertia* (I)	*Distance to the farthest point* (y)	*Section modulus* $Z = \dfrac{I}{y}$	*Radius of gyration* $k = \sqrt{\dfrac{I}{A}}$
(1)	(2)	(3)	(4)	(5)
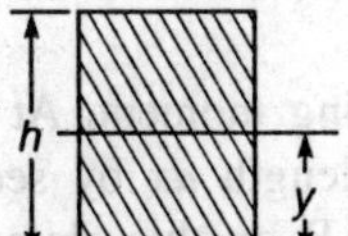	$\dfrac{bh^3}{12}$	$\dfrac{h}{2}$	$\dfrac{bh^2}{6}$	$0.289h$
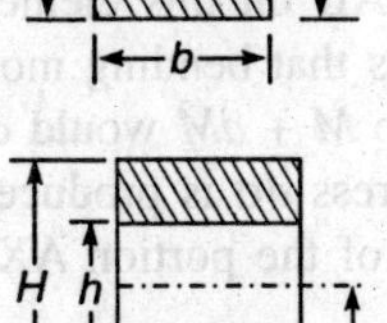	$\dfrac{b}{12}(H^3 - h^3)$	$\dfrac{H}{2}$	$\dfrac{b(H^3 - h^3)}{6H}$	$\sqrt{\dfrac{H^3 - h^3}{12(H-h)}}$
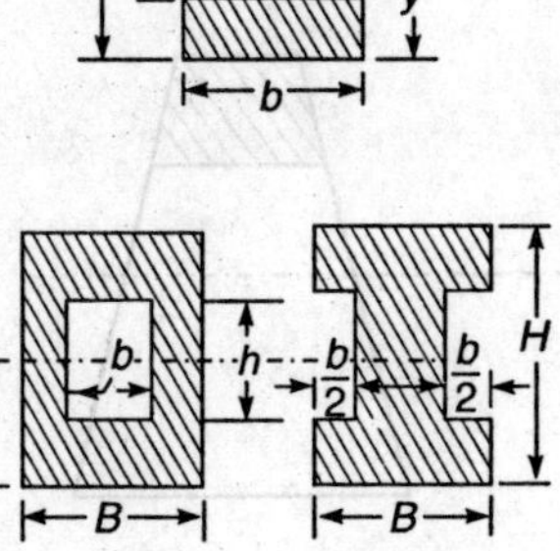	$\dfrac{BH^3 - bh^3}{12}$	$\dfrac{H}{2}$	$\dfrac{BH^3 - bh^3}{6H}$	$\sqrt{\dfrac{BH^3 - bh^3}{12(BH - bh)}}$

(contd.)

Table 4.1 (*contd.*)

(1)	(2)	(3)	(4)	(5)
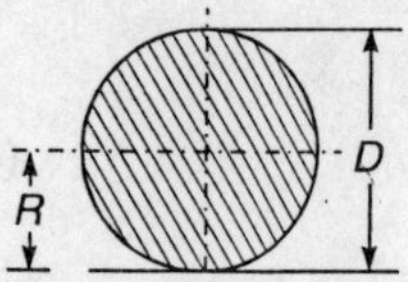	$\dfrac{\pi D^4}{64}$	$\dfrac{D}{2}$	$\dfrac{\pi D^3}{32}$	$\dfrac{D}{4}$
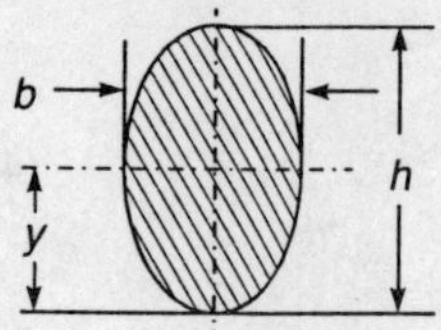	$\dfrac{\pi b h^3}{64}$	$\dfrac{h}{2}$	$\dfrac{\pi b h^2}{32}$	$\dfrac{h}{4}$

4.8 SHEAR STRESS IN BEAMS

In Section 4.7, while discussing the theory of bending we neglected the presence of the shear stress and distortion of the plane section caused by it. This does not imply that shear stress can be neglected altogether, but in practice a combined effect of shear force and bending moment acts on the beam.

A beam of constant cross-section, when subjected to shear force F and bending moment M, the general relation between moment M and shear force F can be written as

$$F = \frac{dM}{dx} \tag{4.26}$$

Consider that a beam of symmetrical section is subjected to bending moment. At some distance along the length of the beam, we cut a transverse section of length dx by sections X-Y and X′-Y′ as shown in Figure 4.14. Let us also consider a fibre AB on this element at distance y from the neutral axis. The study of the element AXX′B reveals that bending moment changes as we move along the length. This change in moment from M to $M + dM$ would cause a change in bending stress from σ to $\sigma + d\sigma$. Therefore, the resultant stress $d\sigma$ is produced on top surface of the element which is creating imbalance. For equilibrium of the portion AXX′B,

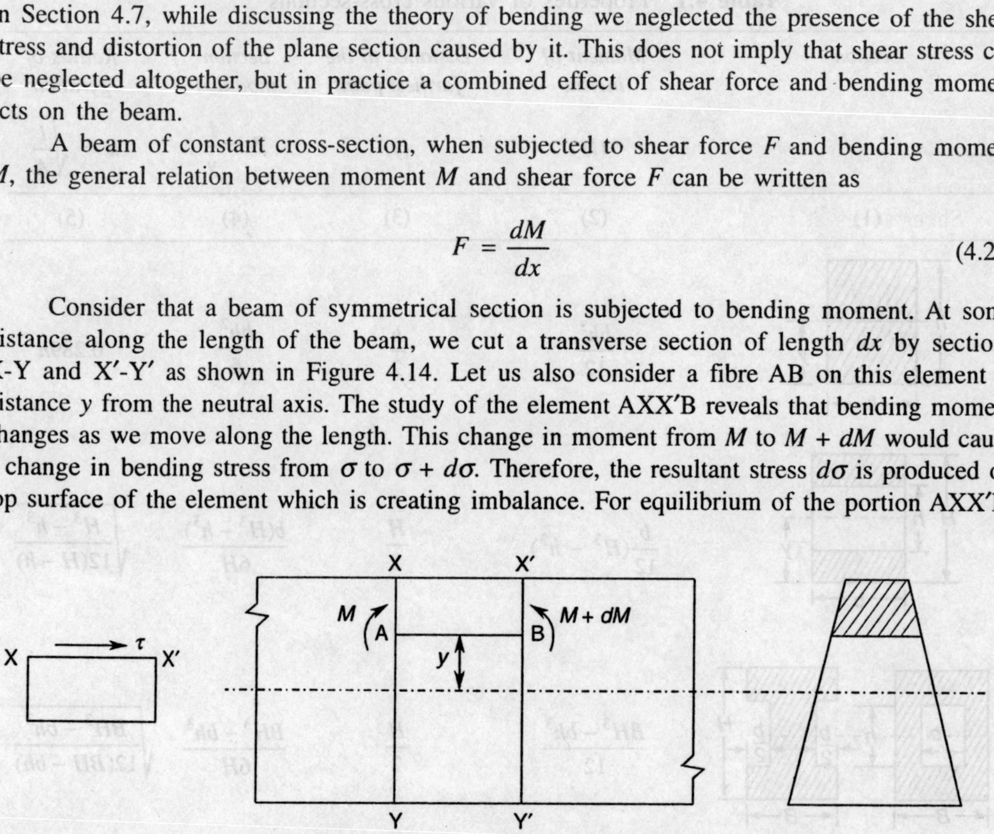

Figure 4.14 Beam under shear stress.

a horizontal force at AB must act to counterbalance the resultant stress $d\sigma$. This counterbalancing force is termed shear force and its intensity is called the shear stress. The shear stress τ for the above beam can be calculated by the following formula,

$$\tau = \frac{F}{Ib} \times A\bar{y} \tag{4.27}$$

where

F is the shear force

I is the moment of inertia of the section

b is the width of the section at a particular distance y from the neutral axis

$A\bar{y}$ is the moment of the area above AB about the neutral axis.

For a rectangular section (Figure 4.15), let us consider any layer AB at height y from the neutral axis. The shear stress is given by

$$\tau = \frac{F}{Ib} \times A\bar{y}$$

$$= \frac{F}{Ib}\left[\left\{b\left(\frac{t}{2} - y\right)\right\}\left\{\frac{1}{2}\left(\frac{t}{2} - y\right) + y\right\}\right] \tag{4.28}$$

$$= \frac{F}{2I}\left(\frac{t^2}{4} - y^2\right)$$

The above equation of shear stress distribution is an equation of the parabola, therefore, the variation of shear stress is parabolic as shown in Figure 4.15. Shear stress at $y = \pm\, t/2$ is 0 (zero)

Shear stress at $y = 0$ is given by

$$\tau = \frac{F}{2I}\frac{t^2}{4}$$

$$= \frac{3F}{2bt} \tag{4.29}$$

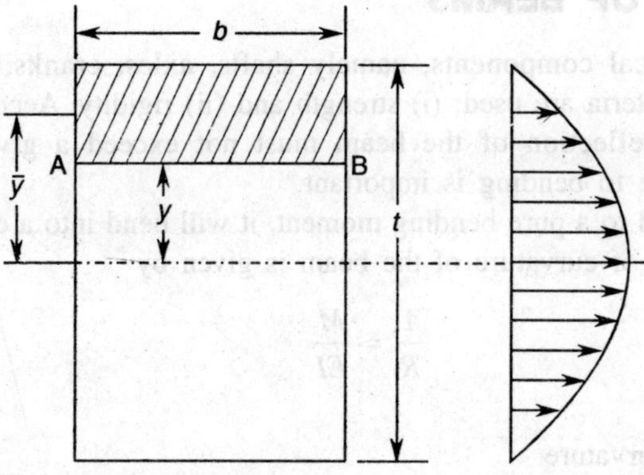

Figure 4.15 Shear stress in a rectangular section.

Thus, the maximum shear stress at the neutral axis is 50 per cent more than the mean value. The shear-stress distributions of various sections are shown in Figure 4.16.

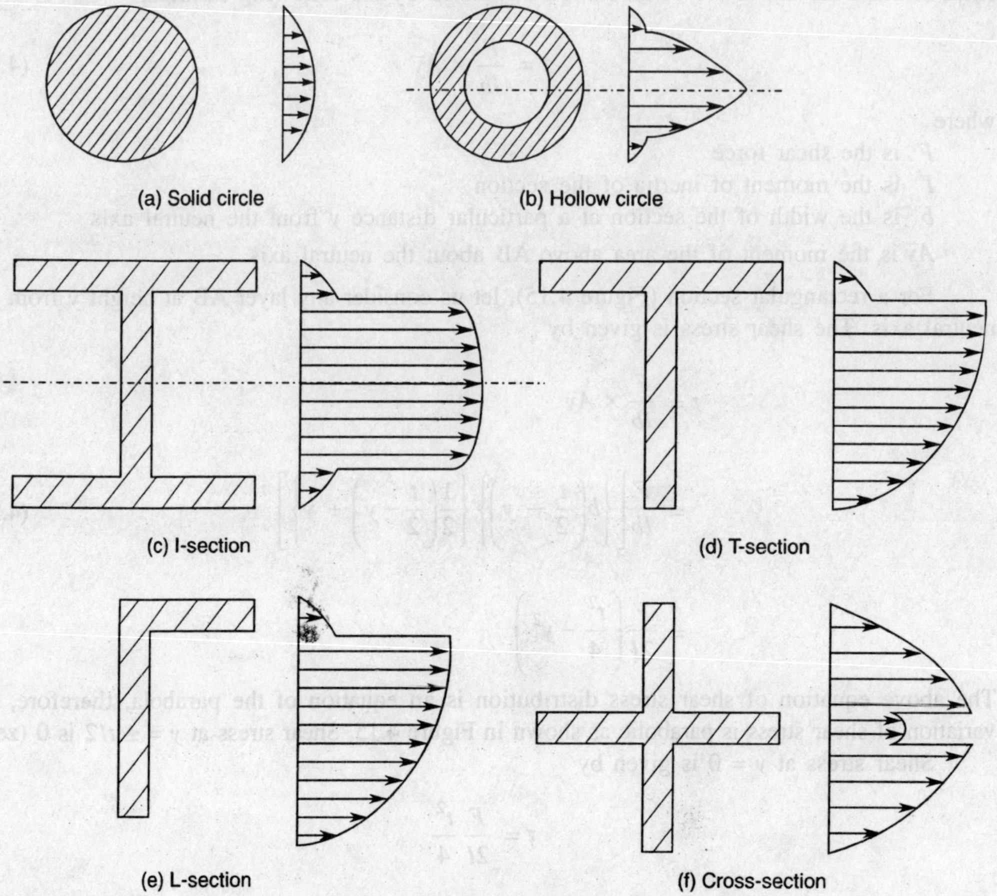

Figure 4.16 Shear stress distribution for some typical sections.

4.9 DEFLECTION OF BEAMS

In the design of mechanical components, namely shafts, axles, cranks, levers, springs and brackets, generally, two criteria are used: (i) strength and (ii) rigidity. According to the rigidity criterion, the maximum deflection of the beam must not exceed a given limit. Thus, the deflection of the beam due to bending is important.

If a beam is subjected to a pure bending moment, it will bend into a circular arc as shown in Figure 4.17. The radius of curvature of the beam is given by

$$\frac{1}{R} = \frac{M}{EI} \tag{4.30}$$

where

R is the radius of curvature

EI is the flexural properties

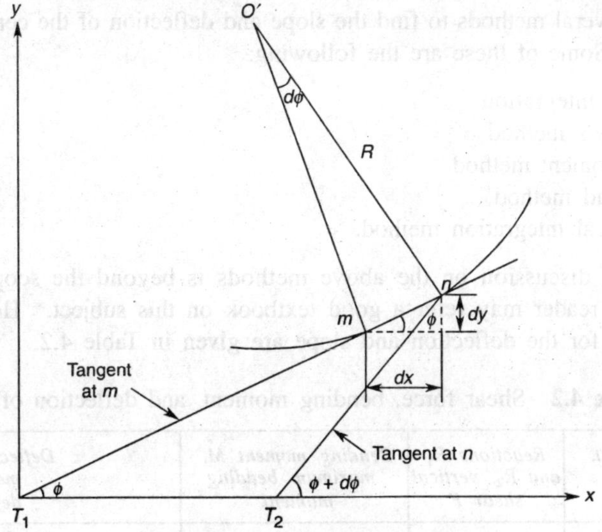

Figure 4.17 Deflection of a beam.

From studies in mathematics (refer Figure 4.17), we know that the curvature of a plane curve is given by the following equation:

$$\frac{1}{R} = \frac{d^2y/dx^2}{\left[1+\left(\dfrac{dy}{dx}\right)^2\right]^{3/2}} \tag{4.31}$$

where

dy/dx is the slope of the beam at any point

y is the deflection of the beam at any point along its length.

Equating Eqs. (4.30) and (4.31), we get

$$\frac{M}{EI} = \frac{d^2y/dx^2}{\left[1+\left(\dfrac{dy}{dx}\right)^2\right]^{3/2}} \tag{4.32}$$

or

$$EI\frac{d^2y}{dx^2} = M\left[1+\left(\frac{dy}{dx}\right)^2\right]^{3/2}$$

or

$$EI\frac{d^2y}{dx^2} \approx M \tag{4.33}$$

The slope (dy/dx) being small, the quantity $\left[1+\left(\dfrac{dy}{dx}\right)^2\right]^{3/2} \approx 1$

There are several methods to find the slope and deflection of the centre line of a beam at a certain section. Some of these are the following:

(i) Double integration
(ii) Maculay's method
(iii) Area moment method
(iv) Unit load method
(v) Numerical integration method.

The detailed discussion on the above methods is beyond the scope of this book. For further details the reader may refer a good textbook on this subject.* However, a few useful standard solutions for the deflection and slope are given in Table 4.2.

Table 4.2 Shear force, bending moment, and deflection of beams

Loading, support, and reference number	*Reactions R_1 and R_2, vertical shear F*	*Bending moment M, maximum bending moment*	*Deflection y and maximum deflection*
1. Cantilever, intermediate load 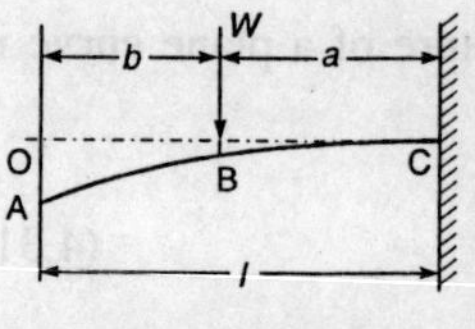	$R_2 = +W$ A to B: $F = 0$ B to C: $F = -W$	A to B: $M = 0$ B to C: $M = -W(x - b)$ $M_{max} = -Wa$ at C	A to B: $y = -\frac{1}{6}\frac{W}{EI}(-a^3 + 3a^2 l - 3a^2 x)$ B to C: $y = -\frac{1}{6}\frac{W}{EI}[(x-b)^3 - 3a^2(x-b) + 2a^3]$ $y_{max} = -\frac{1}{6}\frac{W}{EI}(3a^2 l - a^3)$
2. Cantilever, uniform load 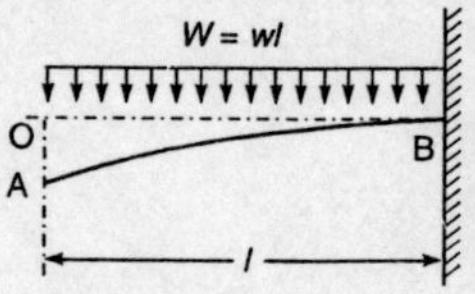	$R_2 = +W$ $F = -\frac{W}{l}x$	$M = -\frac{1}{2}\frac{W}{l}x^2$ $M_{max} = -\frac{1}{2}Wl$ at C	$y = -\frac{1}{24}\frac{W}{EIl}(x^4 - 4l^3 x + 3l^4)$ $y_{max} = -\frac{1}{8}\frac{Wl^3}{EI}$
3. End supports, intermediate load 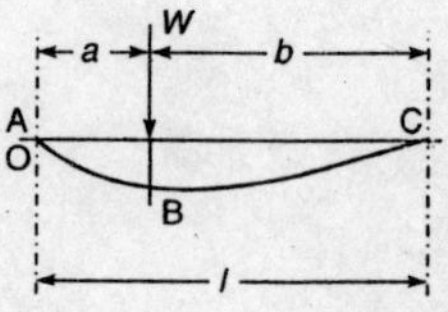	$R_1 = +W\frac{b}{l}$ $R_2 = +W\frac{a}{l}$ A to B: $F = +W\frac{b}{l}$ B to C: $F = -W\frac{a}{l}$	A to B: $M = +W\frac{b}{l}x$ B to C: $M = +W\frac{a}{l}(l - x)$ $M_{max} = +W\frac{ab}{l}$ at B	A to B: $y = -\frac{Wbx}{6EIl}[(2l(l-x) - b^2 - (l-x)^2]$ B to C: $y = -\frac{Wa(l-x)}{6EIl}[2lb - b^2 - (l-x)^2]$ $y_{max} = -\frac{Wab}{27EIl}(a+2b)\sqrt{3a(a+2b)}$ at $x = \sqrt{\frac{1}{3}a(a+2b)}$ when $a > b$
4. End supports, uniform load 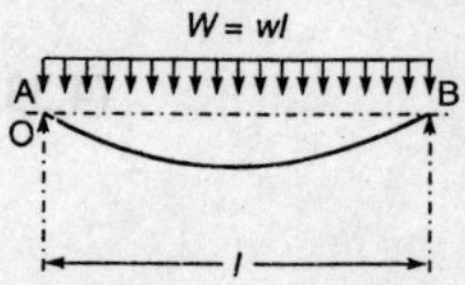	$R_1 = +\frac{1}{2}W'$ $R_2 = +\frac{1}{2}W$ $F = \frac{1}{2}W\left(1 - \frac{2x}{l}\right)$	$M = \frac{1}{2}W\left(x - \frac{x^2}{l}\right)$ $M_{max} = +\frac{1}{8}Wl$ at $x = \frac{l}{2}$	$y = -\frac{1}{24}\frac{Wx}{EIl}(l^3 - 2lx^2 + x^3)$ $y_{max} = -\frac{5}{384}\frac{Wl^3}{EI}$ at $x = \frac{l}{2}$

* *Engineering Mechanics of Solids* by E.P. Popov, 2nd ed., Prentice-Hall of India, New Delhi.

4.10 TORSION

The moment applied in the vertical plane, perpendicular to the longitudinal axis of the beam or shaft as shown in Figure 4.18, is called the *twisting moment* or *torque*. The externally applied twisting moment causes the distortion of the shaft by shifting the radial line from OA to OA_1 and on the surface of the shaft, the line CA [see Figure 4.18(b)] will be distorted to CA_1 through an angle ϕ.

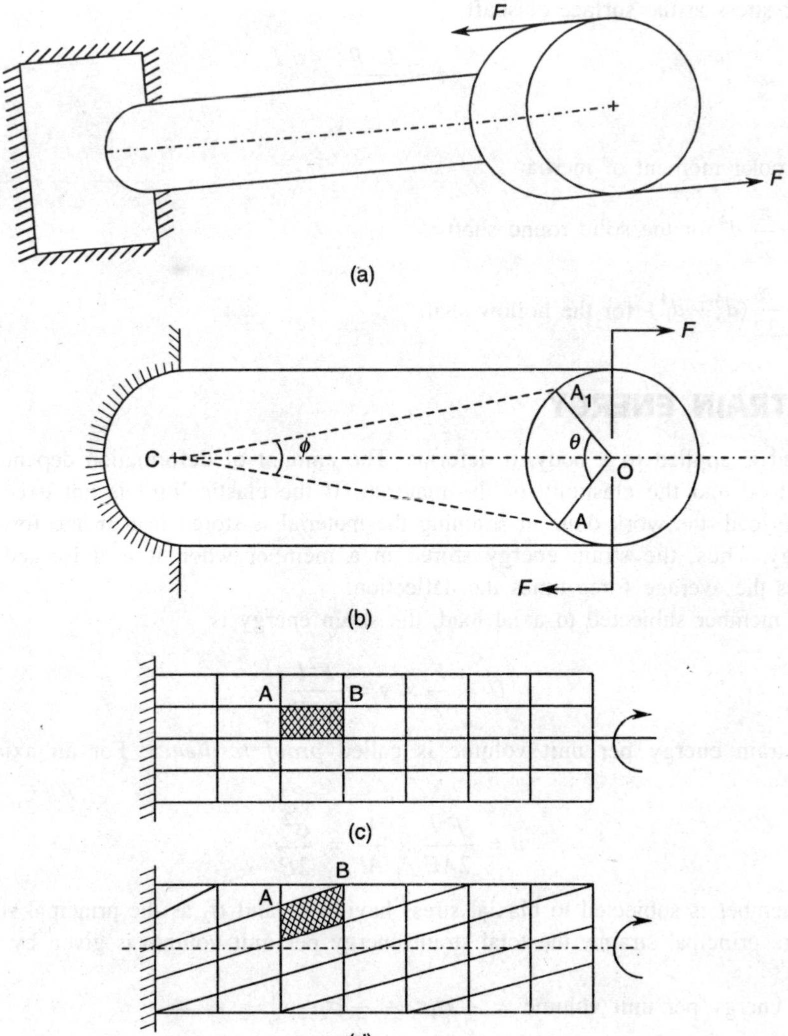

Figure 4.18 Shaft subjected to torsion.

When a shaft is subjected to twisting moment, a set of rectangular grids drawn on the surface turns into a set of parallelograms as shown in Figure 4.18(d). This shows the presence of shear stress on the surface of the shaft.

The shear stress at any point in the cross-section of the shaft subjected to pure torsion is proportional to its distance from the centre. The intensity of the shear stress is zero at the centre and maximum at the outermost surface.

From the condition of the equilibrium, the torque equation is given as

$$\frac{\tau}{R} = \frac{T}{J} = \frac{G\theta}{L} \tag{4.34}$$

or the shear stress at the surface of shaft

$$\tau = \frac{T \cdot R}{J} \tag{4.35}$$

where

J = polar moment of inertia

$= \frac{\pi}{32} d^4$ for the solid round shaft

$= \frac{\pi}{32}(d_o^4 - d_i^4)$ for the hollow shaft

4.11 STRAIN ENERGY

When a load is applied to a body, it deforms. The amount of deformation depends upon the amount of load and the elasticity of the material. If the elastic limit is not exceeded while applying the load, the work done in straining the material is stored in it in the form of elastic strain energy. Thus, the strain energy stored in a member when it is deformed through a distance y is the average force times the deflection.

For a member subjected to axial load, the strain energy is

$$U = \frac{F}{2} \times y = \frac{F^2 l}{2AE} \tag{4.36}$$

The strain energy per unit volume is called *proof resilience.* For an axially-loaded member

$$u = \frac{F^2 l}{2AE} \times \frac{1}{Al} = \frac{\sigma^2}{2E}$$

If a member is subjected to biaxial stress having σ_1 and σ_2 as the principal stresses, and ε_1 and ε_2 are principal strains, the total strain energy per unit volume is given by

$$\text{Total strain energy per unit volume} = \frac{1}{2}\sigma_1\varepsilon_1 + \frac{1}{2}\sigma_2\varepsilon_2$$

$$= \frac{1}{2E}\left(\sigma_1^2 + \sigma_2^2 - 2\mu\sigma_1\sigma_2\right) \tag{4.37}$$

Castigliano theorem. Castigliano theorem states that when forces operate on an elastic system, the displacement corresponding to any force may be found by obtaining the partial derivative of the total strain energy per unit volume with respect to that force. Mathematically,

$$\delta_i = \frac{\partial u}{\partial F_i} \tag{4.38}$$

4.12 ECCENTRIC LOADING

If the line of action of the external force passes through the centroid of the cross-section, it is said to be axial force and the stress can be computed by dividing the load by the area of cross-section (F/A). However, in practice, it is extremely difficult to apply loads with lines of action coincident with the centroid of the member. Further, there are certain mechanical members which are assigned a particular shape for the functional requirement. Under such circumstances, the loads have to be applied at locations other than the c.g. of the cross-section. The load which does not pass through the centroid of the cross-section is called the *eccentric load.* This type of load when acted upon the machine member can be replaced by a parallel force passing through the centroid along with a couple as shown in Figure 4.19. The resultant stresses at the cross-section are obtained by the principle of superimposition, i.e.

$$\sigma = \frac{F}{A} + \frac{F \cdot e}{Z} \tag{4.39}$$

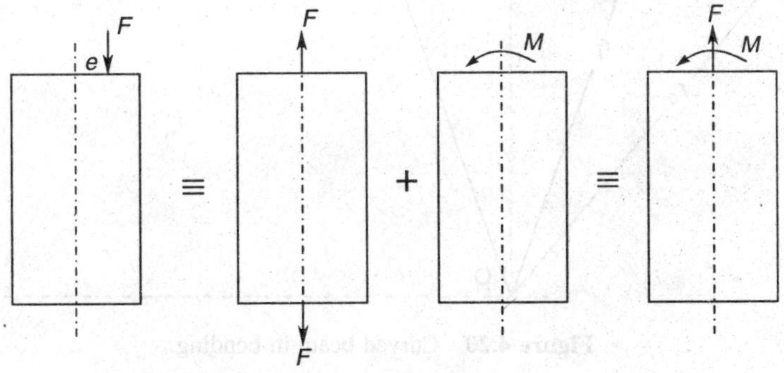

Figure 4.19 Eccentric loading on a component.

4.13 CURVED BEAM

In the design of some mechanical components, namely curved hooks, C-clamps, chain links or any bend-shape element, the simple theory of bending is not applicable because the fundamental assumption of the simple beam bending theory (i.e. the neutral axis and the centroid axis are the same), is not valid.

Consider a portion of the curved beam subtending an angle θ, as shown in Figure 4.20, which is subjected to a bending moment M. Assuming that the cross-section of the beam has an axis of symmetry in a plane along the length of the beam and the plane cross-section remains plane after bending, the general equation for stress at any fibre at a distance y from the neutral axis is given by the following relation,

$$\sigma = \frac{M}{Ae} \cdot \frac{y}{(r_n + y)} \tag{4.40}$$

where

M is the bending moment
A is the area of cross-section
e is the distance, $r - r_n$, between the neutral axis and the centroidal axis
y is the distance of any fibre from the neutral axis
r_n is the radius of curvature of the neutral axis
r is the radius of curvature of the centroidal axis.

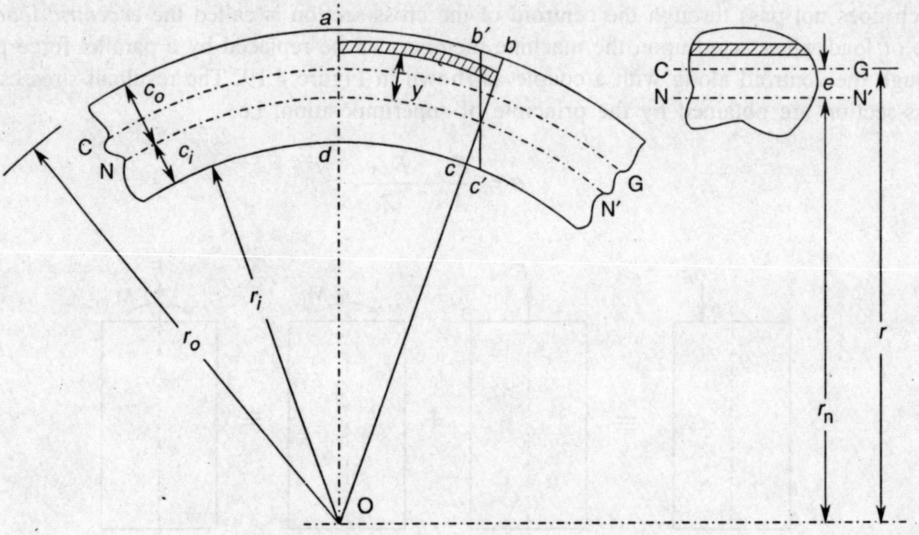

Figure 4.20 Curved beam in bending.

The location of the neutral axis can be found by the following relation:

$$r_n = \frac{A}{\int \frac{dA}{r_n + y}} \tag{4.41}$$

The values of the location of the neutral axis and the centroidal axis for various types of cross-sections are given in Table 4 3.

Table 4.3 Formulae for curved beam

Type of section	*Radius of neutral axis*
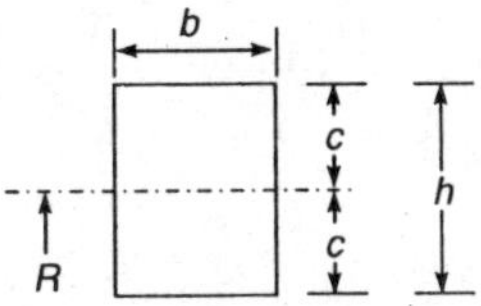	$r_n = \dfrac{h}{\log_e\left(\dfrac{R+c}{R-c}\right)}$
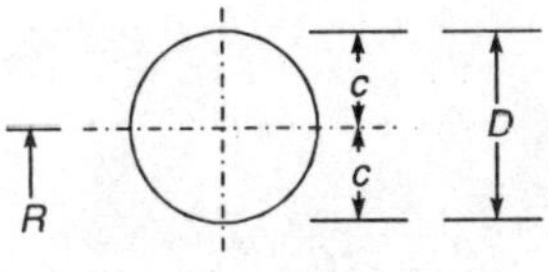	$r_n = \dfrac{0.5c^2}{R - \sqrt{R^2 - c^2}}$
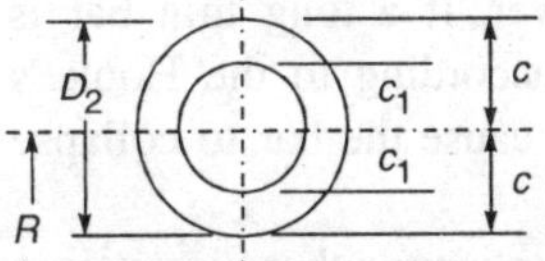	$r_n = \dfrac{0.5(c^2 - c_1^2)}{\sqrt{R^2 - c_1^2} - \sqrt{R^2 - c^2}}$
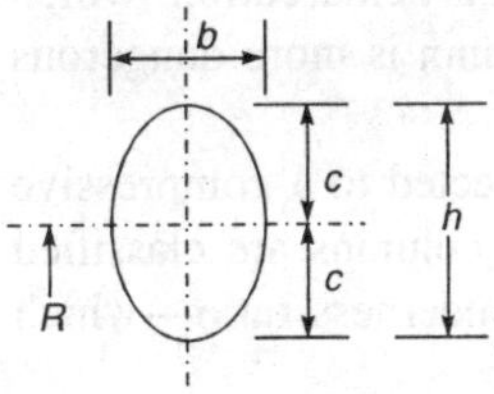	$r_n = \dfrac{0.5c^2}{R - \sqrt{R^2 - c^2}}$
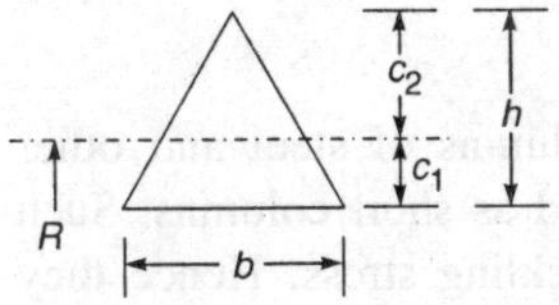	$r_n = \dfrac{0.5h^2}{(R+c_2)\log_e\left(\dfrac{R+c_2}{R-c_1}\right) - h}$
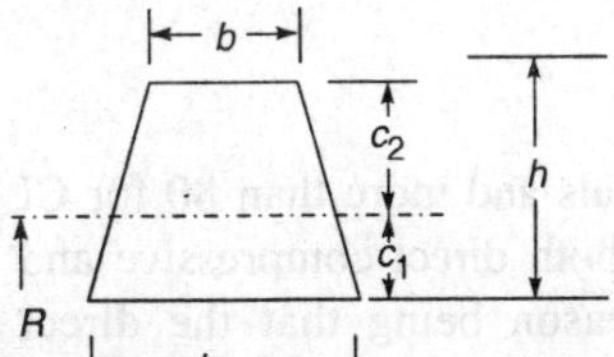	$r_n = \dfrac{A}{\dfrac{b_1(R+c_2) - b(R-c_1)}{h}\log_e\left(\dfrac{R+c_2}{R-c_1}\right) - (b_1 - b)}$ where A is the area of the section.

The distribution of stress in the curved beam is hyperbolic in nature. Thus, the maximum stresses which occur at the inner and outer fibres are:

Stress at the inner fibre

$$\sigma_i = \frac{Mc_i}{Aer_i} \tag{4.42}$$

Stress at the outer fibre

$$\sigma_o = \frac{Mc_o}{Aer_o} \tag{4.43}$$

where

r_i is the radius of curvature at the inner fibre

r_o is the radius of curvature at the outer fibre.

$c_i = r_i - r_n$

$c_o = r_o - r_n$

4.14 STRESS IN COLUMNS

When a short bar is subjected to a compression force along the centroidal axis, it will be shortened according to Hooke's law and ultimately squeezed. However, if a long thin bar is subjected to a pure compression force, the bar will initially behave according to the Hooke's law and a stage would be reached where a slight increase in load may cause the bar to collapse by buckling.

A compression member which fails by buckling is called the *column;* otherwise it is a simple compression member. Unfortunately, there is no clear line of demarcation which distinguishes a column from the compression member. The failure of column is more dangerous than the compression member because there is no early warning.

A member which is in any position other than the vertical and subjected to a compressive load is called the *struct.* Its one or both ends may be hinged. Generally, columns are classified into two groups, short column and long column, depending upon the slenderness ratio—which is defined as the ratio of equivalent length and radius of gyration.

4.14.1 Short Columns

Cast iron columns with slenderness ratio (l/k) less than 80 and columns of steel and other ductile material for which the (l/k) ratio is less than 100 are classified as short columns. Such types of columns are subjected to direct compressive stress and buckling stress. Hence they always fail under the combined effect of both types of stresses.

4.14.2 Long Columns

Those columns whose (l/k) ratio is more than 100 for ductile materials and more than 80 for CI are called long columns. Although these columns are subjected to both direct compressive and buckling stresses, they fail mainly due to buckling; the only reason being that the direct stresses are very small compared to the buckling stresses.

4.14.3 Critical Load

The critical load is defined as the smallest compressive load capable of causing the straight form of the column to become unstable because of a relatively large lateral deflection. The critical load for a column depends upon the column material, its geometry, and the end conditions. The different ideal types of ends of columns and the approximate curves assumed by the neutral axis under critical load are shown in Figure 4.21. The following formulae have been suggested to determine the critical load P_c for columns with various types of ends.

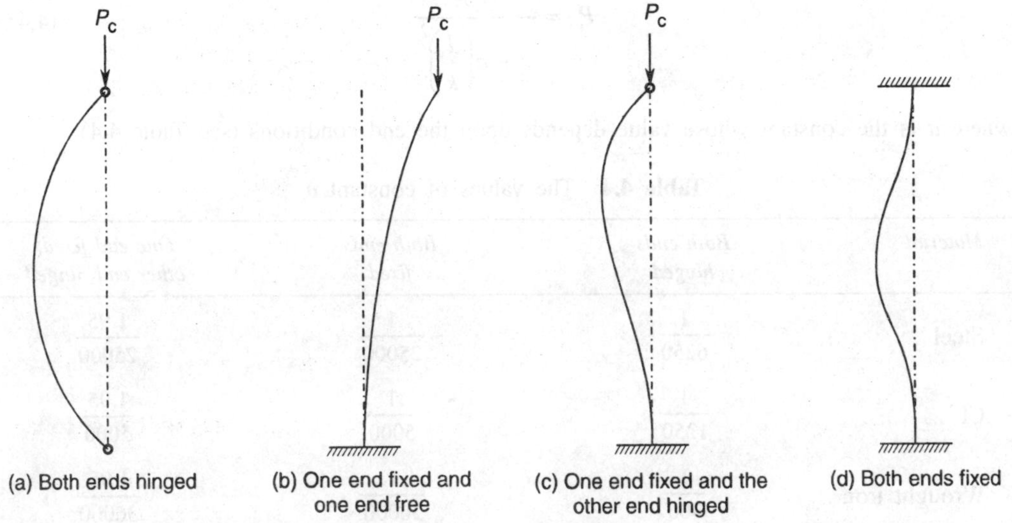

Figure 4.21 Different types of end conditions for a column.

Euler's formula

Euler's formula for determining the critical load is applicable to very long columns, i.e. those columns whose (l/k) ratio is greater than 100. The critical load for columns is very small; the effect of the direct stress is negligible as compared to that of the buckling stress. The Euler's formula is

$$P_c = \frac{n\pi^2 EI}{l^2} \tag{4.44}$$

where

I is the moment of inertia
l is the length of the column
E is the elastic modulus
n = constant for end conditions
= 1 for both ends hinged
= 0.25 for one end free and the other end fixed
= 2.0 for one end hinged and the other end fixed
= 4.0 for both ends fixed.

Rankine formula

Euler's formula has little practical value for the design of the mechanical members. Further, it is useful only for long columns. In the case of short columns, the failure will occur by both direct and buckling stresses. Therefore, to find the critical load for such columns, Rankine proposed an empirical relation. This formula is so ingenious that it covers all cases ranging from short columns to very long columns. The critical load is given by the following relation:

$$P_c = \frac{\sigma_c A}{1 + a\left(\dfrac{l}{k}\right)^2} \tag{4.45}$$

where a is the constant whose value depends upon the end conditions (see Table 4.4).

Table 4.4 The values of constant a

Material	*Both ends hinged*	*Both ends fixed*	*One end fixed, other end hinged*
Steel	$\frac{1}{6250}$	$\frac{1}{25000}$	$\frac{1.95}{25000}$
CI	$\frac{1}{1250}$	$\frac{1}{5000}$	$\frac{1.95}{5000}$
Wrought iron	$\frac{1}{900}$	$\frac{1}{36000}$	$\frac{1.95}{36000}$

Example 4.1* A cantilever beam of I-section supports an electric motor weighing 1000 N at a distance of 400 mm from the fixed end. If the allowable strength of the beam material is 100 N/mm^2, determine the section of the beam. The proportions of I-section are $B = 4t$ and $H = 6t$, where t is the thickness of the flange as well as that of the web.

Solution Load, $W = 1000$ N

Beam length, $l = 400$ mm

Allowable strength = 100 N/mm^2

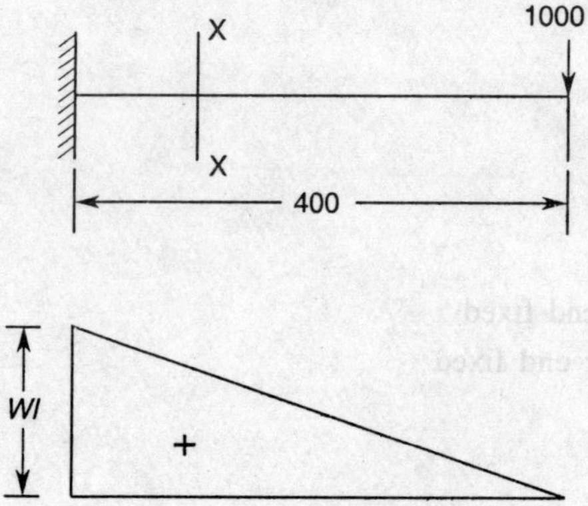

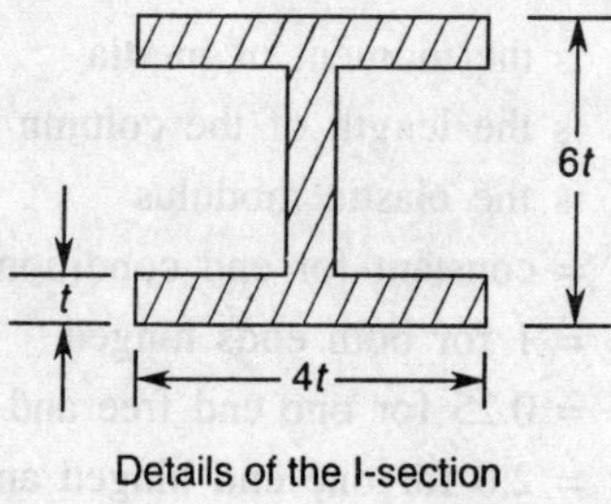

Details of the I-section

* All dimensions on the figures are in mm, unless specified otherwise. This practice is adopted throughout the book.

The maximum bending moment at the fixed end

$$M = W \times l = 1000 \times 0.4 = 400 \text{ N} \cdot \text{m}$$

$$\text{Section modulus, } Z = \frac{BH^3 - bh^3}{6H}$$

or

$$Z = \frac{4t \times (6t)^3 - 3t \times (4t)^3}{6 \times 6t}$$

$$= 18.667t^3$$

$$\text{Bending stress, } \sigma_b = \frac{M}{Z} = \frac{400 \times 1000}{18.667t^3}$$

For safe design, the induced bending stress should be less than the allowable strength, i.e.

$$\frac{400 \times 1000}{18.667\,t^3} \le 100 \text{ N/mm}^2$$

or

$$t = 5.98 \text{ mm} \quad \text{(Let us adopt } t = 6 \text{ mm)}$$

Beam section, $B = 4t = 24$ mm

$H = 6t = 36$ mm

Web/flange thickness, $t = 6$ mm

Example 4.2 A point in an elastic material under strain is subjected to normal stresses 50 N/mm^2 (tensile) along the x-axis and 30 N/mm^2 (compressive) along the y-axis. The element is also subjected to shear stress of magnitude 25 N/mm^2. Find the principal stresses and the position of the principal planes. Also, find the maximum shear stress and its plane.

Solution $\sigma_x = 50$ N/mm^2 (tensile)

$\sigma_y = -30$ N/mm^2 (compressive)

$\tau = 25$ N/mm^2

Analytical solution

Maximum and minimum principal stress

$$\sigma_{1,2} = \frac{\sigma_x + \sigma_y}{2} \pm \sqrt{\left(\frac{\sigma_x - \sigma_y}{2}\right)^2 + \tau^2}$$

$$= \frac{50 - 30}{2} \pm \sqrt{\left(\frac{50 + 30}{2}\right)^2 + 25^2} = 10 \pm 47.1$$

The maximum principal stress, $\sigma_1 = 57.1$ N/mm^2 (tensile). The minimum principal stress, $\sigma_2 = -37.1$ N/mm^2 (compressive). The plane of the principal stress is located at

$$\tan 2\theta = \frac{2\tau}{\sigma_x - \sigma_y}$$

or

$$2\theta = \tan^{-1}\left(\frac{2 \times 25}{50 + 30}\right)$$

or

$$2\theta = 32°$$

or

$$\theta_1 = 16° \text{ and } \theta_2 = 90 + 16 = 106°$$

$$\text{Maximum shear stress, } \tau_{max} = \frac{\sigma_1 - \sigma_2}{2}$$

$$= \frac{57.1 - (-37.1)}{2} = 47.1 \text{ N/mm}^2$$

and the plane of maximum shear stress = $45° + \theta_1 = 61°$ inclination with σ_x.

Graphical method: Mohr's circle

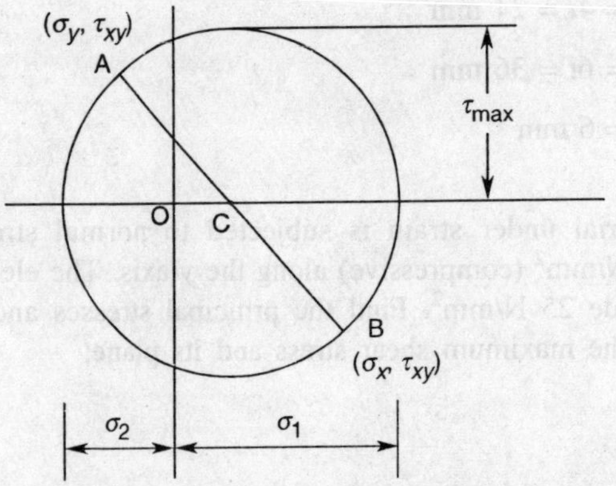

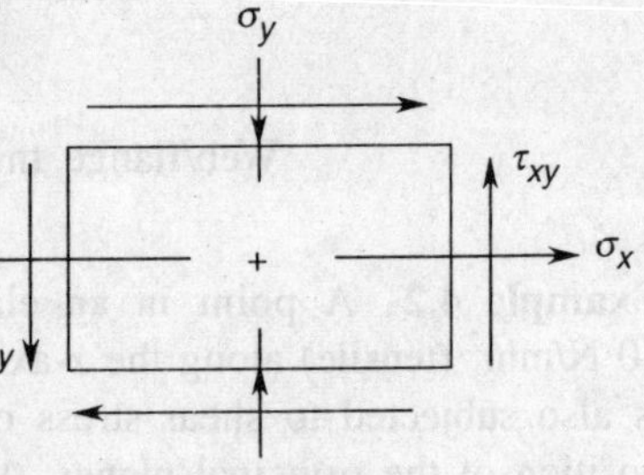

Stressed element

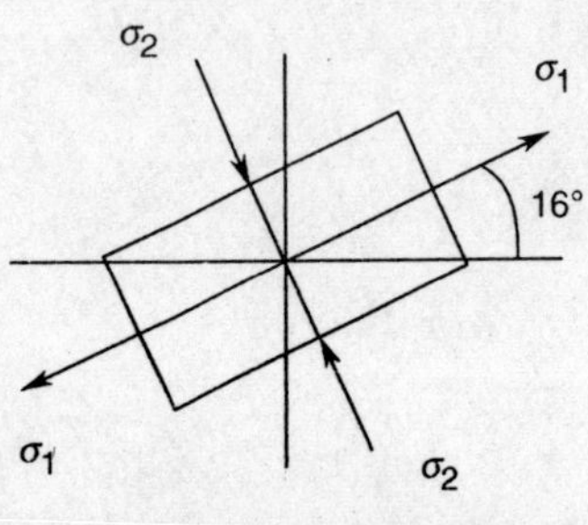

Stressed element subjected to principal stresses

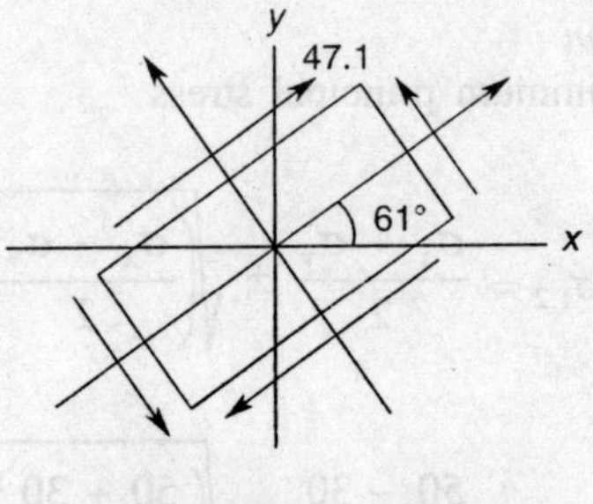

Element under maximum shear stress condition

Example 4.3 The shaft of an overhang crank is subjected to a force F of 1.5 kN as shown in the figure below. The shaft is made of 30 Mn2 steel having the allowable shear strength equal to 80 N/mm^2. Determine the diameter of the shaft.

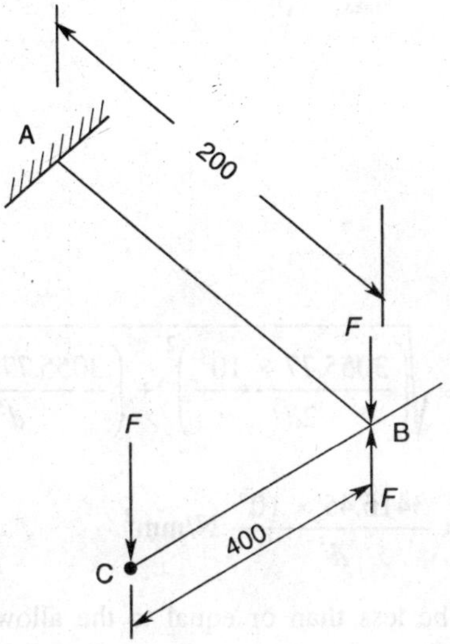

Solution The stresses are critical at point A, the reason being that at this point combined bending and torsional moments are applied. Assume that two equal and opposite forces of magnitude F are applied at point B. The force F at a distance of 200 mm from the support is creating bending moment, i.e.

$$M = F \times 200$$

$$= 1.5 \times 200 = 300 \text{ N} \cdot \text{m}$$

The remaining two parallel but opposite forces produce torsion in the shaft. Therefore, torque

$$T = 1.5 \times 400 = 600 \text{ N} \cdot \text{m}$$

Let the diameter of the shaft be d mm.

$$\text{Bending stress, } \sigma_b = \frac{M}{Z} = \frac{300 \times 10^3}{\dfrac{\pi}{32} \times d^3} = \frac{3055.77 \times 10^3}{d^3} \text{ N/mm}^2$$

$$\text{Shear stress, } \tau = \frac{Tr}{J} = \frac{600 \times 10^3 \times d/2}{\dfrac{\pi}{32} d^4}$$

$$= \frac{3055.77 \times 10^3}{d^3} \text{ N/mm}^2$$

Maximum shear stress at a point, according to Mohr's circle

$$\tau_{max} = \sqrt{\left(\frac{\sigma_x - \sigma_y}{2}\right)^2 + \tau_{xy}^2}$$

where

$\sigma_x = \sigma_b$

$\sigma_y = 0$

$\tau_{xy} = \tau$

Thus,

$$\tau_{max} = \sqrt{\left(\frac{3055.77 \times 10^3}{2d^3}\right)^2 + \left(\frac{3055.77 \times 10^3}{d^3}\right)^2}$$

$$= \frac{3416.45 \times 10^3}{d^3} \text{ N/mm}^2$$

For safe design, τ_{max} should be less than or equal to the allowable shear strength. That is,

$$\frac{3416.45 \times 10^3}{d^3} \leq 80$$

or

$$d \geq 34.95 \text{ mm}$$

Let us adopt 35 mm as the shaft diameter.

Example 4.4 A link of a machine is subjected to a force of 20 kN as shown in the figure below. The link is made of grey cast iron having the allowable strength of 100 N/mm^2. Determine the dimension of the cross-section.

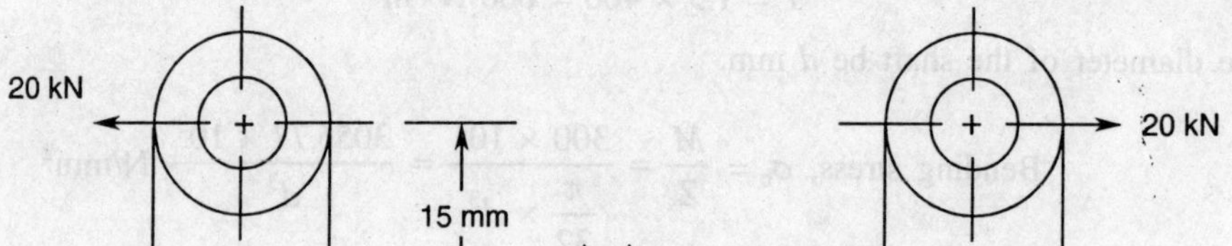

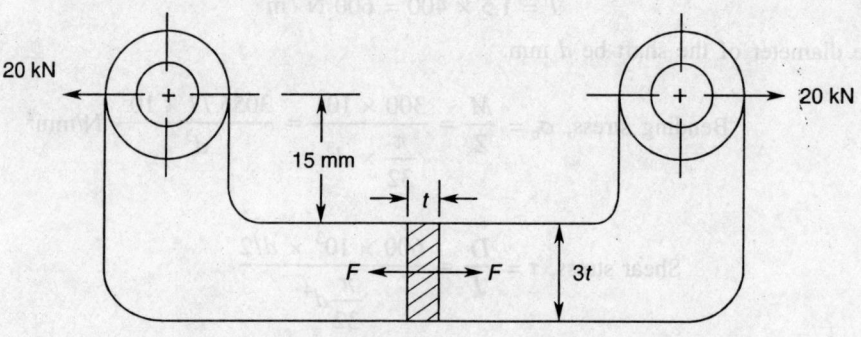

Solution $F = 20$ kN

$\sigma_{allowable} = 100$ N/mm^2

The cross-section of the member link is subjected to a combined direct stress and bending stress. Thus,

$$\sigma = \frac{F}{A} + \frac{M}{Z}$$

Moment, $M = 20 \times 10^3(15 + 1.5t)$

$$\sigma = \frac{20 \times 10^3}{3t \times t} + \frac{20 \times 10^3(15 + 1.5t)}{\frac{1}{6}t \times (3t)^2} \le \sigma_{allowable}$$

$$= \frac{6666.67}{t^2} + \frac{13,333.34(15 + 1.5t)}{t^3} \le 100$$

or

$$t^3 - 266.67t - 2000 = 0$$

By trial and error, $t = 19.5$ mm. Let us adopt $t = 20$ mm.
Depth of the link section, $h = 3t = 60$ mm.

Example 4.5 A C-clamp as shown in the figure below carries a load of 25 kN. The cross-section of the clamp at section X-X is rectangular having a width equal to twice the thickness. Assuming that the clamp is made of steel having allowable strength of 100 N/mm², find its dimensions. Also, determine the stresses at sections Y-Y and Z-Z.

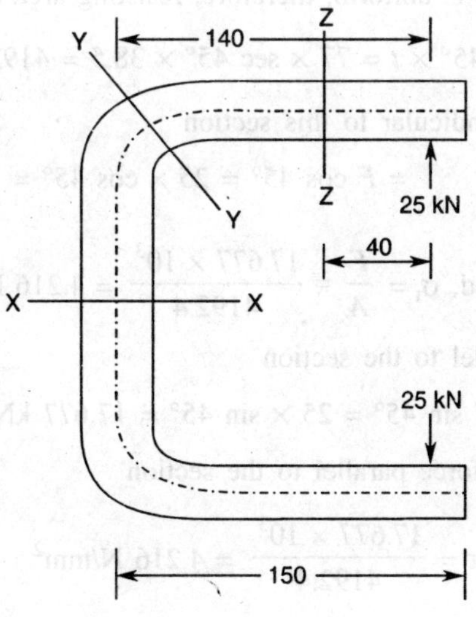

Solution $F = 25$ kN, $\sigma = 100$ N/mm², $b = 2t$, $e = 140$ mm

Cross-section area, $A = b \times t = 2t^2$

$$\text{Direct stress, } \sigma_x = \frac{F}{A} = \frac{25 \times 10^3}{2t^2} = \frac{12.5 \times 10^3}{t^2} \text{ N/mm}^2$$

Bending moment due to force F, $M = F \times e$

$$= 25 \times 140 = 3500 \text{ N} \cdot \text{m}$$

$$\text{Section modulus, } Z = \frac{1}{6} tb^2 = \frac{4t^3}{6}$$

$$\text{Bending stress, } \sigma_b = \frac{M}{Z} = \frac{3500 \times 10^3}{\dfrac{4t^3}{6}} = \frac{5250 \times 10^3}{t^3} \text{ N/mm}^2$$

Induced stresses at section X-X should not exceed the permissible strength, i.e.

$$\frac{12.5 \times 10^3}{t^2} + \frac{5250 \times 10^3}{t^3} \le 100$$

or

$$t^3 - 125t - 52500 = 0$$

Solving by trial and error, $t = 38.5$ mm

width, $b = 2t = 77$ mm

Stresses at section Y-Y

The cross-section of the frame is uniform, therefore, resisting area at section Y-Y is

$$A = b \sec 45° \times t = 77 \times \sec 45° \times 38.5 = 4192.4 \text{ mm}^2$$

Component of the load perpendicular to this section

$$= F \cos 45° = 25 \times \cos 45° = 17.677 \text{ kN}$$

$$\text{Tensile stress produced, } \sigma_t = \frac{F}{A} = \frac{17.677 \times 10^3}{4192.4} = 4.216 \text{ N/mm}^2$$

Component of the force parallel to the section

$$= F \sin 45° = 25 \times \sin 45° = 17.677 \text{ kN}$$

Shear stress produced by the force parallel to the section

$$\tau = \frac{17.677 \times 10^3}{4192.4} = 4.216 \text{ N/mm}^2$$

Bending moment at section Y-Y due to load

$$M = 25 \times 140 = 3500 \text{ N} \cdot \text{m}$$

$$\text{Section modulus, } Z = \frac{t \times (b \sec 45°)^2}{6} = 76{,}088.8 \text{ mm}^3$$

$$\text{Bending stress, } \sigma_b = \frac{M}{Z} = \frac{3500 \times 10^3}{76{,}088.8} = 46 \text{ N/mm}^2$$

Due to bending, the maximum tensile stress is produced at the inner corner and the maximum compressive stress at the outer corner.

Maximum tensile stress = 4.216 + 46 = 50.216 N/mm^2

Maximum compressive stress = 4.216 – 46 = – 41.784 N/mm^2

Since the shear stress acts perpendicular to the tensile stress, therefore, it is a case of biaxial loading. Thus, maximum principal stress

$$\sigma_1 = \frac{\sigma_t}{2} + \sqrt{\left(\frac{\sigma_t}{2}\right)^2 + \tau^2}$$

$$= \frac{50.216}{2} + \sqrt{\left(\frac{50.216}{2}\right)^2 + (4.216)^2}$$

$$= 50.56 \text{ N/mm}^2$$

and maximum shear stress

$$\tau_{max} = \sqrt{\left(\frac{\sigma_t}{2}\right)^2 + \tau^2}$$

$$= \sqrt{\left(\frac{50.216}{2}\right)^2 + 4.216^2} = 25.49 \text{ N/mm}^2$$

Stresses at section Z-Z

Bending moment at section Z-Z, $M = 25 \times 40 = 1000$ N · m

$$\text{Section modulus, } Z = \frac{tb^2}{6} = \frac{38.5 \times 77^2}{6} = 38{,}044.4 \text{ mm}^3$$

$$\text{Bending stress} = \frac{M}{Z} = \frac{1000 \times 10^3}{38{,}044.4} = 26.28 \text{ N/mm}^2$$

Example 4.6 Determine the maximum stress in the frame of a 50 kN punch press as shown in the figure here.

Solution The c.g. of the section from the inner edge is

$$\overline{X} = \frac{A_1X_1 + A_2X_2}{A_1 + A_2} = \frac{300 \times 100 \times 50 + 100 \times 200 \times 200}{300 \times 100 + 100 \times 200} = 110 \text{ mm}$$

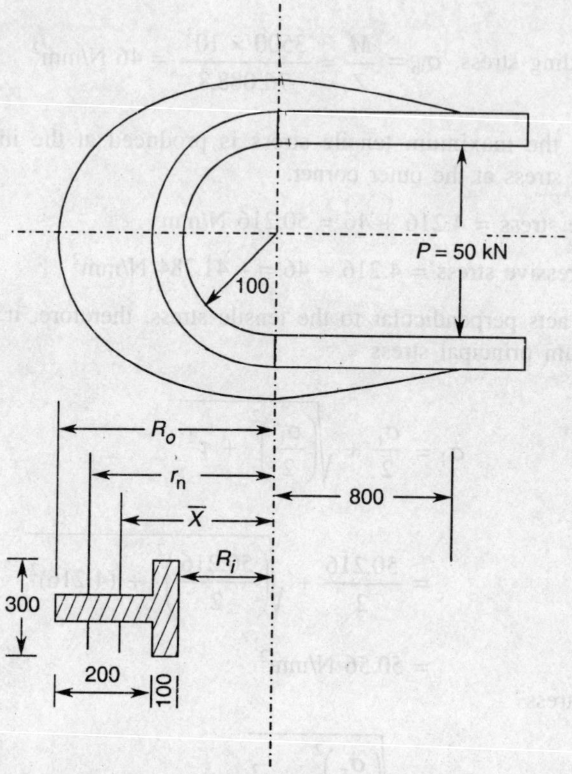

The radius of curvature of the neutral axis

$$r_n = \frac{t(b-t) + tb}{(b-t)\log_e\left(\dfrac{R_i + t}{R_i}\right) + t\log_e\left(\dfrac{R_o}{R_i}\right)}$$

$$= \frac{(300-100)\times 100 + 100 \times 300}{(300-100)\log_e\left(\dfrac{100+100}{100}\right) + 100 \times \log_e\left(\dfrac{400}{100}\right)} = 178.5 \text{ mm}$$

Distance between the c.g. and the neutral axis

$$e = 110 - (178.5 - 100) = 31.5 \text{ mm}$$

Eccentricity of load P

$$= 800 + 178.5 = 978.5 \text{ mm}$$

Bending moment, $M = 50 \times 978.5 = 48{,}925$ N · m

Direct tensile stress, $\sigma_t = \dfrac{P}{A} = \dfrac{50{,}000}{50{,}000} = 1 \text{ N/mm}^2$

Maximum bending stress at the inner fibre

$$\sigma_b = \frac{Mc_i}{AeR_i} = \frac{48{,}925 \times 10^3 (178.5 - 100)}{50{,}000 \times 31.5 \times 100} = 24.38 \text{ N/mm}^2$$

Resultant maximum stress at the inner surface

$$\sigma = \sigma_t + \sigma_b = 1 + 24.38 = 25.38 \text{ N/mm}^2$$

Example 4.7 A central horizontal section of a hook is a symmetrical trapezium 60 mm deep, the inner width being 60 mm and the outer width being 30 mm. Estimate the extreme intensities of stress when the hook carries a load of 20 kN. The load line passes at 40 mm from the inside edge of the section and the centre of curvature lies in the load line.

Solution Section of the hook is trapezoidal, load carried is 20 kN.

The layout of the hook-section is shown in the figure below.

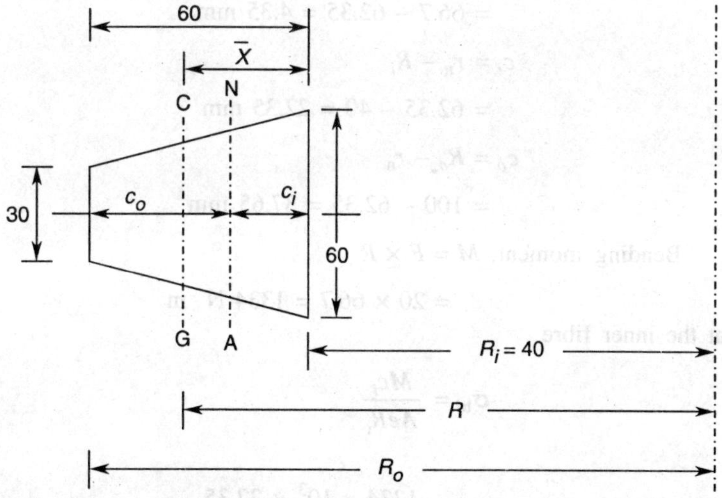

The distance $\overline{X}$ of the centroidal axis from the right side

$$\overline{X} = \frac{60 + 2 \times 30}{60 + 30} \times \frac{60}{3} = 26.7 \text{ mm}$$

$$R_i = 40 \text{ mm}$$

$$R_o = 100 \text{ mm}$$

$$R = R_i + \overline{X} = 40 + 26.7 = 66.7 \text{ mm}$$

$$\text{Area of section, } A = (B + b) \times \frac{h}{2} = (60 + 30) \times \frac{60}{2} = 2700 \text{ mm}^2$$

The radius of curvature of the neutral axis

$$r_n = \frac{\dfrac{B+b}{2} \times h}{\left(\dfrac{BR_o - b \times R_i}{h}\right) \log_e\left(\dfrac{R_o}{R_i}\right) - (B-b)}$$

$$= \frac{\dfrac{60+30}{2} \times 60}{\left(\dfrac{60 \times 100 - 30 \times 40}{60}\right) \log_e\left(\dfrac{100}{40}\right) - (60-30)} = 62.35 \text{ mm}$$

Eccentricity between the centroidal axis and the neutral axis

$$e = R - r_n$$

$$= 66.7 - 62.35 = 4.35 \text{ mm}$$

$$c_i = r_n - R_i$$

$$= 62.35 - 40 = 22.35 \text{ mm}$$

$$c_o = R_o - r_n$$

$$= 100 - 62.35 = 37.65 \text{ mm}$$

Bending moment, $M = F \times R$

$$= 20 \times 66.7 = 1334 \text{ N} \cdot \text{m}$$

Bending stress at the inner fibre

$$\sigma_{bi} = \frac{Mc_i}{AeR_i}$$

$$= \frac{1334 \times 10^3 \times 22.35}{2700 \times 4.35 \times 40} = 63.46 \text{ N/mm}^2$$

Bending stress at the outer fibre

$$\sigma_{bo} = \frac{Mc_o}{AeR_o} = \frac{1334 \times 10^3 \times 37.65}{2700 \times 4.35 \times 100} = 42.76 \text{ N/mm}^2$$

EXERCISES

1. Define the following terms:
   (a) Nominal stress (b) Yield stress
   (c) Proof stress (d) Modulus of elasticity
   (e) Elastic strain
2. What do you mean by the principal plane and the principal stresses?

3. What do you mean by a column? What is the effect of end conditions on the crippling load capacity of a column?
4. Define the Castingliano's theorem.
5. Explain how the bending of a curved beam is different from that of a simple beam.
6. What is the significance of deflection of a beam in the design of a component?
7. A steel pin of 22 mm diameter is fastened in a steel plate of cross-section (100 × 8) mm^2 to a wall. Determine the tensile stress in the plate and the shearing and crushing stresses in the pin if a load of 40 kN is applied in the plane of the plate.
8. Determine the maximum thickness of a steel sheet into which a hole of 20 mm size can be punched, if the ultimate strength of the steel in shear is 400 N/mm^2 and the allowable bearing stress in the hardened end of the steel punch is 2000 N/mm^2.
9. Find the ratio of weight of three beams of the same length subjected to the same bending moment and same stress, having as cross-sections, respectively, a circle, a square, and a rectangle.
10. A simply supported beam of rectangular section having depth three times the width, is subjected to a point load 20 kN at 300 mm from the left support. If the span of the beam is 700 mm, draw the SFD and the BMD of the beam. Also, determine the dimensions of the section if the allowable strength of the material is 200 N/mm^2.
11. A power press has a frame of T-section with flange dimensions as 75 mm × 25 mm and web dimensions as 75 mm × 25 mm. Plot the stress distribution across the section if a load of 20 kN acts vertically downwards along a line at a distance of 35 mm from the inner flange of the section.
12. Find the width of the member section at A-A if the limiting value of stresses to be induced is 100 N/mm^2.

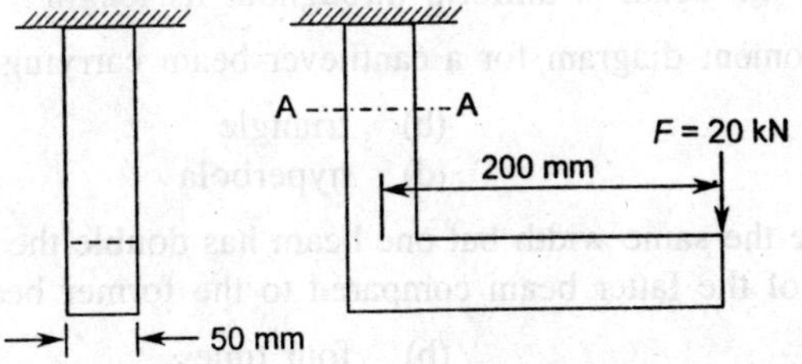

13. A link shown in the figure below is subjected to a tensile load of 30 kN with $h = 2t$. If the maximum elongation is 0.13 mm, determine the section size. Allowable strength of the material is 300 N/mm^2.

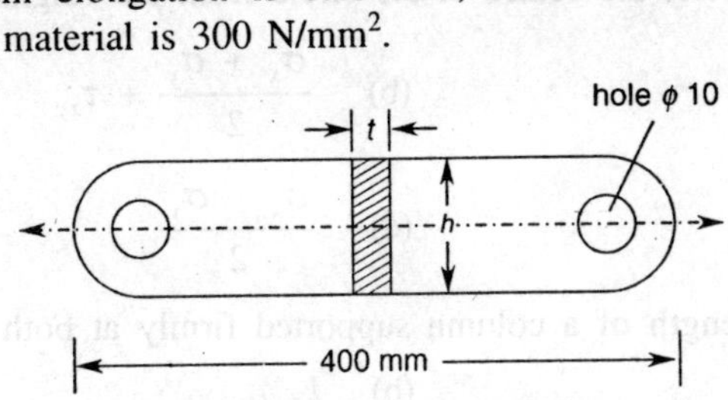

14. If a slot of 10 mm width is cut in a link of Exercise 13, determine the revised dimensions.

15. A propeller shaft in a ship is of 400 mm diameter. The allowable shear stress is 50 N/mm^2 and the allowable twist is 1° per 10 m length. Determine the maximum torque that can be transmitted by the shaft. If the shaft is hollow with the inner to outer diameter ratio as 0.8, determine the revised dimensions of the shaft.

MULTIPLE CHOICE QUESTIONS

1. Hooke's law holds good up to the
   (a) yield point (b) elastic limit
   (c) plastic limit (d) fracture point
2. A beam is loaded as cantilever. If the load at the end is increased, the failure will occur
   (a) at the point of load (b) at the middle of the beam
   (c) at the fixed end (d) at the free end
3. The moment of inertia of an area is always least with respect to the
   (a) bottom axis (b) centroidal axis
   (c) topmost axis (d) radius of gyration
4. The bending moment on a beam is maximum where the shear force is
   (a) maximum (b) minimum
   (c) equal (d) zero
5. A beam is said to be of uniform strength, if the
   (a) bending moment is the same throughout its length
   (b) bending stress is the same throughout its length
   (c) deflection is the same throughout its length
   (d) section of the beam is uniform throughout its length
6. The bending moment diagram for a cantilever beam carrying udl will be a
   (a) rectangle (b) triangle
   (c) parabola (d) hyperbola
7. Two beams have the same width but one beam has double the depth of the other. The elastic strength of the latter beam compared to the former beam will be
   (a) double (b) four times
   (c) half (d) equal
8. In the Mohr's circle, the centre of the circle from the origin is taken as:
   (a) $\sigma_x + \sigma_y$ (b) $\dfrac{\sigma_x + \sigma_y}{2} + \tau_{xy}$
   (c) $\dfrac{\sigma_x + \sigma_y}{2}$ (d) $\dfrac{\sigma_x - \sigma_y}{2}$
9. The equivalent length of a column supported firmly at both ends is
   (a) $2l$ (b) l
   (c) $0.5l$ (d) $0.7l$

Chapter 5

Fundamentals of Machine Design

5.1 INTRODUCTION

The design of machine parts is concerned with the determination of dimensions of the parts. In this process a designer should make sure that the stresses developed in a part, are less than the maximum allowable strength of the material. This maximum allowable stress is referred as the *design stress* and its value depends upon many factors, namely the types of load, the service conditions, the life cycle, and the safety considerations.

The objective of this chapter is to develop relations between the mechanics of machine elements and the material strength in order to achieve reasonable geometry of the part, which will not fail during its service period.

5.2 TYPES OF LOAD

The external forces grouped together constitute what is called the *load* acting on the body. The nature of the load may vary from dead weight to varying load.

The load acting on a member can be classified as under:

Static load. A static or steady load is a stationary force, moment, or torque acting on the member. It does not change its magnitude, point of application, and direction.

Impact or shock load. When a component is subjected to a sudden load or when the duration of loading is less than about half the natural period of vibration, such loading is called impact or shock loading.

Fatigue load. The alternating or repeated loading, whose duration is more than half but less than three times the natural period of vibration is called the *fatigue load*. In cyclic loading, the load fluctuates between the maximum and minimum levels of loading. The loads may further be classified as (a) reversed, (b) repeated, and (c) completely reversed.

5.3 FACTOR OF SAFETY

A designer who designs a component must ensure that the part will not fail during its service period. Thus, there is a need for the designer to be well-versed in the knowledge as to how that part will fail during the service and how to specify a margin of safety to account for uncertainties. If we analyze different failed parts, even though these may have been designed by experienced designers, the failure may be attributed to one or more of the following causes:

(i) Uncertainty about the properties of the material
(ii) Uncertainty about the magnitude of the load, its point of application and direction
(iii) Uncertainty about the accuracy of the mathematical model which represents the predicted behaviour of the machine part
(iv) Uncertainty about the operating conditions and reliability requirements

In machine design to account for all these uncertainties, we provide a sufficient margin of safety. This is usually expressed in the form of a factor of safety (FoS), defined as:

$$\text{Factor of safety, } n = \frac{\text{failure load}}{\text{working load}}$$

$$n_u = \frac{\text{ultimate strength}}{\text{working stress}} \tag{5.1}$$

$$n_y = \frac{\text{yield strength}}{\text{working stress}} \tag{5.2}$$

In the design of mechanical elements, the factor of safety based on the yield strength is quite often used because a mechanical system would become non-functional if some vital component has deformed or yielded permanently.

From its very concept, the value of the factor of safety is more than one. But what should be the factor of safety for a component is something that is very often difficult to decide. With a small value of FoS, the designer is taking a greater risk while with a large value, the component will be uneconomical. Thus, the selection of the factor of safety is more or less based on experience.

Joscepth P. Vidosic has suggested that the following factors of safety can be reasonable. These are based on the yield strength.

(i) n = 1.25–1.5 — for exceptionally reliable materials used under controlled conditions and subjected to load and stresses that can be determined with certainty.
(ii) n = 1.5–2.0 — For well-known materials under reasonable constant conditions and subjected to load or stresses that can be determined readily.
(iii) n = 2–2.5 — For average materials, operated in ordinary environment, load and stress conditions.
(iv) n = 2.5–3.0 — For less tried or brittle materials subjected to average service conditions.
(v) n = 3–4 — Untried materials with average load and stress conditions, or better known materials with uncertain loads.
(vi) n = 1–6 — For repeated loads; must be applied to endurance limit.

5.4 THEORIES OF FAILURE

The material behaviour varies from extreme brittleness to good ductility. The failure of a part made of brittle material may occur by fracture, while the ductile material may fail due to excessive yielding. Therefore, it is always important to know the cause of failure of mechanical components. The particular action subscribing to failure is termed *criterion of failure.* The formal strategy which predicts the failure of a component is called the *theory of failure.* Several theories of failure, based on different criteria of failure, have been proposed by various designers.

Some of the most commonly used theories are given below:

(i) The maximum normal stress or Rankine theory
(ii) The maximum shear stress or Tresca theory
(iii) The maximum distortion energy or von-Mises theory

A brief discussion on these theories of failure is given below:

The maximum normal stress theory

The maximum normal stress theory is based on failure in tension or compression. It may be applied to those materials which are relatively strong in shear but weak in tension or compression. The theory states that the maximum normal stress or the maximum principal stress in combined loading must be less than or equal to the yield strength of material, i.e.

$$\sigma_1 \text{ or } \sigma_{max} \leq \frac{\sigma_y}{n} \tag{5.3}$$

where n is the factor of safety.

This theory is considered to be reasonably satisfactory for those components which are made of brittle material, such as cast iron.

The maximum shear stress theory

The maximum shear stress theory is also known as the Tresca's theory. This theory is easy to use and is always conservative when applied to ductile materials. According to the Tresca's theory, the material will yield when the maximum shear stress in any mechanical element reaches a certain value and this value is the shear stress at the instant of failure in a tensile test. Thus,

$$\sigma_1 - \sigma_2 = 2\tau_{max} = \sigma_y \tag{5.4}$$

where τ_{max} is the radius of the Mohr's circle, i.e. the maximum shear stress in the cycle. For a triaxial stress element the equation of maximum shear stress theory is given as

$$\frac{\sigma_y}{n} \geq \text{the largest of } \begin{Bmatrix} \sigma_1 - \sigma_2 \\ \sigma_2 - \sigma_3 \\ \sigma_1 - \sigma_3 \end{Bmatrix} \tag{5.5}$$

The maximum distortion energy theory

The maximum distortion energy theory is credited to von-Mises and Henky. This theory states that inelastic action at any point in a body under any combination of stresses begins only when the strain energy of distortion per unit volume absorbed at the point is equal to the strain energy of distortion absorbed per unit volume at any point in a bar stressed to elastic limit under a state of uniaxial stress as would occur in a simple tensile test.

According to von-Mises, the equation of maximum distortion energy theory of failure for a triaxial stress element is given as under:

$$(\sigma_1 - \sigma_2)^2 + (\sigma_2 - \sigma_3)^2 + (\sigma_3 - \sigma_1)^2 = 2\sigma_y{}^2 \tag{5.6}$$

In the case of a biaxial stress element with $\sigma_3 = 0$, Eq. (5.6) reduces to

$$\sigma_1{}^2 + \sigma_2{}^2 - \sigma_1\sigma_2 = \left(\frac{\sigma_y}{n}\right)^2 \tag{5.7}$$

Equations (5.6) and (5.7) define the beginning of yield for triaxial and biaxial stress states, respectively.

The graphical representation of the above three theories of failure is shown in Figure 5.1. According to the maximum normal stress theory, this diagram is divided into four quadrants forming a square ABCD. The material will reach its elastic limit when the point (σ_1, σ_2) falls outside the square ABCD.

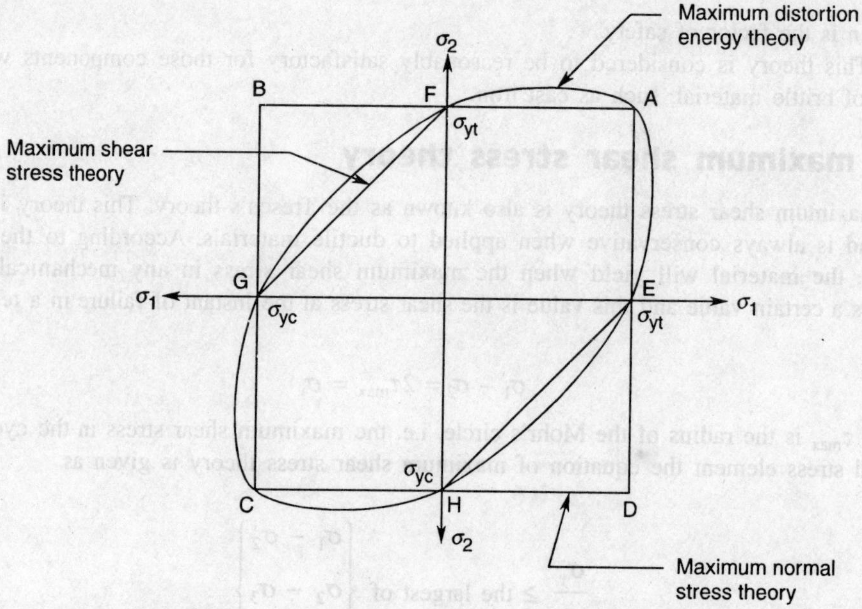

Figure 5.1 Comparison of different theories of failure.

According to the maximum shear stress theory, the shear stress is given by $0.5(\sigma_1 - \sigma_2)$ and the failure in the fourth and second quadrants is represented by

$$\sigma_1 - \sigma_2 = \frac{\sigma_y}{n} \tag{5.8a}$$

or

$$\sigma_2 - \sigma_1 = \frac{\sigma_y}{n} \tag{5.8b}$$

Graphically, these equations are represented by parallel lines EH and GF as shown in Figure 5.1.

The maximum distortion energy theory for biaxial stress state is represented by the following equation:

$$\sigma_1^2 + \sigma_2^2 - \sigma_1\sigma_2 \leq \left(\frac{\sigma_y}{n}\right)^2 \tag{5.9}$$

This is the equation of an ellipse with centre as the origin and axes inclined at 45° as shown in Figure 5.1.

A close look at Figure 5.1 and well documented experimental results indicates that the maximum distortion energy theory predicts yielding with greater accuracy in all the four quadrants whereas the maximum shear stress theory gives results on the conservative side as its graph is inside the ellipse of distortion energy.

The maximum normal stress theory is the same as the maximum shear stress theory in the first and third quadrants. However, it is outside the distortion energy ellipse in the second and fourth quadrants. Thus, it would be totally unsafe to use the maximum normal stress theory here because it might predict certain margins of safety whereas in fact no safety exists.

5.5 STRESS CONCENTRATION

In the development of the basic stress equations for various types of loading, it was assumed that there was no irregularity in the cross-section of the member under consideration. However, most machine elements have some forms of discontinuities, namely a sudden change in section, grooves, holes, keyways, and other changes in sections. These discontinuities in a machine element alter the stress distribution in their neighbourhood so that the elementary stress equations no longer describe the actual state of stress in the part. Such discontinuities are called *stress raisers* and the region in which these occur is called the *area of stress-concentration*. Internal cracks and flaws, cavities in welds, blowholes, pressure at certain points are other common examples of stress raisers. Stress-concentration in a plate with a circular hole is shown in Figure 5.2.

The stress-concentration factor is used to relate the actual maximum stress at discontinuity to the nominal stress. It is a ratio of the maximum stress at the discontinuity to the nominal stress, i.e.

$$K_t = \frac{\text{maximum value of actual stress at discontinuity}}{\text{nominal stress}} = \frac{\sigma_{max}}{\sigma_0} \tag{5.10}$$

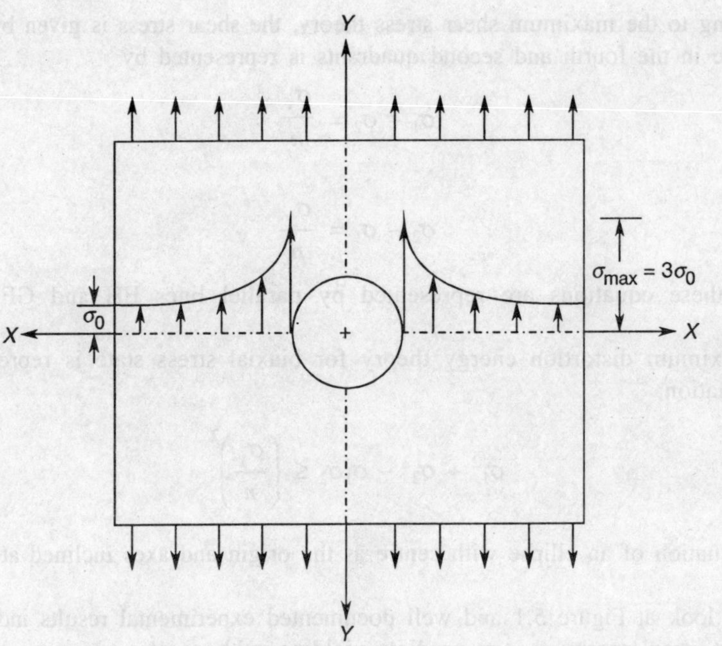

Figure 5.2 Stress-concentration in a plate with a circular hole.

and the stress-concentration factor for torsion

$$K_{ts} = \frac{\tau_{max}}{\tau_0} \tag{5.11}$$

The stress-concentration factor in a mechanical element primarily depends on three main characteristics: (i) the type and size of the discontinuity, (ii) the material of the part, and (iii) the character of the load. The magnitude of stress-concentration is determined by the following methods:

(i) Mathematical analysis using theory of elasticity
(ii) Finite element method
(iii) Experimental methods such as:

(a) Photo elasticity method
(b) Brittle coating method
(c) Soap film method

By nature the theoretical stress-concentration factor depends only on the shape and size of the discontinuity. Therefore, it is also called the form stress-concentration factor.

Most engineering materials are elastic in nature and they have the ability to yield when excessive stress is applied. This characteristic spreads the localized high-stress region into lower intensity. As a result, the actual value of the stress-concentration factor is always lower than that of the form stress factor K_t. Or in other words, the form stress-concentration factor may be considered as the highest limit of stress-concentration under the most adverse condition. The numerical values of the form stress-concentration factors for various types of discontinuities are given in the form of a curve. A few typical such curves are given in Figures 5.3 to 5.7. For the detailed treatment of the subject, the reader may consult the work of Paterson.*

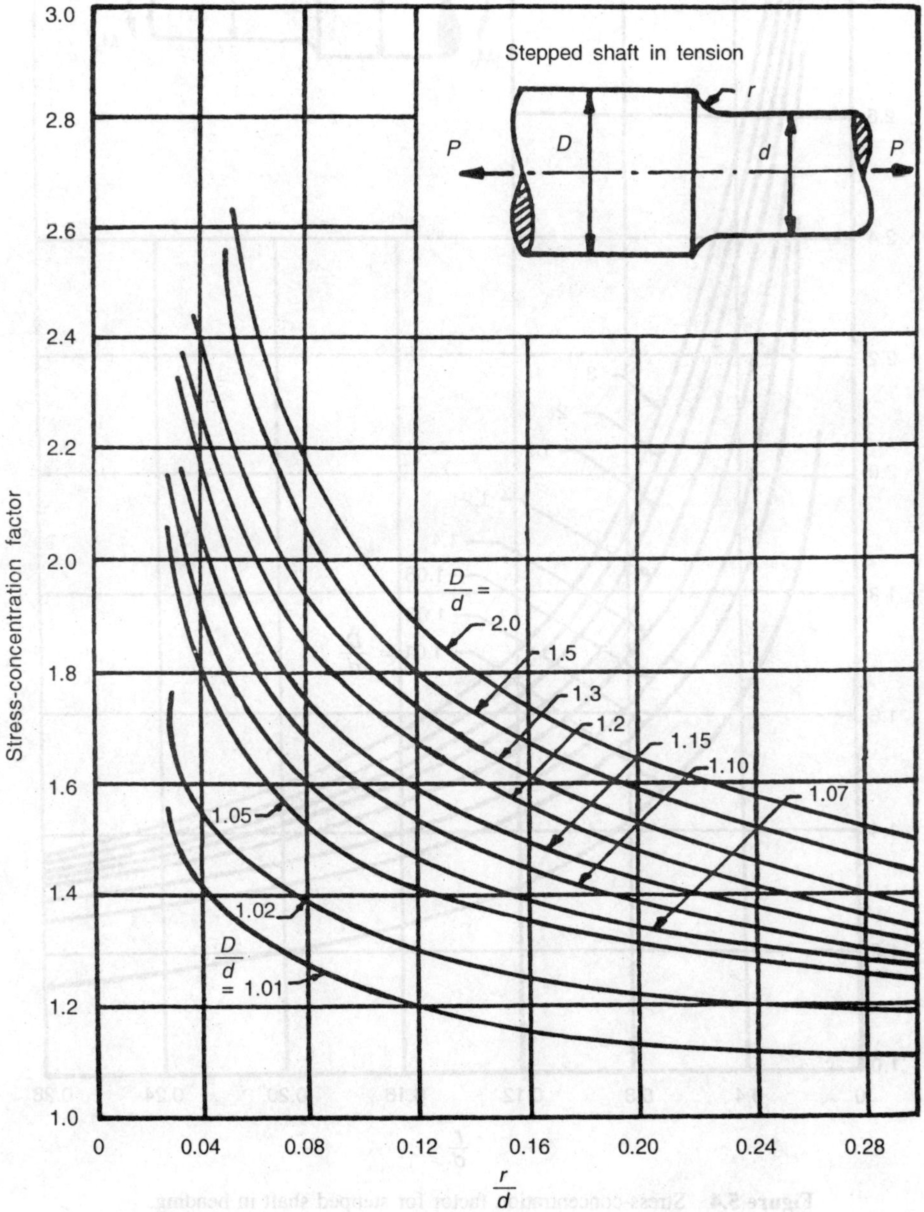

Figure 5.3 Stress-concentration factor for stepped shaft in tension.

---

* R.E. Paterson, *Stress Concentration Factors*, John Wiley and Sons, Inc., N.Y.

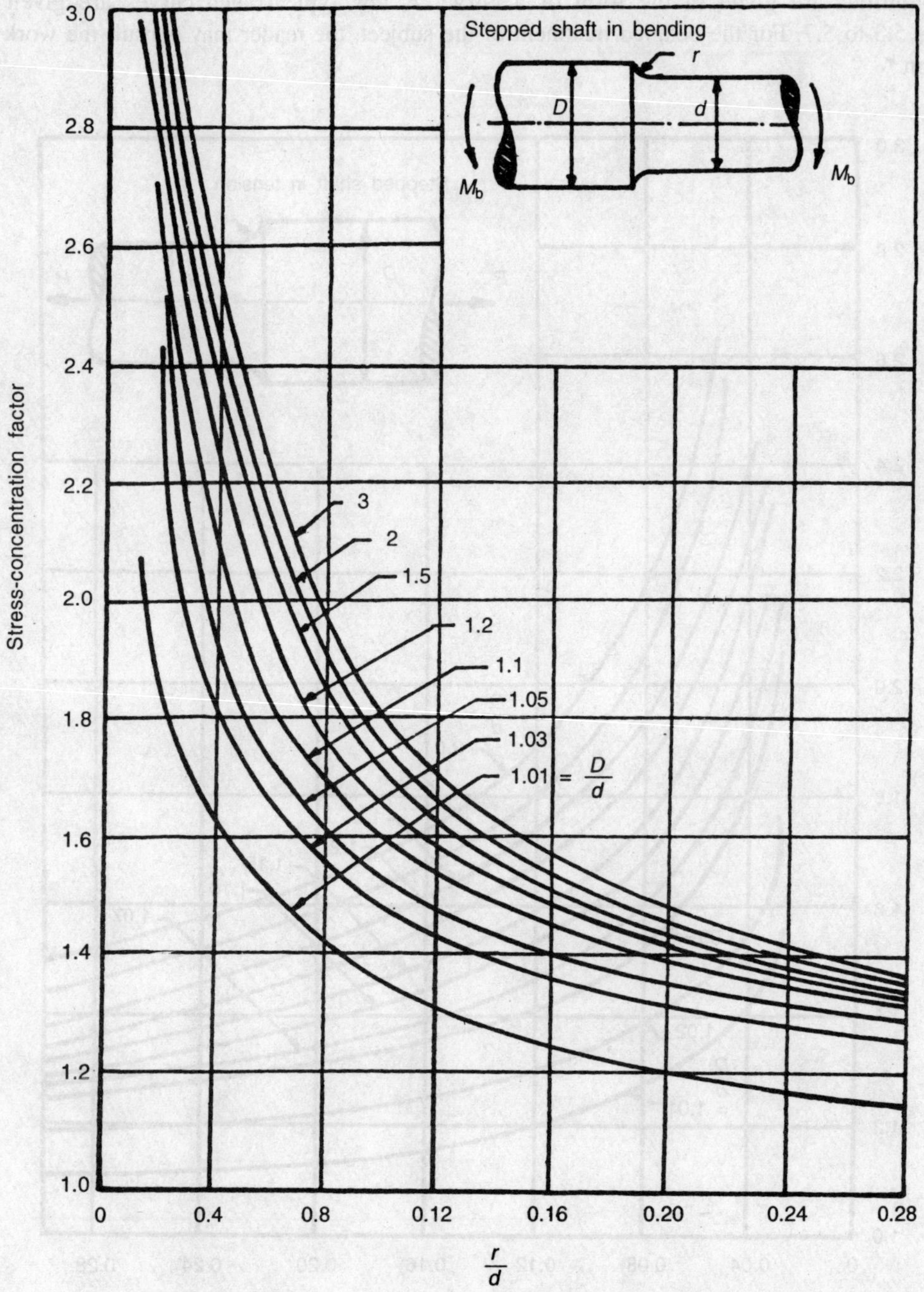

Figure 5.4 Stress-concentration factor for stepped shaft in bending.

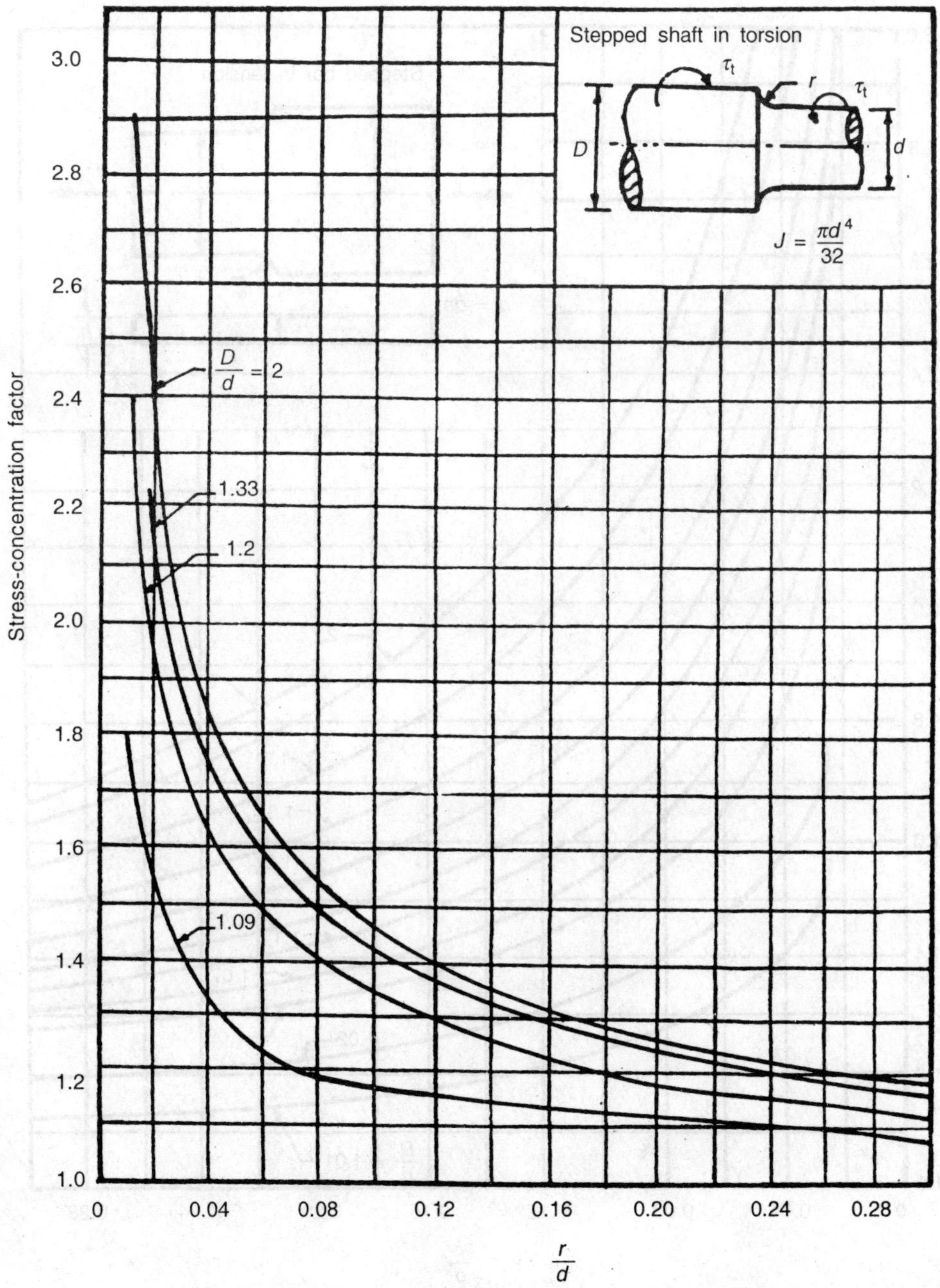

Figure 5.5 Stress-concentration factor for stepped shaft in torsion.

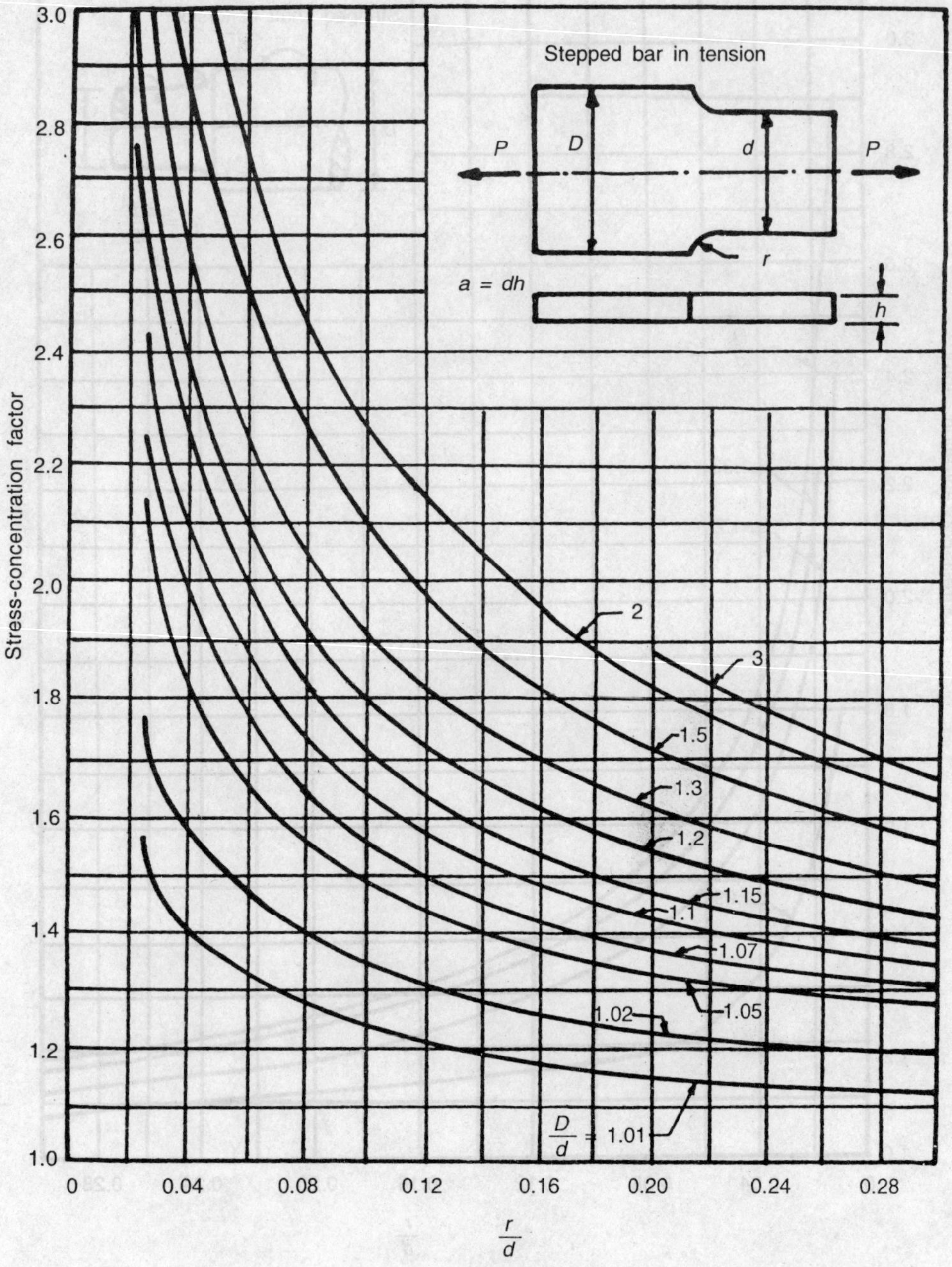

Figure 5.6 Stress-concentration factor for stepped bar in tension.

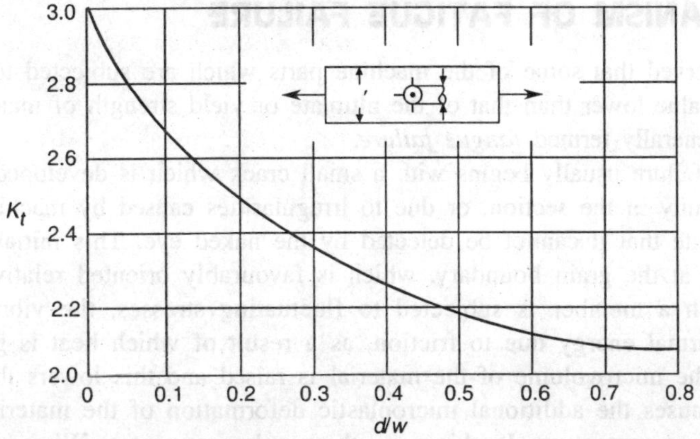

Figure 5.7 Stress-concentration factor for a plate with circular hole.

In the design of most mechanical elements, it is not possible to avoid the discontinuities or abrupt changes in the cross-section. However, some steps can be taken during designing to reduce the value of stress-concentration. For this, the designer must have a feel for stress-concentration and should know when and where it exists so that the same can be checked or reduced. A few examples in which original designs were altered to obtain lower value of stress-concentration are shown in Figure 5.8.

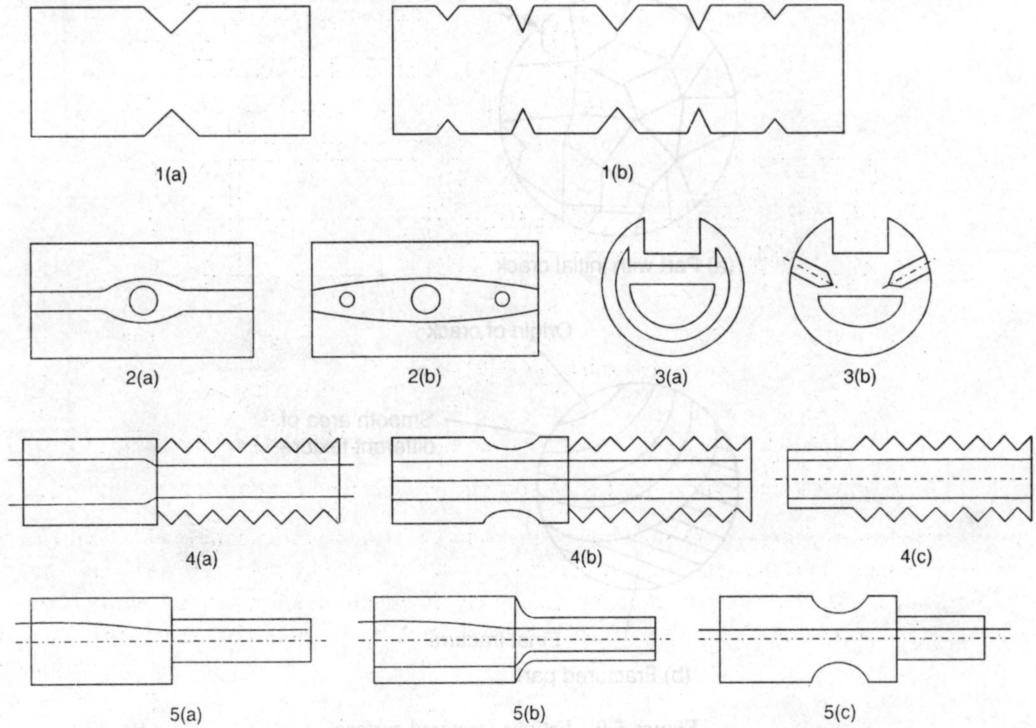

Figure 5.8 Different methods to reduce the effect of stress-concentration.

5.6 MECHANISM OF FATIGUE FAILURE

It has been observed that some of the machine parts which are subjected to variable loading fail at a stress value lower than that of the ultimate or yield strength of material. Such failure of the part is generally termed *fatigue failure.*

A fatigue failure usually begins with a small crack which is developed at a point either due to discontinuity in the section, or due to irregularities caused by machining. Initially the crack is so minute that it cannot be detected by the naked eye. This initial crack shears the material located at the grain boundary, which is favourably oriented relative to the external load. When such a member is subjected to fluctuating stresses, the vibrational energy is converted to thermal energy due to friction, as a result of which heat is generated and the temperature of the microvolume of the material is raised and this lowers the strength of the material. This causes the additional microplastic deformation of the material, which in turn further raises the temperature. In this way, the crack propogates. When the crack extends beyond the grain boundaries, it gets widened and starts to penetrate into the depth of the material along the material's weakened region. Finally, the propagation of the crack results into a small net area of cross-section which resists the load. This ultimately culminates in sudden brittle failure of the part without early warning. A close examination of fatigue fractured part reveals that it has two distinct zones, one with dull smooth surface and the other one is final fracture as shown in Figure 5.9.

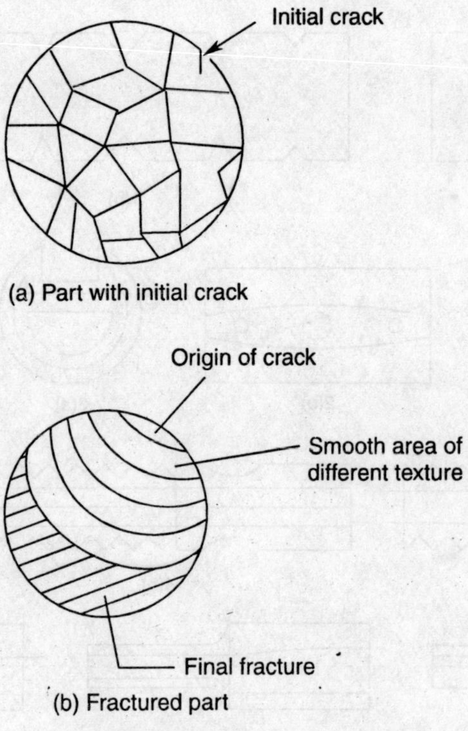

Figure 5.9 Fatigue fractured surface.

Some of the other important facts about fatigue are summarized as under:

(i) The crack generally starts at the surface of the material though metallurgical defects can initiate a crack beneath the surface.
(ii) In a part subjected to axial stress, the crack would usually occur on one side of the cross-section only.
(iii) In a part subjected to reverse bending or axial load combined with bending, the crack would usually occur on both sides of the cross-section.
(iv) In a part subjected to rotating bending stress, the crack is initiated from two opposite directions.

5.6.1 Fatigue Curve

In order to establish the fatigue strength of the material, a large number of test specimens, as shown in Figure 5.10, are carefully machined and polished. These specimens are then subjected to repeated forces of specific magnitude and at the same time the number of stress reversal cycles are counted till the fracture takes place. In this process, the first few specimens are

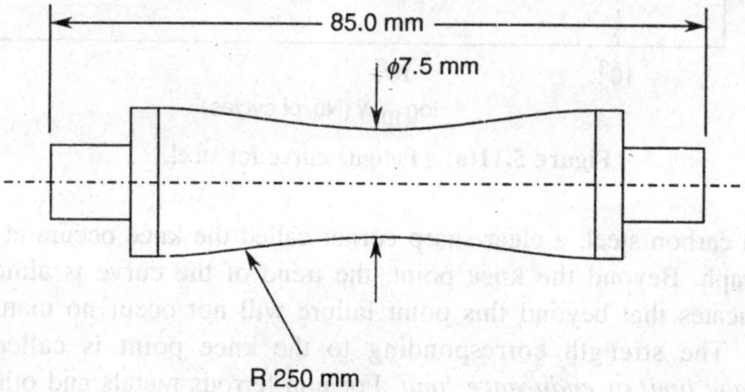

Figure 5.10 Typical test specimen.

loaded at a stress lower than the ultimate strength of the material and the number of cycles they survive, before the fracture takes place, are counted. In the second test, specimens are tested with a stress slightly less than that used for first test and again the number of cycles before the fracture takes place is counted. This process is continued and the results are plotted between stress and the number of cycles on the log-log graph as shown in Figure 5.11(a). This curve is called the *fatigue curve*.

A close look at the curve reveals that the fatigue curve can be broadly divided into two parts. One in which the specimens survived for less than 1000 cycles. It is called the *low cycle area* or *infant mortality period*. The specimens failing before 1000 cycles are generally of no value in the design of the mechanical components. However, the specimens which fail after attaining at least 1000 cycles are classified as *high cycle fatigue*. It is this portion of the diagram which has practical value for designing of parts under fatigue loading.

Since fatigue failure is statistical in nature and depends upon the type of stress and the material of the specimen, it is generally found that σ–N curves of different shapes are obtained.

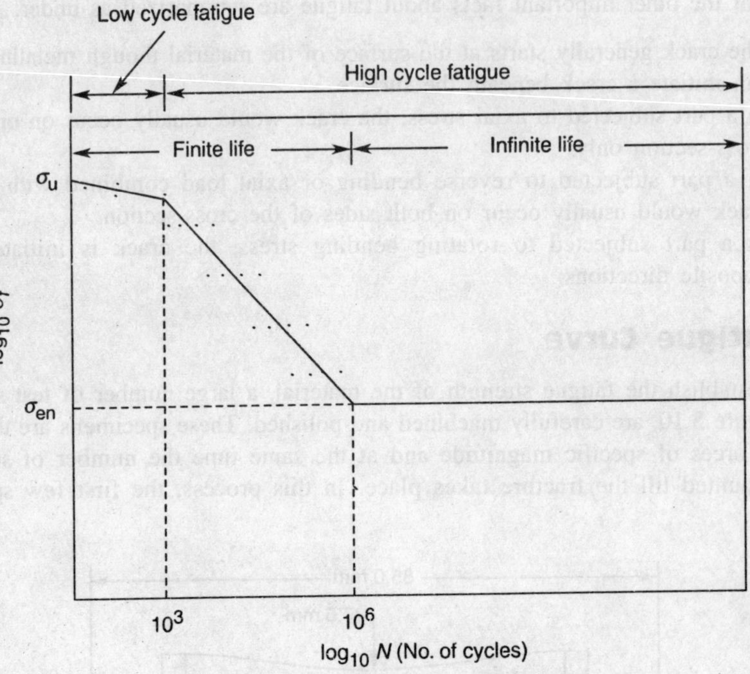

Figure 5.11(a) Fatigue curve for steel.

In the case of plain carbon steel, a clear sharp corner called the knee occurs at about 10^6 cycles as shown in the graph. Beyond the knee point, the trend of the curve is almost asymptotic to abscissa which indicates that beyond this point failure will not occur no matter how great the number of cycles. The strength corresponding to the knee point is called the *endurance strength* σ_{en} or *fatigue limit* or *endurance limit.* For non-ferrous metals and other ferrous metals such as cast iron this curve never becomes horizontal [see Figure 5.11(b)]. Hence, these materials do not have any specific endurance limit.

As mentioned earlier, the endurance limit depend upon the type of load, the stress, and the material of the component. Different materials when subjected to different types of load

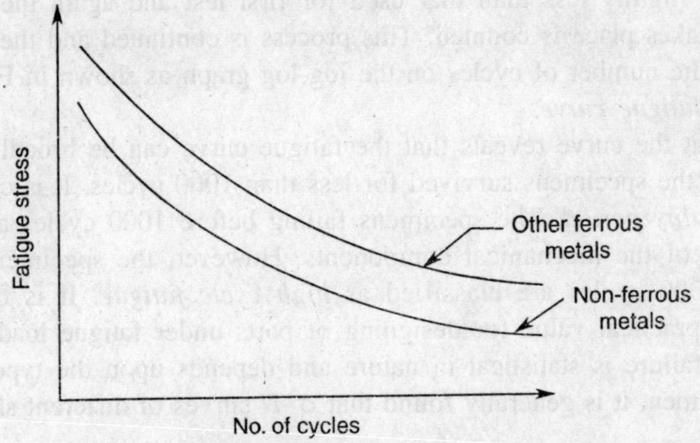

Figure 5.11(b) Fatigue curve.

exhibit different endurance limits. Generally, it is observed that the materials which have better static properties are also good under fatigue loading, although no exact relationship between the ultimate strength and the endurance limit exists. For preliminary design calculations, the following relations are quite useful.

1. Material considerations:
   (a) For steel $\sigma'_{en} = 0.5\sigma_u$ or 700 MPa maximum
   (b) For CI $\sigma'_{en} = 0.4\sigma_u$ or 275 MPa maximum
   (c) For non-ferrous metals $\sigma'_{en} = (0.25\text{–}0.3)\sigma_u$
2. When a component is subjected to direct stress:

$$\sigma_{en-a} = (0.7\text{–}1.0)\sigma'_{en} \text{ for steel, CI and non-ferrous metals}$$

3. For cyclic torsion load:
   (a) $\tau_{en} = (0.5\text{–}0.6)\sigma'_{en}$ for steel
   (b) $\tau_{en} = 0.8\sigma'_{en}$ for CI
   (c) $\tau_{en} = 0.2\sigma_u$ for non-ferrous metals

Besides the type of stresses and materials, the endurance limit also depends upon various other factors as discussed in Section 5.6.2.

5.6.2 Factors Affecting Endurance Limit

While designing a mechanical component it is unrealistic to expect that its endurance limit will match the values obtained in the laboratory under standard test and controlled conditions. Thus, the laboratory derived endurance strength needs a correction. Martin has classified some of the factors which have influence on the endurance limit. According to him, the endurance strength of a mechanical part can be derived using the expression

$$\sigma_{en} = K_a\, K_b\, K_c\, K_d\, K_e\, \sigma'_{en} \tag{5.12}$$

where

σ_{en} is the endurance limit of the mechanical part

σ'_{en} is the endurance limit of the specimen

K_a is the surface finish factor

K_b is the size factor

K_c is the reliability factor

K_d is the temperature factor

K_e is the modifying factor to account for stress concentration

A brief description of the endurance limit modifying factors is given below:

Surface finish factor (K_a)

The surface conditions of machine parts vary with the type of machining or shaping operations that are performed upon them. Experiments have shown that parts with poor surface finish have reduced fatigue limit compared to parts with smooth surface finish. The reduction in fatigue limit is due to signs of toolmarks left on the surface. The degree of smoothness of the surface and the endurance limit reduce in the following order of operations: (i) polishing, (ii) grinding,

(iii) fine turning, and (iv) rough turning. Figure 5.12 gives the value of the surface finish factor for a known value of the ultimate strength and the type of surface finishing operation.

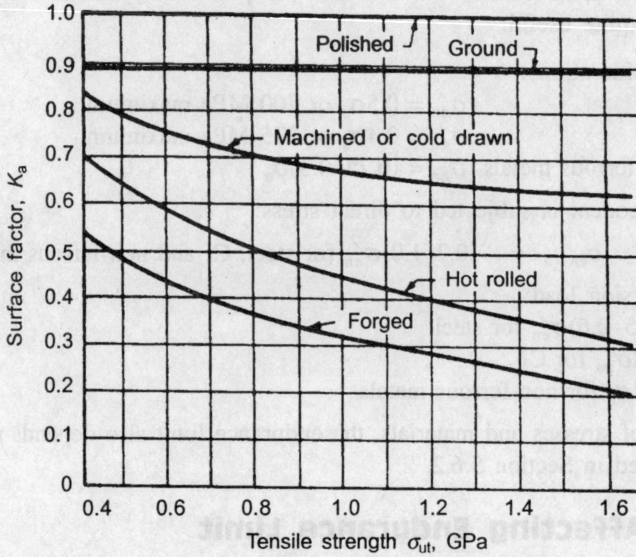

Figure 5.12 Effect of surface finish.

Size factor (K_b)

The endurance limit of the component is found to decrease with the increasing size of the specimen when it is subjected to bending and torsion loading, because an increase in the size of a specimen is apt to have more internal defects. However, for axial loading, tensile tests indicate that no size correction is necessary.

The value of the size factor K_b for a round component is calculated by the following empirical relations:

$$K_b = 1 \quad \text{for } d \le 8 \text{ mm} \tag{5.13}$$

$$= 1.189d^{-0.097} \quad \text{for } 8 \text{ mm} \le d \le 250 \text{ mm}$$

where d is the diameter of the specimen.

In rectangular components, first an equivalent diameter is calculated by the constant volume approach and then the corresponding size factor is determined.

Reliability factor (K_c)

The design of a mechanical element subjected to fatigue load should be such that it will last for any desired life. Since the fatigue failure is statistical in nature and also that the fatigue curve (σ–N) is not a straight line but a scattered band, the life of a component at a particular reliability may constitute a more effective method of measuring design performance.

The component requiring reliability greater than 50 per cent needs a correction in endurance strength. The values of reliability factors at different levels of reliability are given in Table 5.1.

Table 5.1 Reliability factor

Reliability	*Reliability factor*
0.5	1.0
0.9	0.897
0.95	0.868
0.99	0.814
0.999	0.753

Temperature factor (K_d)

When the temperature of a component changes, the properties of the material change too. If the temperature of a component subjected to stresses is increased, it will induce creep in the material which in turn will change the nature of the fatigue curve and the endurance limit. The following guidelines can be used for selecting the temperature factor for correcting the endurance limit.

$K_d = 1$ for temperatures ≤ 300°C
$= 0.5$ for temperatures > 300°C

Stress concentration (K_e)

If a part subjected to fatigue loading has stress-raising notches, it causes more damage than that caused by the ordinary stress raisers. In fatigue design, a factor called the *fatigue stress-concentration factor* is used, instead of the *form stress-concentration factor*, which is defined as the ratio of endurance limit of the notch-free specimen to endurance limit of the notched specimen. That is,

$$K_f = \frac{\text{endurance limit of the notch-free specimen}}{\text{endurance limit of the notched specimen}}$$

The fatigue stress-concentration factor is related to the form stress concentration factor by a term called the *notch sensitivity factor q*, which is defined as the ratio of increase in actual stress over the nominal to the increase in theoretical stress value over the nominal stress. Thus,

$$q = \frac{K_f\sigma_0 - \sigma_0}{K_t\sigma_0 - \sigma_0} = \frac{K_f - 1}{K_t - 1} \tag{5.14}$$

or

$$K_f = 1 + q(K_t - 1) \tag{5.15}$$

The notch sensitivity factor primarily depends upon the notch geometry, the notch radius, the size of the section, the material, heat treatment, the manufacturing process, and so on. The notch sensitivity factor for a given material and notch radius can be found from Figure 5.13.

The value of fatigue stress concentration factor K_f is always greater than one. Therefore for modifying the endurance limit to account for stress concentration it is related to K_e as follows:

$$K_e = \frac{1}{K_f} \tag{5.16}$$

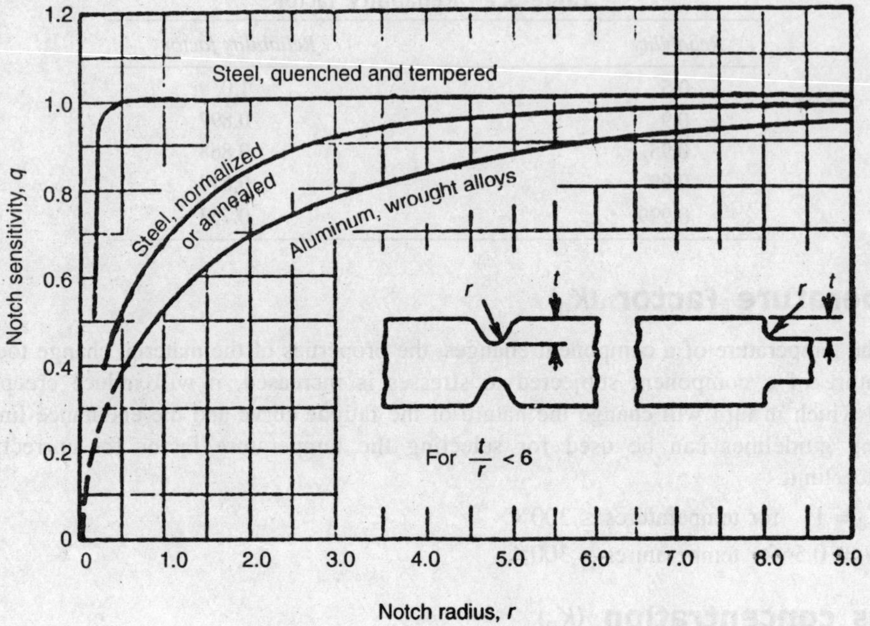

Figure 5.13 Notch sensitivity factor.

5.7 DESIGN FOR FATIGUE

The design of a mechanical element for fatigue failure is classified into two groups: (i) design for infinite life and (ii) design for finite life. When a component is required to be designed for infinite life, the endurance limit becomes the basic criterion of failure. The stress induced in such a component should be less than or equal to the endurance limit so that the component can survive an infinite number of cycles. In other words, the level of stress on the component should be lower than the knee point of the fatigue curve. The following equations are used for the design of these components:

$$\sigma \le \frac{\sigma_{en}}{n} \tag{5.17}$$

and

$$\tau \le \frac{\tau_{en}}{n} \tag{5.18}$$

where σ and τ are induced direct and shear stresses and σ_{en} and τ_{en} are endurance strengths in direct load and torsion, respectively.

5.7.1 Design for Finite Life

In a high cycle fatigue, the region between the 10^3–10^6 cycles is generally called the *finite life region.* If a component is required to be designed for finite life, it is called the *design for finite life.* The fatigue curve is as shown in Figure 5.14. It consists of a straight line AB drawn on the

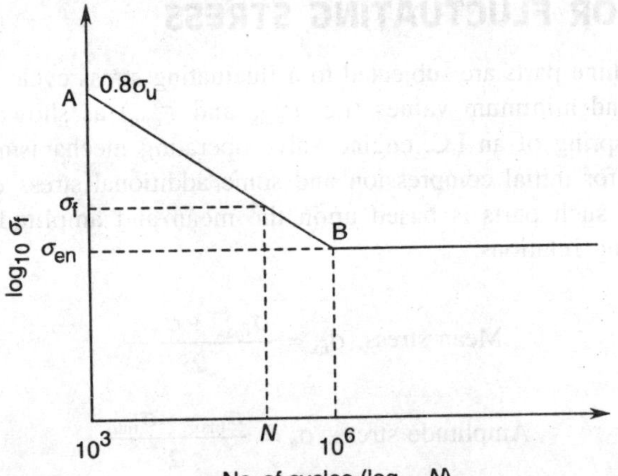

Figure 5.14 Fatigue curve.

σ–N log-log graph. A study of large volume of experimental data on fatigue strength of steel at 10^3 cycles of life indicates that the mean fatigue strength is 0.8 time the ultimate strength. Thus, point A is located at $0.8\sigma_u$ for 10^3 cycles. The point B is located at σ_{en} for 10^6 cycles and points A and B are joined by a straight line. The equation for finite life is therefore given by

$$\log_{10} \sigma_f = b \log_{10} N + c \tag{5.19}$$

where

σ_f is the fatigue strength corresponding to life of N cycles
b, c are constants.

The values of constants b and c can be found by applying the end conditions:

At $N = 10^3$ cycles, $\sigma_f = 0.8\sigma_u$

At $N = 10^6$ cycles, $\sigma_f = \sigma_{en}$

Solving Eq. (5.19) for these conditions, we get

$$b = -\frac{1}{3}\log_{10}\left(\frac{0.8\sigma_u}{\sigma_{en}}\right) \tag{5.20}$$

and

$$c = \log_{10}\frac{(0.8\sigma_u)^2}{\sigma_{en}} \tag{5.21}$$

Substituting the values of constants b and c in Eq. (5.19), we get

$$\text{Fatigue strength, } \sigma_f = 10^c \times N^b \tag{5.22}$$

Alternatively, the expression for the desired life can be given as

$$N = 10^{-c/b} \, \sigma_f^{1/b} \tag{5.23}$$

5.8 DESIGN FOR FLUCTUATING STRESS

In practice, most machine parts are subjected to a fluctuating stress cycle in which stresses vary between maximum and minimum values (i.e. σ_{max} and σ_{min}) as shown in Figure 5.15. For example, the helical spring of an I.C. engine valve operating mechanism is subjected to some minimum stress σ_{min} for initial compression and some additional stress σ_{max} for operating the valve. The design of such parts is based upon the mean and amplitude stresses, which are calculated by following relations:

$$\text{Mean stress, } \sigma_m = \frac{\sigma_{max} + \sigma_{min}}{2} \tag{5.24}$$

$$\text{Amplitude stress, } \sigma_a = \frac{\sigma_{max} - \sigma_{min}}{2} \tag{5.25}$$

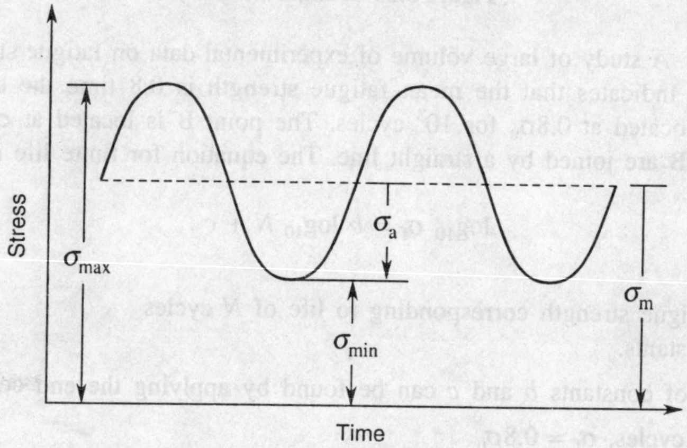

Figure 5.15 Sinusoidal fluctuating stress cycle.

In some typical case where $\sigma_{max} = -\sigma_{min}$, the stress cycle is called completely reversed stress cycle.

For designing a machine part under these stress cycles, generally, two approaches called (i) the Goodman's approach and (ii) the Soderberg's approach are most commonly used. The Goodman's approach is based on ultimate srength of material whereas the Soderberg's approach uses yield strength. Experimentally it is established that in fatigue design of the part for fluctuating stresses, the Goodman's approach is more accurate while the Soderberg's approach is more conservative.

A brief discussion on these design approaches is given below:

Goodman's approach

According to the Goodman's approach, the design of a component_subjected to fluctuating stress can be carried out by the following steps (Figure 5.16).

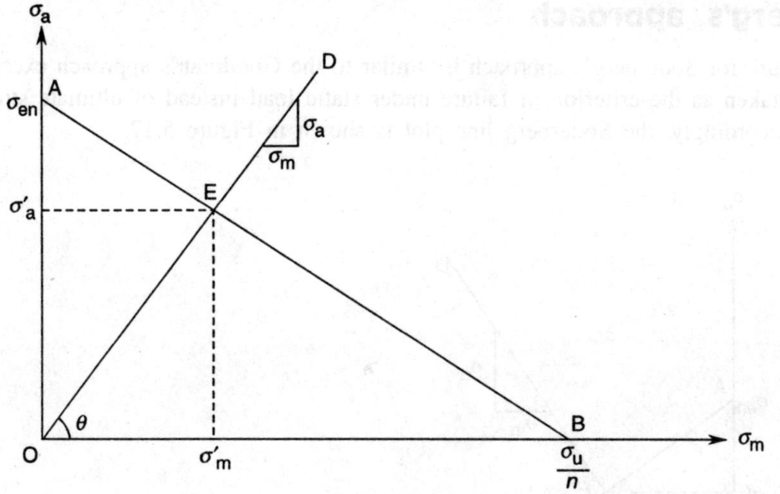

Figure 5.16 Goodman's line.

(a) Determine the endurance strength of the component as explained in Section 5.6.2.
(b) Find the ultimate strength and choose the factor of safety *n*.
(c) On some scale, mark OA equal to σ_{en} along the ordinate line and OB equal to σ_u/n along the abscissa.
(d) Join points A and B by a straight line. This line is called the Goodman's line.
(e) For a given problem, draw a line OD with slope θ as shown in Figure 5.16.

where

$$\tan\theta = \frac{\sigma_a}{\sigma_m} \tag{5.26}$$

(f) The point of intersection of lines AB and OD at point E indicates the dividing line between the safe region and the region of failure. The coordinates of point E (σ_m', σ_a') represent the permissible values of stresses. Therefore,

$$\sigma_a = \frac{\sigma_a'}{n} \tag{5.27}$$

and

$$\sigma_m = \frac{\sigma_m'}{n} \tag{5.28}$$

Mathematically, the equation of the straight line in the intercept form for Goodman's approach can be written as

$$\frac{\sigma_m}{\sigma_u} + \frac{\sigma_a}{\sigma_{en}} = \frac{1}{n} \tag{5.29}$$

Soderberg's approach

The procedure for Soderberg's approach is similar to the Goodman's approach except that yield strength is taken as the criterion of failure under static load instead of ultimate strength of the material. Accordingly, the Soderberg line plot is shown in Figure 5.17.

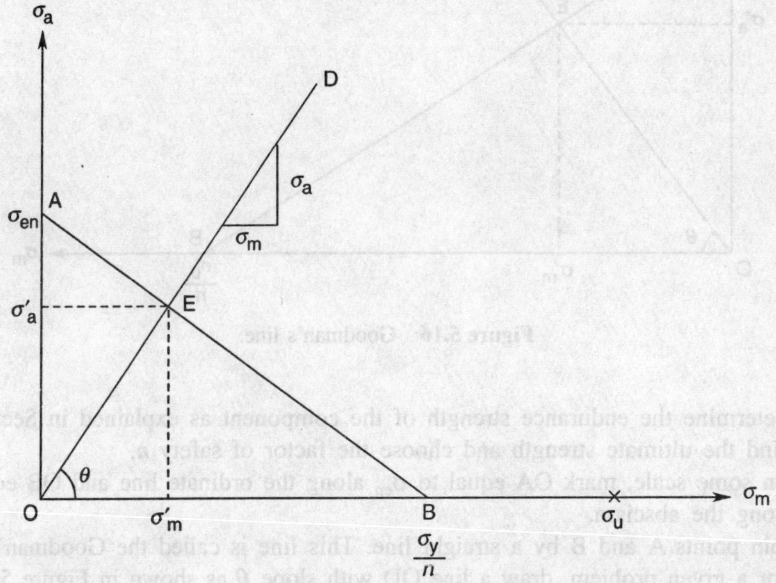

Figure 5.17 Soderberg line.

The mathematical equation resulting in the failure criterion is given as

$$\frac{\sigma_m}{\sigma_y} + \frac{\sigma_a}{\sigma_{en}} = \frac{1}{n} \tag{5.30}$$

In fatigue load design, sometimes a fictitious static stress σ_{st} is used which is equivalent to the fluctuating load. Graphically, the distance between the points O and B is called the static stress. From Eq. (5.30), we get

$$\sigma_{st} = \sigma_m + \frac{\sigma_y}{\sigma_{en}} \cdot \sigma_a \tag{5.31}$$

5.9 DESIGN FOR COMBINED LOADS

Most of the machine parts are subjected to combined stresses, e.g. a rotating shaft subjected to both torque and bending moment. In such cases, it is recommended to first compute the equivalent static stresses as described in Section 5.8 and then use them in a particular theory of failure for designing. For example, when a shaft is subjected to fluctuating bending moment as well as torque, we first compute the mean and amplitude components of stresses. We then translate them into equivalent static stresses. If

σ_m is the mean bending stress

σ_a is the amplitude bending stress

τ_m is the mean torsional stress

τ_a is the amplitude torsional stress

then the equivalent static stresses for bending and torsion are

$$\sigma_{st} = \sigma_m + \frac{\sigma_y}{\sigma_{en}} \cdot \sigma_a \tag{5.32}$$

and

$$\tau_{st} = \tau_m + \frac{\tau_y}{\tau_{en}} \cdot \tau_a \tag{5.33}$$

Now using the maximum shear stress theory we can determine the equivalent maximum shear stress by the following equation:

$$\tau_{max} = \sqrt{\left(\frac{\sigma_{st}}{2}\right)^2 + \tau_{st}^2} \le \frac{\tau_y}{n} \tag{5.34}$$

A similar approach can be adopted for the von-Mises and other theories of failure.

5.10 CUMMULATIVE FATIGUE DAMAGE

The design procedure described earlier assumes that the machine part is subjected to regular stress for the entire life period. However, in practice there are many cases where the machine parts operate with different stress levels and irregular alternating cycles. Sometimes the machines are also subjected to short duration high stress cycles. For example, if a part is subjected to σ_1 stress for n_1 cycles and σ_2 stress for n_2 cycles, it is obvious that the fatigue life prediction under the regular cyclic loading cannot be applied to such an irregular cycle condition. The determination of fatigue life of a part subjected to an irregular cycle is commonly known as *cumulative fatigue damage.*

In order to estimate the damage to the fatigue life, the following two methods are used:

Miner's rule

Let a machine part be subjected to stress σ_1 for n_1 cycles and stress σ_2 for n_2 cycles. Further, let N_1 and N_2 be the number of cycles to failure at σ_1 and σ_2 stresses, respectively.

Suppose one cycle of stress σ_1 causes damage equal to $1/N_1$. Thus for n_1 cycles, the damage will be n_1/N_1 of the full damage. According to the Miner's rule for a given set of stresses the cumulative damage can be written as

$$\frac{n_1}{N_1} + \frac{n_2}{N_2} + \cdots = 1 \tag{5.35}$$

The modified fatigue curve to predict the endurance limit of the overstressed part is shown in Figure 5.18. The detailed procedure to determine the endurance limit is illustrated in Example 5.7.

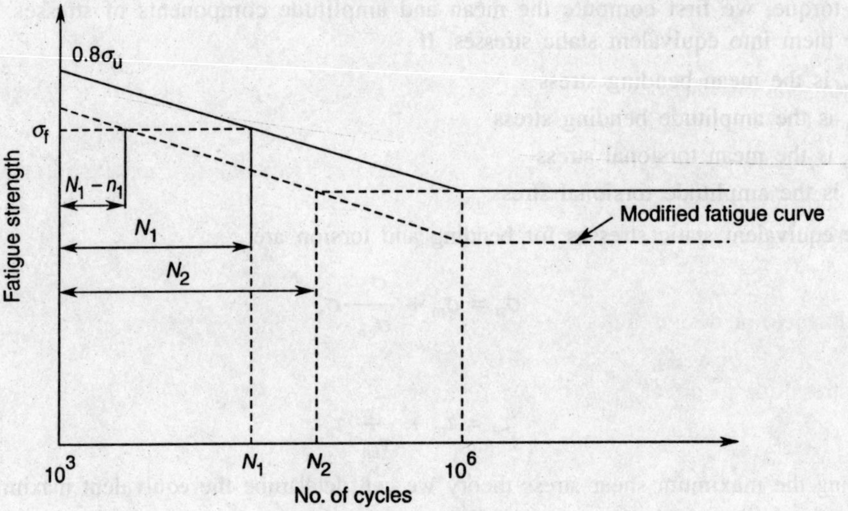

Figure 5.18 Miner's rule.

Though the Miner's rule is quite general, it fails on two counts to conform to experimental evidence.

(i) The theory states that the ultimate strength σ_u decreases because of application of overstress σ_1. However, it cannot be proved experimentally.

(ii) The Miner's rule does not account for the order in which the stresses are applied; hence it ignores any stress less than σ_{en}.

Manson's rule

The Manson's rule overcomes both the deficiencies of the Miner's rule. The cumulative fatigue damage equation is also applicable to Manson's rule. The modified fatigue curve starts with the same starting point as shown in Figure 5.19. The procedure to predict the endurance limit

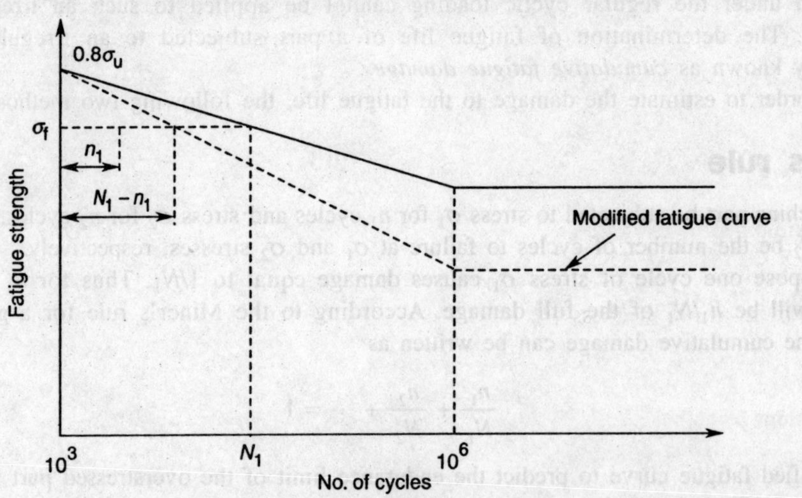

Figure 5.19 Manson's rule.

of a material that has been overstressed for a finite number of cycles is explained in Example 5.7.

Example 5.1 A bolt is subjected to a direct load of 25 kN and shear load of 15 kN. Considering various theories of failure, determine a suitable size of the bolt, if the material of the bolt is C15 having 200 N/mm^2 yield strength.

Solution $F = 25$ kN, $F_s = 15$ kN, and $\sigma_y = 200$ N/mm^2. Although the material properties may be well-known, there may be some change in the magnitude of the load. Thus to account for uncertainties we take the factor of safety, $n = 2.0$. Therefore,

$$\text{Allowable or design stress, } \sigma_d = \frac{\sigma_y}{n} = \frac{200}{2} = 100 \text{ N/mm}^2$$

Tensile stress due to direct load, if A is the resisting area of cross-section

$$\sigma_t = \frac{25}{A} \text{ kN/mm}^2$$

$$\text{Shear stress, } \tau = \frac{15}{A} \text{ kN/mm}^2$$

Since it is a case of biaxial stress condition, the principal stresses are:

$$\sigma_{1,2} = \frac{\sigma_x + \sigma_y}{2} \pm \sqrt{\left(\frac{\sigma_x - \sigma_y}{2}\right)^2 + \tau^2}$$

In this problem $\sigma_x = \sigma_t$ and $\sigma_y = 0$. Thus,

$$\sigma_{1,2} = \frac{\sigma_t}{2} \pm \sqrt{\left(\frac{\sigma_t}{2}\right)^2 + \tau^2}$$

$$= \frac{25}{2A} \pm \sqrt{\left(\frac{25}{2A}\right)^2 + \left(\frac{15}{A}\right)^2}$$

or

$$\text{Maximum principal stress, } \sigma_1 = \frac{32.02}{A} \text{ kN/mm}^2$$

$$\text{Minimum principal stress, } \sigma_2 = -\frac{7.02}{A} \text{ kN/mm}^2$$

(i) *Maximum normal stress theory.* According to the maximum normal stress theory, the maximum principal stress should be less than or equal to the design stress, i.e.

$$\sigma_1 \le \sigma_d$$

or

$$\frac{32.02}{A} \times 1000 \leq 100$$

or

$$A = 320.2 \text{ mm}^2$$

if d_c is the root diameter of the bolt, then

$$\frac{\pi}{4} d_c^2 = 320.2 \quad \text{or} \quad d_c = 20.2 \text{ mm}$$

(ii) *Maximum shear stress theory.* According to the maximum shear stress theory, the failure criterion is

$$(\sigma_1 - \sigma_2) \leq \frac{\sigma_y}{n}$$

or

$$\left(\frac{32.02}{A} + \frac{7.02}{A}\right) \times 10^3 \leq 100$$

or

$$A = 390.4 \text{ mm}^2$$

Therefore, the root diameter of the bolt,

$$d_c = 22.3 \text{ mm}$$

(iii) *von-Mises theory.* According to the von-Mises criterion of failure, the equation of failure is

$$\sigma_1^2 - \sigma_1\sigma_2 + \sigma_2^2 \leq \left(\frac{\sigma_y}{n}\right)^2$$

or

$$10^6 \times \left[\left(\frac{32.02}{A}\right)^2 + \frac{32.02}{A} \times \frac{7.02}{A} + \left(\frac{7.02}{A}\right)^2\right] \leq 100^2$$

or

$$A = 360.4 \text{ mm}^2$$

Therefore,

$$\text{root diameter, } d_c = 21.4 \text{ mm}$$

A comparison of these three theories of failure reveals that maximum shear stress theory is more conservative which has predicted 22.3 mm root diameter whereas von-Mises has predicted 21.4 mm diameter.

Example 5.2 A tension member as shown in Figure (a) below supports an axial load of P newton. It is required to be replaced by another member as shown in Figure (b). Determine the thickness and fillet radius of the second member so that the maximum stress in it does not exceed that in member (a). All dimensions are in mm.

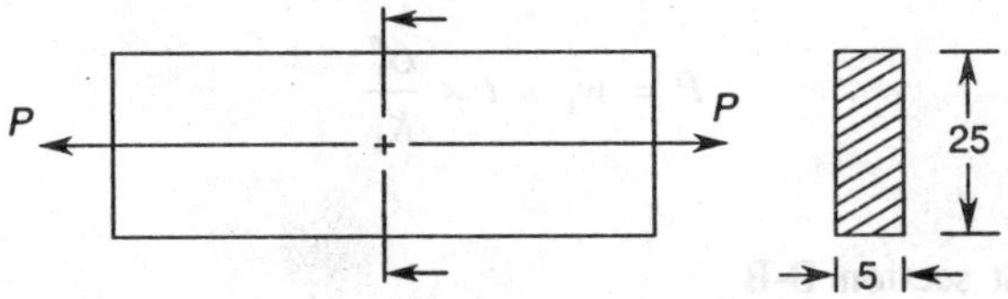

(a)

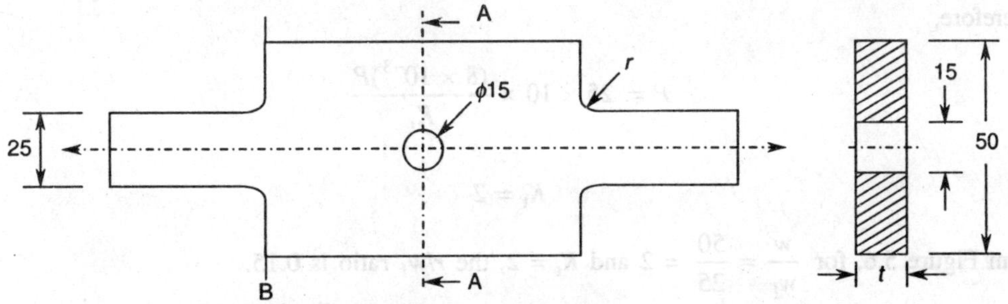

(b)

Solution When a member as shown in Figure (a) above is subjected to tensile load, the maximum tensile stress developed is

$$\sigma_t = \frac{P}{A} = \frac{P}{25 \times 5} = (8 \times 10^{-3})P \text{ N/mm}^2$$

When the member (a) is replaced by the member (b), it may fail at two sections, i.e. at A-A and B-B.

Failure at section A-A

Hole diameter, d = 15 mm

Plate width, w = 50 mm

The stress-concentration factor for $\dfrac{d}{w}$ = 0.3, from Figure 5.7, is 2.37.

$$\text{Resisting force, } P = (w - d) \times t \times \frac{\sigma_t'}{K_t}$$

where σ_t' is the stress induced at section A-A. For equal strength, let $\sigma_t' = \sigma_t$. Therefore,

$$P = (50 - 15) \times t \times \frac{(8 \times 10^{-3})P}{2.37}$$

or

$$t = 8.46 \text{ mm, say, } 10 \text{ mm}$$

Failure at section B-B

The load carrying capacity of the member at section B-B is

$$P = w_1 \times t \times \frac{\sigma_t''}{K_t}$$

where

w_1 is the width at section B-B

σ_t'' is the stress induced at section B-B, which is equal to σ_t for equal strength.

Therefore,

$$P = 25 \times 10 \times \frac{(8 \times 10^{-3})P}{K_t}$$

or

$$K_t = 2$$

From Figure 5.6, for $\dfrac{w}{w_1} = \dfrac{50}{25} = 2$ and $K_t = 2$, the r/w_1 ratio is 0.15.

Thus, fillet radius, $r = 0.15 \times w_1 = 0.15 \times 25 = 3.75$ mm

Hence the required thickness $t = 10$ mm, and the fillet radius $r = 3.75$ mm.

Example 5.3 The figure below shows a rotating shaft supported in ball bearings at supports A and B and loaded by a non-rotating force 10 kN at the mid-span. Estimate the life of the shaft if it is made of 20Mn2 having ultimate tensile strength of 550 N/mm². All dimensions are in mm.

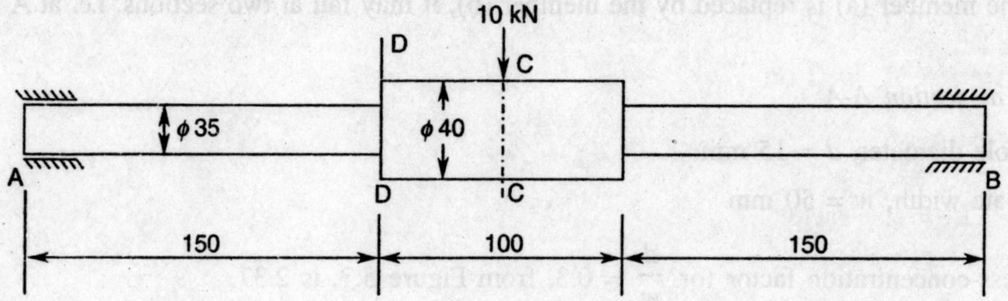

Solution Load = 10 kN

$$\sigma_{ut} = 550 \text{ N/mm}^2$$

For a simply supported beam with a point load at the mid-span, the maximum bending moment is at section C-C. Thus,

$$M_C = 5 \times 10^3 \times 0.2 = 1000 \text{ N} \cdot \text{m}$$

Moment at section D-D

$$M_D = 5 \times 10^3 \times 0.15 = 750 \text{ N} \cdot \text{m}$$

Looking at bending moment and shaft construction it is predicted that the shaft may fail either at section C-C or at section D-D.

Section modulus at section C-C

$$Z_C = \frac{\pi}{32} d^3 = \frac{\pi}{32} 40^3 = 6283.18 \text{ mm}^3$$

At section D-D

$$Z_D = \frac{\pi}{32} 35^3 = 4209.24 \text{ mm}^3.$$

Bending stress at section C-C

$$\sigma_C = \frac{M_C}{Z_C} = \frac{1000 \times 10^3}{6283.18} = 159.15 \text{ N/mm}^2$$

At section D-D

$$\sigma_D = \frac{M_D}{Z_D} = \frac{750 \times 10^3}{4209.24} = 178.18 \text{ N/mm}^2$$

Bending stress at section D-D is more than that at section C-C, further at section D-D there is also a chance of stress-concentration. Thus, section D-D is most critical. For 20Mn2 steel, $\sigma_u = 550$ N/mm^2.

Endurance strength, $\sigma'_{en} = 0.5\sigma_u = 275$ N/mm^2

Modified endurance limit of the shaft is

$$\sigma_{en} = \sigma'_{en} \times K_a \times K_b \times K_c \times K_d \times K_e$$

where

K_a = surface finish factor

= 0.85 for rough surface

K_b = size factor

= $1.189d^{-0.097}$ = $1.189 \times (35)^{-0.097}$ = 0.84

K_c = reliability factor

= 1.0 for 50 per cent reliability

K_d = temperature factor

= 1.0 for temperature less than 300°C

K_e = stress-concentration modifying factor

= $1/K_f$

K_f = fatigue stress-concentration factor

$= 1 + q(K_t - 1)$

with

K_t = form stress-concentration factor; for $\dfrac{D}{d} = \dfrac{40}{35} = 1.143$ and $\dfrac{r}{d} = \dfrac{3}{35} = 0.085$, from Figure 5.4, $K_t = 1.65$.

q = notch sensitivity factor, for notch radius $r = 3$ mm and normalized steel, notch sensitivity factor from Figure 5.13 is 0.9.

$K_f = 1 + 0.9(1.65 - 1) = 1.585$

$K_e = 1/K_f = 0.63$

Endurance strength of the shaft, therefore, is

$$\sigma_{en} = 275 \times 0.85 \times 0.84 \times 1.0 \times 1.0 \times 0.63$$

$$= 123.7 \text{ N/mm}^2$$

Since endurance limit ($\sigma_{en} = 123.7$ N/mm^2) is less than the stress produced at section D-D, i.e. 178.18 N/mm^2, hence the shaft will survive for finite life only.

The finite life at the operating stress level is

$$N = (10)^{-c/b} \times (\sigma_f)^{1/b}$$

where b and c are constants. Now,

$$b = -\frac{1}{3}\log_{10}\left(\frac{0.8\sigma_u}{\sigma_{en}}\right) = -\frac{1}{3}\log_{10}\left(\frac{0.8 \times 550}{123.7}\right)$$

$$= -0.1837$$

$$c = \log_{10}\frac{(0.8\sigma_u)^2}{\sigma_{en}} = \log_{10}\left[\frac{(0.8 \times 550)^2}{123.7}\right]$$

$$= 3.194$$

Thus, the life of the shaft at $\sigma_f = 178.18$ N/mm^2 is

$$N = (10)^{-3.194/(-0.1837)} \times (178.18)^{-1/0.1837}$$

or

$$N = 136{,}639 \text{ cycles}$$

Example 5.4 A bar is subjected to a completely reversed axial load of 150 kN. Determine the size of the bar if it is made of plain carbon steel having ultimate strength equal to 600 N/mm^2.

Solution $F_{max} = 150$ kN

$F_{min} = -150$ kN

The average and amplitude loads on the bar are

$$F_m = \frac{F_{max} + F_{min}}{2} = \frac{150 - 150}{2} = 0$$

$$F_a = \frac{F_{max} - F_{min}}{2} = \frac{150 + 150}{2} = 150 \text{ kN}$$

Endurance limit of the material $\sigma'_{en} = 0.5\sigma_{ut}$

$$= 0.5 \times 600 = 300 \text{ N/mm}^2$$

Endurance strength for axial loading

$$\sigma_{en\text{-}a} = 0.7\sigma'_{en} = 0.7 \times 300 = 210 \text{ N/mm}^2$$

Endurance strength of the bar

$$\sigma_{en} = \sigma_{en\text{-}a} \times K_a \times K_b \times K_c \times K_d \times K_e$$

where

K_a = surface finish factor

= 0.8 for turning

K_b = size factor

= 0.85 (assumed)

K_c = 1.0 for 50 per cent reliability

K_d = 1.0 for $T \leq 300°C$

K_e = stress-concentration modifying factor. Although the bar is straight in cross-section but it may have some toolmarks etc; so let us assume, $K_e = 0.9$.

Modified endurance strength, $\sigma_{en} = 210 \times 0.8 \times 0.85 \times 1.0 \times 1.0 \times 0.9 = 128.52 \text{ N/mm}^2$

$$\text{Amplitude stress, } \sigma_a = \frac{F_a}{A} = \frac{150 \times 10^3 \times 4}{\pi d^2} = \frac{190{,}985.9}{d^2} \text{ N/mm}^2$$

where d is the diameter of the bar.

For a bar to survive for infinite life the induced amplitude stress should be less than the endurance strength of the bar. Thus,

$$\frac{190{,}985.9}{d^2} \leq 128.52$$

or

$$d = 38.55 \text{ mm, say, } 40 \text{ mm}$$

Example 5.5 A rotating shaft loaded by a force F varying from 0 to 50 kN is shown in the figure below. If the shaft material properties are $\sigma_u = 800 \text{ N/mm}^2$ and $\sigma_y = 600 \text{ N/mm}^2$ and the shaft is machine-finished, for a factor of safety 1.5 and 90 per cent reliability, determine the diameter of the shaft. All dimensions are in mm.

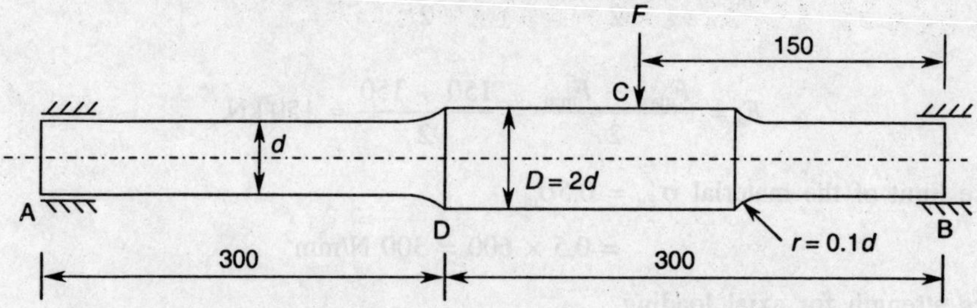

Solution $F_{max} = 50$ kN

$F_{min} = 0$ kN; $n = 1.5$

$\sigma_u = 800$ N/mm^2 and $\sigma_y = 600$ N/mm^2

For machined surface, $K_a = 0.85$

Size factor $K_b = 0.8$ (assumed)

Reliability factor, $K_c = 0.897$ for 90 per cent reliability

Temperature factor, $K_d = 1.0$

Stress concentration factor, $K_t = 1.75$ for $\dfrac{D}{d} = 2$ and $\dfrac{r}{d} = 0.1$

Let us assume notch sensitivity factor, $q = 0.67$

Fatigue stress-concentration factor

$$K_f = 1 + 0.67(1.75 - 1) = 1.5$$

and

$$K_e = \frac{1}{K_f} = \frac{1}{1.5} = 0.667$$

Modified endurance strength

$$\sigma_{en} = 0.5 \times 800 \times 0.85 \times 0.8 \times 0.897 \times 1.0 \times 0.667$$

$$= 162.73 \text{ N/mm}^2$$

The reaction at supports

$$R_B = 37.5 \text{ kN}$$

$$R_A = 12.5 \text{ kN}$$

Bending moment at C

$$M_C = 37.5 \times 150$$

$$= 5625 \text{ N} \cdot \text{m}$$

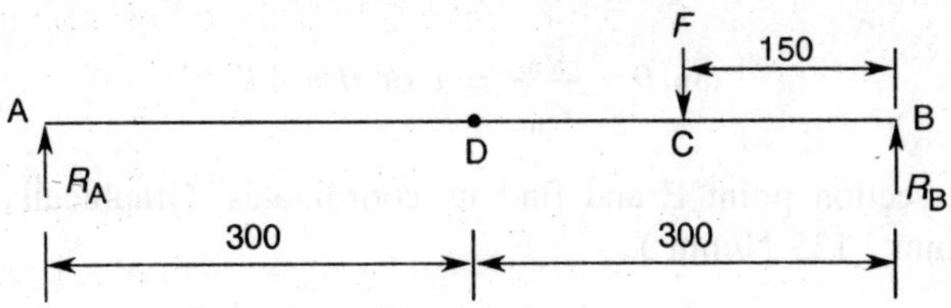

and bending moment at D

$$M_D = 12.5 \times 300.0$$
$$= 3750 \text{ N} \cdot \text{m}$$

At point C, the stress-concentration factor is 1.0 as there is no change in section. However at point D, due to stress concentration the shaft is more prone to failure. The maximum moment at point D

$$M_{max} = 3750 \text{ N} \cdot \text{m}$$
$$M_{min} = 0$$

Therefore,

$$M_m = \frac{M_{max} + M_{min}}{2} = \frac{3750 + 0}{2} = 1875 \text{ N} \cdot \text{m}$$

Amplitude,

$$M_a = \frac{3750 - 0}{2} = 1875 \text{ N} \cdot \text{m}$$

If the section modulus of the shaft at point D is $Z = \dfrac{\pi}{32} d^3$, then the bending stress

$$\sigma_m = \frac{M_m}{Z} = \frac{1875 \times 10^3 \times 32}{\pi d^3} = \frac{19.0986 \times 10^6}{d^3} \text{ N/mm}^2$$

Amplitude stress $\sigma_a = \sigma_m$

Now plot the Goodman's diagram as shown in the figure below. The slope of line OD is

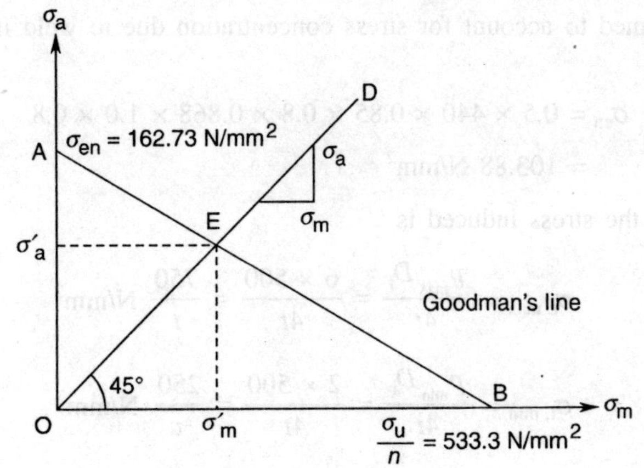

$$\tan \theta = \frac{\sigma_a}{\sigma_m} = 1 \text{ or } \theta = 45°$$

Select the intersection point E and find its coordinates. Graphically, the coordinates of point E are (125 N/mm^2, 125 N/mm^2).

$$\text{For safe design, } \sigma_m = \sigma_a \le \sigma'_m$$

or

$$\frac{19.0986 \times 10^6}{d^3} \le 125.0$$

or

$$d = 53.46 \text{ mm, say, } 55 \text{ mm}$$

Therefore, $D = 2d = 110$ mm

$$\text{Fillet radius, } r = 0.1d = 5.5 \text{ mm}$$

Example 5.6 A spherical pressure vessel with a 500 mm inner diameter is welded from steel plates of cold drawn C20 steel of ultimate strength 440 N/mm^2. The vessel is subjected to internal pressure which varies from 2 N/mm^2 to 6 N/mm^2. If the reliability of the vessel is 95 per cent and the required factor of safety is 3, design the vessel for an infinite life period.

Solution $D_i = 500$ mm

$\sigma_u = 440$ N/mm^2

$p_{max} = 6$ N/mm^2; $p_{min} = 2$ N/mm^2

reliability = 0.95; $n = 3$

Modified endurance strength

$$\sigma_{en} = 0.5\sigma_u \times K_a \times K_b \times K_c \times K_d \times K_e$$

where

K_a = surface finish factor = 0.85 (assumed)

K_b = size factor = 0.8 (assumed)

K_c = reliability factor, for 95 per cent reliability = 0.868

K_d = temperature factor = 1.0

K_e = 0.8 (assumed to account for stress concentration due to weld irregularities)

Therefore,

$$\sigma_{en} = 0.5 \times 440 \times 0.85 \times 0.8 \times 0.868 \times 1.0 \times 0.8$$

$$= 103.88 \text{ N/mm}^2$$

For spherical vessel, the stress induced is

$$\sigma_{t,\,max} = \frac{p_{max} D_i}{4t} = \frac{6 \times 500}{4t} = \frac{750}{t} \text{ N/mm}^2$$

$$\sigma_{t,\,min} = \frac{p_{min} D_i}{4t} = \frac{2 \times 500}{4t} = \frac{250}{t} \text{ N/mm}^2$$

Mean and amplitude stresses are:

$$\sigma_m = \frac{\sigma_{t,max} + \sigma_{t,min}}{2} = \frac{\frac{750}{t} + \frac{250}{t}}{2} = \frac{500}{t} \text{ N/mm}^2$$

$$\sigma_a = \frac{\sigma_{t,max} - \sigma_{t,min}}{2} = \frac{\frac{750}{t} - \frac{250}{t}}{2} = \frac{250}{t} \text{ N/mm}^2$$

Using the Soderberg equation for fatigue failure

$$\frac{\sigma_m}{\sigma_y} + \frac{\sigma_a}{\sigma_{en}} = \frac{1}{n}$$

For C20 steel, the yield strength, $\sigma_y = 240$ N/mm^2. Therefore,

$$\frac{500}{240t} + \frac{250}{103.88t} = \frac{1}{3.0}$$

or

Thickness of vessel plate, $t = 13.47$ mm, say, 15 mm

Example 5.7 A machine member is made of plain carbon steel of ultimate strength 620 N/mm^2 and endurance limit 276 N/mm^2. If this member is subjected to overstress tuned to 413 N/mm^2 for 3000 cycles, determine the revised fatigue limit using (i) the Miner's Rule and (ii) the Manson's rule.

Solution $\sigma_u = 620$ N/mm^2; $\sigma_{en} = 276$ N/mm^2

$\sigma_1 = 413$ N/mm^2 for $n_1 = 3000$ cycles

Using the fatigue life equation, $N = (10)^{-c/b}\,(\sigma_f)^{1/b}$
where b, c are constants.

$$b = -\frac{1}{3}\log_{10}\left(\frac{0.8\sigma_u}{\sigma_{en}}\right)$$

$$= -\frac{1}{3}\log_{10}\left(\frac{0.8 \times 620}{276}\right) = -0.0848$$

$$c = \log_{10}\left(\frac{(0.8\sigma_u)^2}{\sigma_{en}}\right)$$

$$= \log_{10}\left(\frac{(0.8 \times 620)^2}{276}\right) = 2.950$$

Thus for fatigue stress, $\sigma_1 = 413$ N/mm^2, the finite life,

$$N_1 = (10)^{2.950/0.0848} \times (413)^{-1.0/0.0848} = 8697 \text{ cycles}$$

Using the Miner's rule, $\frac{n_1}{N_1} + \frac{n_2}{N_2} = 1$

or

$$n_2 = \left(1 - \frac{n_1}{N_1}\right) N_2 = \left(1 - \frac{3000}{8697}\right) \times 10^6 = 0.6550 \times 10^6 \text{ cycles}$$

Miner's rule

Plot the σ–N curve on the log-log paper and locate point B″ at 413 N/mm² stress on the y-axis and at $(N_1 - n_1)$ distance on the x-axis. Locate another point B′ on the horizontal line passing through point B at distance n_2 cycles. Join points B″ and B′ by a straight line. Extend this line up to point B‴, passing through the line between B and 10^6 cycles. The value of the ordinate corresponding to B‴ point is the revised endurance limit of the overstressed machine part, which is

$$\sigma'''_{en} = 262 \text{ N/mm}^2$$

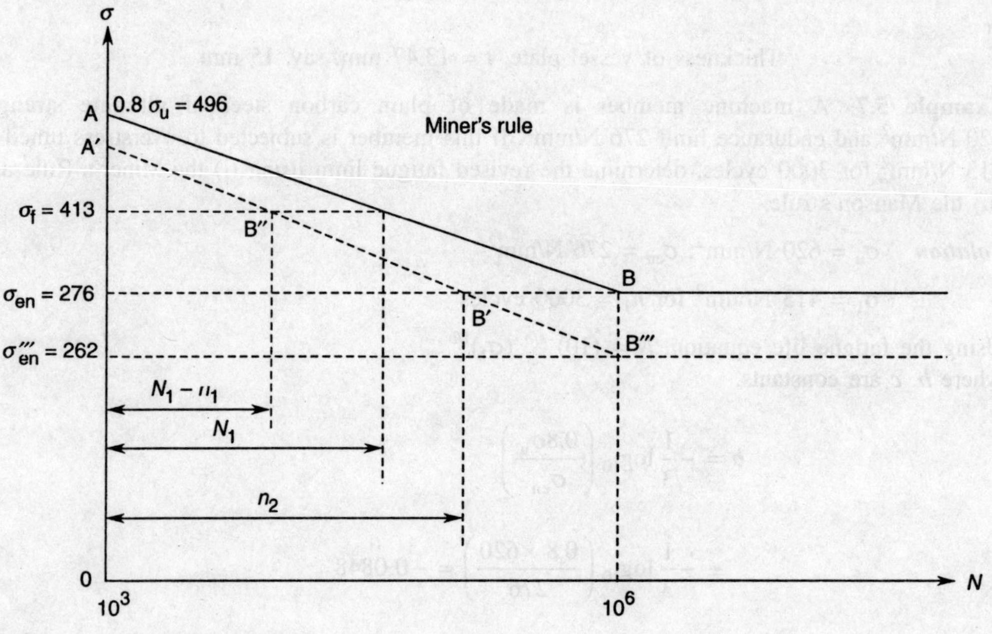

Manson's rule

The rated finite life of the machine part at fatigue stress σ_1 = 413 N/mm² is N_1 = 8697 cycles. Calculate $N_1 - n_1$ = 8697 – 3000 = 5697 cycles.

(i) Locate point A on the ordinate at $0.8\sigma_u$ and point B at coordinates (10^6 cycles, σ_{en}) and join these two points to get the original fatigue curve.

(ii) Locate point A′ at coordinates (σ_1, $N_1 - n_1$), i.e. (413 N/mm², 5697 cycles).

(iii) Join points A and A′ and extend this line till it intersects at B′ with a line drawn between points B and 10^6 cycles.

(iv) The value of the ordinate corresponding to B′ will be the modified endurance limit for the overstressed part, which is

$$\sigma'_{en} = 230 \text{ N/mm}^2$$

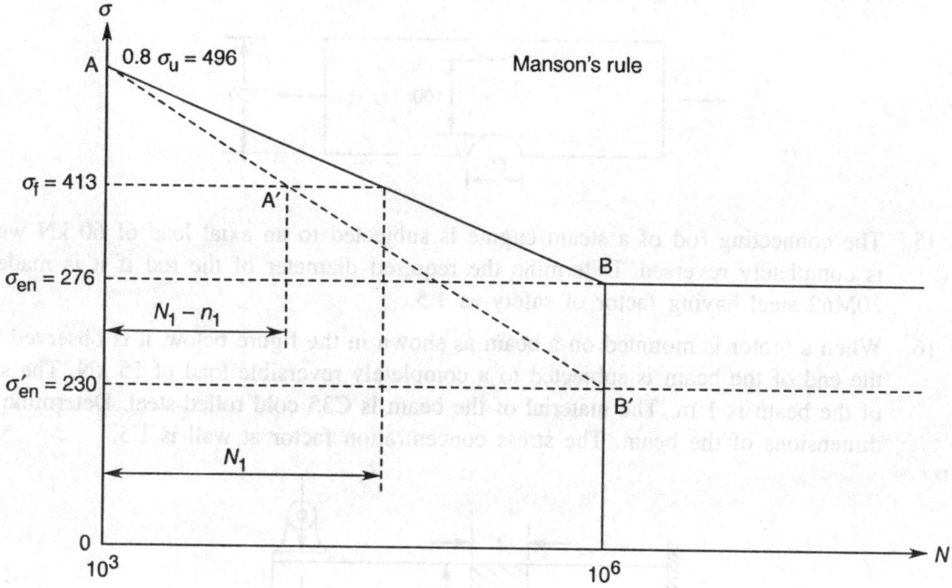

EXERCISES

1. Define load. How are loads classified?
2. Define factor of safety and margin of safety.
3. What is the importance of factor of safety and on what parameters does it depend?
4. What are the modes of failure of a component?
5. What is the significance of theories of failure? Discuss the most commonly used theories.
6. What are the common drawbacks of the maximum shear stress theory compared to the maximum distortion energy theory?
7. What do you mean by stress concentration? How is it accounted for in the design of machine elements?
8. Discuss the mechanism of fatigue failure.
9. Draw the fatigue curve and discuss its importance in the design of a machine element.
10. What are the factors which affect the endurance strength of a component?
11. Explain the procedure for the design of a component for finite life.
12. What do you understand by cumulative fatigue damage?

13. Describe (i) the Miner's rule and (ii) the Manson's rule for computing fatigue damage caused by overstressing of a component.

14. A flat bar, as shown in the figure below, is subjected to an axial load of 40 kN. Assuming that the stress in the bar is limited to 150 N/mm², determine the thickness of the bar. All dimensions are in mm.

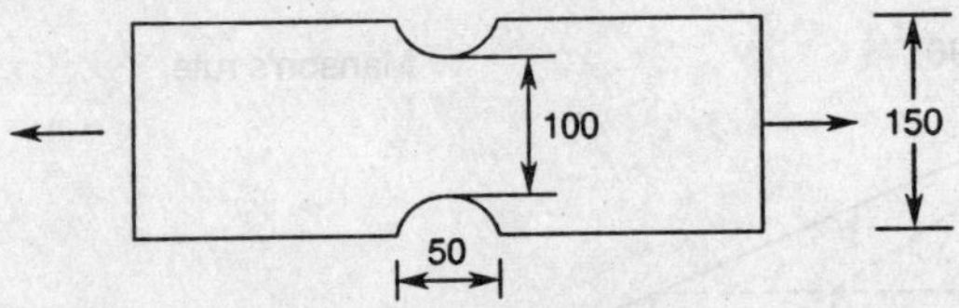

15. The connecting rod of a steam engine is subjected to an axial load of 60 kN which is completely reversed. Determine the required diameter of the rod if it is made of 20Mn2 steel having factor of safety as 1.5.

16. When a motor is mounted on a beam as shown in the figure below, it is observed that the end of the beam is subjected to a completely reversible load of 15 kN. The span of the beam is 1 m. The material of the beam is C35 cold rolled steel. Determine the dimensions of the beam. The stress concentration factor at wall is 1.5.

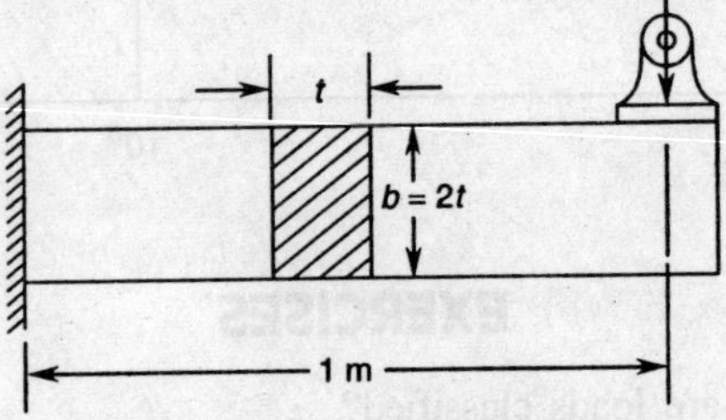

17. A cantilever beam, made of cold drawn steel C30, of circular cross-section as shown in the figure below is subjected to a load which varies from $-F$ to $+3F$. Determine the maximum value of F such that the member can withstand a factor of safety of 2. All dimensions are in mm.

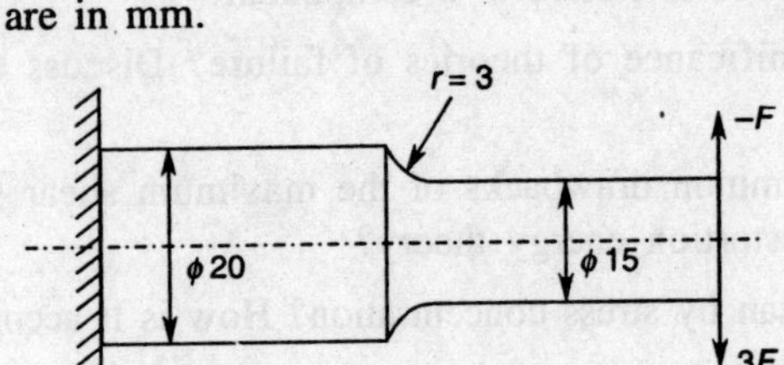

18. A commercial cold rolled C30 steel shaft is required to transmit variable torque ranging from 1 kN · m to 2 kN · m. Specify the diameter of the shaft.

19. A machine component of square cross-section is subjected to a direct tensile load of 20 kN and a shear load of 10 kN. Considering the maximum normal stress and

distortion energy theory of failure, calculate the dimensions of the component and discuss the same.

20. A flat bar as shown in the figure below is subjected to an axial load F equal to 500 kN. Assuming that the stress in the bar is limited to 200 N/mm^2, determine the thickness of the bar. All dimensions are in mm.

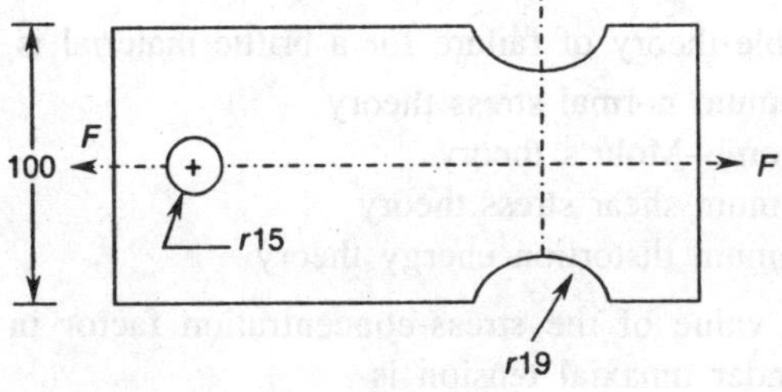

21. A shaft is subjected to a mean torque of 3 MN · mm and is superimposed with variable torque of 3 MN · mm. It is also subjected to a mean bending moment of value zero and a superimposed variable of 6 MN · mm. Design the shaft to support the load for an infinite number of cycles. The shaft material is C20 steel.

22. A circular shaft of an electric motor is 750 mm from bearing to bearing. The weight of the rotor is 7 kN. Determine the diameter of the shaft if the ultimate strength of steel is 650 N/mm^2 and reliability is 0.9.

23. The figure below shows a shaft with load varying from 5 kN to 14 kN. Determine the dimensions of the shaft if it is made of C25 steel. All dimensions are in mm.

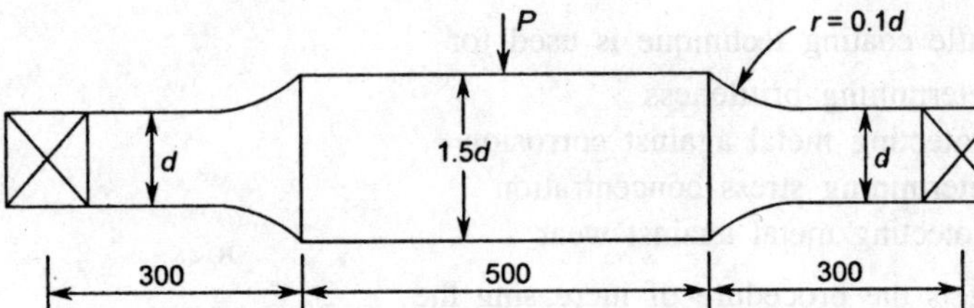

24. A solid round shaft is loaded with a torque which fluctuates between zero and T_{max}. The strengths of the material are σ_u = 1190 MN/m^2 and σ_y = 800 MN/m^2. The critical point is a fillet, ground to 2 mm radius at a square shoulder between sections of 12.5 mm and 10 mm diameter. Estimate the value of maximum torque T_{max} which would cause failure in 50,000 cycles.

25. A steel shaft is subjected to a completely reversed bending moment of 100 kN · m. The shaft transmits 500 kW at 100 rpm. The torque varies over a range of ± 40 per cent. Determine the diameter of the shaft. Take shaft material to be 40Cr1 steel.

26. Determine the wall thickness of a cylindrical pressure vessel of 1200 mm mean diameter subjected to an internal pressure that fluctuates between 5 N/mm^2 and 10 N/mm^2. Take σ_y = 200 N/mm^2 and FoS = 2.

MULTIPLE CHOICE QUESTIONS

1. The most suitable theory of failure for a ductile material is
   (a) The maximum normal stress theory
   (b) The Coulomb–Mohr's theory
   (c) The maximum shear stress theory
   (d) The maximum distortion energy theory

2. The most suitable theory of failure for a brittle material is
   (a) The maximum normal stress theory
   (b) The Coulomb–Mohr's theory
   (c) The maximum shear stress theory
   (d) The maximum distortion energy theory

3. The maximum value of the stress-concentration factor in an infinite plate with a circular hole under uniaxial tension is
   (a) 2 (b) 3 (c) 4 (d) 5

4. The endurance limit of a material can be improved by
   (a) polishing (b) heat treatment
   (c) knurling (d) introducing residual stresses

5. An infinite life of a machine member is generally taken as
   (a) 10^3 cycles (b) 10^4 cycles
   (c) 10^5 cycles (d) 10^6 cycles

6. The resistance to fatigue failure of a material is measured by its
   (a) Young's modulus (b) elastic limit
   (c) ultimate strength (d) endurance limit

7. The brittle coating technique is used for
   (a) determining brittleness
   (b) protecting metal against corrosion
   (c) determining stress concentration
   (d) protecting metal against wear

8. Coaxing is the procedure of increasing the
   (a) metal strength by an alloying element
   (b) hardness by surface treatment
   (c) resistance to corrosion by coating
   (d) fatigue limit by overstressing the metal by successively increasing the load

9. The cold working process
   (a) increases the fatigue strength
   (b) decreases the fatigue strength
   (c) has no influence on the fatigue strength
   (d) has unpredictable influence on the fatigue strength

10. The cumulative fatigue damage can be determined by the
    (a) Miner's rule (b) Soap film method
    (c) Goodman's approach (d) Soderberg's approach

Chapter 6 Manufacturing and Other Aspects in Design

6.1 INTRODUCTION

The development of a machine on the drawing board is only a part of the overall task of design. If the designer's creation is to leave the drawing board and become a physical piece of hardware, it must be manufacturable. In other words, the design of all the parts of a machine should be such that they can be produced by some manufacturing methods and then assembled at competitive cost. The designer should, therefore, have a thorough knowledge of the capabilities and limitations of the manufacturing methods. Only then can he properly design parts, select the materials and manufacturing methods, specify tolerances, consider assembly procedures, specify the reliability of the machine and incorporate human aspect in the design.

In this chapter, we will briefly discuss the following: (i) standardization, (ii) limits, fits and tolerances, (iii) design considerations in manufacturing processes, (iv) reliability, and (v) ergonomic aspects of machine design.

6.2 STANDARDIZATION

Modern systems are increasingly becoming more and more complex. A large number of mechanical components, controls, computers and communication subsystems are found interconnected in a complex system. Such a complex system can have many sources of errors. While some errors may be predictable, others are not due to their random nature. Predictable errors are those which can be foreseen based on the mathematical description of the system's dynamics. In order to minimize predictable errors, the use of standards is advocated.

The main purpose of standardization is to establish mandatory or obligatory norms for the design and production of machines so as to reduce variation in their types and grades and to achieve quality characteristics in raw materials, semi-finished and finished products. Standardization, therefore, provides the following benefits:

(a) Better product quality, reliability and longer service life.
(b) Mass production of components at low cost.
(c) Easy availability of parts for replacement and maintenance.
(d) Less time and effort required to manufacture.
(e) Reduction in variation in size and grades of an article.

The Bureau of Indian Standards (BIS) has standardized a number of items for the benefit of designers and users. In the area of machine design, items of the following categories are standardized and this process is an ongoing one.

(i) Engineering materials, their composition, properties, and methods of testing
(ii) Rules of preparing drawings and use of symbols
(iii) Fits and tolerances for various parts from assembly considerations
(iv) Dimensions and preferred sizes for various machine components, namely rivets, bolts, nuts, keys, couplings, ball and roller bearings, and so on.

In standardization, the concept of preferred numbers helps to reduce unnecessary variation in sizes and grades of an article. Experience has shown that the general requirements of such a grading are mostly satisfied when it follows a geometrical series. Thus, preferred numbers are arranged in geometrical series.

The preferred number contains the power of 10, namely 1, 10, 100, 1000, and so forth. The intermediate numbers are obtained by dividing each number range (1 to 10, 10 to 100, . . .) into 5, 10, 20, and 40 geometrically equal steps. The series thus obtained is known as the *basic series* and designated as R5, R10, R20, and R40, respectively.

The preferred numbers of the basic series are given in IS:1076–1967. As an application of preferred numbers, the power of agriculture tractors is established by the R5 series consisting of 10, 16, 25, 40, and 63 hp. Similarly, the thickness of sheet metal and the diameter of wires are based on the R10 series. The wire diameters of helical springs are in the R20 series.

6.3 LIMITS AND FITS

The ideas of an engineer are expressed on paper in the form of a dimensioned drawing. On the shop-floor, the workers shape the raw material according to the drawing. But unfortunately, it is impossible to shape anything to exact size. Therefore, the designer has to specify the amount of variation that can be tolerated in each dimension. This allowable variation for a given dimension is called *tolerance* and the maximum and minimum permissible sizes are called *limits.*

Depending upon the actual dimensions of the mating parts, an assembly may have a specific kind of fit. Generally, three types of fits, namely (i) clearance, (ii) transition, and (iii) interference are more common in practice.

6.3.1 Definitions

In limits and fits parlance, the following terms are most commonly used (see Figure 6.1).

Basic size. The dimensions obtained by design calculations are called the basic sizes of components. The limits of a size are fixed with reference to the basic size.

Actual size. The size of a manufactured part found by measurement is called its actual size.

Limits. The two extreme sizes between which an actual size is contained are known as limits. The maximum and minimum permissible sizes are called maximum (or higher) limit and minimum (or lower) limit, respectively.

Deviation. It is defined as the algebraic difference between the actual size and the corresponding basic size.

Upper deviation. The algebraic difference between the maximum limit of size and the corresponding basic size is called the upper deviation.

Lower deviation. The algebraic difference between the minimum limit of size and the corresponding basic size is called the lower deviation.

Zero line. A straight line to which the deviations are referred. The zero line is the line of zero deviation and it represents the basic size.

Tolerance. It is the algebraic difference between the upper and lower deviations and has an absolute value without sign. A zone bounded by two limits of the size of the part in relation to the zero line is known as the tolerance zone. If all the tolerance is allowed on one side of the basic size, e.g. $25^{+0.3}_{-0.0}$, the system is said to be unilateral; however, if it is divided, e.g. $25^{+0.2}_{-0.1}$, the system is known to have bilateral tolerance.

Fundamental deviation. One of the two deviations chosen to define the position of the tolerance zone in relation to the zero line is called the fundamental deviation.

Basic hole. A hole whose lower deviation is zero.

Basic shaft. A shaft whose upper deviation is zero.

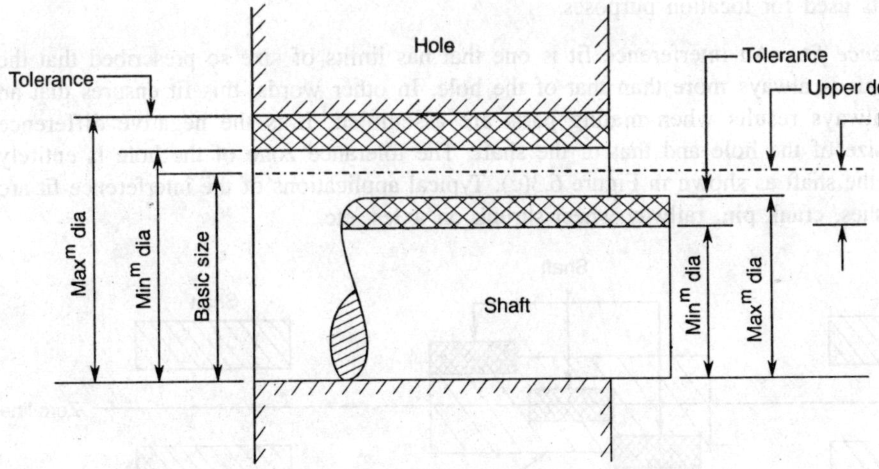

Figure 6.1 Diagram illustrating definitions

6.3.2 Types of Fits

When two machine parts are required to be assembled together, the resulting relation between them due to the difference in their sizes before assembly is called the *fit*. It represents a range of looseness or tightness which may result on account of the application of some specific tolerances. According to the relation due to the difference in the sizes of hole and shaft, a fit between the mating parts may be of three types:

Clearance fit. A clearance fit is one that has limits of size so prescribed that there is always a clearance between the size of the hole and that of the shaft (Figure 6.2). Hence the

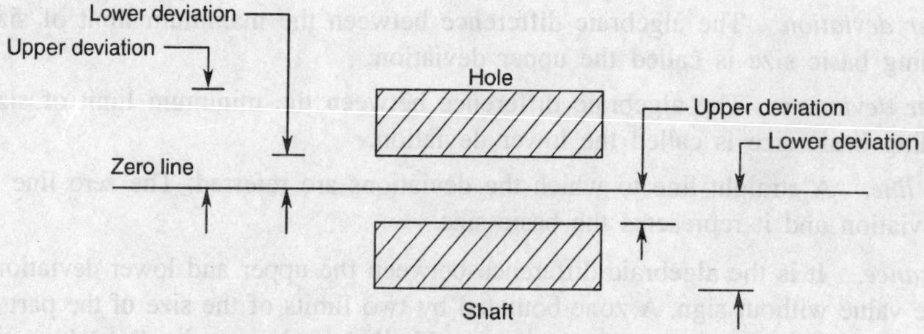

Figure 6.2 Schematic diagram of clearance fit.

shaft is always smaller than the hole into which it fits. The tolerance zone of the hole is entirely above that of the shaft as shown in Figure 6.3(a). Typical applications of the clearance fit are rotating shaft, loose pulley, cross-head slides, etc.

Transition fit. A transition fit is one that has limits of size so prescribed that it may produce either a clearance or an interference. Hence the shaft may be bigger, smaller, or of the same size as that of the hole into which it fits. For this type of fit, the specification of the sizes is such that the tolerance zones for the shaft and the hole overlap each other as shown in Figure 6.3(b). Typical applications of the transition fit are bushes, spigots, fasteners, pins, keys, and other parts used for location purposes.

Interference fit. An interference fit is one that has limits of size so prescribed that the size of the shaft is always more than that of the hole. In other words, this fit ensures that an interference always results when mating parts are assembled. It is the negative difference between the size of the hole and that of the shaft. The tolerance zone of the hole is entirely below that of the shaft as shown in Figure 6.3(c). Typical applications of the interference fit are pressed-in-bushes, crank pin, railway wheel shrunk on axles, etc.

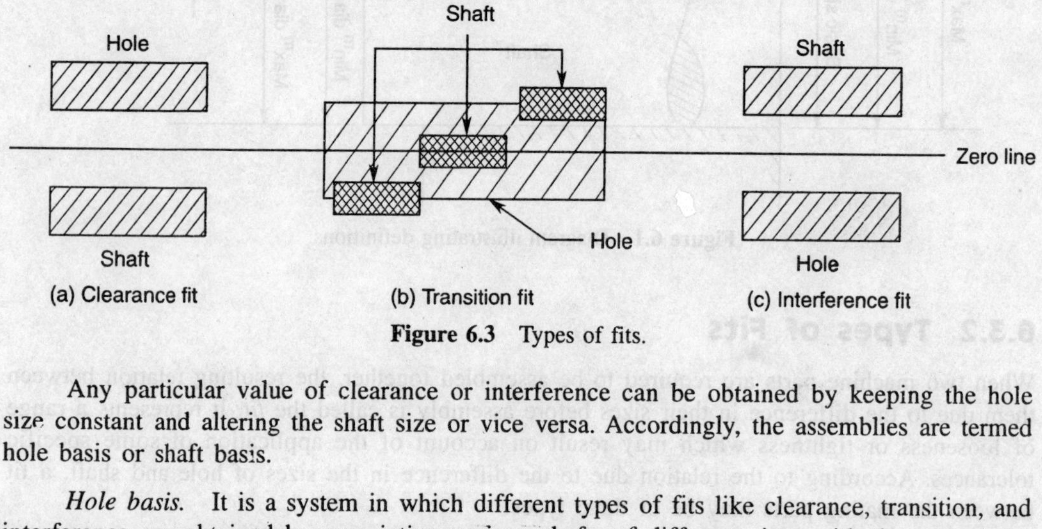

Figure 6.3 Types of fits.

Any particular value of clearance or interference can be obtained by keeping the hole size constant and altering the shaft size or vice versa. Accordingly, the assemblies are termed hole basis or shaft basis.

Hole basis. It is a system in which different types of fits like clearance, transition, and interference are obtained by associating various shafts of different sizes with the same hole. The lower deviation of the hole is zero and the minimum limit of the hole size is the basic size. Such a hole is designated by the symbol 'H' (Figure 6.4).

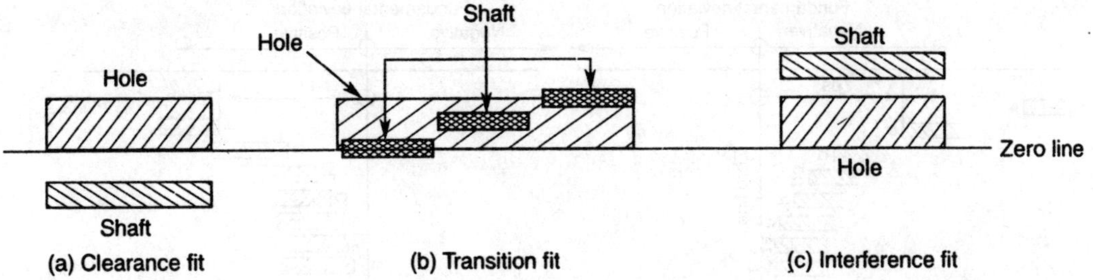

Figure 6.4 Hole-basis system.

Shaft basis. In this system, different types of fits are obtained by associating different size holes with a single shaft. The upper deviation of shaft is zero and the maximum limit of the shaft is the basic size. Such a shaft is designated by the symbol 'h' (Figure 6.5).

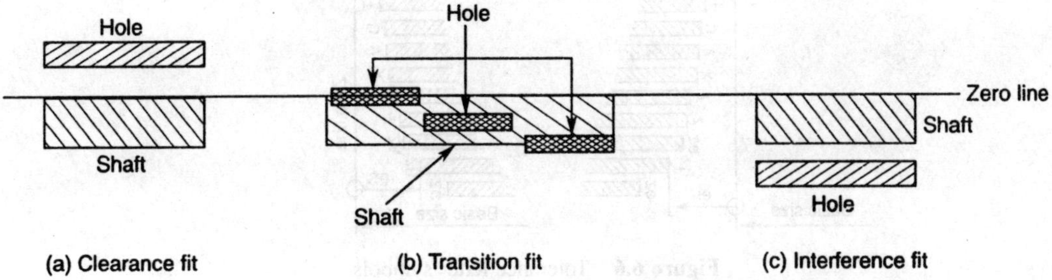

Figure 6.5 Shaft-basis system.

In practice, the hole-basis system is most commonly used. The BIS has also recommended this system, the reason being that due to fixed characteristics of the hole producing tools it is difficult to produce holes of different odd sizes. The shaft being turned and ground to the required size can be produced easily to any given dimension. However, a shaft-basis system is used where a single shaft accommodates a large number of pulleys, bearings, collars, etc.

The fits recommended for different functional requirements are specified in IS: 2790–1964.

6.3.3 BIS System of Limits and Fits

The BIS, under IS:919–1963, has recommended 18 grades of fundamental tolerances which cover a wide range of manufacturing processes and 25 types of fundamental deviations indicated by letter symbols for both holes and shafts. The capital letters A to ZC are reserved for holes and small letters a to zc for shafts, as shown in Figure 6.6. By suitable combinations of fundamental tolerances and fundamental deviations, a large number of fits ranging from extreme clearance to interference can be obtained.

For shafts 'a' to 'g', the upper deviation is below the zero line. For shaft 'h' the upper deviation is zero as it lies on the zero line and for the remaining shafts, i.e. j to zc, it is above the zero line. Similarly, for holes A to G, the lower deviation is above the zero line and for J to ZC it is below the zero line. For hole 'H' the lower deviation is at the zero line as shown in

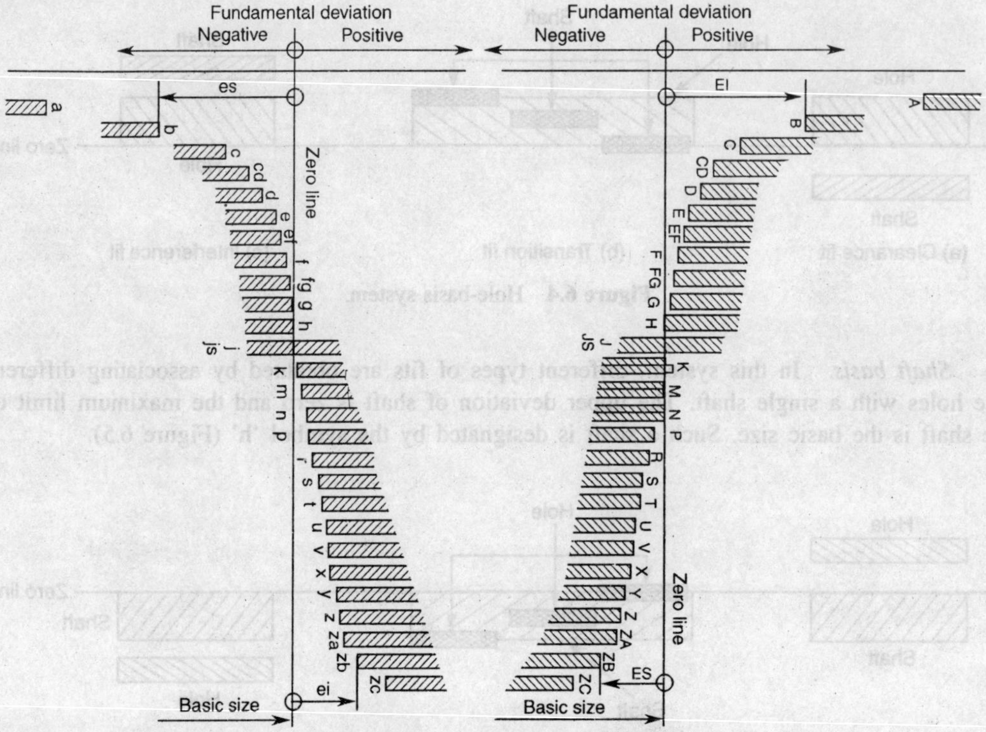

Figure 6.6 Tolerance letter symbols.

Figure 6.4. The upper deviation for shaft and hole is represented by es and ES (Ecart superieur, a French term), respectively. Similarly, the lower deviation for shaft and hole is represented by ei and EI (Ecart inferieur), respectively.

The 18 grades of tolerances are designated as IT01, IT0, IT1 to IT16. These are known as standard tolerances. The values of tolerance for grade IT5 to IT16, in terms of standard tolerance 'i' are listed in Table 6.1. The equation of standard tolerance is

$$i = 0.45 \times \sqrt[3]{D} + 0.001D \ \mu\text{m} \tag{6.1}$$

where D is the geometric mean diameter in mm for a particular range. The various diameter ranges as per IS: 919–1963 are 10–14, 14–18, 18–24, 24–30, 30–50, 50–65, 65–80, and so on.

Table 6.1 Magnitude of tolerance of grades IT5 to IT16

Grade	IT5	IT6	IT7	IT8	IT9	IT10	IT11	IT12	IT13	IT14	IT15	IT16
Value	7i	10i	16i	26i	40i	64i	100i	160i	250i	400i	640i	1000i

The fundamental deviations for the shaft can be calculated with the help of the formulae given in the BIS specification and reported in Table 6.2. The other deviations can be evaluated directly by using the absolute value of tolerance IT. Thus, for the shaft:

Upper deviation, es = ei + IT
Lower deviation, ei = es − IT

Table 6.2 Formulae for fundamental shaft deviation

Upper deviation (es)		*Lower deviation* (ei)	
Shaft designation	es in μm (for D in mm)	Shaft designation	ei in μm (for D in mm)
a	$-(265 + 1.3D)$ for $D \le 120$	j5 to j8	—
	$-3.5D$ for $D > 120$	k4 to k7	$+0.6D^{0.33}$
b	$-(140 + 0.850)$ for $D \le 160$		
	$-1.8D$ for $D > 160$	m	+ (IT7–IT6)
c	$-52D^{0.2}$ for $D \le 40$		
	$-(95 + 0.8D)$ for $D > 40$	n	$+5D^{0.34}$
d	$-16D^{0.44}$	p	+ IT7 + (0–5)
e	$-11D^{0.41}$	r	geometric mean of values of
f	$-5.5D^{0.41}$		ei for p and s
g	$-2.5D^{0.34}$	s	IT8 + 1 to 4 for $D \le 50$
h	0		IT7 + $0.4D$ for $D > 50$
		t	IT7 + $0.63D$
		u	IT7 + D
		v	IT7 + $1.25D$
		x	IT7 + $1.6D$
		y	IT7 + $2D$
		z	IT7 + $2.5D$
		za	IT8 + $3.15D$
		zb	IT9 + $4D$
		zc	IT10 + $5D$

The selection of tolerance on the component is not a random process. It requires to be selected carefully based upon the design calculations, production facility, and cost. Tolerances should be as large as possible because they determine the method of manufacture, which in turn, affects the cost of manufacturing. The cost of production rises exponentially as the tolerance decreases. Therefore, it is essential that while selecting a tolerance the designer should be aware of the accuracies obtainable from various machines and manufacturing processes such as casting, forging, machining, grinding, and the like. Figure 6.7 depicts the attainable grade of tolerances by various manufacturing processes.

6.3.4 Surface Roughness

On every machined surface, some marks of imperfection are bound to be there which take the form of hills and valleys that vary both in height and spacing. The resulting texture of the surface is called *surface roughness*. The characteristics of surface roughness depend upon the method of machining.

Generally, two methods are used to find the value of surface roughness: (i) root mean square (RMS) value and the (ii) centre line average (CLA) value. The RMS value is the square root of the arithmetic mean of the squares of the irregularity heights. The CLA value is defined

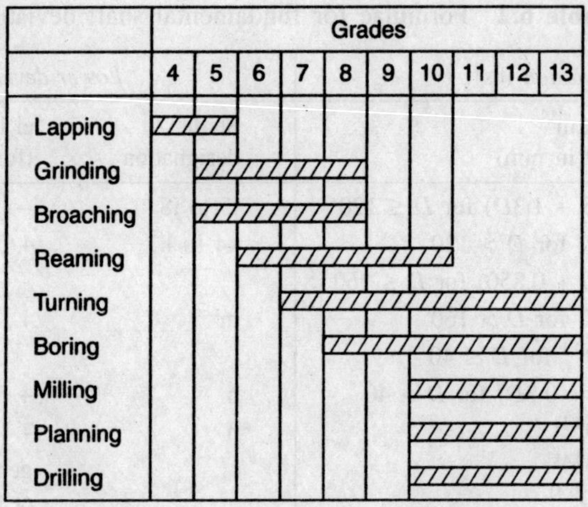

Figure 6.7 Attainable grades of tolerance.

as the sum of all the areas between the surface and the centre line both above and below it and divided by the sampling length (Figure 6.8). The RMS and CLA values are measured in micrometre (μm).

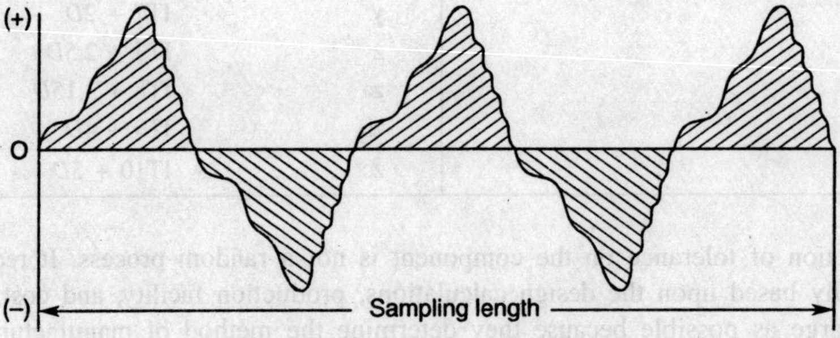

Figure 6.8 Sampling length.

Graphically, the surface roughness of a machined surface is represented by the following information along with the symbol shown in Figure 6.9.

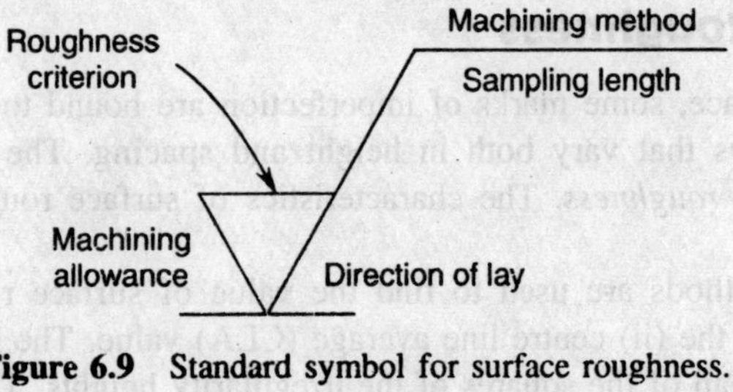

Figure 6.9 Standard symbol for surface roughness.

(i) Surface roughness criterion and value
(ii) Machining method
(iii) Sampling length
(iv) Machining allowance
(v) Direction of lay—it is the direction of the predominant surface pattern and is determined by the manufacturing method. The surface roughness is measured across the direction of lay. The most commonly used symbols for representing the direction of lay are as follows:

=	parallel
⊥	perpendicular
×	cross or angular
M	Multi-directional
C	Circular, relative to the centre of the surface
R	Radial, relative to the centre of the surface

Surface finish on the drawing is indicated by triangles (equilateral) representing the specific values of surface roughness as given below:

Symbol	*Roughness value* (CLA)
∇	8–25 μm
∇∇	1.6–8 μm
∇∇∇	0.025–1.6 μm
∇∇∇∇	< 0.025 μm

Example 6.1 Calculate the fundamental deviation and tolerances and hence obtain the limits of the size for the hole and the shaft in the following fit: 50 H7 f7, a close running fit used for an electric motor, a pump set, etc.

Solution Tolerance $i = (0.45 \times \sqrt[3]{D} + 0.001D)$ μm
where D is the geometrical mean of the diameter step. Diameter step = 50–65

Therefore,

$$D = \sqrt{50 \times 65} = 57 \text{ mm}$$

Hence

$$i = 0.45 \times (57)^{1/3} + 0.001 \times 57$$
$$= 1.7888 \text{ μm} = 0.0017888 \text{ mm}$$

Fundamental tolerance for hole H7 = $16i$ = 0.0286 mm (see Table 6.1)

Tolerance for shaft f7 grade = $16i$ = 0.0286 mm

Fundamental deviation for hole 'H' is zero

Fundamental deviation for shaft f (see Table 6.2)

$$= -5.5D^{0.41} \text{ μm}$$
$$= -5.5 \times (57)^{0.41} = -28.86 \text{ μm} = -0.02886 \text{ mm}$$

Hole

(i) Lower limit of hole size = 50 mm

(ii) Upper limit of hole size = 50 + 16i = 50.0286 mm

Shaft

(i) Upper limit of shaft size = basic size – fundamental deviation

= 50 – 0.02886 = 49.97114 mm

(ii) Lower limit of shaft size = upper limit – tolerance

= 49.97114 – 0.0286 = 49.9425 mm

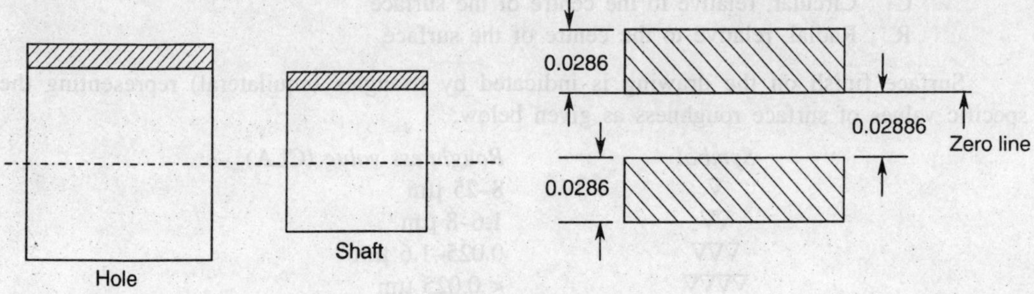

6.4 MANUFACTURING CONSIDERATIONS

The preliminary step in the design of a machine part is to prepare a layout and draw the dimensioned shape of that part. In order to be able to draw this shape, the designer must know in advance how the raw material can be converted into the proposed shape by manufacturing methods. Since each manufacturing method has limitations with regard to its ability to produce a shape, the knowledge of manufacturing methods helps in deciding the shape of the part. The manufacturing processes may be classified into two groups: (i) primary manufacturing methods and (ii) secondary manufacturing methods. The primary manufacturing methods consist of casting, welding, riveting, and forging. In these methods, the raw material is given a shape by application of heat. The secondary manufacturing methods give the piece the exact dimensions required and produce the surface conditions necessary for its functioning. The main methods in this group are turning, drilling, boring, milling, planning, shaping, broaching, grinding, honing, and lapping.

A good knowledge of various manufacturing methods is very helpful to a machine designer. The designer must know and keep in mind these possibilities. The main features of some of the methods, from the standpoint of machine design, are discussed in the following sections.

6.4.1 Design Considerations for Castings

Casting is a basic manufacturing process in which the metal is first liquefied and after properly heating it in a suitable furnace, the liquid or molten metal is poured into a previously prepared mould cavity where it is allowed to solidify. Subsequently, the product is taken out of the mould cavity, trimmed and cleaned to shape. Casting is suitable for a wide range of shapes including complex geometrical shapes. The most important casting processes for iron are sand moulding, permanent moulding, shell moulding, and centrifugal casting. For non-ferrous metals the die-casting process is the most widely used. Die-casting gives good surface finish and better tolerances, sharp corners, and thin sections. The typical values of minimum wall thickness for ferrous and non-ferrous materials are given in Table 6.3.

Table 6.3 Minimum wall thickness

Metal	*Type of casting*	*Minimum wall thickness* (mm)
Steel	Sand	4–5
Grey CI	Sand	3
Malleable CI	Sand	3
Aluminium	Sand	3
Aluminium alloys	Die	1.5
Zinc alloys	Die	0.5

Among iron-casting, the most widely used material is grey cast iron. Experience has shown that many casting problems are related to the design. Therefore, the designer should follow the following recommendations to minimize the casting-related problems.

(i) Cast iron has more strength in compression than in tension. Therefore, the casted parts should always be subjected to compression as shown in Figure 6.10.

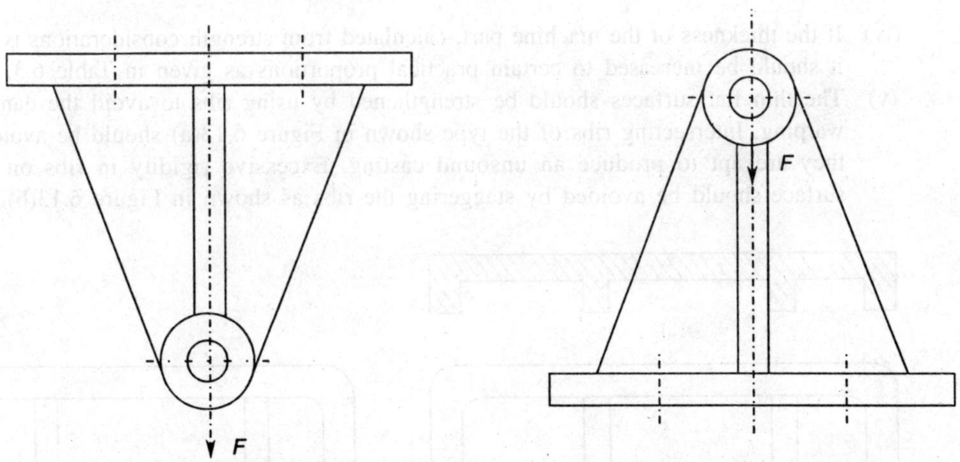

(a) Incorrect (part in tension)

(b) Correct (part in compression)

Figure 6.10 Casted member in compression.

(ii) The section thickness should preferably be uniform and compatible with the overall design considerations. Any abrupt change in the cross-section should be avoided. (Figure 6.11).

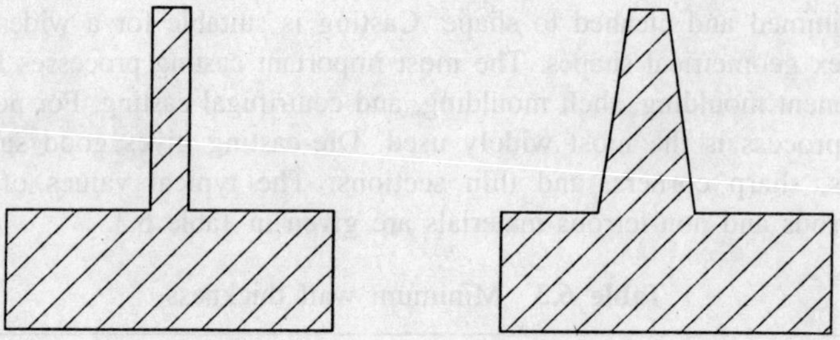

Figure 6.11 Blending of light and heavy sections.

(iii) Avoid concentration of metal at the junction and round all external corners by suitable fillet radii (Figure 6.12).

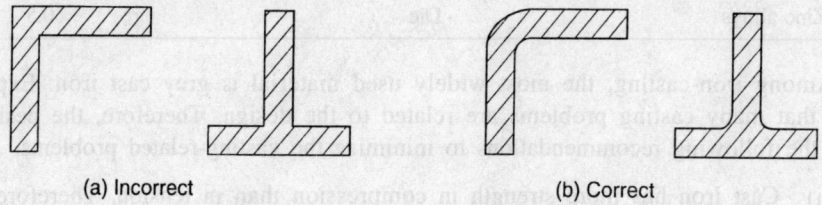

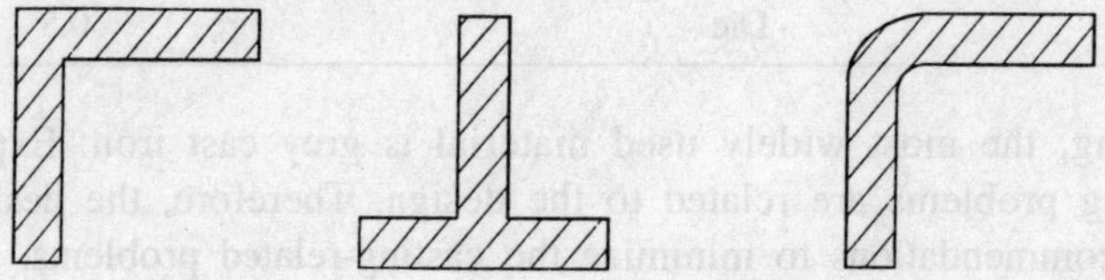

Figure 6.12 Application of fillet radius.

(iv) If the thickness of the machine part, calculated from strength considerations is small, it should be increased to certain practical proportions as given in Table 6.3.

(v) The thin flat surfaces should be strengthened by using ribs to avoid the danger of warping. Intersecting ribs of the type shown in Figure 6.13(a) should be avoided as they are apt to produce an unsound casting. Excessive rigidity in ribs on a flat surface should be avoided by staggering the ribs as shown in Figure 6.13(b).

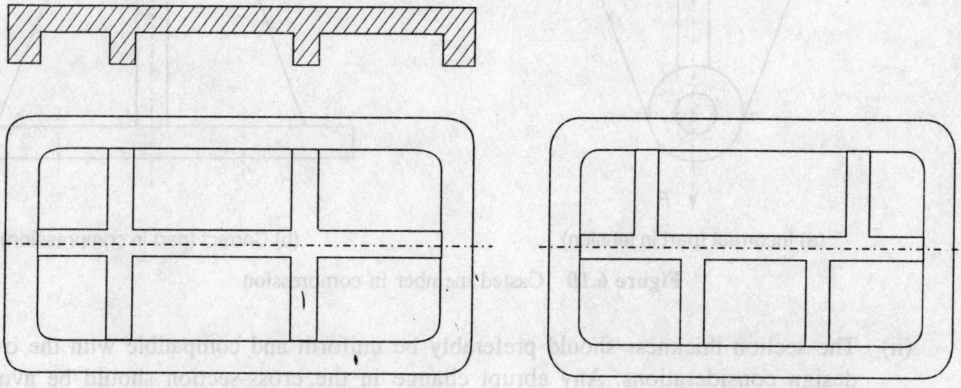

Figure 6.13 Ribbed thin section.

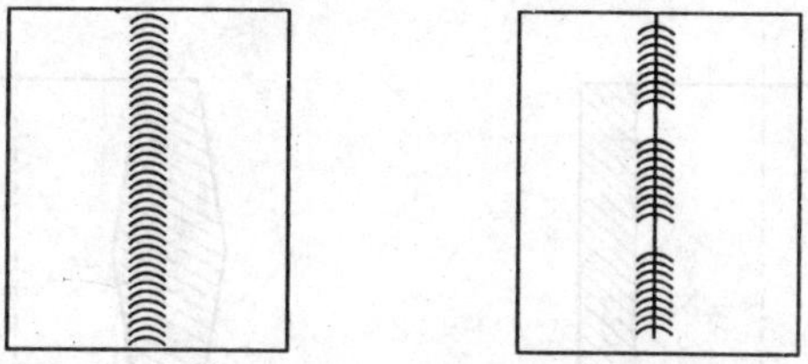

Figure 6.17 Weld profile.

6.4.3 Design Considerations for Forgings

In a forging process, the metal is heated to a plastic state rather than the molten state, and is then subjected to an external force which converts it into the desired complex shape. Forging methods may be classified into four categories, namely (i) blacksmith or hand forging, (ii) drop forging or die forging, (iii) press forging, and (iv) upset forging. The drop forging is extensively used and almost 70–80 per cent of the forged parts are produced by this method. In spite of the wide use of the forging methods, comparatively very little information is available to the designer. The main reason is that with a certain skill the blacksmith can produce almost any shape without much trouble.

A forged component can bear fluctuating stresses and impact loading. However, the limiting factor is the high cost of forging which is due to equipment and tooling. Forging is economical when components are produced on a mass scale. In order to produce sound defect-free forged products, the following general practice should be adopted while designing.

1. The part should be shaped in such a way that the fibre lines remain parallel to the tensile force and perpendicular to the shear force.
2. In the forged part, the deep cuts of machining should be avoided to prevent the fibre lines from being broken. Broken fibre lines result into a weak part.
3. In drop forging, a die should be provided with a sufficient draft angle, usually 7–15°, as shown in Figure 6.18.
4. The parting line of die forgings should be a straight line. All parting surfaces, preferably, should be in one line.
5. The recesses in the forged part should be simple in shape and wherever possible, deep recesses should be avoided. The sharp corners in the recesses should be rounded by proper filleting, which gives a cheaper and long-lasting die. The fillet radii may vary from 1.5 mm to 3.5 mm for a depth up to 50 mm.
6. Any abrupt change in the thickness of the forged components should be avoided. Further, very thin sections should be avoided in forgings to prevent the occurrence of quench cracks. The minimum thickness recommended for moderate-size drop forgings is 3 mm.
7. Where finishing to accurate dimensions is necessary, sufficient material should be allowed for machining. The machining allowance ranges from 1.5 mm to 7.5 mm.

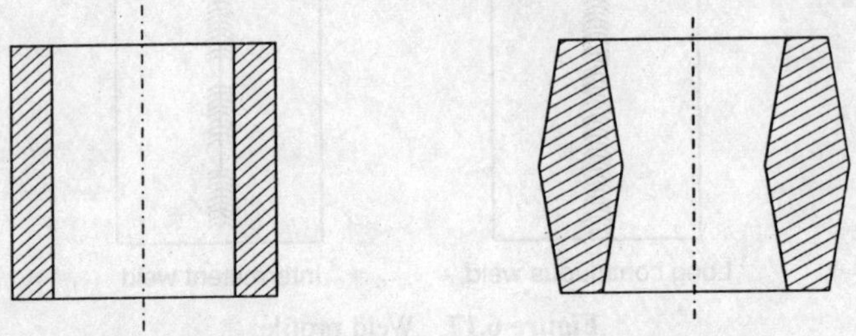

Original product Modified product

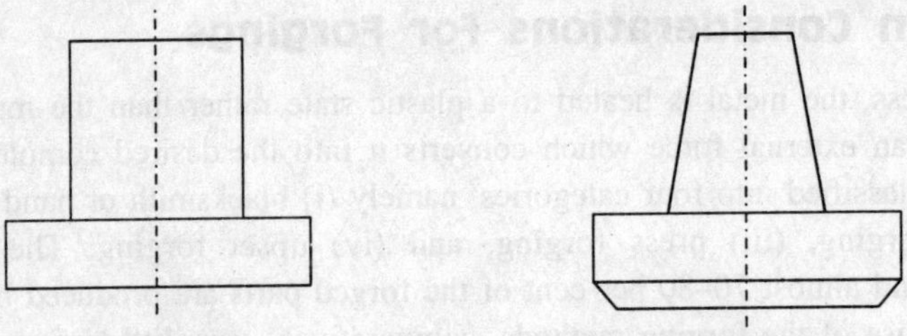

Original product Modified product

Figure 6.18 Forging with suitable draft.

6.4.4 Design Considerations for Machining

The primary manufacturing process helps to give a basic shape to the metal but often the parts are required to be machined to get the required size, surface finish, and geometrical configuration of the surfaces. Therefore, it is necessary for the designer to know the limitations of various secondary manufacturing processes. The cost of manufacturing can be greatly reduced if proper planning for the machining operations is done. The designer should take care of the following points:

1. As far as possible, the geometrical configuration of the machined surface should be plane or circular because any other type of surface would require the use of special tools and fixtures.
2. Adequate tolerances should be prescribed for various surfaces. Tight tolerances would increase the cost of production.
3. A sufficiently large gripping surface should be provided on parts for machining purposes.
4. A run-out clearance for the cutting tool travel should be provided. This is very essential, particularly, if grinding or threading is to be done near a shoulder.
5. The design should involve minimum machining to reduce the cost.
6. A designer should always take into account the provision required for location, holding, and fixing of jobs including the provision for jig and fixture for proper machining.
7. Wherever possible, the sharp corners and shoulders should be avoided. Adequate chamfer or fillet radii should be provided.
8. For the proper location of a drilling hole, it is essential to have a flat and spotted surface to start drilling. The surface should be nearly normal to the axis of the hole.

6.5 RELIABILITY

Reliability is defined as the probability that a given machine or its part will perform its function adequately for its intended period of life under specified operating conditions. In other words, it is the probability of successful performance of any machine or its part. In the present age, a very high degree of reliability is essential as there is too much at stake in terms of cost and human lives. We cannot take any risk with a device which may not function properly when needed most. The most common measures of reliability are: (i) failure rate, (ii) mean time between failures (MTBF), and (iii) probability of survival expressed in percentage.

Mathematically, if in a test, N components are tested, out of which the number of components that survived during time t is $N_S(t)$ and the number of failures that occurred during the same time t is $N_F(t)$, then the reliability can be expressed as the probability of survival, i.e.

$$R(t) = \frac{N_S(t)}{N} = 1 - \frac{N_F(t)}{N} \tag{6.2}$$

Another measure to know the reliability is the rate at which the components in a population fail. This is called the *failure rate* or *hazard rate*. The failure rate for most components follows the bath-tub curve as shown in Figure 6.19. In this curve, three types of failures, namely (i) the quality-related failure, (ii) the stress-related failure, and (iii) the wear-out failure are considered. The sum total of these failures gives the overall failure rate of a component. The failure rate curve has three distinct periods. The failure of components in early age is called *infant mortality*. These failures may be attributed to quality-related aspects such as defective material, poor machining, or poor quality control, and the like. The chances of these types of failure can be eliminated by proper quality control and in-plant testing. The middle portion of the figure which represents design failure, and is mainly stress related, is relatively constant. The third portion indicates old-age failure, i.e. as the product grows old it reaches a wear-out phase which increases the failure rate.

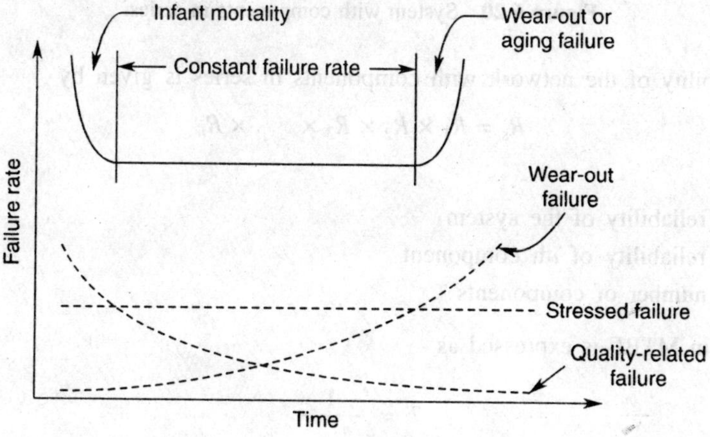

Figure 6.19 Failure bath-tub curve.

Mathematically, the design reliability of a component at constant failure rate is expressed as

$$R = e^{-\lambda t} \tag{6.3}$$

where

λ = failure rate

$$= \frac{\text{number of failures}}{\text{time period during which all components were exposed to failure}}$$

The reciprocal of the failure rate, $1/\lambda$, is called the *mean time between failures* (MTBF). The larger the value of MTBF, the greater the reliability.

6.5.1 System Reliability

During the mechanical design process, the reliability of a system is assessed by the system reliability model. This analysis is usually based on the block diagram representation of the system. The reliability block diagram is constructed from an engineering analysis of the system taking into account the modes of failure and the relationships between various parts.

A system of various blocks may form different combinations, namely the series, the parallel, and k out-of-n units. Thus the reliability of a system can be found by a network of these blocks. These networks are described as under:

Series network

The block diagram of an n unit series network is shown in Figure 6.20. Each block represents a subsystem or a component. If any subsystem or component fails the complete system fails, thus all of the series components must work sucessfully for the system to succeed.

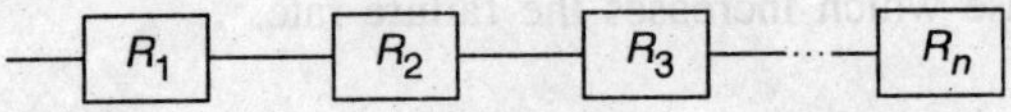

Figure 6.20 System with components in series.

The reliability of the network with components in series is given by

$$R_s = R_1 \times R_2 \times R_3 \times \ldots \times R_n \tag{6.4}$$

where

R_s is the reliability of the system

R_i is the reliability of ith component

n is the number of components

The system MTBF is expressed as

$$T = \frac{1}{\sum_{i=1}^{n} \lambda_i} \tag{6.5}$$

where λ_i is the constant failure rate of the ith component.

Parallel network

When a certain number of blocks, each representing a unit or a component, are connected in parallel, the resulting network is called the parallel network. This system is also called the redundant system. In this system all the components are assumed to be active and at least one component must function normally for the system to succeed. This type of configuration is used to improve the reliability of a system (Figure 6.21).

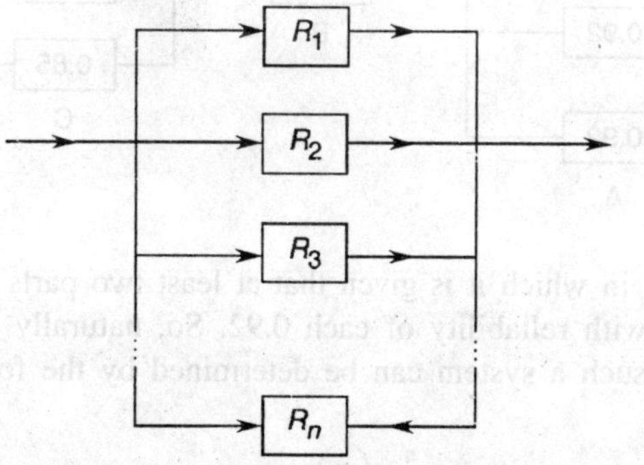

Figure 6.21 System with components in parallel.

The reliability of the parallel network is given as

$$R_p = [1 - (1 - R_1)(1 - R_2) \ldots (1 - R_n)] \tag{6.6}$$

k out-of-*n* units

This arrangement is basically a parallel network with the condition that at least *k* units out of *n* units must function normally for the system to succeed. This network is sometimes referred to as *partially redundant network.* For example, in Boeing 747 aircraft if a condition is imposed that at least three out-of-four engines must operate normally for the aircraft to fly successfully then this system becomes a special case of *k* out-of-*n* units. For independent and identical units, the reliability of *k* out-of-*n* units network is given by

$$R = \sum_{i=k}^{n} \binom{n}{i} R^i (1 - R)^{n-i} \tag{6.7}$$

where

$$\binom{n}{i} = \frac{n!}{i!\,(n-i)!} \tag{6.8}$$

Example 6.2 In a system, three components A, B, and C are assembled in series. In order to improve the reliability of the system, component A is used with 2 out-of-4 unit system, i.e. at least two parts of component A must function out of four. Component C has three parts in parallel. The reliability of the individual parts is written inside each box as shown in the figure below. Determine the reliability of the overall system.

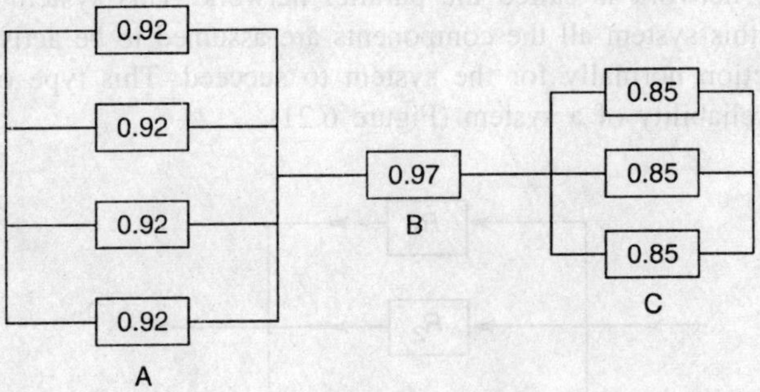

Solution Consider unit A in which it is given that at least two parts must function, out of the four mounted in parallel, with reliability of each 0.92. So, naturally it is a case of k out-of-n system. The reliability of such a system can be determined by the following formula.

$$R_A = \sum_{i=k}^{n} \binom{n}{i} R^i (1-R)^{n-i}$$

where

$$\binom{n}{i} = \frac{n!}{i!(n-1)!}$$

Therefore,

$$R_A = \binom{4}{2} R^2(1-R)^2 + \binom{4}{3} R^3(1-R) + \binom{4}{4} R^4$$

where

$$\binom{4}{2} = \frac{4!}{2!(4-2)!} = 6$$

$$\binom{4}{3} = \frac{4!}{3!(4-3)!} = 4$$

and

$$\binom{4}{4} = 1$$

Hence,

$$R_A = 6R^2(1 - R)^2 + 4R^3(1 - R) + 1\ R^4$$
$$= 6 \times 0.92^2(1 - 0.92)^2 + 4 \times 0.92^3\ (1 - 0.92) + 0.92^4 = 0.998$$

Component B—reliability $R_B = 0.97$

Component C—reliability of components in parallel

$$R_C = 1 - (1 - R_1)(1 - R_2)(1 - R_3)$$

$$= 1 - (1 - 0.85)(1 - 0.85)(1 - 0.85) = 0.9966$$

The combination of components A, B, and C forms a series system as shown in the figure below.

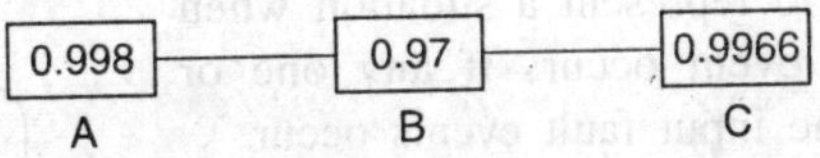

Therefore, the reliability of the overall system

$$R = R_A \times R_B \times R_C$$

$$= 0.998 \times 0.97 \times 0.9966$$

$$= 0.965$$

6.6 DESIGNING FOR RELIABILITY

The starting point for a reliability-based design is the design specification. In this phase, all reliability needs are entrenched into the design specification, such as mean time between failures (MTBF), mean time to repair (MTTR), failure rate, test or demonstration procedure to be used, and so on.

During the design phase of a product, various types of reliability analyses can be performed, namely reliability evaluation and modelling, reliability allocation, reliability testing, and life-cycle costing. In addition, some of the design improvement strategies followed are: (i) zero-failure design, (ii) fault-tolerant design, (iii) built-in testing, (iv) derating, (v) design for damage detection, (vi) modular design, and (vii) maintenance-free design.

There are many reliability analysis techniques and methods available to the designer for use during the design phase. The most popular techniques include: (i) design fault tree analysis and (ii) failure mode and effect analysis (FMEA). A brief description of these techniques is given below:

6.6.1 Design Fault Tree Method

A design fault tree method of determining the reliability of a system is so called because it arranges fault events in a tree-shaped diagram. This technique is well suited for determining the combined effect of multiple failures.

The design fault tree analysis begins by identifying an event which can lead to failure. It is called the top event. The fault events that could cause the occurrence of the top event are generated in a successive manner until the fault events need not be developed any further. These events, called the primary events, are connected by logic gates such as AND, OR, etc. The reliability analysis at each primary event is performed to find the weak link in the system and suitable correction or redesign is performed to improve the reliability of weak links so that the overall reliability of the system is improved.

Figure 6.22 shows the basic symbols used to draw a design fault tree diagram.

Circle — It is used to denote a basic fault event.

Rectangle — It is used to represent an event which needs further analysis.

AND gate — It represents a situation when an output event occurs if all the input fault events occur.

OR gate — It is used to represent a situation when an output event occurs if any one or more of the input fault events occur.

One example of the fault free is shown in Figure 6.23. Once a fault tree is constructed and the reliability of the individual events determined, the system reliability can be found by the system reliability model explained earlier.

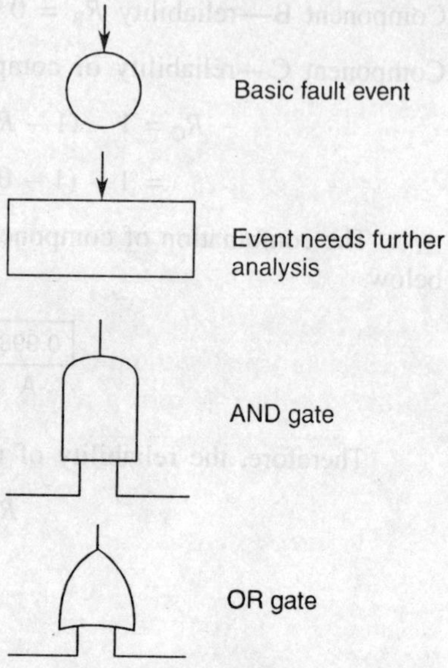

Figure 6.22 Symbols used to draw a design fault tree diagram.

6.6.2 Failure Mode and Effect Analysis

Failure mode and effect analysis (FMEA) is the most widely used tool in industry for evaluating a system design from the point of view of reliability. It was developed in the 1950s to evaluate the design of various flight control systems.

The FMEA is mostly performed for the following purposes:

(i) To identify the design weaknesses
(ii) To help in choosing alternatives during design stages
(iii) To recommend design changes
(iv) To understand all conceivable failure modes and their associated effects
(v) To establish correct action priorities

The major advantage of FMEA is that it helps to identify system weaknesses early enough in the design stage. Therefore, suitable remedial measures can be taken immediately during the design stage itself. However, the major drawback is that it is a single failure analysis and therefore it is not suitable for determining the combined effect of multiple failures.

There are many standards written on FMEA which describe the procedure for performing FMEA. The most common among these is the US Department of Defense Standard MIL-STD 1629 and the Institute of Electrical and Electronics Engineers (IEEE) Standard ANSI N 41.4.

A generalized procedure for performing FMEA is composed of four steps as follows:

Establishing the analysis scope. In the first stage the system boundaries and the scope of the analysis is decided. The basic information regarding the system is gathered. The scope of FMEA depends upon the timing of its conduction, namely the conceptual design stage or the detailed design stage. The scope of FMEA is generally broader for the detailed design stage than for the conceptual design stage.

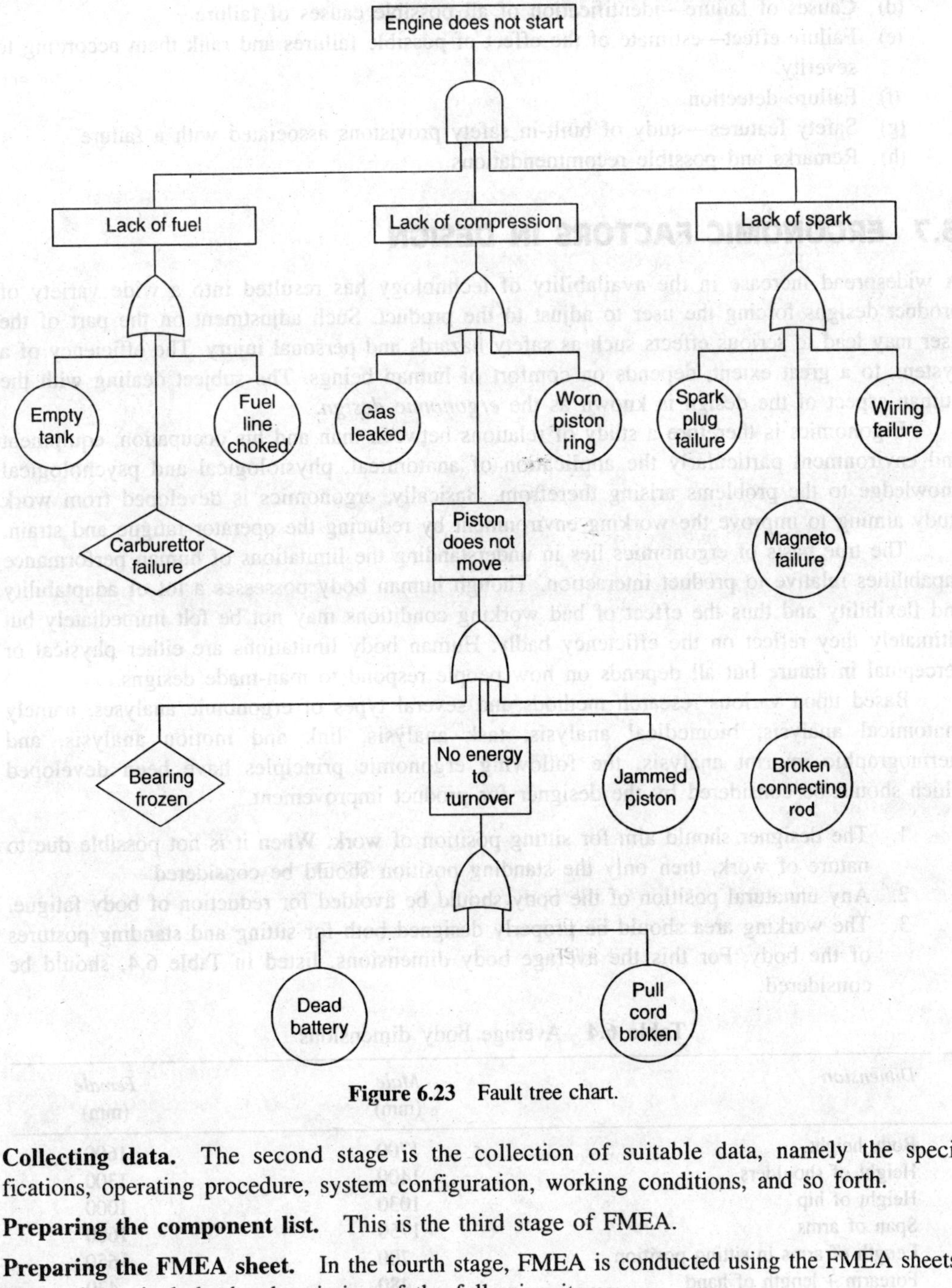

Figure 6.23 Fault tree chart.

Collecting data. The second stage is the collection of suitable data, namely the specifications, operating procedure, system configuration, working conditions, and so forth.

Preparing the component list. This is the third stage of FMEA.

Preparing the FMEA sheet. In the fourth stage, FMEA is conducted using the FMEA sheets. These sheets include the description of the following items:

(a) Part identification and description
(b) Function of the part in different operational modes and the failure frequency
(c) Failure mode—determination of all the possible failure modes associated with a part

(d) Causes of failure—identification of all possible causes of failure
(e) Failure effect—estimate of the effect of possible failures and rank them according to severity.
(f) Failure detection
(g) Safety features—study of built-in safety provisions associated with a failure
(h) Remarks and possible recommendations

6.7 ERGONOMIC FACTORS IN DESIGN

A widespread increase in the availability of technology has resulted into a wide variety of product designs forcing the user to adjust to the product. Such adjustment on the part of the user may lead to serious effects such as safety hazards and personal injury. The efficiency of a system, to a great extent, depends on comfort of human beings. The subject dealing with the human aspect of the design is known as the *ergonomic design.*

Ergonomics is therefore a study of relations between man and his occupation, equipment and environment particularly the application of anatomical, physiological and psychological knowledge to the problems arising therefrom. Basically, ergonomics is developed from work study aiming to improve the working environment by reducing the operator fatigue and strain.

The true basis of ergonomics lies in understanding the limitations of human performance capabilities relative to product interaction. Though human body possesses a lot of adaptability and flexibility and thus the effect of bad working conditions may not be felt immediately but ultimately they reflect on the efficiency badly. Human body limitations are either physical or perceptual in nature but all depends on how people respond to man-made designs.

Based upon various research methods and several types of ergonomic analyses, namely anatomical analysis, biomedical analysis, task analysis, link and motion analysis, and thermographic imprint analysis, the following ergonomic principles have been developed which should be considered by the designer for product improvement.

1. The designer should aim for sitting position of work. When it is not possible due to nature of work, then only the standing position should be considered.
2. Any unnatural position of the body should be avoided for reduction of body fatigue.
3. The working area should be properly designed both for sitting and standing postures of the body. For this the average body dimensions, listed in Table 6.4, should be considered.

Table 6.4 Average body dimensions

Dimension	*Male* (mm)	*Female* (mm)
Body height	1700	1600
Height of shoulders	1400	1300
Height of hip	1030	1000
Span of arms	1750	1600
Length of arms in sitting position	700	650
Forearm + length of hand	480	430
Length of the upper arm	360	325
Height of the knee	525	475
Back of knee sole	460	375

For the standing position, the work table should be:

(a) 50 mm below the elbow height for precision work
(b) 100 mm below the elbow height for light work
(c) 150–300 mm below the elbow height for heavy work.

For the sitting position:

(a) The work table height should be between 680–780 mm.
(b) The height between the ground and seat should be 450 mm.
(c) The height between the seat and table top should be 250–280 mm.
(d) The backrest should be at inclination of 95°–105° from the seat.
(e) The angle between the thigh and leg while sitting should be 105°–110°.

4. According to the anatomical analysis performed by kinesiologist, the most frequent movement of the arms should be as close to the body as possible so that a person can use implements without stretching arms. The anatomical analysis suggests that a human wrist can take a right turn by 40° and left turn by 20°. Both hands can take a right turn by about 40°, whereas the left hand can be turned by 180–200° in the anticlockwise direction. The normal working radius, keeping shoulder as centre, is about 350–450 mm whereas in extended form this radius can be 550–650 mm. Thus, the working table should be designed accordingly. One typical working table design is shown in Figure 6.24 in which the inner curve represents the optimum grasp distance, while the outer curve shows the maximum grasp distance.

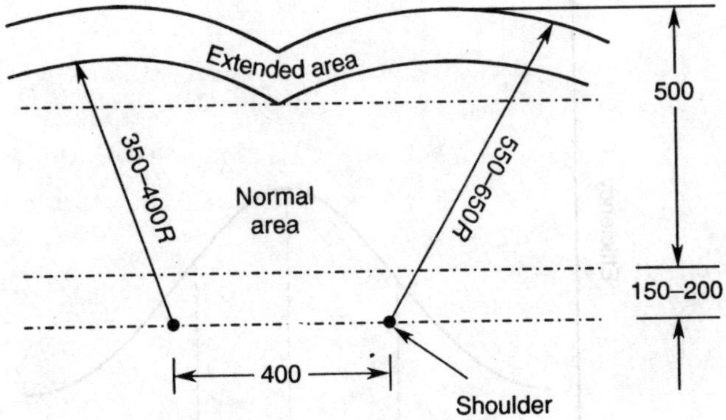

Figure 6.24 Working table layout.

5. The system requiring the use of knobs, levers, hand grips and push buttons should be properly designed and located for efficient control as they directly influence the efficiency of the operator.
6. Dials are generally provided to read the signals so that the operator knows the status of the operating parameters of the system. Hence, proper size and location of dials having clear and easy readable graduations should be designed.
7. The operating controls should be arranged in logical sequence.

8. The principles of consistency of motion should be strictly followed in the design of the system. For example, in a process controlled by turning the wheel clockwise, rotation means an increase and anticlockwise motion means a decrease in the value of the process parameter.
9. The operations requiring constant visual control should be placed in such a way that a comfortable head position of the operator can be maintained. Anatomical investigations have revealed that the most comfortable head position is one in which the line of sight is 38° ± 2° below the horizontal position in the sitting posture and 30° ± 3° in the standing position.
10. The working environment should be provided with adequate glare-free lighting. The intensity level of 160 lux is considered minimum in all workplaces. In the machine shop and precision work areas, the intensity of light should be at least 225 lux.
11. The efficiency of an operator improves if the temperature, humidity, and air flow rate are controlled in surroundings in which the operator has to work. Thus, the design of appropriate heating, cooling, and ventilation systems need to be considered.
12. Noise is the most disturbing factor and it affects the efficiency of the operator. Industrial noise should be less than 70 dB for fatigue-free working of the operator.
13. Human performance is affected by stress. Figure 6.25 shows the relationship between the efficiency of working and stress. At high stress levels, the efficiency is obviously low and chances of making mistakes are high. If the stress level is very low, the working conditions become dull and unchallenging and boredom which sets in reduces efficiency.

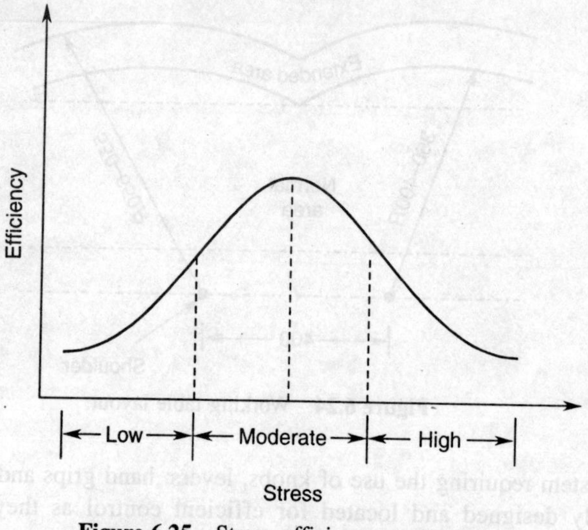

Figure 6.25 Stress–efficiency curve.

The consideration of the above facts at the equipment design stage is essential to produce an acceptable product. If these aspects are considered properly at the design stage, the majority of personnel performance related errors can be reduced.

EXERCISES

1. What do you mean by standardization? What is its importance in the design?
2. What is the meaning of R5 series? Explain the difference between the basic and derived series.
3. Explain the following terms:

   (a) Basic size (b) Actual size (c) Limits
   (d) Tolerance (e) Upper deviation (f) Lower deviation
   (g) Basic hole (h) Basic shaft

4. What is tolerance zone?
5. Explain the different types of fits.
6. Explain the terms CLA and RMS.
7. Explain the importance of manufacturing considerations in the design process.
8. What important considerations are required to be taken into account while designing (a) cast parts, (b) welded parts, and (c) forged parts?
9. What is the importance of reliability in design? Explain the bath-tub curve for reliability of a component.
10. Explain the fault tree method of design.
11. What is failure mode and effect analysis? How is it performed to improve the reliability of a component?
12. Explain the importance of ergonomic factors in design.
13. List the important ergonomic considerations required to be taken into account while designing a component.

MULTIPLE CHOICE QUESTIONS

1. Manufacturing aspects in design mean:

   (a) Economy of manufacture (b) Lightness in weight
   (c) Standardization (d) Manufacturing facility

2. Preferred numbers are arranged

   (a) arithmetically (b) in geometric series
   (c) logarithmically (d) randomly

3. According to BIS, the total number of grades of tolerances are

   (a) 6 (b) 12 (c) 18 (d) 24

4. The basic hole is one whose

   (a) lower deviation is zero (b) upper deviation is zero
   (c) lower deviation is positive (d) upper deviation is negative

5. The basic shaft is one whose

   (a) lower deviation is zero (b) upper deviation is zero
   (c) lower deviation is negative (d) upper deviation is positive

6. The minimum section thickness for grey sand casting is
   (a) 1 mm (b) 2 mm (c) 3 mm (d) 5 mm
7. When a part is cast, it should always be subjected to
   (a) tension (b) compression (c) torsion (d) any load
8. The strength of a forged part compared to a one that is cast is
   (a) greater (b) smaller (c) equal (d) unpredictable
9. The noise level in a workplace should be kept
   (a) below 70 dB (b) above 70 dB
   (c) up to 100 dB (d) at any level
10. The designer should aim for
    (a) standing position of work (b) sitting position of work
    (c) any position of work
11. The average intensity of light level for normal working conditions should be
    (a) below 100 lux (b) 100 lux
    (c) 160 lux (d) 250 lux
12. Performance-related errors can be reduced by
    (a) the use of appropriate mathematical models
    (b) taking into consideration appropriate manufacturing considerations
    (c) improving the reliability of the product
    (d) taking into consideration appropriate ergonomic considerations

Chapter 7

Design of Joints

7.1 INTRODUCTION

A machine is built by joining together several parts either by sliding joints or by fixed joints. The requirement of sliding joints primarily depends upon the kinematic considerations. However, in engineering practice, fixed joints are more important because they are required to carry load.

The fixed joints used in most machines may be either permanent or temporary. The selection of a type of joint is dictated by the functional requirements of the joint in a particular machine. The classification of joints is shown in Figure 7.1.

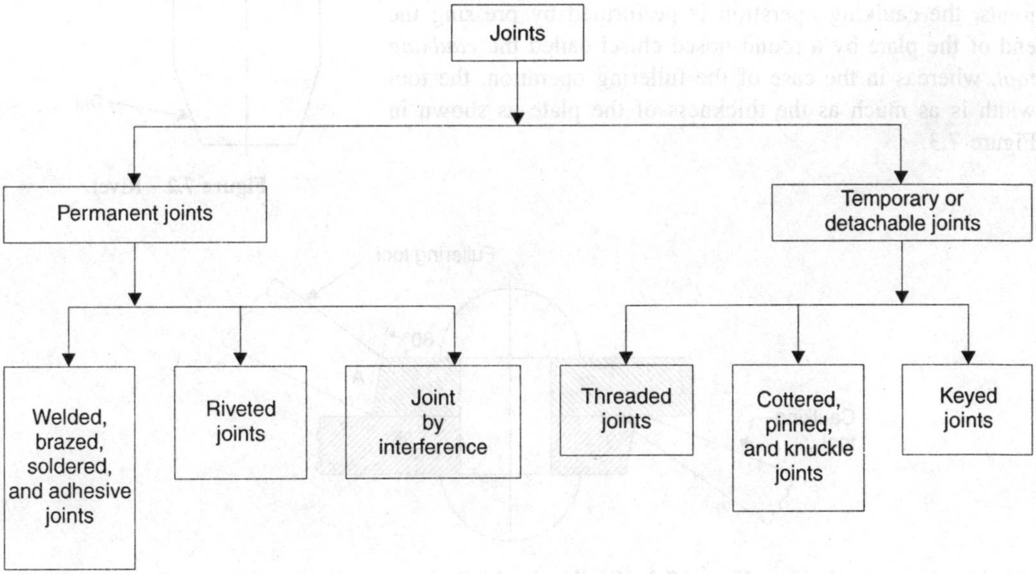

Figure 7.1 Types of joints.

Permanent joints cannot be disassembled without fracture of the connecting parts. These joints are made by welding, brazing, riveting, and adhesive bonding. Soldering and brazing are used where low strength joints will do, for instance, joints in electric and electronic circuits, and tool tip bonding to tool holder, and the like.

Temporary fasteners or detachable joints allow the disassembly of a machine without damaging the elements of assembly. The most commonly used detachable joints are formed by threaded members, cotter, knuckle pin, and keyed members. In this chapter, the detailed designing procedure of riveted joints, welded joints, cotter joints, and pinned joints will be discussed.

7.2 RIVETED JOINTS

A rivet is a round bar consisting of an upset end called the *head* and a long part called the *shank,* as shown in Figure 7.2. In the riveting process, the shank passes through a cylindrical hole made in the plates which are to be joined together and then a second head, called the *point*, is formed on the tail of the rivet. If the point is prepared by the cold working process, then it is called *cold riveting*. However, if the tail end of the rivet is first heated and then the point is formed by upset forging, the process is termed *hot riveting*. Cold riveting is generally used for structural fabrication, whereas hot riveting is employed in pressure vessels or leakproof joints.

In the riveted joints for pressure vessels and hulls of ships, where security against leakage has to be considered, the ends of the plates are bevelled at an angle of about 80° for fullering and caulking operations. To obtain leakproof joints, the caulking operation is performed by pressing the end of the plate by a round-nosed chisel called the *caulking tool,* whereas in the case of the fullering operation, the tool width is as much as the thickness of the plate as shown in Figure 7.3.

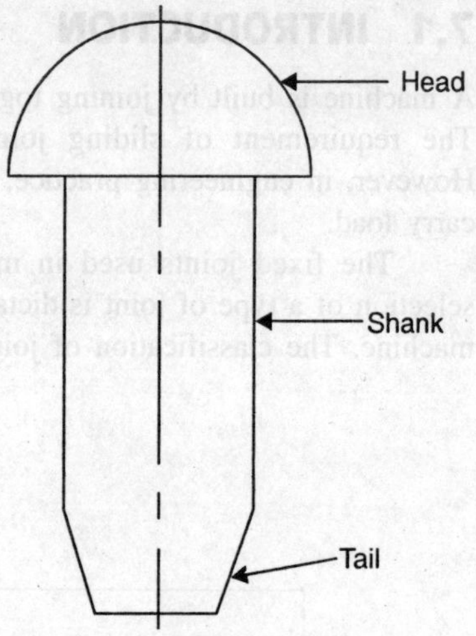

Figure 7.2 Rivet.

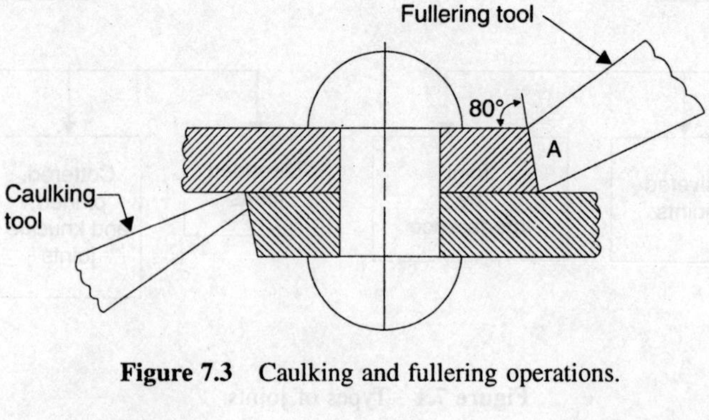

Figure 7.3 Caulking and fullering operations.

7.2.1 Rivet Materials

Rivets are manufactured from the materials conforming to IS:1148–1982 and IS:1149–1982 for structural work and to IS:1990–1973 for pressure vessels. Rivets are usually made from tough and ductile materials, namely low carbon steel C15, nickel alloy steel and wrought iron. Rivets are also made from non-ferrous materials such as copper, aluminium alloys, and brass for anti-corrosive properties where strength is not a major requirement. According to BIS, the rivet material should have tensile strength more than 350 N/mm^2 and elongation not less than 20 per cent.

The BIS has also recommended standard tests. According to IS:1928–1961, the samples of manufactured rivets are subjected to bending and flattening tests. In the bending test, a rivet shank is bent cold and hammered until the two ends of the shank are turned to touch each other without cracking. In the flattening test, the rivet head is flattened until its diameter is 2.5 times the diameter of shank and yet it does not crack at the edges (Figure 7.4).

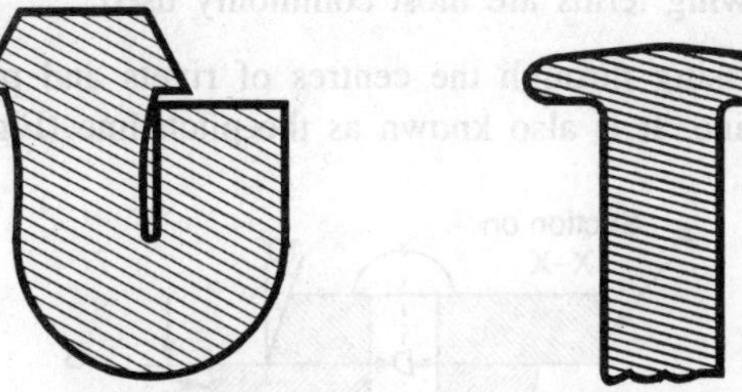

(a) Bending test (b) Flattening test

Figure 7.4 Bending and flattening tests.

7.2.2 Rivet Heads

The rivets with different types of heads are shown in Figure 7.5. The dimensions of a rivet are specified in terms of shank diameter *D*. A snap-head rivet is most widely used in structural

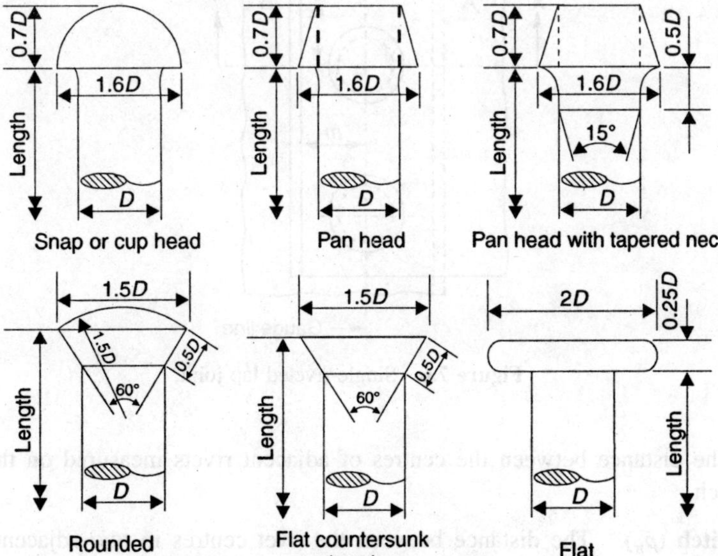

Figure 7.5 General purpose rivet heads.

work and machine riveting, while pan and conical-head rivets with taper neck are used in pressure vessels and leakproof joints. The countersunk head rivet is used for obtaining a smooth surface, e.g. in shipbuilding. Tubular and semi-tubular rivets are used extensively for quick assembly operations. These rivets are used for connecting sheet metal and die-cast components where only the shear stresses are induced in the shank owing to external loads.

The standard diameters of general-purpose rivets are 6, 8, 10, 12, 14, 16, 18, 20, 24, 27, 30, 33, 36, 39, 42, and 48 mm. The diameter of the rivet hole is kept slightly larger than the rivet shank diameter to accommodate the increase in rivet size due to upset forging.

7.2.3 Rivet Terminology

In riveted joints, the following terms are most commonly used.

Gauge line. The line passing through the centres of rivets and parallel to the edge of the plate is called the gauge line. It is also known as the pitch line (Figure 7.6).

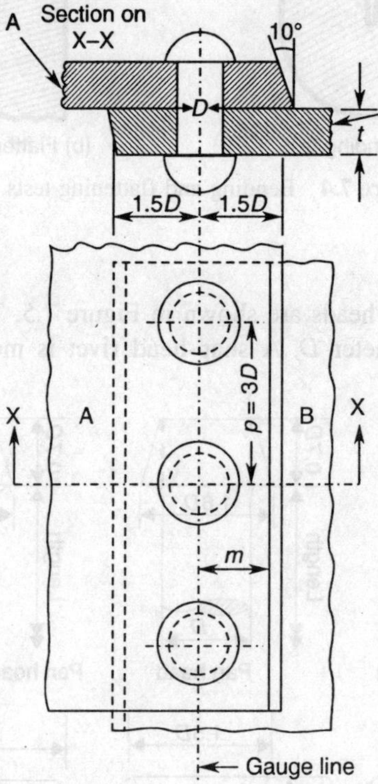

Figure 7.6 Single-riveted lap joint.

Pitch (p). The distance between the centres of adjacent rivets measured on the gauge line is called the pitch.

Transverse pitch (p_{tr}). The distance between the rivet centres in two adjacent gauge lines is called the back pitch or transverse pitch (Figure 7.7).

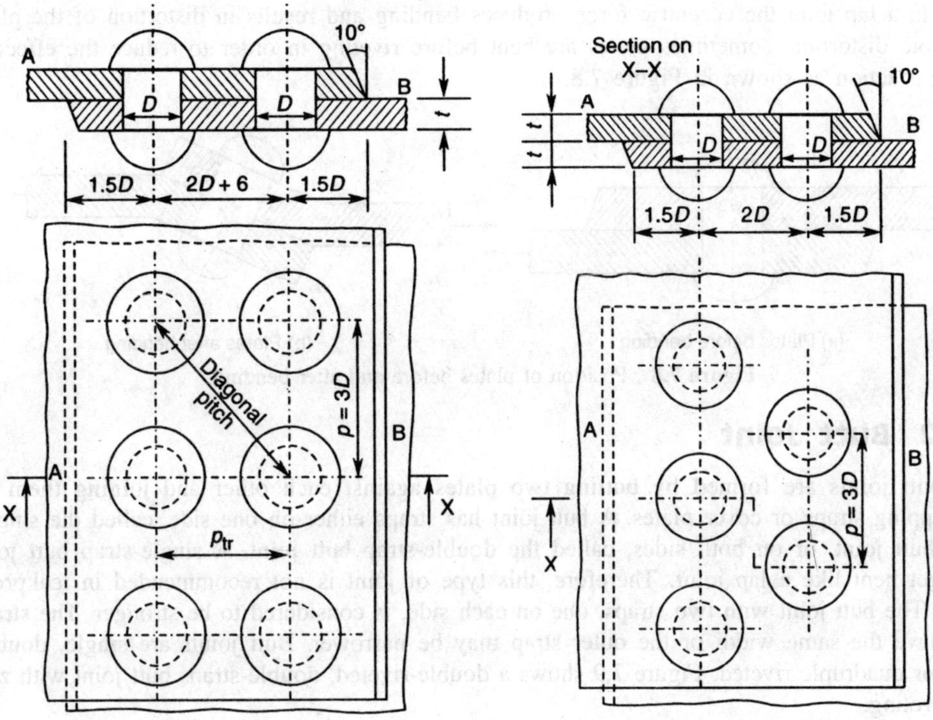

(a) Chain riveting (b) Zig-zag riveting

Figure 7.7 Double-riveted lap joint.

Diagonal pitch (p_d). The distance between the adjacent rivet centres on adjacent gauge lines in zig-zag or chain riveting is called the diagonal pitch.

Chain/zig-zag riveting. If the rivets in adjacent rows are placed opposite to each other, the arrangement is termed *chain riveting*. Otherwise, it is called *zig-zag riveting*.

Margin (m). The distance between the outermost gauge line and the edge of the plate is called the *margin* or *margin pitch*.

7.3 TYPES OF RIVETED JOINTS

The riveted joints can be broadly classified into two categories, namely (i) lap joint and (ii) butt joint. A brief description of these is given below.

7.3.1 Lap Joint

A lap joint consists of two overlapping plates held together by one or more rows of rivets. A single-riveted lap joint is shown in Figure 7.6. A double-riveted lap joint can have rivets either staggered (zig-zag riveting) or in-line (chain riveting) as shown in Figure 7.7.

In a lap joint the eccentric force produces bending and results in distortion of the plate. To avoid distortion, sometimes plates are bent before riveting in order to reduce the effect of bending action as shown in Figure 7.8.

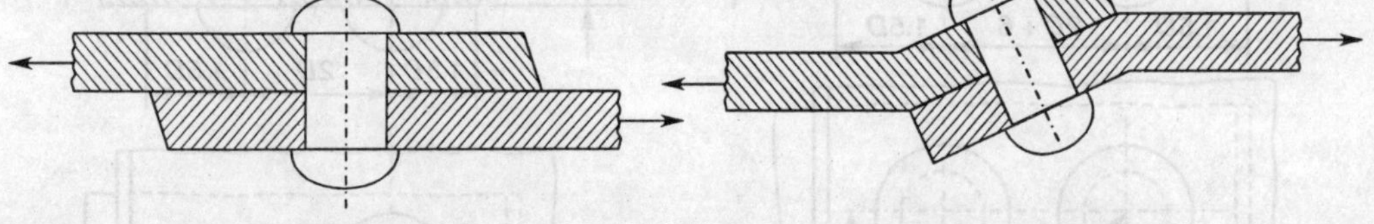

Figure 7.8 Position of plates before and after bending.

7.3.2 Butt Joint

The butt joints are formed by butting two plates against each other and joining them by overlapping straps or cover plates. A butt joint has straps either on one side, called the single-strap butt joint, or on both sides, called the double-strap butt joint. A single-strap butt joint may get bent like a lap joint. Therefore, this type of joint is not recommended in leakproof joints. The butt joint with two straps, one on each side, is considered to be stronger. The straps may have the same width or the outer strap may be narrower. Butt joints are single, double, triple or quadruple riveted. Figure 7.9 shows a double-riveted, double-straps butt joint with zig-zag riveting.

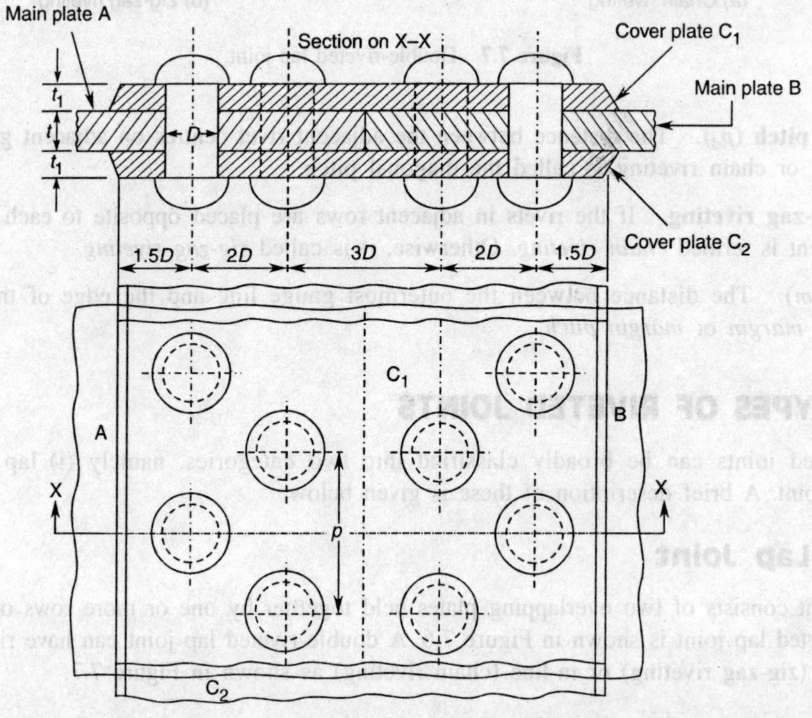

Figure 7.9 Double-riveted butt joint with double cover plates (straps) and zig-zag riveting.

7.4 ANALYSIS OF A RIVETED JOINT

In a riveted joint, the rivets should always be placed at right angle to the direction of the applied force so that maximum stress induced is either shear or crushing. The analysis of a riveted joint is based upon the conventional assumption that the load distribution is equal among all the rivets. However, the actual load distribution in the riveted joint is very complex and depends upon so many factors that no mathematical model can represent the true stress distribution.

In general, the riveted joint is analyzed on the basis of the following assumptions:

(i) Rivets are loaded in shear and the load is distributed in proportion to the shear area of the rivets.
(ii) There are no bending or direct stresses in rivets.
(iii) Rivet holes in the plate do not weaken the plate in compression.
(iv) Rivets after assembly completely fill the holes.
(v) Friction between adjacent surfaces does not affect the strength of joints; however the actual stress produced decreases.
(vi) When a rivet is subjected to double shear, the shear force is equally distributed between the two areas in shear.

According to conventional theory, the failure of a joint may occur in one of the following modes.

Tearing of plate in front of the rivet

The riveted joint may fail due to tearing of the plate in front of the rivet, although such type of failure occurs rarely when the distance between the edge and the nearest row of the rivet, called *margin,* is very small. Generally, this margin distance is kept 1.5 times the diameter of the rivet shank to avoid such failures. This failure mode is shown in Figure 7.10.

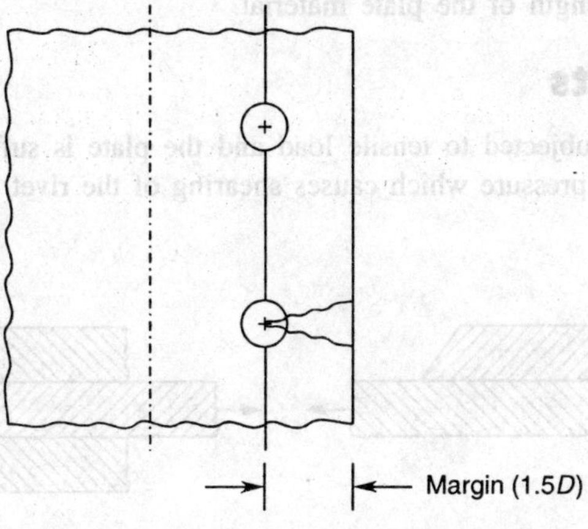

Figure 7.10 Tearing of plate in front of a rivet.

Tearing of plate along the gauge line

In a riveted joint, because of the drilled hole, the plate between the holes is considered to be the weakest along the gauge line. Therefore, the plate can tear off across the pitch line as shown in Figure 7.11.

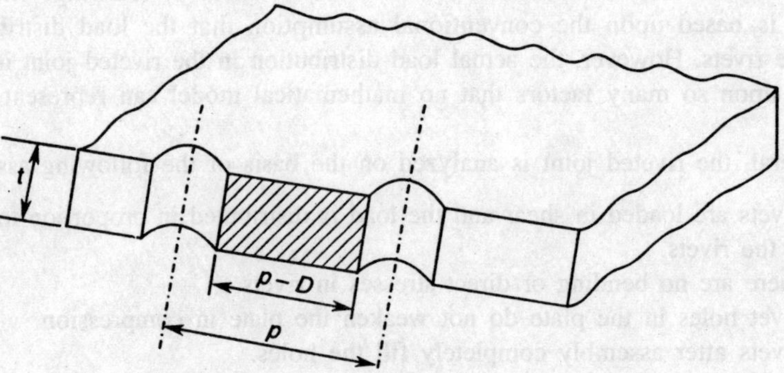

Figure 7.11 Tearing failure of the plate.

The area resistant to tearing failure of the plate per pitch length is

$$A = (p - D)t$$

and tearing resistance of the plate for one pitch length is

$$F = (p - D)t\sigma_t \quad (7.1)$$

where

p is the pitch
D is the diameter of the rivet
t is the thickness of the plate
σ_t is the tensile strength of the plate material.

Shearing of rivets

When a riveted joint is subjected to tensile load and the plate is sufficiently strong, the rivet gets subjected to a radial pressure which causes shearing of the rivet as shown in Figure 7.12.

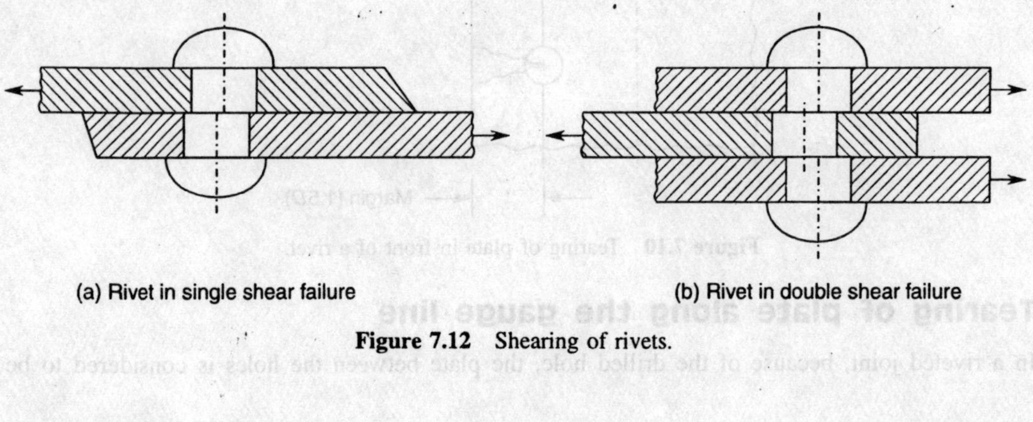

(a) Rivet in single shear failure

(b) Rivet in double shear failure

Figure 7.12 Shearing of rivets.

The shear strength of a riveted joint is given by the following equations:

(i) In single shear: $F_s = \frac{\pi}{4} D^2 \tau_{max}$ (7.2)

(ii) In double shear: $F_s = 1.875 \times \frac{\pi}{4} D^2 \tau_{max}$ (7.3)

where τ_{max} is the shear strength of the rivet material.

In a multi-row riveted joint, the number of rivets in shear is equal to the number of rivets contained between the lines drawn at a pitch distance apart.

Crushing of rivets

When both plates and rivets of a joint are stronger in their respective modes of failure as discussed above, either the plate or the rivet or both may get crushed by compression between the plate and the rivet. Such a failure is called *crushing failure* (Figure 7.13). Sometimes when a joint is subjected to a large force to be resisted by rivets and plates, the plate hole may get elongated and become oval in shape. Hence, the joint becomes loose. Such a failure is termed *bearing failure* of the joint.

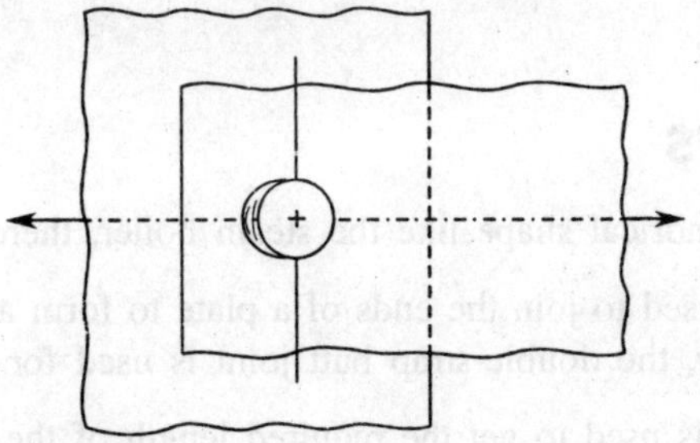

Figure 7.13 Rivet under crushing.

The area resistant to such failure is the projected area of the hole or rivet. Therefore, the crushing strength of the joint is given by

$$F_{cr} = nDt\sigma_{cr} \tag{7.4}$$

where

n is the number of rivets under crushing

σ_{cr} is the crushing strength of a rivet.

Generally, the diameter of the rivet D in mm is computed purely from the empirical formula. Unwin suggested that if the thickness of plate is more than 8 mm, then the diameter of the rivet can be computed by the following relation:

$$D = 6.1\sqrt{t} \tag{7.5}$$

where t is the thickness of the plate in mm. If the thickness of the plate is less than 8 mm, then the shearing strength of the joint must be equated to the crushing strength for determination of the rivet diameter. The pitch length of the joint can be found by equating the shear strength and tearing strength of the joint for one pitch length. That is,

$$(n_1 + 1.875n_2)\frac{\pi}{4}D^2\tau_{max} = (p - D)t\sigma_t \tag{7.6}$$

where

n_1 is the number of rivets in single shear
n_2 is the number of rivets in double shear.

7.5 EFFICIENCY OF A RIVETED JOINT

As discussed earlier, a riveted joint may fail in the following three modes of failure, namely (i) tearing of plate, (ii) shearing of rivets, and (iii) crushing of rivets. The efficiency of a riveted joint is defined as the ratio of the strength of the joint at the weakest section in the weakest mode of failure to the strength of the unpunched plate in one pitch length of the joint.

$$\eta = \frac{\text{strength of the joint in the weakest mode}}{\text{strength of the unpunched plate}}$$

7.6 BOILER JOINTS

In a pressure vessel of cylindrical shape like the steam boiler, there are two types of joints:

Longitudinal joint. It is used to join the ends of a plate to form a cylindrical shell, as shown in Figure 7.14(a). Generally, the double-strap butt joint is used for this purpose.

Circumferential joint. It is used to get the required length of the shell and to close its ends. A lap joint is most widely used for this purpose [Figure 7.14(b)].

7.6.1 Longitudinal Joint Design

The longitudinal joint for a pressure vessel is made by butt joint. Generally, the butt joint with two straps is preferred to make a leakproof joint. The number of rivets per row and the number of rows are largely decided by experience. For design purposes, it is sufficient to consider one pitch length of the joint.

According to IBR, the following steps should be followed for the efficient design of a longitudinal joint.

1. Select a suitable butt joint. The selection primarily depends upon the diameter of the shell and the desired efficiency of the joint. Table 7.1 gives the recommended types of joints and their efficiencies.
2. Sketch the layout of the selected boiler joint.
3. The thickness of the boiler shell plate is determined by the following formula:

$$t = \frac{p_S d}{2\sigma_t \eta_t} + 1 \text{ to } 2 \text{ mm as corrosion allowance} \tag{7.7}$$

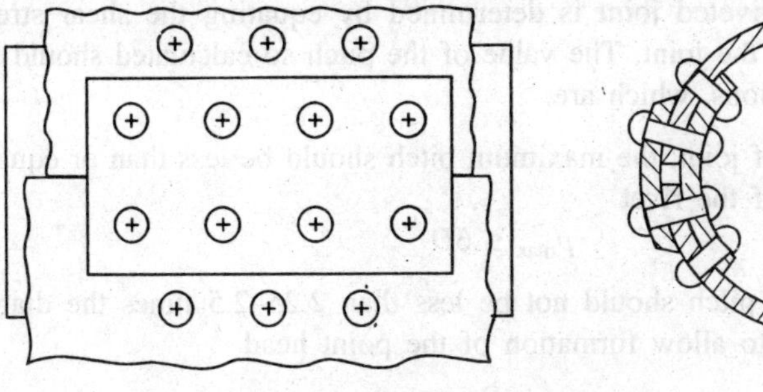

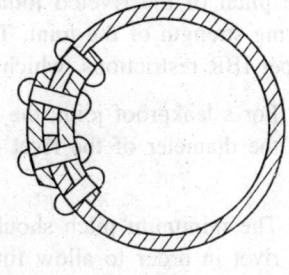

(a) Butt joint for longitudinal joint

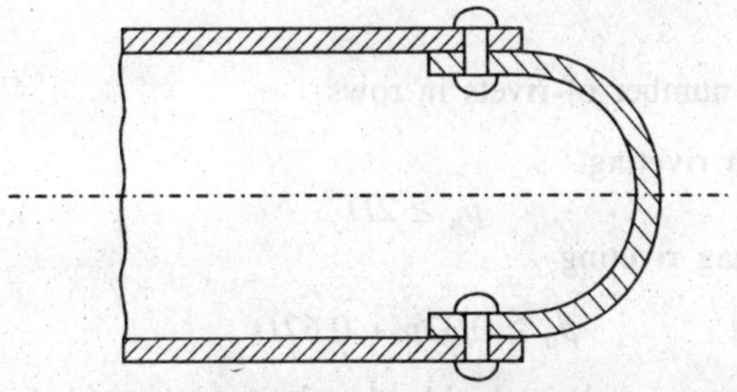

(b) Lap joint for circumferential joint

Figure 7.14 Boiler joints.

where

p_S is the steam pressure, N/mm^2
d is the shell diameter, mm
σ_t is the allowable tensile strength of the material, N/mm^2
η_t is the efficiency of the joint (refer Table 7.1)

4. Calculate the the diameter of the rivet by the Unwin's formula

$$D = 6.1\sqrt{t} \tag{7.8}$$

Table 7.1 Efficiency of commercial boiler joints

Type of joint	*Diameter of shell* (mm)	*Efficiency* (%)
Single riveted lap joint	—	40–60
Double riveted lap joint	—	60–72
Triple riveted lap joint	—	72–82
Double riveted butt joint	600–1800	72–82
Triple riveted butt joint	900–2000	80–90
Quadruple riveted butt joint	1500–2500	85–95

5. The pitch of the riveted joint is determined by equating the shear strength to the tearing strength of the joint. The value of the pitch so calculated should be modified as per IBR restrictions, which are:

   (i) For a leakproof joint, the maximum pitch should be less than or equal to 6 times the diameter of the rivet

$$p_{max} \leq 6D \tag{7.9}$$

   (ii) The minimum pitch should not be less than 2.25–2.5 times the diameter of the rivet in order to allow formation of the point head

$$p_{min} \geq (2.25\text{–}2.5)D \tag{7.10}$$

6. The transverse pitch for a particular joint may be selected from the following relations:

   (i) For an equal number of rivets in rows

   (a) For chain riveting

$$p_{tr} \geq 2D \tag{7.11a}$$

   (b) For zig-zag riveting

$$p_{tr} \geq 0.33p + 0.67D \tag{7.11b}$$

   (ii) Where the longer pitch is double the shorter one and the inner rows are chain riveted

   (a) Transverse pitch between the outer and the inner rows

$$p_{tr} \geq 2D \quad \text{or} \quad p_{tr} > 0.33p + 0.67D \tag{7.12a}$$

   (b) Transverse pitch between the successive inner rows

$$p_{tr} \geq 2D \tag{7.12b}$$

   (iii) Where the longer pitch is double the shorter one and the inner rows are zig-zag riveted

   (a) Transverse pitch between the outer and the inner rows

$$p_{tr} \geq 0.2p + 1.15D \tag{7.13a}$$

   (b) Transverse pitch between the successive inner rows

$$p_{tr} \geq 0.165p + 0.67D \tag{7.13b}$$

7. The thickness of the butt strap should never be less than 10 mm, however, it can be computed by the following relations:

   (i) The thickness of the single-strap butt joint having ordinary riveting is given by

$$t_1 = 1.125t \tag{7.14}$$

   where t is the thickness of the plate.

   (ii) The thickness of the single-strap butt joint having alternate rivets in the outer row omitted, is given by

$$t_1 = 1.125t \frac{p - D}{p - 2D} \tag{7.15}$$

(iii) The thicknesses of the double butt straps of equal width, having alternate rivets in the outer row omitted, are given by

$$t_1 = t_2 = 0.625t \frac{p - D}{p - 2D} \tag{7.16}$$

(iv) The thicknesses of the double butt straps of equal width having ordinary riveting are given by

$$t_1 = t_2 = 0.625t \tag{7.17}$$

(v) The thicknesses of the double butt straps of unequal width, either having ordinary riveting or having every alternate rivet in the outer rows omitted, are given by

$$t_1 = 0.75t \tag{7.18a}$$

$$t_2 = 0.625t \tag{7.18b}$$

8. The margin pitch

$$m = 1.5D \tag{7.19}$$

9. In the zig-zag riveted joint, the diagonal pitch should be

$$p_d \geq \frac{p + D}{2}$$

or

$$p_d = (2.25\text{–}2.5)D \tag{7.20}$$

7.6.2 Circumferential Joint Design

The multi-row lap joint is commonly used for circumferential joints of the boiler. According to IBR, the following steps should be followed for economic design of lap joints:

1. The diameter of the rivet should be the same as that of rivets used for the longitudinal joint.
2. To find the total number of rivets required for the circumferential joint, the shear strength of the rivet is equated to the shearing load

$$N \times \frac{\pi}{4} D^2 \tau_{max} = \frac{\pi}{4} d^2 p_S \tag{7.21}$$

where N is the total number of rivets required for the joint.

3. The pitch of the circumferential joint is calculated from the tearing efficiency of the joint

$$\eta_{tearing} = \frac{p - D}{p} \tag{7.22}$$

where $\eta_{tearing}$ is the tearing efficiency of the circumferential joint. Generally, it is taken as 50–65 per cent of the tearing efficiency of the longitudinal joint.

4. Once the pitch is decided, the number of rivets per row can be calculated from the circumference of the shell

$$n_r = \frac{\pi(d + t)}{p} \tag{7.23}$$

where n_r is the number of rivets per row.

The number of rows required can be computed as

$$\frac{N}{n_r} \tag{7.24}$$

If the calculated value of the number of rows is not a whole number, then it should be increased to the next higher number and all the rivets should be distributed in a revised number of rows. In this process, the pitch of the rivet will be changed. This recalculated pitch should be such that the joint is leakproof.

5. The transverse pitch is selected in a similar way as in the design of the longitudinal joint.

Example 7.1 Two plates of 6 mm thickness are to be joined by a double-riveted zig-zag lap joint. Design the joint, if the allowable strengths of mild steel are

$$\sigma_t = 100 \text{ N/mm}^2,\ \tau = 70 \text{ N/mm}^2, \text{ and } \sigma_{cr} = 130 \text{ N/mm}^2$$

Solution The thickness of the plate is less than 8 mm, therefore, the diameter of the rivet shall be calculated by equating shearing strength to the crushing strength.

$$\text{Shear strength/pitch length} = 2 \times \frac{\pi}{4} D^2\tau \tag{i}$$

(In a double-riveted lap joint, two rivets are in single shear per pitch length)

$$\text{Crushing strength} = 2Dt\sigma_{cr} \tag{ii}$$

Equating (i) and (ii)

$$2 \times \frac{\pi}{4} D^2\tau = 2Dt\,\sigma_{cr}$$

or

$$D = \frac{4t\,\sigma_{cr}}{\pi\tau} = \frac{4 \times 6 \times 130}{\pi \times 70} = 14.18 \text{ mm, say, 16 mm}$$

Pitch of the joint can be found by equating the tearing strength to the shearing strength. Therefore,

$$(p - D)t\sigma_t = 2 \times \frac{\pi}{4} D^2\tau$$

$$p = \frac{2 \times \frac{\pi}{4} 16^2 \times 70}{6 \times 100} + 16 = 62.9 \text{ mm, say, 63 mm}$$

This value of p also satisfies the condition of minimum pitch, i.e. $p \geq 2.5D$. Transverse pitch for equal number of rivets in the row for zig-zag riveting is given by

$$p_{tr} \geq 0.33p + 0.67D$$

$$= 0.33 \times 63 + 0.67 \times 16 = 31.5 \text{ mm, say, } 32 \text{ mm}$$

$$\text{Margin, } m = 1.5D = 1.5 \times 16 = 24 \text{ mm}$$

Efficiency of the joint:

(i) Tearing efficiency

$$\eta_{\text{tearing}} = \frac{p - D}{p} = \frac{63 - 16}{63} = 74.6\%$$

(ii) Shearing efficiency

$$\eta_{\text{shearing}} = \frac{\text{shearing strength}}{\text{strength of unpunched plate}}$$

$$= \frac{2 \times \frac{\pi}{4} D^2 \times \tau}{pt\sigma_t}$$

$$= \frac{2 \times \frac{\pi}{4} 16^2 \times 70}{63 \times 6 \times 100} = 74.4\%$$

(iii) Crushing efficiency

$$\eta_{\text{crushing}} = \frac{\text{crushing strength}}{\text{strength of unpunched plate}}$$

$$= \frac{2Dt\sigma_{cr}}{pt\sigma_t}$$

$$= \frac{2 \times 16 \times 6 \times 130}{63 \times 6 \times 100} = 66\%$$

Among all the three efficiencies, the crushing efficiency is the lowest. Hence, the efficiency of the joint is 66%.

Example 7.2 A penstock of 0.75 m diameter, made of C20 steel plate, operates under 25 m of water head. If the penstock is made by a single riveted lap joint, determine the main dimensions of the joint.

Solution Tensile strength of C20 steel, $\sigma_{ut} = 500$ N/mm^2

Let us assume factor of safety = 4

$$\text{Allowable tensile strength, } \sigma_t = \frac{500}{4} = 125 \text{ N/mm}^2$$

$$\text{Shear strength, } \tau = 0.6\sigma_t = 75 \text{ N/mm}^2$$
$$\text{Crushing strength, } \sigma_{cr} = 1.28\sigma_t = 160 \text{ N/mm}^2$$

Let us assume that the efficiency of the joint is 60% and corrosion allowance of plate thickness is 2 mm. Pressure in the penstock

$$P = \rho H = \frac{1}{100} \times 25 = 0.25 \text{ N/mm}^2$$

According to IBR, the thickness of the plate

$$t = \frac{Pd}{2\sigma_t \eta_t} + 2$$

$$= \frac{0.25 \times 750}{2 \times 125 \times 0.6} + 2 = 3.25 \text{ mm, say, } 4 \text{ mm}$$

Equating shearing and crushing strengths

$$\frac{\pi}{4} D^2 \tau = D t \sigma_{cr}$$

or

$$D = \frac{4t\sigma_{cr}}{\pi\tau} = \frac{4 \times 4 \times 160}{\pi \times 75} = 10.86 \text{ mm, say, } 12 \text{ mm}$$

For pitch p, equating shear strength with tearing strength

$$(p - D)t\sigma_t = \frac{\pi}{4} D^2 \tau$$

or

$$(p - 12) \times 4 \times 125 = \frac{\pi}{4} \times 12^2 \times 75$$

or

$$p = 28.9 \text{ mm, say, } 30 \text{ mm}$$

For a fluid-tight joint, the pitch should be between $2.5D$ and $6D$, which is satisfied, hence the joint is leakproof.

$$\text{Margin pitch, } m = 1.5D = 18 \text{ mm}$$

Example 7.3 Design riveted joints for the longitudinal and circumferential seams of a boiler having 1.25 m diameter to withstand maximum pressure of 2.5 N/mm^2.

Solution Let us assume that the material of the shell plate and rivet is C20 steel having allowable strengths as

$$\sigma_t = 86 \text{ N/mm}^2 \qquad \tau = 52 \text{ N/mm}^2 \qquad \sigma_{cr} = 129 \text{ N/mm}^2$$

Since the diameter of the shell is more than 1.0 m, we should select a triple-riveted butt joint with upper plate having two rows of rivets on each side and one rivet on outer row with alternate rivets omitted, as shown in the figure below. The efficiency of such a longitudinal joint should be approximately 85% (refer Table 7.1).

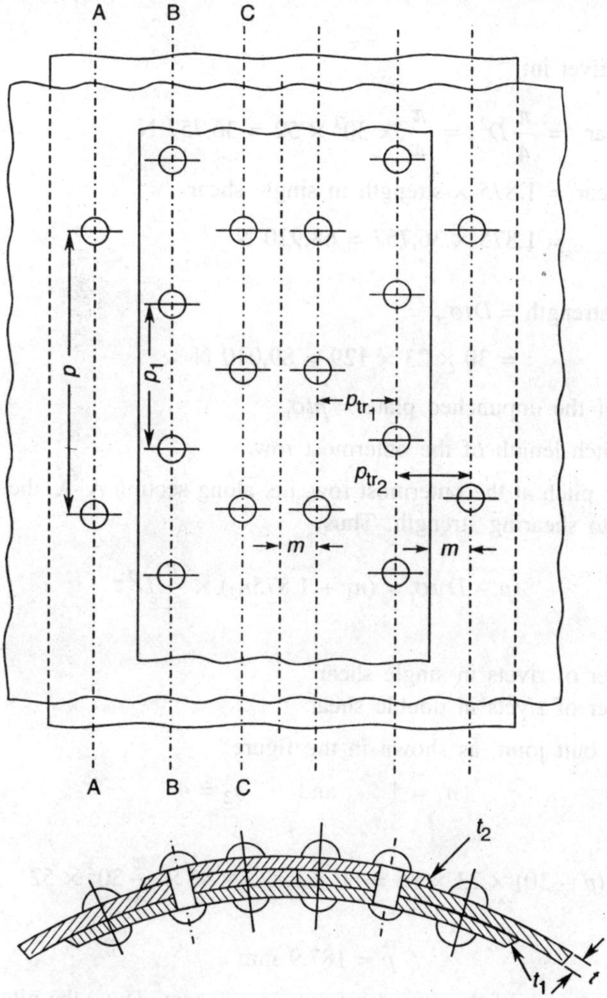

According to IBR, the thickness of the shell plate

$$t = \frac{Pd}{2\sigma_t \eta_t} + 1.0 \text{ (as corrosion allowance)}$$

$$= \frac{2.5 \times 1250}{2 \times 86 \times 0.85} + 1.0 = 22.37 \text{ mm, say, 23 mm}$$

Diameter of the rivet, according to the Unwin's formula

$$D = 6.1\sqrt{t} = 6.1\sqrt{23} = 29.25 \text{ mm}$$

Let us adopt rivet diameter of 29 mm and hole size of 30 mm. After the rivet is driven, its diameter will be 30 mm.

Longitudinal joint

The strength of one rivet in:

(i) Single shear $= \dfrac{\pi}{4} D^2 \tau = \dfrac{\pi}{4} \times 30^2 \times 52 = 36{,}757$ N

(ii) Double shear = 1.875 × strength in single shear

$= 1.875 \times 36{,}757 = 68{,}920$ N

Also,

Crushing strength $= Dt\sigma_{cr}$

$= 30 \times 23 \times 129 = 89{,}010$ N

Strength of the unpunched plate $= pt\sigma_t$

p is the pitch length of the outermost row.

To calculate the pitch at the outermost row, i.e. along section A–A, the tearing strength of the plate is equated to shearing strength. Thus,

$$(p - D)t\sigma_t = (n_1 + 1.875n_2) \times \frac{\pi}{4} D^2 \tau$$

where

n_1 is the number of rivets in single shear
n_2 is the number of rivets in double shear.

For the triple-riveted butt joint, as shown in the figure.

$$n_1 = 1 \quad \text{and} \quad n_2 = 4$$

Therefore,

$$(p - 30) \times 23 \times 86 = (1 + 1.875 \times 4) \times \frac{\pi}{4} 30^2 \times 52$$

or

$$p = 187.9 \text{ mm}$$

Let us adopt the pitch length of the outermost row as 190 mm. Thus, the pitch of the inner row, $p_1 = p/2 = 95$ mm.

For a leakproof joint, as per IBR the following conditions should be satisfied:

$$2.5D \le p_1 \le 6D$$

95 mm pitch of the inner row satisfies the above condition; hence it is safe.

Thickness of straps, $t_2 = 0.625t = 15$ mm

$t_1 = 0.75t = 18$ mm

The strength of the riveted joint at all the sections is required to determine the efficiency. Thus,

(i) Tearing strength along section A–A

$= (p - D)t\sigma_t$

$= (190 - 30) \times 23 \times 86 = 316.48$ kN

(ii) Strength of the joint along section B–B

= tearing strength at B–B + shearing strength at A–A

$= (p - 2D)t\sigma_t + \frac{\pi}{4}D^2\tau$

$= (190 - 2 \times 30) \times 23 \times 86 + 36{,}757$

$= 293.89$ kN

(iii) Strength of the joint along section C–C

= tearing strength at C–C + shearing strength at B–B + shearing strength at A–A

$= (p - 2D)t\sigma_t + 2 \times 1.875 \times \frac{\pi}{4}D^2\tau + \frac{\pi}{4}D^2\tau$

$= (190 - 2 \times 30) \times 23 \times 86 + (1 + 2 \times 1.875) \times 36{,}757$

$= 431.74$ kN

(iv) Shearing strength of all the rivets

$= (n_1 + 1.875n_2) \times \frac{\pi}{4}D^2\tau$

$= (1 + 1.875 \times 4) \times 36{,}757$

$= 312.43$ kN

(v) Crushing strength of all the rivets

$= D(n_1t_1 + n_2t)\sigma_{cr}$

$= 30(1 \times 18 + 4 \times 23) \times 129$

$= 425.7$ kN

The efficiency of the joint is the ratio of the strength at the weakest section to the strength of the unpunched plate. Thus,

$$\eta = \frac{\text{strength at section B–B}}{\text{strength of unpunched plate}}$$

$$= \frac{293.89 \times 1000}{23 \times 190 \times 86}$$

= 0.782 or 78.2% which is reasonable for a triple-riveted butt joint.

Transverse pitch. In a joint where the longer pitch, p, is double the shorter pitch p_1 and the inner rows are zig-zag riveted

(i) Transverse pitch between the outer row and the inner row

$$p_{tr1} \geq 0.2p + 1.15D$$

$$= 0.2 \times 190 + 1.15 \times 30 = 72.5 \text{ mm, say, } 75 \text{ mm}$$

(ii) Transverse pitch between the successive inner rows

$$p_{tr2} \geq 0.165p + 0.67D$$

$$= 0.165 \times 190 + 0.67 \times 30 = 51.45 \text{ mm, say, } 55 \text{ mm}$$

$$\text{Margin, } m = 1.5D = 45 \text{ mm}$$

Circumferential joint

The type of joint used for plate thickness (23 mm) and rivet diameter (30 mm) is the lap joint.

In a lap joint having all the rivets in single shear, the shear strength of one rivet

$$= \frac{\pi}{4} 30^2 \times 52 \text{ N} = 36{,}757 \text{ N}$$

Shear force on the rivets due to fluid pressure

$$= \frac{\pi}{4} d^2 \times P$$

$$= \frac{\pi}{4} \times 1.25^2 \times 2.5 \times 10^6$$

$$= 3.07 \times 10^6 \text{ N}$$

Number of rivets required to withstand the shear load

$$N = \frac{\text{total shear force}}{\text{shear strength of one rivet}}$$

$$= \frac{3.07 \times 10^6}{36{,}757} = 83.5, \text{ say, } 84$$

The length of circumference of the shell

$$= \pi(d + t)$$

$$= \pi(1250 + 23) = 3999.24, \text{ say, } 4000 \text{ mm}$$

Let us select the double-riveted lap joint. Therefore,

$$\text{Number of rivets per row} = \frac{84}{2} = 42$$

$$\text{Pitch length} = \frac{\text{circumference of the shell}}{\text{number of rivets/row}}$$

$$= \frac{4000}{42} = 95.2 \text{ mm}$$

This pitch satisfies the condition of the leakproof joint, i.e. $2.5D \leq p \leq 6D$; hence the joint is leakproof. The efficiency of the joint

$$\eta = \frac{p - D}{p} = \frac{95.2 - 30}{95.2} = 0.6848$$

The 68.48% efficiency is reasonably good for the double-row lap joint. The layout sketch of the double-riveted lap joint is similar to that shown in Figure 7.7.

Example 7.4 Two lengths of mild steel tie rods having width 200 mm and thickness 12.5 mm are to be connected by means of a butt joint with equal straps. Design the Lozenge* joint if the permissible working stresses in plates and rivet material are: $\sigma_t = 80$ N/mm^2, $\tau = 50$ N/mm^2, and $\sigma_{cr} = 150$ N/mm^2.

Solution Plate width, $b = 200$ mm

Plate thickness, $t = 12.5$ mm

According to the Unwin's formula, the rivet diameter

$$D = 6.1\sqrt{t} = 6.1\sqrt{12.5} = 21.56 \text{ mm, say, } 22 \text{ mm}$$

Thickness of the double strap

$$t_1 = t_2 = 0.75t = 0.75 \times 12.5 = 9.375 \text{ mm, say, } 10 \text{ mm}$$

Strength of the joint in different modes

(i) Shear strength of one rivet in double shear

$$= 1.875 \times \frac{\pi}{4} D^2\tau$$

$$= 1.875 \times \frac{\pi}{4} 22^2 \times 50$$

$$= 35.637 \text{ kN}$$

(ii) Crushing strength of one rivet

$$= Dt\,\sigma_{cr}$$

$$= 22 \times 12.5 \times 150$$

$$= 41.25 \text{ kN}$$

(iii) Tearing strength of the plate at outer row

$$= (b - D)t\sigma_t$$

$$= (200 - 22) \times 12.5 \times 80$$

$$= 178 \text{ kN}$$

Minimum number of rivets required

$$N = \frac{\text{tearing strength of the plate}}{\text{lowest strength of one rivet}}$$

$$= \frac{178}{35.637} = 4.99, \text{ say, } 5$$

The layout of the riveted joint is shown in the figure below.

---

*A Lozenge joint is a special type of double cover plate butt joint. It is mainly used in structural joints, namely bridges, girders, and so forth.

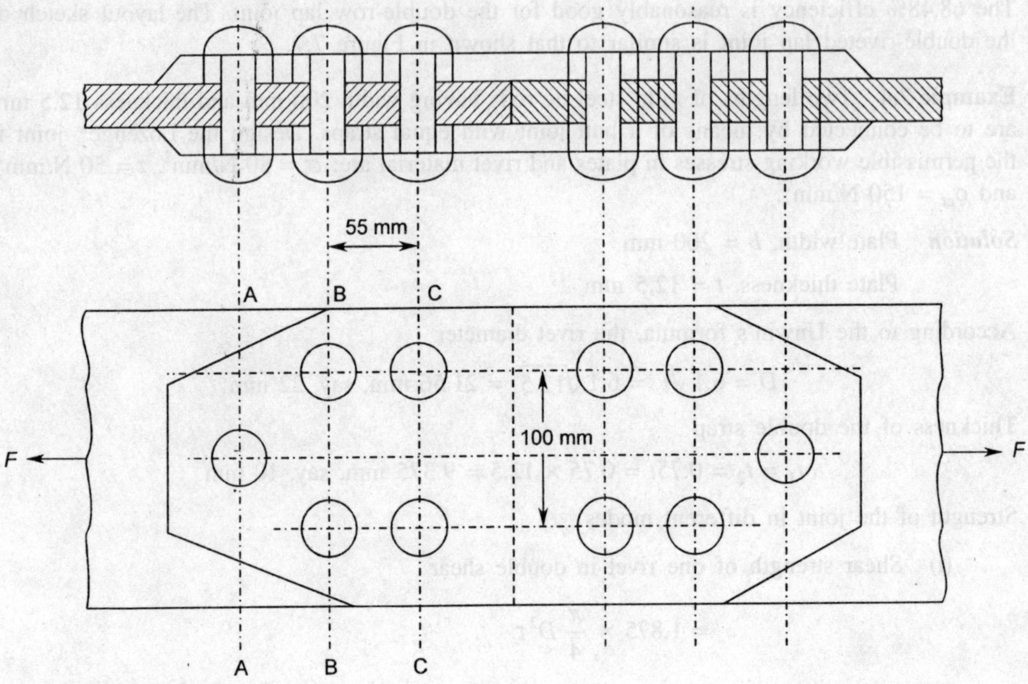

Strength of the Lozenge joint in different modes of failure is:

(i) Tearing of plate along section A–A

$= (b - D)t\sigma_t$

$= (200 - 22) \times 12.5 \times 80$

$= 178$ kN

(ii) Tearing of plate along section B–B + shearing of one rivet at section A–A

$= (b - 2D)t\sigma_t + 1.875 \times \frac{\pi}{4} D^2\tau$

$= (200 - 2 \times 22) \times 12.5 \times 80 + 35{,}637$

$= 191.637$ kN

(iii) Tearing of plate along section C–C + shearing of three rivets

$= (b - 2D)t\sigma_t + 3 \times 1.875 \times \frac{\pi}{4} D^2\tau$

$= (200 - 2 \times 22) \times 12.5 \times 80 + 3 \times 35{,}637$

$= 262.91$ kN

(iv) Shearing strength of all the rivets

$= 5 \times 1.875 \times \frac{\pi}{4} D^2\tau$

$= 5 \times 35{,}637$

$= 178.185$ kN

(v) Crushing strength of all the rivets

$= 5 \times Dt\sigma_{cr}$

$= 5 \times 41.25$

$= 206.25$ kN

$$\text{Efficiency of the joint} = \frac{\text{lowest strength of the joint}}{\text{strength of the unpunched plate}}$$

$$= \frac{178 \times 10^3}{200 \times 12.5 \times 80} = 0.89$$

Thus efficiency of the Lozenge joint is 89%.

Margin pitch, $m = 1.5D = 33$ mm

Pitch along section B–B and C–C $= \frac{b}{2} = 100$ mm

Transverse pitch, $p_{tr} = 2.5D = 55$ mm

7.7 ECCENTRICALLY LOADED RIVETED JOINT

The design of riveted joints is based upon the fundamental assumption that all rivets of the joint are subjected to shear force, i.e. the force line passes through the centre of gravity (c.g.) of the joint. However, in some cases, like in structural joints, the force line does not pass through the c.g. of the joint and has an offset distance e. Structural joints are, therefore, commonly known as eccentrically-loaded riveted joints.

In these joints, the force is shifted by an eccentricity e which produces an additional shear force on the rivet system that tries to rotate the structural joint as shown in Figure 7.15.

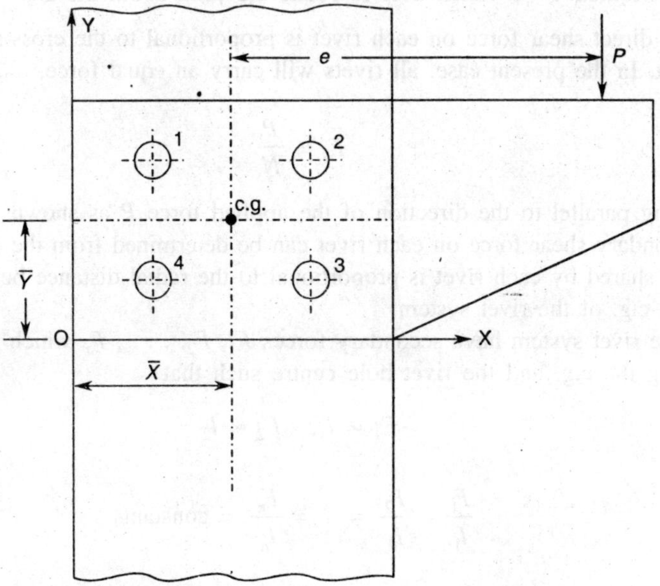

Figure 7.15 Eccentrically-loaded riveted joint.

The torsional force, called the *secondary shear force,* therefore, acts on the rivet in addition to the direct shear force. In such a case, a joint must be designed by considering the effect of both forces.

For force analysis of the above system, the following procedure should be adopted.

1. Determine the centre of gravity (c.g.) of the joint by taking moment about OX and OY axes. Let the rivets in the joint be designated as 1, 2, 3, . . . , etc. with all rivets having the same area of cross-section, i.e. $A_1 = A_2 = \cdots = A_n$. These rivets are situated at $X_1, X_2, \ldots, X_n$ and $Y_1, Y_2, \ldots, Y_n$ from the OX and OY axis, respectively. Then, taking moment about axes.

$$\overline{X} = \frac{X_1 + X_2 + \cdots + X_n}{N} \tag{7.25a}$$

and

$$\overline{Y} = \frac{Y_1 + Y_2 + \cdots + Y_n}{N} \tag{7.25b}$$

where

$\overline{X}$ is the distance of the c.g. from the OX-axis

$\overline{Y}$ is the distance of the c.g. from the OY-axis

N is the number of rivets

2. Let us assume that two forces of equal magnitude (equal to eccentric force) act at the c.g. of the system in opposite directions, which does not disturb the equilibrium of the system. From this assumption, we can conclude that the joint is subjected to two types of forces:

   (i) Direct shear force due to external force P

   (ii) Moment $P \cdot e$ which tries to rotate the joint about the c.g.

   The direct shear force on each rivet is proportional to the cross-sectional area of the rivet. In the present case, all rivets will carry an equal force,

$$F_s = \frac{P}{N} \tag{7.26}$$

acting parallel to the direction of the applied force P as shown in Figure 7.16.

The secondary shear force on each rivet can be determined from the fact that the amount of shear force shared by each rivet is proportional to the radial distance between the rivet hole centre and the c.g. of the rivet system.

Let the rivet system have secondary forces $F_1, F_2, \ldots, F_n$ which act perpendicular to the line joining the c.g. and the rivet hole centre such that

$$F_1 \propto l_1, \quad F_2 \propto l_2$$

or

$$\frac{F_1}{l_1} = \frac{F_2}{l_2} = \cdots = \frac{F_n}{l_n} = \text{constant} \tag{7.27}$$

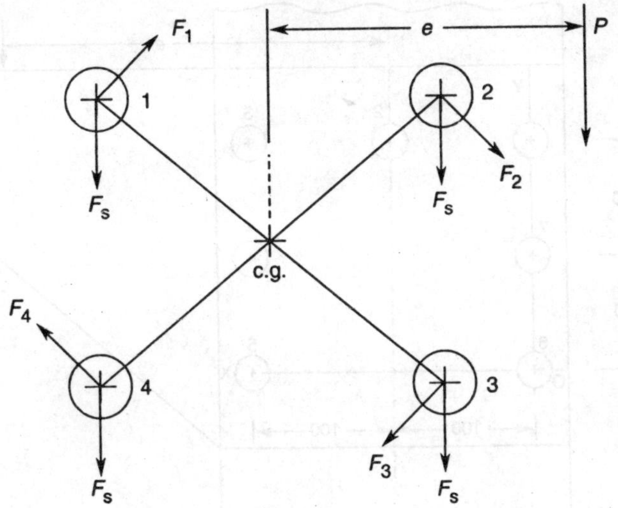

Figure 7.16 Distribution of direct and secondary shear force.

Also, for equilibrium of the system the sum of the moments should be equal to zero. Thus,

$$P \times e = F_1 l_1 + F_2 l_2 + \cdots + F_n l_n \tag{7.28}$$

From Eqs. (7.27) and (7.28), we get

$$P \times e = \frac{F_1}{l_1}(l_1^2 + l_2^2 + \cdots + l_n^2) \tag{7.29}$$

Using Eq. (7.29), we can determine the secondary forces acting on all rivets. The direction of these forces is normal to lines joining rivet centres with the c.g. of the riveted joint as shown in Figure 7.16.

3. Calculate the resultant force on all rivets. The resultant force on the nth rivet

$$R_n = (F_n^2 + F_s^2 + 2F_n F_s \cos \theta_n)^{0.5} \tag{7.30}$$

where θ_n is the angle between the direct and the secondary shear force at the nth rivet. The value of R_n is maximum when θ_n is minimum and maximum resultant force $R_{n\max}$ shall be the design force.

4. Determine the diameter of the rivet by considering the shear failure of the rivet.

Example 7.5 Design an eccentrically loaded lap riveted joint as shown in the figure below. The bracket plate is 25 mm thick. All rivets are to be of the same size. The load on the bracket is 30 kN. The rivet spacing is 100 mm and eccentricity is 400 mm. Permissible shear strength of the rivet is 60 N/mm^2.

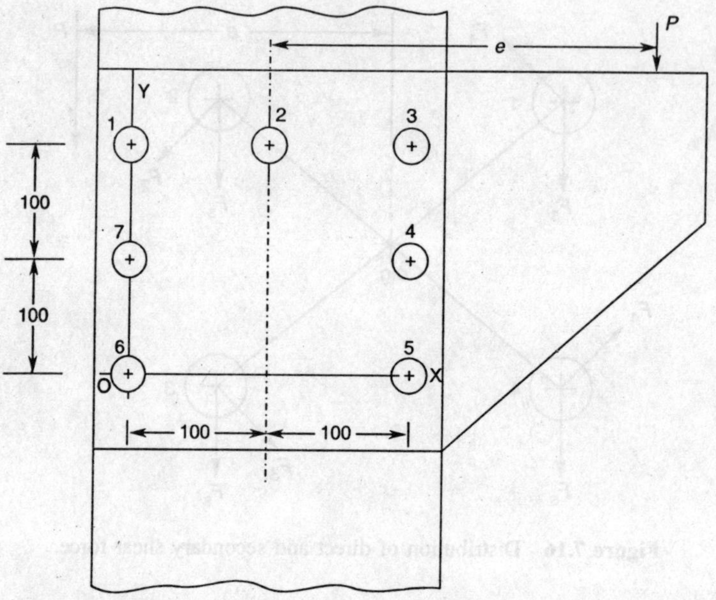

Solution First let us find the c.g. of the rivet system. Let

$\overline{X}$ be the distance of the c.g. from the OX-axis

$\overline{Y}$ be the distance of the c.g. from the OY-axis

$X_1, X_2, \ldots$ be the distances of the rivet centre from the OX-axis

$Y_1, Y_2, \ldots$ be the distances of the rivet centre from the OY-axis

$$\overline{X} = \frac{X_1 + X_2 + \cdots + X_7}{N}$$

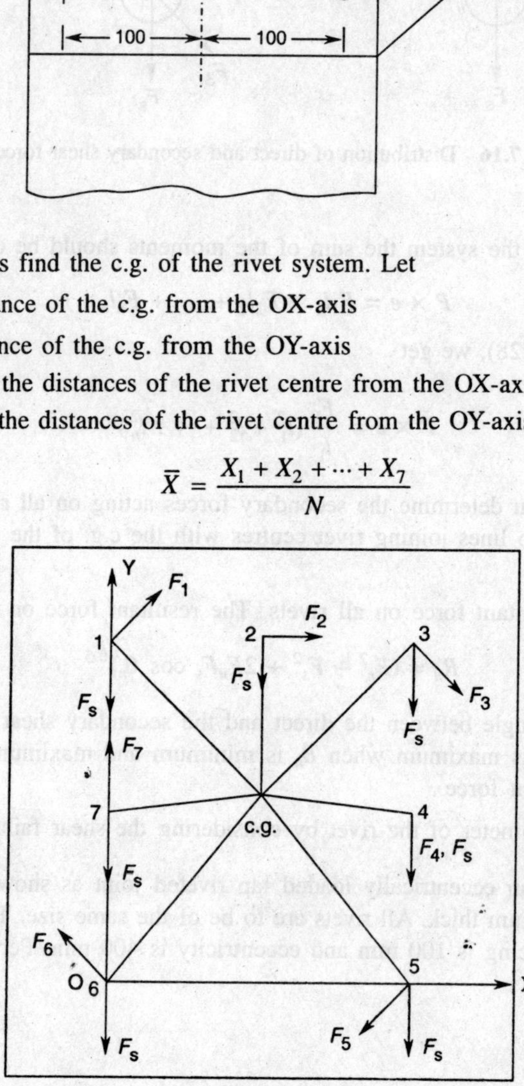

Since $X_1 = X_6 = X_7 = 0$, $X_2 = 100$, $X_3 = X_4 = X_5 = 200$

$$\overline{X} = \frac{100 + 200 + 200 + 200}{7} = 100 \text{ mm}$$

$$\overline{Y} = \frac{Y_1 + Y_2 + \cdots + Y_7}{n}$$

Since $Y_5 = Y_6 = 0$, $Y_1 = Y_2 = Y_3 = 200$, $Y_4 = Y_7 = 100$

$$\overline{Y} = \frac{200 + 200 + 200 + 100 + 100}{7} = 114.3 \text{ mm}$$

Therefore, the c.g. of the rivet system from the centre of the rivet 6 is $\overline{X} = 100$ mm and $\overline{Y} = 14.3$ mm.

Direct shear force on each rivet acting vertically downwards is

$$F_s = \frac{P}{N} = \frac{30}{7} = 4.286 \text{ kN}$$

Turning moment $= P \cdot e = 30 \times 400 = 12{,}000$ kN · mm. Let $F_1, F_2, \ldots, F_7$ be the secondary forces on the rivets 1, 2, . . ., 7 at distances $l_1, l_2, \ldots, l_7$, respectively. From the c.g. of the rivet system

$$l_1 = l_3 = \sqrt{100^2 + (200 - 114.3)^2} = 131.7 \text{ mm}$$

$$l_2 = 200 - 114.3 = 85.7 \text{ mm}$$

$$l_4 = l_7 = \sqrt{100^2 + (114.3 - 100)^2} = 101 \text{ mm}$$

$$l_5 = l_6 = \sqrt{100^2 + 114.3^2} = 152 \text{ mm}$$

Now using the moment equation

$$P \times e = \frac{F_1}{l_1}(l_1^2 + l_2^2 + \cdots + l_7^2)$$

$$= \frac{F_1}{l_1}(2l_1^2 + l_2^2 + 2l_4^2 + 2l_5^2)$$

or

$$30 \times 400 = \frac{F_1}{131.7}(2 \times 131.7^2 + 85.7^2 + 2 \times 101^2 + 2 \times 152^2)$$

or

$$F_1 = 14.54 \text{ kN}$$

$$F_2 = F_1 \frac{l_2}{l_1} = 14.54 \times \frac{85.7}{131.7} = 9.461 \text{ kN}$$

$$F_3 = F_1 = 14.54 \text{ kN}$$

$$F_7 = F_4 = F_1 \frac{l_4}{l_1} = 14.54 \times \frac{101}{131.7} = 11.15 \text{ kN}$$

$$F_6 = F_5 = F_1 \frac{l_5}{l_1} = 14.54 \times \frac{152}{131.7} = 16.78 \text{ kN}$$

By drawing the direct and secondary shear forces on each rivet, as shown in the figure on page 196, we can conclude that rivet 5 is heavily loaded. The angle formed between the direct and secondary shear forces for this rivet is

$$\cos\theta_5 = \frac{100}{l_5} = \frac{100}{152} = 0.658$$

Resultant force on rivet 5

$$R_5 = \sqrt{F_s^2 + F_5^2 + 2F_sF_5\cos\theta_5}$$

$$= \sqrt{4.286^2 + 16.78^2 + 2\times 4.286\times 16.78\times 0.658} = 19.86 \text{ kN}$$

Considering shear failure of the rivet, the diameter of the rivet can be determined as

$$D = \left(\frac{4R_5}{\pi\tau}\right)^{0.5}$$

$$= \left(\frac{4\times 19.86\times 10^3}{\pi\times 60}\right)^{0.5} = 20.5 \text{ mm, say, 21 mm}$$

7.8 COMPUTER-AIDED DESIGN OF RIVETED JOINTS

The application of the computer in the design of riveted joints is enormous. It ranges from boiler joint, pressure vessels, fluid carrying pipes to structural joints. In this section, the use of the computer for rivet force analysis of eccentrically loaded structural riveted joints is demonstrated. The listing of the computer program is given in Program 7.1. The variables used are mostly discussed in theory. Program asks for the total number of rivets and geometrical positions of all the rivets. The user is advised to designate any rivet as rivet number 1 and assign its coordinates (0, 0). The coordinates of other rivets should be supplied in relation to the reference rivet number 1. The secondary shear force and its angle of inclination are computed using the given geometry. The secondary shear force is resolved into horizontal and vertical components. The direct shear force is added to the vertical component and the revised resultant force is computed. Comparing these resultant forces, the maximum resultant force and its location are computed to determine the rivet size. A sample run of this program is demonstrated.

Program 7.1

```
//    A PROGRAM TO DESIGN RIVETED JOINT FOR ECCENTRIC LOADING
#     include <iostream.h>
#     include <math.h>
#     include <conio.h>
#     define PI 3.1415

void main ( )
{
      int n,i,imax;
      double x[20], y[20], length[20], force[20], r[20];
      double p,e,sigmay,fos,sumx,sumy,xbar,ybar,sumlsqr;
      double ratio,fx,fy,fxtotal,fytotal,rmax,d;
//    input of design data
//    if the number of rivets are more than 20 then dimensions may be
//    changed.
      cout <<"\n DESIGN INPUT DATA" << endl;
      cout <<"\n";
      cout <<"\n Number of Rivets in the System:";
      cin >> n;
      cout <<"\n NOTE: COORDINATES OF RIVET NO. 1 SHOULD BE (0, 0) ";
      cout <<"\n AND COORDINATES OF OTHER RIVETS SHOULD";
      cout <<"\n BE SUPPLIED WITH REFERENCE TO RIVET NO. 1" <<endl;
      cout <<"\n";
      for (i=1; i<=n; i=i+1)
      {
      cout <<"\n Coordinates (X, Y) of Rivet No. "<< i <<"= ";
      cin >> x[i] >> y[i];
      }
      cout <<"\n Eccentric Load in kN= ";
      cin >> p;
      cout <<"\n Eccentricity of load in (mm) = ";
      cin >> e;
      cout <<"\n Yield Strength (N/mm2) = ";
      cin >>sigmay;
```

```
        cout <<"\n Factor of Safety (fos) = ";
        cin >> fos;
        sigmay=sigmay/fos;
//      calculate c.g. of the joint
        sumx=0.0;
        sumy=0.0;
        for (i=1; i<=n; i=i+1)
        {
            sumx=sumx+x [i];
            sumy=sumy+y [i];
        }
        xbar=sumx/n;
        ybar=sumy/n;
//      calculate distances from c.g.
        for (i=1; i<=n; i=i+1)
        {
                    length [i] =sqrt (pow ((x[i] -xbar), 2) +pow ((y[i] -ybar), 2));
        }
        sumlsqr=0.0;
        for (i=1; i<=n; i=i+1)
        {
            sumlsqr=sumlsqr+(length [i] *length [i]);
        }
//      calculate secondary, primary and resultant forces
        for (i=1; i<=n; i=i+1)
        {
            force[i] =p*e*length[i]/sumlsqr;
            if (x[i] = =xbar)
            {
                fx=force[i];
                fy=0.0;
            }
            else
            {
            ratio=(y[i] -ybar) / (x[i] -xbar);
            fy=force[i] / (sqrt (1+ratio*ratio));
            fx=fy*ratio;
            }
            fxtotal=fx;
            fytotal=fy+p/n;
//          cout << fxtotal << fytotal <<endl;
            r[i] =sqrt (fxtotal*fxtotal+fytotal*fytotal);
        }
//      find maximum resultant force
        rmax=0.0;
        imax=0;
        for (i=1; i<=n; i=i+1)
```

```
{
        if (rmax < r[i])
        {
            rmax=r[i];
            imax=i;
        }
    }
    d=sqrt (4.0*rmax*1000.0/(PI*sigmay));
    d=int(d+1.0);
    cout <<"\n DESIGN RESULTS" <<endl;
    cout <<"\n RESULTANT FORCE= "<<rmax <<"kN";
    cout <<"\n AT RIVET NO. = "<<imax;
    cout <<"\n RIVET DIAMETER ="<<d <<"mm" <<endl;
```

Sample Run of Program 7.1

DESIGN INPUT DATA
Number of Rivets in the System = 6
NOTE: COORDINATES OF RIVET NO. 1 SHOULD BE(0, 0) AND COORDINATES OF OTHER RIVETS SHOULD BE SUPPLIED WITH REFERENCE TO RIVET NO. 1
Coordinates(X,Y) of Rivet No. 1 = 0 0
Coordinates(X,Y) of Rivet No. 2 = 0 -75
Coordinates(X,Y) of Rivet No. 3 = 0 -150
Coordinates(X,Y) of Rivet No. 4 = -100 -150
Coordinates(X,Y) of Rivet No. 5 = -100 -75
Coordinates(X,Y) of Rivet No. 6 = -100 0
Eccentric Load(kN) =30
Eccentricity of Load =400
Shear Strength of Material(N/mm2)=350
Factor of Safety =5

DESIGN RESULTS

RESULTANT FORCE = 17.691kN
AT RIVET NO. = 1
RIVET DIAMETER = 18 mm

7.9 WELDED JOINTS

Welding is a process of joining two similar or dissimilar pieces of metal by heating to a suitable temperature with or without the application of pressure. Welded joints so formed are permanent joints and they cannot be separated without effecting a fracture. In the welded joint, a metallic bond is established between two pieces which have the same mechanical and physical properties. Welding may be classified into three main types: (i) forge welding, (ii) electric resistance, and (iii) fusion welding. In fusion welding, which is most widely used, the two pieces to be joined are heated to the fusion temperature of the metal; additional filler material is usually applied in the corner of the joint by melting a filler rod of suitable composition and then the joint is allowed to cool. The medium of heating can be an electric arc or an oxyacetylene gas flame.

The general applications of welding are the following:

1. Welding can be used to substitute casting and forging manufacturing methods.
2. Welding can be used as a fabrication medium to join parts permanently and to form built-up parts.
3. Welded joints can substitute riveted joints.

The welded joints are lighter in weight and have higher joining efficiency. These joints are leakproof and economical both from the point of view of cost of material and labour. However, these joints have poor vibration damping characteristics and thermal distortion due to localized heating which may cause residual stresses.

7.9.1 Types of Welded Joints

Welded joints are classified according to the relative positions of the two parts to be joined. There are five basic types of welded joints.

Butt joint. It is a joint between two plates lying in the same plane. The edges of the plates may be bevelled depending upon their thickness. Generally, a plate of 6 mm thickness is not bevelled and such joints are called *square butt joints.* Plates of thickness from 6–20 mm are bevelled to form a single V shape. However, plates of thickness more than 20 mm are welded from both sides of the plates and such joints are called double V-butt joint. Figure 7.17 shows various types of butt joints.

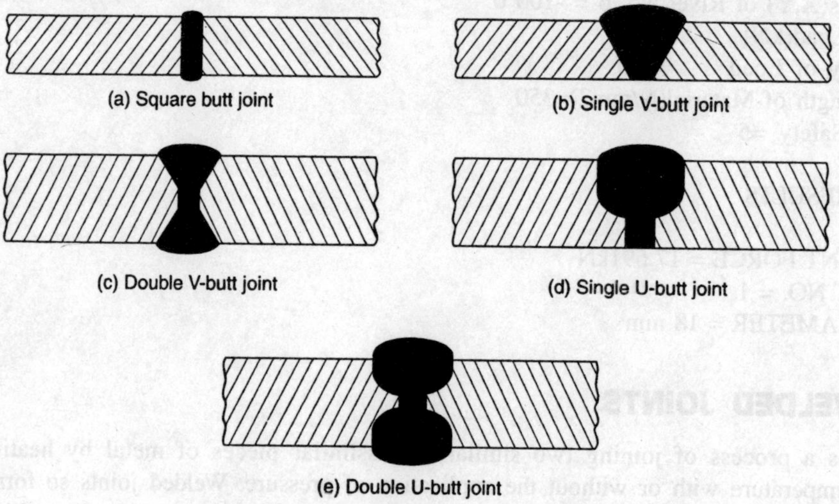

Figure 7.17 Various types of butt joints.

A butt welded joint may be flushed, reinforced on one side or both sides. For fluctuating load conditions, flush butt weld is preferred over to the reinforced one, as the latter type creates discontinuity and gives rise to stress concentration.

Lap joint. In a lap joint, the two plates overlap each other for a certain length and the right angle recess formed between the two plates is filled with the weld metal. Such a weld is also called *fillet weld.* Figure 7.18 shows a standard full fillet weld section of a triangle in which the two sides of the right angle are equal to the thickness of the plate.

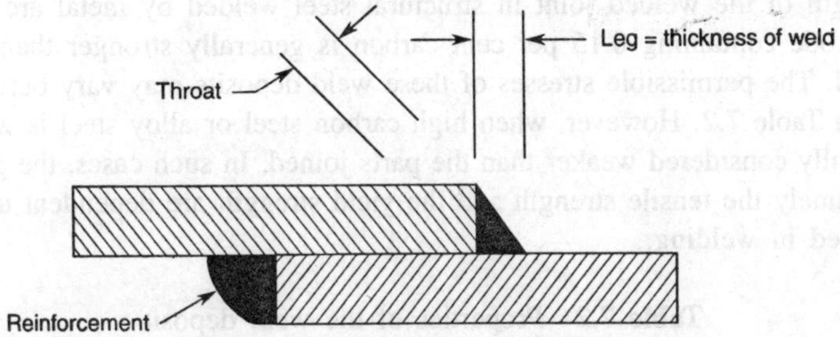

Figure 7.18 Lap joint with fillet weld.

Edge weld. For plates of thickness less than 6 mm, the ends of the overlaping plates can be directly welded at the edges, as shown in Figure 7.19. Such joints are called *edge weld* which can be subjected to light load only.

Corner weld. A corner joint is a joint between two plates which are at right angles to each other in the form of a corner, as shown in Figure 7.20. In a corner joint, the throat of the weld is of the order of 1.35 times the thickness of the plate.

Tee-weld. It is a joint between two plates located at right angles to each other in the form of T, as shown in Figure 7.21. In such joints, the end face of one plate is welded to the sides of the other plate by fillet weld. Generally, both sides of the plate are welded.

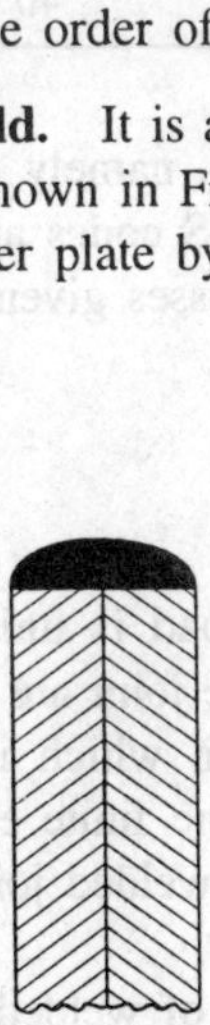

Figure 7.19 Edge weld.

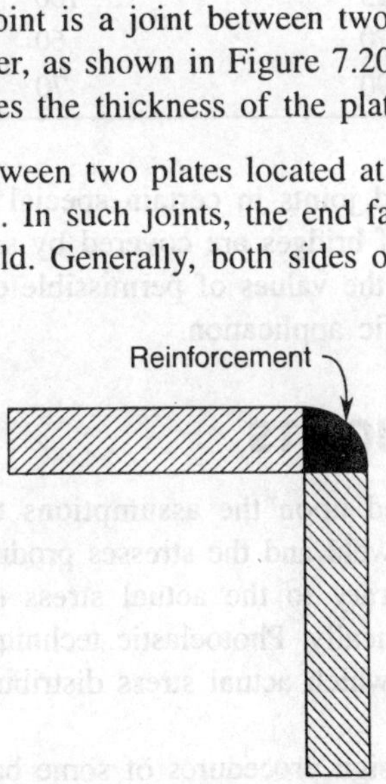

Figure 7.20 Corner weld.

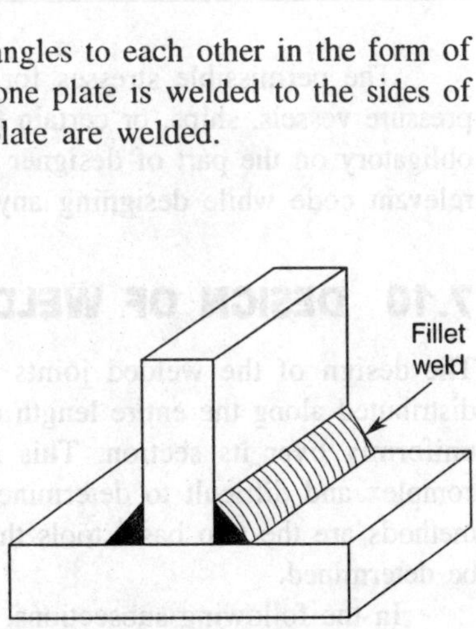

Figure 7.21 Tee-weld.

7.9.2 Weld Material

Practically all commonly used engineering metals, namely plain carbon steel, cast iron, stainless steel, copper, bronze, nickel, etc. can be welded by some welding process. However, some metals can be readily welded while others offer some degree of difficulty. For example, plain carbon steel, except high carbon steel, can be satisfactorily welded but cast iron requires a proper preparation, e.g. pre-heating of the parts to be joined and a special welding process which performs welding under controlled conditions. Similarly, metals containing higher percentages of lead, tin, zinc, aluminium, and molybdenum also offer difficulty in welding due to the fact that vaporization of some of the ingredients and formation of oxides act as insulators which interfere with the flow of current.

The strength of the welded joint in structural steel welded by metal arc welding with a mild steel electrode containing 0.15 per cent carbon is generally stronger than the strength of the plates joined. The permissible stresses of these weld deposits may vary between 35–125 N/mm^2 as given in Table 7.2. However, when high carbon steel or alloy steel is welded, the weld deposit is generally considered weaker than the parts joined. In such cases, the properties of the weld deposit, namely the tensile strength and the yield strength are dependent upon the type of the electrode used in welding.

Table 7.2 Properties of the weld deposit

Types of weld and stress	*Permissible stress* (N/mm^2)		
	Static load	*Fluctuating load*	*Reversed load*
Butt joint			
(a) Tension	120	80	55
(b) Compression	125	100	55
(c) Shear	70	60	35
Fillet weld	90	70	40

The permissible stresses for welded joints in certain special structures, namely unfired pressure vessels, ships, or certain types of bridges are covered by separate BIS codes and it is obligatory on the part of designer to use the values of permissible design stresses given in the relevant code while designing any specific application.

7.10 DESIGN OF WELDED JOINTS

The design of the welded joints is based upon the assumptions that the load is uniformly distributed along the entire length of the weld and the stresses produced at the joint are spread uniformly over its section. This is contrary to the actual stress distribution which is very complex and difficult to determine analytically. Photoelastic techniques and the finite element methods are the two basic tools through which actual stress distribution in a welded joint can be determined.

In the following subsections, the design procedures of some basic types of welded joints are discussed.

7.10.1 Butt Weld

A butt-jointed weld is designed for tension or compressive forces. The type of butt joint formed between two plates and the minimum weld size depend upon the thickness of the plate. Table 7.3 gives the guidelines for the selection of the minimum weld size of butt joints.

Table 7.3 Recommended weld size of butt joints

Plate thickness (mm)	*Minimum weld size* (mm)	*Plate thickness* (mm)	*Minimum weld size* (mm)
3–5	3	18–24	10
6–8	5	26–35	15
10–16	6	Over 38	20

Let us consider a single V-butt joint, as shown in Figure 7.22(a). In the case of a butt joint, the length of the leg or size is equal to the throat thickness which in turn is equal to the thickness of plate t. Therefore, the tensile strength of a single V- or square-butt joint is given by the following equation:

$$P = tl\sigma_t \tag{7.31}$$

where

P is the applied load on the weld
t is the weld size
l is the length of the weld, which is equal to the width of the plate
σ_t is allowable working stress in the weld.

The tensile strength of the double V-butt joint, Figure 7.22(b), is given as

$$P = (t_1 + t_2)l\sigma_t \tag{7.32}$$

where

t_1 is the throat thickness at the top
t_2 is the throat thickness at the bottom.

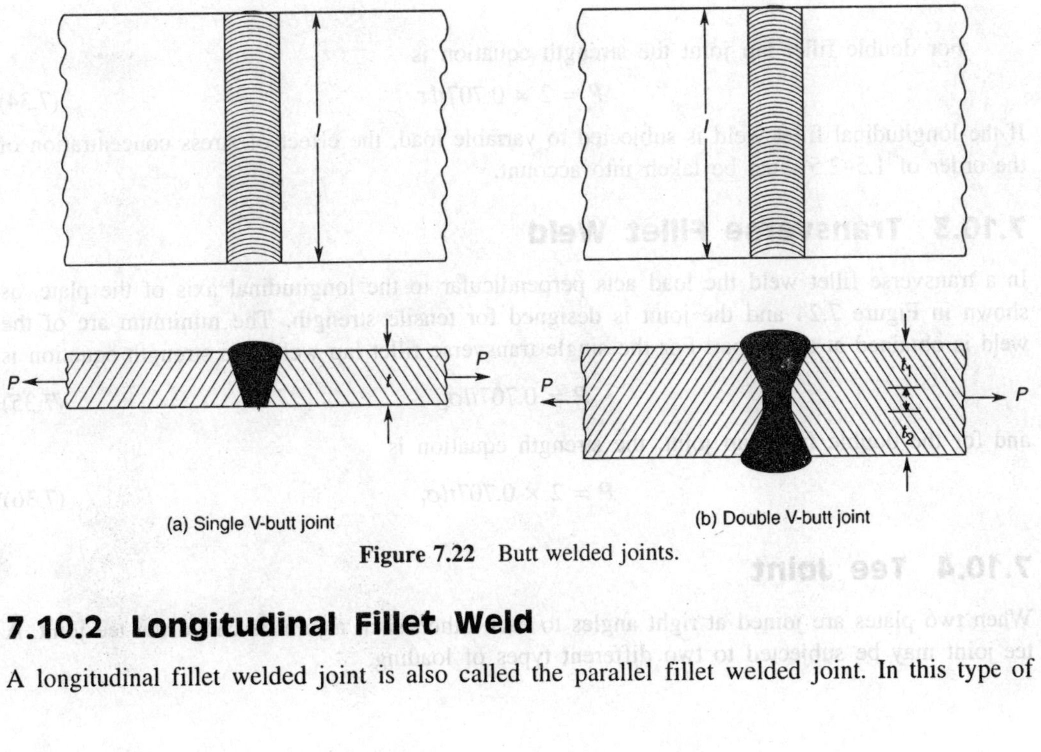

Figure 7.22 Butt welded joints.

7.10.2 Longitudinal Fillet Weld

A longitudinal fillet welded joint is also called the parallel fillet welded joint. In this type of joint, two plates are placed over each other, overlapping, and fillet welding is done along the longitudinal axis (along the length). Such a welded joint, as shown in Figure 7.23 is subjected to either axial tension or compression forces. It is designed for shear strength because the mode of failure is shear. The size of the weld is taken as the thickness of the throat, because the failure of the joint often across more occurs the throat. The throat is the minimum cross-section of the weld located at 45° to the leg (t).

$$\text{Throat thickness} = OC = t \cos 45° = 0.707t$$

$$\text{Area resistant to shear failure} = OC \times l = 0.707tl$$

$$\text{Shear strength of joint, } P = 0.707tl\tau \tag{7.33}$$

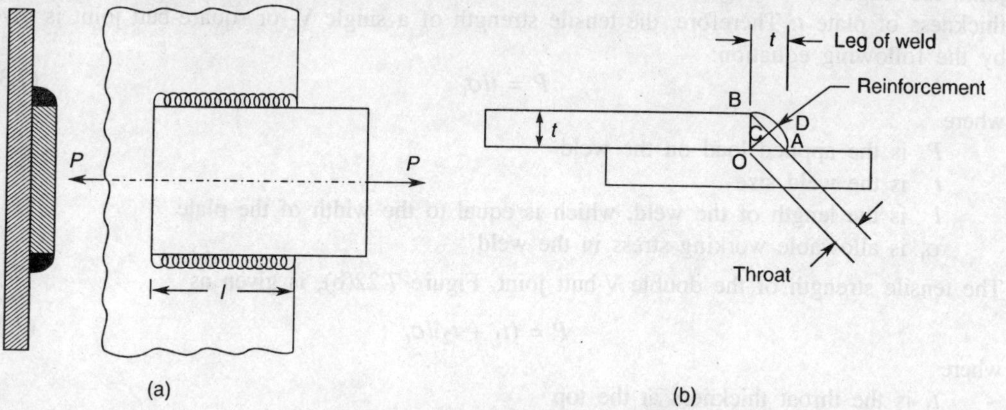

Figure 7.23 Longitudinal fillet weld.

For double fillet lap joint the strength equation is

$$P = 2 \times 0.707tl\tau \tag{7.34}$$

If the longitudinal fillet weld is subjected to variable load, the effect of stress concentration of the order of 1.5–2.5 must be taken into account.

7.10.3 Transverse Fillet Weld

In a transverse fillet weld the load acts perpendicular to the longitudinal axis of the plate, as shown in Figure 7.24 and the joint is designed for tensile strength. The minimum arc of the weld is obtained at the throat. For the single transverse fillet lap weld, the strength equation is

$$P = 0.707tl\sigma_t \tag{7.35}$$

and for the double fillet lap joint, the strength equation is

$$P = 2 \times 0.707tl\sigma_t \tag{7.36}$$

7.10.4 Tee Joint

When two plates are joined at right angles to each other, such a joint is called the tee joint. A tee joint may be subjected to two different types of loading.

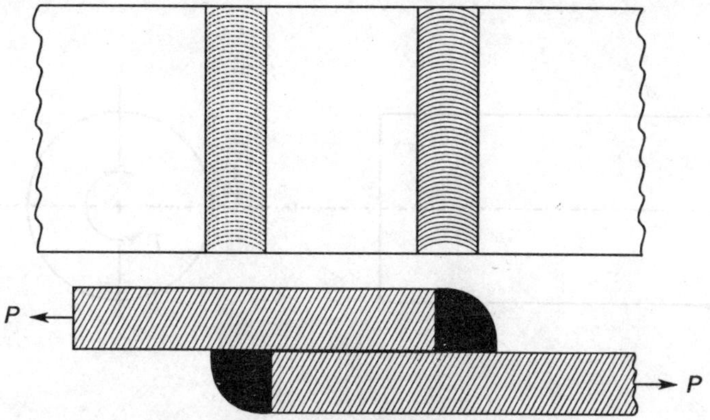

Figure 7.24 Transverse fillet weld.

(i) When an axial tensile force acts perpendicular to the plane of the weld as shown in Figure 7.25(a). For this type of joint the weld design procedure is the same as that for the transverse fillet weld and the governing strength equation for the double fillet welded joint is

$$P = 1.414tl\sigma_t \tag{7.37}$$

(ii) In another case, the tee joint is subjected to a force which is parallel to the plane of the weld with small eccentricity, as shown in Figure 7.25(b). The weld design procedure for this type of joint will be the same as that for the parallel fillet weld. The governing strength equation for the double fillet weld joint is

$$P = 1.414tl\tau \tag{7.38}$$

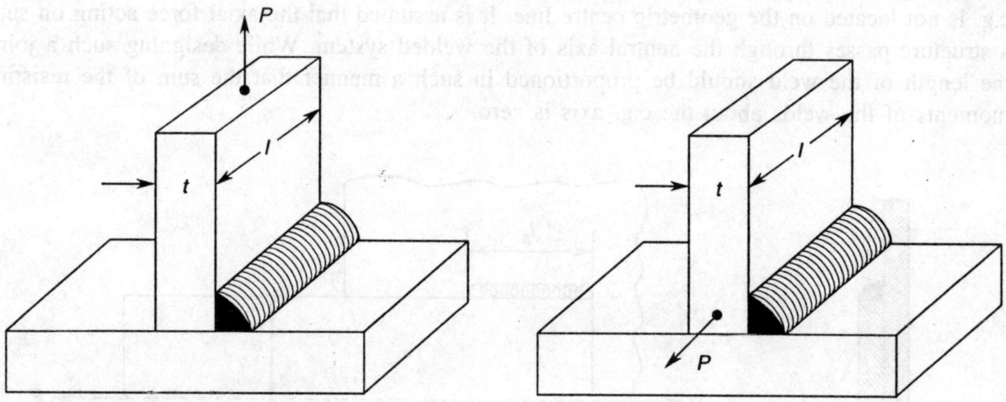

(a) Tee joint with force perpendicular to the plane of weld (b) Tee joint with force parallel to the plane of weld

Figure 7.25

7.10.5 Torsion of Circular Rod with Fillet Weld

A circular shaft welded to a flat plate using fillet welding is shown in Figure 7.26. When the shaft is subjected to torque *T*, the weld will be under stress. If the length of the weld is equal

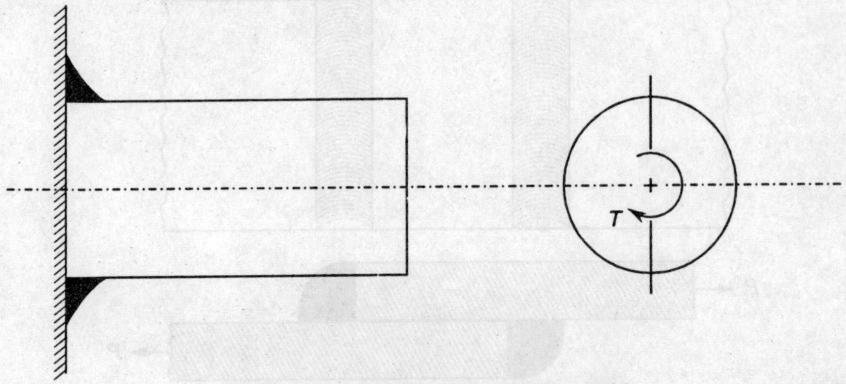

Figure 7.26 Circular rod with fillet weld.

to the circumference of the shaft (πd) and the size of the weld is t, then the torsional strength equation of the welded shaft, considering the throat area as the weakest section, is given by

$$T = 0.707 t\pi d \times d/2 \times \tau$$

or

$$\text{Induced shear stress, } \tau = \frac{2T}{0.707 t\pi d^2} \tag{7.39}$$

7.11 WELD JOINT DESIGN FOR UNSYMMETRICAL SECTIONS

Sometimes unsymmetrical sections such as angles, channels, tee sections, etc. are welded on the flange edge and these are loaded axially as shown in Figure 7.27. In these types of joints, the c.g. is not located on the geometric centre line. It is assumed that the axial force acting on such a structure passes through the neutral axis of the welded system. While designing such a joint, the length of the weld should be proportioned in such a manner that the sum of the resisting moments of the welds about the c.g. axis is zero.

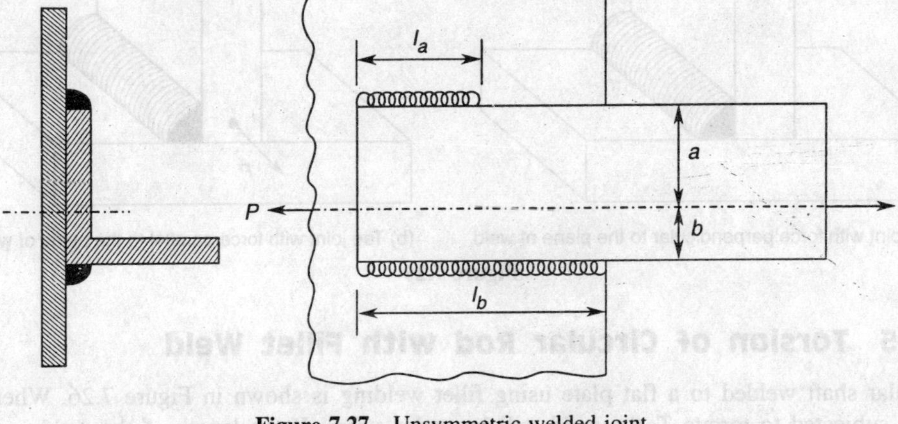

Figure 7.27 Unsymmetric welded joint.

Let us consider an angle-section welded to the flange as shown in Figure 7.27. Let

l_a be the length of the weld at the top
l_b be the length of weld at the bottom
l be the total length $(= l_a + l_b)$
P be the axial force
a be the distance of the top weld from the c.g. axis
b be the distance of the bottom weld from the c.g. axis.
S be the resistance offered by the weld per unit length

The moment of the top weld about the c.g. axis is

$$l_a \times S \times a$$

The moment of the bottom weld about the c.g. axis is

$$l_b \times S \times b$$

Since the sum of the moments about the c.g. axis must be zero, therefore,

$$l_a \times S \times a - l_b \times S \times b = 0$$

or

$$l_a = \frac{l \times b}{a + b} \tag{7.40}$$

and

$$l_b = \frac{l \times a}{a + b} \tag{7.41}$$

Therefore, the total length of the weld l must be divided into these two proportions [Eqs. (7.40) and (7.41)] and provided at the desired locations of the section to ensure zero moment on the weld.

7.12 WELDED JOINT SUBJECTED TO ECCENTRIC LOAD

In the previous sections, the stress analysis of various types of welded joints subjected to direct force has been presented. However there are many practical situations, namely structural joints, where the external applied force may not pass through the geometric centre. Such joints are called eccentrically loaded joints. In eccentric loading of welded joints, two types of stresses are produced. One is the direct shear stress, while the other may be bending or torsional shear stress. A few typical cases of design of eccentrically loaded welded joints are discussed below.

Case I

Consider a cantilever beam as shown in Figure 7.28(a) whose one end is welded to a fixed bracket and the load P acts at the free end. In other words, it is a tee joint which is subjected to a force parallel to the fillet weld at certain eccentricity e as shown in Figure 7.28(b). The strength of this type of welded joint can be calculated by considering the combined effect of bending moment and shear force.

Let us assume that two forces of the same magnitude P act opposite to each other near the fillet weld as shown in Figure 7.28(b). The forces acting on the joint are:

(i) Direct shear force due to force P
(ii) Couple formed between the two equal, parallel and opposite forces, i.e. Pe

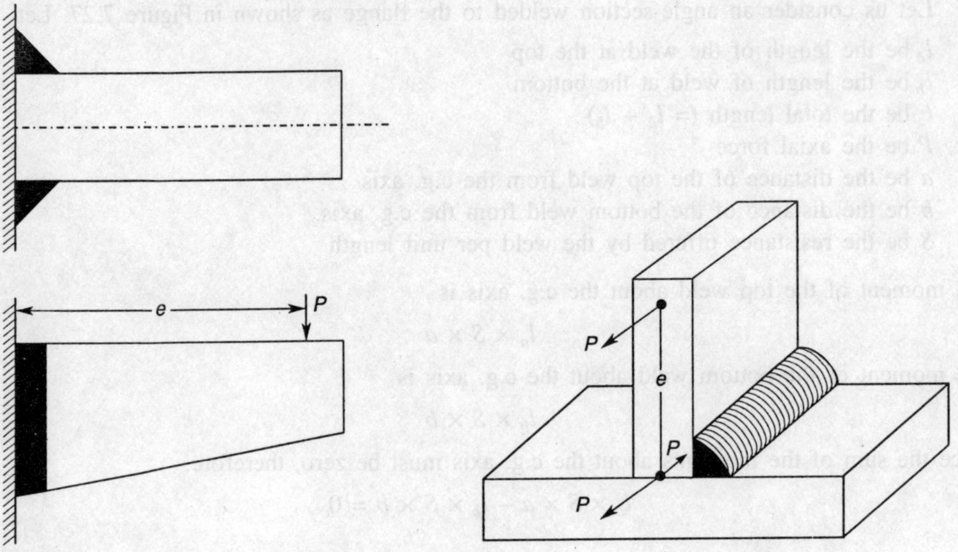

Figure 7.28 Joint under bending load.

On account of the above two types of forces, two different types of stresses are produced.

1. The stress produced due to direct shear force will be shearing in nature, i.e. the same as discussed in Section 7.10.2 for a parallel fillet weld (both sides welded). That is,

$$\tau = \frac{P}{2 \times 0.707\, tl} = \frac{P}{1.414\, tl} \tag{7.42}$$

2. The couple acting on the member will produce bending stress, which is given by the following relation:

$$\sigma_b = \frac{M}{Z} = \frac{Pe}{Z}$$

where Z is the section modulus.

Assuming that failure of the weld may take place at the minimum resisting area, i.e. at the weld throat which is double welded,

$$Z = \frac{1}{6} \times \text{throat thickness} \times l^2$$

where the throat thickness = $0.707t$. Now, bending stress

$$\sigma_b = \frac{Pe}{2 \times \frac{1}{6} \times 0.707 tl^2}$$

or

$$\sigma_b = \frac{4.243\, Pe}{tl^2} \tag{7.43}$$

Since it is a case of biaxial loading, therefore, according to maximum shear stress theory,

$$\tau_{max} = 0.5(\sigma_b^2 + 4\tau^2)^{0.5} \leq \tau_a \tag{7.44}$$

where τ_a is the allowable shear strength.

Case II

Let us consider a generalized case of eccentric loading as shown in Figure 7.29 in which a plate is welded at three sides having different weld lengths.

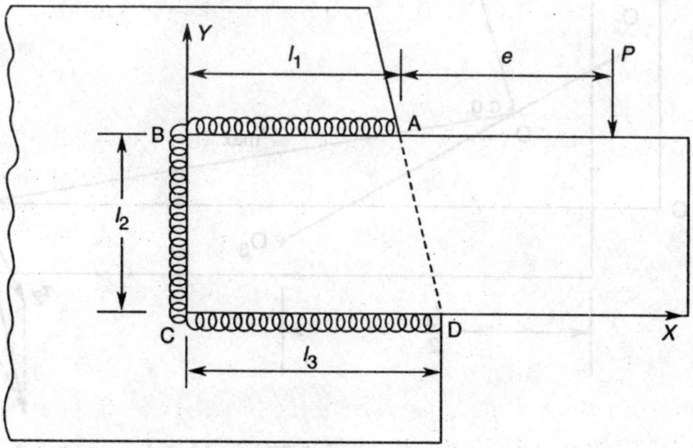

Figure 7.29 Compound fillet weld under eccentric load.

Let

size of weld = t

throat thickness, $a = 0.707t$

throat area of weld AB = $A_1 = 0.707tl_1$

throat area of weld BC = $A_2 = 0.707tl_2$

throat area of weld CD = $A_3 = 0.707tl_3$

and total area $A = 0.707t(l_1 + l_2 + l_3)$

To locate the c.g. of the weld, let us assume that point C is the origin and CD acts as the X-axis while CB acts as the Y-axis (Figure 7.29).

Taking moments about C, the location of the c.g. from the X-axis is given by

$$\overline{X} = \frac{A_i x_1 + A_2 x_2 + \cdots}{A}$$

and the location of the c.g. from the Y-axis is

$$\overline{Y} = \frac{A_1 y_1 + A_2 y_2 + \cdots}{A}$$

The turning moment, $$T = P(e + l_1 - \overline{X}) \tag{7.45}$$

Now let us calculate the polar moment of inertia about the c.g. of the weld (Figure 7.30).

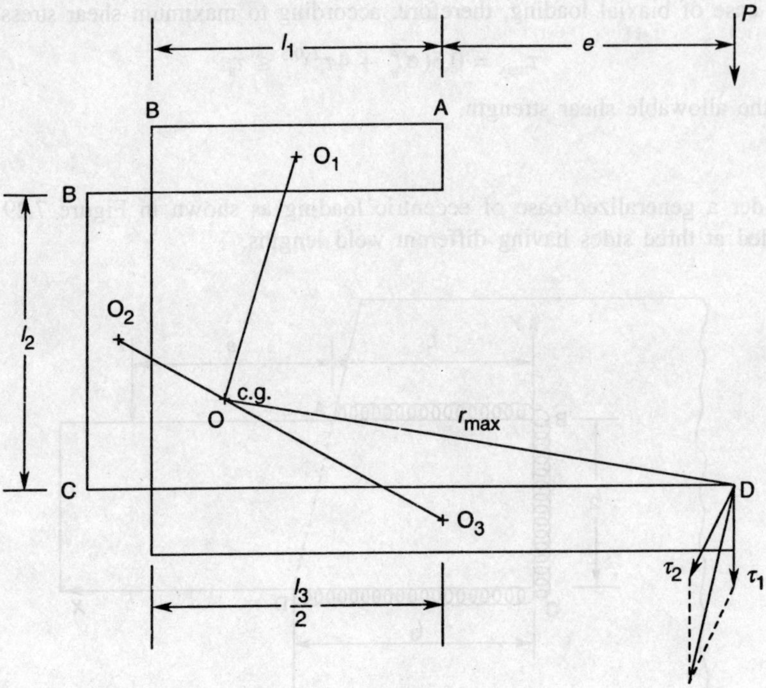

Figure 7.30 Stresses in a compound fillet weld under eccentric load.

Polar moment of inertia of the weld AB about its c.g. is

$$J_1 = \frac{1}{12} a l_1^3$$

The polar moment of inertia about the c.g. of the weld system, according to the parallel axis theorem, is

$$J'_1 = J_1 + A_1 k_1^2$$

where

k_1 = distance between the c.g. point O and O_1

$$= \sqrt{(l_1/2 - \overline{X})^2 + (l_2 - \overline{Y})^2}$$

A_1 = area of the weld = $0.707tl_1$

Therefore,

$$J'_1 = \frac{0.707tl_1^3}{12} + 0.707tl_1k_1^2$$

or

$$J'_1 = (0.707tl_1/12)(l_1^2 + 12k_1^2) \tag{7.46}$$

Similarly, the polar moments of inertia of welds BC and CD about the c.g. of the weld system can be computed and are given below:

$$J_2' = (0.707tl_2/12)(l_2^2 + 12k_2^2) \tag{7.47}$$

and

$$J_3' = (0.707tl_3/12)(l_3^2 + 12k_3^2) \tag{7.48}$$

Total polar moment of inertia of the weld

$$J = J_1' + J_2' + J_3'$$

Therefore, maximum secondary shear stress is given as

$$\tau_2 = \frac{T \cdot r_{max}}{J} \tag{7.49}$$

where r_{max} = distance OD

$$= \sqrt{(l_3 - \overline{X})^2 + \overline{Y}^2}$$

and direct shear stress,
$$\tau_1 = \frac{P}{0.707t(l_1 + l_2 + l_3)} \tag{7.50}$$

Angle between the direct and secondary shear stress

$$\cos\theta = \frac{\overline{Y}}{r_{max}}$$

maximum shear stress

$$\tau_{max} = \sqrt{\tau_1^2 + \tau_2^2 + 2\tau_1\tau_2 \cos(90 - \theta)} \tag{7.51}$$

which should be less than or equal to the design stress.

Example 7.6 A plate 100 mm wide and 10 mm thick is welded to another plate by a single transverse fillet weld and a double parallel fillet weld as shown in the following figure. The maximum working tensile and shear stresses are 75 N/mm² and 55 N/mm², respectively. Find the length of the respective welds. Assume overtravel length equal to 12.5 mm.

Solution Plate width, b = 100 mm

Plate thickness, t = 10 mm

Allowable tensile strength, σ_t = 75 N/mm²

Allowable shear strength, τ = 55 N/mm²

Length of the transverse fillet weld, $l_2 = b$ – overtravel length

$= 100 - 12.5 = 87.5$ mm

Maximum load, $P = bt\sigma_t = 100 \times 10 \times 75 = 75{,}000$ N

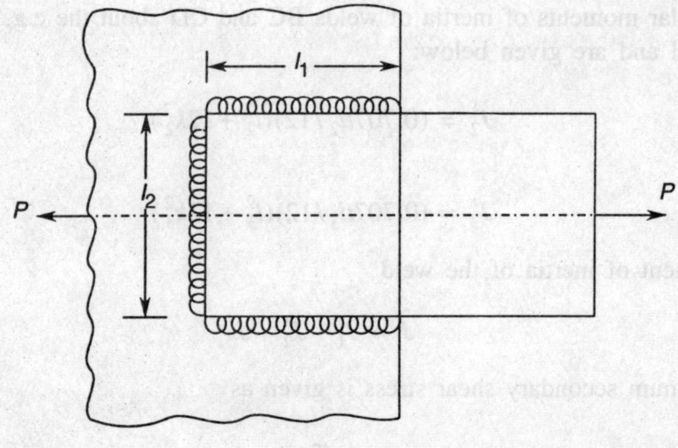

The load carrying capacity of a single transverse weld can be determined from the strength equation

$$P_2 = 0.707tl_2\sigma_t$$

$$= 0.707 \times 10 \times 87.5 \times 75$$

$$= 46,396.8 \text{ N}$$

Therefore, the remaining load will be shared by the double parallel fillet weld, which is

$$P_1 = P - P_2$$

$$= 75,000 - 46,396.8 = 28,603.2 \text{ N}$$

Strength of the parallel double weld in shear failure is

$$P_1 = 1.414tl_1\tau$$

or

$$l_1 = \frac{P_1}{1.414t\tau} = \frac{28,603.2}{1.414 \times 10 \times 55} = 36.78 \text{ mm}$$

Adding 12.5 mm for overtravel, we get

$$l_1 = 36.78 + 12.5 = 49.28 \text{ mm, say, } 50 \text{ mm}$$

Example 7.7 A 50 mm diameter shaft is welded to a flat plate by fillet weld. Determine the size of the weld if the shaft is required to transmit a torque of 1300 N · m. The permissible working shear strength of the weld material is 60 N/mm^2.

Solution Shaft diameter, d = 50 mm

Torque, T = 1300 N · m

Allowable shear strength, τ = 60 N/mm^2

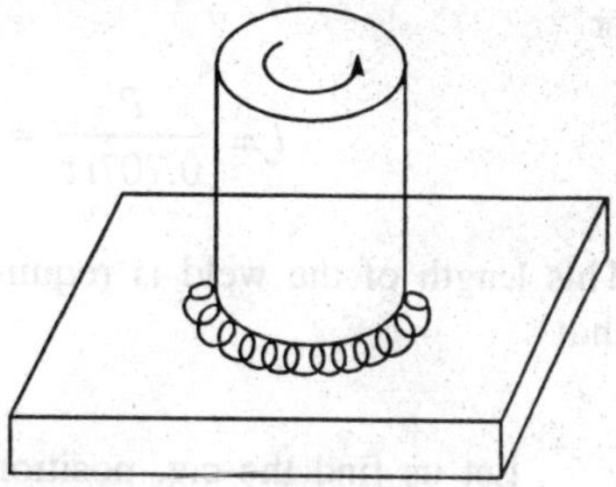

If the weld size is t, the torque transmission capacity of the weld at the throat is given by

$$T = 0.707t\frac{\pi}{2}d^2\tau$$

or

$$t = \frac{2T}{0.707\pi d^2\tau} = \frac{2\times 1300\times 10^3}{0.707\times\pi\times 50^2\times 60}$$

$$= 7.8 \text{ mm, say, } 8 \text{ mm}$$

Example 7.8 An angle of size 200 mm × 150 mm × 10 mm is required to be welded to a steel plate by a fillet weld as shown in the following figure. If the angle is subjected to a static load of 100 kN, determine the top and bottom weld lengths. The allowable shear strength for static loading may be taken as 65 N/mm².

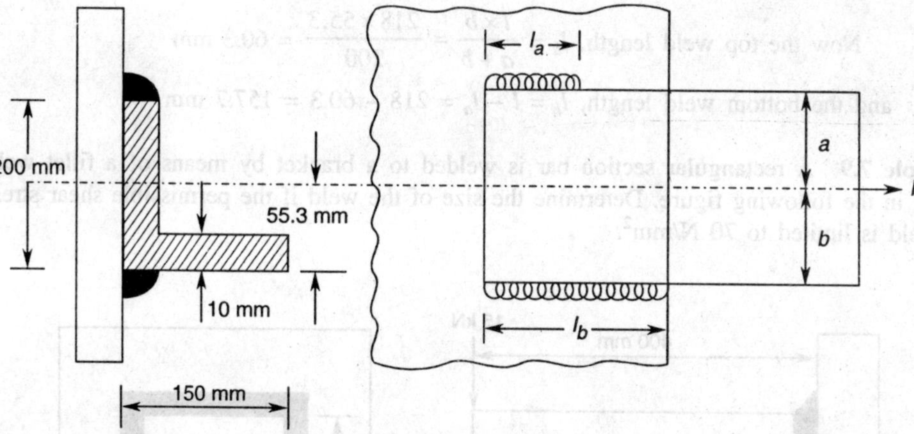

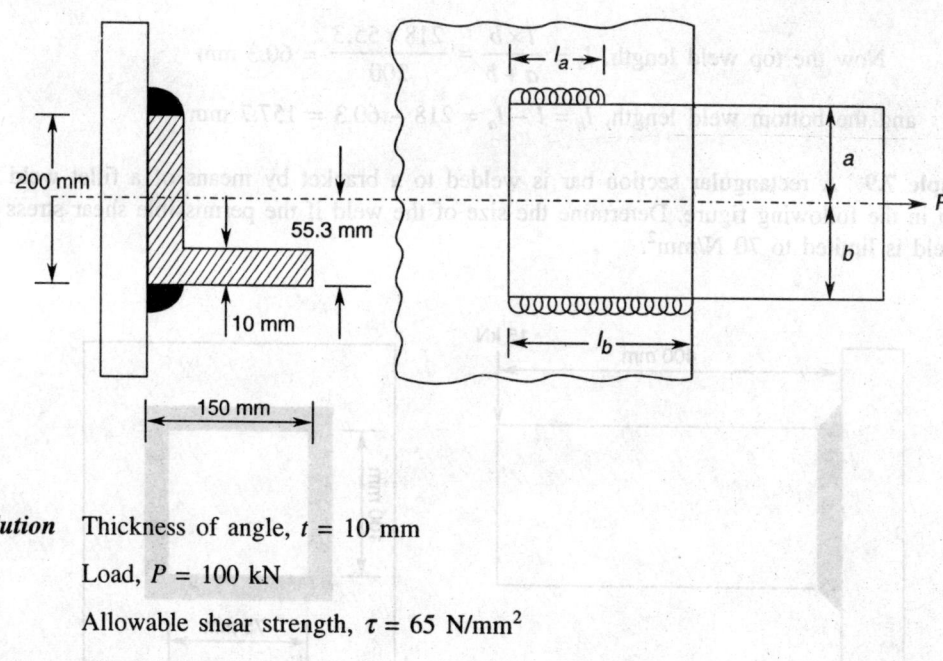

Solution Thickness of angle, t = 10 mm

Load, P = 100 kN

Allowable shear strength, τ = 65 N/mm²

Let

l_a be the length of the weld at top

l_b be the length of the weld at bottom

l be the total length (= $l_a + l_b$)

Suppose the entire load is taken by a single parallel fillet weld. Then the required weld length can be found from the shear strength equation,

$$P = 0.707tl\tau$$

or

$$l = \frac{P}{0.707t\tau} = \frac{100,000}{0.707 \times 10 \times 65} = 217.6 \text{ mm, say, } 218 \text{ mm}$$

This length of the weld is required to be divided into top and bottom lengths, l_a and l_b, such that

$$l = l_a + l_b = 218 \text{ mm}$$

Let us find the c.g. position of the weld system.

Let b be the distance of the c.g. axis from the bottom edge of the angle. Then,

$$b = \frac{(200 - 10) \times 10 \times 95 + 150 \times 10 \times 5}{190 \times 10 + 150 \times 10} = 55.3 \text{ mm}$$

Let a be the distance from the c.g. to the top of the angle. Then,

$$a = 200 - 55.3 = 144.7 \text{ mm}$$

Now the top weld length, $l_a = \frac{l \times b}{a + b} = \frac{218 \times 55.3}{200} = 60.3$ mm

and the bottom weld length, $l_b = l - l_a = 218 - 60.3 = 157.7$ mm

Example 7.9 A rectangular section bar is welded to a bracket by means of a fillet weld as shown in the following figure. Determine the size of the weld if the permissible shear stress in the weld is limited to 70 N/mm^2.

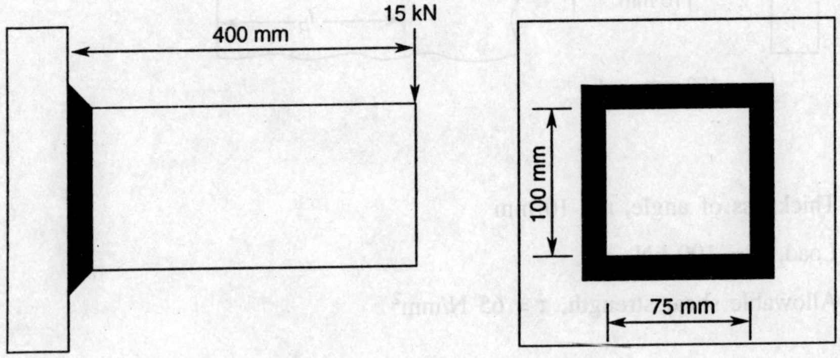

Solution Applied load, $P = 15$ kN

Eccentricity, $e = 400$ mm

Cross-section = (100×75) mm^2

This type of the welded joint is subjected to the following two types of forces:

(i) Direct shear force, P

(ii) Couple Pe which creates bending moment

Let

t be the weld size

a be the throat size.

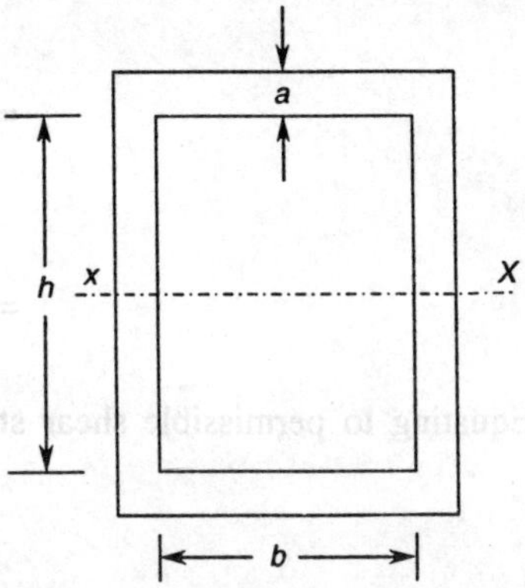

Area of the weld, $A = 2(100 + 75)a = 350a \text{ mm}^2$

Direct shear stress, $\tau = \dfrac{P}{A} = \dfrac{15,000}{350a} = \dfrac{42.85}{a} \text{ N/mm}^2$

Bending stress, $\sigma_b = \dfrac{Pe}{Z}$

Bending moment, $Pe = 15 \times 400 = 6000 \text{ kN} \cdot \text{mm}$

Section modulus, $Z = \dfrac{I_{XX}}{y}$

I_{XX} = moment of inertia of four weld beads about the X-X axis

$$= 2\left[\frac{ba^3}{12} + ba \times \frac{h^2}{2} + \frac{2ah^3}{12}\right]$$

Assuming that dimensions b and h are very large compared to throat dimension a, the term containing a^3 can be neglected. After simplification, we have

$$I_{XX} = a\left[\frac{bh^2}{2} + \frac{h^3}{6}\right]$$

$$= a\left[\frac{75 \times 100^2}{2} + \frac{100^3}{6}\right] = 541,666.67a \text{ mm}^4$$

and

y = distance of the outer fibre from the c.g.

$$= \frac{h + 2a}{2} \approx \frac{h}{2} = 50 \text{ mm}$$

Section modulus, $Z = \dfrac{I_{XX}}{y} = \dfrac{541,666.67a}{50} = 10,833.3a \text{ mm}^3$

Bending stress, $\sigma_b = \dfrac{6000 \times 10^3}{10,833.3a} = \dfrac{553.84}{a} \text{ N/mm}^2$

According to maximum shear stress theory,

$$\tau_{max} = \sqrt{\left(\frac{\sigma_b}{2}\right)^2 + \tau^2}$$

$$= \left[\left(\frac{553.84}{2a}\right)^2 + \left(\frac{42.85}{a}\right)^2\right]^{0.5}$$

$$= \frac{280.2}{a} \text{ N/mm}^2$$

Equating to permissible shear strength, we have

$$\tau_{max} = \frac{280.2}{a} = 70 \text{ N/mm}^2$$

or throat thickness,

$$a = \frac{280.2}{70} = 4 \text{ mm}$$

and weld size,

$$t = \frac{a}{0.707} = \frac{4}{0.707} = 5.65 \text{ mm, say, 6 mm}$$

Example 7.10 A bracket carrying a load of 20 kN is to be welded as shown in the following figure. Calculate the size of the weld if the working shear stress is not to exceed 70 N/mm^2.

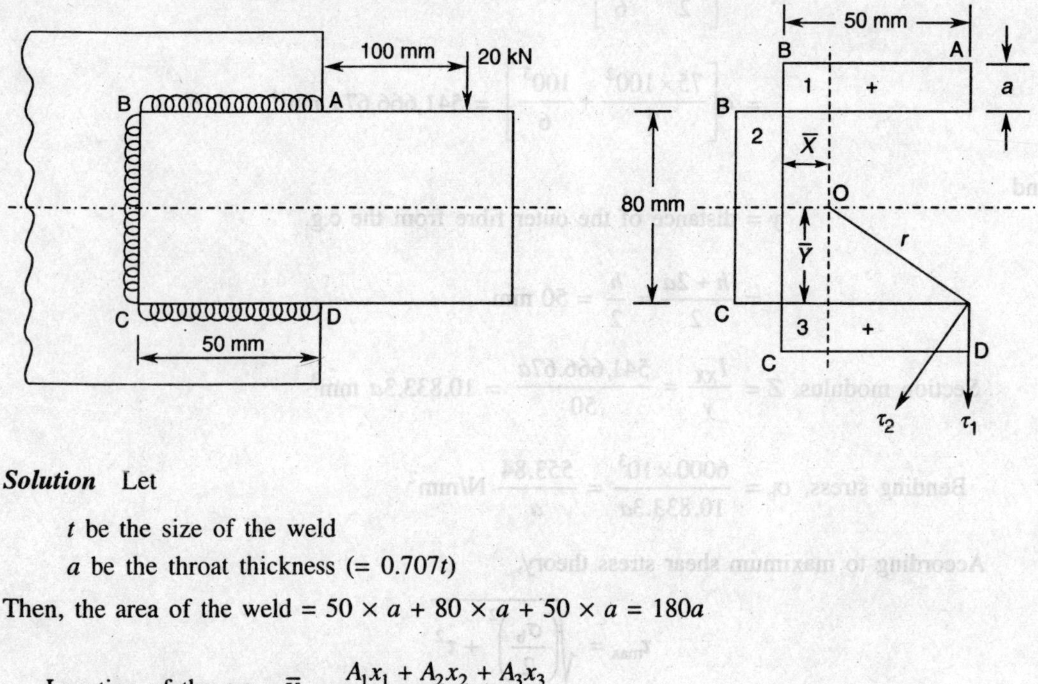

Solution Let

t be the size of the weld

a be the throat thickness (= 0.707*t*)

Then, the area of the weld = 50 × *a* + 80 × *a* + 50 × *a* = 180*a*

$$\text{Location of the c.g., } \bar{X} = \frac{A_1x_1 + A_2x_2 + A_3x_3}{A}$$

$$= \frac{50 \times a \times 25 + 80 \times a \times 0 + 50 \times a \times 25}{180a}$$

$= 13.9$ mm

$\bar{Y} = 40$ mm (by symmetry)

Turning moment, $T = P(100 + 50 - 13.9)$

$= 20 \times 136.1 = 2722$ kN · mm

Polar moment of inertia of the weld about the c.g. at O is

$$J = J'_1 + J'_2 + J'_3$$

Here

$$J'_1 = J'_3$$

Therefore,

$$J = 2J'_1 + J'_2$$

Now,

$$J'_1 = (0.707tl_1/12)(l_1^3 + 12k_1^2)$$

where

$$k_1^2 = (l_1/2 - \bar{X})^2 + (l_2 - \bar{Y})^2$$

$$= (25 - 13.9)^2 + (80 - 40)^2 = 1723.21 \text{ mm}^2$$

Therefore,

$$J'_1 = (0.707t \times 50/12)\ (50^3 + 12 \times 1723.21)$$

$$= 429{,}144.6t \text{ mm}^4$$

Similarly,

$$J'_2 = (0.707tl_2/12)\ (l_2^3 + 12k_2^2)$$

where

$$k_2^2 = \left(\frac{l_2}{2} - \bar{Y}\right)^2 + \bar{X}^2 = \bar{X}^2 \quad \left(\text{as } \frac{l_2}{2} = \bar{Y}\right)$$

$$= 13.9^2 = 193.21 \text{ mm}^2$$

Therefore,

$$J'_2 = (0.707t \times 80/12)(80^3 + 12 \times 193.21)$$

$$= 2{,}424{,}154.6t \text{ mm}^4$$

Total polar moment of inertia

$$J = 2 \times 429{,}144.6t + 2{,}424{,}154.6t$$

$$= 3{,}282{,}443.8t \text{ mm}^4$$

Secondary shear stress, $\tau_2 = \dfrac{Tr}{J}$

where

r = distance from c.g. to the farthest point on the weld

$= [(50 - \bar{X})^2 + \bar{Y}^2]^{0.5}$

$= [(50 - 13.9)^2 + 40^2]^{0.5} = 53.88$ mm

$$\tau_2 = \frac{2722 \times 10^3 \times 53.88}{3,282,443.8\, t} = \frac{44.68}{t} \text{ N/mm}^2$$

and direct shear stress, $\tau_1 = \dfrac{P}{0.707\, t\, (2l_1 + l_2)}$

$$= \frac{20 \times 1000}{0.707t(2 \times 50 + 80)} = \frac{157.16}{t} \text{ N/mm}^2$$

Angle between the direct shear stress τ_1 and the secondary shear stress τ_2

$$\cos\theta = \frac{\bar{Y}}{r} = \frac{40}{53.88} = 0.7424$$

or

$$\theta = 42.06°$$

Resultant shear stress

$$\tau_{max} = \left[\tau_1^2 + \tau_2^2 + 2\tau_1\tau_2 \cos(90 - \theta)\right]^{0.5}$$

$$= \left[\left(\frac{157.16}{t}\right)^2 + \left(\frac{44.68}{t}\right)^2 + 2 \times \frac{157.16}{t} \times \frac{44.68}{t} \times \cos(90 - 42.06)\right]^{0.5}$$

$$= \frac{190}{t} \text{ N/mm}^2$$

Equating τ_{max} to allowable shear strength, we have

$$\frac{190}{t} = 70$$

or

$$t = \frac{190}{70} = 2.7 \text{ mm, say, 3 mm}$$

7.13 COTTER JOINT

A cotter joint is used to rigidly connect two rods making a temporary fastening. The jointed rods may carry tensile or compressive forces. The cotter joint as shown in Figure 7.31 is made up of three parts: the rod end, also called the spigot end, the socket end, and the cotter that fits into the tapered slot. A cotter is a flat wedge-shaped piece of rectangular cross-section having tapering width. The taper on the width varies from 1:48 to 1:8. The taper of the slot as well as that on the cotter is usually on one side as shown in Figure 7.31. The clearance between the cotter and the slot in the 'rod end' and socket allows the driven cotter to draw together the two parts of the joint until the 'socket end' comes in contact with the collar of the 'rod end'. Further driving of the cotter will bend it, which helps in keeping the joint tight under the action of a variable load. Usually the clearance of the cotter is limited to 3 mm.

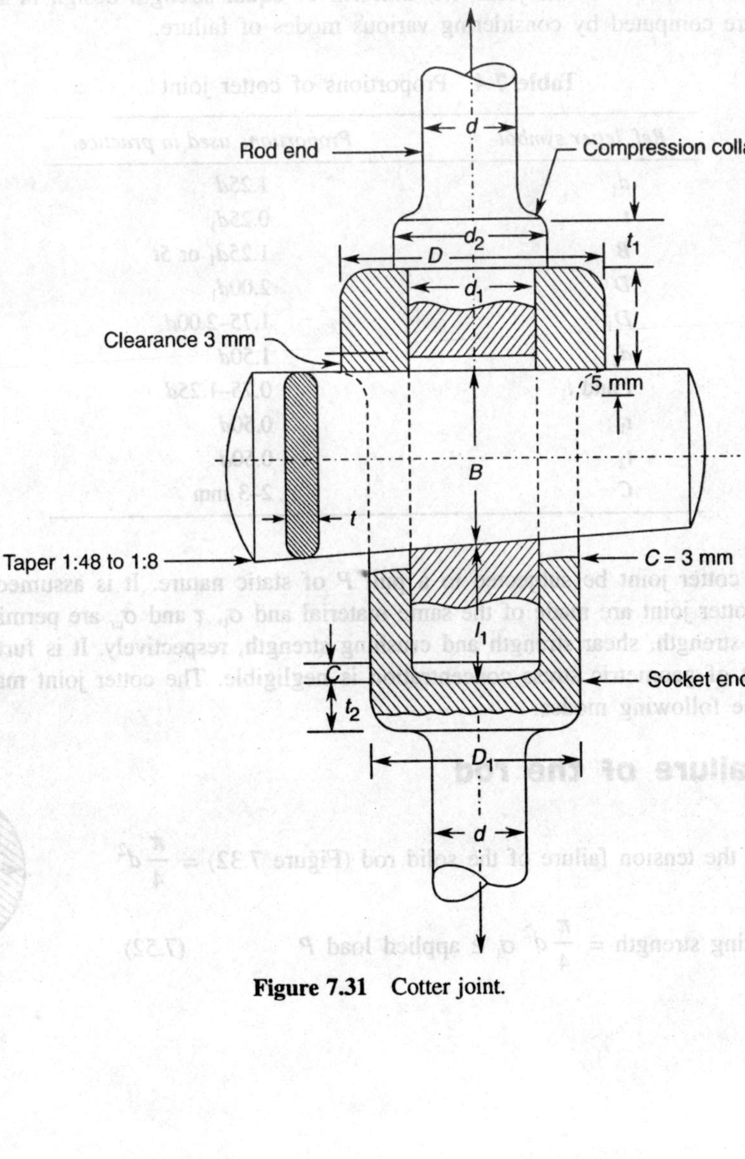

Figure 7.31 Cotter joint.

Cotter joints are generally made of plain carbon steel with carbon percentage varying from 0.2 to 0.5%. IS:1570–1979 has recommended certain grades of steel for both cotter and knuckle joints.

7.13.1 Design of Cotter Joints

A cotter-jointed fastening can work satisfactorily only if the cotter has been sufficiently tightened to an extent that the initial force F_i set up by the wedge action is greater than the external force P. Generally, the initial force F_i should be approximately 1.25 times the external force.

The design of a cotter joint can be accomplished in two ways: (i) design by empirical relations, where all dimensions of various sections are computed using empirical relations defined in terms of the rod diameter, listed in Table 7.4. Later these dimensions are checked in various modes of failure of the joint; (ii) uniform or equal strength design in which various dimensions are computed by considering various modes of failure.

Table 7.4 Proportions of cotter joint

Ref. letter symbol	*Proportions used in practice*
d_1	$1.25d$
t	$0.25d_1$
B	$1.25d_1$ or $5t$
D	$2.00d_1$
D_1	$1.75–2.00d$
d_2	$1.50d$
l and l_1	$0.75–1.25d$
t_1	$0.50d$
t_2	$0.50d$
C	2–3 mm

Let the cotter joint be subjected to a pull P of static nature. It is assumed that all the parts of the cotter joint are made of the same material and σ_t, τ and σ_{cr} are permissible values of the tensile strength, shear strength and crushing strength, respectively. It is further assumed that the effect of geometric stress concentration is negligible. The cotter joint may fail under any one of the following modes:

Tensile failure of the rod

Area resisting the tension failure of the solid rod (Figure 7.32) $= \dfrac{\pi}{4} d^2$

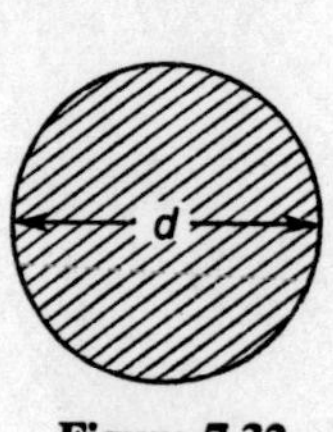

Figure 7.32

$$\text{Resisting strength} = \frac{\pi}{4} d^2 \sigma_t \geq \text{applied load } P \qquad (7.52)$$

Tensile failure of the rod across the slot

Area resisting the tension failure of the rod across the slot at section X–X (Figure 7.33) is

$$A = \left(\frac{\pi}{4} d_1^2 - d_1 t\right)$$

$$\text{Resisting strength} = \left(\frac{\pi}{4} d_1^2 - d_1 t\right)\sigma_t \geq P \qquad (7.53)$$

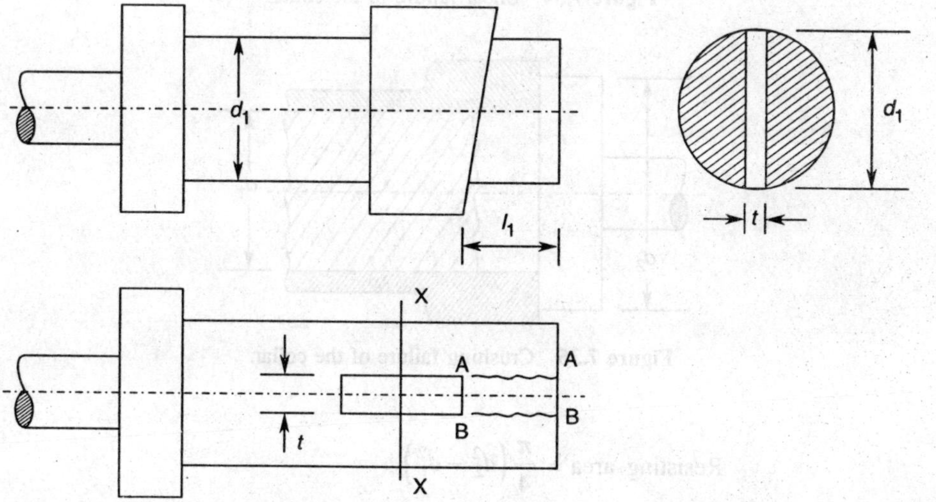

Figure 7.33 Tensile failure of the rod.

Shear failure of the rod end

The rod end when subjected to load may fail in double shear. The cotter may try to shear off a strip of length l_1 at sections A–A, B–B as shown in Figure 7.33.

$$\text{Resisting shear area} = 2d_1 l_1$$

$$\text{Resisting strength} = 2d_1 l_1 \tau \geq P \qquad (7.54)$$

Shear failure of the collar

The collar of the rod may shear as shown in Figure 7.34 at the shear plane X–X.

$$\text{Resisting area} = \pi d_1 t_1$$

$$\text{Resisting strength} = \pi d_1 t_1 \tau \geq P \qquad (7.55)$$

Crushing failure of the collar

When the collar of the rod is sufficiently strong to resist shear failure, it may get crushed against the socket end, as shown in Figure 7.35.

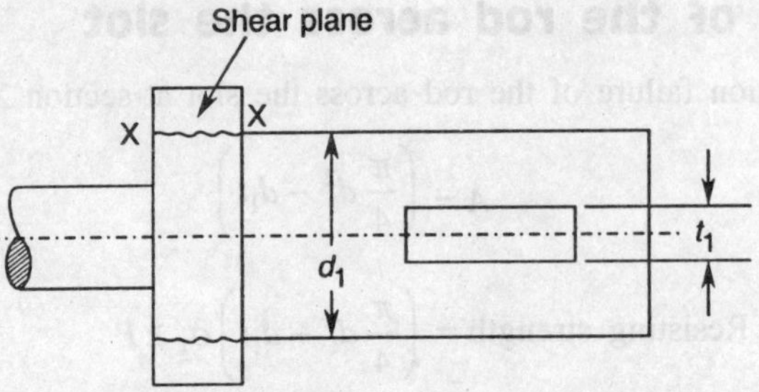

Figure 7.34 Shear failure of the collar.

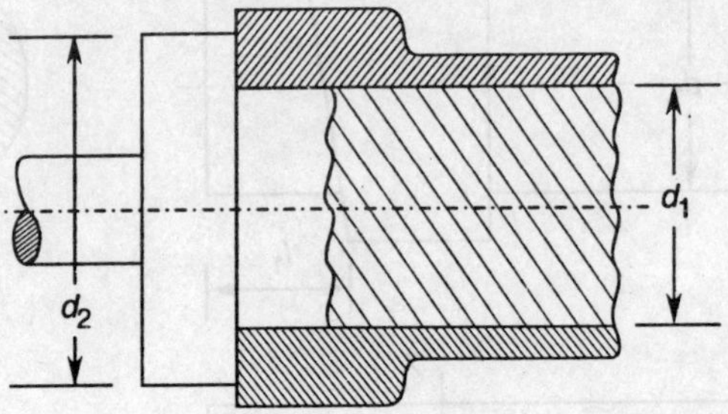

Figure 7.35 Crushing failure of the collar.

$$\text{Resisting area} = \frac{\pi}{4}\left(d_2^2 - d_1^2\right)$$

$$\text{Resisting strength} = \frac{\pi}{4}\left(d_2^2 - d_1^2\right)\sigma_{cr} \geq P \tag{7.56}$$

Crushing failure of the cotter-rod end

The compression between the cotter and rod end may cause crushing failure of either the cotter or the rod end.

$$\text{Resisting area} = d_1 t$$

$$\text{Resisting strength} = d_1 t \sigma_{cr} \geq P \tag{7.57}$$

Tension failure of the socket across the slot

The socket may fail under tension across the slot as shown in Figure 7.36.

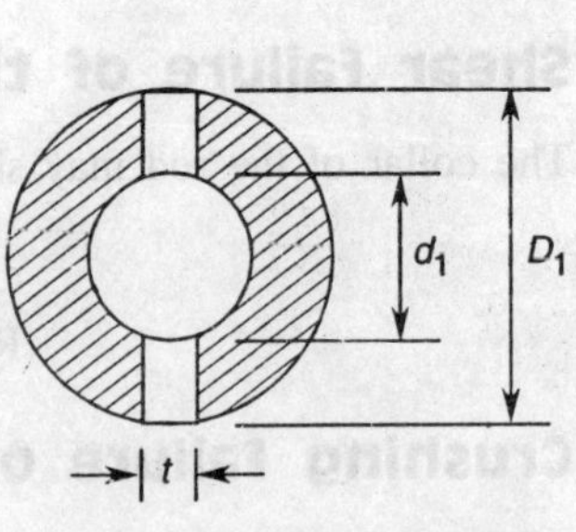

Figure 7.36 Tension failure of the socket.

$$\text{Resisting area} = \frac{\pi}{4}\left(D_1^2 - d_1^2\right) - \left(D_1 - d_1\right)t$$

$$\text{Resisting strength} = \left[\frac{\pi}{4}\left(D_1^2 - d_1^2\right) - \left(D_1 - d_1\right)t\right]\sigma_t \geq P \tag{7.58}$$

Shear failure of the socket end

The socket end in front of the collar may fail due to shear as shown in Figure 7.37.

$$\text{Resisting area} = 2(D - d_1)l$$

$$\text{Resisting strength} = 2(D - d_1)l\tau \geq P \qquad (7.59)$$

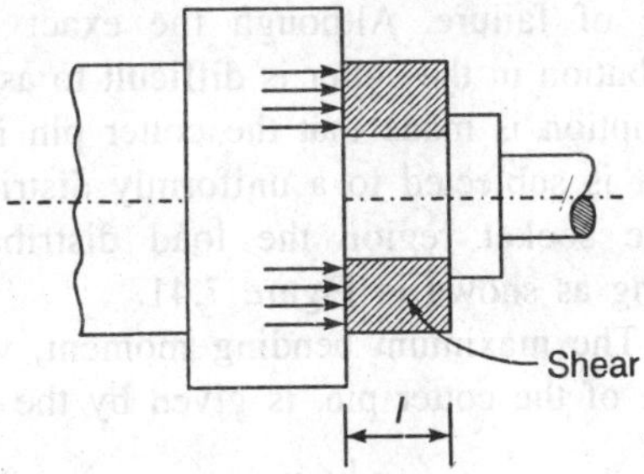

Figure 7.37 Shear failure of the socket end.

Shear failure of the rod–socket connection

The rod–socket connection may shear off as shown in Figure 7.38.

$$\text{Resisting area} = \pi d t_2$$

$$\text{Resisting strength} = \pi d t_2 \tau \geq P \qquad (7.60)$$

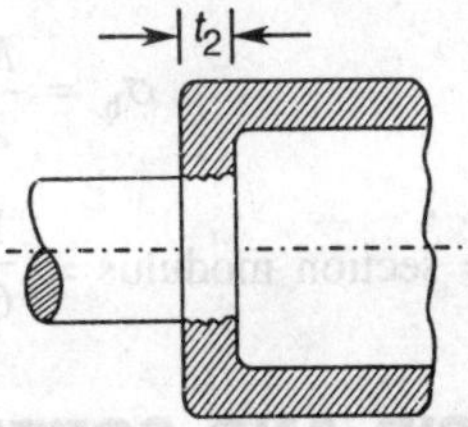

Figure 7.38 Shear failure of the rod–socket connection.

Crushing failure of the socket or the cotter

Either the socket or the cotter may fail under crushing as shown in Figure 7.39.

$$\text{Resisting strength} = (D - d_1)t\sigma_{cr} \geq P \qquad (7.61)$$

Shear failure of the cotter

The cotter pin may fail in double shear across the planes X–X and Y–Y as shown in Figure 7.40. If the average width of the cotter is B,

$$\text{Resisting area} = 2Bt$$

$$\text{Resisting strength} = 2Bt\tau \geq P \qquad (7.62)$$

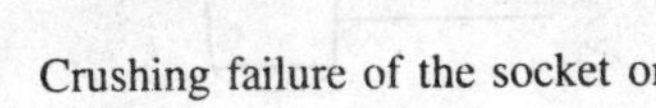

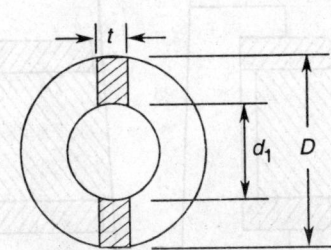

Figure 7.39 Crushing failure of the socket or the cotter.

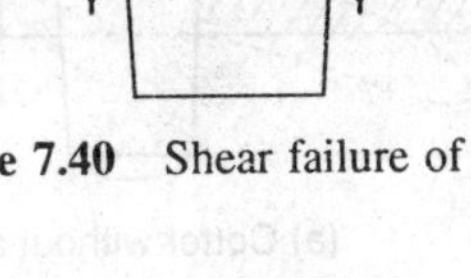

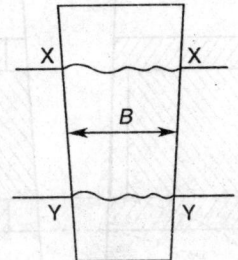

Figure 7.40 Shear failure of the cotter.

Bending force on the cotter

The force acting on the cotter joint when transferred to the cotter pin, acts as the bending force and tries to bend the cotter pin. Therefore, it is important to check the cotter in the bending mode of failure. Although the exact nature of the load distribution in the cotter is difficult to ascertain, a simplified assumption is made that the cotter pin in the region of the spigot is subjected to a uniformly distributed load whereas in the socket region the load distribution is uniformly varying as shown in Figure 7.41.

The maximum bending moment, which occurs at the centre of the cotter pin, is given by the following equation:

$$M = \frac{P}{2}\left[\left(\frac{D-d_1}{6}+\frac{d_1}{2}\right)-\frac{d_1}{4}\right] \tag{7.63}$$

Bending stress is given by

$$\sigma_b = \frac{M}{Z} \tag{7.64}$$

where Z = section modulus = $\frac{1}{6}B^2t$

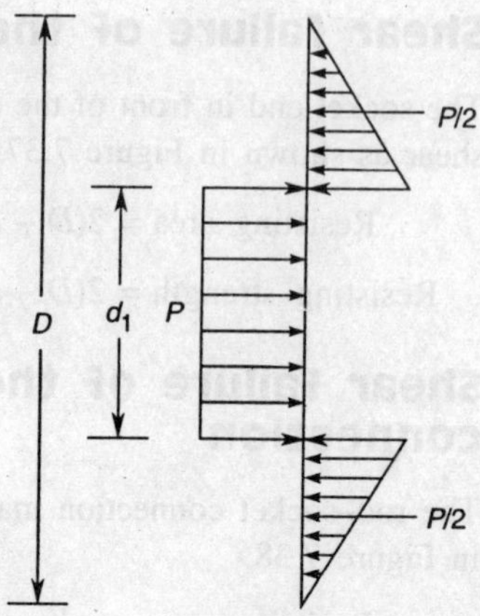

Figure 7.41 Bending force on the cotter.

7.14 GIB AND COTTER JOINT

In certain applications the cotter joint is used to connect rectangular section rods where one of the rods is made into a strap, like the connection of a connecting rod of a steam engine or a marine engine. In such a joint, if a cotter is driven in to make the connection, the frictional force between the cotter and the strap causes the strap to open out as shown in Figure 7.42. In order to prevent this opening out of the strap, it is generally recommended that a gib be used along with a cotter. The gib also provides a larger bearing surface for the cotter to slide on, such that the tendency of the cotter to become loose is considerably decreased. The design procedure of the gib and cotter joint is similar to that of the socket–spigot type cotter joint.

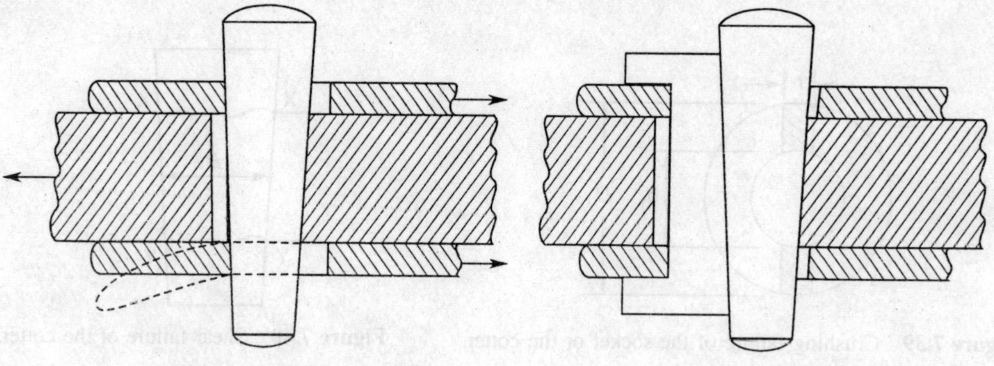

(a) Cotter without a gib

(b) Cotter with a gib

Figure 7.42 Cotter joint of rectangular section.

7.15 KNUCKLE JOINT

Knuckle joints or pinned joints are used to connect two rods that are under the action of tensile forces, though such a joint can also support a compressive force provided it is properly guided. These joints are also called *forked* joints as one of the rods has a forked end while the other has an eye end; the knuckle pin connects these two elements as shown in Figure 7.43. Therefore, the connecting pin acts as the fulcrum and permits angular movement in one plane. The pin is held in place by a small cotter pin or split pin.

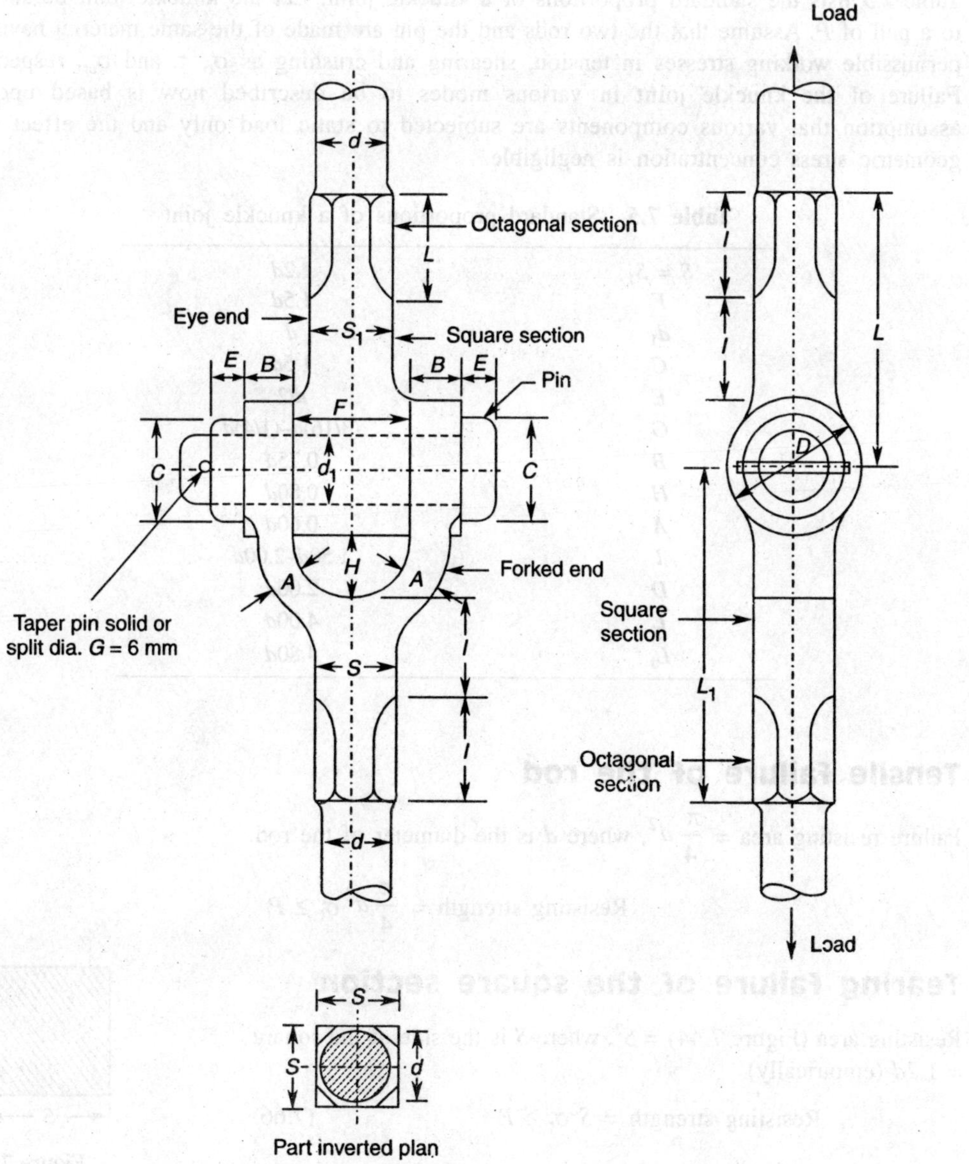

Figure 7.43 Knuckle joint.

A knuckle joint forms a temporary connection between two rods. Therefore, it can be readily assembled or dissembled for repairs, etc. This type of joint finds a wide range of applications in machines, namely the connection of the D-slide valve and the eccentric rod, actuating levers of valves, brake mechanism, and so on.

A knuckle joint can be made from mild steel or wrought iron. However, the IS:1570–1979 code recommends certain grades of steel having 0.3–0.5 per cent carbon.

The design of a knuckle joint can also be carried out in two ways similar to the cotter joint, that is, design by empirical relations and design by uniform or equal strength criteria. Table 7.5 lists the standard proportions of a knuckle joint. Let the knuckle joint be subjected to a pull of P. Assume that the two rods and the pin are made of the same material having the permissible working stresses in tension, shearing and crushing as σ_t, τ, and σ_{cr}, respectively. Failure of the knuckle joint in various modes to be described now is based upon the assumption that various components are subjected to static load only and the effect of the geometric stress concentration is negligible.

Table 7.5 Standard proportions of a knuckle joint

$S = S_1$	$1.2d$
F	$1.5d$
d_1	d
C	$1.5d$
E	$d/2$
G	$(3/16)d$–$(1/4)d$
B	$0.75d$
H	$0.80d$
A	$0.60d$
l	$1.50d$–$2.00d$
D	$2.00d$
L	$4.00d$
L_1	$4.80d$

Tensile failure of the rod

Failure resisting area = $\frac{\pi}{4}d^2$, where d is the diameter of the rod.

$$\text{Resisting strength} = \frac{\pi}{4}d^2\,\sigma_t \geq P \tag{7.65}$$

Tearing failure of the square section

Resisting area (Figure 7.44) = S^2, where S is the side of the square = $1.2d$ (empirically)

$$\text{Resisting strength} = S^2\sigma_t \geq P \tag{7.66}$$

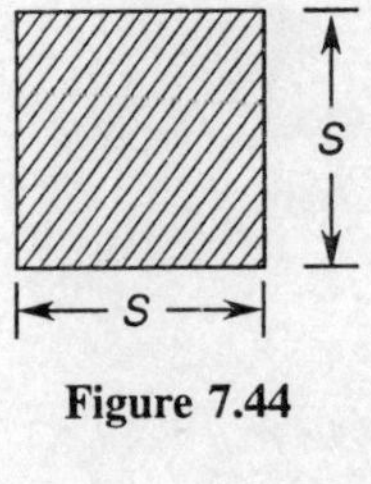

Figure 7.44

Tension failure of the eye end

The tearing resistance of the section across the eye end along a plane passing through the pin axis, as shown in Figure 7.45 is given by

$$F(D - d_1)\sigma_t \geq P \tag{7.67}$$

Shear failure of the eye end

The shear failure of the eye end can take place as shown in Figure 7.46.

The resisting strength of the eye is given by

$$(D - d_1)F\tau \geq P \tag{7.68}$$

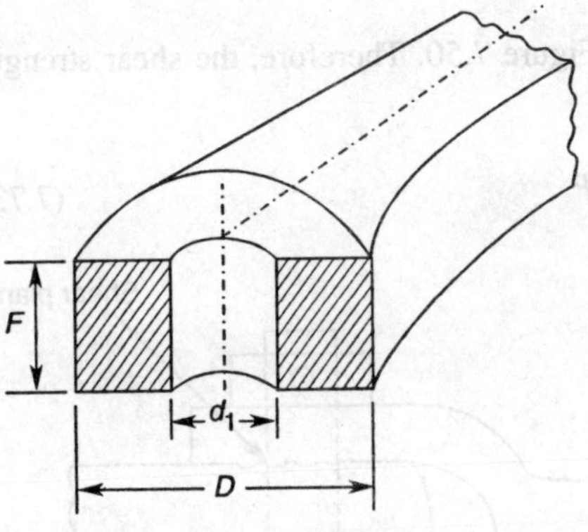

Figure 7.45 Tension failure of the eye end.

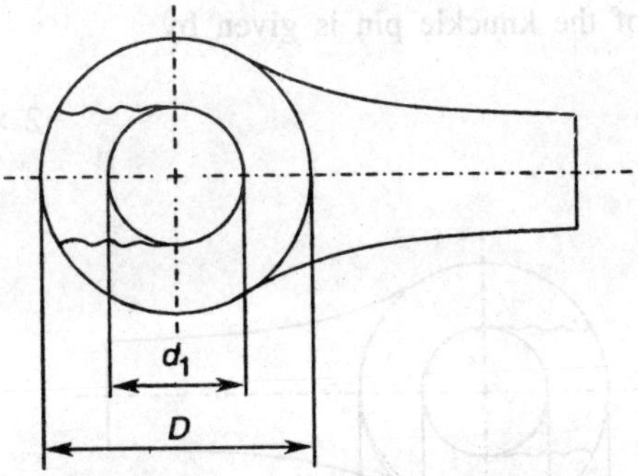

Figure 7.46 Shear failure of the eye end.

Crushing failure of the pin or the eye

Compression of the eye hole or the pin owing to excessive force can cause crushing of the pin or the eye as shown in Figure 7.47.

The resisting strength is computed on the basis of the projected area and is given by

$$d_1 F\sigma_{cr} \geq P \tag{7.69}$$

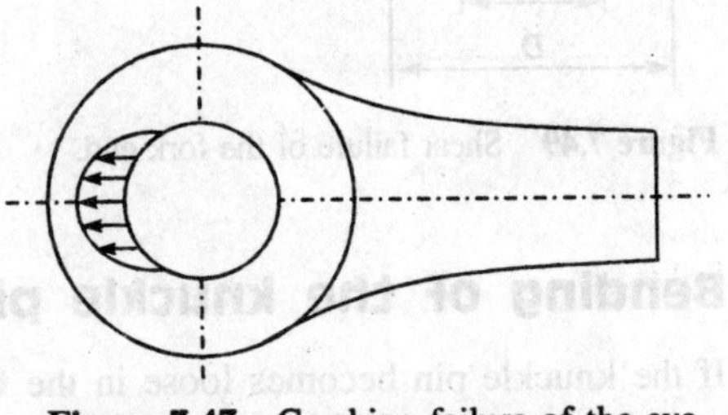

Figure 7.47 Crushing failure of the eye.

Tension failure of the fork end

Tension failure of the fork end across the pin hole may take place as shown in Figure 7.48. The resisting strength is given by

$$2B(D - d_1)\sigma_t \geq P \tag{7.70}$$

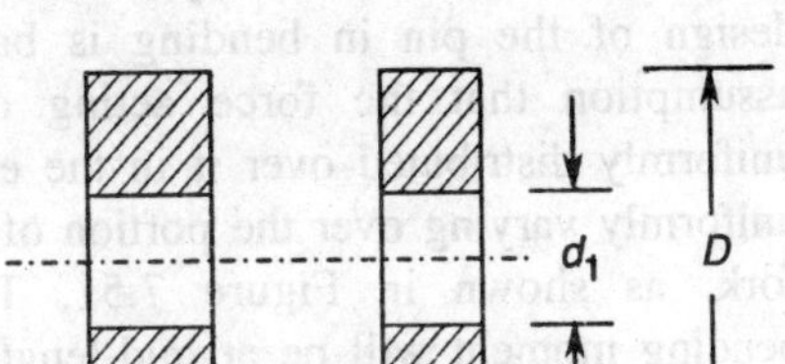

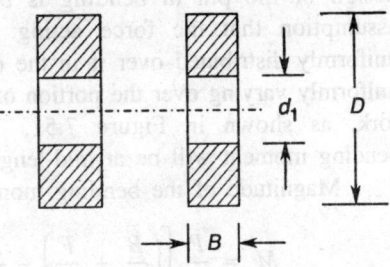

Figure 7.48 Tension failure of the fork end.

Shear failure of the fork end

The fork end may fail under shear action as shown in Figure 7.49. The resisting strength is given by

$$2(D - d_1)B\tau \geq P \tag{7.71}$$

Crushing failure of the fork end

If the fork end and the pin are strong enough in other modes of failure, the fork end may fail due to crushing. The crushing strength of the fork end is given by

$$2\, d_1 B \sigma_{cr} \geq P \tag{7.72}$$

Shear failure of the knuckle pin

The knuckle pin may fail in double shear as shown in Figure 7.50. Therefore, the shear strength of the knuckle pin is given by

$$2 \times \frac{\pi}{4} d_1^2 \tau \geq P \tag{7.73}$$

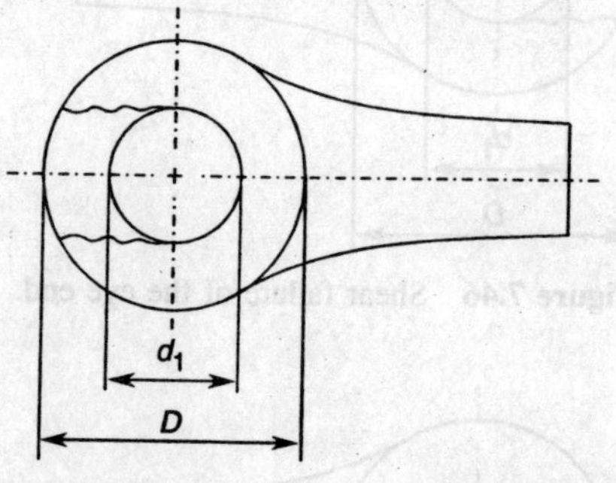

Figure 7.49 Shear failure of the fork end.

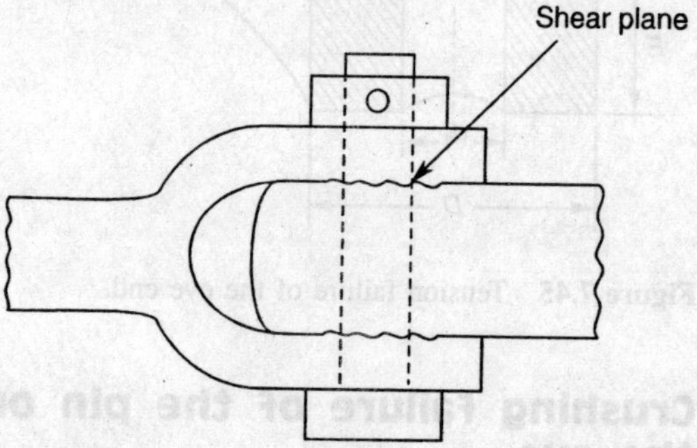

Figure 7.50 Double shear failure of the knuckle pin.

Bending of the knuckle pin

If the knuckle pin becomes loose in the fork, which is quite common, it is subjected to bending. The design of the pin in bending is based upon the assumption that the force acting on the pin is uniformly distributed over it in the eye and is also uniformly varying over the portion of the pin in the fork, as shown in Figure 7.51. The maximum bending moment will be at mid-length of the pin.

Magnitude of the bending moment,

$$M = \frac{P}{2}\left[\left(\frac{B}{3} + \frac{F}{2}\right) - \frac{F}{4}\right] \tag{7.74}$$

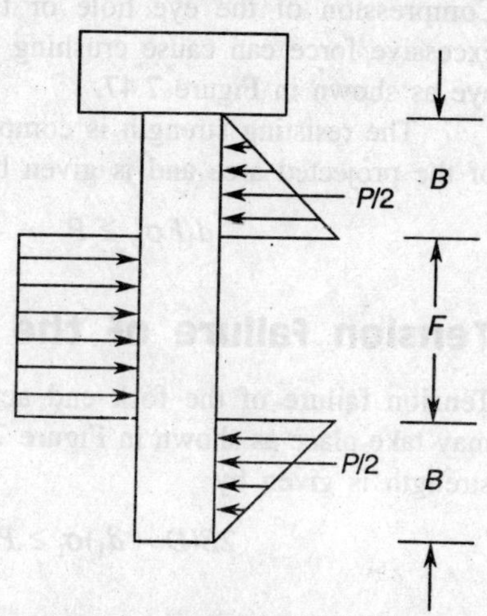

Figure 7.51 Bending forces on the knuckle pin.

and the bending stress

$$\sigma_b = \frac{M}{Z} \tag{7.75}$$

where Z = section modulus = $\dfrac{\pi}{32} d_1^3$

Example 7.11 Design a cotter joint to connect two mild steel rods. The joint is subjected to a 20 kN tensile force. The allowable limits of tensile, shear, and crushing strengths are 60 N/mm², 40 N/mm², and 75 N/mm², respectively.

Solution Design load, P = 20 kN

Allowable strengths: σ_t = 60 N/mm², τ = 40 N/mm², and σ_{cr} = 75 N/mm². To find various dimensions of the cotter joint, we will consider its different modes of failure. (See Figure 7.31 of the cotter joint and the text describing its different modes of failure.)

Tensile failure of the rod

$$P = \frac{\pi}{4} d^2 \sigma_t$$

or

$$d = \sqrt{\frac{4P}{\pi \sigma_t}} = \sqrt{\frac{4 \times 20{,}000}{\pi \times 60}} = 20.6 \text{ mm, say, 21 mm}$$

Tensile failure of the rod across the slot

$$P = \left(\frac{\pi}{4} d_1^2 - d_1 \tau\right) \sigma_t$$

Here both d_1 and t are unknown, so let us use the empirical relation $t = 0.25d_1$. Therefore,

$$\left(\frac{\pi}{4} d_1^2 - 0.25 d_1^2\right) \sigma_t = P$$

or

$$d_1 = \sqrt{\frac{P}{0.5353\, \sigma_t}} = \sqrt{\frac{20{,}000}{0.5353 \times 60}} = 24.9 \text{ mm, say, 25 mm}$$

Thickness of the cotter, $t = 0.25d_1 = 0.25 \times 25 = 6.25$ mm, say, 7 mm

Crushing failure of the cotter-rod end

$$P = d_1 t \sigma_{cr}$$

or

$$\sigma_{cr} = \frac{P}{d_1 t} = \frac{20{,}000}{25 \times 7} = 114.28 \text{ N/mm}^2$$

which is larger than the given permissible value. So let us redesign the spigot diameter d_1 and the cotter thickness t on the basis of the crushing strength. Let

$$t = 0.25d_1$$

then,

$$P = d_1 t \sigma_{cr}$$

or

$$d_1 = \sqrt{\frac{P}{0.25\sigma_{cr}}} = \sqrt{\frac{20,000}{0.25 \times 75}} = 32.6 \text{ mm, say, } 33 \text{ mm}$$

and

$$t = 0.25 \times 33 \approx 9 \text{ mm}$$

Now, we must check these dimensions in tensile failure. Thus,

$$P = \left(\frac{\pi}{4} d_1^2 - d_1 t\right)\sigma_t$$

$$= \left(\frac{\pi}{4} \times 33^2 - 33 \times 9\right)60$$

$= 33.497$ kN, which is larger than the design load and hence it is safe.

Shear failure of the rod end

$$\text{Shear strength, } P = 2d_1 l_1 \tau$$

or

$$l_1 = \frac{P}{2d_1\tau} = \frac{20,000}{2 \times 33 \times 40} = 7.57 \text{ mm, say, } 8 \text{ mm}$$

Shear failure of the collar

$$\text{Shear strength of the collar, } P = \pi d_1 t_1 \tau$$

or

$$\text{collar thickness, } t_1 = \frac{P}{\pi d_1 \tau} = \frac{20,000}{\pi \times 33 \times 40} = 4.8 \text{ mm, say, } 5 \text{ mm}$$

Tension failure of the socket across the slot

$$\text{Tearing strength, } P = \left[\frac{\pi}{4}(D_1^2 - d_1^2) - (D_1 - d_1)t\right]\sigma_t$$

Empirically, let us assume $D_1 = 2d = 2 \times 21 = 42$ mm

Therefore,

$$P = \left[\frac{\pi}{4}(42^2 - 33^2) - (42 - 33) \times 9\right]60$$

$= 25.868$ kN, which is larger than the design load and hence it is safe.

Crushing failure of the cotter or *the socket*

$$\text{Crushing strength, } P = (D - d_1)t\sigma_{cr}$$

or

$$D = \frac{P}{t\sigma_{cr}} + d_1$$

$$= \frac{20{,}000}{9 \times 75} + 33 = 62.6 \text{ mm, say, } 63 \text{ mm}$$

Shear failure of the socket end in front of the cotter

$$P = 2(D - d_1)l\tau$$

or

$$l = \frac{P}{2(D - d_1)\tau} = \frac{20{,}000}{2(63 - 33)40} = 8.3 \text{ mm, say, } 9 \text{ mm}$$

Shear failure of the rod–socket connection

$$P = \pi d t_2 \times \tau$$

or

$$t_2 = \frac{P}{\pi d\tau} = \frac{20{,}000}{\pi \times 21 \times 40} = 7.58 \text{ mm, say, } 8 \text{ mm}$$

Shear failure of the cotter

$$P = 2Bt\tau$$

or

$$B = \frac{P}{2t\tau} = \frac{20{,}000}{2 \times 9 \times 40} = 27.78 \text{ mm}$$

Empirically, the width of the collar should be five times its thickness, i.e. $B = 5t = 5 \times 9 = 45$ mm

Bending failure of the cotter

$$\text{Bending moment, } M = \frac{P}{2}\left[\frac{D - d_1}{6} + \frac{d_1}{4}\right]$$

$$= \frac{20{,}000}{2}\left[\frac{63 - 33}{6} + \frac{33}{4}\right]$$

$$= 132{,}500 \text{ N} \cdot \text{mm}$$

$$\text{Section modulus, } Z = \frac{1}{6}tB^2 = \frac{1}{6} \times 9 \times 45^2 = 3037.5 \text{ mm}^2$$

$$\text{Bending stress, } \sigma_b = \frac{M}{Z} = \frac{132{,}500}{3037.5} = 43.62 \text{ N/mm}^2$$

which is less than the allowable strength. Hence the cotter is safe in bending.

Example 7.12 Design a cottered foundation bolt, as shown in the following figure, to carry an axial pull of 15 kN. The bolt and the cotter are made of the same material having the following permissible stresses:

$$\sigma_t = 40 \text{ N/mm}^2, \tau = 25.0 \text{ N/mm}^2, \text{ and } \sigma_{cr} = 75.0 \text{ N/mm}^2$$

Neglect the effect of stress-concentration.

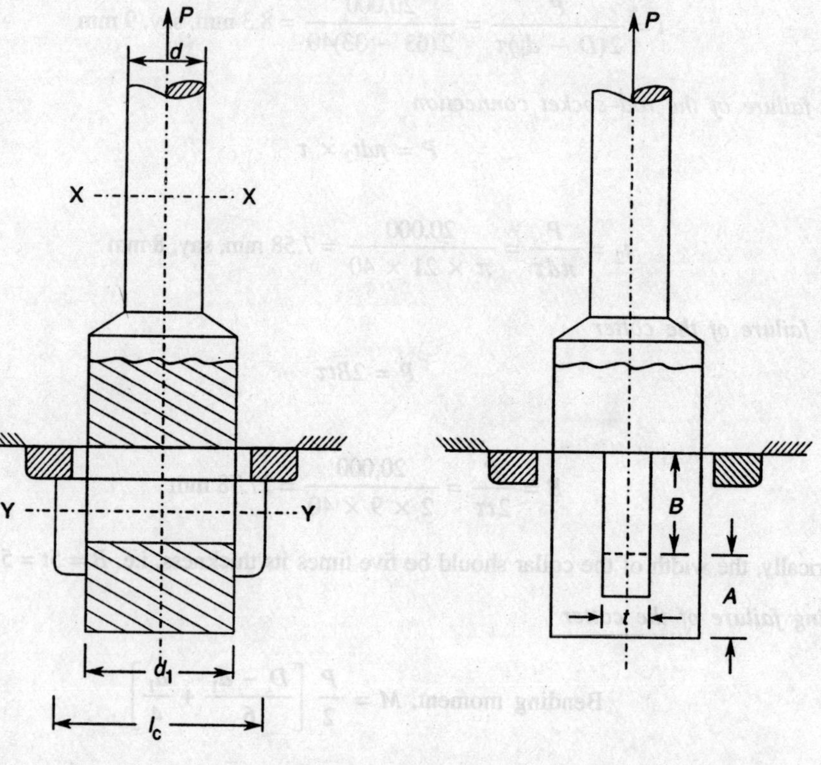

Solution (i) Tension failure of the bolt at section X–X gives the nominal diameter of the bolt as

$$d = \sqrt{\frac{4P}{\pi\sigma_t}} = \sqrt{\frac{4 \times 15{,}000}{\pi \times 40}} = 21.85 \text{ mm, say, 22 mm}$$

(ii) For tension failure of the bolt at section Y–Y

$$P = \left(\frac{\pi}{4}d_1^2 - d_1 t\right)\sigma_t$$

where

d_1 = diameter of the extended portion of the bolt

t = thickness of the cotter

= $d_1/4$ (empirical relation)

Therefore,

$$P = \left(\frac{\pi}{4}d_1^2 - 0.25\, d_1^2\right)\sigma_t$$

or

$$d_1 = \sqrt{\frac{P}{\left(\frac{\pi}{4} - 0.25\right)\sigma_t}} = \sqrt{\frac{15{,}000}{\left(\frac{\pi}{4} - 0.25\right)40}}$$

= 26.4 mm, say, 30 mm

and cotter thickness, $t = d_1/4 = 7.5$ mm, say, $t = 8$ mm

(iii) Check for crushing stress between the cotter and the bolt

$$\sigma_{cr} = \frac{P}{d_1 t} = \frac{15{,}000}{30 \times 8} = 62.5 \text{ N/mm}^2$$

which is less than the permissible strength.

(iv) For shear failure of the end of the enlarged portion of the bolt

$$P = 2d_1 A\tau$$

or

$$A = \frac{P}{2d_1\tau} = \frac{15{,}000}{2 \times 30 \times 25} = 10 \text{ mm}$$

(v) For shear failure of the cotter

$$P = 2Bt\tau \text{ (considering double shear)}$$

or

$$B = \frac{P}{2t\tau} = \frac{15{,}000}{2 \times 8 \times 25} = 37.5 \text{ mm}$$

Empirically the width B should be $5t$. That is,

$$B = 5t = 40 \text{ mm}$$

(vi) For crushing failure between the cotter and the base plate

$$P = l_c t \sigma_{cr}$$

or

$$l_c = \frac{P}{t\sigma_{cr}} = \frac{15,000}{8 \times 75} = 25 \text{ mm}$$

Since cotter length should be larger than the diameter d_1, empirically it should be $1.25d_1$, i.e.

$$l_c = 1.25 \times 30 = 37.5 \text{ mm, say, } 38 \text{ mm}$$

Example 7.13 Design a knuckle joint to connect two mild steel rods which transmit a tensile force of 25 kN. The safe working stresses for tension, shear, and crushing are 100 N/mm^2, 60 N/mm^2, and 160 N/mm^2, respectively.

Solution In order to find all the dimensions of the knuckle joint, let us consider various modes of failure. (For the diagram of the joint and figures of various failure modes, refer Figure 7.43 and the relevant text material.)

Tension failure of the rod

$$\text{Resisting strength, } P = \frac{\pi}{4} d^2 \sigma_t$$

or

$$d = \sqrt{\frac{4P}{\pi\sigma_t}} = \sqrt{\frac{4 \times 25,000}{\pi \times 100}} = 17.8 \text{ mm, say, } 18 \text{ mm}$$

Tearing failure of the square section

$$\text{Resisting strength} = S^2\sigma_t = P$$

or

$$S = \sqrt{\frac{25,000}{100}} = 15.8 \text{ mm}$$

According to the empirical relation the square section should be at least $1.2d$. Therefore,

$$S = 1.2d = 1.2 \times 18 = 21.6 \text{ mm, say, } 22 \text{ mm}$$

Shear failure of the knuckle pin

The knuckle pin may fail in double shear, therefore, the resisting strength

$$2 \times \frac{\pi}{4} d_1^2 \times \tau = P$$

or

$$d_1 = \sqrt{\frac{4P}{2\pi\tau}} = \sqrt{\frac{4 \times 25,000}{2 \times \pi \times 60}} = 16.28 \text{ mm, say, } 17 \text{ mm}$$

Tension failure of the eye end

$$\text{Resisting strength} = F(D - d_1)\sigma_t = P$$

$$\text{Empirically, } F = 1.5d = 1.5 \times 18 = 27 \text{ mm}$$

and

$$D = 2d = 2 \times 18 = 36 \text{ mm}$$

$$\text{Tearing resistance} = 27(36 - 17) \times 100 = 513{,}00 \text{ N}$$

which is larger than the design load; hence the design is safe.

Crushing failure of the pin or the eye

$$\text{Resisting strength, } P = d_1 F \sigma_{cr}$$

or

$$\sigma_{cr} = \frac{P}{d_1 F} = \frac{25{,}000}{17 \times 27} = 54.46 \text{ N/mm}^2$$

which is reasonably smaller than the allowable strength.

Shear failure of the eye end

$$\text{Resisting strength, } P = (D - d_1)F\tau$$

or

$$\tau = \frac{P}{(D - d_1)F} = \frac{25{,}000}{(36 - 17)27} = 48.73 \text{ N/mm}^2$$

which is smaller than the allowable strength; hence the design is safe.

Tension failure of the fork end

$$\text{Resisting strength, } P = 2(D - d_1)B\sigma_t$$

or

$$B = \frac{P}{2(D - d_1)\sigma_t} = \frac{25{,}000}{2(36 - 17)100} = 6.57 \text{ mm}$$

Empirically $B = 0.75d = 13.5$ mm, say, 14 mm

Shear failure of the fork end

$$\text{Resisting strength, } P = 2(D - d_1)B\tau$$

or

$$\tau = \frac{P}{2(D - d_1)B} = \frac{25{,}000}{2(36 - 17)14} = 47 \text{ N/mm}^2$$

which is within the permissible limit.

Crushing failure of the fork end

$$\text{Resisting strength, } P = 2d_1 B\sigma_{cr}$$

or

$$\sigma_{cr} = \frac{P}{2d_1 B} = \frac{25{,}000}{2 \times 17 \times 14} = 52..5 \text{ N/mm}^2$$

which is less than the allowable limit; hence the design is safe.

Bending failure of the pin

$$\text{Bending moment, } M = \frac{P}{2}\left[\frac{B}{3} + \frac{F}{4}\right]$$

$$= \frac{25,000}{2}\left[\frac{14}{3} + \frac{27}{4}\right] = 142,708.3 \text{ N}\cdot\text{mm}$$

$$\text{Bending stess, } \sigma_b = \frac{M}{Z} = \frac{142,708.3}{\frac{\pi}{32}\times 17^3} = 295.8 \text{ N/mm}^2$$

which is too high. Therefore, we have to redesign the pin on the basis of the bending failure.

$$\text{Revised pin diameter, } d_1 = \left[\frac{32 \times 142,708.3}{\pi \times 100}\right]^{1/3}, = 24.4 \text{ mm, say, } 25 \text{ mm}$$

The revision of the pin diameter will affect the strength in the following modes of failure, which have to be rechecked.

(a) *Shear strength of the fork*

$$P = 2(D - d_1)(B \times \tau)$$

or

$$25,000 = 2(D - 25)(14 \times 60)$$

or

$$D = 39.88 \text{ mm, say, } 40 \text{ mm}$$

(b) *Tension failure across the eye*

$$P = F(D - d_1)\sigma_t$$

or

$$\sigma_t = \frac{P}{F(D - d_1)} = \frac{25,000}{27(40 - 25)} = 61.73 \text{ N/mm}^2$$

which is within the permissible limit; hence the design is safe.

(c) *Tension failure of the fork*

$$P = 2(D - d_1)B\sigma_t$$

or

$$\sigma_t = \frac{P}{2(D - d_1)B} = \frac{25,000}{2(40 - 25)14} = 59.52 \text{ N/mm}^2$$

which is less than the permissible limit; hence the design is safe.

The remaining dimensions can be found from the empirical relations as follows:

$$C = 1.5d_1 = 1.5 \times 25 \approx 37.5 \text{ mm, say } 38 \text{ mm}$$

$$E = d_1/2 \approx 13 \text{ mm}$$

$$H = 0.8d \approx 15 \text{ mm}$$

$$\text{Diameter of the taper pin} = 0.2d \approx 4 \text{ mm}$$

EXERCISES

1. What are the basic requirements of a rivet material?
2. Describe the tests that are performed on rivets.
3. What is the limitation of the single-strap butt joint?
4. Show by neat sketches the various modes of failure of riveted joints.
5. Explain the term efficiency of a riveted joint.
6. What type of joint is used for the longitudinal joint of pressure vessels?
7. Explain the procedure used to determine the number of rivets in single and double shear.
8. What do you mean by a uniform strength riveted joint?
9. What do you understand by caulking and fullering? Explain with suitable sketches?
10. Why is a lap joint (welded) not recommended for tensile force?
11. What is throat thickness of a fillet weld?
12. What is a reinforced weld?
13. Why are welded joints preferred over riveted joints?
14. Define a cotter. Why is it tapered?
15. What is uniform strength concept used in the design of a cotter joint?
16. Why is gib used in gib type cotter joints?
17. Explain the bending failure of cotter and knuckle pin.
18. Design and draw the longitudinal joint for a Cornish boiler with shell plate thickness of 23 mm. The joint is triple-riveted, double-cover plate butt joint with straps of unequal width. The rivets and plates are made of mild steel having ultimate strength in tension, crushing, and shear as 60 N/mm^2, 100 N/mm^2, and 45 N/mm^2, respectively. The efficiency of the joint is 80 per cent.
19. Select and design the boiler joints which should conform to IBR. The maximum steam pressure is 12 bar (gauge). The diameter of the boiler shell is 2 m. Make a neat dimensioned sketch of the joint.
20. Design a lap joint for a C20 steel flat tie-bar of size 250 mm × 10 mm.
21. Two mild steel flats in a bridge structure are to be joined by means of a butt joint. The flats are 200 mm wide and 10 mm thick. Design the joint.
22. Design a diamond (Lozenge) joint to connect two C20 steel flats of 10 mm thickness. The maximum load on the joint is 400 kN.
23. Determine the suitable diameter of the rivet for the bracket shown below. The rivets and plates are made of C20 steel.

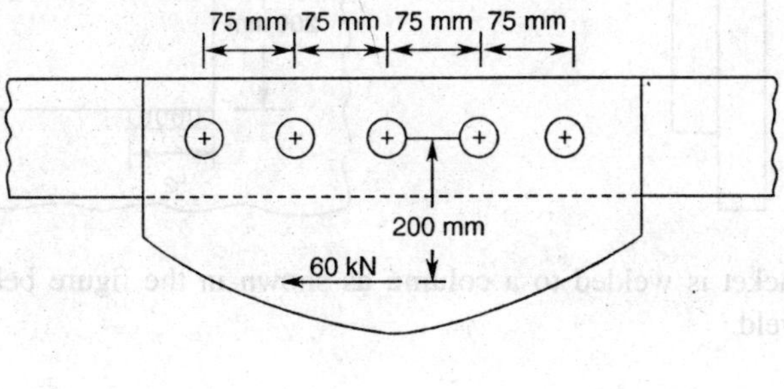

24. Determine the forces acting on all the rivets and the size of the rivet for the structural joint shown below.

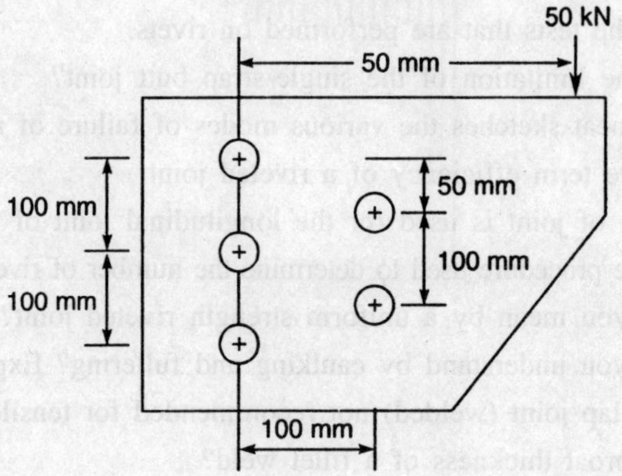

25. Find the suitable size of the weld shown below. The maximum load is 25 kN. The permissible shear strength is 80 MPa.

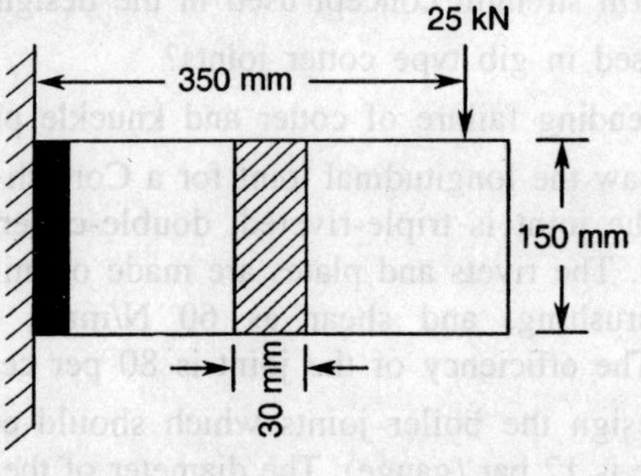

26. An angle of size 200 mm × 150 mm × 20 mm is welded to a flat plate with long side of the angle along the length of the plate, as shown in the figure below. Design the joint.

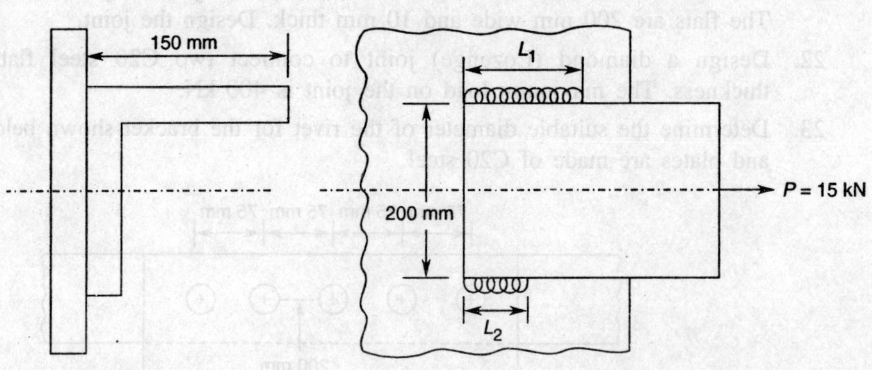

27. A bracket is welded to a column as shown in the figure below. Calculate the size of the weld.

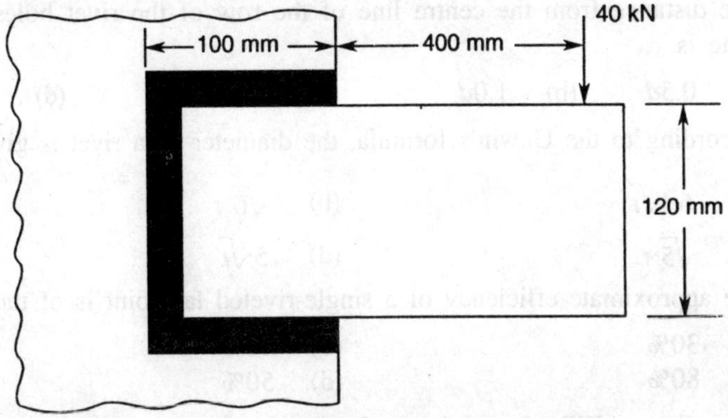

28. Design a cotter joint to support a load of 50 kN which is subjected to a slow reversal of direction. Assume a suitable material and factor of safety.
29. Design a gib and cotter joint of rectangular section. The joint is subjected to 30 kN. It is made of C30 steel.
30. A knuckle joint is required to connect two rods which are subjected to 60 kN tensile force. Design the joint and also give a dimensioned sketch.
31. A plain carbon steel C15 rod is hinged to a CI bracket by a steel pin as shown in the figure below. Design the joint.

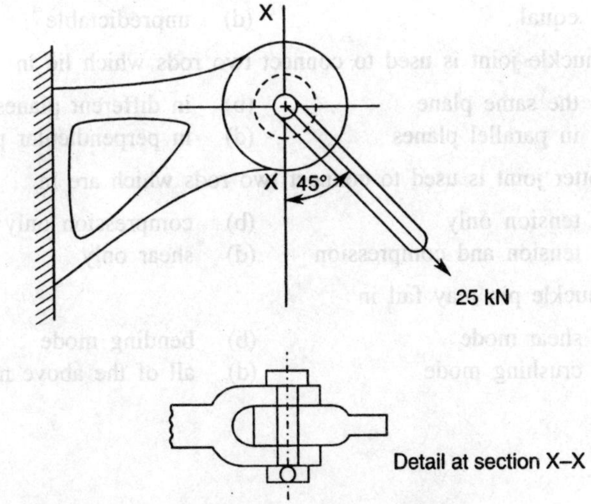

MULTIPLE CHOICE QUESTIONS

1. For riveted joints the type of joint preferred is

(a) lap joint	(b) butt joint with two straps
(c) butt joint with one strap	(d) any of the above

2. The distance from the centre line of the row of the rivet holes to the edge of the plate is

   (a) $0.5d$ (b) $1.0d$ (c) $1.5d$ (d) $2.0d$

3. According to the Unwin's formula, the diameter of a rivet is given by

   (a) $6.1\sqrt{t}$ (b) $\sqrt{6}\,t$

   (c) $\sqrt{5}\,t$ (d) $5\sqrt{t}$

4. The approximate efficiency of a single-riveted lap joint is of the order of

   (a) 30% (b) 40%

   (c) 80% (d) 50%

5. According to IBR, the minimum factor of safety needs to be

   (a) 2 (b) 3 (c) 4 (d) 6

6. The thickness of a boiler plate is 25 mm. The rivet diameter will be

   (a) 20 mm (b) 30 mm (c) 40 mm (d) 15 mm

7. In a reinforced fillet weld, the throat thickness is

   (a) $0.707t$ (b) $0.75t$

   (c) $0.8t$ (d) $0.85t$

8. The efficiency of a welded joint, compared to a riveted joint is

   (a) more (b) less

   (c) equal (d) unpredictable

9. A knuckle joint is used to connect two rods which lie in

   (a) the same plane (b) in different planes

   (c) in parallel planes (d) in perpendicular planes

10. A cotter joint is used to connect two rods which are in

    (a) tension only (b) compression only

    (c) tension and compression (d) shear only

11. A knuckle pin may fail in

    (a) shear mode (b) bending mode

    (c) crushing mode (d) all of the above modes

Chapter 8 Screw Fastenings and Power Screws

8.1 INTRODUCTION

When two or more machine members need to be held together the mechanical fastening most frequently used is the screw fastening. In screw fastening, a screw thread is formed by cutting a continuous helical groove on a cylindrical surface. When such a screw engages with a corresponding threaded hole, it forms an inclined plane contact. When torque is applied, it results in motion in the axial direction.

Screwed connections are highly reliable, convenient to assemble or dissemble. They are available in wide ranges and relatively cheap to produce owing to standardization and specialized manufacturing processes. The main disadvantage of the screw connection is the high stress concentration near the sharp edges of the thread which may damage the thread under variable load conditions.

Screws are also useful for transmission of power such as lead screws of lathe machines, presses, jacks, and other similar devices. A typical screw connection using bolt and nut is shown in Figure 8.1(a). The terminology used in bolt and nut is explained in Figures 8.1(b) and (c).

8.2 FORMS OF SCREW THREAD

The basic form of a screw thread is V-shaped. However, due to high stress concentration at sharp edges, screw threads are either chamfered or filleted at edges. The most popular forms of threads are (i) British Standard Whitworth (BSW), and (ii) American national thread (Figure 8.2). The British Standard Whitworth thread is superseded by the ISO metric thread.

The basic profile of the ISO metric thread is V-form with included angle 60°, as shown in Figure 8.3. In these threads the root, which is chamfered in the American national thread, is rounded at minor diameter of the external thread. In the case of internal threads, the root is rounded at the major diameter to avoid sharp corners.

The standard proportions of the ISO metric thread of coarse series are given in IS: 4218 (Part III)—1967. See also Table 8.1. The tensile stress area, which is the area of an imaginary circle whose diameter is the mean of the pitch and minor diameters, is used to compute the tensile strength of the bolt. According to BIS, an ISO metric threaded bolt is designated by a letter M followed by basic diameter and pitch, Example: M10 × 1.5—IS: 4218 (Part III)—1967 represents an ISO metric threaded bolt having basic diameter 10 mm and pitch 1.5 mm.

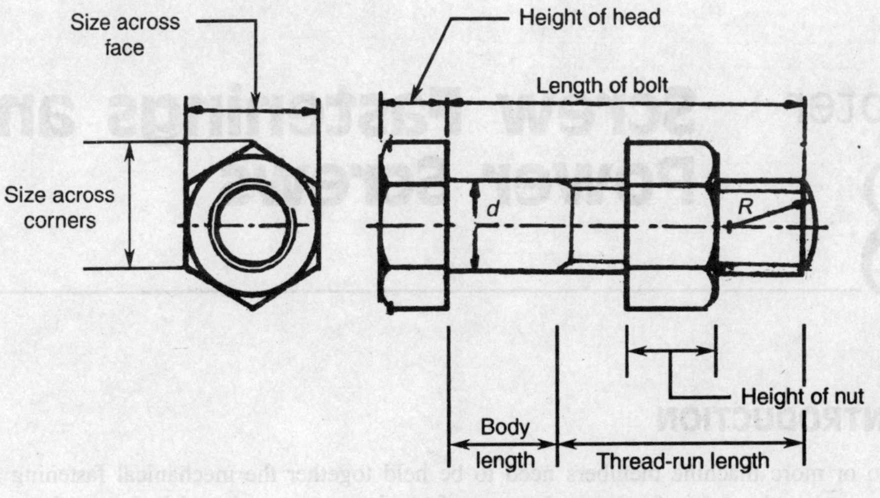

(a) Parts of bolt and nut

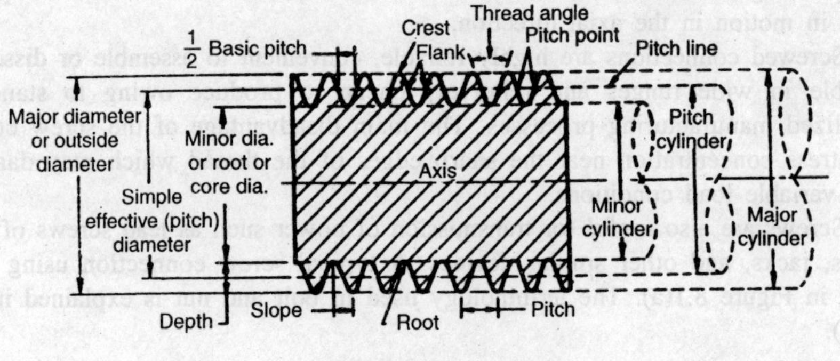

(b) External thread

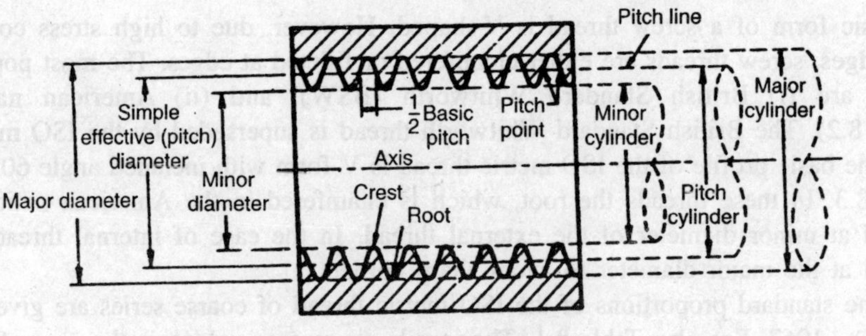

(c) Internal thread

Figure 8.1 A typical screw connection.

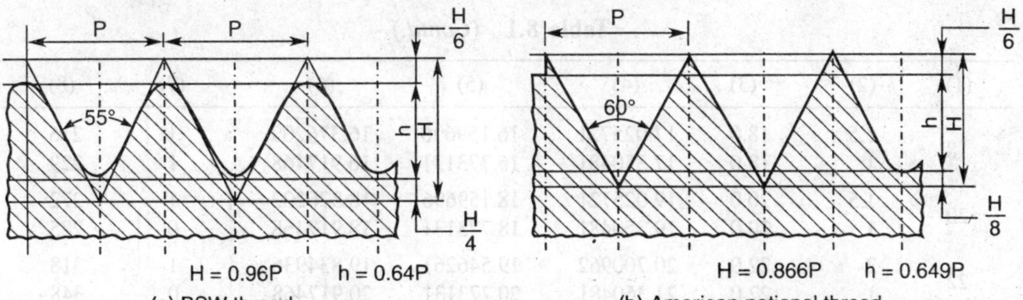

Figure 8.2 Forms of screw thread.

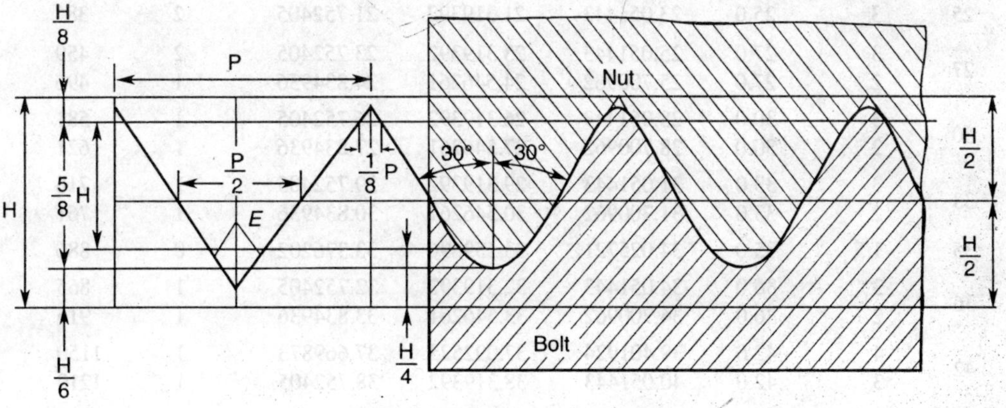

Figure 8.3 ISO metric thread.

Table 8.1 Basic dimensions of ISO metric screw threads IS : 4218 (Part III)—1967

Basic diameter (mm)	*Pitch* (mm)	*Major diameter* (mm)	*Pitch diameter* (mm)	*Minor diameter* (mm) *External threads*	*Minor diameter* (mm) *Internal threads*	*Lead angle* (θ°)	*Tensile stress* (mm²)
(1)	(2)	(3)	(4)	(5)	(6)	(7)	(8)
10	1.5	10.0	9.025721	8.159696	8.376202	3	58.0
	1	10.0	9.350481	8.773131	8.917468	1	64.5
12	1.5	12.0	11.025721	10.159696	10.376202	2	88.1
	1	12.0	11.350481	10.773131	10.917468	1	96.1
14	1.5	14.0	13.025721	12.159696	12.376202	2	125
	1	14.0	13.350481	12.773131	12.917468	1	134
16	1.5	16.0	15.025721	14.159696	14.376202	1	167
	1	16.0	15.350481	14.773131	14.917468	1	178

(*Contd.*)

Table 8.1 (*Contd.*)

(1)	(2)	(3)	(4)	(5)	(6)	(7)	(8)
18	1.5	18.0	17.025721	16.159696	16.376202	1	216
	1	18.0	17.350481	16.773131	16.917468	1	229
20	1.5	20.0	19.025721	18.159696	18.376202	1	272
	1	20.0	19.350481	18.773131	18.917468	0	285
22	2	22.0	20.700962	19.546261	19.834936	1	318
	1	22.0	21.350481	20.773131	20.917468	0	348
24	2	24.0	22.700962	21.546261	21.834936	1	384
	1	24.0	23.350481	22.773131	22.917468	0	418
25	3	25.0	23.051443	21.319392	21.752405	2	385
27	3	27.0	25.051443	23.319392	23.752405	2	459
	2	27.0	25.700962	24.546261	24.834936	1	496
30	3	30.0	28.051443	26.319392	26.752405	1	581
	2	30.0	28.700962	27.546261	27.834936	1	621
33	3	33.0	31.051443	29.319392	29.752405	1	716
	2	33.0	31.700962	30.546261	30.834936	1	761
35	1.5	35.0	34.025721	33.159696	33.376202	0	886
36	3	36.0	34.051443	32.319392	32.752405	1	865
	2	36.0	34.700962	33.546261	33.834936	1	915
42	4	42.0	39.401924	37.092523	37.669873	1	1150
	3	42.0	40.051443	38.319392	38.752405	1	1210
45	4	45.0	42.401924	40.092523	40.669873	1	1340
	3	45.0	43.051443	41.319392	41.752405	1	1400
48	4	48.0	45.401924	43.092523	43.569873	1	1540
	3	48.0	46.051443	44.319392	44.752405	1	1600
52	4	52.0	49.401924	47.092523	47.669873	1	1830
	3	52.0	50.051443	48,319392	48.752405	1	1900
56	4	56.0	53.401924	51.092523	51.669873	1	2140
	3	56.0	54.051443	52.319392	52.752405	1	2220
60	4	60.0	57.401924	55.092523	55.669873	1	2490
	3	60.0	58.051443	56.319392	56.752405	0	2570

8.3 TYPES OF SCREW FASTENINGS

Some of the common types of screw fastenings used as mechanical fasteners are (i) through bolt, (ii) tap bolt, (iii) stud, (iv) machine screw, (v) set screw, and (vi) washer. A brief description of some of these fasteners is given below.

Through bolt. It is commonly called the bolt. It is a round bar, one end of which is threaded and fitted with a nut while the other end is upset to form a head. Depending upon the shape of the head, these bolts are called hexagonal bolt or square bolt. These bolts are used where both the head and the nut can stay easily accessible. Figure 8.4 shows some typical types of bolts. The standard proportions of the ISO metric thread bolt are given in Table 8.1.

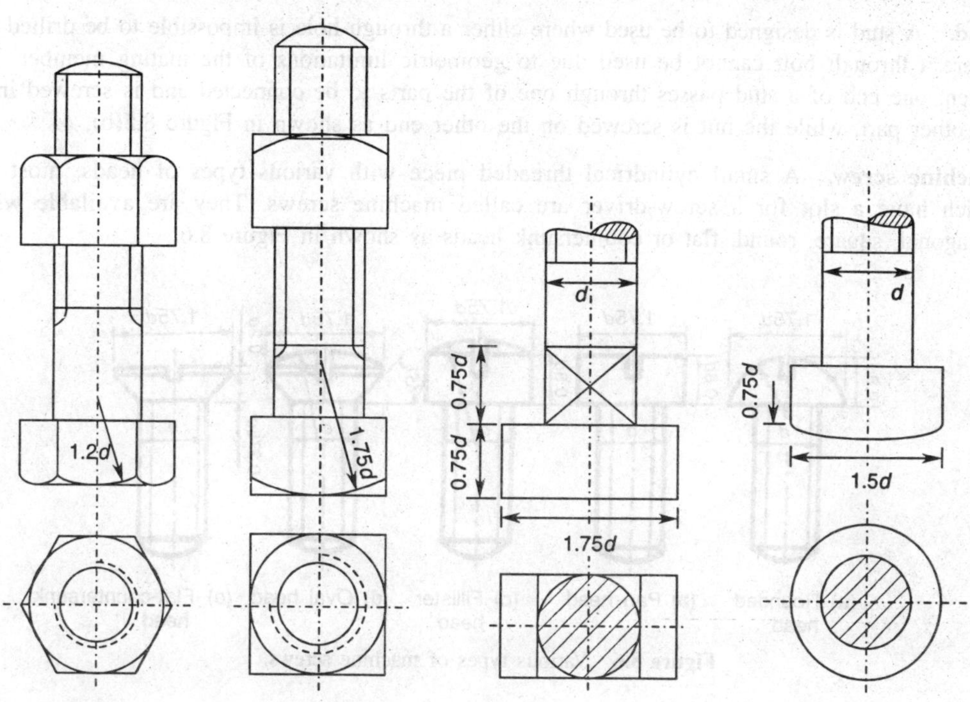

(a) Hexagon bolt (b) Square-head bolt (c) Tee-head bolt (d) Cheese-head bolt

Figure 8.4 Different types of bolts.

Tap bolt. This bolt is the through bolt without the nut; the internal threading is done into the mating member itself, as shown in Figure 8.5(c). However, this bolt should not be used when frequent unscrewing of a machine member is required as it may damage the threads in the machine member. In order to ensure a good fastening, the depth of the threaded hole should be at least 1.5–2.0 times the diameter of the bolt.

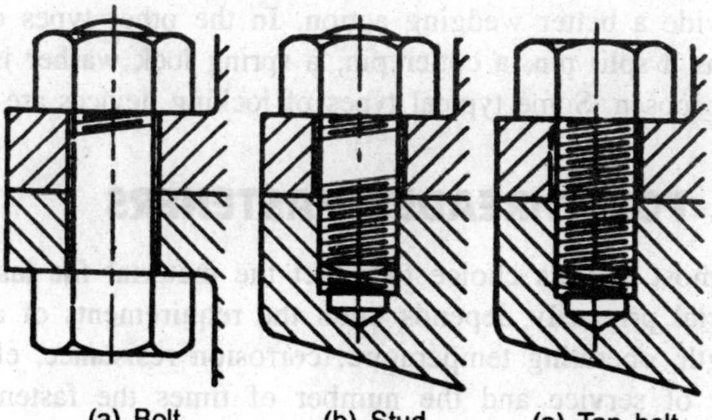

(a) Bolt (b) Stud (a) Tap bolt

Figure 8.5 Use of bolt, stud and tap bolt.

Stud. A stud is designed to be used where either a through hole is impossible to be drilled or where a through bolt cannot be used due to geometric limitations of the mating member. To fasten, one end of a stud passes through one of the parts to be connected and is screwed into the other part, while the nut is screwed on the other end as shown in Figure 8.5(b).

Machine screw. A small cylindrical threaded piece with various types of heads, most of which have a slot for a screw-driver are called machine screws. They are available with hexagonal, square, round, flat or countersunk heads as shown in Figure 8.6.

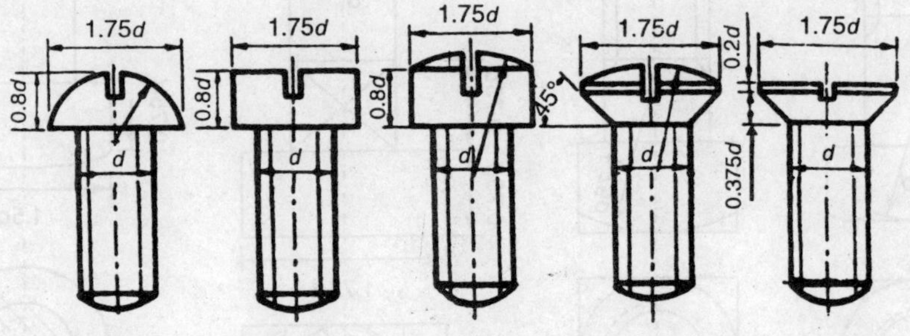

Figure 8.6 Various types of machine screws.

8.4 LOCKING DEVICES

In a screw fastening, the external force tends to unscrew the fastening. However, the frictional force between the wedging threads tends to resist the unscrewing motion. When a bolted joint is subjected to cyclic loading or vibrations, there is tendency of the nut to become loose. Thus, the nut-bolt fastening requires some additional means of restraining the nut. The devices used for this purpose are called the locking devices. A large number of locking devices are available, among them the *lock nut* is the most common locking device. In this device, a nut is tightened in the ordinary way and then the lock nut is tightened over it almost to its limit. Holding the upper nut by a spanner, the lower nut is now turned back by another spanner. The threads of the two nuts will now be in contact as shown in Figure 8.7(a) with opposite side clearances which provide a better wedging action. In the other types of locking devices, a physical device such as a split pin, a cotter pin, a spring lock washer is used to prevent the tendency of the nut to loosen. Some typical types of locking devices are shown in Figure 8.7.

8.5 MATERIAL FOR THREADED FASTENERS

A designer has an almost endless choice to select the material for fasteners. However, the selection of the material primarily depends upon the requirements of a particular fastening system, namely strength, operating temperature, corrosion resistance, electrical, thermal and other properties, type of service and the number of times the fastener is required to be detached. The most commonly used materials for general purpose fasteners are plain carbon steels and free cutting steels having carbon percentage between 0.1–0.4. Alloy steels with

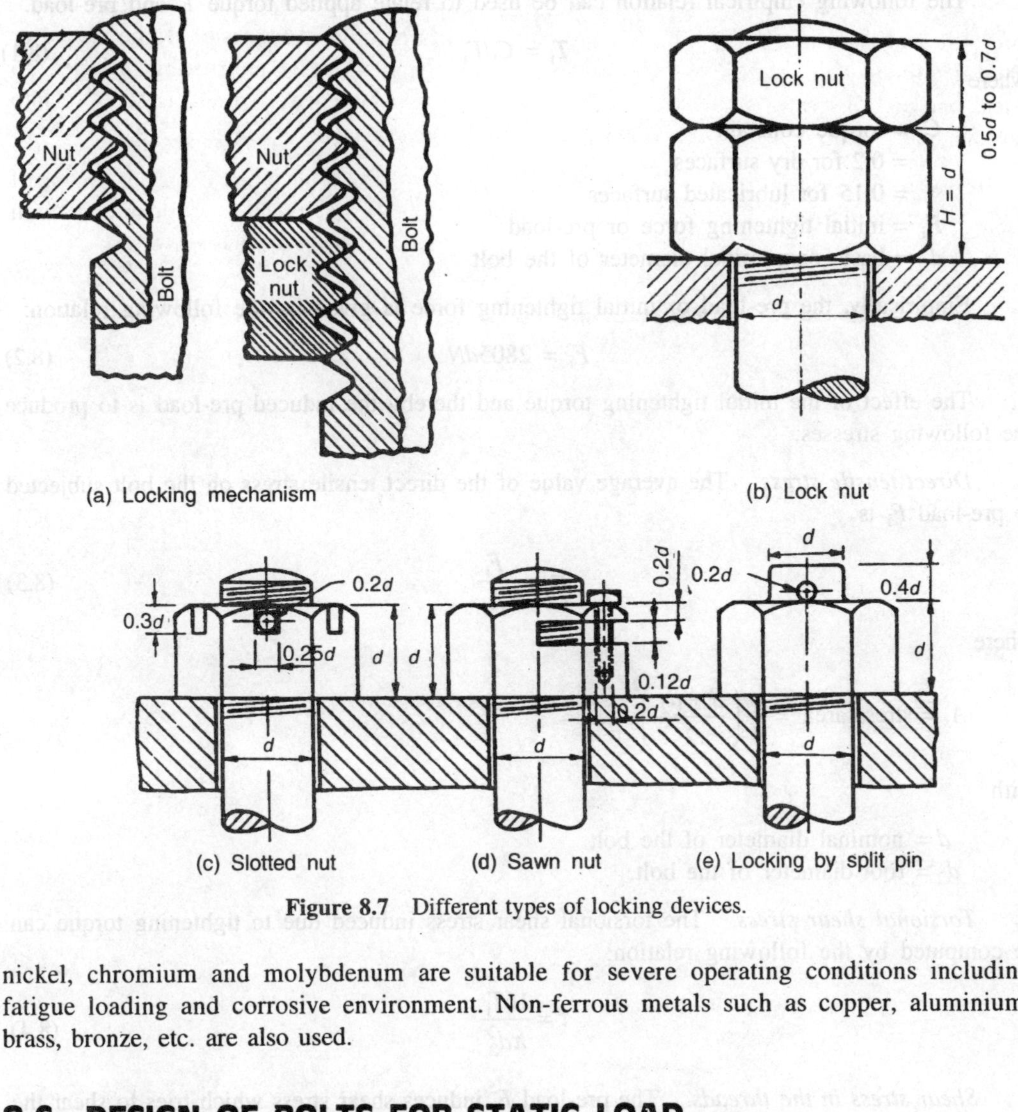

Figure 8.7 Different types of locking devices.

nickel, chromium and molybdenum are suitable for severe operating conditions including fatigue loading and corrosive environment. Non-ferrous metals such as copper, aluminium, brass, bronze, etc. are also used.

8.6 DESIGN OF BOLTS FOR STATIC LOAD

The main purpose of a bolt is to fasten the given mechanical components. A bolt is inserted in the hole drilled through parts and a nut is tightened. This tightening of the nut produces a clamping force, generally called the initial tightening force or pre-load. Besides, the bolted assembly is also subjected to external forces. In a bolted assembly, the outer surfaces of the parts, where the head or nut rests, should be normal to the axis of the bolt; otherwise the bolt will be subjected to bending stresses.

8.6.1 Initial Tightening Force

The initial tightening force on a bolt is necessary, as the strength of the joint depends upon it. The amount of pre-load on the bolt depends upon the torque applied to tighten the bolt.

The following empirical relation can be used to relate applied torque T_i and pre-load.

$$T_i = CdF_i \tag{8.1}$$

where

C = torque constant
= 0.2 for dry surfaces
= 0.15 for lubricated surfaces
F_i = initial tightening force or pre-load
d = basic or nominal diameter of the bolt

Empirically, the pre-load or initial tightening force is given by the following relation:

$$F_i = 2805dN \tag{8.2}$$

The effect of the initial tightening torque and thereby the induced pre-load is to produce the following stresses:

Direct tensile stress. The average value of the direct tensile stress on the bolt subjected to pre-load F_i is

$$\sigma_t = \frac{F_i}{A_s} \tag{8.3}$$

where

$$A_s = \text{stress area} = \frac{\pi}{4}\left(\frac{d + d_2}{2}\right)^2$$

with

d = nominal diameter of the bolt
d_2 = root diameter of the bolt.

Torsional shear stress. The torsional shear stress induced due to tightening torque can be computed by the following relation:

$$\tau = \frac{16T_i}{\pi d_2^3} \tag{8.4}$$

Shear stress in the threads. The pre-load F_i induces shear stress which tries to shear the threads. The induced shear stresses in the threads of bolt and nut are given by the following relations (Figure 8.8):

$$\tau = \frac{F_i}{\pi d_2 N b} \quad \text{in the bolt} \tag{8.5}$$

$$= \frac{F_i}{\pi d_1 N b} \quad \text{in the nut} \tag{8.6}$$

where d_1 and d_2 are the root diameters of the nut and the bolt, respectively, b is width of the threaded section at the root, and N is number of threads.

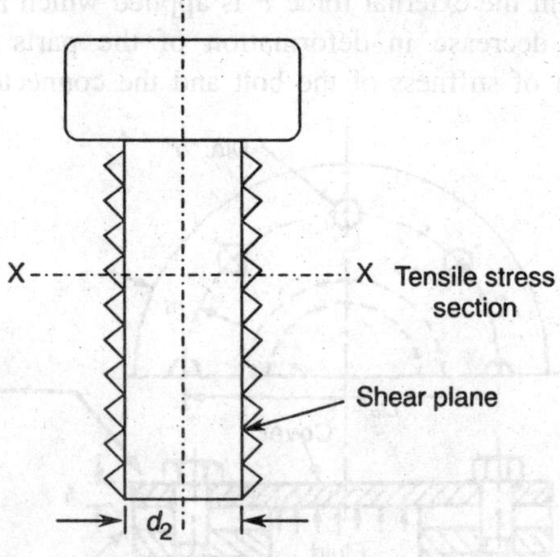

Figure 8.8 Threads under stress.

Compression stress. The threads of the screw connection are subjected to compressive stress which can be computed by the following relation:

$$\sigma_{cr} = \frac{F_i}{\frac{\pi}{4}\left(d_1^2 - d_2^2\right)N} \tag{8.7}$$

8.6.2 External Force

The bolted joints are often required to carry an external force. Therefore, while designing a bolt it is necessary to find what portion of the externally applied force is taken up by the bolt and what portion by the connecting parts.

Let us consider a bolted assembly as shown in Figure 8.9. It is assumed that the pre-load F_i has been applied and both the bolt and the connected parts have gone through deformation

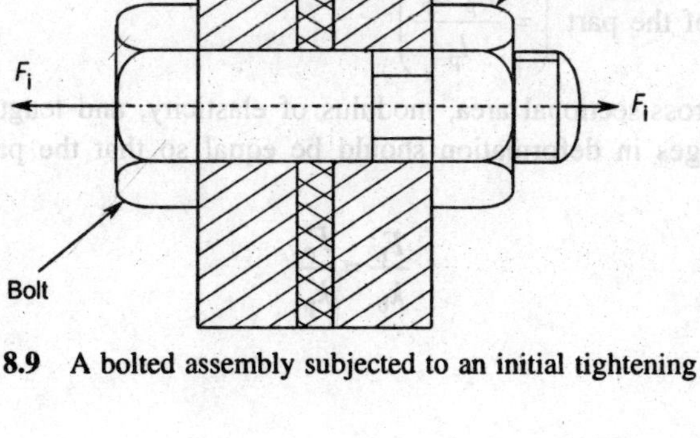

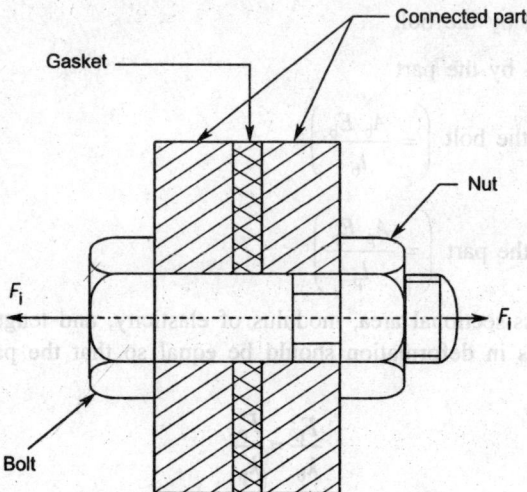

Figure 8.9 A bolted assembly subjected to an initial tightening force.

due to this load. At this point the external force F is applied which results in an increase in the length of the bolt and a decrease in deformation of the parts (see Figure 8.10). These deformations are a function of stiffness of the bolt and the connected parts.

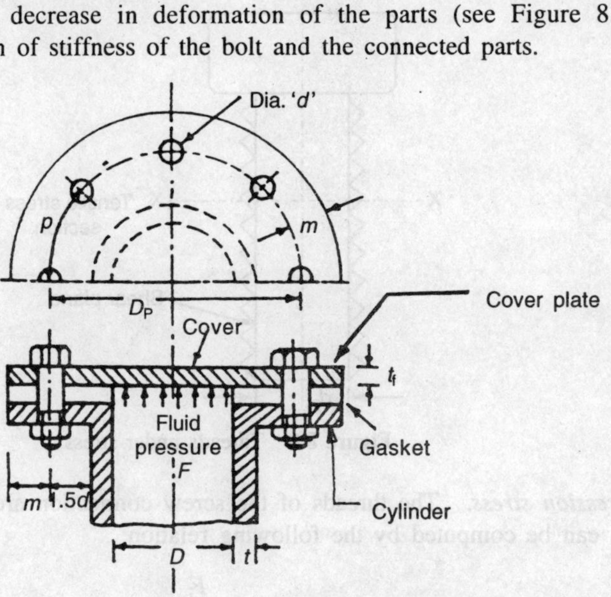

Figure 8.10 Bolted assembly subjected to external force.

Let

$$\text{increase in length of bolt} = \frac{F_b}{k_b} \tag{8.8}$$

and

$$\text{deformation of part} = \frac{F_p}{k_p} \tag{8.9}$$

where

F_b = force shared by the bolt

F_p = force shared by the part

$$k_b = \text{stiffness of the bolt} \left(= \frac{A_b E_b}{l_b} \right) \tag{8.10}$$

$$k_p = \text{stiffness of the part} \left(= \frac{A_p E_p}{l_p} \right) \tag{8.11}$$

and A, E, and l are cross-sectional area, modulus of elasticity, and length, respectively.

These two changes in deformation should be equal so that the parts are not separated. Therefore,

$$\frac{F_b}{k_b} = \frac{F_p}{k_p} \tag{8.12}$$

and the total external force is sum of the two individual forces

$$F = F_b + F_p \tag{8.13}$$

Solving Eqs. (8.12) and (8.13), we get

$$\text{Force shared by the bolt, } F_b = \frac{k_b}{k_b + k_p} F \tag{8.14}$$

and

$$\text{Force shared by the part, } F_p = \frac{k_p}{k_b + k_p} F \tag{8.15}$$

Thus, the total force on the bolt

$$F_{b,t} = F_b + F_i$$

$$= \frac{k_b}{k_b + k_p} F + F_i$$

or

$$F_{b,t} = kF + F_i \tag{8.16}$$

The term k is called the gasket factor and may have value between 0 and 1. Typical values of gasket factor for various types of gaskets are given in Table 8.2.

Table 8.2 Typical values of gasket factor (k)

Connection and type of gasket	k
Soft packing with studs	1.0
Soft packing with through bolt	0.75
Asbestos	0.60
Soft copper gasket with through bolt	0.50
Hard copper gasket with through bolt	0.25
Metal to metal joint	0.00

In many practical applications, the bolted members, as shown in Figure 8.11, are

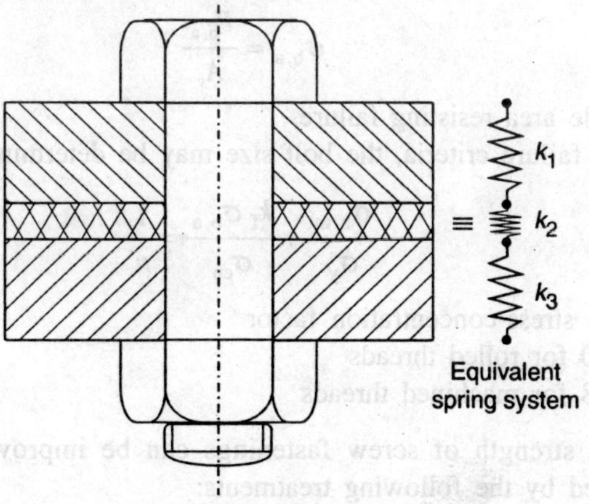

Figure 8.11 Bolted assembly and its equivalent spring system.

composed of two or more different types of materials. For example, a copper gasket is used between two steel plates. In such cases, the stiffness of part k_p is given by the relation

$$\frac{1}{k_p} = \frac{1}{k_1} + \frac{1}{k_2} + \ldots \tag{8.17}$$

where k_1, k_2, ... etc. are stiffnesses of the individual components.

8.7 BOLTS SUBJECTED TO VARIABLE LOAD

When a bolted joint is subjected to an external variable load, as observed in the connecting rod and cylinder head bolts of an I.C. engine, the Soderberg equation of fatigue failure can be used.

Let us assume that the bolt is subjected to a load varying from F_{max} to F_{min}. The total load on the bolt will be

$$F_{b,\,max} = kF_{max} + F_i$$

$$F_{b,\,min} = kF_{min} + F_i$$

The mean and amplitude forces on the bolt will be

$$F_{b,\,m} = \frac{F_{b,\,max} + F_{b,\,min}}{2}$$

$$F_{b,\,a} = \frac{F_{b,\,max} - F_{b,\,min}}{2}$$

The mean stress induced in the bolt

$$\sigma_{b,\,m} = \frac{F_{b,\,m}}{A_s} \tag{8.18}$$

and the amplitude stress induced in the bolt

$$\sigma_{b,\,a} = \frac{F_{b,\,a}}{A_s} \tag{8.19}$$

where A_s is the tensile area resisting failure.
Using the Soderberg failure criteria, the bolt size may be determined by the equation

$$\frac{\sigma_{b,\,m}}{\sigma_y} + \frac{k_f\,\sigma_{b,\,a}}{\sigma_{en}} = \frac{1}{n} \tag{8.20}$$

where k_f = fatigue stress-concentration factor
= 2.2–3.0 for rolled threads
= 2.8–3.8 for machined threads

The endurance strength of screw fastenings can be improved and chances of fatigue failure can be reduced by the following treatments:

1. The high initial or pre-load on the bolt improves the endurance strength.

2. The stress concentration in the bolt should be reduced by the following treatments to improve the fatigue strength.
   (a) A large fillet radius should be provided between the shank and head of the bolt.
   (b) Threads should be rolled instead of cut.
   (c) A round circumferential groove immediately after the thread should be provided.
   (d) The shank diameter should be equal to the root diameter of the thread.
3. Nitriding of the bolt also improves the endurance strength.
4. A corrosion resisting coating is also helpful in improving the fatigue strength of the bolt.

8.8 OTHER DESIGN CONSIDERATIONS

The following considerations should be taken into account while designing bolts.

1. Bolts are designed as tensile members and they should not be subjected to shear force. However if shear force is unavoidable, as found in the case of flange coupling, the bolts should be fitted into reamed holes.
2. In order to permit the use of the wrench for screwing or unscrewing, generally the pitch and margin should be decided by the following relations:

$$\text{pitch: } p \geq 3.25d + 2 \text{ mm} \tag{8.21}$$

$$\text{margin: } m \geq 1.625d + 2 \text{ mm} \tag{8.22}$$

3. In the case of a fluid-tight joint, the pitch should not exceed $30\sqrt{d_1}$ so that uniform contact pressure on gasket can be maintained.
4. The nut height should be at least equal to the diameter of the bolt.

Example 8.1 The cylinder head of a steam engine is held in position by 12 studs. The cylinder bore is 500 mm and the maximum pressure is 1.2 N/mm². A copper gasket is used to make the joint steam-tight. Select a suitable size of the stud. Assume appropriate initial tightening force, material of the studs and factor of safety.

Solution Number of studs, $N = 12$

Cylinder bore, $D = 500$ mm

Maximum pressure, $p_{max} = 1.2$ N/mm^2

For hard copper gasket, $k = 0.25$

Maximum external force on 12 studs due to steam pressure

$$P = \frac{\pi}{4} \times 500^2 \times 1.2 = 235.62 \text{ kN}$$

Load on each stud, $F = \dfrac{P}{N} = \dfrac{235.62}{12} = 19.635$ kN

Assuming initial tightening force, $F_i = 2.805d$ kN

$$\text{Total force on the bolt} = kF + F_i$$
$$= 0.25 \times 19.635 + 2.805d$$
$$= (4.9088 + 2.805d)\ \text{kN}$$

If the material of the bolt is C25 steel for which the yield strength is 275 N/mm^2 and the factor of safety is 2, then the allowable strength

$$\sigma_t = \frac{\sigma_y}{\text{FoS}} = \frac{275}{2} = 137.5\ \text{N/mm}^2$$

$$\text{Induced stress} = \frac{(4.9088 + 2.805d) \times 10^3}{A_s} \le 137.5\ \text{N/mm}^2$$

By hit-and-trial method, we select the M36 × 2 size stud having stress area $A_s = 915$ mm^2 (see Table 8.1).

Example 8.2 A steam engine cylinder has an effective diameter of 350 mm and the maximum steam pressure acting on the cylinder cover is 1.25 N/mm^2. Calculate the number and the size of studs required to fix the cylinder cover. Assume the permissible stress in the studs to be 70 N/mm^2.

Solution Effective cylinder dia, $D = 350$ mm

Steam pressure, $p_{max} = 1.25$ N/mm^2

$\sigma_t = 70$ N/mm^2

$$\text{Steam force} = \frac{\pi}{4} \times D^2 \times p_{max}$$
$$= \frac{\pi}{4} \times 350^2 \times 1.25 = 120.26\ \text{kN}$$

$$\text{Resisting area} = \frac{\text{steam force}}{\sigma_t} = \frac{120.26 \times 10^3}{70} = 1718\ \text{mm}^2$$

From Table 8.1, we choose M18 × 1.5 bolt having stress area of 216 mm^2. Therefore,

$$\text{Number of studs} = \frac{\text{resisting area}}{\text{tensile area of one bolt}} = \frac{1718}{216} = 7.95, \text{ say } 8$$

Let us take stud hole size, d_1 = stud size + 1 mm = 19 mm

$$\text{Pitch circle diameter, } d_p = D + 3d_1$$
$$= 350 + 3 \times 19 = 407\ \text{mm}$$

$$\text{Pitch of the stud layout} = \frac{\pi d_p}{N}$$
$$= \frac{\pi \times 407}{8} = 159.8\ \text{mm}$$

However for a leakproof joint, the pitch of the stud layout should be less than $30\sqrt{d_1}$, i.e.

130.76 mm. The calculated pitch 159.8 mm is more than the maximum limit. Hence there is a chance of leakage.

Let us increase the number of studs to 10, though from strength considerations 8 studs are sufficient.

$$\text{Revised pitch} = \frac{\pi \times 407}{10} = 127.8 \text{ mm}$$

which is reasonable for a leakproof joint. Hence the design specifications are:

(i) Stud size = M18 × 1.5
(ii) Number of studs = 10
(iii) Pitch of the stud layout = 127.8 mm

Example 8.3 An air compressor cylinder of effective diameter 300 mm is subjected to air pressure of 1.5 N/mm². The cylinder head is connected by means of 8 bolts having yield strength of 350 N/mm² and endurance limit of 240 N/mm². The bolts are tightened with an initial pre-load force of 1.5 times that of the external force. A copper gasket is used to make the joint leakproof. Assume stress concentration factor of 2.5 and factor of safety of 2. Determine the required size of the bolt.

Solution Cylinder diameter, D = 300 mm
Gas (air) pressure, p_{max} = 1.5 N/mm²
Number of bolts, N = 8
σ_y = 350 N/mm²
σ_{en} = 240 N/mm²
$F_i = 1.5F$, where F is the gas force.
k_f = 2.5

Factor of safety, n = 2

For hard copper gasket, k = 0.25

$$\text{Maximum gas force} = \frac{\pi}{4} \times D^2 \times p_{max}$$

$$= \frac{\pi}{4} \times 300^2 \times 1.5 = 106.03 \text{ kN}$$

$$\text{External force per bolt, } F = \frac{106.03}{8} = 13.26 \text{ kN}$$

Initial tension, $F_i = 1.5F = 1.5 \times 13.26 = 19.89$ kN

External force on the bolt varies from 0 to 13.26 kN, or in other words,

$$F_{max} = 13.26 \text{ kN} \quad \text{and} \quad F_{min} = 0 \text{ kN}$$

$$\text{Maximum force on a bolt, } F_{b,\,max} = kF_{max} + F_i$$

$$= 0.25 \times 13.26 + 19.89 = 23.205 \text{ kN}$$

$$\text{Minimum force on a bolt, } F_{b,\,min} = kF_{min} + F_i$$

$$= 0.25 \times 0 + 19.89 = 19.89 \text{ kN}$$

$$\text{Mean force on a bolt, } F_{b,m} = \frac{F_{b,max} + F_{b,min}}{2}$$

$$= \frac{23.205 + 19.89}{2} = 21.5475 \text{ kN}$$

$$\text{Amplitude force on a bolt, } F_{b,a} = \frac{F_{b,max} - F_{b,min}}{2}$$

$$= \frac{23.205 - 19.89}{2} = 1.6575 \text{ kN}$$

Using the Soderberg equation for the fluctuating force

$$\frac{\sigma_{b,m}}{\sigma_y} + \frac{k_f\ \sigma_{b,a}}{\sigma_{en}} = \frac{1}{n}$$

where

$\sigma_{b,m}$ is the mean stress $\left(= \dfrac{F_{b,m}}{A_s} \right)$

$\sigma_{b,a}$ is the amplitude stress $\left(= \dfrac{F_{b,a}}{A_s} \right)$

A_s is the stress area

Substituting the values of the mean and amplitude stresses in the Soderberg equation and simplifying, we get

$$A_s = \frac{F_{b,m} + \left(\dfrac{\sigma_y}{\sigma_{en}} \right) \times k_f \times F_{b,a}}{(\sigma_y / n)}$$

$$= \frac{21547.5 + \left(\dfrac{350}{240} \right) \times 2.5 \times 1657.5}{350/2}$$

$$= 157.6 \text{ mm}^2$$

For tensile stress area of 157.6 mm^2, the M16 × 1.5 bolt having stress area of 167 mm^2 is suitable (see Table 8.1).

8.9 ECCENTRIC LOADING

In many engineering applications the bolted joints are subjected to eccentric loading, e.g. a bracket attached to the wall, a pillar crane, steel structural members, and so forth. In general, there are three different ways by which a bolted joint may be subjected to eccentric loading. These are:

(i) Force acting parallel to the bolt axis

(ii) Force acting perpendicular to the bolt axis
(iii) Force acting in the plane containing the bolts.

These three cases of eccentric loading are discussed below

8.9.1 Force Acting Parallel to the Bolt Axis

The bracket shown in Figure 8.12 is a typical example of an eccentrically loaded bolted joint in which the force acts parallel to the bolt axis. The applied force tends to turn the bracket about the edge X-X, thus stretching each bolt to a varying degree which depends on its distance from the tilting edge.

Let us consider a bracket shown in Figure 8.12 which carries a force F acting at a distance L from the edge of the bracket. Let F_1 and F_2 be the forces acting on the bolts situated at distances l_1 and l_2 respectively from the tilting edge X-X. Since the stresses induced in the bolts are directly proportional to the elongation which in turn is proportional to the distances of the bolts from the tilting edge X-X, the force acting on the bolt can be expressed as

$$F_2 = F_1\left(\frac{l_2}{l_1}\right) \tag{8.23}$$

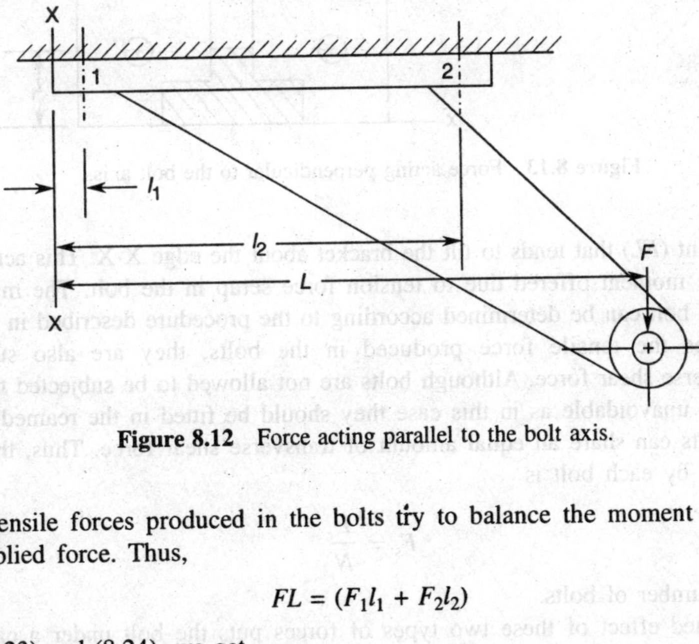

Figure 8.12 Force acting parallel to the bolt axis.

These tensile forces produced in the bolts try to balance the moment produced by the externally applied force. Thus,

$$FL = (F_1 l_1 + F_2 l_2) \tag{8.24}$$

From Eqs. (8.23) and (8.24), we get

$$F_1 = \frac{FLl_1}{l_1^2 + l_2^2} \tag{8.25}$$

$$F_2 = \frac{FLl_2}{l_1^2 + l_2^2} \tag{8.26}$$

The maximum force out of F_1 and F_2 should be considered as the external design force on the bolt.

8.9.2 Force Acting Perpendicular to the Bolt Axis

In certain applications, namely a bracket attached to the wall by means of bolts as shown in Figure 8.13, the external force F acts perpendicular to the axis of the bolt. In such a case, the external force F creates two effects:

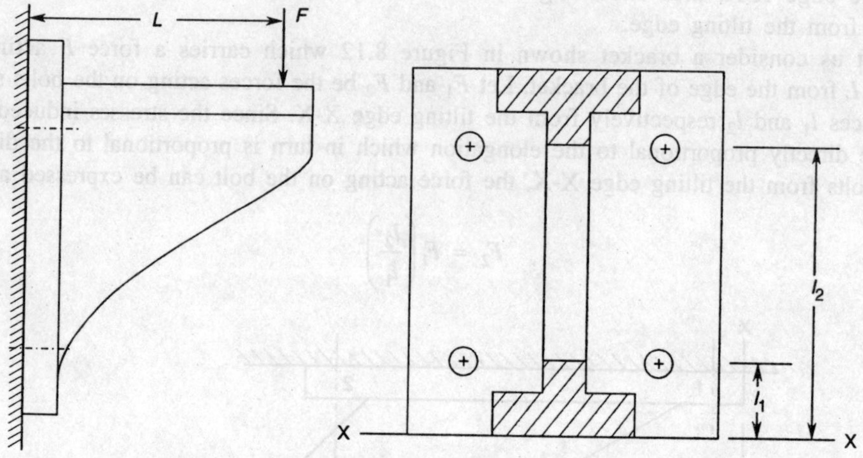

Figure 8.13 Force acting perpendicular to the bolt axis.

(a) Moment (FL) that tends to tilt the bracket about the edge X-X. This action is resisted by the moment offered due to tension force setup in the bolt. The maximum force on the bolt can be determined according to the procedure described in Section 8.9.1.

(b) Besides the tensile force produced in the bolts, they are also subjected to a transverse shear force. Although bolts are not allowed to be subjected to shear force, if it is unavoidable as in this case they should be fitted in the reamed holes so that all bolts can share an equal amount of transverse shear force. Thus, the shear force shared by each bolt is

$$F_s = \frac{F}{N} \tag{8.27}$$

where N is the number of bolts.

The combined effect of these two types of forces puts the bolt under a case of biaxial loading. Thus, according to the maximum principal stress theory of failure, the equivalent tensile force can be computed by the relation,

$$F_{eq} = \frac{1}{2}\left[F_i + \sqrt{F_i^2 + 4F_s^2}\right] \tag{8.28}$$

where F_i is the maximum tensile force in the ith number bolt.

The equivalent force F_{eq} should be taken as the design force to determine the size of the bolt. All bolts should be of the same size for good interchangeability.

8.9.3 Force Acting in the Plane Containing the Bolts

A plate carrying external force F, fastened to a vertical column by means of certain number of bolts, as shown in Figure 8.14, is a typical example of force acting in the plane containing the bolts. In such a case, the force taken by each bolt can be determined by the procedure described in Section 7.7, for eccentrically loaded riveted joints. The reader is advised to refer to this section to determine the maximum design force on the bolt. The computer program given in Chapter 7 (Program 7.1) may be used for these types of joints as well.

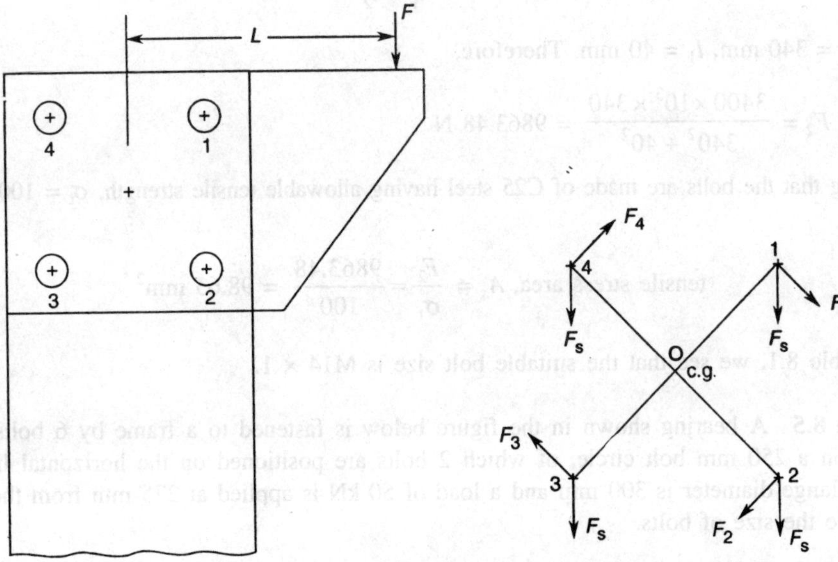

Figure 8.14 Force acting in the plane containing the bolts.

Example 8.4 Find a suitable size of the bolt to be used for fixing the J hanger bracket to the ceiling as shown in the figure below. The maximum external force is 10 kN.

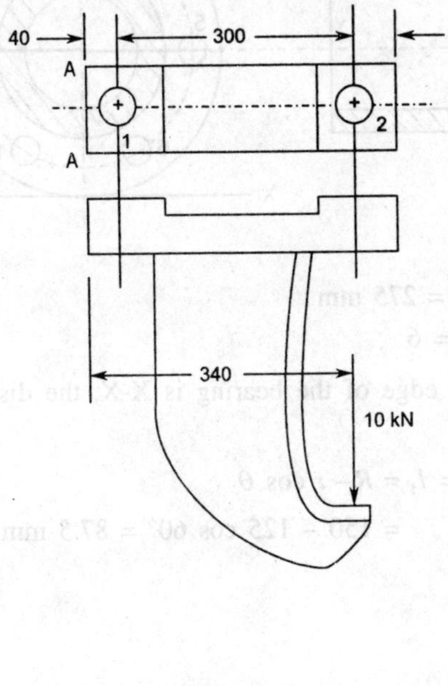

Solution Moment about the tilting edge A-A

$$M = FL$$

$$= 10 \times 340 = 3400 \text{ kN} \cdot \text{mm}$$

If F_2 is the load in the bolt number 2 located farther from the tilting edge, then

$$F_2 = \frac{FLl_2}{l_1^2 + l_2^2}$$

where l_2 = 340 mm, l_1 = 40 mm. Therefore,

$$F_2 = \frac{3400 \times 10^3 \times 340}{340^2 + 40^2} = 9863.48 \text{ N}$$

Assuming that the bolts are made of C25 steel having allowable tensile strength, σ_t = 100 N/mm², then

$$\text{tensile stress area, } A_s = \frac{F_2}{\sigma_t} = \frac{9863.48}{100} = 98.63 \text{ mm}^2$$

From Table 8.1, we see that the suitable bolt size is M14 × 1.

Example 8.5 A bearing shown in the figure below is fastened to a frame by 6 bolts spaced equally on a 250 mm bolt circle, of which 2 bolts are positioned on the horizontal line. The bearing flange diameter is 300 mm and a load of 50 kN is applied at 275 mm from the frame. Determine the size of bolts.

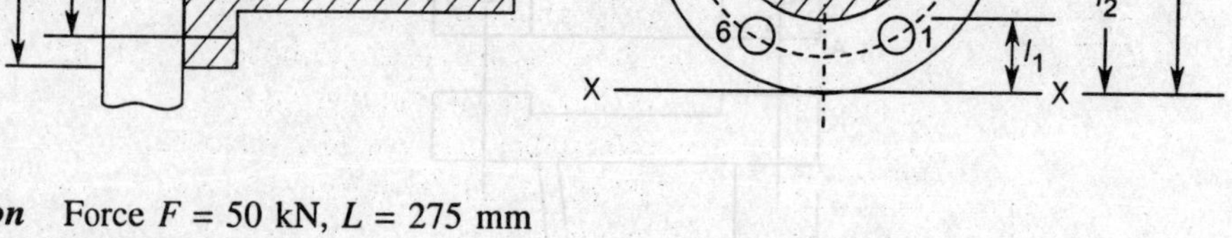

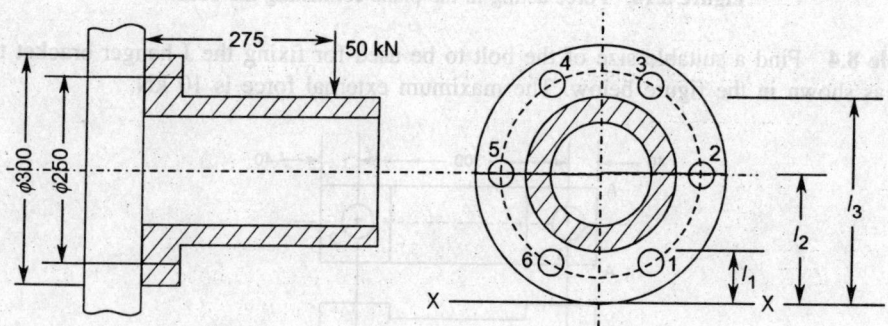

Solution Force F = 50 kN, L = 275 mm

Number of bolts, N = 6

Assuming that the tilting edge of the bearing is X-X, the distances l_1, l_2, . . ., etc. are expressed as

$$l_6 = l_1 = R - r \cos \theta$$

$$= 150 - 125 \cos 60° = 87.5 \text{ mm}$$

$$l_5 = l_2 = R = 150 \text{ mm}$$

$$l_4 = l_3 = R + r\cos\theta$$

$$= 150 + 125\cos 60° = 212.5 \text{ mm}$$

Since the load shared by the bolt is proportional to its distance from the tilting edge, therefore, the maximum force will be experienced by bolt number 3 and 4

$$F_4 = F_3 = \frac{FLl_3}{2\left(l_1^2 + l_2^2 + l_3^2\right)}$$

$$= \frac{50 \times 275 \times 212.5}{2(87.5^2 + 150^2 + 212.5^2)} = 19.4 \text{ kN}$$

Assuming that the bolt is made of C20 steel having yield strength of 245 N/mm^2 and factor of safety 3, the allowable strength

$$\sigma_t = \frac{\sigma_y}{\text{FoS}} = \frac{245}{3} = 81.67 \text{ N/mm}^2$$

$$\text{Tensile stress area} = \frac{F_3}{\sigma_t} = \frac{19.4 \times 1000}{81.67} = 237.54 \text{ mm}^2$$

From Table 8.1, we see that the suitable bolt size is M20 × 1.5.

Example 8.6 A steel bracket, as shown in the figure below, is secured to a wall by means of four steel bolts. The load on the bracket is 15 kN which acts at a distance of 175 mm from the wall. Determine the size of the bolt.

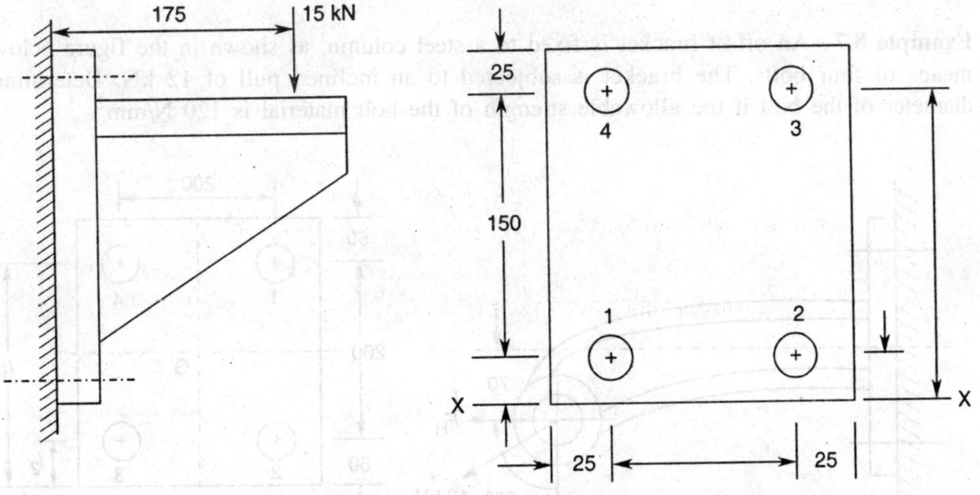

Solution Force $F = 15$ kN

Moment arm length, $L = 175$ mm

The bracket tends to tilt about the X-X edge. Maximum force will be shared by the bolts numbered 3 and 4 as they are farthest from the tilting edge.

Direct shear force, $F_s = \dfrac{F}{N} = \dfrac{15}{4} = 3.75 \text{ kN}$

The distance from the tilting edge X-X to:

bolt numbered 1 and 2, $l_1 = l_2 = 25$ mm

bolt numbered 3 and 4, $l_4 = l_3 = 175$ mm

Tensile force in bolts numbered 3 and 4

$$F_4 = F_3 = \frac{FL \times l_3}{2(l_1^2 + l_3^2)}$$

$$= \frac{15 \times 175 \times 175}{2(25^2 + 175^2)} = 7.35 \text{ kN}$$

$$\text{Equivalent force, } F_{eq} = \frac{1}{2}\left[F_3 + \sqrt{F_3^2 + 4\,F_s^2}\right]$$

$$= 0.5\left[7.35 + \sqrt{7.35^2 + 4 \times 3.75^2}\right] = 8.925 \text{ kN}$$

Assuming that the bolts are made of C35 steel having allowable strength, $\sigma_t = 120$ N/mm^2

$$\text{Tensile stress area, } A_s = \frac{8.925 \times 10^3}{120} = 74.37 \text{ mm}^2$$

From Table 8.1, we see that M12 size bolt with 1.5 mm coarse pitch is suitable.

Example 8.7 An offset bracket is fixed to a steel column, as shown in the figure below, by means of four bolts. The bracket is subjected to an inclined pull of 12 kN. Determine the diameter of the bolt if the allowable strength of the bolt material is 120 N/mm^2.

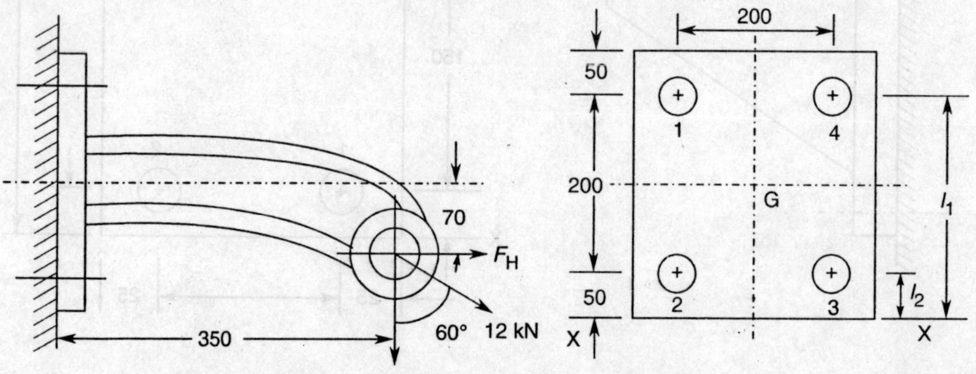

Solution Load F = 12 kN inclined at 60° to vertical

Number of bolts, $N = 4$

Offset of load centre = 70 mm

Moment arm length, L = 350 mm

Since bolts are arranged symmetrically, the geometric centre G is at middle of the length and width of the plate.

The inclined force F has two components:

Horizontal force, $F_H = F \sin 60° = 12 \sin 60° = 10.39$ kN

Vertical force, $F_V = F \cos 60° = 12 \cos 60° = 6$ kN

The horizontal component acts parallel to the axis of the bolt but is eccentric, therefore, it will produce a direct tensile force and a turning moment in the anticlockwise direction. Thus,

$$\text{Direct tensile force, } F_{H,\,t} = \frac{10.39}{4} = 2.5975 \text{ kN}$$

and

$$\text{Turning moment, } M_H = 10.39 \times 70 = 727.3 \text{ kN} \cdot \text{mm}$$

Similarly, the vertical component produces a transverse shear force and a turning moment in the clockwise direction. Thus,

$$\text{Direct shear force, } F_{V,\,s} = \frac{6}{4} = 1.5 \text{ kN}$$

$$\text{Turning moment, } M_V = 6 \times 350 = 2100 \text{ kN} \cdot \text{mm}$$

Therefore,

$$\text{Net turning moment, } M = 2100 - 727.3 = 1372.7 \text{ kN} \cdot \text{mm}$$

The bracket will tend to tilt about the edge X-X due to this net turning moment. The bolts numbered 1 and 4 will be subjected to maximum force as they are farthest from the edge. The resisting force in the bolt

$$F_4 = F_1 = \frac{M l_1}{2(l_1^2 + l_2^2)}$$

where $l_1 = 50 + 200 = 250$ mm and $l_2 = 50$ mm. Thus,

$$F_1 = \frac{1372.7 \times 10^3 \times 250}{2(250^2 + 50^2)} = 2639.8 \text{ N}$$

Total tensile force on the bolt

$$F_{\text{total}} = F_1 + F_{H,\,t} = 2639.8 + 2597.5 = 5237.3 \text{ N}$$

$$\text{Now equivalent load, } F_{\text{eq}} = 0.5\left[F_{\text{total}} + \sqrt{F_{\text{total}}^2 + 4F_s^2}\right]$$

$$= 0.5\left[5237.3 + \sqrt{5237.3^2 + 4 \times 1500^2}\right]$$

$$= 5636.48 \text{ N}$$

$$\text{Tensile stress area} = \frac{F_{eq}}{\sigma_t} = \frac{5636.48}{120} = 46.97 \text{ mm}^2$$

From Table 8.1, we observe that the suitable bolt size is M10 with 1.5 mm pitch.

8.10 POWER SCREWS

Power screws are also called *translation screws.* They are used to transmit motion or power by the use of helical translatory motion of the screw threads. Power screws may also be used as linear actuators which transform rotary motion into linear motion. The general kinematics of the power screw are the same as those of the ordinary nut and bolt, the only difference being the geometry of the threads. Typical applications of power screws are: (i) automotive vehicle jack, (ii) screw type press, (iii) lead screw of lathe machine, (iv) valve stems, and (v) nuclear reactor control device.

The power screws generally have an efficiency of the order of 40–70 per cent depending upon the helix angle of the thread and the coefficient of sliding friction μ between the nuts and screws. However, high transmission efficiency, usually 90 per cent or more, can be obtained by ball screws.

8.10.1 Forms of Thread

The most commonly used forms of thread for power screws are: (i) Acme thread, (ii) square thread, (iii) modified square thread, and (iv) buttress thread. The specifications for these threads are given in IS: 4694–1968 and IS: 7008 (Parts I–IV)–1973.

Acme thread. Acme threads are the oldest type of power screw threads. Acme thread is a trapezoidal-shaped thread with a slight slope given to its sides which lowers its efficiency and also produces radial pressure on the nut. However, it increases the base area which improves its shear strength. The Vee angle of the acme thread is 29° which is large enough to permit the use of an adjustable split nut for wear compensation while not so large as to decrease the efficiency of the drive. These threads can be produced economically by the use of taps, dies, thread milling, and rolling. The standard proportions of acme thread are shown in Figure 8.15(a).

Square thread. A thread with sides perpendicular to the thread axis is known as the *square thread.* These threads have 0° included angle which results into maximum transmission efficiency and minimum radial pressure on the nut. However, it has comparatively poor mechanical advantage; it is difficult to machine and cannot be easily compensated for wear by the use of the split nut. Square threads are useful in all cases where the load is applied in both directions of the axis as in the case of the screw jack. Standard proportions are shown in Figure 8.15(b). The basic dimensions of square threads, coarse series, are listed in Table 8.3.

Modified square thread. The modified square threaded screw is a trapezoidal threaded screw with 10° included angle. Since its included angle is small, its transmission efficiency for most practical purposes is equivalent to that of the true square thread without the disadvantage of the square thread. The standard proportions are shown in Figure 8.15(c).

Buttress thread. The buttress thread is useful in those applications which have to bear a large force acting along the axis in one direction only. A nearly perpendicular thrust surface results in low radial pressure and the transmission efficiency is approximately equivalent to that of the square thread. The standard proportions of the buttress thread are shown in Figure 8.15(d).

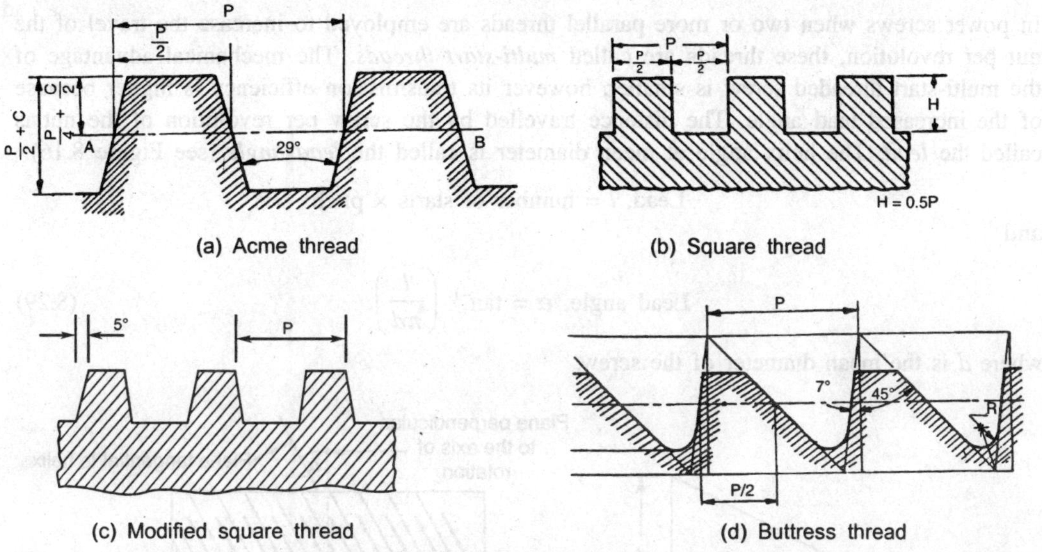

(a) Acme thread (b) Square thread

(c) Modified square thread (d) Buttress thread

Figure 8.15 Forms of power screw threads.

Table 8.3 Dimensions of square threads [IS: 4694–1968]

Nominal diameter	*Major diameter* *Bolt* d	*Nut* D	*Minor diameter* d_2	*Pitch* p
22	22	22.5	14	
24	24	24.5	16	8
26	26	26.5	18	
28	28	28.5	20	
30	30	30.5	20	
32	32	32.5	22	10
36	36	36.5	26	
40	40	40.5	28	
44	44	44.5	32	
48	48	48.5	36	12
50	50	50.5	38	
52	52	52.5	40	
55	55	55.5	41	14
60	60	60.5	46	
65	65	65.5	49	
70	70	70.5	54	
75	75	75.5	59	16
80	80	80.5	64	
85	85	85.5	67	
90	90	90.5	72	
95	95	95.5	77	18
100	100	100.5	80	

In power screws when two or more parallel threads are employed to increase the travel of the nut per revolution, these threads are called ***multi-start threads.*** The mechanical advantage of the multi-start threaded screw is smaller, however its transmission efficiency is higher because of the increased lead angle. The distance travelled by the screw per revolution of the nut is called the *lead.* The helix angle at mean diameter is called the *lead angle* (see Figure 8.16).

$$\text{Lead, } l = \text{number of starts} \times \text{pitch}$$

and

$$\text{Lead angle, } \alpha = \tan^{-1}\left(\frac{l}{\pi d}\right) \tag{8.29}$$

where d is the mean diameter of the screw.

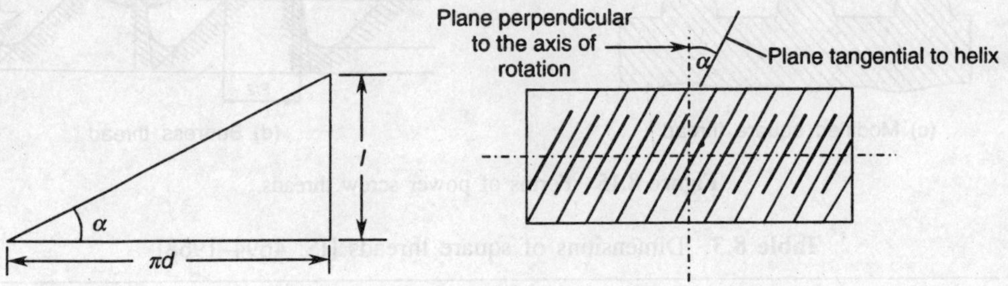

Figure 8.16 Lead angle of thread.

8.11 TORQUE AND EFFICIENCY OF POWER SCREW

In the power screw, the movement between the nut and the screw against the applied axial load is analogous to the movement of a weight on an inclined plane as shown in Figure 8.17. The load on the screw, when transferred to the nut, acts as a distributed load on the surface of the thread in contact. For simplicity it is assumed that the distributed load is concentrated at a point on the mean circumference of the thread

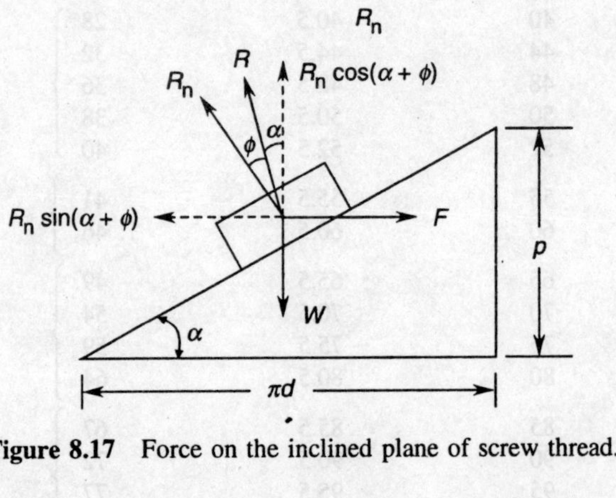

Figure 8.17 Force on the inclined plane of screw thread.

Let

α be the helix or lead angle of the thread
p be the pitch of the thread
d be the mean diameter of the screw
μ be the coefficient of friction
ϕ be the angle of friction ($\tan^{-1}(\mu)$).

Referring to Figure 8.17, under static load conditions, the direction of force R on the thread will be normal to the thread surface. However, when the screw rotates to lift the load, the line of action of force R will be rotated through the angle of friction ϕ, i.e. force R_n is ϕ degrees away from the original position of force R.

For equilibrium of forces, the component of R_n along the weight W is

$$W = R_n \cos(\alpha + \phi) \tag{8.30}$$

The component of R_n at right angles to the axis of the screw which is required to raise the load is

$$F = R_n \sin(\alpha + \phi) \tag{8.31}$$

or

$$F = W \tan(\alpha + \phi) \tag{8.32}$$

The frictional torque required to lift the load at mean radius is

$$T = \frac{Wd}{2} \tan(\alpha + \phi) \tag{8.33}$$

Similarly, the force required to lower the load is

$$F = W \tan(\phi - \alpha) \tag{8.34}$$

Now let T_{ideal} be the ideal torque required to lift the load, i.e. when the coefficient of friction $\mu = 0$ (no friction is present). Therefore,

$$T_{\text{ideal}} = \frac{Wd}{2} \tan\alpha \tag{8.35}$$

Thus the transmission efficiency of the power screw can be expressed as

$$\eta = \frac{\text{ideal torque}}{\text{frictional torque}} = \frac{T_{\text{ideal}}}{T}$$

$$= \frac{\dfrac{Wd}{2}\tan\alpha}{\dfrac{Wd}{2}\tan(\phi + \alpha)}$$

$$= \frac{\tan\alpha}{\tan(\phi + \alpha)} \tag{8.36}$$

Equation (8.36) can be used for the acme thread by substituting ϕ by $\tan^{-1}(\mu/\cos\beta)$, where β is the semi-thread angle of the acme thread. It is obvious from Eq. (8.36) that the efficiency of the square-threaded screw depends on the helix angle α and friction angle ϕ. For maximum efficiency, $d\eta/d\alpha$ must be equal to zero, which is possible if $\alpha = 45° - \phi/2$. Hence the maximum efficiency of the square-threaded screw is given by

$$\eta_{max} = \frac{1 - \sin\phi}{1 + \sin\phi} \tag{8.37}$$

Figure 8.18 shows two curves plotted between efficiency and helix angle, α for different values of coefficient of friction. It is noticed from this figure that the transmission efficiency increases rapidly up to 20° helix angle and then the increment rate is low. The transmission efficiency is more or less maximum at approximately 30–45° helix angle. Thus, in a power screw not much is gained by increasing the helix angle above 30° and moreover it becomes difficult to cut threads with large helix angles.

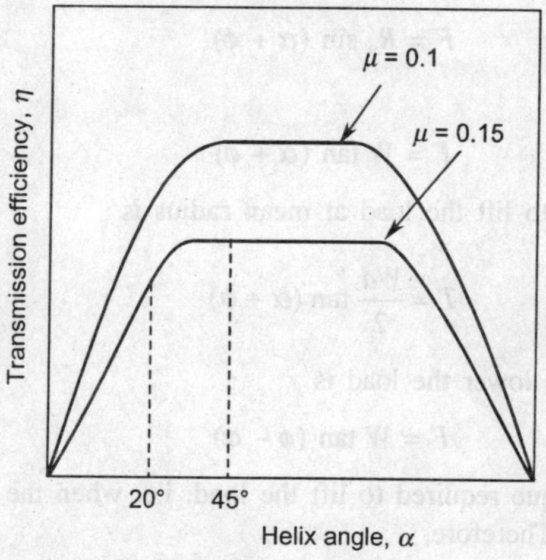

Figure 8.18 Transmission efficiency vs. helix angle.

8.11.1 Self-locking Screws

The torque required to lower the load is given as

$$T = \frac{Wd}{2} \tan(\phi - \alpha) \tag{8.38}$$

If in the above equation the coefficient of friction between the nut and screw is small or the helix angle is so large that the condition $\phi < \alpha$ is satisfied, then the torque required to lower the load will be negative, i.e. effort will be required to resist the tendency of the load to descend. Such a condition is known as *overhauling* of the screw. On the other hand if $\phi > \alpha$, then the torque required to lower the load will be positive. Such a screw is known as the *self-locking* screw.

8.11.2 Collar Friction

In the power screw, the axial force required to be raised is also resisted by a collar along with screw threads as shown in Figure 8.19. Thus, an additional torque is required to overcome the friction produced by the thrust collar. The frictional torque offered by the collar can be determined by assuming that the frictional force acts at the mean diameter of the collar and is given by the relation

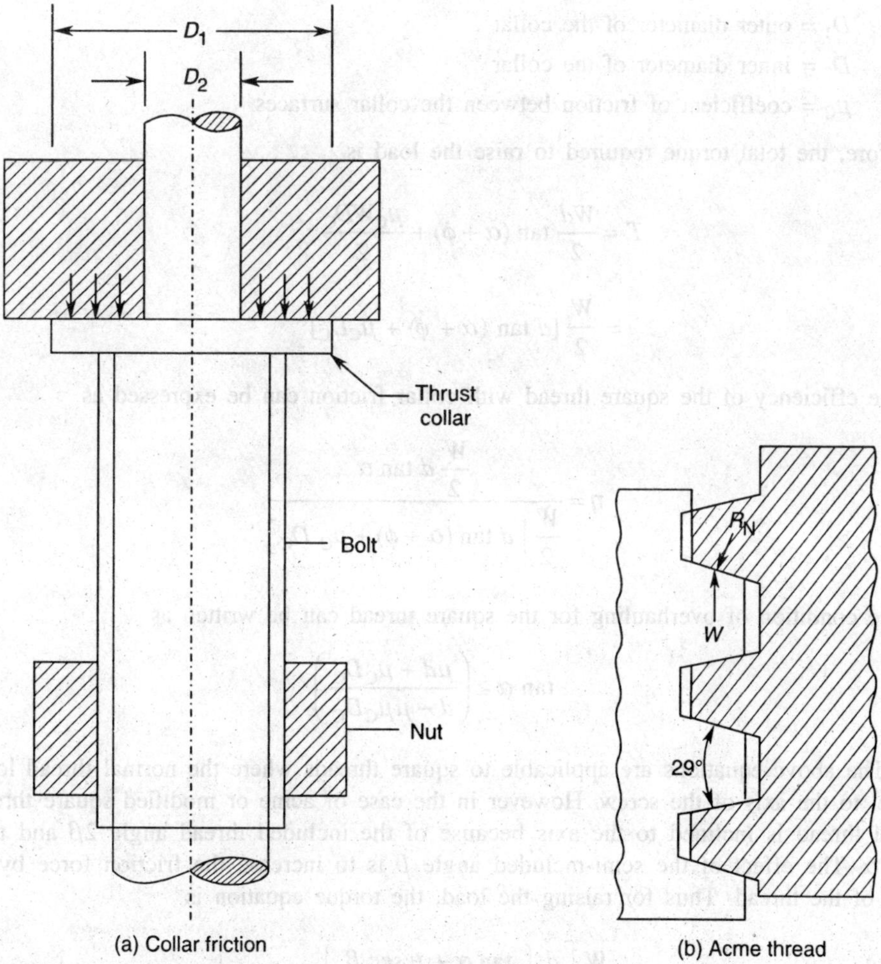

Figure 8.19 Collar friction in acme thread.

$$T_C = \frac{\mu_C W D_C}{2} \tag{8.39}$$

where

D_C = mean diameter of the thrust collar

$$= \frac{D_1 + D_2}{2} \quad \text{for the uniform wear rate theory}$$

$$= \frac{2}{3}\left(\frac{D_1^3 - D_2^3}{D_1^2 - D_2^2}\right) \quad \text{for the constant pressure theory}$$

with

D_1 = outer diameter of the collar

D_2 = inner diameter of the collar

μ_C = coefficient of friction between the collar surfaces.

Therefore, the total torque required to raise the load is

$$T = \frac{Wd}{2}\tan(\alpha + \phi) + \frac{\mu_C W D_C}{2}$$

$$= \frac{W}{2}[d \tan(\alpha + \phi) + \mu_C D_C] \tag{8.40}$$

and the efficiency of the square thread with collar friction can be expressed as

$$\eta = \frac{\frac{W}{2} d \tan \alpha}{\frac{W}{2}\left[d \tan(\alpha + \phi) + \mu_C D_C\right]} \tag{8.41}$$

and the condition of overhauling for the square thread can be written as

$$\tan \alpha \geq \left(\frac{\mu d + \mu_C D_C}{d - \mu \mu_C D_C}\right) \tag{8.42}$$

The above equations are applicable to square threads where the normal thread loads are parallel to the axis of the screw. However in the case of acme or modified square thread, the normal thread is inclined to the axis because of the included thread angle 2β and the lead angle α. The effect of the semi-included angle β is to increase the friction force by wedge action of the thread. Thus for raising the load, the torque equation is

$$T = \frac{W}{2}\left[\frac{d}{2}\left\{\frac{\tan \alpha + \mu \sec \beta}{1 - \mu \sec \beta \tan \alpha}\right\} + \mu_C D_C\right] \tag{8.43}$$

and the efficiency equation for the acme thread is given as

$$\eta = \frac{d \tan \alpha}{\left[d\left\{\frac{\tan \alpha + \mu \sec \beta}{1 - \mu \sec \beta \tan \alpha}\right\} + \mu_C D_C\right]} \tag{8.44}$$

8.12 STRESS ANALYSIS OF POWER SCREW

The stress analysis of power screw and nut is based upon the following assumptions:

1. It is commonly assumed that the load carried by the screw and the nut is uniformly distributed throughout the thread engagement. However, in actual practice Goodier has shown that due to elastic deflection of threads only the first few threads in engagement carry the major portion of the load. The remaining threads carry a small portion of the load depending upon the elastic deformation of threads.
2. It is also assumed that the bearing pressure is uniformly distributed throughout the engagement, however, this is possible only if proper lubrication is provided.
3. The effect of additional factors such as fillet radii, surface finish, class of fit, etc. on actual stress distribution is neglected.

The various stresses produced in a power screw are discussed below:

Direct stress. When a power screw is subjected to an external load W, the direct tensile or compressive stresses may be produced which depend upon the type of service and method of mounting. The magnitude of the direct stress is equal to the load divided by the area at the root of the thread. In the absence of fillets in the threads, the stress concentration factor may be assumed to be 1.5 to 1.75.

Bearing pressure. The bearing pressure between the surfaces of the screw threads and the contacting surface of the nut is basically the crushing stress between the lubricated surfaces (Figure 8.20). Although it is not uniform, such an assumption simplifies the design and is therefore generally made. The inaccuracy due to this assumption is taken into account by using low values for bearing pressure. The required number of threads N can be found from the relation

$$N = \frac{4W}{p_b \pi \left(d_1^2 - d_2^2\right)} \tag{8.45}$$

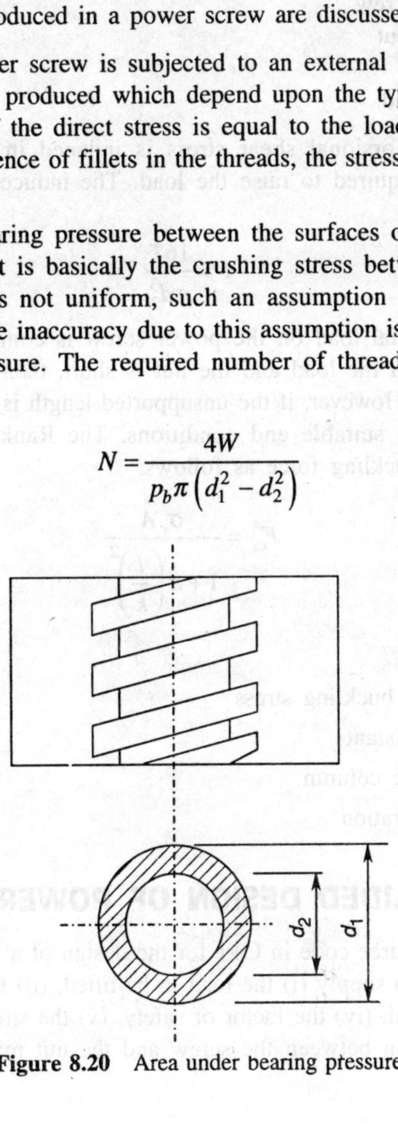

Figure 8.20 Area under bearing pressure.

where

d_1 is the major diameter
d_2 is the minor diameter
p_b is the bearing pressure (5–25 N/mm²).

Transverse shear stress. In power screws, the threads of both the screw and the nut experience a transverse shearing stress. For the rectangular cross-section, the transverse shear stress is given as

$$\tau = \frac{3W}{2A} \tag{8.46}$$

where

A = shear area
= $N\pi d_2 b$ for the screw
= $N\pi d_1 b$ for the nut

with

b = width of the thread.

Torsional shear stress. Torsional shear stress is induced in the screw by the externally applied torque, which is required to raise the load. The induced stress can be found by the relation

$$\tau = \frac{16T}{\pi d_2^3} \tag{8.47}$$

Buckling stress. If the axial load on the power screw is compressive and the unsupported length of the screw between the load and the nut is short, then the design may be based on average compressive stress. However, if the unsupported length is large, then the screw must be designed as a column with suitable end conditions. The Rankine formula may be used to compute the crippling or buckling force as follows:

$$F_{\text{cr}} = \frac{\sigma_c A}{1 + a\left(\frac{l}{k}\right)^2} \tag{8.48}$$

where

σ_c is the crippling or buckling stress
a is the Rankine constant
l is the length of the column
k is the radius of gyration

8.13 COMPUTER-AIDED DESIGN OF POWER SCREW

The Program 8.1 lists the source code in C++ for the design of a bottle-type screw jack. In this program, the user is asked to supply (i) the load to be lifted, (ii) the required lift, (iii) the yield strength of the screw material, (iv) the factor of safety, (v) the strength of the nut material, and (vi) the coefficient of friction between the screw and the nut material. The program assumes that the jack will be operated by one person who can apply 400 N load. The algorithm of the program is illustrated in Example 8.8. The program is well commented. The sample run of the program on the data given in Example 8.8 is demonstrated.

Program 8.1

```
// A PROGRAM TO DESIGN A SCREW JACK (BOTTLE TYPE)
# include <iostream.h>
# include <math.h>
# include <conio.h>
# define PI 3.1415
   double major_dia (double d1);
   double pitch (double p,double d1);
void main ( )
{
   double w, lift, sigmascy, fos, sigmanuty, mu, d1, d2, p, d;
   double alpha, phi, dd, ptch, ts, D1, D2, DC, muc, tr, sigma, dr, h;
   double lever_length, tau, lscrew, a, k, sigmac, moment, sigmab;
   double sigmamax, n, h1, d3, d4, d5, d6, t1, t2, tc, tn, height;
   double sigmad, sigm;
// input design data
   cout << "\n    DESIGN INPUT DATA" << endl;
   cout << "\n    Load to be lifted(kN) = ";
   cin >> w;
   cout << "\n    Lift required (mm) = ";
   cin >> lift;
   cout << "\n    Yield strength of the screw material(N/mm2) = ";
   cin >> sigmascy;
   cout << "\n    Factor of safety(fos) = ";
   cin >> fos;
   cout << "\n    Yield strength of the nut material(N/mm2) = ";
   cin >> sigmanuty;
   cout << "\n    Coefficient of friction(mu) = ";
   cin >> mu;
   sigmad=sigmascy/fos;
   d1=22.0;
   p=5.0
repeat: d2=d1-p;
      sigm=4000.0*w/(PI*d2*d2);
      if (sigm>sigmad)
      {
         dd=major_dia (d1);
         d1=dd;
```

```
            ptch=pitch(p,d1);
            p=ptch;
            goto repeat;
        }
    d=(d1+d2)/2.0;
    alpha=atan(p/(PI*d));
    phi=atan(mu);
// check friction angle is greater than helix angle
    if (alpha > phi)
        {
            dd=major_dia (d1);
            d1=dd;
            ptch=pitch (p,d1);
            p=ptch;
            goto repeat;
        }
// calculate friction torque
    ts=w*d*tan(alpha+phi)/2.0;
// dimensions of collars D1 and D2
    D1=int (1.6*d1+1.0);
    D2=int (0.5*d1+1.0);
    DC=(pow(D1,3)–pow(D2,3))/(pow(D1,2)–pow(D2,2))*(2.0/3.0);
    muc=0.15; //user may change mu between collars
    tc=0.5*muc*w*DC;
    tr=ts+tc;
// assuming allowable strength of the operating lever material
    sigma=150.0; // user may change for different materials
    dr=pow((32.0*tr*1000.0/(PI*sigma)), 0.333);
    dr=int (dr+2.0); //dia of the operating rod
    h=2.0*dr; // height of the screw head
// calculate length of the lever
    lever_length=tr*1000.0/400.0; // assume 400N load can be applied
// one person with 100.0mm for the grip allowance
    lever_length=10.*int((lever_length+5.0)/10.0)+100.0;
// check stress in screw
    tau=16.0*ts*1000.0/(PI*d2*d2*d2); // torsional shear stress
    lscrew=lift+h; // length of the screw
    a=1.95/25000; // Rankine Constant
    k=0.25*d2; // radius of gyration
// buckling stress
    sigmac=w*4000*(1+a*pow((lscrew/k),2.0))/(PI*d2*d2);
// bending stress for load eccentricity of 5.0mm
    moment=w*5.0;
    sigmab=32.0*moment*1000/(PI*d2*d2*d2);
    sigma=sigmac+sigmab;
```

```
   sigmamax=0.5*(sigma+sqrt (sigma*sigma+4.0*tau*tau));
   fos=sigmascy/sigmamax;
   if (fos <2.0)
   {
        dd=major_dia (d1);
     d1=dd;
     ptch=pitch (p,d1);
        p=ptch;
        goto repeat;
   }
// design of nut; find the number of threads using bearing
// consideration. Assume bearing pressure 12.0N/mm2
   n=4.0*w/(PI*(d1*d1-d2*d2)*12.0);
   n=int(n+1.0);
   h1=n*p;
   if (h1<2.0*d1)
   {
        h1=2.0*d1;
        n=int (h1/p+0.5);
        h1=n*p;
   }
// transverse shear stress
   tau=w/(PI*d1*p*n);
   while (tau>25.0)
   {
        n=n+1;
        tau=w/(PI*d1*p*n);
   }
   h1=n*p;
// body dimensions of the nut
   d3=int(1.5*d1+0.5); // outside dia of the nut
   d4=int(1.5*d3+0.5); // top dia of the nut
   tn=int(d3/3.0+1); // thickness of the nut collar
// body dimensions of the bottle
   d5=10*int((1.5*d4+5)/10.0); // top dia of the body
   d6=4*d5; // base dia of the body
   t1=int(0.3*d1+1); // thickness of the body
   t2=2.0*t1; //thickness of the body at the base
   height=lift+25.0; // height of the body
// design report
   cout <<"\n    DESIGN REPORT" << endl;
   cout <<"\n    Major Diameter of the Screw          = "<<d1 <<"mm";
   cout <<"\n    Minor Diameter of the Screw          = "<<d2 <<"mm";
```

```
    cout <<"\n   Mean Diameter of the Screw              = "<<d <<"mm";
    cout <<"\n   Pitch of the Screw                      = "<<p <<"mm";
    cout <<"\n   Screw Head Diameter                     = "<<D1 <<"mm";
    cout <<"\n   Minor Diameter of the Collar            = "<<D2 <<"mm";
    cout <<"\n   Diameter of the Handle Bar              = "<<dr <<"mm";
    cout <<"\n   Height of the Screw Head                = "<<h <<"mm";
    cout <<"\n   Length of the Handle Bar                = "<<lever_length <<"mm";
    cout <<"\n   Top Diameter of the Nut                 = "<<d4 <<"mm";
    cout <<"\n   Outside Diameter of the Nut             = "<<d3 <<"mm";
    cout <<"\n   Thickness of the Nut                    = "<<tn <<"mm";
    cout <<"\n   Number of Threads                       = "<<n;
    cout <<"\n   Height of the Nut                       = "<< h1 <<"mm";
    cout <<"\n   Top Diameter of the Body                = "<<d5 <<"mm";
    cout <<"\n   Base Diameter of the Body               = "<<d6 <<"mm";
    cout <<"\n   Height of the Body                      = "<<height <<"mm";
    cout <<"\n   Thickness of the Body                   = "<<t1 <<"mm";
    cout <<"\n   Thickness of the Body at the Base       = "<<t2 <<"mm"; << endl;
}
// function to find the major diameter of the screw
double major_dia (double dia)
{
    if (dia <32.0)
        dia=dia+2;
    else
        if (dia <48.0)
            dia=dia+4.0;
        else
            if (dia <52.0)
                dia=dia+2.0;
            else
                if (dia <110)
                    dia=dia+5.0;
                else
                    if (dia <170)
                        dia=dia+10.0;
                    else
                        goto stop;
    return (dia);
stop:
        cout <<"=n   Required diameter is more than 170 mm";
    return (dia);
} //end function
//     function to select pitch
double pitch (double pp, double dia)
```

```
{
  if (dia <=28)
     pp=5.0;
  else
     if (dia <=36)
        pp=6.0;
     else
        if (dia <=44)
           pp=7.0;
        else
           if (dia <=52)
              pp=8;
           else
              if (dia <=60)
                 pp=9;
              else
                 if (dia <=80)
                    pp=10;
                 else
                    if (dia <=110)
                       pp=12;
                    else
                       if (dia <=140)
                          pp=14;
                       else
                          if (dia <=170)
                             pp=16;

  return (pp);
} //end of function
```

Sample Run of Program 8.1

DESIGN INPUT DATA

Load to be lifted(kN)= 20
Lift required(mm)= 200
Yield strength of the screw material(N/mm2)=350
Factor of safety(fos)= 5
Yield strength of the nut material(N/mm2)=35
Coefficient of friction(mu)= 0.15

DESIGN REPORT

Major Diameter of the Screw = 28 mm
Minor Diameter of the Screw = 23 mm
Mean Diameter of the Screw = 25.5 mm

Pitch of the Screw	= 5 mm
Screw Head Diameter	= 45 mm
Minor Diameter of the Collar	= 15 mm
Diameter of the Handle Bar	= 21 mm
Height of the Screw Head	= 42 mm
Length of the Handle Bar	= 360 mm
Top Diameter of the Nut	= 63 mm
Outside Diameter of the Nut	= 42 mm
Thickness of the Nut	= 15 mm
Number of Threads	= 11
Height of the Nut	= 55 mm
Top Diameter of the Body	= 90 mm
Base Diameter of the Body	= 360 mm
Height of the Body	= 225 mm
Thickness of the Body	= 9 mm
Thickness of the Body at the Base	= 18 mm

Example 8.8 Design a screw jack for lifting a load of 20 kN through a distance of 200 mm.

Solution We assume that the power screw of the screw jack is made of plain carbon steel C45 having yield strength of 350 N/mm^2 and the nut is made of phosphor bronze. Since the screw jack is subjected to various stresses, namely compression, torsional, shear, bending and buckling, therefore, in order to account for these stresses we consider a high factor of safety, say, 5.

$$\text{Allowable tensile strength, } \sigma = \frac{\sigma_{y,t}}{\text{FoS}} = \frac{350}{5} = 70 \text{ N/mm}^2$$

Design of the screw

Considering the compression failure of the screw at the root diameter d_2

$$d_2 = \sqrt{\frac{4W}{\pi\sigma}} = \sqrt{\frac{4 \times 20,000}{\pi \times 70}} = 19.07 \text{ mm}$$

We select square thread of coarse series as per IS: 4694–1968. Thus,

Major diameter, d_1 = 32 mm
Minor diameter, d_2 = 22 mm
Pitch, p = 10 mm
Mean diameter, d = 27 mm

Assuming single start thread for which the average helix angle

$$\alpha = \tan^{-1}\left(\frac{p}{\pi d}\right) = \tan^{-1}\left(\frac{10}{\pi \times 27}\right) = 6.72°$$

For steel screw and bronze nut, the coefficient of friction $\mu = 0.15$. Therefore,

$$\text{Friction angle, } \phi = \tan^{-1} \mu = \tan^{-1} 0.15$$

or

$$\phi = 8.53°$$

Since the friction angle is greater than the helix angle ($\phi > \alpha$), the screw satisfies the condition of self-locking. The torque required to drive the screw against friction between the screw and the nut during lifting of a load is

$$T = \frac{Wd}{2} \tan(\phi + \alpha)$$

$$= \frac{20{,}000 \times 27}{2 \times 1000} \tan(8.53 + 6.72) = 73.67 \text{ N} \cdot \text{m}$$

Besides the frictional torque between threads, an additional torque will be needed to overcome the frictional resistance between collars. We assume that the diameter of the screw head and cup seating is kept between (1.5–1.8) times the major diameter of the screw d_1 and the cup is attached to head by means of a pin which is approximately $0.5d_1$, as shown in the figure below.

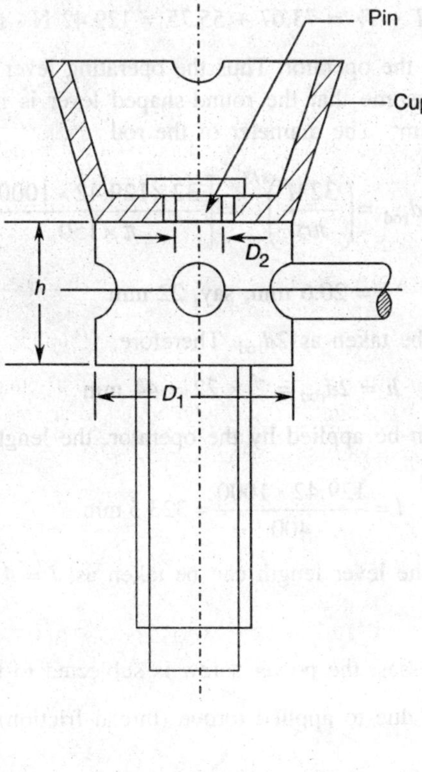

Let

$$D_1 = 1.6d_1 = 1.6 \times 32 = 51.2 \text{ mm, say, } 52 \text{ mm}$$

$$D_2 = 0.5d_1 = 16 \text{ mm}$$

Assuming a uniform pressure distribution between the cup and top of the screw head, the mean diameter of the screw head and cup assembly is

$$D_C = \frac{2}{3}\left(\frac{D_1^3 - D_2^3}{D_1^2 - D_2^2}\right)$$

$$= \frac{2}{3}\left(\frac{52^3 - 16^3}{52^2 - 16^2}\right) = 37.17 \text{ mm}$$

If the coefficient of friction between the cup and the screw head is 0.15, the frictional torque due to collar action between the cup and the screw head is

$$T_C = \frac{1}{2}\mu_C W D_C$$

$$= \frac{1}{2} \times 0.15 \times 20{,}000 \times \frac{37.17}{1000} = 55.75 \text{ N}\cdot\text{m}$$

Total torque required to drive the screw is

$$T_R = T + T_C = 73.67 + 55.75 = 129.42 \text{ N}\cdot\text{m}$$

This total torque is applied by the operator. Thus the operating lever is subjected to a moment of the same magnitude. We assume that the round-shaped lever is made of C20 steel having allowable strength of 150 N/mm^2. The diameter of the rod

$$d_{rod} = \left(\frac{32M}{\pi\sigma}\right)^{1/3} = \left[\frac{32 \times 129.42 \times 1000}{\pi \times 150}\right]^{1/3}$$

$$= 20.6 \text{ mm, say, } 22 \text{ mm}$$

Height of the screw head can be taken as $2d_{rod}$. Therefore,

$$h = 2d_{rod} = 2 \times 22 = 44 \text{ mm}$$

Assuming that 400 N force can be applied by the operator, the length of the lever

$$l = \frac{129.42 \times 1000}{400} = 323.5 \text{ mm}$$

With suitable grip allowance, the lever length can be taken as, l = 425 mm.

Stresses in screw

Besides direct compression stresses, the power screw is subjected to the following stresses:

(i) Induced shear stress due to applied torque (thread friction)

$$\tau = \frac{16T}{\pi d_2^3} = \frac{16 \times 73.67 \times 1000}{\pi \times 22^3} = 35.2 \text{ N/mm}^2$$

(ii) When the screw is at its full lifted condition, it may be treated as column with lower end fixed.

Length of the screw = maximum lift + height of the screw head

= 200 + 44 = 244 mm

Using the Rankine column theory, the crippling stresses can be computed by the relation

$$\sigma_{cr} = \frac{W}{A}\left[1 + a\left(\frac{l}{k}\right)^2\right]$$

where

a = Rankine constant = $\frac{1.95}{25,000}$ for one end free and the other end fixed

k = radius of gyration = $\sqrt{I/A}$ = $0.25d_2$ = 0.25 × 22 = 5.5 mm

$$\text{Crippling stress, } \sigma_{cr} = \frac{20,000}{\frac{\pi}{4} \times 22^2}\left[1 + \frac{1.95}{25,000}\left(\frac{244}{5.5}\right)^2\right] = 60.69 \text{ N/mm}^2$$

(iii) Let there be the possibility that load may act eccentric. We assume eccentricity, e = 5 mm. Then, the bending moment

$$M = 20,000 \times \frac{5}{1000} = 100 \text{ N} \cdot \text{m}$$

$$\text{Bending stress, } \sigma_b = \frac{M}{Z} = \frac{32M}{\pi d_2^3}$$

$$= \frac{32 \times 100 \times 1000}{\pi \times 22^3} = 95.66 \text{ N/mm}^2$$

Combining these stresses for maximum principal stress theory

$$\sigma_{max} = \frac{1}{2}\left[(\sigma_{cr} + \sigma_b) + \sqrt{(\sigma_{cr} + \sigma_b)^2 + 4\tau^2}\right]$$

$$= \frac{1}{2}\left[(60.69 + 95.66) + \sqrt{(60.69 + 95.66)^2 + 4 \times 35.2^2}\right]$$

$$= 163.9 \text{ N/mm}^2$$

Thus factor of safety, $n = \dfrac{\sigma_{y,t}}{\text{induced stresses}}$

$$= \frac{350}{163.9} = 2.14 \text{ (which is satisfactory)}$$

Design of nut

The limiting pressure between the mating threads of the screw and the nut should be less than 12 N/mm^2. Thus, the number of threads required to keep bearing pressure up to 12 N/mm^2

$$N = \frac{4W}{\pi\left(d_1^2 - d_2^2\right)p_b} = \frac{4 \times 20{,}000}{\pi\left(32^2 - 22^2\right) \times 12} = 3.93, \text{ say, 4 threads}$$

Height of the nut, $h_{nut} = N \times \text{pitch} = 40$ mm; however, the height of the nut should be at least 1.5 to 2.0 times the screw diameter. So, we adopt $h_{nut} = 60$ mm, having 6 threads.
Threads of the nut may fail in shear, therefore, the induced shear stress

$$\tau = \frac{W}{\pi d_1 b N} = \frac{20{,}000}{\pi \times 32 \times 5 \times 6} = 6.63 \text{ N/mm}^2$$

which is satisfactory for phosphor bronze having yield strength of 35 N/mm^2.
The design of the nut is shown in the figure below.

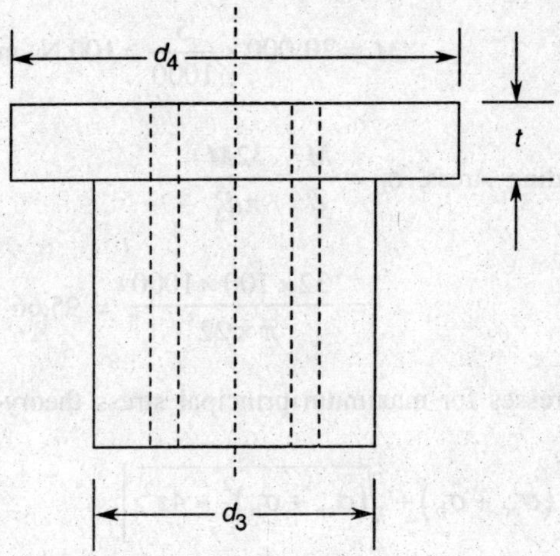

The nut body may come under tensile load if not properly fitted in the jack. Now

$$\sigma_t = \frac{4W}{\pi\left(d_3^2 - d_1^2\right)}$$

or

$$35 = \frac{4 \times 20,000}{\pi \left(d_3^2 - 32^2\right)}$$

or

$$d_3 = 41.85 \text{ mm, say, } 50 \text{ mm}$$

Empirically, let $d_4 = 1.5d_3 = 75$ mm

Crushing stress between the collar of the nut and the body

$$\sigma_{cr} = \frac{4W}{\pi\left(d_4^2 - d_3^2\right)} = \frac{4 \times 20,000}{\pi\left(75^2 - 50^2\right)}$$

$$= 8.15 \text{ N/mm}^2 \text{ (which is satisfactory)}$$

Shearing of collar

Thickness of the collar, $t = \dfrac{W}{\pi d_3 \times \tau}$

where τ = allowable shear strength of the nut material

$= 0.56 \times 35 \approx 20$ N/mm^2

Therefore,

$$t = \frac{20,000}{\pi \times 50 \times 20} = 6.36 \text{ mm, say, 7 mm}$$

Body design

The design of the body is shown in the following figure. Various dimensions are selected empirically as follows:

(i) Diameter of the body at the top = $1.5d_4$ = 1.5 × 75 = 112.5, say, 115 mm
(ii) Thickness of the body, t_1 = (0.25–0.4)d_1 = 0.3 × 32 = 9.6 mm, say, 10 mm
(iii) Thickness at the base flange, $t_2 = 2t_1$ = 2 × 10 = 20 mm
(iv) For stability of the body, it should be tapered towards the base by 10°.
(v) Height of the body = lift + h_{nut} + bottom clearance = 200 + 60 + 20 = 280 mm

$$\text{Mechanical advantage} = \frac{\text{load to be lifted}}{\text{effort}}$$

$$= \frac{20,000}{400} = 50$$

$$\text{Efficiency of the drive} = \frac{\text{load} \times \text{distance per turn}}{2\pi T}$$

$$= \frac{20,000 \times 10}{2\pi \times 129.42 \times 1000} \times 100 = 24.6\%$$

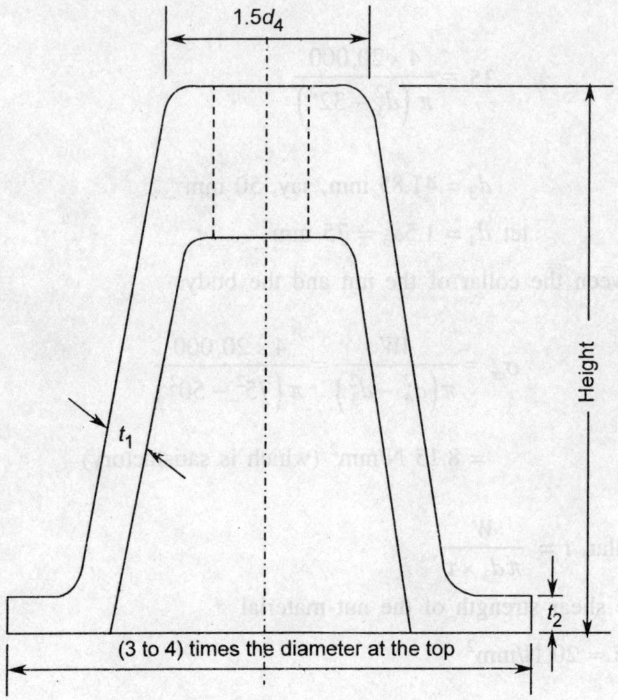

Example 8.9 A tool carriage of a lathe machine is driven by a lead screw revolving at 30 rpm. To drive the carriage, the maximum axial force exerted by the lead screw is 3 kN. The length of the screw is 1.8 m. Design the lead screw.

Solution We assume that the lead screw is made of C40 steel having allowable tensile strength of 80 N/mm^2 and shear strength of 45 N/mm^2.

Assuming that the screw is subjected to direct stresses the core diameter of the screw

$$d_2 = \sqrt{\frac{4W}{\pi\sigma}} = \sqrt{\frac{4 \times 3000}{\pi \times 80}} = 6.9 \text{ mm}$$

To account for the torsional shear stress and the buckling load, we adopt the acme thread with the following proportions:

Major diameter, d_1 = 22 mm
Minor diameter, d_2 = 17 mm
Pitch, p = 5 mm
Included angle, 2β = 29°
Mean diameter, d = 19.5 mm

For single-start screw, the lead angle

$$\alpha = \tan^{-1}\left(\frac{p}{\pi d}\right) = \tan^{-1}\left(\frac{5}{\pi \times 19.5}\right) = 4.66°$$

Assume that the nut is made of cast iron, the coefficient of friction is 0.14. Since the acme thread is Vee thread, therefore, the effective friction coefficient

$$\mu_1 = \frac{\mu}{\cos\beta} = \frac{0.14}{\cos 14.5°} = 0.1446$$

$$\text{Frictional angle, } \phi = \tan^{-1}\mu_1 = \tan^{-1}(0.1446) = 8.23°$$

Since the friction angle ϕ is greater than the lead angle, hence it satisfies the self-locking condition.

Frictional torque between the screw and the nut

$$T_{\text{screw}} = \frac{Wd}{2}\tan(\phi + \alpha)$$

$$= \frac{3000 \times 19.5}{2000} \times \tan(8.23 + 4.66) = 6.69 \text{ N} \cdot \text{m}$$

Stresses induced in screw are:

(i) Direct stress, $\sigma = \sqrt{\dfrac{4W}{\pi d_2^2}} = \sqrt{\dfrac{4 \times 3000}{\pi \times 17^2}} = 3.63 \text{ N/mm}^2$

(ii) Torsional shear stress

$$\tau = \frac{16T_{\text{screw}}}{\pi d_2^3} = \frac{16 \times 6.69 \times 10^3}{\pi \times 17^3} = 6.93 \text{ N/mm}^2$$

Maximum shear stress induced for biaxial loading

$$\tau_{\max} = \sqrt{\left(\frac{\sigma}{2}\right)^2 + \tau^2}$$

$$= \sqrt{\left(\frac{3.63}{2}\right)^2 + 6.93^2} = 7.16 \text{ N/mm}^2$$

which is less than the allowable strength. Hence the design of the screw is satisfactory.

Design of nut

In order to prevent wearing of the nut, the bearing pressure for the CI–steel combination should be limited to 8 N/mm^2. Therefore, the required number of threads

$$N = \frac{4W}{\pi(d_1^2 - d_2^2) \times p_b}$$

$$= \frac{4 \times 3000}{\pi(22^2 - 17^2) \times 8}$$

$$= 2.44, \text{ say, 3 threads}$$

However, from the manufacturing point of view, the height of the split nut should be about three times the mean diameter. Therefore,

$$h_{nut} = 3 \times 19.5 = 58.5, \text{ say, } 60 \text{ mm}$$

Therefore, the number of threads, $N = \dfrac{h_{nut}}{p} = \dfrac{60}{5} = 12$

Generally, the thrust is carried on a collar. The empirical relations for the collar are:

Inner diameter, $D_1 = (1.1\text{–}1.2)d_1$, say, $1.1 \times 22 \approx 25$ mm

Outer diameter, $D_2 = 2D_1 = 50$ mm

$$\text{For uniform pressure, } D_{collar} = \frac{2}{3}\left(\frac{50^3 - 25^3}{50^2 - 25^2}\right) = 38.89 \text{ mm}$$

$$\text{Frictional torque at collar, } T_{collar} = \frac{W\,\mu_{collar}\,D_{collar}}{2}$$

$$= \frac{3000 \times 0.13 \times 38.89}{2 \times 1000} = 7.583 \text{ N}\cdot\text{m}$$

$$\text{Total frictional torque, } T = T_{screw} + T_{collar}$$

$$= 6.69 + 7.583 = 14.273 \text{ N}\cdot\text{m}$$

$$\text{Power loss} = \frac{2\pi NT}{60} = \frac{2\pi \times 30 \times 14.273}{60}$$

$$= 44.8 \text{ W}$$

$$\text{Transmission efficiency, } \eta = \frac{\tan\alpha}{\tan(\alpha + \phi)}$$

$$= \frac{\tan 4.66}{\tan(4.66 + 8.23)} \times 100 = 35.6\%$$

Example 8.10 A screw press is required to exert a force of 40 kN. The unsupported length is 400 mm. The screw is made of steel and the nut of cast iron. The allowable strengths of steel screw are as follows:

Allowable yield strength, $\sigma_{y,t} = 100$ N/mm²

Allowable shear strength, $\tau = 60$ N/mm²

Solution

Design of screw

Considering that the screw is subjected to direct stress, the root diameter of the screw is

$$d_2 = \sqrt{\frac{4W}{\pi\sigma_{y,t}}} = \sqrt{\frac{4 \times 40 \times 10^3}{\pi \times 100}} = 22.56 \text{ mm}$$

To account for buckling and torsional shear stress we adopt a square-threaded power screw of the following dimensions:

Major diameter, d_1 = 36 mm

Minor diameter, d_2 = 30 mm

Pitch, p = 6 mm

Mean diameter, d = 33 mm

For single-start screw thread, the helix angle

$$\alpha = \tan^{-1}\left(\frac{p}{\pi d}\right) = \tan^{-1}\left(\frac{6}{\pi \times 33}\right) = 3.31°$$

Assuming coefficient of friction $\mu = 0.13$ between the steel screw and the CI nut, the friction angle

$$\phi = \tan^{-1}(0.13) = 7.4°$$

Since $\phi > \alpha$ condition is satisfied, hence the screw is self-locked.

The frictional torque between the screw threads

$$T_{\text{screw}} = \frac{Wd}{2} \tan(\phi + \alpha)$$

$$= \frac{40,000 \times 33}{2000} \tan(7.4 + 3.31) = 124.82 \text{ N} \cdot \text{m}$$

(i) Torsional shear stress

$$\tau = \frac{16 T_{\text{screw}}}{\pi d_2^3} = \frac{16 \times 124.82 \times 10^3}{\pi \times 30^3} = 23.54 \text{ N/mm}^2$$

(ii) The power screw under load conditions is subjected to buckling stress. Using the Rankine formula, the crippling stress

$$\sigma_{\text{cr}} = \frac{W}{\frac{\pi}{4} d_2^2}\left[1 + a\left(\frac{l}{k}\right)^2\right]$$

$$= \frac{40,000}{\frac{\pi}{4} \times 30^2}\left[1 + \frac{1}{25,000}\left(\frac{400}{0.25 \times 30}\right)^2\right]$$

$$= 63 \text{ N/mm}^2$$

$$\text{FoS} = \frac{\sigma_{y,t}}{\sigma_{cr}} = \frac{100}{63} = 1.58, \text{ which is satisfactory.}$$

(iii) Direct compression stress

$$\sigma = \frac{4W}{\pi d_2^2} = \frac{4 \times 40,000}{\pi \times 30^2} = 56.59 \text{ N/mm}^2$$

According to maximum shear stress theory

$$\tau_{max} = \sqrt{\left(\frac{\sigma}{2}\right)^2 + \tau^2}$$

$$= \sqrt{\left(\frac{56.59}{2}\right)^2 + 23.54^2} = 36.8 \text{ N/mm}^2$$

$$\text{FoS} = \frac{\tau_{allowable}}{\tau_{max}} = \frac{60}{36.8} = 1.63, \text{ which is satisfactory.}$$

Design of nut

Considering bearing pressure between steel–CI as 12 N/mm^2, the required number of threads

$$N = \frac{4W}{\pi\left(d_1^2 - d_2^2\right)p_b} = \frac{4 \times 40,000}{\pi\left(36^2 - 30^2\right) \times 12}$$

$$= 10.7, \text{ say, } 11$$

Length of the nut, $h_{nut} = N \times p = 11 \times 6 = 66$ mm, however, the length of the nut for a screw press should be between 3–5 times the diameter of the screw. We adopt

$$h_{nut} = 4d_1 = 4 \times 36 = 144 \text{ mm}$$

So, the number of threads, $N = \dfrac{h_{nut}}{p} = \dfrac{144}{6} = 24$

Shearing of thread

$$\text{Shear stress, } \tau = \frac{W}{\pi d_1 b N}$$

$$= \frac{40,000}{\pi \times 36 \times 3 \times 24} = 4.9 \text{ N/mm}^2$$

which is less than the permissible strength.

Diameter of handwheel

$$\text{Torque, } T_{screw} = \text{Force applied by the operator} \times \frac{D}{2}$$

where D is the diameter of the handwheel. Let the force applied by the operator be 600 N (assumed). Thus,

$$124.82 \times 10^3 = 600 \times \frac{D}{2}$$

or

$$D = 416.0, \text{ say, } 450 \text{ mm}$$

Example 8.11 Design a toggle jack, operated by a lever, to lift a load of 5 kN. It utilizes eight symmetrical links each 110 mm long as shown in the figure below. The distance between the nuts corresponding to the top-most position and the bottom-most position is 50 mm and 210 mm, respectively. The link pins at the base are 30 mm apart. Assume that links, screw and nuts are made of plain carbon steel having allowable strength σ_t = 100 N/mm².

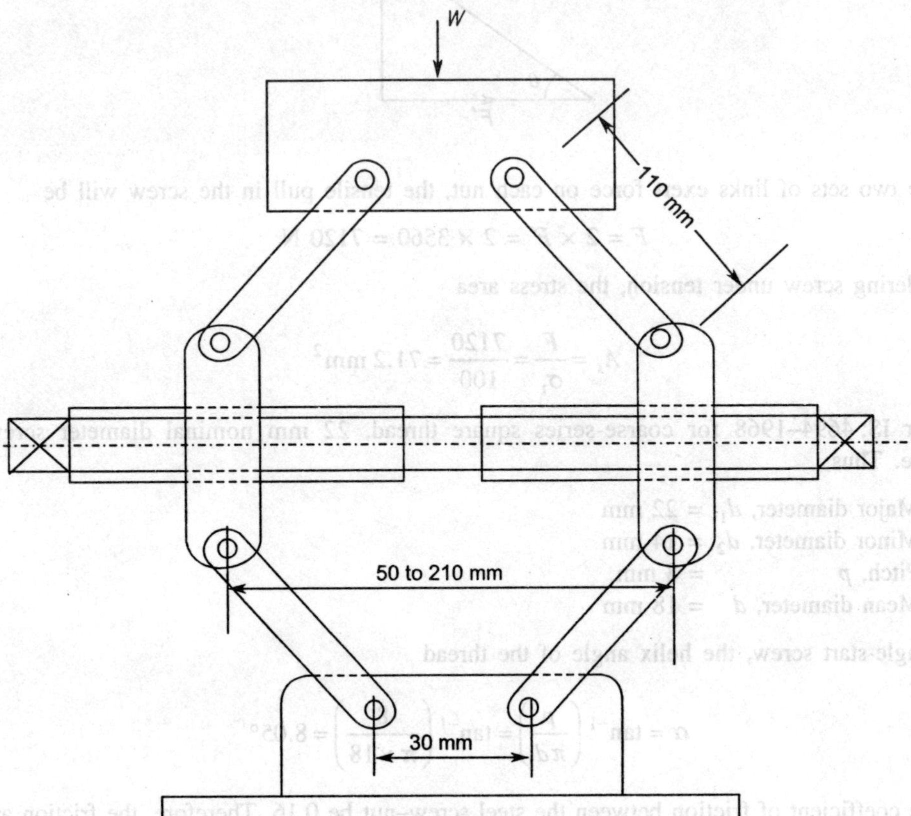

Solution

Design of screw

The screw will experience the maximum pull when the jack is at the bottom-most position. In this position, the link will be inclined at an angle of

$$\theta = \cos^{-1}\left(\frac{210-30}{2\times 110}\right) = 35.1° \text{ with respect to the screw axis}$$

The component of the force along the screw (see the figure below)

$$F' = \frac{5000}{2\times \tan 35.1°} = 3557 \text{ N, say, } 3560 \text{ N}$$

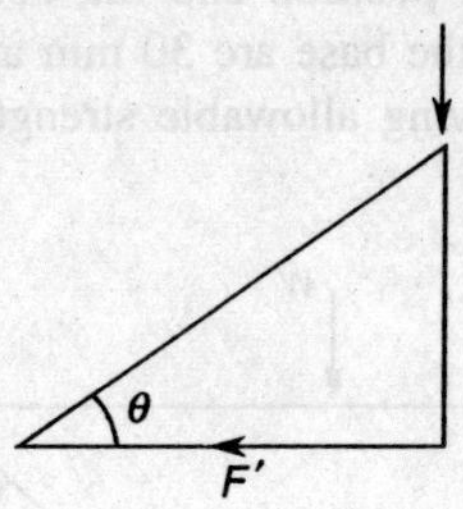

As the two sets of links exert force on each nut, the tensile pull in the screw will be

$$F = 2 \times F' = 2 \times 3560 = 7120 \text{ N}$$

Considering screw under tension, the stress area

$$A_s = \frac{F}{\sigma_t} = \frac{7120}{100} = 71.2 \text{ mm}^2$$

As per IS: 4694–1968 for coarse-series square thread, 22 mm nominal diameter screw is suitable. Thus,

Major diameter, d_1 = 22 mm
Minor diameter, d_2 = 14 mm
Pitch, p = 8 mm
Mean diameter, d = 18 mm

For single-start screw, the helix angle of the thread

$$\alpha = \tan^{-1}\left(\frac{p}{\pi d}\right) = \tan^{-1}\left(\frac{8}{\pi \times 18}\right) = 8.05°$$

Let the coefficient of friction between the steel screw–nut be 0.16. Therefore, the friction angle

$$\phi = \tan^{-1}(0.16) = 9.09°$$

The torque required to overcome friction between the screw and the nut threads

$$T = \frac{Fd}{2}\tan(\phi + \alpha)$$

$$= \frac{7120 \times 18}{2000} \times \tan(9.09 + 8.05) = 19.7 \text{ N . m}$$

Stresses induced in screw

(i) Torsional shear stress

$$\tau = \frac{16T}{\pi d_2^3} = \frac{16 \times 19.7 \times 10^3}{\pi \times 14^3} = 36.56 \text{ N/mm}^2$$

(ii) Direct tensile stress

$$\sigma = \frac{4F}{\pi d_2^2} = \frac{4 \times 7120}{\pi \times 14^2} = 46.25 \text{ N/mm}^2$$

Maximum principal stress

$$\sigma_1 = \frac{1}{2}\left[\sigma + \sqrt{\sigma^2 + 4\tau^2}\right]$$

$$= \frac{1}{2}\left[46.25 + \sqrt{46.25^2 + 4 \times 36.56^2}\right] = 66.38 \text{ N/mm}^2$$

which is less than the permissible value of 100 N/mm^2.

According to maximum shear stress theory

$$\tau_{max} = \sqrt{\left(\frac{\sigma}{2}\right)^2 + \tau^2}$$

$$= \sqrt{\left(\frac{46.25}{2}\right)^2 + 36.56^2} = 43.26 \text{ N/mm}^2$$

which is less than the permissible value of 56 N/mm^2. Hence the design is safe.

Design of nut

Let N be the number of threads in contact. Assuming that the total force of 7120 N is distributed uniformly over the cross-section area of the nut

$$N = \frac{4F}{\pi\left(d_1^2 - d_2^2\right) \times p_b}$$

Assuming safe bearing pressure between the contacting threads to be 15 N/mm^2,

$$N = \frac{4 \times 7120}{\pi \left(22^2 - 14^2\right) \times 15} = 2.09, \text{ say, } 3$$

In order to have good stability and also to prevent rocking of the screw in the nut, we adopt 4 threads. Therefore,

Thickness of the nut, $h_{nut} = N \times p = 4 \times 8 = 32$ mm

Width of the nut, $b = (1.5\text{–}2.0)d_1$

Let us adopt $b = 2d_1 = 2 \times 22 = 44$, say, 45 mm

Length of the threaded portion of the screw

= 210 + 2 × thickness of the nut + clearance + some length at centre which is kept equal to core diameter + length of two square ends

= 210 + 2 × 32 + 15 + 25 + 2 × 15 = 344 mm

Design of links

$$\text{Load on each link} = \frac{1}{2} \times \frac{W}{2 \sin \theta}$$

$$= \frac{5000}{4 \times \sin 35.1°} = 2173.89 \text{ N}$$

Due to this load, the link may buckle in any of the two planes at right angle to each other. Let the thickness of the link be t_1. Then

Width of the link, $b_1 = 3t_1$

Cross-section, $A = 3t_1^2$

Moment of inertia, $I = \frac{1}{12} t_1 b_1^3 = 2.25\, t_1^4$

Also,

$$I = Ak^2$$

where k is the radius of gyration. Thus,

$$k^2 = \frac{I}{A} = \frac{2.25\, t_1^4}{3t_1^2} = 0.75\, t_1^2$$

For buckling in the plane of links, the ends are considered to be hinged.

Equivalent length of link, $l = 110$ mm

Buckling load = load on link × FoS = 2173.89 × 5 = 10,869.45 N

Using the Rankine formula

$$10{,}869.45 = \frac{\sigma_{cr} A}{1 + a\left(\frac{l}{k}\right)^2} = \frac{100 \times 3t_1^2}{1 + \frac{1}{6250}\left(\frac{110^2}{0.75\, t_1^2}\right)}$$

By trial and error, $t_1 = 6.5$ mm, say, 7 mm. Thus, width $b_1 = 21$ mm.

Considering buckling in a plane perpendicular to the plane of the link, both ends are fixed.

$$k^2 = \frac{I}{A} = \frac{\frac{1}{12} \times 21 \times 7^3}{21 \times 7} = 4.08$$

$$\text{Crippling load} = \frac{100 \times 3 \times 7^2}{1 + \frac{1}{25000} \times \left(\frac{110^2}{4.08}\right)} = 13{,}141 \text{ N}$$

$$\text{Factor of safety} = \frac{\text{crippling load}}{\text{load on link}}$$

$$= \frac{13{,}141}{2173.8} = 6$$

which is satisfactory.

Design of link pins

$$\text{Load on pin} = F_1 = \frac{W/2}{\sin\theta} = \frac{5000}{2 \times \sin 35.1°} = 4347.78 \text{ N}$$

The diameter of the pin based on bearing pressure

$$d_{\text{pin}} = \frac{F_1}{l \times p_b}$$

Let the length of the pin be, $l = 2.5d_{\text{pin}}$ and $p_b = 15$ N/mm^2, then the diameter of the pin

$$d_{\text{pin}} = \sqrt{\frac{4347.78}{2.5 \times 15}} = 10.76 \text{ mm, say, 12 mm}$$

which is less than the link width. So, there is no chance of tearing failure of the link.

The pin may fail in double shear. Therefore, the shear stress induced

$$\tau = \frac{4F_1}{2\pi d_{\text{pin}}^2} = \frac{4 \times 4347.78}{2 \times \pi \times 12^2} = 19.22 \text{ N/mm}^2$$

which is satisfactory.

EXERCISES

1. Define the following terms related to screw fastenings:

(a)	Stress area	(b)	Major diameter	(c)	Minor diameter
(d)	Pitch	(e)	Helix angle	(f)	Multi start

2. How is an ISO metric screw designated?
3. What is the effect of initial tightening?

4. What is the effect of gasket on bolt load?
5. What do you mean by a bolt of uniform strength?
6. Explain the mechanism of the lock nut.
7. What is the difference between the forms of threads used for fastening and those used for power screw purposes?
8. Why is the modified square thread preferred to square thread?
9. Why should helix angle be not more than 30° for power screws?
10. Explain the phenomenon of self-locking.
11. Why is the nut of a power screw made of a soft material?
12. Why should the tommy bar in a screw jack be the weakest?
13. A steam engine of effective diameter 300 mm is subjected to a steam pressure of 1.5 N/mm^2. The cylinder head is connected by means of eight bolts having yield strength of 350 N/mm^2 and endurance limit of 250 N/mm^2. The bolts are tightened with an initial preload of 1.5 times that of steam force. A soft copper gasket is used to make a leakproof joint. The stress-concentration factor is 2.8. Determine the size of the bolts.
14. The figure below shows a cap screw subjected to a force of 30 kN. Determine the size of the screw if the allowable tensile and shear stresses are 50 and 35 N/mm^2, respectively.

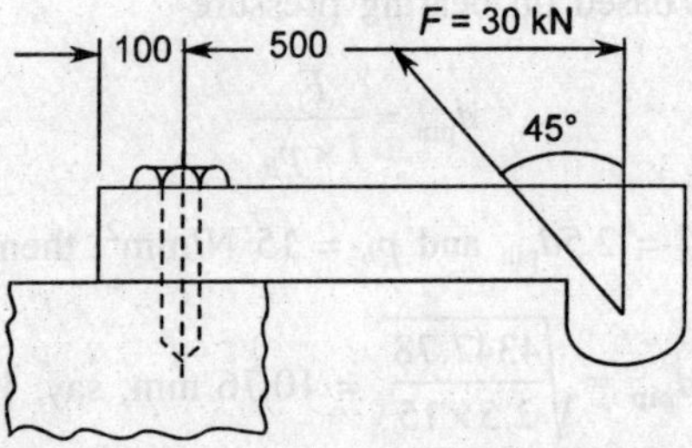

15. What load can the crane runway bracket, shown in the figure below, support if two M25 × 3 bolts are used to fasten the bracket to the roof truss?

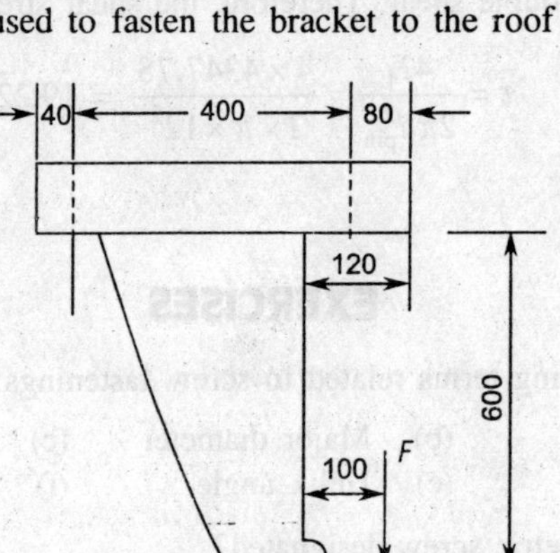

16. A bracket, as shown in the figure below, is fixed to a wall by means of four bolts. Suggest a suitable size of the bolt.

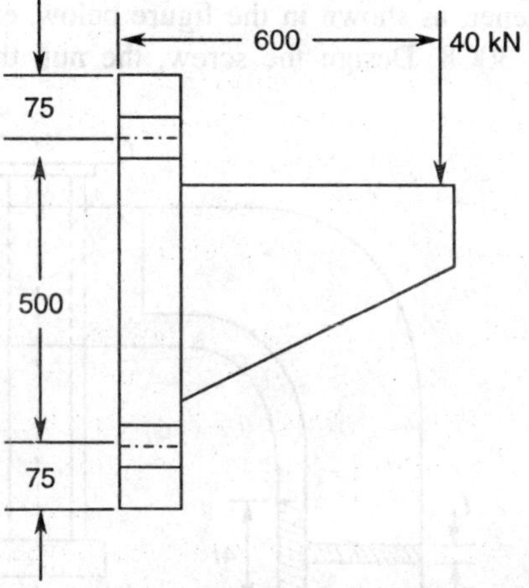

17. A CI bracket carrying a shaft and pulley is shown in the following figure. The bracket is fastened by four bolts. The total force on the pulley is 6000 N. Determine the size of the C20 steel bolts.

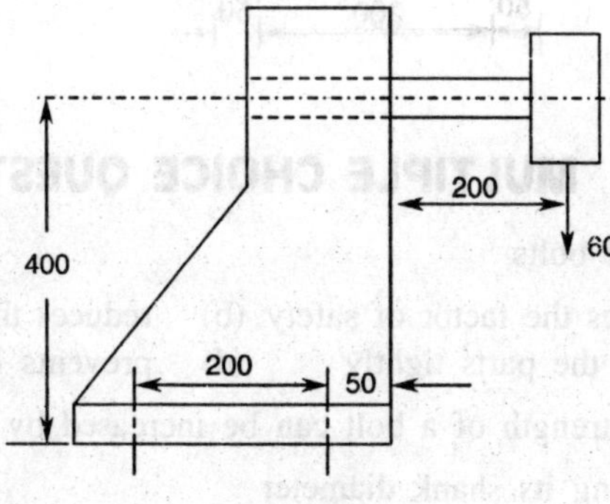

18. Design a screw jack suitable for a maximum load of 50 kN and having a lift of 225 mm. Assume a suitable material and factor of safety.

19. The tool carriage of a lathe is driven by a lead screw revolving at 50 rpm. To drive the carriage the maximum axial force exerted by the lead screw is 2.6 kN. The length of the screw is 1.5 m. The thrust is carried on a collar of 80 mm outer diameter and 40 mm inner diameter. Design the lead screw and determine the transmission efficiency.

20. A lock gate weighing 30 kN is raised or lowered which moves the square-threaded nut mounted on the gate. The maximum fluid resistance is 5 kN. Design the screw and the nut and sketch the arrangement.

21. A punching press is required to punch a maximum hole size of 20 mm diameter in a material having ultimate shear strength of 300 N/mm^2. If the thickness of the sheet is 5 mm, design the screw and the nut.

22. A shaft straightener, as shown in the figure below, exerts a 30 kN force. The material of the screw is 30C8. Design the screw, the nut, the cast iron body, and the fixing bolts.

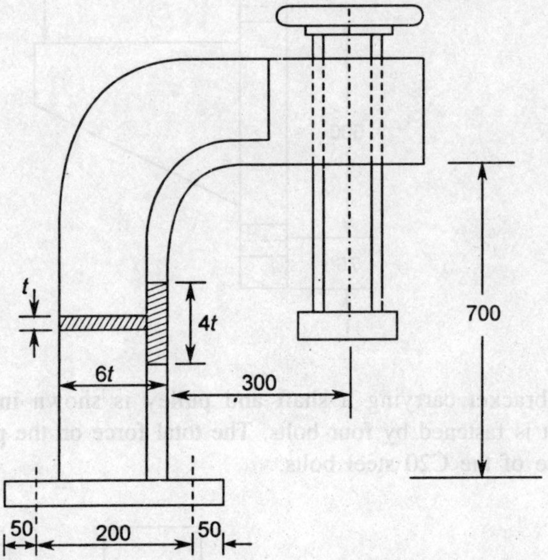

MULTIPLE CHOICE QUESTIONS

1. Preloading of bolts
   (a) improves the factor of safety (b) reduces the factor of safety
   (c) secures the parts tightly (d) prevents leakage

2. The impact strength of a bolt can be increased by
   (a) increasing its shank diameter
   (b) reducing the shank diameter to its core diameter
   (c) having fine threads
   (d) incorporating surface treatment

3. The M10 screw indicates a
   (a) machine screw with 10 mm major diameter
   (b) machine screw with 10 mm root diameter
   (c) metric thread with 10 mm major diameter
   (d) metric thread with 10 mm pitch diameter

4. The efficiency of the screw is given by
   (a) $\tan\phi / \tan\alpha$ (b) $\tan\alpha / (\tan\phi - \alpha)$
   (c) $\tan\alpha / (\tan\phi + \alpha)$ (d) $\tan(\phi - \alpha) / \tan\alpha$

5. The best method for making threads on a bolt is
   (a) forging (b) casting
   (c) machining on a lathe (d) rolling
6. The acme threads are preferred to the square threads because
   (a) their efficiency is high
   (b) they can be manufactured easily
   (c) of their high coefficient of friction
   (d) none of the above are true
7. The condition for self-locking of a power screw is
   (a) $\phi > \alpha$ (b) $\alpha > \phi$
   (c) $\phi = \alpha$ (d) any of the above
8. The material of the nut in a power screw is
   (a) aluminium (b) 30C8 steel
   (c) phosphor bronze (d) alloy steel
9. The transmission efficiency of the square thread is
   (a) minimum (b) maximum
   (c) unpredictable (d) less than the acme thread
10. The buttress thread can take up load in
    (a) one direction (b) two directions
    (c) any direction (d) can't say

Chapter 9 Mechanical Springs

9.1 INTRODUCTION

In mechanical systems wherever flexibility or a relative large deflection under a given load is required, some form of spring is used. A mechanical spring is an elastic member whose primary function is to deflect under load and then to recover its original shape and position when the load is released. Springs are used to connect two parts by a flexible joint to exert force or torque on an element or to absorb energy. Strength and flexibility are two essential requirements of spring design. Springs are usually required to perform the following functions:

(i) To apply force and to control motion by maintaining contact between two elements, namely cam and follower, governor, I.C. engine valve.

(ii) To cushion, absorb, or control energy due to shock and vibration, e.g. automobile springs, aircraft landing gears, railway buffers, and vibration dampers.

(iii) To measure force, e.g. spring balance, meters, and engine indicator.

(iv) To store energy, e.g. clocks, toys, circuit breakers, and starters.

(v) To alter the vibration characteristics of a member, e.g. the flexible mountings of a motor.

Springs can be classified according to their shape and the type of stresses they have to withstand. On the basis of shape, springs may be classified as wire spring, flat spring, and special-shape spring such as disc spring.

The most common types of springs are:

Helical spring. Cylindrical springs having a certain helix angle, usually about 10°, are called helical springs. These springs may sustain either tensile force or compressive force along their axes. Accordingly, they are called helical tension or compression springs. Helical springs are made of either circular, rectangular section or square section wire. These springs are available in a wide range and are easy to manufacture. They provide a more accurate prediction of performance [Figure 9.1(a)].

Conical spring. A conical spring works in compression. It is made of round wire in the shape of cone [Figure 9.1(b)]. It is used either where space limitation does not allow to use cylindrical helical spring or where a variable rate of stiffness is desired with a single spring.

Leaf spring. A leaf spring comprises a flat plate supported at both ends, thus acting as a double cantilever. The major stresses are tensile or compressive. These types of springs may have more than one plate and in that case they are called *laminated leaf springs* [Figure 9.1(c)]. A leaf spring may be of full-elliptical, semi-elliptical, or cantilever type.

Spiral spring. A flat spring of rectangular cross-section when wound in the form of a spiral, i.e. with zero helix angle, results in a spiral spring [Figure 9.1(d)]. These springs are loaded in torsion. Major stresses developed in such springs are tensile and compressive in nature due to bending.

Disc spring. These springs are also known as *Belleville springs*. They are made in the form of a cone disc to carry a high compressive force. In order to improve their load carrying capacity, they may be stacked up together. The major stresses are either tensile or compressive [Figure 9.1(e)].

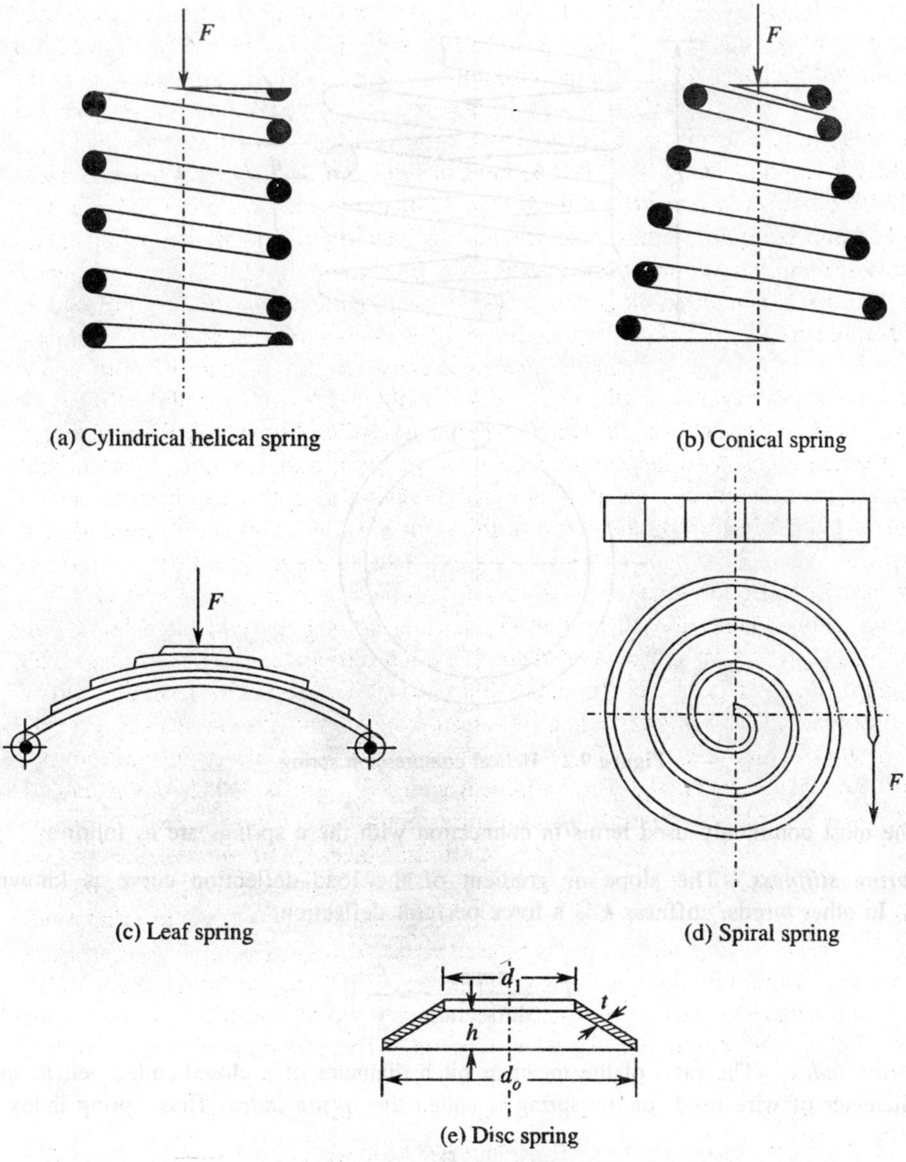

Figure 9.1 Types of springs.

9.2 HELICAL SPRINGS

Helical springs are used to take up forces which tend to either shorten, lengthen or twist them. However, most springs work in compression. The main advantages of these springs are that they are cheaper to manufacture and they continue to function satisfactorily for prolonged periods. Figure 9.2 shows the constructional details of the helical compression springs.

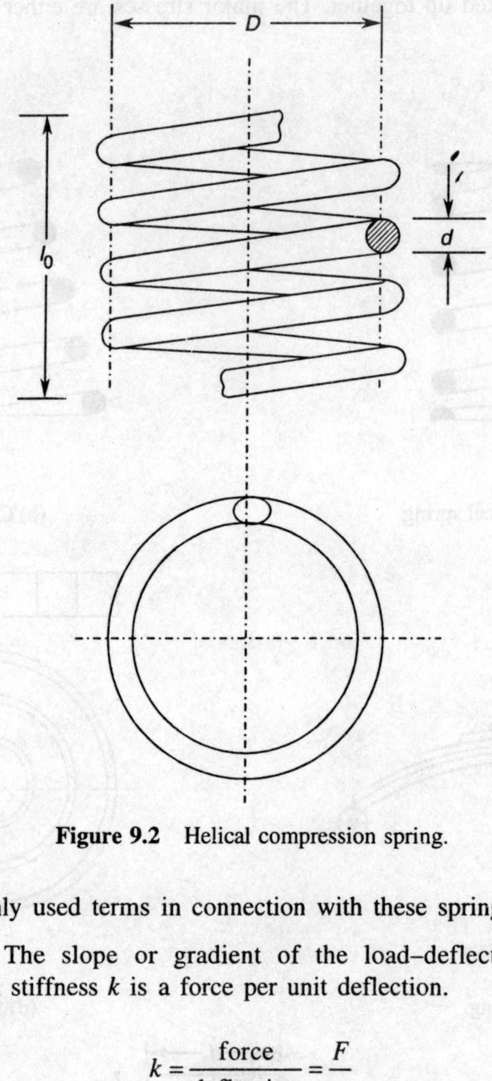

Figure 9.2 Helical compression spring.

The most commonly used terms in connection with these springs are as follows:

Spring stiffness. The slope or gradient of the load–deflection curve is known as *stiffness.* In other words, stiffness k is a force per unit deflection.

$$k = \frac{\text{force}}{\text{deflection}} = \frac{F}{y} \tag{9.1}$$

Spring index. The ratio of the mean or pitch diameter of a closed-coiled helical spring to the diameter of wire used for the spring is called the *spring index*. Thus, spring index

$$C = \frac{D}{d} \tag{9.2}$$

where

D is the mean or pitch diameter of the spring
d is the diameter of the wire.

For the rectangular section wire,

$$C = \frac{D}{b} \tag{9.3}$$

where b is the thickness of the wire in the radial direction.

Active coils. Those coils which are free to deflect under load are called *active coils.*

Inactive coils. The coils which do not take part in deflection of a spring are known as *inactive coils.* The number of inactive coils in a spring depend upon the type of end conditions.

Free length. The full length of a spring under the no-load condition is called its *free length.*

Solid length. The length of a spring under the maximum rated load condition is called its *solid length.*

9.2.1 Stress Analysis

A helical spring is said to be closed coiled, if the plane containing each coil is nearly perpendicular to the axis of the helix. This is true only if the helix angle is less than 10°. The design requirements of these springs are: (i) the maximum stress developed should be within the permissible limits and (ii) the required spring stiffness should be obtained for a given application. These springs may be made of round, rectangular, or square cross-section. Figure 9.3(a) shows a round wire closed coil helical compression spring loaded by axial force F. If this spring is cut at some point and a portion of it is removed then the cut portion would exert two types of forces—torsional moment or torque and direct shear force on the remaining portion of the spring—as shown in Figure 9.3(b).

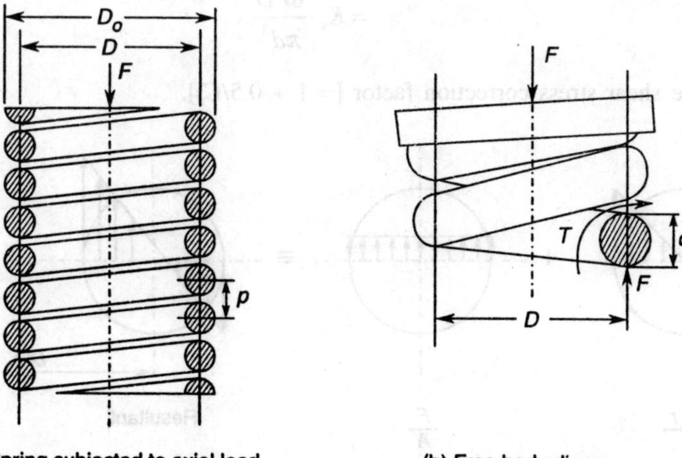

(a) Spring subjected to axial load (b) Free body diagram

Figure 9.3 Closed coiled helical spring.

Torsional moment

The action of applied force F is to twist or turn the wire about its own axis. Thus, flexing of a helical spring creates a torsion in the wire. This torque produces torsional shear stress τ in the wire [see Figure 9.4(a)].

$$\tau = \frac{Tr}{J} = \frac{8FD}{\pi d^3} \tag{9.4}$$

where

T is the torque (= $FD/2$)
J is the polar moment of inertia (= $\pi d^4/32$)
F is the axial force on the spring
D is the mean coil diameter
d is the diameter of wire.

Direct shear stress

The axial force F acts as the shear force on the wire which produces a transverse shear stress in the wire. The intensity of the transverse shear stress

$$\sigma_s = \frac{4F}{\pi d^2} \tag{9.5}$$

is uniform as shown in Figure 9.4(b).

Superimposing the above two shear stresses as shown in Figure 9.4(c), the maximum shear stress in the wire is given by

$$\tau_{max} = \frac{8FD}{\pi d^3} + \frac{4F}{\pi d^2}$$

$$= \left(1 + \frac{0.5}{C}\right)\frac{8FD}{\pi d^3}$$

$$= K_s \frac{8FD}{\pi d^3} \tag{9.6}$$

where K_s is the shear stress correction factor [= $1 + 0.5/C$)].

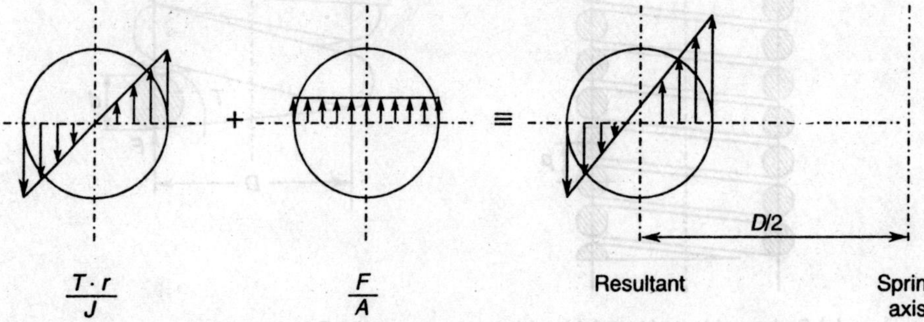

Figure 9.4 Superposition of stresses.

From the manufacturing point of view, the spring index C should not be less than 3. For most springs the value of C ranges between 5 and 12. For industrial springs the value of C should be between 6 and 10. For greater resilience and better utilization of the material, it is advisable to use a spring with a large value of the spring index.

When a wire is bent in the form of a helical coil, the length of the inner fibre of the wire is reduced in comparison to the length of the outer fibre. This results in stress concentration at the inner fibre. The shear stress Eq. (9.6) does not take into account the effect of stress concentration due to curvature of the coil. A.M. Wahl introduced a correction factor which takes into account the effect of curvature as well as the shear stress correction factor. Therefore, the modified form of the maximum shear stress equation is

$$\tau_{\max} = K \cdot \frac{8FD}{\pi d^3} \tag{9.7}$$

where

K is the Wahl's correction factor (= $K_c \cdot K_s$)
K_c is the curvature correction factor.

According to the Wahl's hypothesis, the correction factor K is defined as

$$K = \frac{4C - 1}{4C - 4} + \frac{0.615}{C} \tag{9.8}$$

The relation between K and C is shown in Figure 9.5. The resulting stress distribution on the wire is shown in Figure 9.6.

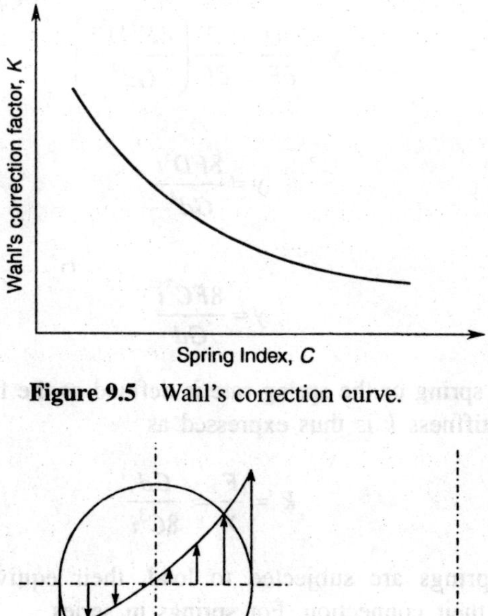

Figure 9.5 Wahl's correction curve.

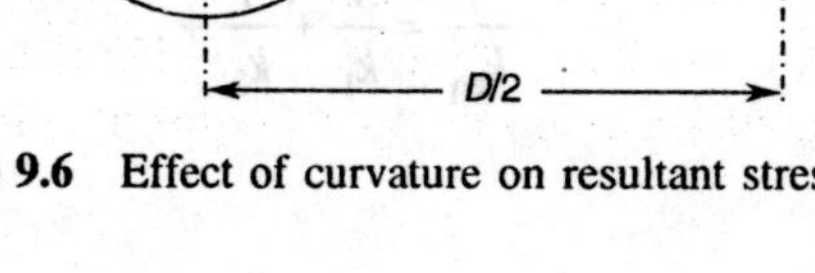

Figure 9.6 Effect of curvature on resultant stress.

9.2.2 Deflection Analysis

In a helical spring the work done by an external axial force F is converted into strain energy and is stored in the spring. The amount of energy absorbed by the spring is called *resilience*, U. Therefore, the resilience or strain energy U of the spring is given by

$$U = \frac{\text{average torque} \times \text{angular displacement}}{2} = \frac{T\theta}{2} \tag{9.9}$$

where

T is the average torque (= $FD/2$)
θ is the angular displacement (= Tl/JG)
l is the length of the spring wire (= πDi)
J is the polar moment of inertia of the wire section
G is the modulus of rigidity
i is the number of active turns in a spring

Thus, the strain energy Eq. (9.9) takes the form

$$U = \frac{4F^2D^3i}{Gd^4} \tag{9.10}$$

According to Castigliano's theorem, the displacement of the spring corresponding to force F can be found by partially differentiating the strain energy relation with respect to force F. Therefore, axial deflection

$$y = \frac{\partial U}{\partial F} = \frac{\partial}{\partial F}\left(\frac{4F^2D^3i}{Gd^4}\right)$$

or

$$y = \frac{8FD^3i}{Gd^4} \tag{9.11}$$

or

$$y = \frac{8FC^3i}{Gd} \tag{9.12}$$

Since the stiffness of a spring or the spring rate is defined as the force required to produce unit deflection, the spring stiffness k is thus expressed as

$$k = \frac{F}{y} = \frac{Gd}{8C^3i} \tag{9.13}$$

When two or more springs are subjected to load, their equivalent stiffness is computed according to nature of their connection. For springs in series

$$\frac{1}{k_{eq}} = \frac{1}{k_1} + \frac{1}{k_2} + \dots \tag{9.14}$$

and for parallel springs

$$k_{eq} = k_1 + k_2 + \ldots \tag{9.15}$$

where

$k_1, k_2, \ldots$ are the stiffnesses of the individual springs
k_{eq} is the stiffness of the equivalent spring.

The length of the spring under the no-load condition, known as the free length l_0, is determined by the following relation:

$$l_0 = \text{solid length} + \text{maximum spring deflection} + \text{clash allowance}$$
$$= Nd + y_{max} + 0.15y_{max} \tag{9.16}$$

where N is the total number of turns.

The total number of turns in a spring depends upon the type of the end conditions used for them. It is the sum of the active number of turns and the inactive turns due to end conditions. Generally, four different types of end conditions are used as shown in Figure 9.7.

Plain end. It is most economical to produce. It has poor seating space. Total number of turns $N = i$ (active turns).

Ground end. It has better seating than the plain end condition. Here, $N = i + 1$.

Squared end. It has better seating than the ground end and more close axial loading is possible. Here, $N = i + 2$.

Squared and ground end. It is the closest approach to the axial loading. Highly stressed springs are made with this type of ends. Here, $N = i + 2$.

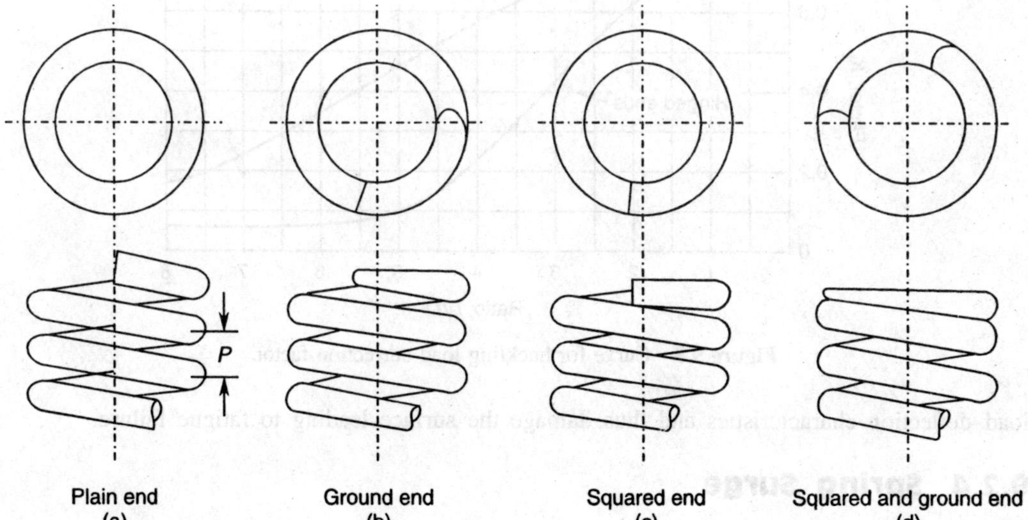

Figure 9.7 Types of end conditions.

9.2.3 Buckling of Springs

The helical compression spring behaves like a column and buckles at a comparative small load when the length of the spring is large compared to its mean coil diameter. Experimentally it has been found that a compression spring may buckle if the following condition is not satisfied.

$$\frac{l_0}{D} < \frac{2.6}{\text{const.}} \tag{9.17}$$

where

const. = a constant
= 0.5 for both ends fixed
= 0.707 for one end fixed and the other hinged
= 1.0 for both ends hinged

The crippling load, under which a spring can buckle, is computed from the following equation

$$F_{cr} = kK_L l_0 \tag{9.18}$$

where K_L is the load correction factor which can be found from Figure 9.8.

In certain cases, when compression springs cannot be designed buckle-proof, they must be guided either in sleeve or over an arbour or be installed in a limited space. Such guidance is, however, undesirable because the friction between the spring and the guide may change the

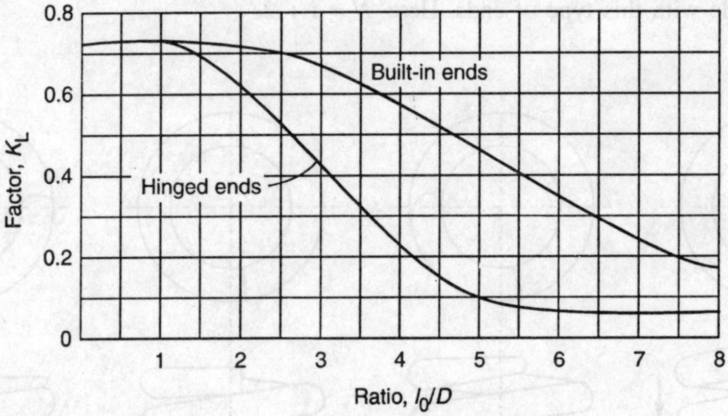

Figure 9.8 Curve for buckling load correction factor.

load–deflection characteristics and thus damage the surface leading to fatigue failure.

9.2.4 Spring Surge

If one end of a compression spring is held against a flat surface and the other end is disturbed or excited, a compression wave is created which travels to and fro from one end to the other. This causes the spring to jump out of the end plate. This effect is called *spring surge.* When helical springs are used in those applications which require rapid motion, e.g. the valve spring of an I.C. engine, special care should be taken to avoid any possible occurrence of spring surge. The designer should ensure that the physical dimensions of the spring are such that the natural frequency of vibration of the spring does not lie in the range of forcing frequency of the applied force, otherwise resonance may occur resulting in setting up of a compression wave that may cause failure of the spring.

The natural frequency of vibration of the spring is given by the following equations:

(i) Spring held between two plates, i.e. both ends are fixed.

$$\omega = \frac{1}{2}\sqrt{\frac{k}{m}} \tag{9.19}$$

(ii) Spring with one end free and the other end fixed by holding it against a flat plate.

$$\omega = \frac{1}{4}\sqrt{\frac{k}{m}} \tag{9.20}$$

where

k is the stiffness of the spring

m is the mass of the spring $\left(= Al\rho = \dfrac{\pi^2 d^2 DN\rho}{4} \right)$ (9.21)

To avoid resonance of the spring with the forcing frequency of the applied force, the spring should be designed in such a way that its natural frequency is significantly higher, i.e. 15–20 times the forcing frequency. This condition can be obtained by increasing the spring stiffness or by decreasing its mass.

9.3 SPRING MATERIALS

Springs are manufactured using either the hot or the cold working process. The selection of the process depends upon the spring index, the type of the material, and the size of the wire. Springs having wire diameters larger than 12 mm are generally wound hot and then they are heat treated to get the required properties. A prehardened wire is not recommended if the spring index is less than 4 or if the diameter of the wire is less than 4 mm. Winding of the springs by the cold wound method induces residual stress due to bending. These stresses are relieved by mild heat treatment or by shot peening method.

Springs are used to absorb energy which is proportional to the square of the stress induced. Therefore, it is always advantageous to use those materials which permit operation at high stresses. A wide variety of spring materials are available to the designer. The most commonly used materials are plain carbon steel, music wire, alloy steel, stainless steel, and non-ferrous metals such as brass, bronze, copper, and berilium, etc.

The BIS has recommended four basic varieties of steels for various applications.

(i) Patented and cold drawn steel
(ii) Oil hardened unalloyed steel
(iii) Oil hardened alloy steel
(iv) Stainless steel.

A brief description of these materials is given below.

Patented and cold drawn steel

It is the toughest and most widely used spring material. It has high strength and can withstand fatigue loading. There are four grades of this material. Grade-1 is useful for springs subjected to static loading. Grade-2 is useful for a moderate static load cycle. Grade-3 is used for highly stressed springs or those springs which are subjected to moderate dynamic loads. Grade-4 is recommended for springs subjected to severe operating conditions. The chemical composition and minimum tensile strength of these materials are covered by IS: 4454 (Part I)—1981 (Revised).

Oil hardened unalloyed steel

It is a general-purpose spring steel which can be used for springs subjected to static and fatigue loads. It is available in two grades—SW and VW. The SW grade is suitable for springs subjected to moderate fatigue loads whereas the VW grade, which is the valve spring grade, is recommended for those springs which are subjected to a high magnitude of fluctuating stress. The maximum operating temperature is 180°C. The tensile strength of these two grades is covered by IS: 4454 (Part II)—1975.

Oil hardened alloy steel

These steels contain carbon (0.45–0.55 per cent), manganese (0.6–0.8 per cent), chromium (0.8–1.2 per cent), and vanadium (0.1–0.3 per cent). These steels can be used for high shock and fatigue load conditions. The maximum operating temperature for these steels should not exceed 250°C. The tensile strength of these steels may vary from 1470 to 2060 N/mm^2 depending upon the wire diameter. More details are given in IS: 4454 (Part III)—1975.

Stainless steel

It is nickel–chrome steel mainly used where resistance to corrosion and creep at high temperatures is required. Austenitic type stainless steels retain high stress at moderate temperatures, i.e. 350°–400°C. The typical stainless steel is 18-8 steel having 18 per cent chromium, 8 per cent nickel, and 0.7 per cent carbon. The tensile strength of this steel ranges from 1270 to 2000 N/mm^2. IS: 4454 (Part IV)—1975 provides further details about strengths of stainless steels at various diameters.

Table 9.1 gives the values of constants used to estimate the tensile strength of some types of steels. The tensile strength is given by the equation

$$\sigma_{ut} = \frac{A}{d^m} \tag{9.22}$$

where

σ_{ut} is the ultimate tensile strength (N/mm^2)
A is the constant
m is the exponent
d is the wire diameter.

Most of the compiled data on various spring steels give the values of ultimate tensile strength. However, little information regarding shear strength is known. In the absence of this information, the designer may use the following relations.

$$\text{Yield strength, } \sigma_y = 0.75\sigma_{ut}$$
$$\text{Shear strength, } \tau = 0.577\sigma_y$$

The standard sizes of wires are listed in Table 9.2.

Table 9.1 Constants used to estimate the tensile strength of spring steel

Material	*Exponent (m)*	*Constant (A)*
Patented cold drawn steel		
Grade-1	0.251	1630
Grade-2	0.191	1720
Grade-3	0.179	1980
Grade-4	0.145	2160
Oil tempered unalloyed steel	0.186	1880
Chrome–vanadium steel	0.167	2000
Chrome–silicon steel	0.112	2000

Table 9.2 Standard sizes of wires (mm)

Cold drawn steel wire unalloyed	*Hardened and tempered spring steel wire and valve spring wire*	*Stainless steel wire for normal corrosion resistance*
increment	increment	increment
0.07 to 0.12–0.01	1.00 to 1.10–0.05	0.10, 0.11, 0.125
0.14 to 0.22–0.02	1.2, 1.25	0.14 to 0.22–0.02
0.25	1.30 to 2.10–0.10	0.25
0.28 to 0.40–0.02	2.25	0.28 to 0.40–0.02
0.43, 0.45, 0.48	2.40 to 2.60–0.10	0.43, 0.45, 0.48, 0.50
0.50, 0.53, 0.56	2.80 to 4.00–0.20	0.53, 0.56, 0.60, 0.63
0.60, 0.63	4.25 to 5.00–0.25	0.65 to 1.30–0.05
0.65 to 1.30–0.05	5.30, 5.60, 6.00, 6.30	1.40 to 2.10–0.10
1.40 to 2.10–0.10	6.50 to 11.0–0.50	2.25; 2.40, 2.50, 2.60
2.25, 2.40, 2.50	12.0, 12.5, 13.0	2.80, 3.00, 3.15
2.60, 2.80, 3.00	14.0	3.20 to 4.00–0.20
3.20 to 4.00–0.20		4.25 to 5.00–0.25
4.25 to 5.00–0.25		5.30, 5.60, 6.00, 6.30
5.30, 5.60, 6.00, 6.30		6.50 to 10.00–0.50
6.50 to 11.0–0.50		
12.0, 12.50		
13.00 to 17.00–1.00		

9.4 DESIGN PROCEDURE

The design of a helical spring that is subjected to static loading involves a trial-and-error solution methodology. While designing such a spring, the designer has to consider the following factors.

(a) It should be able to carry the designed load.
(b) It should have the required load–deflection characteristics.
(c) It should not buckle under load.
(d) It should also satisfy the given set of constraints, namely the space limitation, the minimum height, the desired life, the specific vibrational characteristics, etc.

Based upon the objective function and the given set of constraints, a designer usually selects the spring design parameters, namely material, size of wire, number of turns, the type of end, free length, and spring rate, etc.

Although no definite set of steps can be given for spring design because of involvement of a large number of interdependent parameters, an algorithm given below can work as a guideline for spring design.

1. For the given load and deflection (if not given directly then compute from the given data) and the purpose for which the spring is to be designed, select a suitable material.
2. For the given space limitation, select the mean coil diameter. If no other condition or space limitation is given, the designer may assume a suitable value of the spring index which normally ranges from 5 to 10.
3. Determine the wire diameter for $K = 1$, if the spring index is not chosen.
4. Find the revised diameter using the value of K computed on the basis of the calculated diameter and the given space limitation.
5. Adopt the nearest standard size of wire.
6. Determine the number of coils for the required deflection. If the calculated number of turns is small, the spring will be too soft. So, reduce the mean diameter of the spring. This change will slightly increase the diameter of wire. If the number of turns is large, the mean coil diameter may be increased.
7. Decide the end conditions and select the number of inactive coils.
8. Calculate the solid length and free length of the spring.
9. Check the spring for buckling.
10. Compute the natural frequency of vibration, avoiding spring surge.
11. Compute the spring rate.

Example 9.1 Design a closed-coil helical spring for a boiler safety valve which is required to blow off steam at the pressure of 1.5 N/mm². The diameter of the valve is 50 mm. The initial compression of the spring is 40 mm and the lift is limited to 20 mm.

Solution

$$\text{Steam force on the valve, } F = \frac{\pi}{4} d_{\text{valve}}^2 \times p$$

$$= \frac{\pi}{4} \times 50^2 \times 1.5 = 2945.2 \text{ N}$$

This force keeps the spring compressed by 40 mm. Thus,

$$\text{Stiffness, } k = \frac{\text{load}}{\text{initial deflection}} = \frac{2945.2}{40} = 73.63 \text{ N/mm}$$

When the valve lifts completely, the total deflection of the spring is the sum of the initial compression and the valve lift. That is,

$$y = 40 + 20 = 60 \text{ mm}$$

The required maximum force on the spring is

$$F = k \times y$$

$$= 73.63 \times 60 = 4417.8 \text{ N}$$

Assuming that a static load of 4417.8 N is applied to the spring, the material chosen for the spring is patented chrome–vanadium steel for which the ultimate tensile strength is

$$\sigma_{ut} = \frac{A}{d^m} = \frac{2000}{d^{0.167}} \qquad \text{(see Table 9.1)}$$

$$\text{Yield strength, } \sigma_y = 0.75\sigma_{ut}$$

and

$$\text{Shear strength, } \tau = 0.577\sigma_y = 0.43\sigma_{ut}$$

Thus, the designed shear stress in the wire

$$\tau_d = 0.43 \times \frac{2000}{d^{0.167}} = \frac{860}{d^{0.167}}$$

Since no space limitation is given, so we choose spring index $C = 6$ (which varies between 5–10).

$$\text{Wahl's stress correction factor, } K = \frac{4C-1}{4C-4} + \frac{0.615}{C}$$

or

$$K = \frac{4 \times 6 - 1}{4 \times 6 - 4} + \frac{0.615}{6} = 1.252$$

Shear stress produced in the wire

$$\tau = K \times \frac{8FD}{\pi d^3} = K \times \frac{8FC}{\pi d^2}$$

$$= \frac{1.252 \times 8 \times 4417.8 \times 6}{\pi d^2} = \frac{84,508.7}{d^2}$$

Trial I: We choose a wire of 12 mm diameter.

$$\text{Shear stress produced, } \tau = \frac{84,508.7}{12^2} = 586.8 \text{ N/mm}^2$$

$$\text{Design shear stress, } \tau_d = \frac{860}{(12)^{0.167}} = 567.9 \text{ N/mm}^2$$

Since $\tau > \tau_d$, so 12 mm diameter is not sufficient. We therefore select the next higher diameter.

Trial II: Let $d = 12.5$ mm

$$\tau = \frac{84508.7}{12.5^2} = 540.8 \text{ N/mm}^2$$

$$\tau_d = \frac{860}{12.5^{0.167}} = 564 \text{ N/mm}^2$$

Since $\tau_d > \tau$, so 12.5 mm diameter wire is safe from the strength point of view.

For maximum deflection of 60 mm, the required number of turns can be determined from the spring stiffness equation

$$k = \frac{Gd}{8C^3 i}$$

Therefore,

$$\text{Number of active turns, } i = \frac{Gd}{8C^3 k}$$

$$= \frac{0.84 \times 10^5 \times 12.5}{8 \times 6^3 \times 73.63} = 8.25, \text{ say, } 9$$

For ground and squared end conditions, the total number of turns, $N = i + 2 = 9 + 2 = 11$

$$\text{Solid length, } l_s = Nd = 11 \times 12.5 = 137.5 \text{ mm}$$

$$\text{Free length, } l_0 = l_s + y + 0.15y$$

$$= 137.5 + 60 + 0.15 \times 60 = 206.5 \text{ mm}$$

Check for buckling of the spring. In order to prevent buckling, the following condition should be satisfied.

$$\frac{l_0}{D} < \frac{2.6}{\text{const.}}$$

For both ends fixed, const. = 0.5. So

$$\frac{l_0}{D} < \frac{2.6}{0.5} = 5.2$$

Mean coil diameter, $D = C \times d = 6 \times 12.5 = 75$ mm

Therefore,

$$\frac{l_0}{D} = \frac{206.5}{75} = 2.75, \text{ i.e.} < 5.2$$

Hence the spring is safe in buckling and does not require mandrill.

Resonant frequency—for both ends fixed

$$\omega = 0.5\sqrt{\frac{k}{m}}$$

Let the density of wire ρ be 7800 kg/m^3. Therefore, mass of the spring

$$m = \frac{\pi^2 d^2 DN\rho}{4} = \frac{\pi^2 \times 0.0125^2 \times 0.075 \times 11 \times 7800}{4} = 2.48 \text{ kg}$$

$$\text{Frequency, } \omega = 0.5\sqrt{73.63 \times 1000/2.48} = 86.15 \text{ rad/s}$$

$$\text{Pitch of the coil, } p = \frac{l_0 - l_s}{N} + d$$

$$= \frac{206.5 - 137.5}{11} + 12.5 = 18.77 \text{ mm}$$

Design specifications of the spring

Material	:	Chrome–Vanadium steel
Wire diameter	:	12.5 mm
Mean coil diameter	:	75 mm
Number of turns (squared and ground end conditions)	:	11
Spring index	:	6
Free length, l_0	:	206.5 mm
Pitch of the coil, p	:	18.77 mm

Example 9.2 From a toy gun, a bullet of 1 N is fired. The bullet travels a distance of 10 m. The compression of the spring, when the gun is loaded, is 100 mm and the bore of the barrel is 20 mm. Design a suitable spring.

Solution When the bullet is fired, it travels a distance of 10 m. The work done by the bullet is $(1 \times 10) = 10$ N · m. To produce this work, the energy stored in the spring should be at least equal to the work done by the bullet, if all other losses are neglected. Thus,

$$\text{Energy stored} = 0.5 \times F \times y = 10$$

where

F is the maximum force

y is the deflection of the spring.

Therefore,

$$F = \frac{2 \times 10}{0.1} = 200 \text{ N}$$

The bore of the barrel is given as 20 mm. It means that maximum coil diameter should be less than 20 mm. Considering this, we select spring index $C = 5.5$.

$$\text{Wahl's correction factor, } K = \frac{4C - 1}{4C - 4} + \frac{0.615}{C}$$

$$= \frac{4 \times 5.5 - 1}{4 \times 5.5 - 4} + \frac{0.615}{5.5} = 1.28$$

Assuming that the spring is made of patented cold drawn steel of grade-3 which can take up moderate shock load, the design shear stress

$$\tau_d = 0.43\,\sigma_{ut} = 0.43 \times \frac{A}{d^m}$$

$$= 0.43 \times \frac{1980}{d^{0.179}} = \frac{851.4}{d^{0.179}} \quad \text{(see Table 9.1)}$$

Shear stress produced in the wire

$$\tau = K \times \frac{8FD}{\pi d^3} = K \times \frac{8FC}{\pi d^2}$$

$$= \frac{1.28 \times 200 \times 8 \times 5.5}{\pi d^2} = \frac{3585.4}{d^2}$$

Trial I: We choose a wire diameter, d = 2 mm

$$\text{Shear stress produced, } \tau = \frac{3585.4}{2^2} = 896.3 \text{ N/mm}^2$$

$$\text{Design shear stress, } \tau_d = \frac{851.4}{2^{0.179}} = 752.0 \text{ N/mm}^2$$

As $\tau > \tau_d$, so we try the next higher diameter.

Trial II: Let d = 2.5 mm

$$\text{Shear stress produced, } \tau = \frac{3585.4}{2.5^2} = 573.6 \text{ N/mm}^2$$

$$\text{Design shear stress, } \tau_d = \frac{851.4}{2.5^{0.179}} = 722.6 \text{ N/mm}^2$$

Since the design shear stress $\tau_d > \tau$ (induced stress), so a standard wire diameter of 2.5 mm is selected.

Outside diameter = mean coil diameter + wire diameter

$$= 5.5 \times 2.5 + 2.5 = 16.25 \text{ mm}$$

which is less than the bore of the barrel. Hence, it satisfies the space limitation.

For a deflection of 100 mm, the number of active coils

$$i = \frac{dG}{8kC^3}, \quad \text{where} \quad k = \frac{\text{force}}{\text{deflection}} = \frac{200}{100} = 2 \text{ N/mm}$$

$$= \frac{2.5 \times 0.84 \times 10^5}{8 \times 2 \times 5.5^3} = 78.88, \text{ say, 79 turns}$$

For squared and ground end conditions, the total number of turns, $N = i + 2$ or N = 81 turns.

Solid length, $l_s = N \times d$ = 81 × 2.5 = 202.5 mm

Free length, $l_0 = l_s + y + 0.15y$

$$= 202.5 + 100 + 0.15 \times 100 = 317.5 \text{ mm}$$

Since l_0/D for this spring is greater than 5.2 for both ends fixed, so the spring may buckle. However, as it is to be mounted in the barrel, the chances of buckling are remote.

Design specifications of the spring

Material	:	Patented steel of grade-3
Wire diameter	:	2.5 mm
Mean coil diameter	:	13.75 mm
Maximum coil diameter	:	16.25 mm
Total number of turns (squared and ground end conditions)	:	81
Free length	:	317.5 mm

9.5 SPRINGS SUBJECTED TO VARIABLE LOAD

Helical compression springs in many applications are subjected to variable or fatigue load. For example, the valve spring of an I.C. engine may be subjected to several millions of stress reversals during its life cycle. Those springs which can carry at least 10 million stress reversal cycles before fatigue failure are called springs with infinite life. Springs subjected to fluctuating stresses are designed on the basis of fatigue failure. However, there is a basic difference between the fatigue failure of a rotating shaft and that of a spring. A helical compression spring is never subjected to a completely reverse loading as observed in a rotating shaft. The load on these springs may be either a variable compressive force, or a pulsating force which varies between zero and maximum compressive force F_{max}. Figure 9.9(a) shows the variable force on a helical compression spring varying between F_{max} and F_{min}, where F_{min} is the force due to initial compression of the helical compression spring. In the worst condition, F_{min} may be zero, as shown in Figure 9.9(b).

Let us now consider a spring subjected to a force varying between F_{max} and F_{min} in a load cycle. The mean and amplitude forces are given as:

$$\text{Mean force, } F_m = \frac{F_{max} + F_{min}}{2} \tag{9.23a}$$

$$\text{Amplitude force, } F_a = \frac{F_{max} - F_{min}}{2} \tag{9.23b}$$

The mean torsional shear stress

$$\tau_m = K_s\left(\frac{8F_m D}{\pi d^3}\right) \tag{9.24}$$

where K_s is the shear stress correction factor [= $(1 + 0.5/C)$]. It is used only for mean shear stress calculation.

The torsional shear stress amplitude

$$\tau_a = K \times \frac{8F_a D}{\pi d^3} \tag{9.25}$$

where K is the Wahl's stress correction factor.

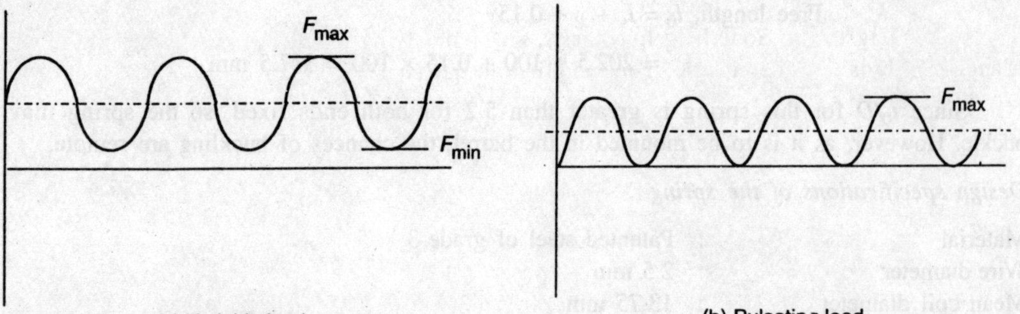

Figure 9.9 Fatigue load on a spring.

The design of a spring under fatigue load conditions is based upon the modified Soderberg line approach as shown in Figure 9.10. Accordingly, the point A can be found by drawing a line OA inclined at 45° with the *X*-axis, which intersects the original Soderberg line at point A. The point A on this diagram indicates the limiting value of stress due to pulsating load condition. The point B on the *X*-axis indicates the limiting value of stress due to static load condition. Thus the line joining points A and B is called the Soderberg line of failure. To take into account the effect of factor of safety (FoS), a line CD is drawn parallel to line AB, from point D on the *X*-axis, where OD = τ_y/FoS. The line CD is called the modified Soderberg line of failure.

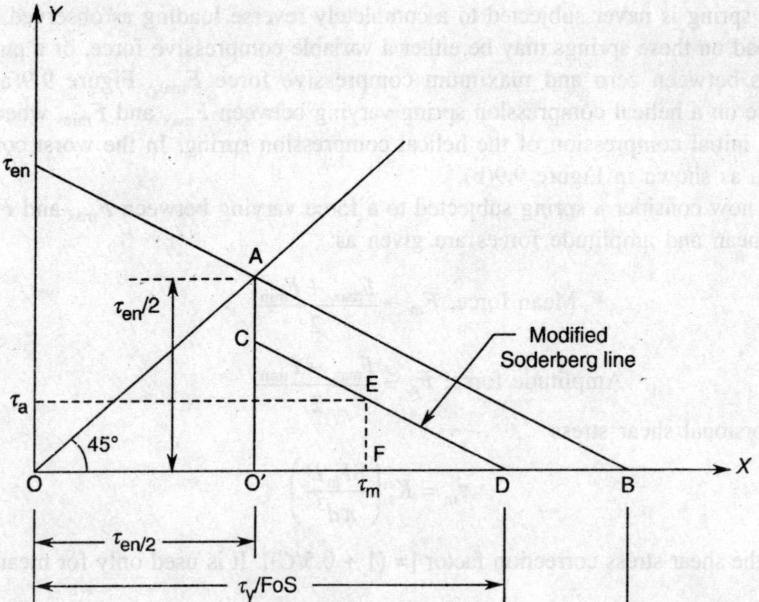

Figure 9.10 Modified Soderberg line.

According to the Soderberg hypothesis, any point lying either on line CD or within the triangle O′CD is considered safe. The modified design equation

$$\frac{1}{\text{FoS}} = \frac{\tau_m - \tau_a}{\tau_y} + \frac{2\tau_a}{\tau_{en}} \tag{9.26}$$

where

τ_{en} = shear endurance strength = $(0.5–0.6)\sigma_{en}$, with $\sigma_{en} = 0.5\sigma_{ut}$.

Example 9.3 A helical valve spring is to be designed for an operating load range of 90 N to 135 N. The 90 N load acts when the valve is closed and the 135 N force acts when the valve is open. The deflection of the spring is 7.5 mm.

Solution The valve spring is subjected to fatigue loading. However, we will first design the valve spring for a static load of 135 N and then check the design for fatigue loading.

Design for static load

We assume that the spring is made of chrome–silicon steel having ultimate strength

$$\sigma_{ut} = \frac{2000}{d^{0.112}} \quad \text{(see Table 9.1)}$$

$$\text{Allowable or design shear strength, } \tau_d = 0.43 \times \frac{2000}{d^{0.112}} = \frac{860}{d^{0.112}} \text{ N/mm}^2$$

Since space limitation is not given, we select spring index $C = 7$.

$$\text{Wahl's correction factor, } K = \frac{4C-1}{4C-4} + \frac{0.615}{C}$$

$$= \frac{4\times 7-1}{4\times 7-4} + \frac{0.615}{7} = 1.213$$

Shear stress produced in wire

$$\tau = K \times \frac{8FD}{\pi d^3} = K \times \frac{8FC}{\pi d^2}$$

$$= \frac{1.213 \times 8 \times 135 \times 7}{\pi d^2} = \frac{2918.99}{d^2} \text{ N/mm}^2$$

We assume wire diameter $d = 2$ mm.

$$\text{Induced shear stress, } \tau = \frac{2918.99}{2^2} = 729.75 \text{ N/mm}^2$$

$$\text{Design shear stress } \tau_d = \frac{860}{d^{0.112}} = \frac{860}{2^{0.112}} = 795.7 \text{ N/mm}^2$$

Since $\tau_d > \tau$, 2 mm diameter of the wire is safe. But as the spring is subjected to fatigue loading, we select a slightly larger diameter, say, $d = 2.5$ mm to account for the fatigue load.

Check for fluctuating load

$$\text{Mean force, } F_m = \frac{F_{max} + F_{min}}{2} = \frac{135 + 90}{2} = 112.5 \text{ N}$$

$$\text{Amplitude force, } F_a = \frac{F_{max} - F_{min}}{2} = \frac{135 - 90}{2} = 22.5 \text{ N}$$

$$\text{Mean shear stress, } \tau_m = K_s \times \frac{8F_m D}{\pi d^3}$$

$$= \left(1 + \frac{0.5}{7}\right) \times \frac{8 \times 112.5 \times 17.5}{\pi \times 2.5^3} = 343.7 \text{ N/mm}^2$$

$$\text{Amplitude stress } \tau_a = K \times \frac{8F_a D}{\pi d^3}$$

$$= \frac{1.213 \times 8 \times 22.5 \times 17.5}{\pi \times 2.5^3} = 77.84 \text{ N/mm}^2$$

For chrome–silicon steel

$$\text{Endurance limit, } \sigma_{en} = 0.5\sigma_{ut}$$

$$= 0.5 \times \frac{2000}{(2.5)^{0.112}} = 902.5 \text{ N/mm}^2$$

$$\text{Shear endurance limit, } \tau_{en} = 0.5\sigma_{en} = 0.5 \times 902.5 = 451.2 \text{ N/mm}^2$$

$$\text{Shear yield strength, } \tau_y = 0.43\sigma_{ut} = 776.1 \text{ N/mm}^2$$

Now applying the modified Soderberg equation

$$\frac{1}{\text{FoS}} = \frac{\tau_m - \tau_a}{\tau_y} + \frac{2\tau_a}{\tau_{en}}$$

$$= \frac{343.7 - 77.84}{776.1} + \frac{2 \times 77.84}{451.2}$$

or FoS = 1.45, which is reasonable for the valve spring. Hence the spring is safe in fluctuating load conditions.

$$\text{Stiffness of spring, } k = \frac{135 - 90}{7.5} = 6 \text{ N/mm}$$

Number of active turns

$$i = \frac{dG}{8kC^3} = \frac{2.5 \times 0.84 \times 10^5}{8 \times 6 \times 7^3}$$

or

$$i = 12.75 \approx 13 \text{ turns}$$

Assuming square and ground end conditions, the total number of turns, $N = i + 2 = 15$

$$\text{Solid length } l_s = N \times d$$

$$= 15 \times 2.5 = 37.5 \text{ mm}$$

$$\text{Free length, } l_0 = l_s + y_{max} + 0.15y_{max}$$

$$= 37.5 + 7.5 + 0.15 \times 7.5 = 46.13 \text{ mm}$$

In order to check for buckling, the following condition must be satisfied

$$\frac{l_0}{D} < \frac{2.6}{\text{const.}}$$

Assuming both ends fixed, const. = 0.5. Thus,

$$\frac{l_0}{D} < \frac{2.6}{0.5} = 5.2$$

Mean coil diameter, $D = C \times d = 7 \times 2.5 = 17.5$ mm

$$\frac{l_0}{D} < \frac{46.13}{17.5} = 2.63 < 5.2$$

so the spring will not buckle under load.

Resonance frequency—for both ends fixed

$$\omega = 0.5\sqrt{\frac{k}{m}}$$

m = mass of the spring

$$= \frac{\pi^2 d^2 DN\rho}{4} = \frac{\pi^2 \times 0.0025^2 \times 0.0175 \times 15 \times 7800}{4}$$

$$= 0.03157 \text{ kg}$$

Therefore,

$$\omega = 0.5\sqrt{6 \times 1000 / 0.03157} = 217.97 \text{ rad/s}$$

$$\text{Resonance frequency, } f = \frac{\omega}{2\pi} = \frac{217.97}{2\pi} = 34.69 \text{ Hz}$$

or

$$34.69 \times 60 = 2081.4 \text{ rpm}$$

Design specifications of the spring

Material	: Chrome–silicon steel
Wire diameter	: 2.5 mm
Coil diameter	: 17.5 mm
Number of turns	: 15, squared and ground ends
Free length	: 46.13 mm
Resonance frequency	: 34.69 Hz
FoS	: 1.45

9.6 RECTANGULAR- OR SQUARE-SECTION HELICAL SPRINGS

A helical-coiled spring made of rectangular- or square-section is used where a strong spring is required to be placed in a limited space. These springs, for a given space restriction, are stronger than the round-wire springs. However, their use is generally not recommended unless necessary, because of the following disadvantages:

(i) The wire is generally not uniform in cross-section.
(ii) The quality of the material is not good; it is produced in smaller quantities than the round wire.
(iii) The stress distribution is not as favourable as that for the round wire.
(iv) The shape of the wire does not remain rectangular or square while forming the coil, resulting in trapezoidal cross-section.
(v) The change in cross-section reduces the energy absorbing capacity.

For rectangular-section wire springs, the design equations are (Figure 9.11):

$$\text{Induced shear stress, } \tau = K \times \frac{FD\left(1.5h + 0.9b\right)}{b^2h^2} \tag{9.27}$$

where

K is the Wahl's stress correction factor $\left(= \dfrac{4C-1}{4C-4} + \dfrac{0.615}{C}\right)$

C is the spring index $\left(= \dfrac{D}{b}\right)$

b is the width of the wire section, measured perpendicular to the axis of the spring

h is the thickness of the wire, measured parallel to the spring axis

F is the axial force

D is the mean coil diameter.

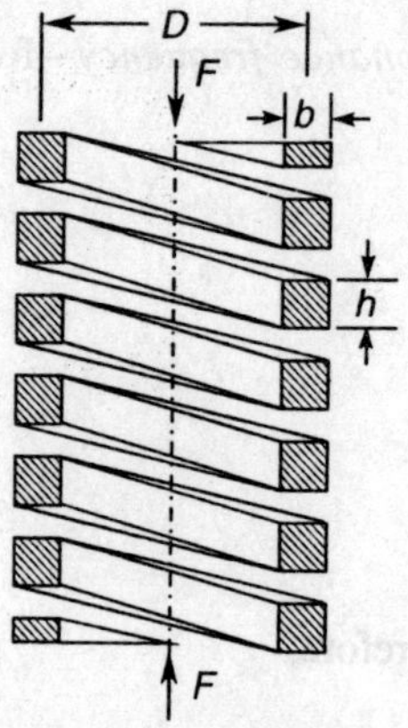

Figure 9.11 Rectangular-section spring.

The axial deflection of rectangular-section spring

$$y = \frac{2.83\, i\, FD^3\left(b^2 + h^2\right)}{b^3h^3G} \tag{9.28}$$

where

i is the number of active coils

G is the modulus of rigidity.

For the rectangular cross-section wire spring, the ratio b/h is usually kept between 0.4 and 1.0. When $b/h = 1$, it results into a square-section wire spring. The total number of turns for the square and round end condition is $i + 2$. The design procedure for the rectangular-section springs is similar to that for round-wire helical springs.

9.7 HELICAL TENSION SPRING

A helical tension spring is also called *extension spring* because when a tensile force is applied, the spring gets extended in length. Generally, these springs have some means of transferring the load from the support to the body of the spring. For this purpose, the ends of springs are converted into hooks which are highly stressed due to a sharp bend between the spring body and the end hook. Figure 9.12 shows a helical tension spring with the end hook.

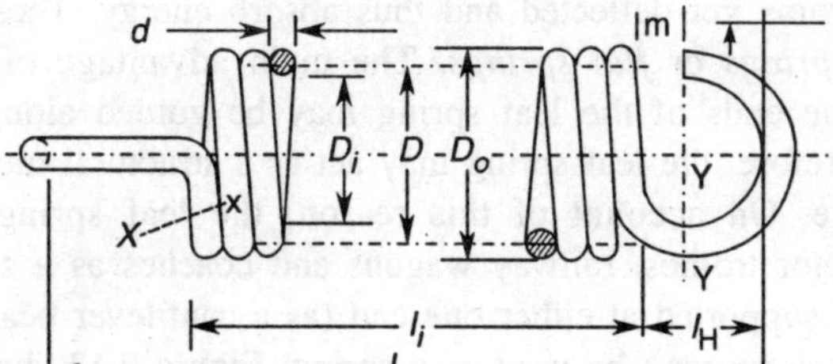

Figure 9.12 Helical tension spring.

The design procedure for helical tension springs is quite similar to that for compression springs except for the design of the end hook. The coils of a tension spring are usually wound tightly together so that there is an initial tension F_i. This force holds the spring accurately. The magnitude of initial tension is about 15–25 per cent of the maximum external load. These springs do not get deflected until the external force exceeds the initial tension.

Thus the deflection is

$$y = \frac{8(F - F_i) D^3 i}{G d^4} \tag{9.29}$$

and the spring stiffness

$$k = \frac{F - F_i}{y} \tag{9.30}$$

Thus the total design load is the sum of the initial load and the load required for the desired extension, i.e.

$$F = F_i + ky \tag{9.31}$$

The stress produced in the wire of the spring, say at section X–X shown in Figure 9.12 is due to torsion and direct shear. Thus the basic design equation of the compression spring is applicable, i.e.

$$\tau = K \times \frac{8FD}{\pi d^3}$$

However, the stresses produced in the hook are due to bending moment (Fr_m) and direct force F. The bend in the end coil causes a complicated stress distribution which may be approximated by using a stress–concentration factor. The stresses at section Y–Y in the hook

$$\sigma_{YY} = K_1 \times \frac{32Fr_m}{\pi d^3} + \frac{4F}{\pi d^2} \quad (9.32)$$

where

K_1 is the stress concentration factor ($= r_m/r_i$)

r_m is the mean radius of the hook end

r_i is the inside radius of the hook end

9.8 LEAF SPRINGS

Simply supported beams and cantilever beams may be used as springs because under a certain amount of load these beams get deflected and thus absorb energy. These types of springs are commonly called *leaf springs* or *flat springs*. The main advantage of leaf springs over the helical springs is that the ends of the leaf spring may be guided along a definite path as it deflects under load. Therefore, the leaf spring may act as a structural member in addition to an energy absorbing device. On account of this reason, the leaf springs are widely used in automotive vehicles, tractor trollies, railway wagons and coaches as a suspension system.

A single thin plate supported at either one end (as a cantilever beam) or at both the ends (as a simply supported beam), may be used as a spring. Figure 9.13 shows a cantilever spring and a simply supported spring, both of uniform cross-sections.

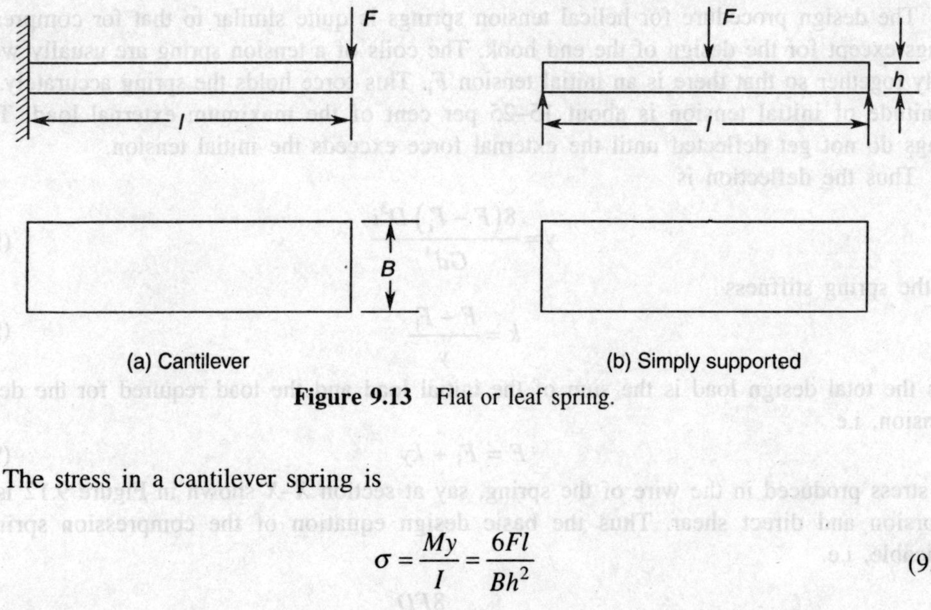

Figure 9.13 Flat or leaf spring.

The stress in a cantilever spring is

$$\sigma = \frac{My}{I} = \frac{6Fl}{Bh^2} \quad (9.33)$$

and the deflection

$$y = \frac{1}{3}\frac{Fl^3}{EI} = \frac{4Fl^3}{EBh^3} \tag{9.34}$$

Similarly, the stress and deflection in a simply supported spring are, respectively,

$$\sigma = \frac{3Fl}{2Bh^2} \tag{9.35}$$

and

$$y = \frac{Fl^3}{4EBh^3} \tag{9.36}$$

where

B is the width of the cross-section
h is the thickness
l is the length of the beam.

The main drawback of the above springs is that they are stressed heavily at one specific location and the other parts of the length are stressed lightly. Therefore, these springs can be made of uniform strength by keeping either a constant thickness or a constant width. Generally the thickness of these springs is kept constant and the width is made variable as shown in Figure 9.14.

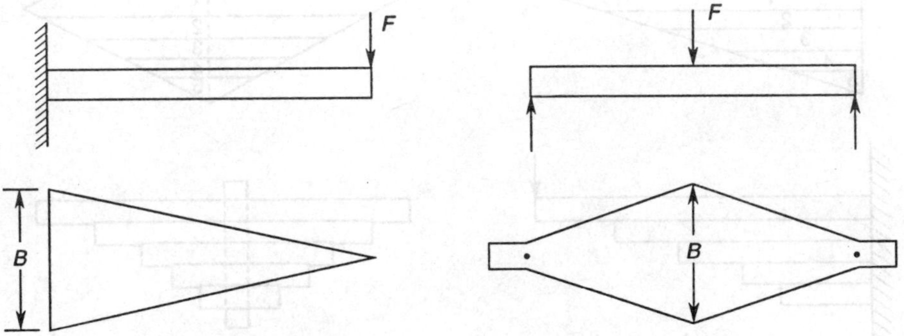

(a) Cantilever spring with varying width (b) Simply supported spring with varying width

Figure 9.14 Leaf springs with varying width.

The stresses and deflections of the spring with varying width are given as

(i) *Cantilever spring*

$$\text{Stress, } \sigma = \frac{6Fl}{Bh^2} \tag{9.37}$$

$$\text{Deflection, } y = \frac{6Fl^3}{EBh^3} \tag{9.38}$$

(ii) *Simply supported spring*

$$\text{Stress, } \sigma = \frac{3Fl}{2Bh^2} \tag{9.39}$$

$$\text{Deflection, } y = \frac{3Fl^3}{8EBh^3} \tag{9.40}$$

9.9 LAMINATED SPRINGS

As observed in Section 9.8 that in order to design a spring of uniform strength, the width of the spring is increased which becomes too large in a single leaf spring. Therefore, in order to decrease the width of the spring the triangular shape plate (in the case of cantilever springs) and the diamond shaped plate (in the case of simply supported springs) can be assumed to be cut in various strips as shown in Figure 9.15 and then assembled with clamps. The width of each leaf is $b = B/N$, where N is number of leaves. Such a spring is called the *laminated spring*. This type of spring is widely used in automobiles, railway carriages, coaches, and so on.

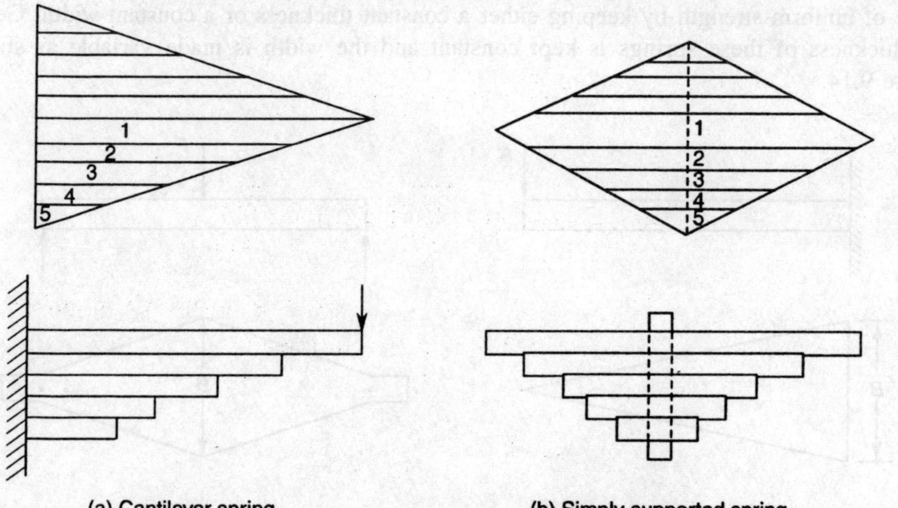

Figure 9.15 Laminated springs.

The maximum bending stress and deflection value for these springs are the same as for the original plate except that width B is replaced by Nb.

For cantilever springs

$$\text{Stress, } \sigma = \frac{6Fl}{Nbh^2} \tag{9.41}$$

$$\text{Deflection, } y = \frac{6Fl^3}{NEbh^3} \tag{9.42}$$

For simply supported springs

$$\sigma = \frac{3Fl}{2Nbh^2} \tag{9.43}$$

$$y = \frac{3Fl^3}{8NEbh^3} \tag{9.44}$$

9.9.1 Semi-elliptical Laminated Springs

The most common type of leaf spring used in automobiles is the semi-elliptical leaf spring as shown in Figure 9.16. It consists of a long curved leaf called the *master leave* and a certain number of small leaves called *graduated leaves*. In some springs there may be more than one

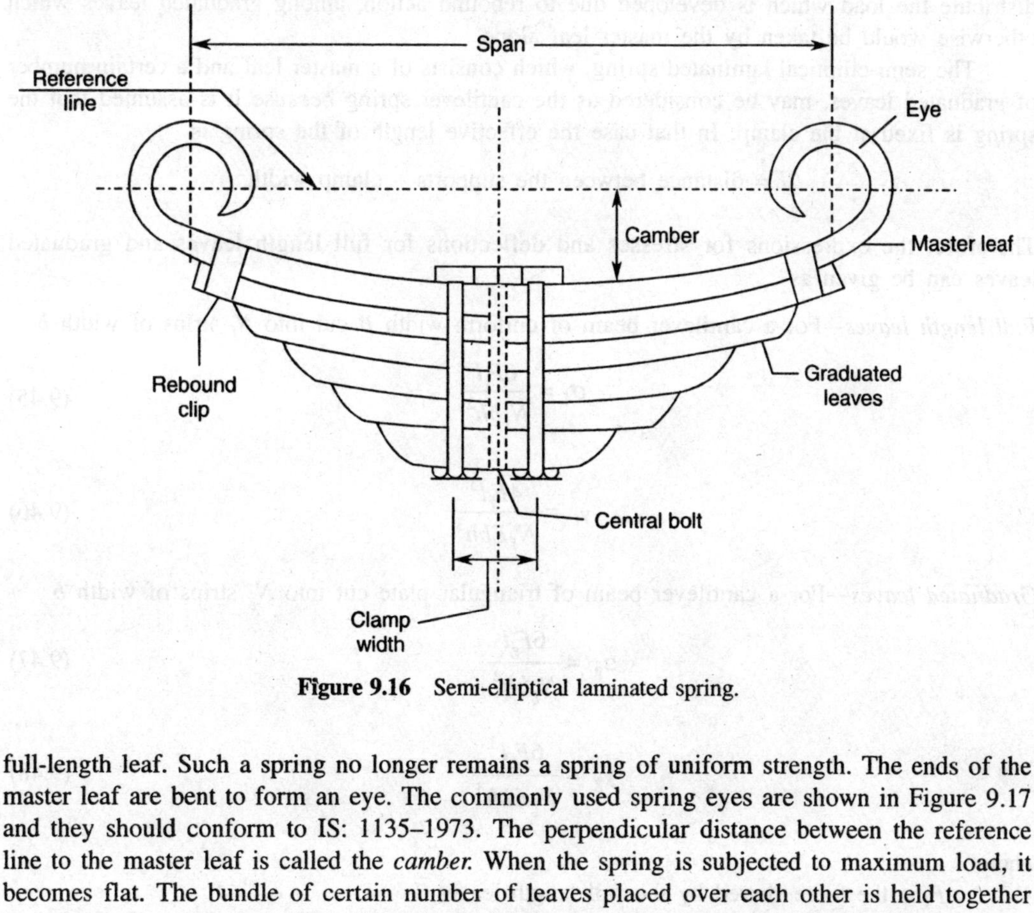

Figure 9.16 Semi-elliptical laminated spring.

full-length leaf. Such a spring no longer remains a spring of uniform strength. The ends of the master leaf are bent to form an eye. The commonly used spring eyes are shown in Figure 9.17 and they should conform to IS: 1135–1973. The perpendicular distance between the reference line to the master leaf is called the *camber.* When the spring is subjected to maximum load, it becomes flat. The bundle of certain number of leaves placed over each other is held together by means of a centre bolt. The dimensions of the centre bolt for leaf spring should conform to IS: 9484—1980. The hole drilled in the plate for the bolt weakens the spring. The pressure exerted by the U-clip, which holds the spring to the seat, reduces the bending stress in the

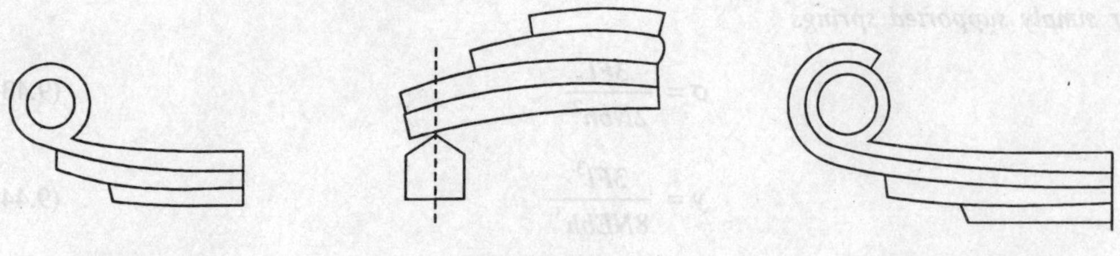

Figure 9.17 Types of leaves eye ends.

central part of the spring. The U-clip creates a stress concentration at the edge of the spring seat. In order to reduce this stress concentration, a soft pad is placed between the leaf and the seat. The ends of shorter or graduated leaves are clamped by the rebound clips. They help to distribute the load which is developed due to rebound action, among graduated leaves which otherwise would be taken by the master leaf alone.

The semi-elliptical laminated spring, which consists of a master leaf and a certain number of graduated leaves, may be considered as the cantilever spring because it is assumed that the spring is fixed at the clamp. In that case the effective length of the spring is

$$2l = \text{distance between the supports} - \text{clamp width}$$

Therefore, the expressions for stresses and deflections for full length leaves and graduated leaves can be given as

Full length leaves—For a cantilever beam of uniform width B cut into N_f strips of width b

$$\sigma_f = \frac{6F_f l}{N_f b h^2} \tag{9.45}$$

$$y_f = \frac{4F_f l^3}{N_f E b h^3} \tag{9.46}$$

Graduated leaves—For a cantilever beam of triangular plate cut into N_g strips of width b

$$s_g = \frac{6F_g l}{N_g b h^2} \tag{9.47}$$

$$y_g = \frac{6F_g l^3}{N_g E b h^3} \tag{9.48}$$

where

F_f is the force shared by the full length leaves

F_g is the force shared by the graduated leaves.

Now,

$$y_f = y_g$$

or

$$\frac{4F_f l^3}{N_f Ebh^3} = \frac{6F_g l^3}{N_g Ebh^3} \tag{9.49}$$

and

$$\text{Total force, } F = F_f + F_g \tag{9.50}$$

From Eqs. (9.49) and (9.50), we get

$$F_g = \frac{2N_g}{3N_f + 2N_g} F \tag{9.51}$$

and

$$F_f = \frac{3N_f}{3N_f + 2N_g} F \tag{9.52}$$

and stresses in full length leaves and graduated leaves are

$$\sigma_f = \frac{18Fl}{bh^2\left(3N_f + 2N_g\right)} \tag{9.53}$$

and

$$\sigma_g = \frac{12Fl}{bh^2\left(3N_f + 2N_g\right)} \tag{9.54}$$

The deflection, which is the same for both full length leaves and graduated leaves, is

$$y = \frac{12Fl^3}{Ebh^3\left(2N_g + 3N_f\right)} \tag{9.55}$$

9.9.2 Nipping of Laminated Springs

It may be seen from Eqs. (9.53) and (9.54) that the stress in the full length leaves is 50% greater than the stresses in the graduated leaves. Therefore, in order to stress all the leaves equally, the full length leaves are given a greater radius of curvature than the graduated leaves and thereafter they should be assembled to form a spring as shown in Figure 9.18. This will create an initial gap (c) between the leaves, called the *nip*. When the centre spring clip is drawn up tight, the top leaf will bend backwards and have an initial stress opposite to that produced by the external service load. Therefore when the service load is applied, the resultant stress in the top leaf will have a lower value than that of without the initial gap. Such a pre-stressing obtained by a difference of radii curvature is known as *nipping*. It is very common in automobile suspension springs where the top leaf has to carry an additional load due to swaying of the car, brake torque, etc.

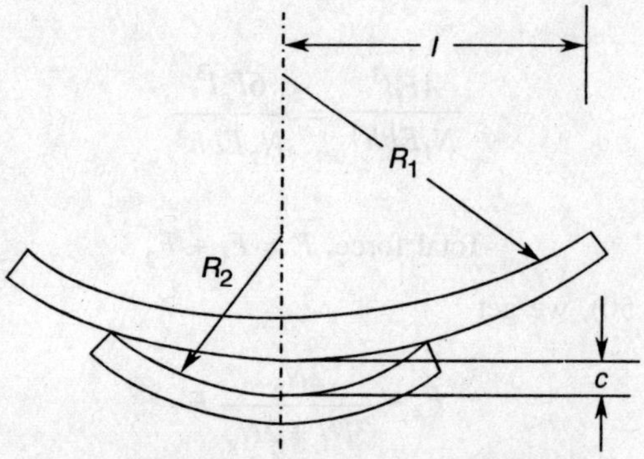

Figure 9.18 Nipping of laminated spring.

Assuming that the above pre-stressing results in stress equalization, by equating Eqs. (9.45) and, we get

$$\sigma_g = \sigma_f$$

or

$$\frac{F_g}{F_f} = \frac{N_g}{N_f} \tag{9.56}$$

Also

$$F = F_g + F_f \tag{9.57}$$

Solving Eqs. (9.56) and (9.57), we get

$$F_g = \frac{N_g}{N_f} F \quad \text{and} \quad F_f = \frac{N_f}{N_g} F$$

Under the maximum force F, the deflection of the graduated leaves will exceed the deflection of the full length leaves by an amount equal to the initial gap c or nip. Therefore,

$$c = \frac{6F_g l^3}{N_g Ebh^3} - \frac{4F_f l^3}{N_f Ebh^3}$$

Substituting the values of F_g and F_f, we get

$$c = \frac{2Fl^3}{NEbh^3} \tag{9.58}$$

where $N = N_f + N_g$.

The load on the clip bolt F_b required to close the gap can be determined by the fact that the total gap

$$c = y_g + y_f$$

$$= \frac{6l^3}{N_g Ebh^3} \times \frac{F_b}{2} + \frac{4l^3}{N_f Ebh^3} \times \frac{F_b}{2} \tag{9.59}$$

From Eqs. (9.58) and (9.59), we get

$$F_b = \frac{2N_g N_f}{N(3N_f + 2N_g)} \times F \tag{9.60}$$

The final stress in the spring leaves will be the stress in the full length leaves due to the applied load minus the initial stress. Thus,

$$\sigma = \frac{6l}{N_f bh^2}\left(F_f - \frac{F_b}{2}\right)$$

Substituting the values of F_f and F_b and simplifying, we get

$$\sigma = \frac{6Fl}{Nbh^2} \tag{9.61}$$

The deflection caused by the maximum force is the same as that caused in the spring without the initial deflection. Therefore,

$$y = \frac{12Fl^3}{Ebh^3(3N_f + 2N_g)} \tag{9.62}$$

9.9.3 Materials for Leaf Springs

Leaf springs are generally made of annealed plain carbon steel having 0.9–1.0 per cent carbon. In certain cases where high toughness and high endurance strength are required, chrome–vanadium and silicon–manganese steels may be used. SAE 6140, 6150, 9250, and 9260 are prominent among this category. The properties of these steels may be taken from a standard handbook. The working stress should be based on endurance limit which should not exceed the yield strength. The endurance limit in reverse bending may be taken as 40 per cent of ultimate strength. In order to take into account the surface defects due to rolling and large variations in stress during service, a factor of safety of the order of 1.5–3.5 may be assumed.

The nominal sizes of leaves, as recommended by IS: 3431–1965 code, are given in Table 9.3.

Table 9.3 Standard sizes of leaves

Width (mm)	25, 30, 35, 40, 45, 50, 55, 60, 65, 70, 75, 80, 90, 100, 120, and 140
Thickness (mm)	2, 3, 4, 5, 6, 7, 8, 10, 12, 14, and 16

9.9.4 Other Design Considerations

Besides computing the cross-section of flats, number of leaves, size of the centre bolt, etc. the following additional design considerations should be taken into account:

1. The camber of the spring may be taken equal to half the maximum deflection.
2. The radius of curvature of a leaf may be computed by the geometric relation shown in Figure 9.19.

$$\text{Curvature radius, } R = \sqrt{CB^2 + OC^2} \tag{9.63}$$

where

OC = R – CD

CD = camber

3. The length of leaves may be computed by the following relation:

$$l_i = \frac{\text{effective length}}{N-1} \times i + \text{ineffective length} \tag{9.64}$$

where

effective length = distance between the spring supports – width of the central band ineffective length = width of the central band

In the case of the master leaf with eye end, the length of the leaf is equal to

$$l_{master} + 2\pi(d + h) \tag{9.65}$$

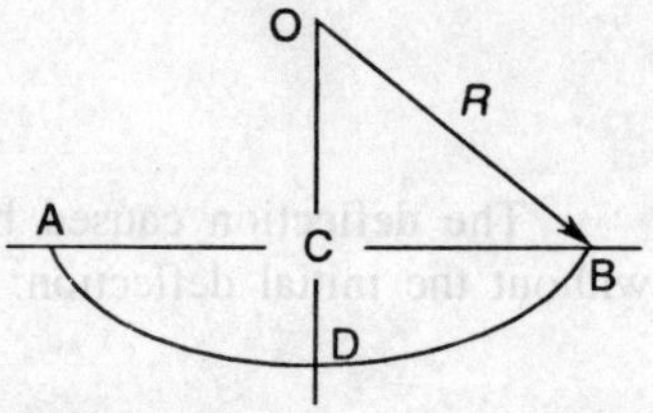

Figure 9.19 Geometry of a leaf.

where

d is the diameter of the eye

h is the thickness of the leaf.

4. While developing the equation for the laminated spring it was assumed that leaves were uniform in thickness and their ends were pointed. However, in automobile springs the ends are usually round and thinned, as shown Figure 9.20. So this condition slightly changes the deflection.
5. The friction between leaves tends to reduce the deflection and makes the spring stiffer. Therefore, in order to reduce the wear and to obtain uniform spring action, separator pads may be used between the ends of adjacent leaves.
6. Instead of the central bolt which increases stress concentration, shrunk bands may be used to avoid stress concentration.

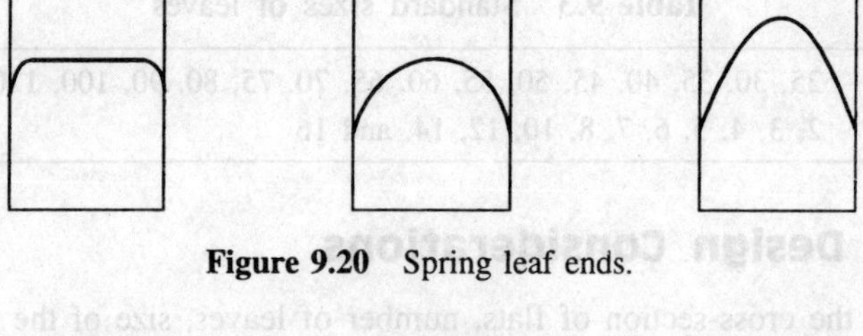

Figure 9.20 Spring leaf ends.

Example 9.4 Design a leaf spring for the rear axle of a tractor trolley. The load on the rear axle of the trolley is 10,000 N. The span is 1200 mm and the width of clamp is 100 mm. In all, 12 leaves are used out of which two are main leaves and the remaining graduated leaves.

Solution Assuming that the 10 kN load is equally shared by the two springs, the load on each spring

$$F = \frac{10{,}000}{2} = 5000\ \text{N}$$

$$\text{Effective length, } 2l = \text{span} - \text{width of the clamp}$$

$$= 1200 - 100 = 1100$$

or

$$l = 550\ \text{mm}$$

We assume that the spring material is chrome–vanadium steel having ultimate strength, σ_{ut} = 1500 N/mm^2 and endurance limit = $0.4\sigma_{ut}$/FoS.

Let FoS = 2 for the spring subjected to fatigue load. Then,

$$\sigma_d = 0.4 \times \frac{1500}{2} = 300\ \text{N/mm}^2$$

The stress produced in the spring

$$\sigma = \frac{6Fl}{Nbh^2}$$

or

$$300 = \frac{6 \times 5000 \times 550}{12 \times bh^2}$$

or

$$bh^2 = 4583\ \text{mm}^3$$

Assuming width of the flat, b = 55 mm

$$\text{Thickness, } h = \sqrt{\frac{4583}{55}} = 9.12\ \text{mm, say, 9 mm}$$

$$\text{Ratio of the total depth to width of the spring} = \frac{N \times h}{b} = \frac{12 \times 9}{55} = 1.96$$

which is reasonable.

$$\text{Maximum deflection, } y = \frac{12Fl^3}{Ebh^3(2N_g + 3N_f)}$$

$$= \frac{12 \times 5000 \times 550^3}{2 \times 10^5 \times 55 \times 9^3(2 \times 10 + 3 \times 2)}$$

$$= 47.88\text{, say, 48 mm}$$

$$\text{Camber} = 0.5 \times \text{maximum deflection}$$

$$= 0.5 \times 48 = 24\ \text{mm}$$

Radius of curvature

$$R = \sqrt{OC^2 + CB^2} \qquad (OC = R - 24;\ CB = 550)\ \text{(Refer Figure 9.19)}$$

$$= \sqrt{(R-24)^2 + 550^2}$$

or

$$R = 6314 \text{ mm}$$

Design of the pin

$$\text{Load on the pin} = \frac{F}{\cos 45°} = \frac{5000}{\cos 45°} = 7071 \text{ N}$$

(i) *Bearing consideration*

Let the bearing pressure p_b be 10 N/mm^2. Therefore,

$$d_p \times p_b \times b = 7071$$

or

$$d_p = \frac{7071}{10 \times 55} = 12.8 \text{ mm}$$

(ii) *Bending consideration*

$$\text{Bending moment, } M = 7071 \times \frac{b'}{4}$$

where

$$b' = \text{moment arm length} = b + 2 \times \text{clearance}$$
$$= 55 + 2 \times 2.5 = 60 \text{ mm}$$

Assuming that the pin is made of plain carbon steel having allowable bending strength of 120 N/mm^2,

$$\text{Diameter of the pin, } d_p = \left(\frac{32M}{\pi\sigma}\right)^{1/3} = \left[\frac{32 \times 7071 \times 60}{\pi \times 120 \times 4}\right]^{1/3} = 20.8 \text{ mm, say, 22 mm}$$

(iii) *Check for double shear failure*

$$\sigma_s = \frac{F}{A} = \frac{7071}{2 \times \frac{\pi}{4} \times 22^2} = 9.3 \text{ N/mm}^2$$

which is very low. Hence the safe value of pin diameter is 22 mm

Length of leaves

$$l_i = \frac{\text{effective length}}{N-1} \times i + \text{ineffective length}$$

$$l_1 = \frac{1100}{11} \times 1 + 100 = 200 \text{ mm}$$

$$l_2 = \frac{1100}{11} \times 2 + 100 = 300 \text{ mm}$$

Similarly, other lengths can be determined, which are

$$l_3 = 400 \text{ mm}, l_4 = 500 \text{ mm}, l_5 = 600 \text{ mm}, l_6 = 700 \text{ mm}$$

$$l_7 = 800 \text{ mm}, l_8 = 900 \text{ mm}, l_9 = 1000 \text{ mm}, l_{10} = 1100 \text{ mm, and } l_{11} = 1200 \text{ mm}$$

Length of the master leaf

$$l_{\text{master}} = 1200 + 2\pi(d_p + h)$$

$$= 1200 + 2\pi(22 + 9)$$

$$= 1394.7 \text{ mm}$$

Load on centre bolt

$$F_b = \frac{2N_g N_f}{N(2N_g + 3N_f)} \times F$$

$$= \frac{2 \times 10 \times 2}{12(2 \times 10 + 3 \times 2)} \times 5000 = 641 \text{ N}$$

Assuming that the bolt is made of mild steel having allowable tensile strength of 100 N/mm^2,

$$\text{Resisting area} = \frac{\text{load}}{\text{stress}} = \frac{641}{100} = 6.41 \text{ mm}^2$$

Let us adopt the M8 × 1.0 bolt.

Design specifications of the spring

Material	: Chrome–vanadium steel
Width of the flat	: 55 mm
Thickness	: 9 mm
Number of leaves	: 12
Full length leaves	: 2
Graduated leaves	: 10
Radius of curvature	: 6314 mm
Camber	: 24 mm
Maximum deflection	: 48 mm
Length of leaves (mm)	: 300, 400, 500, 600, 700, 800, 900, 1000, 1100, 1200
Length of the master leaf	: 1394.7 mm
Size of the bolt	: M8 × 1.0

9.10 CONCENTRIC HELICAL SPRINGS

The concentric spring, as the name suggests, is a composite or cluster spring in which two or more than two coiled helical springs are placed—one inside the other. A typical two-coiled concentric spring is shown in Figure 9.21. In this spring, the adjacent coils are wound in opposite directions. If the materials of both springs are the same, it is desirable to have the same index so that the stress values induced in both the springs are the same.

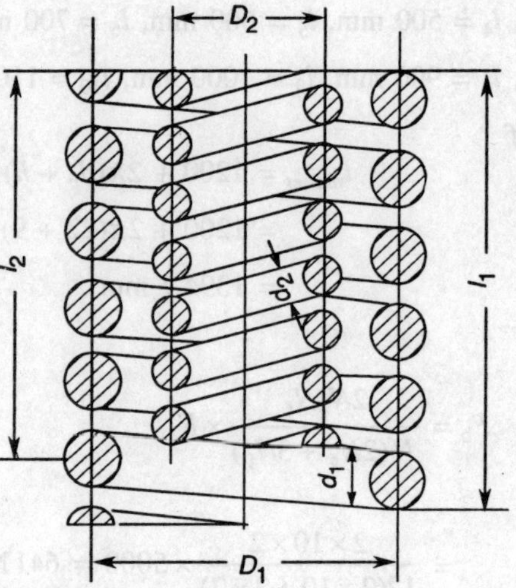

Figure 9.21 Concentric spring.

These springs are used primarily for the following purposes:

(i) To obtain a greater spring force in a limited space.
(ii) To ensure the operation of a mechanism in the event when one spring fails.
(iii) To obtain a variable force–deflection relation.

In order to realize the first two objectives, the concentric springs must have the same free length besides fulfilling the following conditions.

1. Assuming the same material for both the springs the maximum shear stress in both the springs should be the same. Thus,

$$\frac{8F_1D_1}{\pi d_1^3} = \frac{8F_2D_2}{\pi d_2^3}$$

or

$$\frac{F_1}{F_2} = \left(\frac{D_2}{D_1}\right)\left(\frac{d_1}{d_2}\right)^3 \tag{9.66}$$

where

F_1, F_2 is the force shared by springs 1 and 2, respectively ($F = F_1 + F_2$)

D_1, D_2 are the mean coil diameters

d_1, d_2 are the wire diameters

2. The deflection should be the same for both the springs. Thus,

$$\frac{8F_1D_1^3i_1}{G_1d_1^4} = \frac{8F_2D_2^3i_2}{G_2d_2^4}$$

For $G_1 = G_2$ and i_1, i_2 = number of active coils

$$\frac{F_1}{F_2} = \left(\frac{D_2}{D_1}\right)^3 \left(\frac{d_1}{d_2}\right)^4 \left(\frac{i_2}{i_1}\right) \tag{9.67}$$

Dividing Eq. (9.67) by Eq. (9.66), we get

$$1 = \left(\frac{D_2}{D_1}\right)^2 \left(\frac{d_1}{d_2}\right) \left(\frac{i_2}{i_1}\right) \tag{9.68}$$

3. Both springs, when compressed, must have the same solid length. Thus

$$i_1d_1 = i_2d_2 \quad \text{or} \quad \frac{d_1}{d_2} = \frac{i_2}{i_1} \tag{9.69}$$

Using Eqs. (9.68) and (9.69), we get

$$1 = \left(\frac{D_2}{D_1}\right)^2 \left(\frac{d_1}{d_2}\right)^2$$

or

$$\frac{D_1}{d_1} = \frac{D_2}{d_2} = C \tag{9.70}$$

Thus while designing the concentric springs, their proportions should be so chosen that their spring index is the same.

Using Eqs. (9.66) and (9.70), we get

$$\frac{F_1}{F_2} = \frac{d_1^2}{d_2^2} \tag{9.71}$$

Another important requirement of concentric springs is that the net clearance between them be kept $(d_1 - d_2)/2$. Thus,

$$\frac{(D_1 - d_1) - (D_2 + d_2)}{2} = \frac{d_1 - d_2}{2}$$

After simplification, we get

$$\frac{d_1}{d_2} = \frac{C}{C-2} \tag{9.72}$$

The load on each spring can, therefore, be determined by using Eqs. (9.71) and (9.72). Thereafter, the usual design procedure can be used to determine other dimensions of concentric springs.

When concentric springs are to be designed for variable force–deflection ratio, the springs should have different free lengths and different deflections. Figure 9.22 shows the force–deflection relations for three concentric springs. Their lengths are l_1, l_2, and l_3 and maximum deflections are y_1, y_2, and y_3, respectively. The inclined lines 1–1, 2–2, and 3–3 represent the force variation of springs and the curve 1–4–5–6 represents the gradual increase of the spring force.

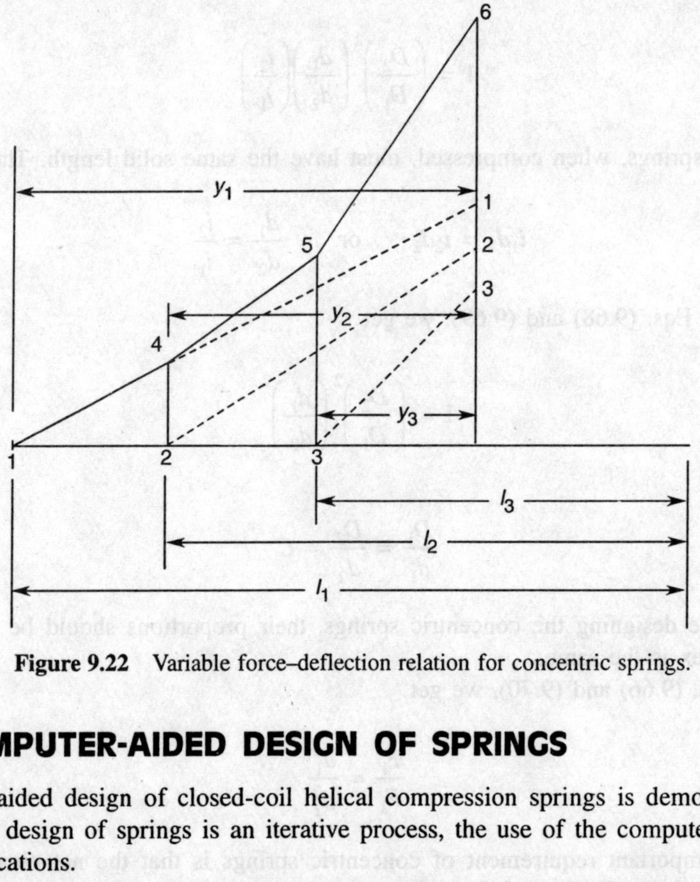

Figure 9.22 Variable force–deflection relation for concentric springs.

9.11 COMPUTER-AIDED DESIGN OF SPRINGS

The computer-aided design of closed-coil helical compression springs is demonstrated in this section. As the design of springs is an iterative process, the use of the computer is best suited for these applications.

The listing of the computer program for the design of closed-coil helical springs is given in Program 9.1. The variables used are mostly discussed in theory. In this program, seven different types of materials are used, however, the user can modify the program to increase the list of materials. The strength of the wire material is calculated by empirical relations reported in *Machine Design* by J.E. Sighley. The selection of the wire size is based on standard sizes available in IS: 4454–1981. In order to perform design, the user is expected to furnish the following input details: (i) maximum force, (ii) deflection, (iii) selection of material, (iv) modulus of rigidity, (v) factor of safety, and (vi) spring index. In this program, square and ground end conditions have been chosen and spring material density is kept 7800 kg/m^3. However, the user may change these values for different applications. A sample run of this program is demonstrated.

Program 9.1

```
//HELICAL SPRING DESIGN PROGRAM IN C++
#include <iostream.h>
#include <math.h>
#define PI 3.1416
void main()
{
     double f, y, sigmau, tauy, g, fos, c;
     double    a, m, tau_d, tau, k, b, d, od, 1solid, 1free, n;
     double    mass, stiff, freq;
     int       mat_code;
     char      another_try;
// input to spring design
COUT << "\N          HELICAL SPRING DESIGN \n";
     cout << "\n     Maximum Spring Force (Newton)          F=";
     cin >> f;
     cout << "\n     Maximum Deflection (mm)                Y = ";
     cin >> y;
     cout << "\n     Select Material: \n";
     cout << "\n          1.   Patented Cold Drawn Steel Grade-1";
     cout << "\n          2.   Patented Cold Drawn Steel Grade-2";
     cout << "\n          3.   Patented Cold Drawn Steel Grade-3";
     cout << "\n          4.   Patented Cold Drawn Steel Grade-4";
     cout << "\n          5.   Oil Tempered Unalloy Steel";
     cout << "\n          6.   Chrome-Vanadium Steel";
     cout << "\n          7.   Chrome-Silicon Steel\n";
     cout << "\n                     Enter Your Selection (1-7):";
     cin >> mat_code;
     cout << "\n     Modulus of Rigidity (N/mm2)     G=";
     cin >> g;
     cout << "\n     Factor of Safety                FoS=";
     cin >> fos;
// input of spring index "C"
retry:
     cout << "\n          Spring Index (D/d)         C=";
     cin >> c;
//   print design specification
     for (int i=1; i<=10; i=i+1)
```

```
        {
            cout <<"\n";
        }
        cout << "\n                    SPRING DESIGN SPECIFICATIONS\n";
//      classify material selection and compute strength
        switch (mat_code)
        {
        case 1:
        cout << "\n            Spring Material: Patented Cold Drawn Steel Grade-1";
        cout << "\n                             as per IS: 4454–1981\n";
            a=1630.0;
            m=0.251;
            break;
        case 2:
        cout << "\n            Spring Material: Patented Cold Drawn Steel Grade-2";
        cout << "\n                             as per IS: 4454–1981\n";
            a=1720.0;
            m=0.191;
            break;
        case 3:
        cout << "\n            Spring Material: Patented Cold Drawn Steel Grade-3";
        cout << "\n                             as per IS: 4454–1981\n";
            a=1980.0;
            m=0.179;
            break;
        case 4:
        cout << "\n            Spring Material: Patented Cold Drawn Steel Grade-4";
        cout << "\n                             as per IS: 4454–1981\n";
            a=2160.0;
            m=0.145;
            break;
        case 5:
        cout << "\n            Spring Material: Oil Tempered Steel";
        cout << "\n                             as per IS: 4454–1981\n";
            a=1880.0;
            m=0.186;
            break;
        case 6:
        cout << "\n            Spring Material: Chrome-Vanadium Steel";
        cout << "\n                             as per IS: 4454–1981\n";
            a=2000.0;
            m=0.167;
            break;
        case 7:
        cout << "\n            Spring Material: Chrome-Silicon Steel";
        cout << "\n                             as per IS: 4454–1981\n";
            a=2000.0;
            m=0.112;
            break;
```

```
      default:
            cout << "Illegal Material selection";
            goto stop;
      }
//    Choose trial value of wire diameter (d)
            d=0.05;
            k=(4.0*c-1.0)/(4*c-4.0) + 0.615/c;
//    calculate design stress
calculate: sigmau=a/pow (d,m);
                  tauy=0.43*sigmau;
                  if (d > 12.0)
                  {
                        b=pow(d,0.25)/1.885;
                  }
                  else
                  {
                        b=1.0;
                  }
                  tau_d=tauy/(b*fos);
                  tau_d=0.01*int(100.0*tau_d);
//    compute induced shear stress
      tau=(8.0*k*f*c)/(PI*d*d);
//    compare value of tau and tau_d and if necessary,
//    select another wire diameter
      if (tau_d < tau)
      {
            if (d<=1.1)
            {
                  d=d+0.05;
                  goto calculate;
            }
            if (d<=2.6)
            {
                  d=d+0.1;
                  goto calculate;
            }
            if (d<=4.0)
            {
                  d=d+0.2;
                  goto calculate;
            }
            if (d<=6.5)
            {
                  d=d+0.25;
                  goto calculate;
            }
            if (d<=11.0)
```

```
        {
                d=d+0.5;
                goto calculate;
        }
        if (d<=17.0)
        {
                d=d+1.0;
                goto calculate;
        }
        else
        {
                cout<< "\n        Wire diameter beyond 17 mm not available";
                goto stop;
        }
}
//      calculate number of turns
        n=y*d*g/(8.0*f*pow(c,3.0));
        n=int(n+0.5);
//      calculate other specification
        od=c*d+d;
        lsolid=(n+2)*d;
        1free=1solid+1.15*y;
//      print design specification
        cout << "\n          Wire Diameter            =" << d << "mm";
        cout << "\n          Mean Coil Diameter       =" << c*d << "mm";
        cout << "\n          Outside Diameter         =" << od << "mm";
        cout << "\n          Spring Index             =" << c;
        cout << "\n          Number of Turns          =" << n+2;
        cout << "\n          Square and Ground Ends";
        cout << "\n          Free Length                    ="<<1free << "mm";
        if (lfree/(c*d)>3.2)
        {
                cout <<" with Mandrill";
        }
        else
        {
                cout << "without Mandrill";
        }
        mass=PI*PI*d*d*c*d*(n+2)*7800.0/4.0e+0.9;
        stiff=f*1000.0/y;
        freq=(0.5*sqrt(stiff/mass))/(2.0*PI);
        cout << "\n Surge Frequency= " << freq << "Hz";
//      Try another solution with different value of spring index
        cout << "\n Try with Another Value of Spring Index <Y/N>:";
        cin >> another_try;
        if (another_try =='Y' || another_try == 'y')
```

```
{
        goto retry;
}
stop: cout << "\n        End of Program!\n";
}
```

Sample Run of Program 9.1

```
                    HELICAL SPRING DESIGN

Maximum Spring Force F(Newton)=3000
Maximum Deflection(mm)          Y=50
Select Material:
        1.  Patented Cold Drawn Steel Grade-1
        2.  Patented Cold Drawn Steel Grade-2
        3.  Patented Cold Drawn Steel Grade-3
        4.  Patented Cold Drawn Steel Grade-4
        5.  Oil Tempered Unalloy Steel
        6.  Chrome-Vanadium Steel
        7.  Chrome-Silicon Steel
            Enter Your Selection(1-7):5
Modulus of Rigidity(N/mm^2)                 G=84000
Factor of Safety                            FoS =1.5
Spring Index (D/d)                          C =5.5

                SPRING DESIGN SPECIFICATIONS

Spring Material: Oil Tempered Steel
                 As per IS: 4454-1981
Wire Diameter          =13mm
Mean Coil Diameter     =71.5mm
Outside Diameter       =84.5mm
Spring Index           =5.5
Number of Turns        =15
Square and Ground Ends
Free Length            =252.0mm with Mandrill
Surge Frequency        =10Hz
Try with Another Value of Spring Index(Y/N):y
Spring Index (D/d)                          C=6

                SPRING DESIGN SPECIFICATIONS

Spring Material: Oil Tempered Steel
                 As per IS:4454-1981
Wire Diameter          =14mm
Mean Coil Diameter     =84mm
Outside Diameter       =98mm
Spring Index           =6
```

Number of Turns =13
Square and Ground Ends
Free Length =239.5mm without Mandrill
Surge Frequency =14Hz
Try with Another Value of Spring Index(Y/N):n

EXERCISES

1. What type of stresses are produced in the wire of a closed-coiled helical spring? Draw the distribution of stresses.
2. What is Wahl's correction factor?
3. What do you mean by buckling of a spring? How can it be prevented?
4. What do you understand by surge in a spring and how can it be eliminated?
5. What are the commonly used materials for springs? What are the factors which govern the choice of a particular material?
6. What are the advantages and disadvantages of square-section helical springs?
7. What is nipping in leaf springs? Discuss its role in spring design.
8. In what ways, can the fatigue resistance of a leaf spring be increased?
9. Why are separator pads used between the ends of adjacent leaves of laminated springs?
10. For what purpose is a concentric helical compression spring used?
11. A railway bogie resting on eight helical springs weighs 240 kN along with its goods. The dynamic load on the wagon due to irregularities on the rail track may be 40 kN. Design the spring, if the required stiffness of the spring is 2 kN/mm. Assume a suitable material and factor of safety.
12. Design a suitable spring for the exhaust valve of a petrol engine. The spring should be capable of exerting a net force of 360 N when the valve is open and 220 N when it is closed. The maximum inside diameter of the spring is 25 mm. The compression in spring is 8 mm.
13. A circular cam 300 mm in diameter rotates with an eccentricity of 35 mm and operates a roller follower that is carried by the arm as shown in the figure below. The follower is kept in contact with the cam by means of an extension spring. The force between the follower and the cam is approximately 300 N at the lowest position and 600 N at the highest position. Design a suitable spring.

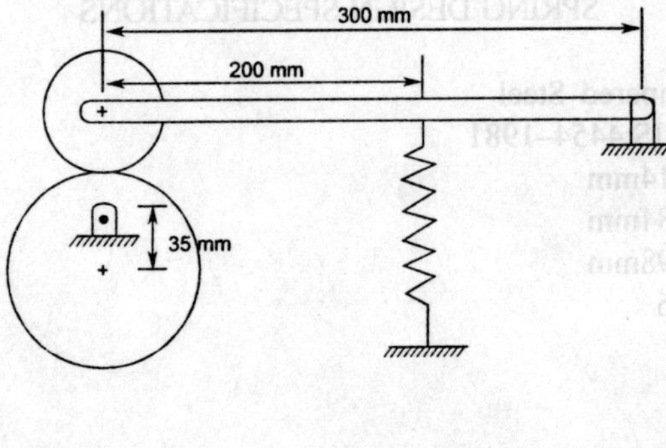

14. An engine indicator has a plunger diameter of 20 mm. The indicator spring attached to this plunger is to be compressed by 12 mm when the steam pressure acting on the plunger is 5 N/mm^2. The mean diameter of the coil is three times that of the wire diameter. Design the spring.

15. A car weighing 15 kN is supported by four semi-elliptical springs with the load equally distributed on front and rear axles. Considering the available space, it is decided to use a spring which is 1400 mm long and 50 mm in width. Determine the number of leaves and the thickness, if the deflection at rest is assumed to be 100 mm.

16. The spring in a truck has 12 leaves, two of which are full length. The spring supports are 1.05 m apart and the central band is 85 mm wide. The central load is 5.4 kN and the permissible stress of the material is 280 N/mm^2. Design the spring.

MULTIPLE CHOICE QUESTIONS

1. Shear stress induced in a closed-coiled helical spring is

(a) $\dfrac{8FD}{\pi d^3}$ (b) $\dfrac{4FD}{\pi d^3}$

(c) $\dfrac{32FD}{\pi d^3}$ (d) $\dfrac{\pi d^3}{8FD}$

2. The Wahl's correction factor accounts for

(a) direct shear stress (b) effect of curvature
(c) both (a) and (b) (d) none of the above

3. The value of the Wahl's stress factor with increase in the value of C

(a) decreases (b) increases
(c) remains the same (d) is unpredictable

4. Concentric helical springs should be wound

(a) in the same helix hand (b) in opposite helix hands
(c) in any direction (d) depending on the load

5. The allowable stress in compression springs for most of the materials with increase in the size of wire will

(a) increase (b) decrease
(c) remain the same (d) be unpredictable

6. The deflection of helical spring is inversely proportional to the

(a) wire diameter (b) (wire diameter)^2
(c) (wire diameter)^3 (d) (wire diameter)^4

7. The type of spring used in a clock is

(a) leaf (b) helical
(c) conical (d) spiral

8. To equalize the stress in leaf springs, the initial radius of curvature of full length leaves should be
   (a) smaller than the graduated leaves
   (b) greater than the graduated leaves
   (c) equal and opposite to the graduated leaves
   (d) equal to the graduated leaves
9. The minimum number of full length leaves should be
   (a) 4 (b) 3
   (c) 2 (d) 1
10. For square and ground end condition of helical springs, the number of inactive turns are
   (a) 1 (b) 2
   (c) 0 (d) 3
11. The stiffness of conical springs is
   (a) constant (b) variable (c) unpredictable

Chapter 10 Levers

10.1 INTRODUCTION

A lever is a simple mechanical device which is either a straight or a curved link or a rigid rod moving about some point called *fulcrum* and works on the principle of moment. It is used to facilitate the application of force in a desired direction and to get mechanical advantage. The basic terms associated the design of levers (Figure 10.1) are defined as:

Load (W). A load which is required to be overcome by effort.

Effort (P). A force applied to a lever against the load to be lifted is known as effort.

Mechanical advantage. The ratio of load to effort is called the mechanical advantage (MA).

$$MA = \frac{\text{load } (W)}{\text{effort } (P)}$$

Leverage. The ratio of the effort arm length l_2 to the load arm length l_1 is known as leverage.

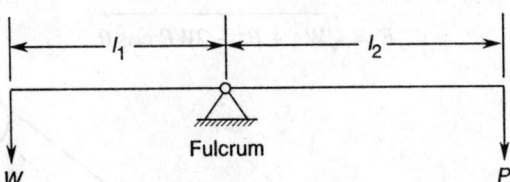

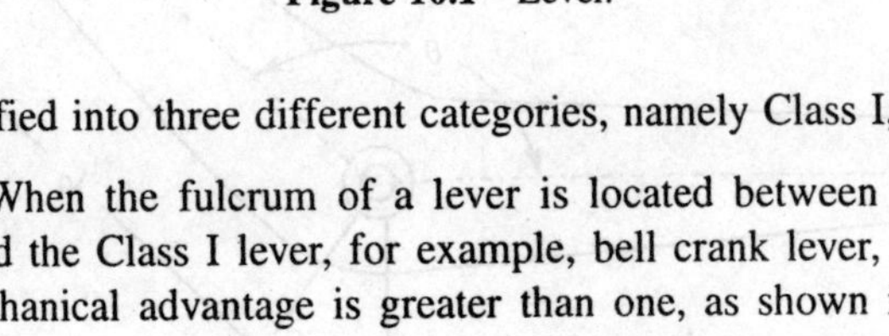

Figure 10.1 Lever.

Levers are classified into three different categories, namely Class I, Class II, and Class III.

Class I lever. When the fulcrum of a lever is located between the load point and the effort point, it is called the Class I lever, for example, bell crank lever, rocker arm, etc. In this type of lever, the mechanical advantage is greater than one, as shown in Figure 10.2(a).

Class II lever. A lever having the load point between the effort point and the fulcrum is known as the Class II lever, for example, the lever used in safety valves. Figure 10.2(b) shows a Class II lever.

Class III lever. A lever having the effort point located between the load point and the fulcrum is classified as Class III lever. The mechanical advantage of this class of levers is always less than one; hence they are seldom used [Figure 10.2(c)].

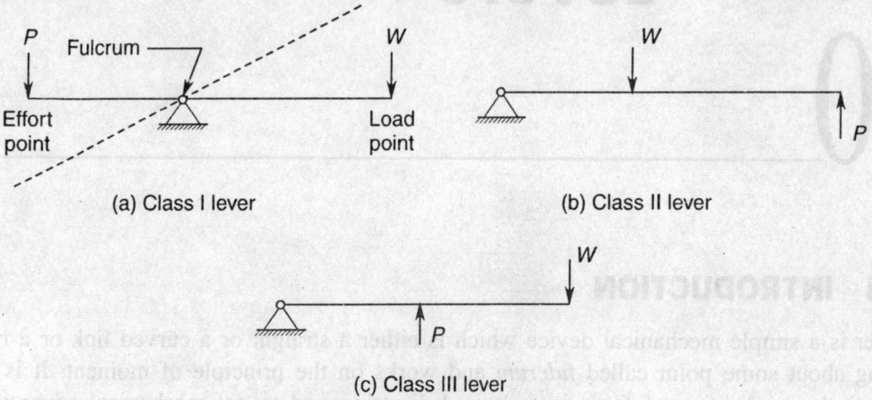

Figure 10.2 Types of levers.

10.2 DESIGN PROCEDURE

The following procedure is generally adopted for the design of a lever.

1. The reaction force at the fulcrum is determined as below:
   (a) When the load and effort are parallel, the resultant reaction force at the fulcrum is the algebraic sum of these two forces.
   (b) When the load and effort arms are inclined to each other at an angle θ as shown in Figure 10.3, the reaction F at the fulcrum can be obtained by the following relation:

$$F = \sqrt{W^2 + P^2 - 2WP \cos\theta} \tag{10.1}$$

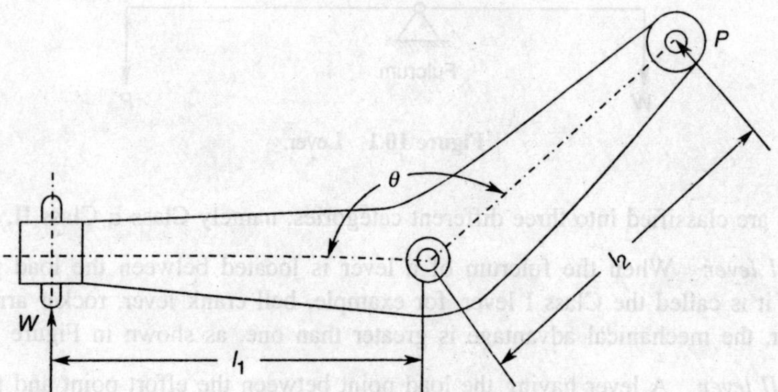

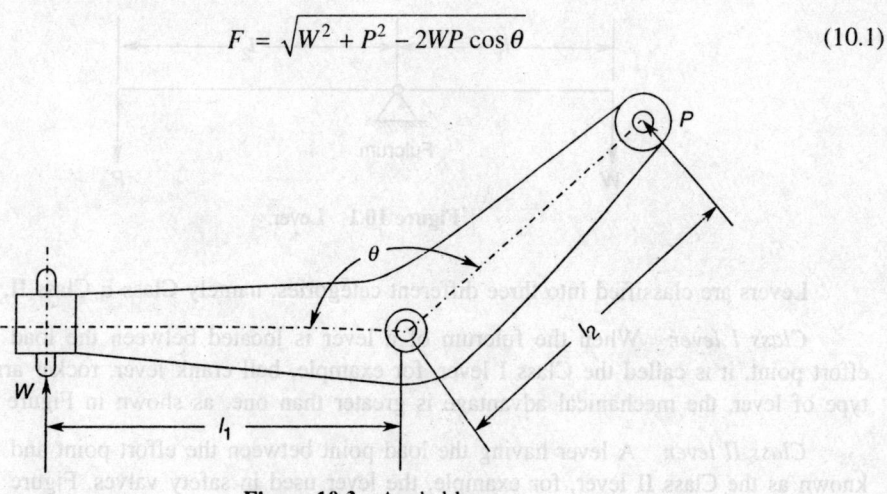

Figure 10.3 Angled lever.

2. The maximum bending moment is determined as illustrated in Figure 10.4.

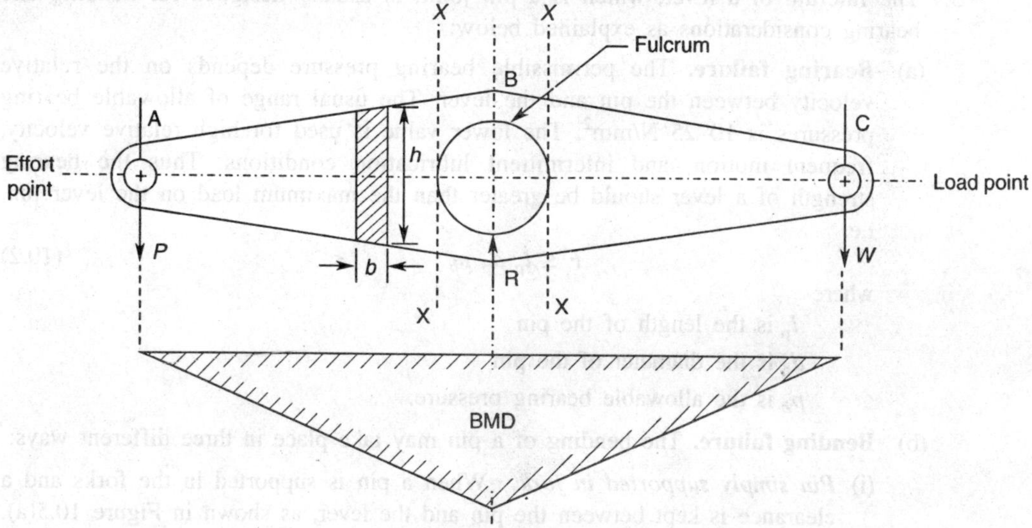

Figure 10.4 Taper lever.

3. Using the theory of bending the size of the lever section is determined. Though the maximum bending moment occurs at the fulcrum point in a Class I lever, the section dimensions calculated on this basis are assumed at section X–X, i.e. adjacent to the boss section as shown in Figure 10.4. The most commonly used sections are rectangular, elliptical, and I-section. The section moduli and usual proportions of these sections are given in Table 10.1.

Table 10.1 Section modulus and usual proportions

	Rectangular	*Elliptical*	*I-section*
Shape	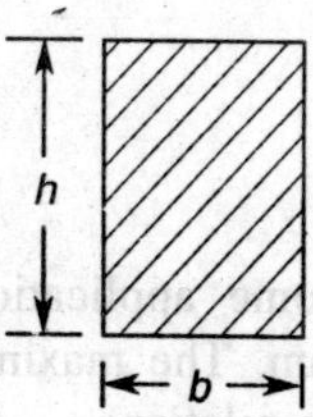	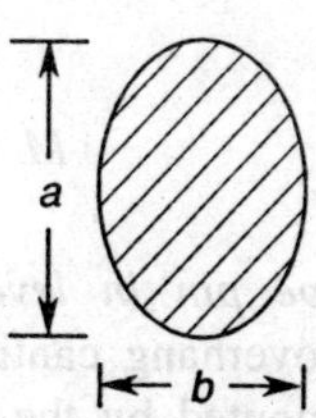	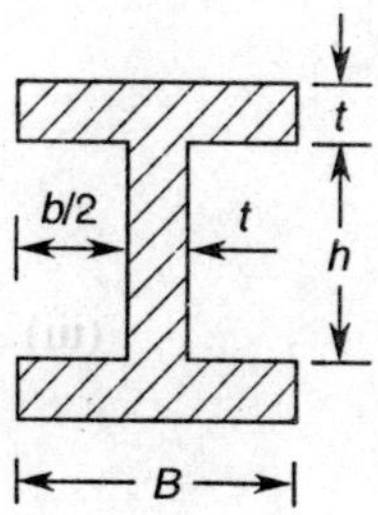
Section modulus (Z)	$\dfrac{bh^2}{6}$	$\dfrac{\pi}{32}a^2b$	$\dfrac{BH^3 - bh^3}{6H}$
Usual proportions	$h = (2 \text{ to } 5)b$	$a = (2 \text{ to } 3)b$	$H = (4 \text{ to } 6)t$ $B = (3 \text{ to } 4)t$

4. If required, the lever should be suitably tapered to obtain a lever of uniform strength.
5. The fulcrum of a lever, which is a pin joint, is usually designed for bending and bearing considerations as explained below:
   (a) **Bearing failure.** The permissible bearing pressure depends on the relative velocity between the pin and the lever. The usual range of allowable bearing pressures is 10–25 N/mm^2. The lower value is used for high relative velocity, frequent motion, and intermittent lubrication conditions. Thus the bearing strength of a lever should be greater than the maximum load on the lever pin, i.e.

$$F \le l_p \, d_p \, p_b \tag{10.2}$$

   where

   l_p is the length of the pin

   d_p is the diameter of the pin

   p_b is the allowable bearing pressure.

   (b) **Bending failure.** The bending of a pin may take place in three different ways:

   (i) *Pin simply supported in fork.* When a pin is supported in the forks and a clearance is kept between the pin and the lever, as shown in Figure 10.5(a), the pin is said to be simply supported and the maximum bending moment occurs at the centre. The force acting on the pin is shown in Figure 10.5(b). The maximum bending moment is computed by the following relation:

$$M = \frac{F}{2}\left(\frac{b}{4} + \frac{t}{3}\right) \tag{10.3}$$

   where F is the maximum force which is equal to reaction at the fulcrum or load W, whichever is greater.

   (ii) *Cantilever type pin in fork.* When clearance is provided between the pin and the fork instead of between the lever and the pin, the pin is treated as fixed at the lever section. Thus it acts as a double cantilever and the maximum bending moment is computed by the following relation (see Figure 10.6).

$$M = \frac{Ft}{4} \tag{10.4}$$

   (iii) *Cantilever type pin in lever.* In some applications, the fulcrum pin is supported as overhang cantilever beam. The maximum bending moment in the pin is computed by the following relation:

$$M = \frac{Fb}{2} \tag{10.5}$$

   The bending stress in the pin computed as per any of the relations (10.3), (10.4), and (10.5) should be less than the allowable strength of the pin material.

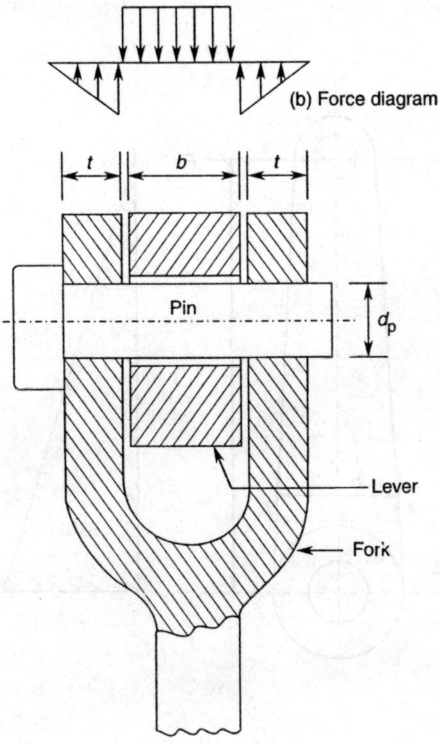

Figure 10.5 Simply supported pin.

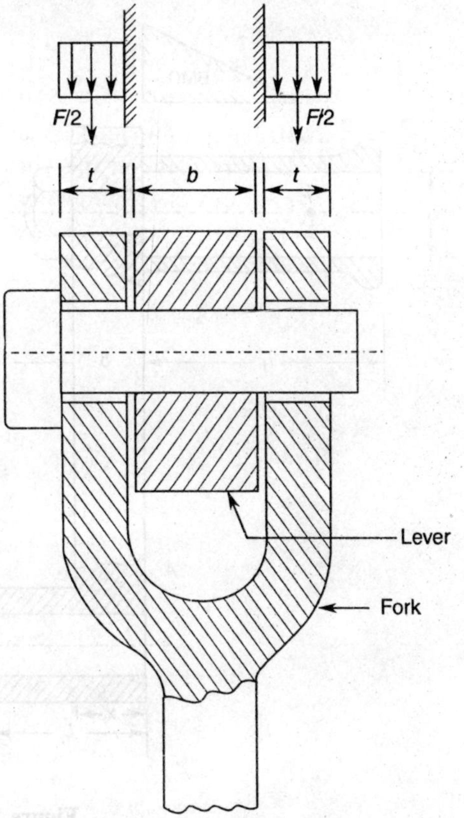

Figure 10.6 Cantilever pin in fork.

10.3 CRANKED LEVERS

The hand-operated crank levers are commonly used for hoisting or lowering loads with winches or for starting a diesel engine. In these levers, the force applied by a human being at the handle is transmitted to the shaft through the lever. Although the force applied on the handle is distributed over its entire length, it is assumed to be applied as point load at two-third of length of the handle away from the lever. The selection of the lever length depends upon the torque capacity and the number of persons operating it. Normally, 300-500 mm length is considered to be sufficient. The length of the lever, i.e. the crank radius is usually taken between 350 mm and 400 mm. Cranked levers are usually made of rectangular or elliptical sections. They are sometimes inclined from the handle by 5–10° to avoid interference with the operator's hand.

1. *Handle*. When a force F is applied by an operator as shown in Figure 10.7, the handle bar of the lever is subjected to bending moment M which equals $2Fl_1/3$. The diameter of the handle bar is computed from the following relation:

$$d = \left(\frac{32M}{\pi\sigma}\right)^{1/3} \tag{10.6}$$

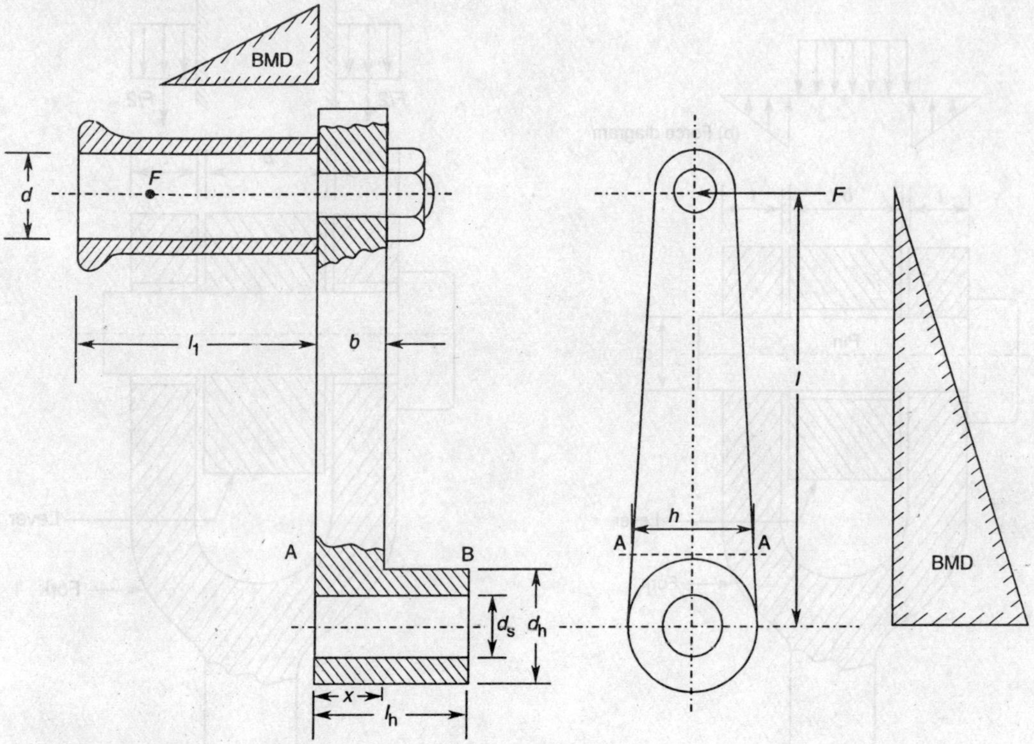

Figure 10.7 Cranked lever.

where σ is the allowable strength of the material. In order to prevent hand scoring and for better grip, the handle bar is slipped into a wooden handle or a metallic pipe.

2. *Lever.* The lever section near the boss is subjected to two types of forces:

(a) Constant twisting moment, $T = \dfrac{2}{3}Fl_1$

(b) Varying bending moment which is maximum at section A–A. However, it is assumed that the length of lever is extended up to the centre of the crank and maximum bending moment is Fl. The equivalent torque in the lever

$$T_{eq} = \sqrt{M^2 + T^2} \tag{10.7}$$

Considering the torsional shear failure,

$$T_{eq} = \tau Z_{polar}$$

where Z_{polar} is the polar-section modulus and is given by

$$= \frac{2}{9} hb^2 \text{ for rectangular sections of } h \times b$$

$$= \frac{\pi}{16} b^2 a \text{ for elliptical sections of major axis } a \text{ and minor axis } b.$$

3. *Shaft journal.* The shaft journal is subjected to (i) twisting moment and (ii) bending moment.

$$\text{Twisting moment, } T' = Fl$$

$$\text{Bending moment, } M' = F\left(\frac{2}{3}l_1 + x\right)$$

where x is the distance from side A of the boss to the centre of journal. Therefore, the equivalent torque

$$T'_{\text{eq}} = \sqrt{M'^2 + T'^2}$$

Considering shear failure, the shaft journal diameter d_s can be found. The following empirical relations are used for the hub diameter and its length.

$$\text{Diameter of the hub, } d_h = (1.5\text{–}2.0)\, d_s$$

$$\text{Length of the hub } l_h = (1\text{–}1.5)\, d_s$$

Example 10.1 Design a simple lever of a safety valve for a boiler having a gauge pressure of 1.5 MN/m². The valve diameter is 90 mm. The lever is 1 m long and the distance between the fulcrum and the valve point is 100 mm as shown in the figure below.

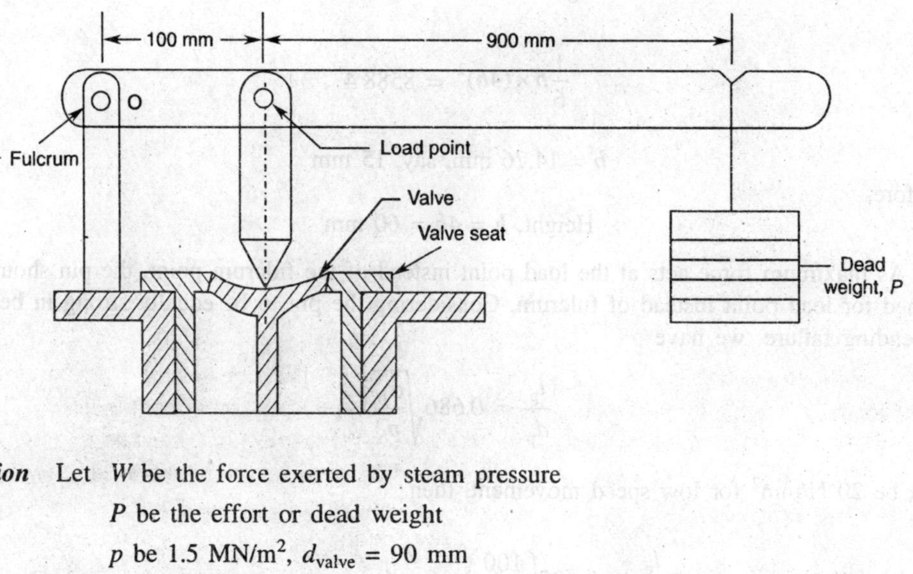

Solution Let W be the force exerted by steam pressure

P be the effort or dead weight

p be 1.5 MN/m², $d_{\text{valve}} = 90$ mm

l_1 be 100 mm

$$\text{Force, } W = \frac{\pi}{4} d_{\text{valve}}^2 p = \frac{\pi}{4} \times 90^2 \times 1.5 = 9542.6 \text{ N}$$

Taking moment about fulcrum O, shown in the figure below, we have

$$Wl_1 = Pl_2$$

or

$$P = \frac{Wl_1}{l_2} = 9542.6 \times \frac{100}{1000} = 954.26 \text{ N}$$

Reaction at fulcrum, $F = W - P = 9542.6 - 954.26 = 8588.34$ N

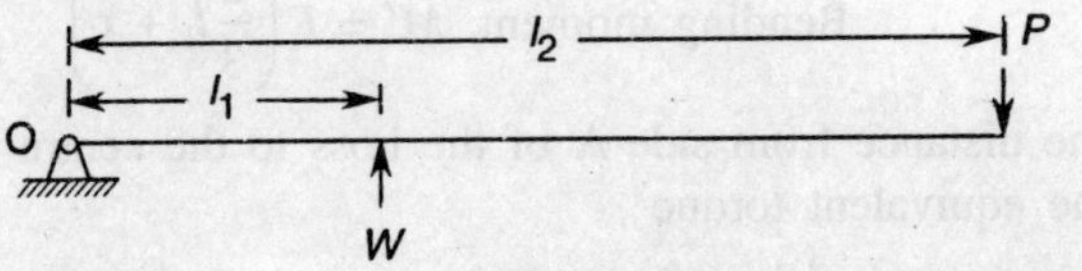

Lever section. Let us assume that the lever section is rectangular having *h*/*b* ratio equal to 4. The lever is made of C20 steel having allowable strength of 100 N/mm^2.

Bending moment at the load point, $M = P(l_2 - l_1) = 954.26\ (1.0 - 0.1) = 858.84$ N · m

Considering bending failure, the section modulus $Z = \dfrac{M}{\sigma_b} = \dfrac{858.84 \times 10^3}{100} = 8588.4 \text{ mm}^3$

or

$$\frac{1}{6} b \times h^2 = 8588.4 \text{ mm}^3$$

or

$$\frac{1}{6} b \times (4b)^2 = 8588.4$$

or

$$b = 14.76 \text{ mm, say, } 15 \text{ mm}$$

Therefore,

$$\text{Height, } h = 4b = 60 \text{ mm}$$

Pin. As maximum force acts at the load point instead of the fulcrum point, the pin should be designed for load point instead of fulcrum. Considering the pin to be equally strong in bearing and bending failure, we have

$$\frac{l_p}{d_p} = 0.686\sqrt{\frac{\sigma_b}{p_b}}$$

Let p_b be 20 N/mm^2 for low speed movement, then

$$\frac{l_p}{d_p} = 0.686\left(\frac{100}{20}\right)^{0.5} = 1.534, \text{ say, } 1.5$$

Bearing force, $F = l_p d_p p_b$

or

$$9542.6 = 1.5\,d_p^2 \times 20$$

or

$$d_p = 17.8 \text{ mm, say, } 18 \text{ mm}$$

Therefore, the length of the pin, $l_p = 1.5d_p = 1.5 \times 18 = 27$ mm

Let

$$\begin{aligned} d_h &= \text{diameter of the hole} \\ &= d_p + 2 \times \text{thickness of the bush} \\ &= 18 + 2 \times 3 = 24 \text{ mm} \end{aligned}$$

The modified section of the lever is shown in the figure below:

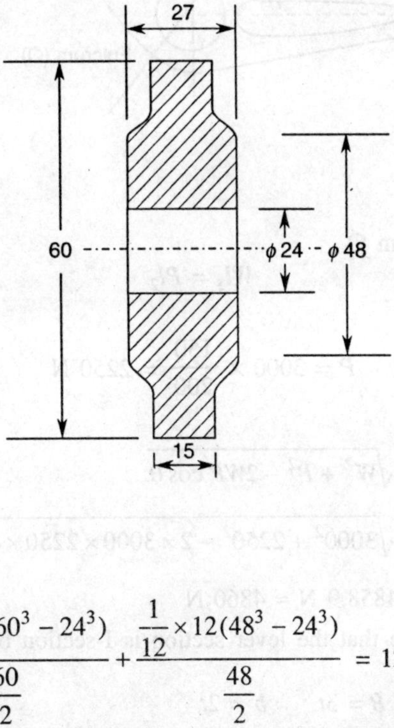

Revised section modulus

$$Z = \frac{\frac{1}{12}\times 15(60^3 - 24^3)}{\frac{60}{2}} + \frac{\frac{1}{12}\times 12(48^3 - 24^3)}{\frac{48}{2}} = 12{,}456 \text{ mm}^3$$

which is larger than the required value 8588.4 mm^3. Hence the lever section is safe.

Example 10.2 A rocker arm lever, as shown in the following figure, is used for opening a valve of an I.C. engine. The lever is supported in the bracket and the maximum load acting on the valve arm end is 3 kN. Design the lever if the allowable tensile strength is 80 N/mm^2 and the allowable shear strength is 45 N/mm^2.

Solution Load, $W = 3000$ N

Load arm, $l_1 = 150$ mm

Effort arm, $l_2 = 200$ mm

Let P be the effort to be applied by the roller.

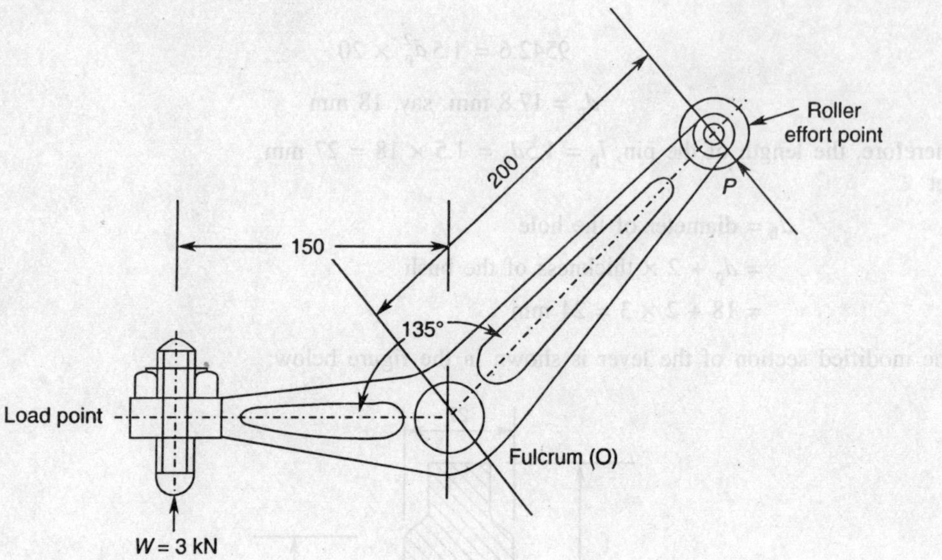

Taking moments about fulcrum O

$$Wl_1 = Pl_2$$

or

$$P = 3000 \times \frac{150}{200} = 2250 \text{ N}$$

The reaction at fulcrum

$$F = \sqrt{W^2 + P^2 - 2WP\cos\theta}$$

$$= \sqrt{3000^2 + 2250^2 - 2 \times 3000 \times 2250 \times \cos 135°}$$

$$= 4858.9 \text{ N} \approx 4860 \text{ N}$$

Lever section. Let us assume that the lever section is I-section of the following proportions. Let

$$B = 3t \qquad b = 2t$$

$$H = 4t \qquad h = 2t$$

Section modulus

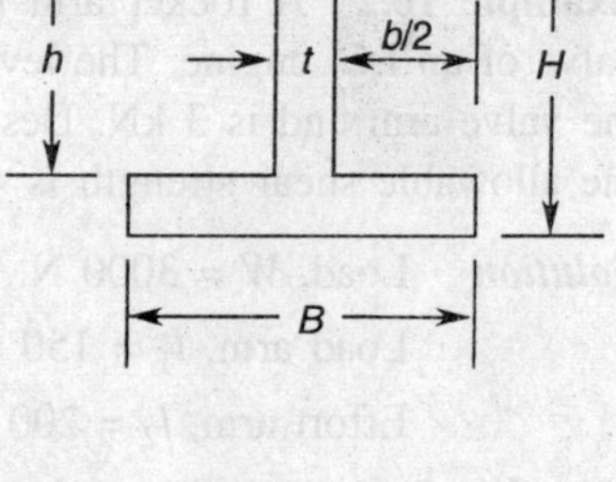

$$Z = \frac{\frac{1}{12}(BH^3 - bh^3)}{H/2}$$

$$= \frac{\frac{1}{12}[3t \times (4t)^3 - 2t \times (2t)^3]}{4t/2}$$

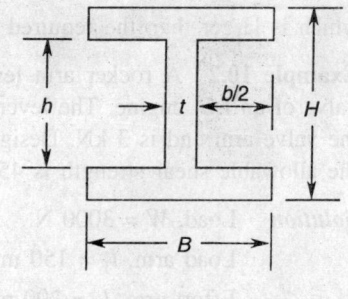

$$= 7.334t^3$$

Bending moment, $M = 3000 \times 0.15 = 450$ N · m

$$\text{Section modulus, } Z = \frac{M}{\sigma} = \frac{450 \times 10^3}{80} = 5625 \text{ mm}^3$$

or

$$7.334t^3 = 5625$$

or

$$t = 9.15, \text{ say, } 9.5 \text{ mm}$$

Therefore,

$$H = 4t = 38 \text{ mm}$$

$$B = 3t = 28.5 \text{ mm}$$

Fulcrum pin. For equal strength in bending and bearing failure

$$\frac{l_p}{d_p} = 0.686\sqrt{\frac{\sigma}{p_b}}$$

Let

p_b = bearing pressure = 10 N/mm^2 for frequent motion

Then,

$$\frac{l_p}{d_p} = 0.686\sqrt{\frac{80}{10}} = 1.94, \text{ say, } 2$$

Considering bearing failure

$$l_p \times d_p \times p_b = F$$

or

$$2d_p^2 \times p_b = F$$

or

$$2d_p^2 \times 10 = 4860$$

or

$$d_p = 15.58 \text{ mm, say, } 16 \text{ mm}$$

Therefore.

$$l_p = 2 \times d_p = 32 \text{ mm}$$

Diameter of the hole = d_p + 2 × thickness of the phosphor bronze bush

$= 16 + 2 \times 2 = 20$ mm

Boss diameter, $d = 2 \times 20 = 40$ mm

The boss section is a rectangular section of 40 mm × 32 mm in which a hole of 20 mm diameter is drilled. The revised section modulus is therefore

$$Z = \frac{\frac{1}{12} \times 32(40^3 - 20^3)}{40/2} = 7466.6 \text{ mm}^3$$

which is larger than the required section modulus. Hence the lever section is safe.

Example 10.3 Design a cross lever to operate a twin-cylinder double-acting pump, as shown in the figure below, which carries the force 4.5 kN acting downwards at pin A and 6.5 kN acting upwards at pin B.

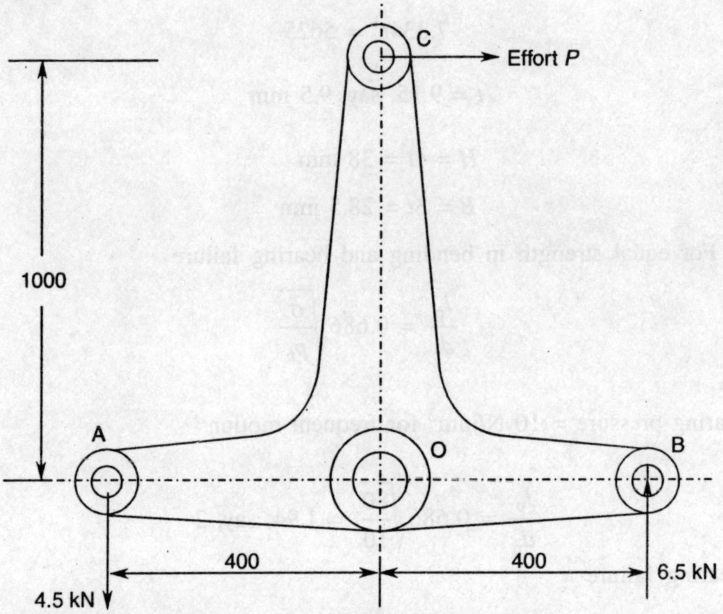

Solution

Case I — When both pumps are working

Let P be the effort required at point C.

Taking moments about fulcrum O

$$P \times 1000 = 6.5 \times 400 + 4.5 \times 400$$

or

$$P = 4.4 \text{ kN}$$

$$\text{Reaction at fulcrum, } F_1 = \sqrt{4.4^2 + (6.5 - 4.5)^2} = 4.834 \text{ kN}$$

Case II — When one pump is working

Let P_1 be the effort required at point C.

Taking moments about fulcrum O

$$P_1 \times 1000 = 6.5 \times 400$$

or

$$P_1 = 2.6 \text{ kN}$$

$$\text{Reaction, } F_2 = \sqrt{2.6^2 + 6.5^2} = 7 \text{ kN}$$

Since reaction F_2 in the second case is greater than F_1, the design of the pin should be on the basis of reaction F_2.

Lever section

(i) *For horizontal arm*

$$\text{Bending moment, } M = 6.5 \times 400 = 2600 \text{ N} \cdot \text{m}$$

Let the cross-section of the arm be rectangular having $h = 4b$.
Let σ_b be the allowable strength; for steel C20, $\sigma_b = 100 \text{ N/mm}^2$
Now,

$$\text{Bending stress, } \sigma_b = \frac{M}{Z}$$

or

$$\text{Section modulus, } Z = \frac{M}{\sigma_b} = \frac{2600 \times 10^3}{100} = 26{,}000 \text{ mm}^3$$

Also,

$$Z = \frac{1}{6}bh^2$$

or

$$26{,}000 = \frac{1}{6} \times b \times (4b)^2$$

or

$$b = 21.36 \text{ mm, say, } 22 \text{ mm}$$

Therefore,

$$\text{height, } h = 4b = 88 \text{ mm}$$

(ii) *For vertical arm*

The maximum bending moment in Case I is

$$M = P \times 1000 = 4.4 \times 1000 = 4400 \text{ N} \cdot \text{m}$$

Let the cross-section be $b \times h_1$ (rectangular with $h_1 = 4b$)

$$\text{Section modulus, } Z = \frac{M}{\sigma_b} = \frac{4400 \times 10^3}{100} = 44{,}000 \text{ mm}^3$$

or

$$\frac{1}{6} b \times h_1^2 = 44{,}000$$

Keeping thickness constant, $b = 22$ mm

$$h_1 = \left(\frac{6 \times 44000}{22}\right)^{0.5} = 109.5 \text{ mm, say, } 110 \text{ mm}$$

Design of the pin. The fulcrum pin is subjected to a reaction of 7 kN load in the worst condition. So let us design it considering the pin to be equally strong in bending and bearing failure. Therefore,

$$\frac{l_p}{d_p} = 0.686\sqrt{\frac{\sigma_b}{p_b}}$$

Let p_b = bearing pressure (10–25 N/mm^2) = 12 N/mm^2 (say)

Therefore,

$$\frac{l_p}{d_p} = 0.686\sqrt{\frac{100}{12}} = 1.98, \text{ say, } 2$$

Considering bearing failure of the pin

$$F_2 = l_p d_p \times p_b$$

or

$$7000 = 2d_p^2 \times 12$$

or

$$d_p = 17 \text{ mm and } l_p = 2d_p = 34 \text{ mm}$$

Diameter of the hole = d_p + 2 × thickness of the bush

= 17 + 2 × 2 = 21 mm

Diameter of the boss = 42 mm

Example 10.4 Design a bell crank lever of the spring-loaded Hartnell governor shown in the figure below. The mass of ball is 2.5 kg and the ball arm attains vertical position corresponding to speed of 300 rpm. The lengths of the ball arm and the sleeve arm are 250 mm

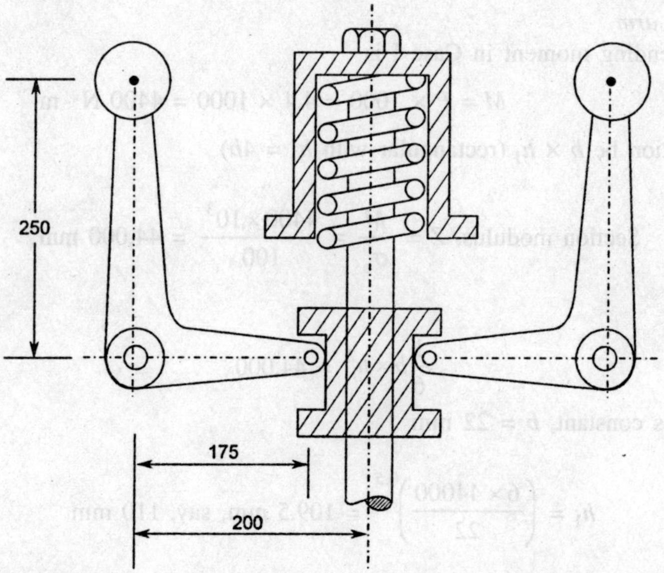

and 175 mm, respectively. The distance of the pivot of each ball crank lever is 200 mm from the axis of revolution. The speed of the governor increases by 3% for a lift of 16 mm of the sleeve.

Solution Let F_C = centrifugal force

$$= m\omega^2 r$$

$$= 2.5 \times \left(\frac{2\pi \times 300}{60}\right)^2 \times 0.2$$

$$= 493.5 \text{ N}$$

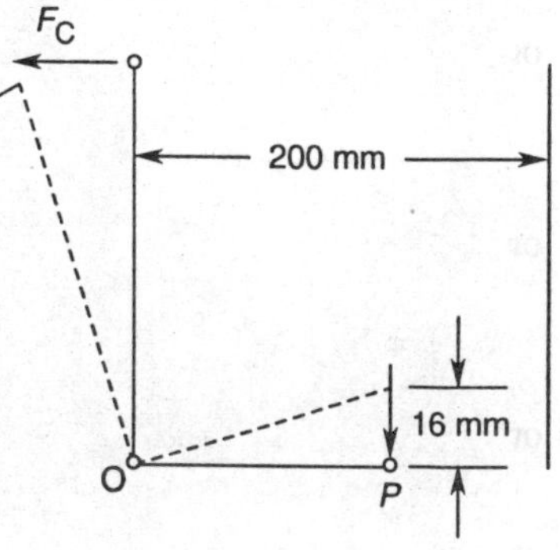

Figure Ex. 10.4(a)

Taking moment about fulcrum O (see the adjoining figure)

$$P = \frac{F_C \times 250}{175} = \frac{493.5 \times 250}{175} = 705 \text{ N}$$

When the speed is increased by 3%

$$N_1 = 300 \times 1.03 = 309 \text{ rpm}$$

$$\text{Radius, } r_1 = r + \text{lift} \times \frac{\text{length of the ball arm}}{\text{length of the sleeve arm}}$$

$$= 200 + 16 \times \frac{250}{175} = 222.8 \text{ mm}$$

$$\text{The centrifugal force, } F_{C1} = 2.5 \times \left(\frac{2\pi \times 309}{60}\right)^2 \times 0.2228 = 583.2 \text{ N}$$

Taking moment about fulcrum O

$$F_{C1} \times 250 = P_1 \times 175$$

or

$$P_1 = 583.2 \times \frac{250}{175} = 833.1 \text{ N}$$

$$\text{Reaction at fulcrum, } F = \sqrt{F_{C1}^2 + P_1^2} = \sqrt{583.2^2 + 833.1^2} = 1016.94 \text{ N}$$

Lever section

The maximum bending moment

$$M = P_1 \times \text{length of the sleeve arm}$$

$$= 833.1 \times \frac{175}{1000} = 145.8 \text{ N} \cdot \text{m}$$

Let the lever cross-section be rectangular with $h = 3b$ and allowable strength σ be 100 N/mm^2.

Considering bending failure

$$\text{Section modulus, } Z = \frac{M}{\sigma} = \frac{145.8 \times 10^3}{100} = 1458 \text{ mm}^3$$

or

$$\frac{1}{6} bh^2 = 1458$$

or

$$\frac{1}{6} \times b \times (3b)^2 = 1458$$

or

$$b = 9.9 \text{ mm, say, } 10 \text{ mm}$$

Therefore,

$$\text{Height, } h = 3b = 30 \text{ mm}$$

Fulcrum pin

$$\text{For equal strength, } \frac{l_p}{d_p} = 0.686\sqrt{\frac{\sigma}{p_b}}$$

Let p_b be 15 N/mm^2. Therefore,

$$\frac{l_p}{d_p} = 0.686\sqrt{\frac{100}{15}} = 1.77, \text{ say, } 1.8$$

Considering bearing failure

$$F = l_p d_p p_b$$

or

$$1016.94 = 1.8 d_p^2 \times 15$$

or

$$d_p = 6.13 \text{ mm, say, } 6 \text{ mm}$$

Hence,

$$l_p = 1.8 \times 6 = 10.8 \text{ mm, say, } 11 \text{ mm}$$

$$\text{Revised thickness, } b = l_p = 11 \text{ mm}$$

$$\text{Hole size} = 6 + 2 \times 2 = 10 \text{ mm}$$

$$\text{Revised section modulus, } Z = \frac{\frac{1}{12}\left[11(30^3 - 10^3)\right]}{30/2} = 1588.9 \text{ mm}^3$$

which is larger than the required value. Hence the design of the lever section is safe.

Example 10.5 Design a crank lever to operate diesel engines. The following data are given:

(i) Length of the lever handle, $l_1 = 300$ mm
(ii) Length of the lever, $l = 400$ mm
(iii) Force on the handle, $F = 400$ N

The lever is operated by one person. It is made of mild steel having allowable tensile strength of 100 N/mm^2 and shear strength of 50 N/mm^2.

Solution

Design of the handle (see Figure 10.7)

The bending moment in the handle

$$M = \frac{2}{3} F l_1 = \frac{2}{3} \times 400 \times \frac{300}{1000} = 80 \text{ N} \cdot \text{m}$$

Considering bending failure, diameter of the circular handle

$$d = \left(\frac{32M}{\pi\sigma}\right)^{1/3} = \left[\frac{32 \times 80 \times 10^3}{\pi \times 100}\right]^{1/3} = 20.12 \text{ mm}$$

Let us adopt $d = 25$ mm with a handle cover of 50 mm outside diameter and 25 mm inside diameter

Lever section

The lever section near the hub is subjected to two types of forces:

(i) Twisting moment, $T = \frac{2}{3} F l_1 = 80 \text{ N} \cdot \text{m}$

(ii) Bending moment, $M = F \times l = 400 \times \frac{400}{1000} = 160 \text{ N} \cdot \text{m}$

The equivalent twisting moment, $T_{eq} = \sqrt{M^2 + T^2} = \sqrt{160^2 + 80^2} = 178.9 \text{ N} \cdot \text{m}$

Considering torsional failure

$$\textbf{Polar-section modulus, } Z_{polar} = \frac{T_{eq}}{\tau} = \frac{178.9 \times 10^3}{50} = 3578 \text{ mm}^3$$

For the rectangular section with $h = 2b$, $Z_{polar} = \frac{2}{9} hb^2 = 3578$

or

$$\frac{2}{9} \times 2b \times b^2 = 3578$$

or

$$b = 20 \text{ mm}$$

$$\text{Height, } h = 2b = 40 \text{ mm}$$

Journal

$$\text{Torque transmitted, } T' = Fl = 400 \times 0.4 = 160 \text{ N} \cdot \text{m}$$

$$\text{Bending moment, } M' = F\left(\frac{2}{3}l_1 + x\right)$$

where x is the distance from the side of the boss to the centre of the journal length [= 40 mm (assumed)]. Therefore,

$$M' = 400\left(\frac{2}{3} \times 300 + 40\right) = 96 \text{ N} \cdot \text{m}$$

$$\text{Equivalent torque, } T'_{eq} = \sqrt{M'^2 + T'^2} = \sqrt{96^2 + 160^2} = 186.6 \text{ N} \cdot \text{m}$$

$$\text{Diameter of the journal shaft, } d_s = \left(\frac{16T_{eq}}{\pi\tau}\right)^{1/3} = \left[\frac{16 \times 186.6 \times 10^3}{\pi \times 50}\right]^{1/3} = 26.7 \text{ mm, say, } 30 \text{ mm}$$

Let

$$\text{Hub diameter, } d_h = 2d_s = 60 \text{ mm; Length of hub, } l_h = 1.8ds = 54 \text{ mm}$$

$$\text{The distance } x = \frac{1}{2} \times \text{length of the hub} = \frac{54}{2} = 37 \text{ mm}$$

which is less than the assumed value. Hence the design of the journal shaft and hub is safe.

EXERCISES

1. Define a lever. How are levers classified?
2. What do you understand by the terms mechanical advantage and leverage?
3. Why are levers usually tapered?
4. Explain the procedure of designing a fulcrum pin.
5. Design a right-angled bell crank lever having arms length of 500 mm and 750 mm respectively. A load of 10 kN is acted upon the small arm length.
6. Design a simple lever for safety valve of a boiler having a gauge pressure of 0.75 MPa. The valve diameter is 60 mm. The dead weight at the end of the lever should not exceed 400 N. The maximum length of the lever is 800 mm.
7. In a spring-loaded Hartnell governor, the weight of each ball is 20 N and the ball arm attains vertical position at the speed of 400 rpm. The length of the ball arm and the sleeve arm are 150 mm and 100 mm respectively. The distance between the pivot of the lever and the axis of rotation is 125 mm. Design the lever if the speed is to be limited to 5% for a lift of 20 mm.
8. Design a spanner to tighten a nut of 10 mm bolt. The distance between the centres is 150 mm and the maximum load applied is 100 N. The section of the spanner may be assumed to be approximately I-section.

Chapter 11 Belt Drives

11.1 INTRODUCTION

The transmission of power from the prime mover shaft to the driven machine shaft can be made either through flexible or through non-flexible drive elements. The flexible drive elements include belt, rope, and chain. These elements greatly simplify machine construction and allow the designer considerable flexibility in the location of the driving and driven machines. The location tolerances are not critical as with positive drive elements such as gears, clutches, and power screws. Flexible drives are simple in construction, run quieter, are suitable for long centre-distances between shafts, involve low maintenance and also cost less. In addition, flexible drives play an important role in absorbing shock loads and in damping out the effects of vibration forces on account of their long length and elastic properties. The main limitations of these drive elements are the low values of velocity ratio which does not remain constant because of slip. They also require regular adjustment of the centre-distance to maintain initial setting of tension in the drive.

11.2 BELT DRIVE

A belt drive consists of a driving pulley, a driven pulley, and a belt which is wrapped around the pulleys with a certain amount of tension as shown in Figure 11.1. It transmits tangential

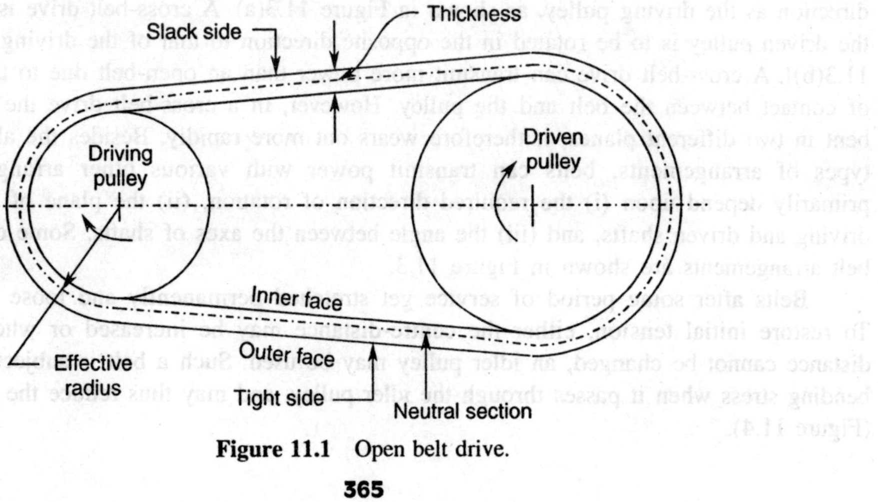

Figure 11.1 Open belt drive.

force from one pulley to another, thereby transmitting power between two shafts. For an unstretched belt mounted on the pulley, the outer and inner faces remain in tension and compression, respectively. In between these two faces, there is a neutral section which is neither under tension nor under compression. Usually, the effective radius of rotation of the pulley is obtained by adding half the thickness of belt to the radius of pulley. A belt in the drive can be of rectangular section, known as flat belt, or of trapezoidal section known as V-belt as shown in Figure 11.2. In the case of flat belt, the rim of the pulley is slightly crowned which helps to keep the belt running centrally on the pulley rim. In the V-belt drive, the groove on the rim of pulley is made deeper to take advantage of the wedging action. The belt does not touch the bottom of the V-groove. Owing to wedging action, a V-belt does not need regular adjustment and it can transmit more power without any slip, as compared to a flat belt.

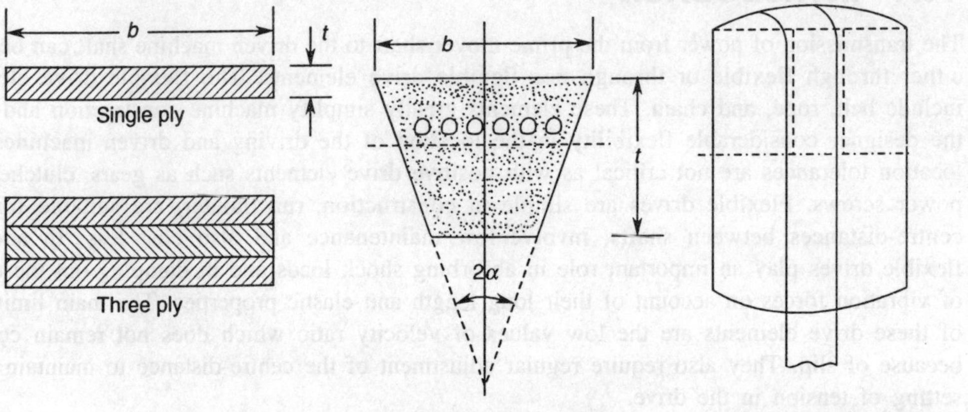

(a) Rectangular section flat belt (b) Trapezoidal section V-belt (c) Effect of pulley crown

Figure 11.2 Types of belt sections and effect of pulley crown.

The arrangement of belt drives is generally of two types: (i) open-belt drive and (ii) cross-belt drive. An open-belt drive is used when the driven pulley is to be rotated in the same direction as the driving pulley, as shown in Figure 11.3(a). A cross-belt drive is adopted when the driven pulley is to be rotated in the opposite direction to that of the driving pulley [Figure 11.3(b)]. A cross-belt drive can transmit more power than an open-belt due to the larger angle of contact between the belt and the pulley. However, in a cross-belt drive the belt has to be bent in two different planes; it therefore wears out more rapidly. Besides the above two basic types of arrangements, belts can transmit power with various other arrangements which primarily depend upon (i) the required direction of rotation, (ii) the plane of rotation of the driving and driven shafts, and (iii) the angle between the axes of shafts. Some of these special belt arrangements are shown in Figure 11.3.

Belts after some period of service get stretched permanently and loose initial tension. To restore initial tension, either the centre-distance may be increased or where the centre-distance cannot be changed, an idler pulley may be used. Such a belt is subjected to reversed bending stress when it passes through the idler pulley, and may thus reduce the life of the belt (Figure 11.4).

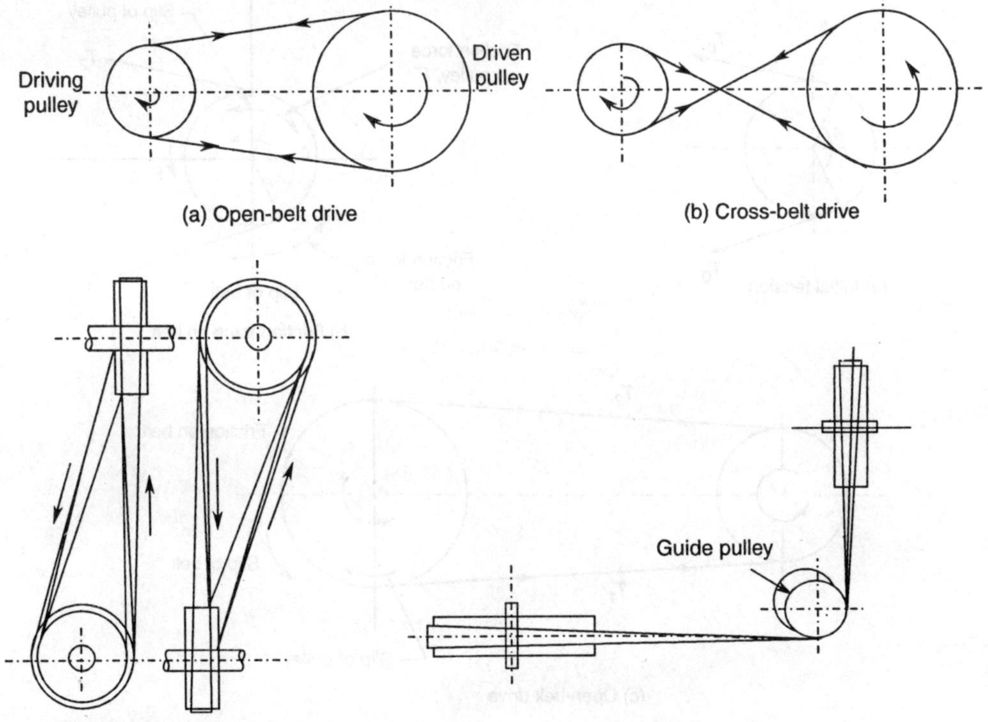

Figure 11.3 Various arrangements of belt drives.

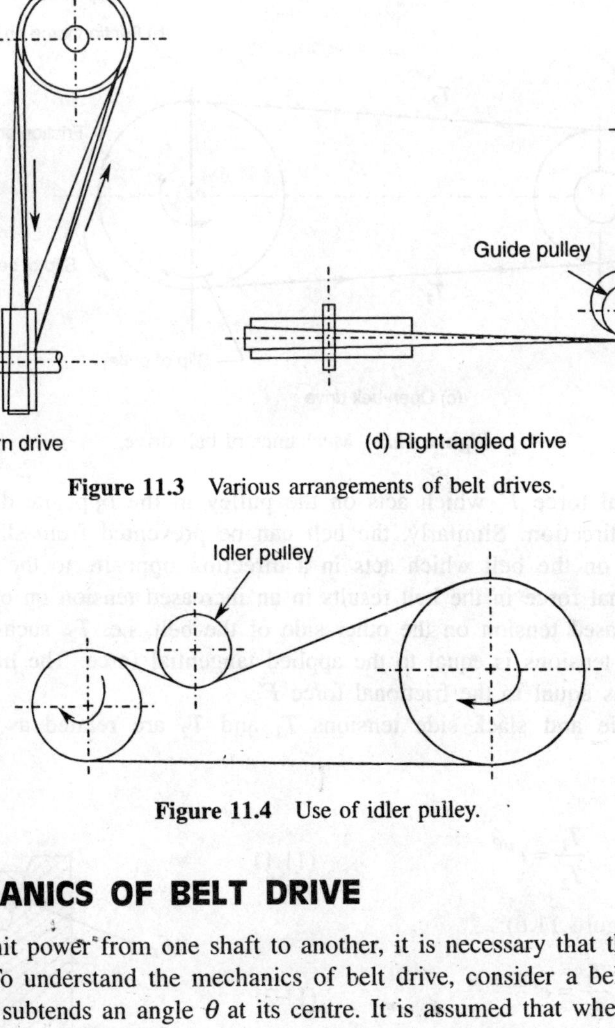

Figure 11.4 Use of idler pulley.

11.3 MECHANICS OF BELT DRIVE

In order to transmit power from one shaft to another, it is necessary that the belt does not slip over the pulley. To understand the mechanics of belt drive, consider a belt wrapped around a pulley and which subtends an angle θ at its centre. It is assumed that when the drive does not operate or when it is idling, there is only an initial tension T_0 on the belt. Now if a tangential force F is applied to the driving pulley in any direction, say in clockwise direction, as shown in Figure 11.5, it tends to rotate the belt with it. If the motion of the belt is resisted by some means, the pulley will try to slip over the belt. In order to prevent the slipping of pulley, there

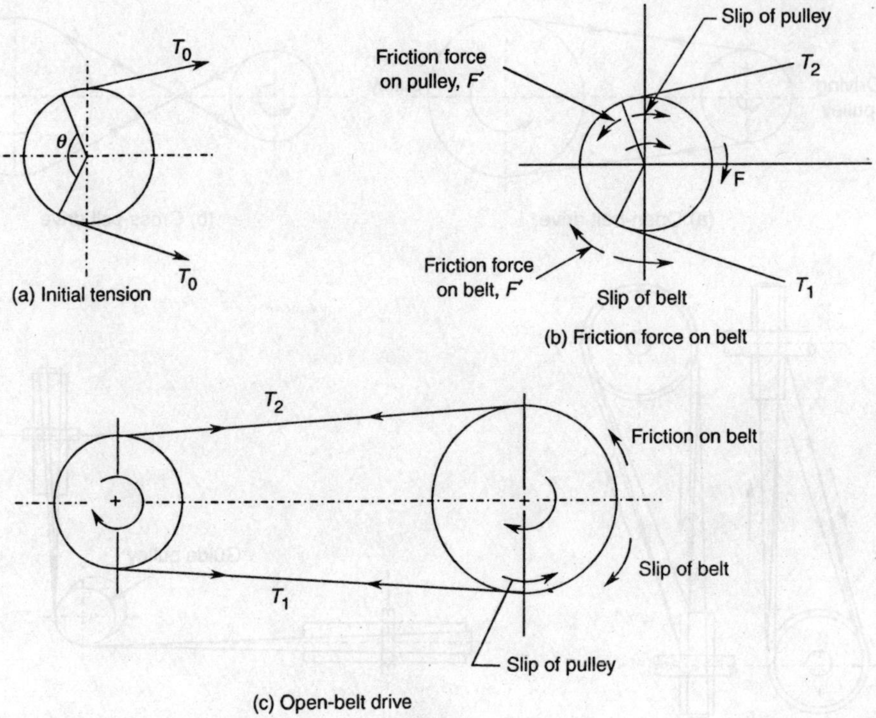

Figure 11.5 Mechanics of belt drive.

must be a frictional force F' which acts on the pulley in the opposite direction, i.e. in the counterclockwise direction. Similarly, the belt can be prevented from slipping if there is a frictional force F' on the belt which acts in a direction opposite to the frictional force on pulley. This frictional force in the belt results in an increased tension on one side of the belt, i.e. T_1 and a decreased tension on the other side of the belt, i.e. T_2, such that the difference between these two tensions is equal to the applied tangential force. The limiting value of the tangential force F is equal to the frictional force F'.

The tight side and slack side tensions T_1 and T_2 are related as per the following equations:

For flat belt:

$$\frac{T_1}{T_2} = e^{\mu\theta} \tag{11.1}$$

For V-belt (refer Figure 11.6):

$$\frac{T_1}{T_2} = e^{\mu\theta/\sin\alpha} \tag{11.2}$$

where the coefficient of friction

$$\mu = 0.54 - \frac{0.712}{2.542 + v} \tag{11.3}$$

α = semi-cone angle of the V-pulley groove

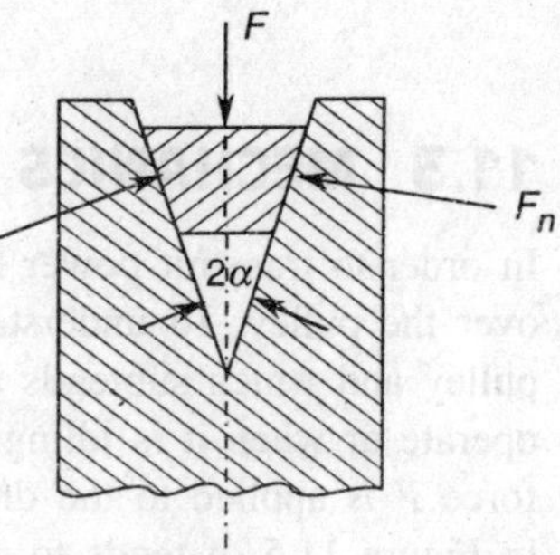

Figure 11.6 Forces on V-belt.

The power transmission capacity of the belt drive is

$$P = \frac{(T_1 - T_2)\,v}{1000} \tag{11.4}$$

where v is the peripheral velocity of the belt in m/s.

Further when the belt transmits certain power, then tensions in the two branches of the belt are different. It is apparent that the tight-side tension would be greater than the initial tension T_0 whereas the slack side tension would be less than the initial tension. This means that branch 1 of the belt is elongated with strain e_1 and branch 2 is contracted with strain e_2 from the original length. Thus, while the longer length of the belt is approaching the driving pulley, the shorter length will be leaving it, as shown in Figure 11.7. This phenomenon is called *elastic creep*. The effect of the creep is to slow down the speed of the belt on the driving pulley and make it less than that of the rim of the pulley and to reduce the rim velocity of the driven pulley. The difference between these two strains is called the belt slip (s). The effect of slip on the velocity ratio (VR) of the belt drive can be incorporated as follows:

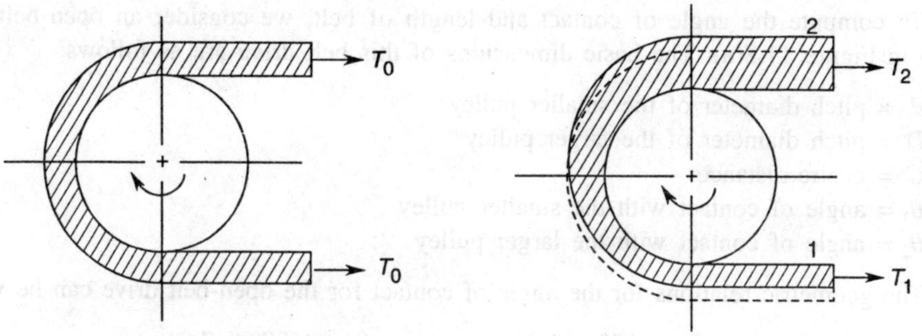

(a) Belt in idle condition (b) Belt in normal running condition

Figure 11.7 Effect of elastic creep.

$$\text{VR} = \frac{N_2}{N_1} = \frac{(d+t)}{(D+t)}\left(\frac{100 - s}{100}\right) \tag{11.5}$$

where

d is the pitch diameter of the smaller pulley
D is the pitch diameter of the larger pulley
t is the thickness of the belt
s is the percentage slip.

It is to be remembered that slip will first occur on the pulley which has the smaller angle of contact. Equation (11.5) shows that the velocity ratio is related not only to the pulley diameters and thickness of the belt but also to the relative slip which depends upon the operating conditions and is not constant. This is a serious shortcoming of belt drives. Therefore, on account of this reason the drive cannot be used where we require constant output speed. The flat belt and V-belt can be used effectively for transmission ratio up to 5 and 15, respectively.

11.3.1 Geometric Factors

In a belt drive, some geometric factors such as length of belt, angle of contact and centre-distance between two pulleys need to be known. It is observed that power transmission capacity of a belt drive largely depends on the angle of contact with the smaller pulley which in turn depends upon the diameters of pulleys and the centre-distance between them. The centre-distance between two shafts is primarily dependent on the layout of the application machines and availability of space. An empirical relation for selection of the optimum centre-distance from the point of view of long service life of the belt is given below:

$$C = (0.07\text{–}0.1)v \tag{11.6a}$$

or

$$C \geq (1.5\text{–}2.0)(d + D) \tag{11.6b}$$

If the centre-distance is too large, the belt whips, i.e. vibrates in a direction perpendicular to the direction of motion. But for a very short centre-distance, the belt may slip.

To compute the angle of contact and length of belt, we consider an open-belt drive as shown in Figure 11.8(a). The basic dimensions of this belt drive are as follows:

d = pitch diameter of the smaller pulley
D = pitch diameter of the larger pulley
C = centre-distance
θ_1 = angle of contact with the smaller pulley
θ_2 = angle of contact with the larger pulley

The geometric relations for the angle of contact for the open-belt drive can be written as

$$\theta_1 = (180 - 2\phi) \quad \text{and} \quad \theta_2 = (180° + 2\phi)$$

where

$$\phi = \sin^{-1}\frac{D - d}{2C} \quad \text{[see Figure 11.8(a)]} \tag{11.7}$$

The length of belt L_o can be determined as

$$L_o = 2(\text{arc CD + DE + arc EF})$$

$$= \frac{\pi(D + d)}{2} + \phi(D - d) + 2C \cos \phi$$

For small values of ϕ, we get

$$L_o = \frac{\pi(d + D)}{2} + \frac{(D - d)^2}{4C} + 2C \tag{11.8}$$

Similarly, the geometric relations for the cross-belt drive, shown in Figure 11.8(b), can be written as

$$\theta_1 = \theta_2 = 180° + 2\phi$$

where

$$\phi = \sin^{-1}\frac{D + d}{2C}$$

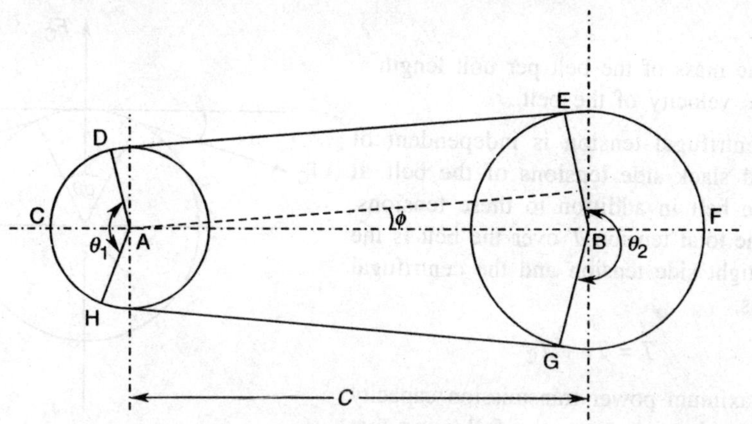

(a) Open-belt drive

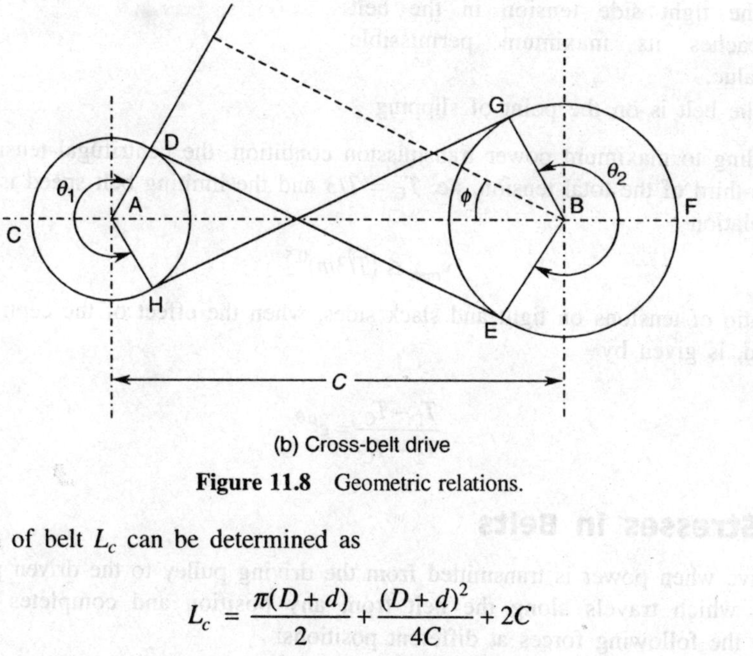

(b) Cross-belt drive

Figure 11.8 Geometric relations.

and the length of belt L_c can be determined as

$$L_c = \frac{\pi(D+d)}{2} + \frac{(D+d)^2}{4C} + 2C \quad (11.9)$$

The belt length calculated by the above formulae is required to be reduced slightly to compensate for the elastic extension of the belt due to initial tension.

11.3.2 Effect of Centrifugal Force

When a belt is operated at a high speed, considerable inertial force acts on the belt. This inertial force is due to centrifugal action of belt weight which tends to lift the belt from the pulley. Consider an element of belt on the pulley as shown in Figure 11.9. The centrifugal tension in the belt is computed by the following relation:

$$T_C = mv^2 \quad (11.10)$$

where

m is the mass of the belt per unit length
v is the velocity of the belt.

The centrifugal tension is independent of the tight and slack side tensions of the belt. It acts over the belt in addition to these tensions. Therefore, the total tension T over the belt is the sum of the tight side tension and the centrifugal tension. Thus,

$$T = T_1 + T_C$$

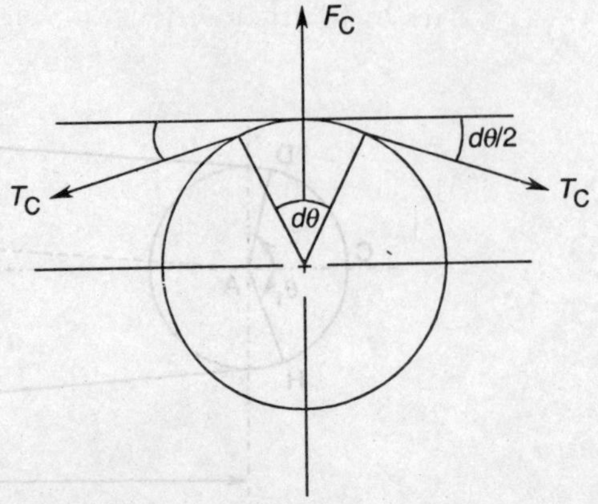

Figure 11.9 Centrifugal tension in belt.

The maximum power transmission capacity of a belt drive depends upon the following two conditions:

(i) The tight side tension in the belt reaches its maximum permissible value.
(ii) The belt is on the point of slipping.

According to maximum power transmission condition, the centrifugal tension should be equal to one-third of the total tension, i.e. $T_C = T/3$ and the limiting belt speed is given by the following relation

$$v_{max} \leq (T/3m)^{0.5} \tag{11.11}$$

The ratio of tensions on tight and slack sides, when the effect of the centrifugal tension is considered, is given by

$$\frac{T_1 - T_C}{T_2 - T_C} = e^{\mu\theta} \tag{11.12}$$

11.3.3 Stresses in Belts

In a belt drive when power is transmitted from the driving pulley to the driven pulley, a fibre on the belt which travels along the belt from any position and completes one cycle is subjected to the following forces at different positions:

(i) Tension on tight side
(ii) Tension on slack side
(iii) Centrifugal force
(iv) Bending of belt over pulley.

These forces in the belt drive mainly produce three types of stresses:

(a) Static stress due to tension on tight side

$$\sigma_1 = \frac{T_1}{bt} \tag{11.13}$$

(b) The centrifugal force produced due to weight of the belt produces centrifugal stresses, which are tensile in nature.

$$\sigma_C = \frac{T_C}{bt} = \frac{\rho v^2}{g \times 10^6} \quad (11.14)$$

(c) Maximum bending stress due to curvature effect of the belt around the smaller pulley is given by the relation

$$\sigma_{b,1} = \frac{Et}{d} \quad (11.15)$$

where E is the elastic modulus of the belt material.

The distribution of these stresses at various points in a belt is shown in Figure 11.10 which indicates that maximum stress occurs at the beginning of the arc of the driving pulley. For safe design, it should be less than or equal to the allowable strength of the belt material, i.e.

$$\sigma_{max} = \sigma_1 + \sigma_C + \sigma_{b,1} \leq \sigma_d \quad (11.16)$$

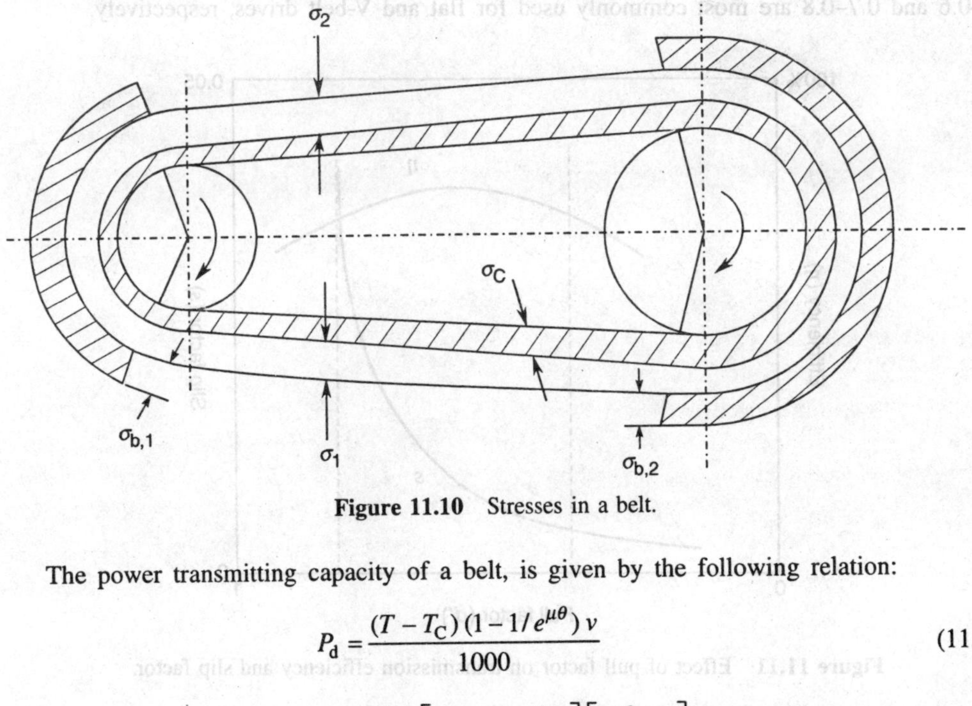

Figure 11.10 Stresses in a belt.

The power transmitting capacity of a belt, is given by the following relation:

$$P_d = \frac{(T - T_C)(1 - 1/e^{\mu\theta})\, v}{1000} \quad (11.17)$$

$$= \frac{btv}{1000}\left[\sigma_d - \frac{\rho v^2}{g \times 10^6}\right]\left[\frac{e^{\mu\theta} - 1}{e^{\mu\theta}}\right] \quad (11.18)$$

where

σ_d is the design stress (= 2.0 MN/m^2 for leather belts and 1.6 MN/m^2 for rubber belts)
ρ is the density of the belt material
bt is the belt cross-sectional area
μ is the coefficient of friction.

11.3.4 Pull Factor

In a belt drive, the ratio of the tangential force to twice the initial tension is called the *pull factor d′*. Thus,

$$d' = \frac{F}{2T_0} = \frac{T_1 - T_2}{T_1 + T_2} = \frac{k-1}{k+1} \tag{11.19}$$

where

$$k = \frac{T_1}{T_2} = e^{\mu\theta}$$

In the design of a belt drive, the pull factor is important for quiet belt operation. The transmission efficiency and the slip factor of the belt are both a function of the pull factor as shown in Figure 11.11. The transmission efficiency is maximum at a particular point. Before this point, i.e. at low values of pull factor, the belt runs very hard and the higher losses reduce the efficiency. After the optimum point, sliding friction increases rapidly and so do the losses which also lower the efficiency. For efficient operation, pull factors in the range between 0.4–0.6 and 0.7–0.8 are most commonly used for flat and V-belt drives, respectively.

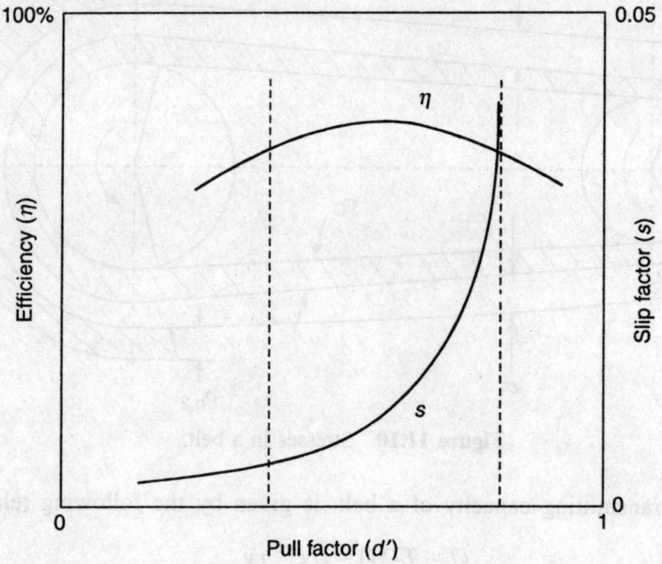

Figure 11.11 Effect of pull factor on transmission efficiency and slip factor.

11.4 BELT MATERIALS

A belt material should have a high coefficient of friction, strength, flexibility and durability. The commonly used materials are the following:

Leather. It is the most widely used material for belts. Leather belts are available in two varieties—oak tanned and chrome tanned. Oak-tanned leather is most commonly used in ordinary applications whereas for special applications involving damp environment, chemical handling machinery and oiled surfaces, the chrome-tanned leather belts are preferred. In order to obtain a reasonable life of a belt, the layers of leather strips are cemented together and these belts are specified according to the number of layers called plys. Properties of various belt materials are listed in Tables 11.1 and 11.2.

Table 11.1 Properties of belt materials

Material	*Design strength* (N/mm²)	*Endurance strength* (N/mm²)	*Elastic modulus* (N/mm²)	*Density* × 10^{-3} (N/m³)	*Maximum velocity* (m/s)	*Constant, m*
Rubber	1.6	6.0	10.0	11–12	30	5
Leather	2.0	5.0	30.0	10–11	30–50	6
Fabric	1.5	3.0	15.0	9–10	20–25	5
Plastic	4.0	6.0	60.0	10–11	60	6

Table 11.2 Leather belt data

Belt grade	*Thickness* (mm) *Single ply*	*Double ply*	*Triple ply*	*Quadruple ply*	*Width-increment* (mm)
Light	3	6	—	—	12–24 by 3 24–102 by 6 102–198 by 12
Medium	4	8	12.5	17.5	200–800 by 25 800–1400 by 50
Heavy	5	10	15	20	1500–2100 by 100

Rubber. Rubber belts are made from cotton duck or canvas impregnated with rubber. These are normally available in three to ten plys. They are vulcanized for use where exposed to oil or sunlight. The strength of the belt is mainly obtained from cotton duck or canvas whereas the rubber lining provides protection against adverse working conditions. Rubber belts have short service life compared to leather belts. These belts are cheaper and most suited for outdoor service. The standard sizes of rubber belts are given in Table 11.3.

Table 11.3 Rubber belt data

Number of plys	*Thickness* (mm)	*Belt width* (mm)
3	3.9	25, 32, 40, 50, 63, 71
4	5.2	40, 50, 63, 71, 80, 90, 100, 112, 125, 140
5	6.5	71, 80, 90, 100, 112, 125, 140, 160
6	7.8	112, 125, 160, 180, 200, 224, 250
7	9.1	160, 180, 200, 224, 250, 280, 315
8	10.4	224, 250, 280, 315, 355, 400, 450, 500

Batta. Batta belts are made from closely woven cotton duck impregnated with batta gum; they need not be vulcanized. Batta does not oxidize and does not age in air or sunlight. It is waterproof, and is not affected by acids, alkalies, and humidity. However, it is seriously affected by mineral oils. Batta belts are 20–40 per cent stronger than the rubber belts.

Fabric. Fabric is another material popularly used as belt material. Fabric belts are made from canvas or cotton ducks in which a number of layers of cotton ducks are closely woven. The woven belt is further treated in linseed oil to make it waterproof. These belts are used for temporary installations and rough service where little attention is needed. The mechanical properties of fabric belts are comparable to those of rubber belts.

Plastics. Plastic strips are used as core material of the belts. Plastic-cored belts are made from nylon canvas or thin plastic sheets with a layer of rubber surrounding the whole belt structure. Plastic-cored belts can be run at high speeds and can be wrapped around very small pulleys. These belts possess high strength which is approximately two times that of leather belts.

11.5 SELECTION OF A PULLEY

The design of a belt drive begins with the determination of diameter of the driving pulley which is critical. This selection is associated with centrifugal stresses on the belt which, if the diameter of the driving pulley is large enough, lead to such values that produce tension on belt greater than the initial tension. There are some empirical relations which are all based on an idea of optimum selection of the driving pulley diameter. One such empirical relation developed by Saverin for the cast iron pulley is most commonly used.

$$d = (525\text{–}630) \times \left(\frac{P}{\omega}\right)^{1/3} \tag{11.20}$$

where

d is the diameter of the driving pulley
P is the power transmitted (kW)
ω is the angular velocity (rad/s)

Based upon the diameter calculated by the above equation, a pulley of standard nominal diameter may be selected from Table 11.4.

Table 11.4 Nominal diameters (mm) of CI pulleys

40, 45, 50, 63, 71, 80, 90, 100, 112, 140, 160, 180, 200, 224, 280, 315, 355, 400, 450, 500, 560, 630, 710, 800, 900, 1000, 1120, 1400, 1600, 1800, 2000

The other proportions of the pulley as shown in Figure 11.12 are given below:

(i) Number of arms (i)
$i = 4$ for $d = 200\text{–}450$ mm
$i = 6$ for $d > 450$ mm

(ii) Width of pulley
$B = (1.1\text{–}1.25)b + 5\text{–}10$ mm
where b is the width of the belt (mm).

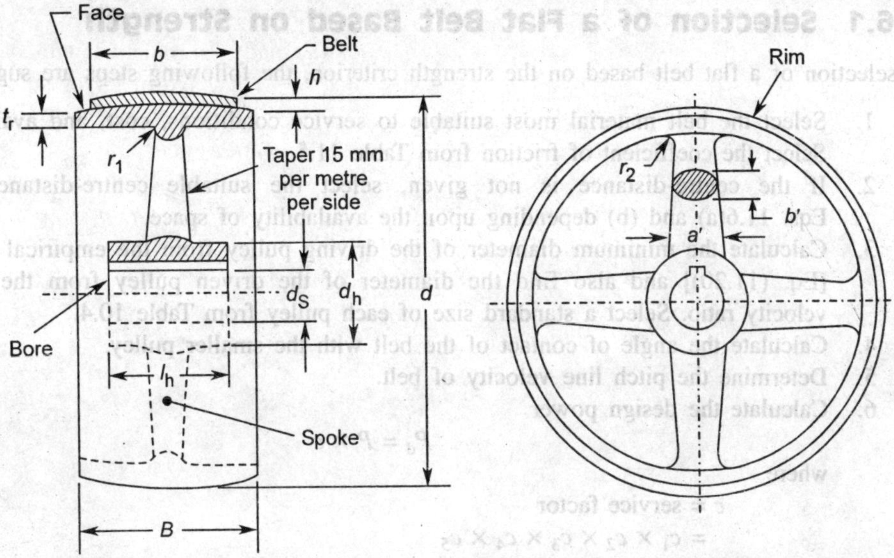

Figure 11.12 Cast iron pulley.

(iii) Thickness of the rim, $t_r = \dfrac{d}{200} + 3.0$ mm

(iv) Crown height, $h = 0.003d$

(v) Length of the hub, $l_h = (1.5\text{–}2.5)d_S$

where d_S is the diameter of the shaft.

(vi) Thickness of the major axis of the elliptical arm near the boss

$$a' = 3.44\left(\frac{T}{i\,\sigma_t}\right)^{1/3}$$

where

T is the bending moment near the boss

σ_t is the allowable tensile strength

i is the number of arms

and minor axis $b' = a'/2$

11.6 SELECTION OF FLAT BELTS

For most applications, the selection of the belt is done from the manufacturer's catalogue. Each manufacturer gives a suggested design procedure. However, to substantiate the manufacturer's procedure, a rational design method based on strength of the belt material should be used. In this section, both these methods of belt selection—one based on strength and the other based upon the manufacturer's catalogue—are presented to give hands-on experience to students.

11.6.1 Selection of a Flat Belt Based on Strength

For selection of a flat belt based on the strength criterion, the following steps are suggested:

1. Select the belt material most suitable to service conditions, cost, and availability. Select the coefficient of friction from Table 11.5.
2. If the centre-distance is not given, select the suitable centre-distance from Eqs. 11.6(a) and (b) depending upon the availability of space.
3. Calculate the minimum diameter of the driving pulley from the empirical relation [Eq. (11.20)] and also find the diameter of the driven pulley from the known velocity ratio. Select a standard size of each pulley from Table 10.4.
4. Calculate the angle of contact of the belt with the smaller pulley.
5. Determine the pitch line velocity of belt.
6. Calculate the design power

$$P_d = P \times c$$

where

$$c = \text{service factor}$$
$$= c_1 \times c_2 \times c_3 \times c_4 \times c_5$$

The values of these factors are given in Table 11.6.

7. Find the required cross-section by using Eq. (11.18).
8. Select the suitable number of plys and the thickness of the belt. Also, calculate the width of the belt from the calculated area of cross-section.
9. Calculate the service life of the belt based on fatigue considerations, discussed later in this section.
10. Determine the length of the belt using Eq. (11.8) or (11.9). Reduce the calculated length to take up initial tension as suggested below:
    (a) Belt of 3 plys— 1.5 per cent of length
    (b) Belt of 4, 5, and 6 plys — 1.0 per cent of length
    (c) Belt of more than 6 plys — 0.5 per cent of length
11. Calculate the pull factor.
12. Calculate the shaft diameter and decide the other dimensions of the pulley.
13. Sketch the layout of the drive.

Table 11.5 Coefficient of friction for belt-pulley material

Belt material	*Pulley material*			
		Cast iron		
	Steel	*Dry*	*Wet*	*Greased*
Oak-tanned leather	0.22	0.25	0.2	0.15
Chrome-tanned leather	0.35	0.4	0.35	0.25
Batata	0.28	0.32	0.20	—
Rubber	0.3	0.35	0.2	—
Cotton woven	0.18	0.22	0.15	0.12

Table 11.6 Service factors

c_1	Overload factor	1.0–1.5
c_2	Environment factor	1.0–1.3
c_3	Continuous operation factor, if fatigue test is not performed	
	(a) 3–4 h/day	1.45
	(b) 8–10 h/day	1.50
	(c) 16–18 h/day	1.90
	(d) 24 h/day	2.0
c_4	Angle of contact factor	

θ	80°	100°	140°	180°
c_4	1.5	1.35	1.1	1.0

c_5	Tension factor	
	(a) Tension with bolts	1.0
	(b) Belt shortening	1.2
	(c) Self-tensioning system	0.8

Service life of belt

Although various fatigue tests on belts have not revealed any specific endurance limit, the point where the deterioration of a belt starts has great uncertainty. It seems that after a number of stress reversals, deterioration starts independently of the magnitude of stress. The service life calculation based on fatigue strength is given by

$$(\sigma_{max})^m N = \text{constant} \tag{11.21}$$

where

σ_{max} is the maximum stress

m is the constant (refer Table 11.1)

N is the number of stress cycles.

If the fatigue life is assumed to be N_e cycles, usually taken as 10^6 cycles, and endurance strength is σ_{en}, then

$$(\sigma_{max})^m N = (\sigma_{en})^m N_e \tag{11.22}$$

or the number of cycles before failure of the belt is

$$N = N_e \left(\frac{\sigma_{en}}{\sigma_{max}}\right)^m \tag{11.23a}$$

$$= 3600\, HZ \frac{v}{L} \tag{11.23b}$$

where

H is the life in hours

Z is the number of bends per one full cycle of the belt

v/L is the number of full revolutions of the belt/second

σ_{en} is the endurance strength of the belt material (refer Table 11.1).

Therefore

$$\text{Service life, } H = \frac{N_e \left(\dfrac{\sigma_{en}}{\sigma_{max}} \right)^m}{3600 \times Z \times \dfrac{v}{L}} \tag{11.24}$$

11.6.2 Selection of a Flat Belt Based on Manufacturer's Catalogue

In practice for most of the applications, a designer usually selects a belt from the manufacturer's catalogue, the reason being to take advantage of the long-term experience of experimental data and assured availability of a standard size belt. Besides this, the procedure discussed in Section 11.6.1 has some inherent limitations such as uncertainty associated with the coefficient of friction, working conditions, type of application, and strength of the belt material. These uncertainties result in a high degree of unreliability even if a large factor of safety is adopted. For the selection of a belt, the following input data are required: (i) power to be transmitted, (ii) transmission ratio, and (iii) centre-distance.

The basic procedure for the selection of a belt is suggested as under:

1. Select a suitable belt material. Usually two different types of belts are available—HI-SPEED duck belting and FORT duck belting impregnated with rubber. The transmission capacities of these belts are as follows:

   (i) HI-SPEED 0.012 kW/mm width/ply
   (ii) FORT 0.015 kW/mm width/ply.

   These values of power transmission capacities are based upon the assumptions that (i) the angle of contact is 180° and (ii) the belt velocity is 5 m/s.

2. Select an optimum belt velocity between 15 m/s and 23 m/s and calculate the diameter of pulleys.
3. Calculate the angle of contact and find the value of angle of contact factor (K_1) from Table 11.7.

Table 11.7 Angle of contact factor (K_1)

Angle of contact	120	130	140	150	160	170	180	190	200	210
Factor (K_1)	1.33	1.26	1.19	1.13	1.08	1.04	1.00	0.97	0.94	0.91

4. Calculate the design power using the following relation

$$P_d = P \times c \tag{11.25}$$

where

P is the power to be transmitted

c is the service factor (refer Table 11.6)

5. Determine the corrected power rating for the belt by the following relation:

$$\text{Corrected power rating, } P_r = \frac{\text{Power rating} \times v \times K_1}{5} \tag{11.26}$$

6. Determine the product of the number of plys and the width of the belt by dividing the design power by the corrected power rating.

$$\text{No. of plys} \times \text{width} = \frac{\text{design power } (P_d)}{\text{corrected power rating } (P_r)}$$

Select a suitable width and the number of plys from Table 11.3.

11.7 SELECTION OF V-BELTS

A V-belt having a trapezoidal section runs over a grooved pulley which may be single- or multiple-grooved. Here the increased coefficient of friction, due to wedging action against the sides of the grooved pulley, permits to transmit more power compared to a flat belt. The advantages of the V-belt drive are: (i) the belt cannot come out of the groove, (ii) the wedging action allows operation at reduced angle of contact with the smaller pulley, (iii) lower initial tension, (iv) short centre-distance, (v) the drive is quieter at high speeds and is capable of absorbing high shock, and (vi) the standardization of V-belts results in better initial installation and facilitates subsequent replacement.

A V-belt consists of a central layer of cords which is located above the neutral axis to carry the load (Figure 11.13). Generally nylon, rayon or cotton fabric materials are used for cords. This layer of cords is surrounded by rubber to transmit the pressure of the cords to the side walls of the belt. Both the rubber and cords are enclosed in an elastic wear resisting cover made of rubber impregnated canvas. These endless belts are made of standard lengths and cross-sections. The BIS under IS: 2494–1974 and IS: 2122 (Part II)—1973 has given five basic symbols A, B, C, D, and E to represent each cross-section. The dimensions of standard cross-sections and the recommended power range are given in Table 11.8.

Table 11.8 Dimensions of V-belts

	V-belt section				
	A	B	C	D	E
Nominal top width, *b* (mm)	13	17	22	32	38
Nominal thickness, *t* (mm)	8	11	14	19	23
Recommended velocity (m/s)	25	25	25	30	30
Recommended power range (kW)	0.4–4.0	1.5–15	10–70	35–150	70–260
Maximum number of strands	6	9	12	14	20
Minimum pitch diameter of pulley	125	200	300	500	630

[IS: 2494–1974]

The V-belts are designated by the symbol of the cross-section followed by the inside length of the belt. The inside length of a belt can be calculated by subtracting a fixed length

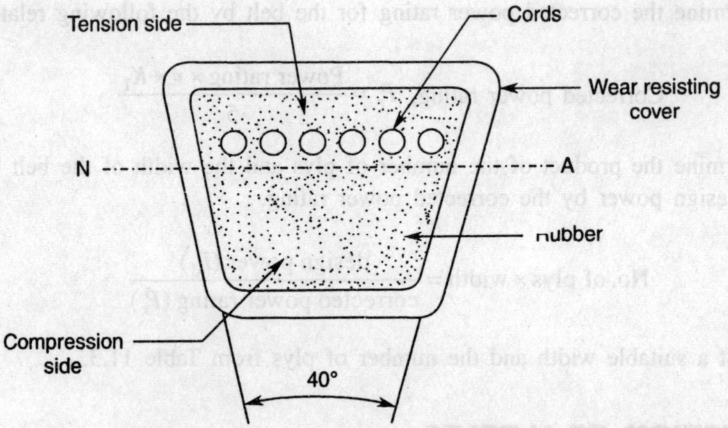

Figure 11.13 Cross-section details of V-belt.

from the pitch length of the belt. The values of pitch length for various sections are given in Table 11.11(a). For example, a V-belt of cross-section A having pitch length of 1051 mm is designated as A-1015-IS: 2494–1974, where the inside length 1015 mm is computed by subtracting 36 mm from the pitch length [see Table 11.11(b)]. The other details of the belt section such as the number of strands used in the cords, their dimensions, minimum pulley diameter and power capacity range for each section are generally given by belt manufacturers.

The power rating of V-belts in kW, based on 180° angle of contact, for various pulley diameters, belt speed and section of belt have been given in a tabular form by BIS under code IS: 2494–1974. However, the empirical formulae for computing the power rating as given by BIS are listed in Table 11.9.

Table 11.9 Empirical formula for power rating of V-belts (IS : 2494–1974)

Belt section	*Power rating* (kW)	*Maximum value* of d_e (mm)
A	$P' = \left(0.61v^{-0.09} - \dfrac{26.68}{d_e} - 1.04 \times 10^{-4} v^2\right) \times 0.7355v$	125
B	$P' = \left(1.08v^{-0.09} - \dfrac{69.68}{d_e} - 1.78 \times 10^{-4} v^2\right) \times 0.7355v$	175
C	$P' = \left(2.01v^{-0.09} - \dfrac{194.8}{d_e} - 3.18 \times 10^{-4} v^2\right) \times 0.7355v$	300
D	$P' = \left(4.29v^{-0.09} - \dfrac{690}{d_e} - 6.48 \times 10^{-4} v^2\right) \times 0.7355v$	425
E	$P' = \left(6.22v^{-0.09} - \dfrac{1294}{d_e} - 9.59 \times 10^{-4} v^2\right) \times 0.7355v$	700

where P' is the power rating (kW)
v is the belt speed (m/s)
d_e is the equivalent diameter = $d \times K_d$
K_d is the small diameter factor (= 1.0 to 1.14)

11.7.1 Selection Procedure

The following procedure may be used for the selection of a V-belt.

1. The selection of the cross-section of a V-belt depends upon two factors, namely the power to be transmitted and the speed of the faster pulley. Belt manufacturers generally supply a curve between these two parameters as shown in Figure 11.14. With the known values of actual power to be transmitted and the speed of the faster pulley, a point can be located on this figure and the corresponding belt-section selected. In the absence of this diagram, the power range of belts recommended by IS : 2494 may be used for selection of the cross-section of belt (refer Table 11.8).

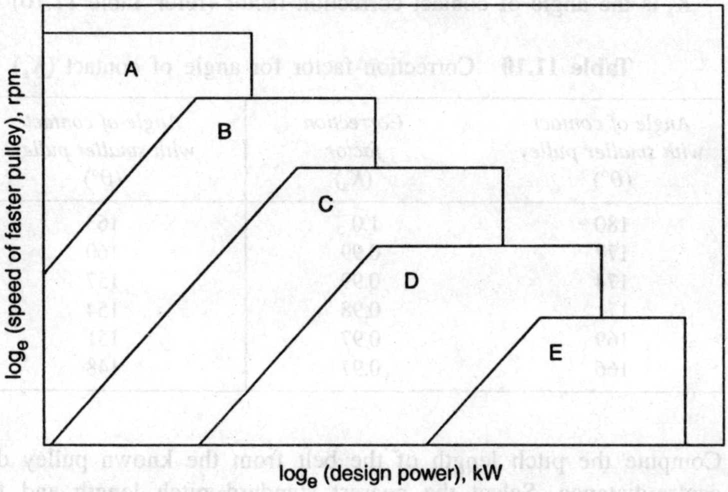

Figure 11.14 Selection of cross-section of V-belt drive.

2. Calculate the pitch diameters of the driven and driving pulleys based upon the given rpm, the velocity ratio, and the pitch line velocity of the belt. If the pitch line velocity is not given, a value in the range 15–25 m/s may be selected. In no case it should be more than 30 m/s. The nominal diameters of pulley as per IS: 3142–1965, are listed in Table 11.4.
3. If the centre-distance is not given, it may be selected empirically as follows:

$$C_{min} = 0.55(d + D) + t$$
$$C_{max} = 2(d + D) \qquad (11.27)$$

4. Compute the power rating in kW for the chosen belt-section using the empirical formula from Table 11.9. The power rating of the belt is based upon certain assumptions that the angle of contact is 180° and the length of the belt has no effect on the service life. The earlier value of the power rating of the belt now needs to be corrected.

5. Determine the number of belts n required for a particular power transmission.

$$n = \frac{P \times K_S}{P' \times K_L \times K_a} \tag{11.28}$$

where

P is the power to be transmitted

P' is the standard power rating

K_S is the service factor (=1.0–2.0) depending upon the type of service

K_L is the correction factor for belt length which takes into account the reduction in life of a short length. It is usually taken between 0.8 and 1.12.

K_a is the angle of contact correction factor (refer Table 11.10)

Table 11.10 Correction factor for angle of contact (K_a)

Angle of contact with smaller pulley ($\theta°$)	*Correction factor* (K_a)	*Angle of contact with smaller pulley* ($\theta°$)	*Correction factor* (K_a)
180	1.0	163	0.96
177	0.99	160	0.95
174	0.99	157	0.94
171	0.98	154	0.93
169	0.97	151	0.93
166	0.97	148	0.92

6. Compute the pitch length of the belt from the known pulley diameters and the centre-distance. Select the nearest standard pitch length and finally the inside length.
7. Adjust the centre-distance to accommodate the standard pitch length

$$C = A + \sqrt{A^2 - B^2} \tag{11.29}$$

where

$$A = \frac{L}{4} - \frac{\pi (D + d)}{8} \tag{11.30a}$$

$$B = \frac{D - d}{8} \tag{11.30b}$$

8. For initial tensions in the belt, reduce the length by 0.5 to 1.0 per cent.
9. Find the shaft diameter, the hub diameter, and dimensions of arms, keys.
10. Sketch the layout.

Table 11.11(a) Nominal pitch length of V-belt sections

Belt section	*Pitch length* (mm)
A	1051, 1102, 1128, 1204, 1255, 1331, 1433, 1458, 1509, 1560, 1636, 1661, 1687, 1763, 1814, 1905, 1981, 2032, 2057, 2159, 2286, 2458, 2464, 2540, 2667, 2845, 3048, 3150, 3251, 3404
B	932, 1008, 1059, 1110, 1212, 1262, 1339, 1415, 1440, 1466, 1567, 1694, 1770, 1824, 1948, 2024, 2101, 2202, 2329, 2507, 2583, 2710, 2888, 3091, 3294, 3701, 4050, 4158
C	1351, 1580, 1783, 1961, 2113, 2215, 2342, 2494, 2723, 2901, 3104, 3205, 3307, 3459, 3713, 4069, 4171, 4450, 4628, 5009, 5390, 6101, 6863, 7625

Table 11.11(b) Conversion of pitch length to inside length

Belt section	A	B	C	D	E
Difference between pitch length and inside length	36	43	56	79	92

Example 11.1 Design a flat-belt drive to transmit 15 kW at 720 rpm to a driven machine operating at 360 rpm. Assume that the belt slips over pulley by 3 per cent.

Solution Since service conditions are not given, we assume that the belt drive operates in damp and chemical environment. So the belt material should be chrome-tanned leather. Pulleys are made of cast iron.

Diameter of pulleys

According to the Saverin formula, the diameter of the driving pulley

$$d = (525\text{–}630)\left(\frac{P}{\omega}\right)^{1/3}, \text{ with } \omega = \frac{2\pi \times 720}{60} = 75.4 \text{ rad/s.}$$

$$= 600 \times \left(\frac{15}{75.4}\right)^{1/3} = 350.2 \text{ mm, say, 355 mm (nearest standard size) (see Table 11.4)}$$

$$\text{Diameter of the driven pulley, } D = \frac{(1-s)d}{VR} \text{ (neglecting belt thickness)}$$

$$= (1 - 0.03) \times 355 \times 2 = 688.7 \text{ mm, say 710 mm (see Table 11.4)}$$

$$\text{Pitch line velocity, } v = \frac{\pi d N_1}{60} = \frac{\pi \times 355 \times 720}{60 \times 1000} = 13.38 \text{ m/s}$$

Centre-distance

No space limitation is given, therefore, the centre-distance can be found empirically.

$$C \geq (1.5\text{–}2.0)(d + D)$$

Let

$$C = 1.75(d + D) = 1.75(355 + 710) = 1863.7 \text{ mm, say, } 1850 \text{ mm}$$

Angle of contact with the smaller pulley, $\theta_1 = 180 - 2\phi$

where $\phi = \sin^{-1} \dfrac{D-d}{2C} = \sin^{-1} \dfrac{710-355}{2 \times 1850} = 5.5°$

Thus,

$$\theta_1 = 180° - 2 \times 5.5° = 169°$$

Assuming open-belt drive, the length of the belt

$$L_o = \frac{\pi (D+d)}{2} + \frac{(D-d)^2}{4C} + 2C$$

$$= \frac{\pi (710+355)}{2} + \frac{(710-355)^2}{4 \times 1850} + 2 \times 1850 = 5389.9 \text{ mm}$$

Let

σ_d = design stress = 2 N/mm^2 (see Table 11.1 for leather)

Coefficient of friction, $\mu = 0.4$ (see Table 11.5 for chrome-tanned leather belt and CI pulley)

density, $\rho = 10 \times 10^3$ N/m^3 (see Table 11.1)

Design power, $P_d = P \times c$

c = constant = service factor = $c_1 \times c_2 \times c_3 \times c_4 \times c_5$ (see Table 11.6)

$= 1.1 \times 1.0 \times 1.5 \times 1.0 \times 1.2 = 1.98$

where

c_1 = overload factor = 1.1; c_2 = environment factor = 1.0; c_3 = operation factor = 1.5; c_4 = angle of contact factor = 1.0; c_5 = tension factor = 1.2.

Therefore,

$$P_d = 15 \times 1.98 = 29.7 \text{ kW}$$

Neglecting bending stresses in the belt, the power transmission capacity is given by the following equation

$$P_d = \frac{bt}{1000}\left[\sigma_d - \frac{\rho v^2}{g \times 10^6}\right]\left[\frac{e^{\mu\theta}-1}{e^{\mu\theta}}\right] v$$

Here,

$$e^{\mu\theta} = e^{0.4 \times \frac{\pi}{180} \times 169} = 3.25$$

Therefore,

$$29.7 = \frac{bt}{1000}\left[2.0 - \frac{10 \times 10^3 \times 13.38^2}{9.81 \times 10^6}\right]\left[\frac{3.25-1}{3.25}\right] \times 13.38$$

or

$$bt = 1764.1 \text{ mm}^2$$

For medium duty, a double-ply belt with thickness t = 8 mm is chosen (see Table 11.2). Therefore,

$$\text{Belt width, } b = \frac{1764.1}{8} = 220.5 \text{ mm, say, } 225 \text{ mm}$$

$$\text{Tight side tension, } T_1 = \frac{P_d}{\left(\frac{e^{\mu\theta}-1}{e^{\mu\theta}}\right)\times v} = \frac{29.7\times 10^3}{\left(\frac{3.25-1}{3.25}\right)\times 13.38} = 3206.2 \text{ N}$$

Direct stress in the belt due to tight side tension

$$\sigma_1 = \frac{T_1}{bt} = \frac{3206.2}{8\times 225} = 1.78 \text{ N/mm}^2$$

Centrifugul stress

$$\sigma_C = \frac{\rho v^2}{g\times 100\times 10^6} = \frac{10\times 10^3\times 13.38^2}{9.81\times 10^6} = 0.18 \text{ N/mm}^2$$

Bending stress

$$\sigma_b = \frac{Et}{d} = \frac{30\times 8}{355} = 0.67 \text{ N/mm}^2$$

$$\text{Maximum stress, } \sigma_{max} = \sigma_1 + \sigma_C + \sigma_b$$

$$= 1.78 + 0.18 + 0.67 = 2.63 \text{ N/mm}^2$$

Fatigue life of belt (H)

Endurance strength, σ_{en} = 6 N/mm² (see Table 11.1)

Constant, m = 6 (see Table 11.1)

Therefore,

$$H = \frac{N_e\left(\frac{\sigma_{en}}{\sigma_{max}}\right)^m}{3600\times Z\times \frac{v}{L}}$$

where, $N_e = 10^6$ cycles and Z = 2. Thus,

$$H = \frac{10^6\times\left(\frac{6}{2.63}\right)^6}{3600\times 2\times\frac{13.38}{5.3899}}$$

$$= 7888 \text{ h}$$

For an average 10 h/day and 300 working days per year, the belt will operate for 2.6 years, which is satisfactory.

Dimensions of the driving pulley

We assume that the shaft is made of plain carbon steel C20 having allowable shear strength $\tau = 65$ N/mm^2.

$$\text{Bending moment, } M = (T_1 + T_2) \times \text{overhang}$$

$$T_1 = 3206.2 \text{ N}; \; T_2 = \frac{T_1}{e^{\mu\theta}} = \frac{3206.2}{3.25} = 986.5 \text{ N}$$

$$\text{Overhang, } l_1 = \frac{1}{2} \times \text{pulley width} + \text{clearance} + d_{S_1}$$

$$\text{Pulley width, } B = 1.1b + 5 \text{ mm} = 1.1 \times 225 + 5 = 252.5 \text{ mm, say, 255 mm}$$

$$l_1 = \frac{255}{2} + 25 + d_{S_1}$$

Let us assume, $l_1 = 200$ mm (a trial value). Then:

$$\text{Moment, } M = (3206.2 + 986.5) \times 0.2 = 838.5 \text{ N} \cdot \text{m}$$

$$\text{Torque, } T = (T_1 - T_2) \times r$$

$$= (3206.2 - 986.5) \times \frac{0.355}{2} = 394 \text{ N} \cdot \text{m}$$

Equivalent torque according to ASME code, with $C_m = 1.5$ and $C_t = 1.0$

$$T_{eq} = \sqrt{(C_m M)^2 + (C_t T)^2}$$

$$= \sqrt{(1.5 \times 838.5)^2 + (394)^2} = 1318 \text{ N} \cdot \text{m}$$

$$\text{Shaft diameter, } d_{S1} = \left(\frac{16 T_{eq}}{\pi \tau}\right)^{1/3} = \left[\frac{16 \times 1318 \times 10^3}{\pi \times 65}\right]^{1/3} = 46.9 \text{ mm, say, 50 mm}$$

Actual overhang $= \dfrac{255}{2} + 25 + 50 = 202.5$ mm, which is approximately equal to that assumed.

Arms-section

Let the number of arms be $i = 4$ and out of which half the numbers actually take load.

$$\text{Section modulus, } Z = \frac{2T}{i\sigma_t}$$

$$\sigma_t = 15 \text{ N/mm}^2 \text{ for ordinary grey CI (FG200)}$$

Therefore,

$$Z = \frac{2 \times 394 \times 10^3}{4 \times 15} = 13{,}133.3 \text{ mm}^3$$

Assuming elliptical cross-section

$$\text{Major axis, } a_1 = \left(\frac{64Z}{\pi}\right)^{1/3} = \left[\frac{64 \times 13133.3}{\pi}\right]^{1/3} = 64.43 \text{ mm, say, } 66 \text{ mm}$$

$$\text{Minor axis, } b_1 = \frac{a_1}{2} = 33 \text{ mm}$$

So elliptical section (66 × 33) mm^2 with 4 mm taper per 100 mm length is recommended.

Specifications of the driving pulley

(i) Pitch diameter $d = 450$ mm

(ii) Shaft diameter $d_S = 50$ mm

(iii) Hub diameter $d_h = 75$ mm

(iv) Length of the hub $l_h = \frac{2}{3} \times B = 150$ mm

(v) Width of the belt $b = 225$ mm

(vi) Thickness of the rim $t_{r_1} = \left(\frac{d}{200} + 3\right)$ mm = 5.25 mm

(vii) Number of arms $i = 4$

(viii) Arms-section = 66 mm × 33 mm (elliptical)

(ix) Keys = 13 mm × 13 mm (square key)

The dimensions of the driven pulley can be found on the lines similar to the driving pulley. It is left as an exercise to the student.

Example 11.2 A 30 kW, 1000 rpm motor transmits power to a stone-crushing machine, which operates at 250 rpm. Select a suitable flat belt.

Solution We select a rubber belt of HI-SPEED quality from a manufacturer's catalogue (partly given in Section 11.6.2). The power transmission capacity of HI-SPEED is

0.012 kW/mm width/ply

For operational economy, the belt speed should be between 15 and 25 m/s. We select, $v = 20$ m/s. Now,

$$\text{Diameter of the driving pulley, } d = \frac{v \times 60}{\pi N_1} = \frac{20 \times 60}{\pi \times 1000} = 0.3819 \text{ m, say, } 400 \text{ mm}$$

(standard size from Table 11.4)

$$\text{Diameter of the driven pulley, } D = \frac{(1 - s)d}{\text{VR}}$$

Let the belt slip by 2 per cent, i.e. $s = 0.02$. Thus,

$$D = (1 - 0.02) \times 400 \times \frac{1000}{250} = 1568 \text{ mm, say, } 1600 \text{ mm (standard size from Table 11.4).}$$

$$\text{Revised belt speed, } v = \frac{\pi d N_1}{60 \times 1000} = \frac{\pi \times 400 \times 1000}{60 \times 1000} = 20.94 \text{ m/s}$$

The centre-distance between the two shafts is not given. However, it should be large enough to protect the motor from stone dust. By the empirical relation

$$C > (1.5\text{–}2.0)(d + D)$$

$$\text{we adopt, } C = 2(d + D)$$
$$= 2(400 + 1600) = 4000 \text{ mm}$$

Angle of contact on smaller pulley

$$\theta_1 = 180° - 2\phi$$

$$\text{where } \phi = \sin^{-1}\frac{D - d}{2C} = \sin^{-1}\frac{1600 - 400}{2 \times 4000} = 8.62°$$

$$\text{Therefore, } \theta_1 = 180° - 2 \times 8.62° = 162.76°$$

Considering open-belt drive, the length of the belt can be computed by the following relation:

$$L = \frac{\pi (D + d)}{2} + \frac{(D - d)^2}{4C} + 2C$$

$$= \frac{\pi (1600 + 400)}{2} + \frac{(1600 - 400)^2}{4 \times 4000} + 2 \times 4000 = 11.231 \text{ m}$$

$$\text{Design power, } P_d = P \times c \text{ (where } c = \text{service factor} = 2.5)$$
$$= 30 \times 2.5 = 75 \text{ kW}$$

$$\text{Corrected power rating, } P_r = \frac{\text{Power rating} \times v \times K_1}{5}$$

where K_1 = angle of contact factor = 1.08 (see Table 11.7).

Therefore,

$$P_r = \frac{0.012 \times 20.94 \times 1.08}{5} = 0.0542 \text{ kW/mm width/ply}$$

$$\text{Number of plys} \times \text{width} = \frac{\text{design power}}{\text{corrected power}} = \frac{P_d}{P_r} = \frac{75}{0.0542} = 1383.7 \text{ mm width} \times \text{ply}$$

Let us adopt a 6-ply belt having thickness, t = 7.8 mm (see Table 11.3). Then, width of belt,

$b = \dfrac{1383.7}{6}$ = 230.6 mm, say, 250 mm (standard size from Table 11.3). Hence the flat belt selected is HI-SPEED rubber belt of 6 ply × 250 mm width.

Example 11.3 A V-belt drive is required to transmit 16 kW power to a compressor. The motor rpm is 1440 and the speed reduction ratio is 3.6. Design the belt drive.

Solution For operational economy, the belt speed should be between 15 and 25 m/s. We adopt belt speed, v = 22.0 m/s.

$$\text{Diameter of the driving pulley, } d = \frac{v \times 60}{\pi N_1} = \frac{22 \times 60}{\pi \times 1440} = 0.291 \text{ m, say, 300 mm}$$

Diameter of the driven pulley, $D = d \times$ speed reduction ratio

$$= 300 \times 3.6 = 1080 \text{ mm, say, 1120 mm}$$

$$\text{Revised belt speed, } v = \frac{\pi d N_1}{60 \times 1000} = \frac{\pi \times 300 \times 1440}{60 \times 1000} = 22.62 \text{ m/s}$$

$$\text{Revised speed reduction ratio} = \frac{D}{d} = \frac{1120}{300} = 3.73$$

Based upon the power to be transmitted, i.e. 16 kW, we select a C-section belt (see Table 11.8). The centre-distance is not given, so we select it by the empirical formula

$$C \geq 0.55(D + d)$$

$$C \leq 2(D + d)$$

We select the centre-distance, $C = D + d = 1120 + 300 = 1420$ mm.

Considering open-belt drive, the length of the belt is

$$L = \frac{\pi (D + d)}{2} + \frac{(D - d)^2}{4C} + 2C$$

$$= \frac{\pi (1120 + 300)}{2} + \frac{(1120 - 300)^2}{4 \times 1420} + 2 \times 1420$$

$$= 5188.9 \text{ mm}$$

We select the nearest pitch length, L = 5390 mm from Table 11.11a. The nominal inside length of belt is 5390 − 56 = 5334 mm (see Table 11.11b).

$$\text{Revised centre-distance, } C = A + \sqrt{A^2 - B^2}$$

where

$$A = \frac{L}{4} - \frac{\pi (D + d)}{8} = \frac{5390}{4} - \frac{\pi (1120 + 300)}{8} = 789.86$$

$$B = \frac{(D - d)}{8} = \frac{(1120 - 300)}{8} = 102.5$$

Therefore,

$$\text{Revised centre-distance, } C = 789.86 + \sqrt{789.86^2 - 102.5^2} = 1573 \text{ mm}$$

Angle of contact on smaller pulley, $\theta = 180° - 2\phi$

where

$$\phi = \sin^{-1} \frac{D - d}{2C} = \sin^{-1} \frac{1120 - 300}{2 \times 1573} = 15.1°$$

Hence,

$$\theta = 180° - 2 \times 15.1° = 149.8°$$

Power rating of the belt as per IS: 2494–1974

$$P'(\text{kW}) = (2.01v^{-0.09} - \frac{194.8}{d_e} - 3.18 \times 10^{-4} v^2) \times 0.7355v$$

where

d_e = equivalent diameter (≤ 300 mm)
$= d \times K_d$
with K_d = small diameter factor = 1.14 (see Table 11.9)

Hence, d_e = 300 × 1.14 = 342 mm. But the maximum value of d_e can only be 300 mm. So we adopt, d_e = 300 mm. Therefore,

$$P' = [2.01 \times (22.62)^{-0.09} - \frac{194.8}{300} - 3.18 \times 10^{-4} \times (22.62)^2] \times 0.7355 \times 22.62 = 11.74 \text{ kW}$$

$$\text{Number of belts, } n = \frac{P \times K_S}{P' K_L K_a}$$

where

K_S = service factor = 1.2
K_L = Length correction factor = 1.08
K_a = angle of contact correction factor = 0.92 (see Table 11.10)

Therefore,

$$n = \frac{16 \times 1.2}{11.74 \times 0.92 \times 1.08} = 1.646$$

We adopt two V-belts of C5334 – IS: 2494 – 1974.

Note: The design of the pulley and the drive elements is left as an exercise to the student.

11.8 COMPUTER-AIDED DESIGN OF BELTS

The application of computer-aided selection of belts is demonstrated in this section. The listing of computer programs for the selection of V-belts is listed in Program 11.1. The program asks for the following inputs:

(i) Power to be transmitted in kW
(ii) Speed of the driving pulley in rpm
(iii) Speed of the driven pulley in rpm
(iv) Service factor
(v) Centre-distance

The user is advised to supply the centre-distance between the minimum and maximum limits suggested by the program. The power rating of the belt is selected using the empirical relation given by IS: 2494–1974. The pitch length is computed by using an open-belt drive. The user is advised to modify the program to incorporaté data file for standard pitch lengths. The correction factor for belt length and the angle of contact are assumed to be constant, which may be modified as per requirement. A sample run of the program is demonstrated.

Program 11.1

```
//    A PROGRAM FOR SELECTION OF V-BELT
# include <iostream.h>
# include <math.h>
# include <conio.h>
# define PI 3.1415
double pitch_dia (double dia);
void main ( )
{
      double power, n1, n2, ks, v, d1, d2, dd, d22, c, pitch_length, n;
      double p_rating, de, k1, ka, dd1;
      char section;
//    input design data
      cout <<"\n          DESIGN INPUT DATA" <<endl;
      cout <<"\n          Power to be Transmitted (kW)      = ";
      cin >> power;
      cout <<"\n          Speed of Driving Pulley (rpm)     = ";
      cin >>n1;
      cout <<"\n          Speed of Driven Pulley (rpm)      = ";
      cin >>n2;
      cout <<"\n          Service Factor (ks)    = ";
      cin >>ks;
//    calculate the diameter of the smaller pulley, assuming v = 20 m/s
      d1 = 40.0;
repeat: v = PI*d1*n1/60000.0;
          if (v < 20.0)
          {
                dd = pitch_dia (d1);
                d1 = dd;
                goto repeat;
          }
      d2 = d1*n1/n2;
      d2 = 10*int (d2/10.0 + 0.5);
//    revise velocity, centre distance and pitch length
      v = PI*d1*n1/60000;
      v = (int (v*100 + 0.5))/100.0;
      cout <<"\n          Centre Distance ("<<0.55* (d1 + d2) <<" to "<< 2* (d1 + d2) <<") = ";
      cin >> c;
      pitch_length = PI* (d1 + d2)/2.0 + pow ((d2 - d1), 2.0)/(4*c) + 2*c;
      pitch_length = 10*int (pitch_length/10 + 0.5);
//    selection of belt section
      if (power < = 2.5)
            section = 'A';
      else
            if (power < = 12.0)
                  section = 'B';
            else
```

```
            if (power < = 50)
                  section = 'C';
            else
                  if (power <=100)
                        section = 'D';
                        else
                              section = 'E';
//     calculate power rating
       if (section = = 'A'
       {
             de = 125.0;
             p_rating = (0.61*pow (v, – 0.09) – 26.68/de – 0.000104*v*v) * 0.7355*v;
       }
       else
       if (section = = 'B')
       {
             de = 175.0;
             p_rating = (1.08*pow (v, – 0.09) – 69.68/de – 0.000178*v*v)*0.7355*v;
       }
       else
if (section = = 'C')
       {
             de = 300.0;
             p_rating = (2.01*pow (v, – 0.09) – 194.8/de – 0.000318*v*v)*0.7355*v;
       }
       else
if (section = = 'D')
       {
             de = 425.0;
             p_rating = (4.29*pow (v, – 0.09) – 690.0/de – 0.000648*v*v)*0.7355*v;
       }
       else
       {
             de = 700.0;
             p_rating = (6.22*pow (v, – 0.09) – 1294/de – 0.000959*v*v)*0.7355*v;
       }
//     some constants K1 = length correction factor, Ka = Arc angle factor
       k1 = 0.92; // user may take between 0.8 and 1.12
       ka = 0.95; // user may change as per table 11.10
//     number of belts
       n=power*ks/(p_rating*k1*ka);
       n = int (n + 1);
//     design results
       cout <<"\n        DESIGN RESULTS" <<endl;
       cout <<"\n        Belt Section                = " << section;
       cout <<"\n        Number of Belts             = " <<n;
```

```
    cout <<"\n      Pitch Length                  = " << pitch_length <<"mm";
    cout <<"\n      Diameter of Driving Pulley    = " << d1 <<"mm";
    cout <<"\n      Diameter of Driven Pulley     = " << d2 << "mm"<<endl;
}
//  function to select standard size of pulley
double pitch_dia (double dia)
{
        if (dia >400)
            goto next;
        else
            if (dia <63)
                dia = int (dia*1.11+1);
        else
            if (dia <90)
                dia = int (dia*1.11+2);
        else
            if (dia <125)
                dia = int (dia*1.11+1);
        else
            if (dia <140)
                dia = int (dia*1.11+2);
        else
            if (dia <160)
                dia = int (dia*1.11+5);
        else
            if (dia <180)
                dia = int (dia*1.11+3);
        else
            if (dia <200)
                dia = int (dia*1.11+1);
        else
            if (dia <280)
                dia = int (dia*1.12);
        else
            if (dia <315)
                dia = int (dia*1.12+2);
        else
            if (dia<400)
                dia = int (dia*1.12+3);
        return (dia);
        next:
            if (dia <630)
                dia = int (dia*1.11+1)*10;
        else
            if (dia <900)
                dia = int (dia*1.11+2)*10;
```

```
else
    if (dia <1250)
        dia = int (dia*1.11+1)*10;
else
    if (dia <1400)
        dia = int (dia*1.11+2)*10;
else
    if (dia <1600)
        dia = int (dia*1.11+5)*10;
else
    if (dia <1800)
        dia = int (dia*1.11+3)*10;
else
    if (dia <2000)
        dia = int (dia*1.11+1)*10;
else
    if (dia <2800)
        dia = int (dia*1.12)*10;
else
    if (dia<3150)
        dia = int (dia*1.12+2)*10;
else
    if (dia < 4000)
        dia = int(dia*1.12+3)*10;
return (dia);
} // end of function
```

Sample Run of Program 11.1

DESIGN INPUT DATA
Power to be Transmitted (kW) = 16
Speed of Driving Pulley (rpm) = 1440
Speed of Driven Pulley (rpm) = 400
Service Factor (ks) = 1.2
Centre Distance (709.5 to 2580) = 1420
DESIGN RESULTS
Belt Section = C
Number of Belts = 2
Pitch Length = 4960 mm
Diameter of Driving Pulley = 280 mm
Diameter of Driven Pulley = 1010 mm

EXERCISES

1. What are the important factors which decide the selection of a drive for power transmission?
2. Compare belt drive with gear drive.

3. For a horizontal belt, which side (tight or slack) of the belt should be on the top and why?
4. What is the importance of the adjustment of belt tension in a belt drive? Explain any two methods used in practice for the adjustment of initial tension.
5. Enlist the merits and demerits of V-belt over the flat-belt drive.
6. What types of stresses are produced in a belt used for power transmission? Show the distribution of these stresses by a suitable sketch.
7. What is the effect of length of belt on its life? How does the pulley size affect the power transmission capacity of the belt?
8. How are the V-belts designated?
9. Sketch the cross-section of the V-belt and label its parts.
10. A 50 kW, 1200 rpm, high torque squirrel-cage motor is used to drive a punch press. The speed of the punch press flywheel is 300 rpm. If the centre-distance is 2.5 m, select a suitable leather belt.
11. The following particulars refer to a belt drive using a cast iron pulley mounted on a shaft:

Power	: 25 kW
Speed of driving shaft	: 300 rpm
Speed of driven shaft	: 100 rpm
Allowable tension	: 10 N/mm width

Select a suitable belt size.

12. A machine is driven at 1440 rpm by means of a flat belt. The pulleys on the motor and machine shafts are of 200 mm and 800 mm diameter, respectively. Design the belt for transmitting 25 kW power.
13. Design a flat-belt drive to transmit 7.5 kW power at 720 rpm to run a compressor at 300 rpm. Sketch the layout.
14. Select a suitable V-belt drive to connect a 7.5 kW, 1440 rpm induction motor to run a fan at approximately 480 rpm for a service of 16 h per day. The space available for centre-distance is 1.0 m.
15. A V-belt drive is to transmit 15 kW to a compressor. The motor runs at 1150 rpm and the compressor is to run at 400 rpm. Determine the size and the number of belts required.
16. Design a V-belt drive to be used from an electric motor to the flywheel of a forging press, given the following data.

Motor power	: 100 kW
Overload factor	: 1.6
Speed of the motor	: 740 rpm
Speed of the flywheel	: 250 rpm
Diameter of the flywheel pulley	: 1500 mm

Centre-distance : 2000 mm approx.

17. A single acting reciprocating pump making 60 strokes per minute is driven from a geared speed reducer having a 10:1 ratio. The reducer is to be driven through a V-belt drive by a 40 kW motor running at 1440 rpm. Select a suitable belt and design the drive. Sketch the layout.
18. The drive from a motor to a machine consists of three V-belts of size B. The motor pulley has a pitch diameter 200 mm and the machine pulley diameter is 500 mm. If the centre-distance is 1.3 m, give the full specification details of the belt.

MULTIPLE CHOICE QUESTIONS

1. A cross belt transmits
   (a) more power than an open belt
   (b) less power than an open belt
   (c) power equal to an open belt
   (d) power that is unpredictable
2. The life of an open belt in comparison to a cross belt is
   (a) more (b) less
   (c) equal (d) unpredictable
3. The crowning of a flat-belt pulley is done to avoid
   (a) slipping of the belt (b) running off the belt
   (c) creep of the belt (d) wearing out of the belt
4. The angle of contact of a flat-belt should not be less than
   (a) 100° (b) 140°
   (c) 155° (d) 165°
5. The angle of contact of a V-belt should not be less than
   (a) 100° (b) 140°
   (c) 155° (d) 165°
6. In a flat-belt drive, the maximum value of tension is
   (a) T_c (b) $2T_c$
   (c) $3T_c$ (d) $4T_c$
7. In a V-belt drive, the belt touches the pulley
   (a) at bottom (b) at side only
   (c) at bottom and sides (d) can't say
8. The standard angle between the sides of a V-belt is
   (a) 25° (b) 30°
   (c) 40° (d) 45°

Chapter 12 Gears

12.1 INTRODUCTION

Gear is a machine element which, by means of progressive engagement of projections called *teeth*, transmits motion and power between two rotating shafts. Gear teeth, in general, have an involute profile which provides a constant pitch line velocity. The action of such mating gear teeth consists of a combination of rolling and sliding motions, thus producing a positive drive. Gears are classified on the following basis:

(i) Relation between axes
(ii) Shape of the solid on which teeth are cut
(iii) Curvature of the tooth profile

Generally, the following types of gears are most commonly used in industry for power transmission purposes.

Spur gears. A gear having straight teeth along the axis is called the spur gear. Spur gears are used to transmit power between two parallel shafts as shown in Figure 12.1(a). A rack is a straight tooth gear which can be thought of as a segment of spur gear of infinite diameter.

Helical gears. They are also used to transmit power between two parallel shafts and teeth are cut on the cylindrical disc. The tooth faces of these gears have a certain degree of helix angle of opposite hand on pinion and gear as shown in Figure 12.1(b). These gears are smooth in operation and therefore can transmit power at a high pitch line velocity.

Bevel gear. When power is to be transmitted between two intersecting shafts, bevel gears are used. The angle of intersection of shafts is called the *shaft angle.* The gear blank is a frustum of cone on which teeth are generated. The teeth are straight but their sides are tapered so that all lines, when extended, meet at a common point called the *apex* of the cone, as shown in Figure 12.1(c).

Worm and worm gears. In this system of gearing, the axes of the power transmitting shafts are neither parallel nor intersecting but the planes containing the axes are generally at right angles to each other. The teeth used are helical. The schematic diagram of a worm gear set is shown in Figure 12.1(d).

In a gear drive, the smaller of the two gears in mesh is called the *pinion* and the larger gear is customarily designated as *gear.* In most of the applications, the pinion is the driving element whereas the gear is the driven element. There are some applications like the epicyclic

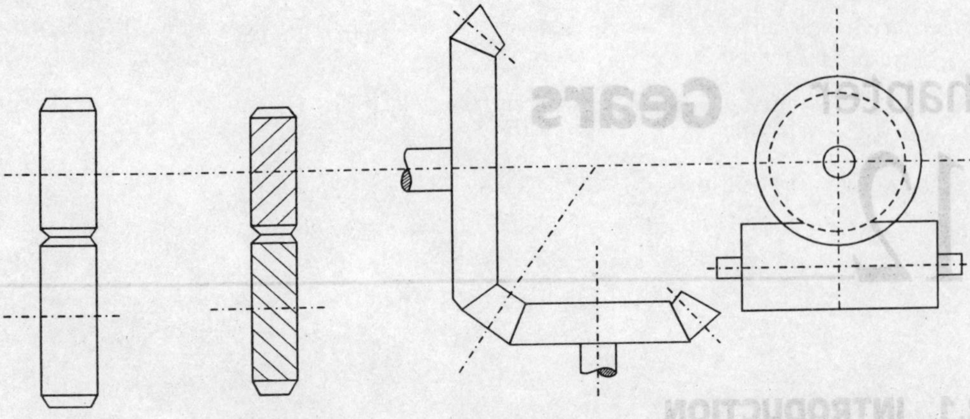

Figure 12.1 Types of gears.

gear train where the gear teeth are cut on the inside of the rim. Such gears are known as *internal* gears.

12.2 GEAR TERMINOLOGY

The terminology and notations for toothed gearing are covered by Bureau of Indian Standards (BIS) in their codes IS: 2458–1965 and IS: 2467–1965. The following definitions of different terms are given with reference to Figure 12.2.

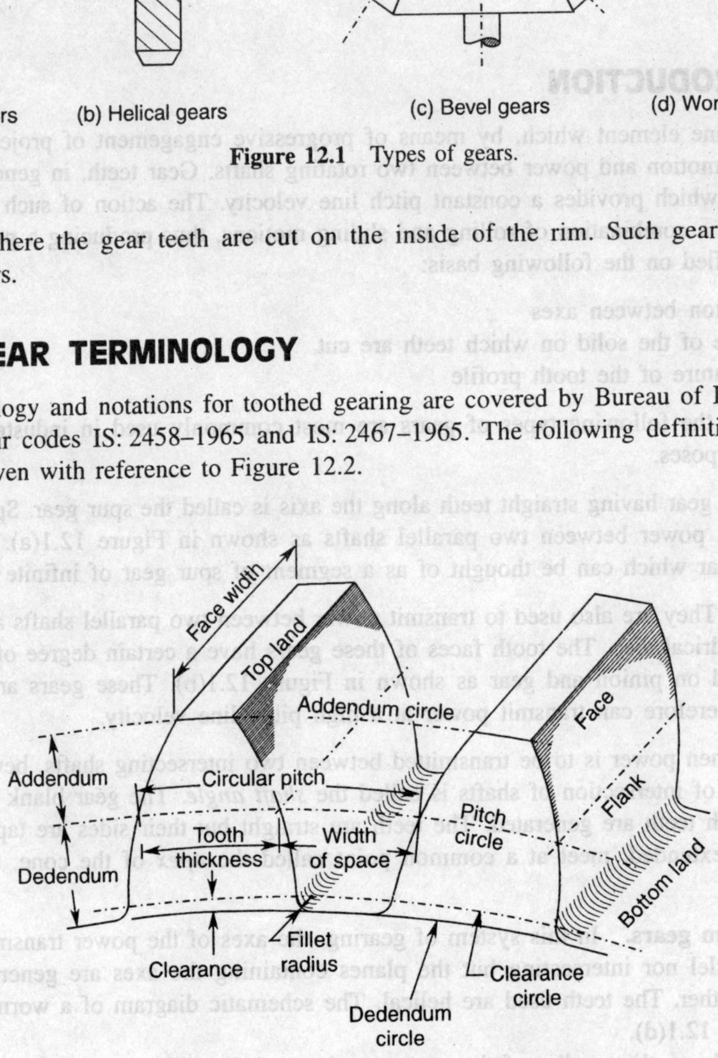

Figure 12.2 Terminologies of spur gear tooth.

Pitch surface. The surface on which teeth are cut to ensure a positive action drive is called the *pitch surface.*

Pitch circle. The intersection of the pitch surface with a plane perpendicular to the axis of rotation is called the *pitch circle.*

Pitch point. The contact point of two pitch circles is called the *pitch point.*

Addendum circle. It is the circle which bounds the outer ends of the teeth. In other words, it is the diameter of a blank on which teeth are cut.

Addendum. The radial distance between the pitch circle and the addendum circle is called the *addendum.* Generally, this distance is kept equal to one module in a 20° full-depth teeth gear and 0.8 time the module in 20° stubbed teeth gear.

Dedendum circle. The circle which bounds the bottom of the teeth.

Dedendum. The radial distance between the pitch circle diameter and the dedendum circle is called the *dedendum.* The value of the dedendum is taken as 1.25 times the module for a full-depth teeth gear and one module for a stubbed teeth gear.

Total depth of tooth. The sum of addendum and dedendum is defined as the total depth of the tooth.

Clearance. The difference between the dedendum and addendum of a mating gear teeth is known as *clearance.* For a spur gear having 20° pressure angle full-depth teeth, its value is 0.25 time the module.

Base circle. A circle from which the tooth profile curve is generated is known as the *base circle.*

Tooth thickness. The chord length measured along the pitch circle between the opposite faces of the same tooth is called *tooth thickness.*

Circular pitch. The distance measured along the pitch circle from a point on one tooth to the corresponding point on the adjacent tooth is called the *circular pitch.* It is denoted by p and given by the following relation:

$$p = \frac{\pi d_1}{Z_1} \tag{12.1a}$$

where

d_1 is the pitch circle diameter of the pinion
Z_1 is the number of teeth on the pinion.

Module. The ratio of the pitch circle diameter to the number of teeth, i.e. the reciprocal of the diametral pitch (DP) is called *module* and is denoted by m.

$$m = \frac{d_1}{Z_1} = \frac{1}{\text{DP}} \tag{12.1b}$$

Diametral pitch. The ratio of the number of teeth to the pitch circle diameter is called the *diametral pitch.*

$$\text{DP} = \frac{Z_1}{d_1} \tag{12.1c}$$

Backlash. The space between two consecutive teeth, measured along the pitch circle, is known as *backlash.*

12.3 KINEMATICS OF GEARING

In a gear drive the action of tooth profile is to transmit motion at a constant angular velocity ratio for which the gears must satisfy the fundamental law of gearing which states that the common normal at the point of action between two teeth must always pass through a fixed point, called the *pitch point.*

When gear tooth profiles are designed to produce a constant angular velocity ratio during meshing, they are said to have *conjugate* action. In order to obtain conjugate action, most gears are cut in the shape of an involute curve between the base circle and the addendum circle. However, there are few gears in existence which are cut in the form of a cycloidal curve.

The gear tooth action is shown in Figure 12.3 in which the pitch circles of mating gears meet at the pitch point P. A line AB which is normal to the line joining the centres O_1, O_2 passes through the pitch point. Further, another line CD, which is tangent to the base circles of gear and pinion passes through the same point and is normal to teeth in contact. Therefore, all the points of contact of the two teeth must lie on this line. For this reason, the line CD is called the *line of action.* The angle between the lines AB and CD is called the *pressure angle* and the normal load that one tooth exerts on the other passes through the pressure line. The pressure line in a gear pair can be located by rotating the line AB through the pressure angle α in the direction opposite to the direction of rotation of the driving gear.

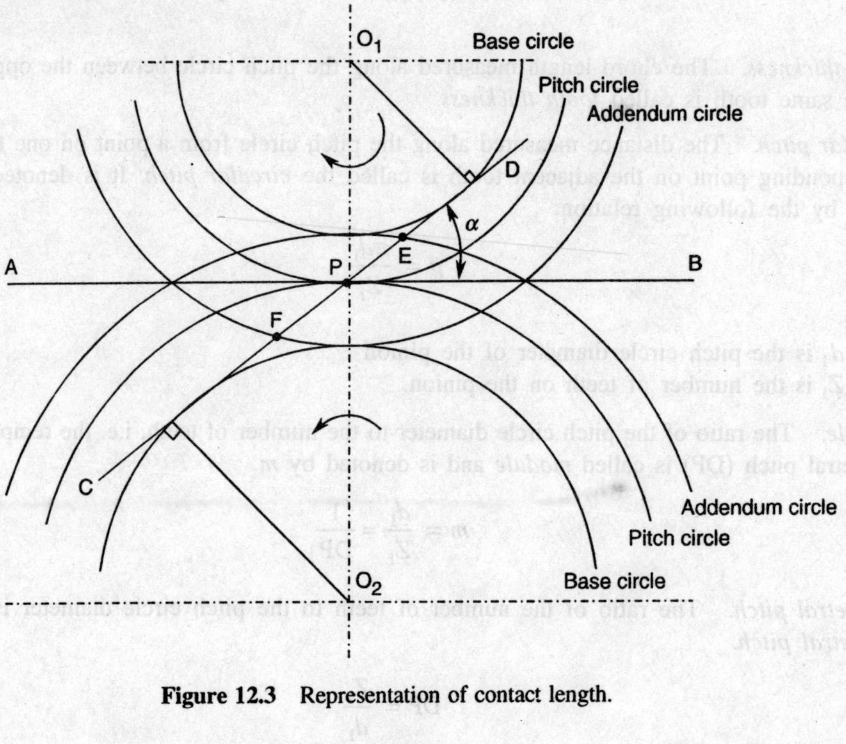

Figure 12.3 Representation of contact length.

When two gears start transmitting motion, the initial contact occurs at a point where the flank of the driving gear comes in contact with the tip of the driven gear and the contact ends when the tip of the driving tooth comes in contact with the flank of the driven tooth. In other words, the contact between the two gears starts when the addendum circle of the driven gear cuts the pressure line at point E and ends where the addendum circle of the driving gear cuts the pressure line at point F, as shown in Figure 12.3. The distance between these two points is called the *contact length* which is useful to determine the contact ratio. Generally, gears are designed for contact ratios from 1.0 to 1.6.

Another important aspect of kinematics of gearing is interference. When a gear tooth tries to dig below the base circle of the mating gear then the gear tooth action shall be non-conjugate and violate the fundamental law of gearing. This non-conjugate action is called *interference*. In other words, the condition of interference arises when contact occurs outside the points C and D as shown in Figure 12.4. Interference can be avoided by employing some preventive measures as given below:

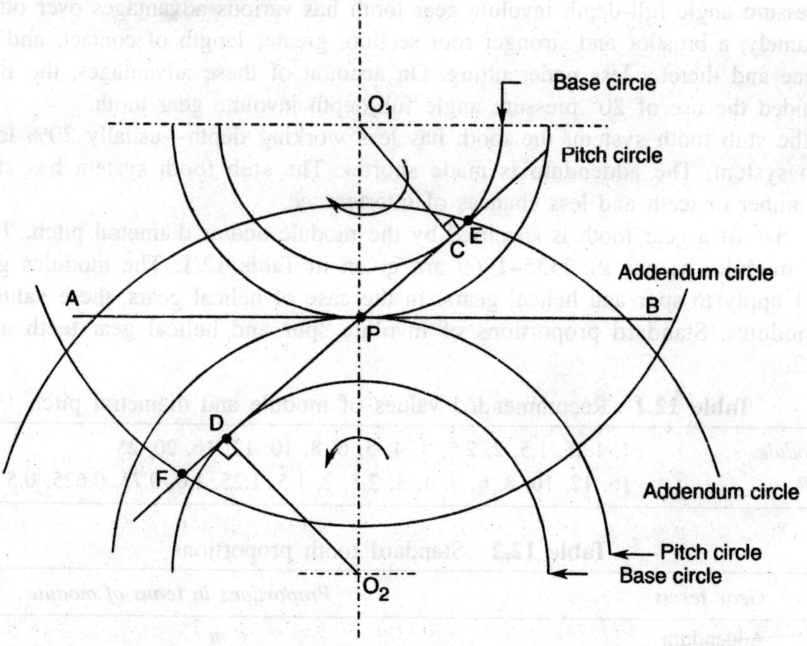

Figure 12.4 Interference of gears.

Undercutting. In this method, a portion of the tooth flank, causing interference, is cut away. Thus there will be no non-conjugate action between teeth. However, by undercutting the actual contact ratio decreases which causes a more noisy and rough gear action. Secondly, it also reduces the thickness of tooth which ultimately reduces the bending strength of the tooth.

Stubbed tooth. When a portion of a tooth near the top is cut away, such a tooth is called stubbed tooth. Such a measure prevents interference, but it also reduces the contact ratio.

Pressure angle. Increasing the pressure angle decreases the base circle diameter of the gear, which means that it increases the involute portion of the tooth profile and hence eliminates the interference. However, it will increase the radial force component which may try to dislodge the gear.

12.4 STANDARD SYSTEM OF GEAR TOOTH

In a gear drive, as mentioned earlier, two types of curves, the cycloidal and the involute, are being used for gear teeth. With regard to efficiency and strength, both forms give practically the same results. An advantage of the cycloidal tooth over the involute one is that a convex surface is always in contact with a concave one; therefore the wear of the cycloidal tooth is not as fast as that of the involute tooth. On the other hand, the involute profile does not affect velocity ratio if the actual centre distance is slightly deviated. The BIS has therefore recommended the use of the involute profile.

In a gear drive, the shape of the tooth depends upon the pressure angle. Gears of involute profile with 14.5°, 20° full-depth and 20° stub pressure angles are most widely used in industry. A 20° pressure angle full-depth involute gear tooth has various advantages over other pressure angles, namely, a broader and stronger root section, greater length of contact, and low risk of interference and thereby less undercutting. On account of these advantages, the BIS has also recommended the use of 20° pressure angle full-depth involute gear tooth.

In the stub tooth system, the tooth has less working depth—usually 20% less than the full-depth system. The addendum is made shorter. The stub tooth system has stronger and smaller number of teeth and less chances of interference.

The size of a gear tooth is specified by the module and/or diametral pitch. The standard values of modules as per IS: 2535–1969 are given in Table 12.1. The modules given in the Table 12.1 apply to spur and helical gears. In the case of helical gears, these values represent normal modules. Standard proportions of involute spur and helical gear teeth are given in Table 12.2.

Table 12.1 Recommended values of module and diametral pitch

Module	1, 1.25, 1.5, 2, 2.5, 3, 4, 5, 6, 8, 10, 12, 16, 20, 25
DP	16, 12, 10, 8, 6, 5, 4, 3, 2.5, 2, 1.5, 1.25, 1.0, 0.75, 0.625, 0.5

Table 12.2 Standard tooth proportions

Gear terms	*Proportions in terms of module*
Addendum	m
Dedendum	$1.25m$
Tooth thickness	$1.5708m$
Tooth space	$1.5708m$
Working depth	$2m$
Total depth	$2.25m$
Clearance	$0.25m$
Pitch diameter	Zm
Outside diameter	$(Z + 2)m$
Root diameter	$(Z - 2.5)m$
Fillet radius	$0.4m$

12.5 GEAR MATERIAL

The gear material should have the following properties:

(i) High tensile strength to prevent failure against static loads
(ii) High endurance strength to withstand dynamic loads
(iii) Good wear resistance to prevent failure due to contact stresses which cause pitting and scoring
(iv) Low coefficient of friction
(v) Good manufacturability

The properties of the gear material primarily depend upon the percentage of carbon, alloying elements, grain size, core and surface hardness and heat-treatment methods.

Gears are made of cast iron, steel, bronze and some non-metallic materials, namely phenolic resin, nylon, and teflon. Cast iron is a very useful material. It has good wear resistance. It is cheaper, easy to cast and machine. It produces less noise and vibrations due to inherent high damping capacity. Generally, large-size gears are made of grey cast iron of grade FG300. However, cast iron has poor strength. The plain carbon steel is another widely used material in gear manufacturing. It offers the best combination of wear resistance, high strength with ductility to absorb shock loads. Generally C45, C50, and 55Mn75 steels are used for light to medium duty applications. For heavy duty applications, the use of alloy steels, namely 40Cr1, 35Ni1Cr60, and 40Ni2Cr1Mo28 is recommended.

In non-ferrous metals, bronze is the most widely used gear material. A tin–bronze alloy containing small percentages of nickel, zinc and lead having 70–80 BHN is most commonly used. The non-metallic gears, made of phenolic resins, teflon, and nylon, can absorb the effect of tooth profile error due to lower elastic modulus. These gears can operate at high speed with marginal boundary lubrication.

12.6 DESIGN OF SPUR GEARS

The basic requirement of a gear drive is to transmit power at a particular velocity ratio for certain service conditions, for example, operating time, nature of load, etc. While designing a gear drive, the following points must be considered.

1. The highest static load acting on the gear tooth due to high starting torque.
2. The dynamic load at normal running conditions due to profile error on the tooth.
3. Wear characteristics of the gear tooth for a long satisfactory life.

Besides the above basic requirements, adequate attention to lubrication of teeth, alignment of gears, stress concentration at the root of the teeth, and deflection of gear teeth and shaft should be given.

12.6.1 Force Analysis

A gear drive is generally specified by power to be transmitted, the speed of the driving shaft, and the velocity ratio. The power is transmitted by means of a force exerted by the tooth of the driving gear on the mating tooth of the driven gear. According to the law of gearing, this exerted force F_n is always normal to the tooth surface and acts along the pressure angle line.

Obviously, a force would be exerted on the driven tooth equal to F_n in magnitude but in the opposite direction. This normal force is designated by two subscripts, for example, F_{12} which means the force exerted by gear 1 against gear 2.

Figure 12.5(a) shows a pair of spur gears in mesh mounted on respective shafts. The driving gear 2 is mounted on shaft 1 and rotates in the clockwise direction. The driven gear 3 is mounted on shaft 4. The free body diagram of the forces acting upon the two gears along the pressure line is shown in Figure 12.5(b). The driving gear 2 exerts a force F_{23} on the driven gear 3. Similarly, the driving gear experiences a reaction force F_{32}.

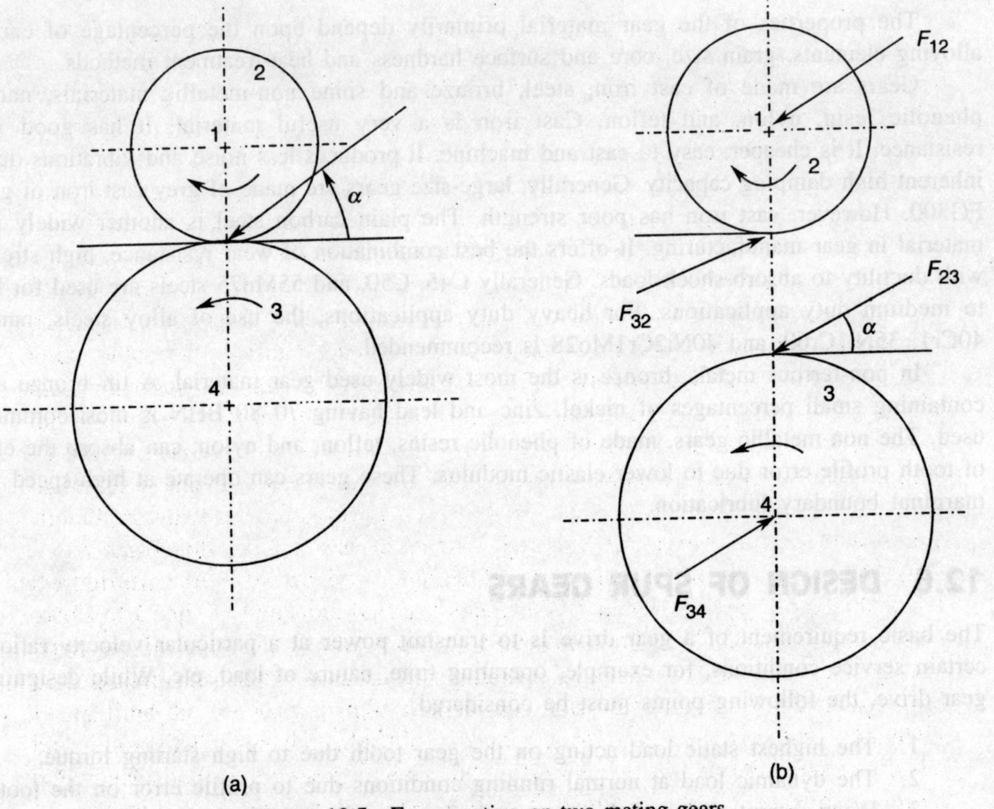

Figure 12.5 Forces acting on two mating gears.

The normal force F_n (or F_{23}), as shown in Figure 12.6, acting along the pressure line, can be resolved into two components, namely the tangential force F_t and the radial force F_r. Thus,

$$F_t = F_n \cos \alpha \tag{12.2a}$$

$$F_r = F_n \sin \alpha = F_t \tan \alpha \tag{12.2b}$$

where α is the pressure angle.

The tangential component F_t is mainly responsible for transmitting torque and consequently the power. The radial force F_r is called the *separating force*, which always acts towards the centre of the gear.

In the force analysis of a gear drive, it is assumed that the tangential force remains constant in magnitude as the contact between two teeth moves from top of the tooth to its bottom. The torque that the tangential force transmits with respect to the centre of the gear is

$$T = \frac{F_t d_1}{2} \quad (12.3)$$

where d_1 is the pitch circle diameter.

Alternatively, the tangential force responsible for transmitting power can be found from the following relation

$$P = F_t \times v \quad (12.4)$$

where

P is the power (kW)
v is the pitch line velocity (m/s)
F_t is the tangential force (kN).

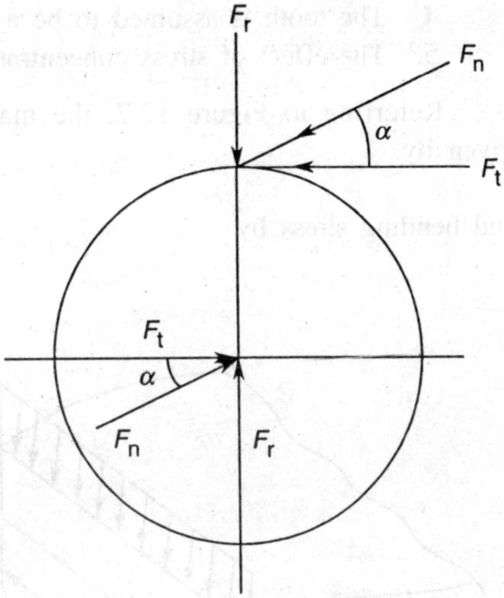

Figure 12.6 Forces on a gear tooth.

12.6.2 Beam Strength of Spur Gear Tooth

The accurate stress analysis of a gear tooth for a particular application is a complex problem because of the following reasons:

1. There is continuous change in the point of application of load on the tooth profile.
2. The magnitude and direction of applied load also change.
3. In addition to static load, the dynamic load due to inaccuracy of the tooth profile, error in machining and mounting, tooth deflection, acceleration and stress concentration also act on the tooth which are all difficult to model mathematically.

Wilfred Lewis, in a paper titled, "The investigation of the strength of gear tooth" published in 1892, derived an equation for determining the approximate stress in a gear tooth by treating it as a cantilever beam of uniform strength. The beam strength calculation is based upon the following assumptions:

1. The tangential component of the force on gear tooth, F_t, is uniformly distributed across the face width; however, in actual force distribution it is found to be non-uniformly distributed. This assumption is valid for small face widths b, i.e.

$$9.5m \leq b \leq 12.5\ m$$

2. The effect of the radial component F_r, which produces direct compressive stress, is neglected.
3. The maximum stress is assumed to occur when the entire load is at the tip of the tooth. This is, however, not true because when more than one pair of teeth are in contact, the load is shared between all of them. Further, as the tooth moves through its path, the magnitude of the force and its moment are changed.

4. The tooth is assumed to be a simple cantilever beam.
5. The effect of stress concentration and manufacturing errors are neglected.

Referring to Figure 12.7, the maximum bending moment at the root section X–X is given by

$$M = F_t \times h$$

and bending stress by

$$\sigma_b = \frac{My}{I}$$

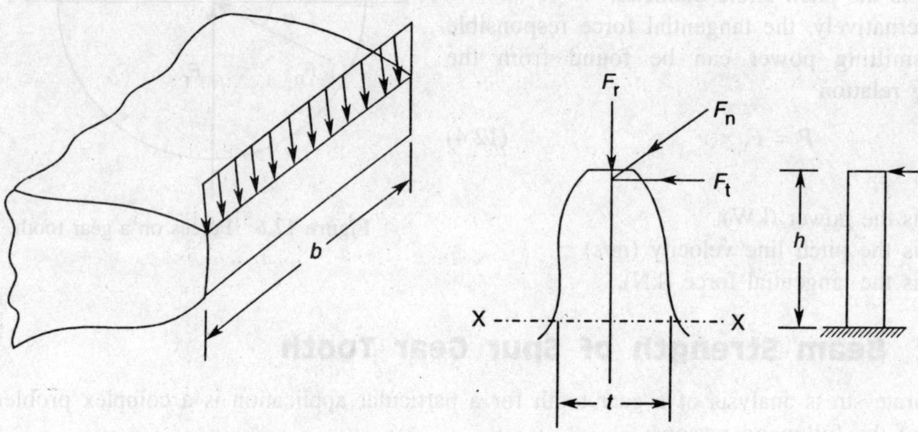

Figure 12.7 Gear tooth as a cantilever beam.

Assuming the tooth section to be of rectangular shape, the moment of inertia $I = bt^3/12$ and $y = t/2$. The bending stress equation can then be written as

$$\sigma_b = \frac{F_t \times h \times t/2}{bt^3/12}$$

or

$$F_t = b\sigma_b \frac{t^2}{6h}$$

Multiplying the numerator and denominator of the right-hand side of above equation by module m,

$$F_t = bm\sigma_b \left(\frac{t^2}{6hm}\right) \tag{12.5}$$

or

$$F_t = bm\sigma_b Y \tag{12.6}$$

where the term $Y = t^2/6hm$ is called the *Lewis form factor*, whose value depends on the number of teeth and pressure angle of the gear drive. The empirical relations for the Lewis form factor are:

For a 20° involute full-depth tooth:

$$Y = \pi\left(0.154 - \frac{0.912}{Z}\right) \tag{12.7a}$$

For a 20° stub tooth:

$$Y = \pi\left(0.175 - \frac{0.95}{Z}\right) \tag{12.7b}$$

For a 14.5° tooth:

$$Y = \pi\left(0.124 - \frac{0.684}{Z}\right) \tag{12.7c}$$

Equation (12.6) gives the relation between the tangential force and the bending stress produced in the gear tooth. When the bending stress reaches the permissible magnitude of bending strength of the material, σ_d, the corresponding force F_t is called the *beam strength* F_{beam}. Thus beam strength is the maximum value of the tangential force that can be resisted by a tooth without bending failure. Therefore, Eq. (12.6) can be rewritten as

$$F_{beam} = bm\sigma_d Y \tag{12.8}$$

In any gear drive, the average tangential force on a gear tooth can be computed from Eq. (12.4). However, in many practical applications the actual force may be greater than the average force on account of the poor service conditions and the dynamic load on the tooth due to profile being inaccurate. In order to include the effect of the above two conditions, two factors, namely the service factor C_S and the velocity factor C_v are introduced and the maximum force between the two mating teeth is computed by the following relation:

$$F_{max} = \frac{C_S\, F_t}{C_v} \tag{12.9}$$

The values of the factors C_S and C_v may be found from Tables 12.3 and 12.4, respectively.

In order to avoid failure of the gear tooth due to bending, the beam strength, F_{beam}, should be greater than the maximum force on the gear tooth. Thus,

$$F_{beam} \geq F_{max} \tag{12.10}$$

Table 12.3 Values of service factor (C_S)

Type of load	*Type of service*		
	Intermittent or 3 h/day	8–10 h/day	Continuous
Steady	1.0	1.0	1.25
Light shock	1.0	1.25	1.5
Medium shock	1.25	1.5	1.8
Heavy shock	1.50	1.8	2.0

Table 12.4 Velocity factor (C_v)

Conditions	C_v
Ordinary cut gear and pitch line velocity up to 8 m/s	$\frac{3.05}{3.05 + v}$
Carefully cut gear with pitch line velocity up to 13 m/s	$\frac{4.58}{4.58 + v}$
Accurately cut gear with pitch line velocity up to 20 m/s	$\frac{6.1}{6.1 + v}$
Hardened, ground and lapped precision gear with speed more than 20 m/s	$\frac{5.56}{5.56 + \sqrt{v}}$

12.6.3 Dynamic Load on Gear

The maximum force acting on a gear tooth during power transmission is called the *dynamic load*, F_{dyn}, which is the sum of the tangential force F_t and incremental force F_i. The reasons behind the incremental force F_i are the following:

1. Inaccuracies of the tooth profile
2. Error in tooth spacing resulting in excessive backlash
3. Elastic deformation of a tooth under load, affecting kinematic perfection
4. Inertia of the rotating masses
5. Misalignment between bearings supporting gears

Buckingham had found that the maximum effect of the above factors occurs when the contact is transferred from one pair to the next pair of teeth. According to him, the dynamic load F_{dyn} is given by

$$F_{dyn} = F_t + F_i \tag{12.11}$$

The increment force F_i for average conditions may be found from the following equation

$$F_i = \frac{K_3 v (cb + F_t)}{K_3 v + \sqrt{cb + F_t}} \tag{12.12}$$

where

$$c \text{ is the dynamic load factor } \left[= \frac{e}{K_1(1/E_1 + 1/E_2)} \right] \tag{12.13}$$

e is the profile error in action between gears (refer Tables 12.5 and 12.6)

K_1 = constant = 9 for 20° full-depth teeth

K_3 = 20.67, a constant.

Table 12.5 Maximum allowable profile error

Velocity (m/s)	*Error* (mm)	*Velocity* (m/s)	*Error* (mm)
1.0	0.096	8.0	0.05
2.0	0.088	10.0	0.039
2.5	0.084	12.0	0.033
3.0	0.078	15.0	0.023
4.0	0.071	20.0	0.016
5.0	0.064	25.0	0.013
6.0	0.059	Over 25.0	0.013

Table 12.6 Maximum error in action between gears

Module, m (mm)	*Error, e* (mm)		
	Class I (industrial or commercial cut gears)	Class II (accurate or carefully cut gears)	Class III (precision gears)
25	0.122	0.061	0.031
20	0.117	0.058	0.029
15	0.108	0.054	0.027
12	0.099	0.049	0.025
10	0.089	0.044	0.022
8	0.078	0.039	0.020
7	0.071	0.036	0.018
6	0.065	0.032	0.017
5	0.055	0.028	0.015
4	0.051	0.025	0.013

Further, it is suggested that for satisfactory design of the gear under dynamic load conditions, the dynamic load should be equal to endurance strength F_{en}.

Endurance strength is computed from the modified Lewis equation as

$$F_{en} = F_{dyn} \times \text{FoS} = bm\, \sigma_{en} Y \tag{12.14}$$

where

σ_{en} is the endurance limit (= 1.75 BHN)
BHN is the Core Brinel hardness number

12.6.4 Wear Strength

Due to rolling and sliding actions of the gear teeth, the following types of surface destructions (wear) may occur:

Abrasive wear. Scratching of the tooth surface due to the presence of foreign materials in the lubricant is called *abrasive wear.*

Corrosive wear. Chemical reactions on the surface of a gear cause *corrosive wear.*

Pitting. Repeated application of the stress cycle, known as *pitting*, causes fatigue failure.

Scoring. Inadequate lubrication between metal-to-metal contact causes *scoring*.

It is observed from the results of various experiments that abrasion, corrosion, and scoring are caused by improper lubrication, whereas pitting usually occurs because of repeated application of Hertz contact stress on the portion of a gear tooth which has relatively little sliding motion compared to rolling motion. Clearly, the spur gear will have pitting near the pitch line where motion is almost all of rolling.

In designing a gear against wear, the material should be able to resist the repeated contact stresses. In order to obtain a formula for actual surface stress that exists between two mating gear teeth, the Hertz equation for contact stress between two cylinders in rolling contact is used (Figure 12.8).

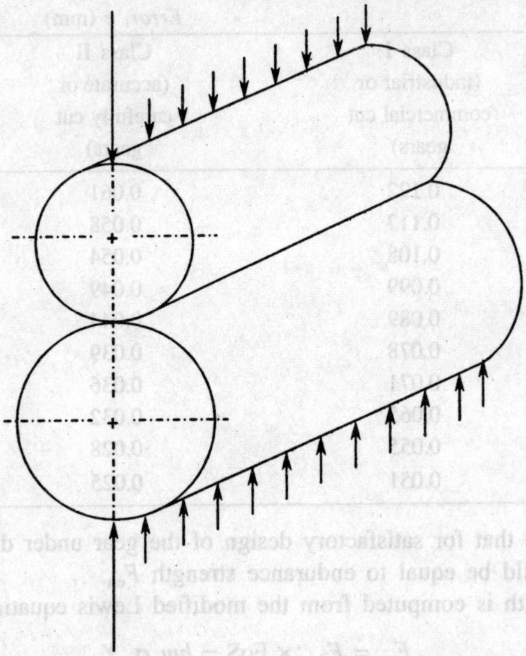

Figure 12.8 Two cylinders in rolling contact.

Referring to Figure 12.8, the Hertz contact stress between two rolling cylinders is given by the equation

$$\sigma = \left[\frac{F\left(\dfrac{1}{r_1}+\dfrac{1}{r_2}\right)}{\pi l\left[\dfrac{1-\mu_1^2}{E_1}+\dfrac{1-\mu_2^2}{E_2}\right]}\right]^{0.5} \tag{12.15}$$

where

σ is the Hertz contact stress

r_1, r_2 are the radii of cylinders

E_1, E_2 are the elastic moduli of cylinder materials

μ_1, μ_2 are the Poisson's ratios

l is the length of contact between two cylinders.

In order to apply the Hertz contact stress equation to spur gears, the force F is replaced by the allowable wear load F_{wear}, radii r_1 and r_2 are considered as pitch radii of pinion and gear and contact stress is replaced by the allowable surface endurance limit σ_{es}.

The limiting wear load equation for spur gear pairs is given as

$$F_{wear} = d_1 bQK \tag{12.16}$$

where

Q is the ratio factor $(= 2Z_2/(Z_1 + Z_2))$

$$K \text{ is the load stress factor } \left[= \frac{\sigma_{es}^2 \sin\alpha}{1.4}\left(\frac{1}{E_1} + \frac{1}{E_2}\right)\right] \tag{12.17}$$

For steel gears, the surface endurance limit may be calculated by the following formula

$$\sigma_{es} = (2.75 \text{ BHN} - 70) \text{ N/mm}^2 \tag{12.18}$$

Buckingham has reported that the allowable wear load as obtained by Eq. (12.16) is clearly a normal force that must be greater than the dynamic load for satisfactory design.

12.6.5 Gear Construction

The dimensions of the parts of a gear are determined largely by empirical relations. The relations for the proportions of tooth in terms of module are listed in Table 12.2. The dimensions of other gear parts such as rim, hub, arm section and solid web, as shown in Figure 12.9, are determined empirically as given below:

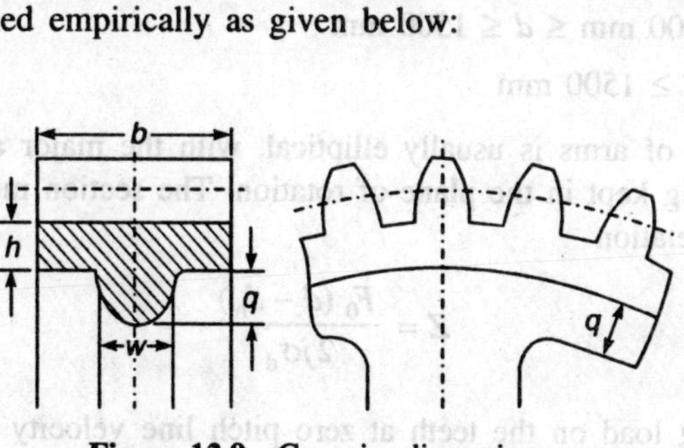

Figure 12.9 Gear rim dimensions.

1. If the pitch diameter $d \leq 14.8\text{m} + 60.0$ mm, the pinion/gear should be of solid disc type.
2. If the pitch diameter $d \geq 23.5\text{m} + 85.0$ mm, the gear to be constructed should be of arm type, else it should be of web type having web thickness 1.6 to 2.0 times the module m (see Figure 12.10).

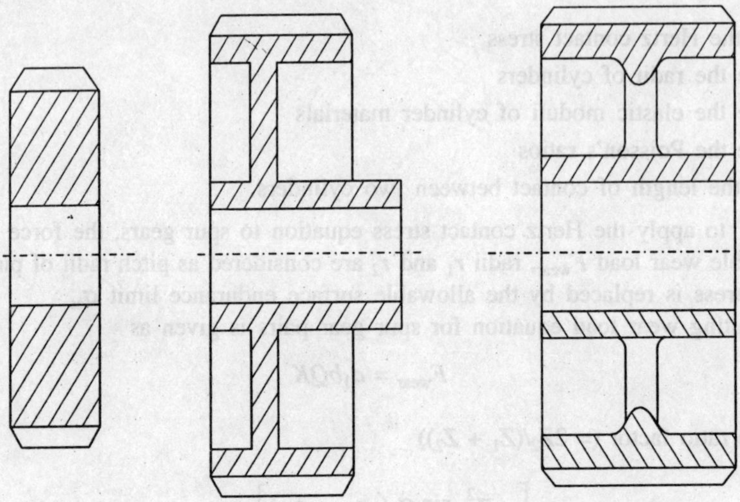

Figure 12.10 Types of gear construction.

3. The thickness of the rim, $h = (2 \text{ to } 4)m$. The rim should be tapered 1 : 5 towards the centre.
4. The thickness of the stiffening rib, $q = (1.0 \text{ to } 1.25)h$.
5. Hub diameter, $d_h = (1.6 \text{ to } 2.0)d_S$.
6. Hub length, $l_h = 2d_S$ or at least equal to face width b.
7. Number of arms (j)

   $j = 4$ if $d \le 500$ mm

   $j = 6$ if 500 mm $\le d \le$ 1500 mm

   $j = 8$ if $d \ge 1500$ mm

The cross-section of arms is usually elliptical, with the major axis twice the minor axis and the major axis being kept in the plane of rotation. The section modulus of the arm section is determined by the relation

$$Z = \frac{F_0 (d - d_h)}{2 j \sigma_d} \tag{12.19}$$

where F_0 is the stalling load on the teeth at zero pitch line velocity ($= bmY\sigma_d$).

Arms are usually tapered towards the rim.

12.6.6 Stress Concentration

In a gear, where the tooth joins the bottom land, there is a concentration of stresses due to the complex geometric shape (Figure 12.11). Though stress concentration is difficult to model

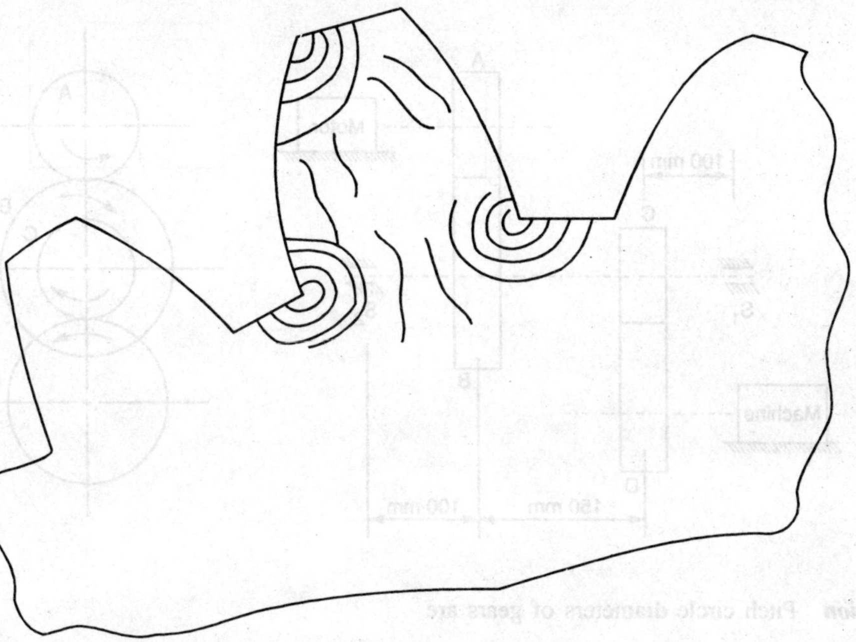

Figure 12.11 Stress concentration zone in a spur gear tooth.

mathematically, the photoelastic method and the finite element analysis are used to determine the empirical relation yielding a reasonable value of it. According to Dolan and Broghamer, the stress-concentration factor is related to tooth thickness *t*, fillet radius *r* and tooth height *h* by the relation

$$K_t = 0.18\left(\frac{t}{r}\right)^{0.15}\left(\frac{t}{h}\right)^{0.45} \quad \text{for } 20° \text{ pressure angle} \tag{12.20}$$

While designing a gear tooth, the static stress-concentration factor K_t is converted to fatigue stress-concentration factor K_f and the beam strength is modified accordingly as below:

$$F_{\text{beam}} = \frac{bm\,\sigma_{\text{d}}Y}{K_f} \tag{12.21}$$

Example 12.1 A gear train transmitting 5 kW at 1440 rpm is shown in the following figure. The number of teeth on gears A, B, C, and D are 25, 100, 30, and 150, respectively. All gears have 5 mm module and a 20° full-depth involute profile gear tooth. Calculate the tangential and radial components of forces between gears A and B and between gears C and D. Also calculate the resultant reactions at the bearing supports S_1 and S_2.

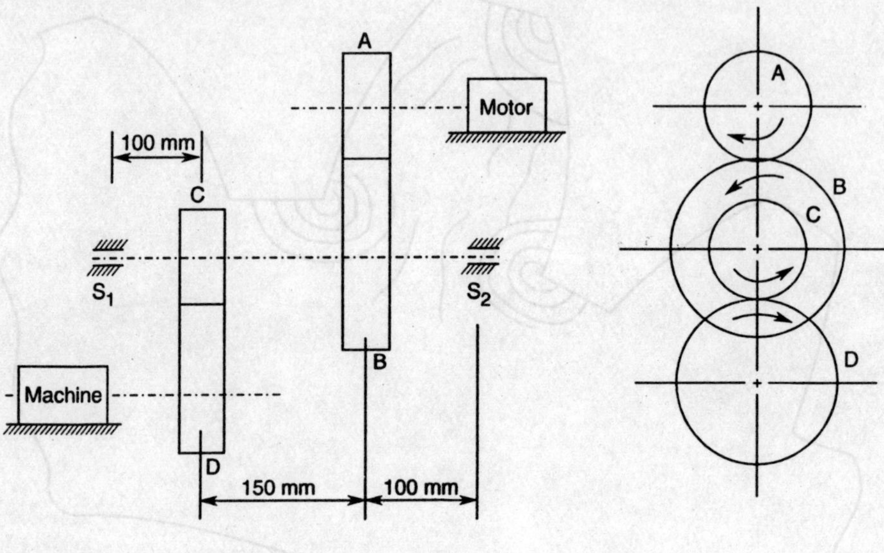

Solution Pitch circle diameters of gears are

$$d_A = mZ_A = 25 \times 5 = 125 \text{ mm}$$

$$d_B = mZ_B = 100 \times 5 = 500 \text{ mm}$$

$$d_C = mZ_C = 30 \times 5 = 150 \text{ mm}$$

$$d_D = mZ_D = 150 \times 5 = 750 \text{ mm}$$

Pitch line velocity between gears A and B

$$v_1 = \frac{\pi d_A N_A}{60 \times 1000} = \frac{\pi \times 125 \times 1440}{60 \times 1000} = 9.42 \text{ m/s}$$

Tangential force between gears A and B

$$F_{t1} = \frac{P}{v} = \frac{5 \times 10^3}{9.42} = 530.78 \text{ N}$$

Radial force between gears A and B

$$F_{r1} = F_{t1} \tan \alpha = 530.78 \times \tan 20° = 193.2 \text{ N}$$

Similarly, for pitch line velocity between gears C and D, rpm of gear C = rpm of gear B

$$N_C = N_B = 360 \text{ rpm}$$

$$\text{Pitch line velocity, } v_2 = \frac{\pi d_C N_C}{60 \times 1000} = \frac{\pi \times 150 \times 360}{60 \times 1000} = 2.83 \text{ m/s}$$

Tangential force between gears C and D

$$F_{t2} = \frac{P}{v_2} = \frac{5 \times 10^3}{2.83} = 1766.78 \text{ N}$$

Radial force between gears C and D

$$F_{r2} = F_{t2} \tan \alpha = 1766.78 \times \tan 20° = 643 \text{ N}$$

The free body diagrams of gears are shown in the figure below.

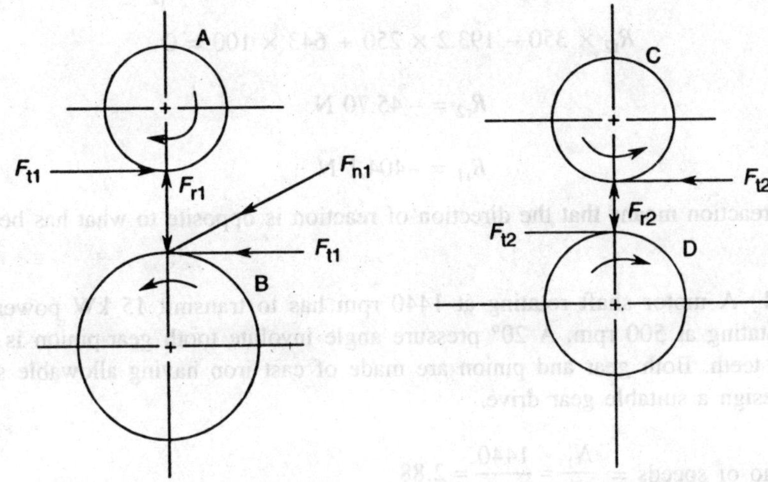

The horizontal forces acting on gears B and C, which are mounted on the shaft S_1–S_2 are shown in the figure below.

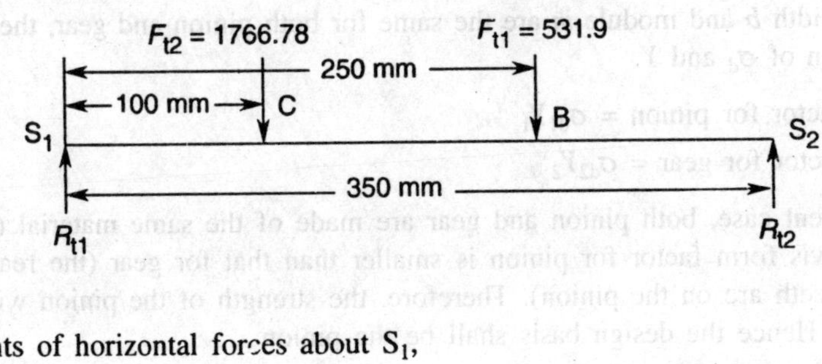

Taking moments of horizontal forces about S_1,

$$R_{t2} \times 350 - 530.78 \times 250 - 1766.78 \times 100 = 0$$

or

$$R_{t2} = 884 \text{ N}$$

Therefore

$$R_{t1} = 1766.78 + 530.78 - 884.0 = 1413.56 \text{ N}$$

The vertical force diagram is shown in the figure below. Taking moments of vertical forces about S_1,

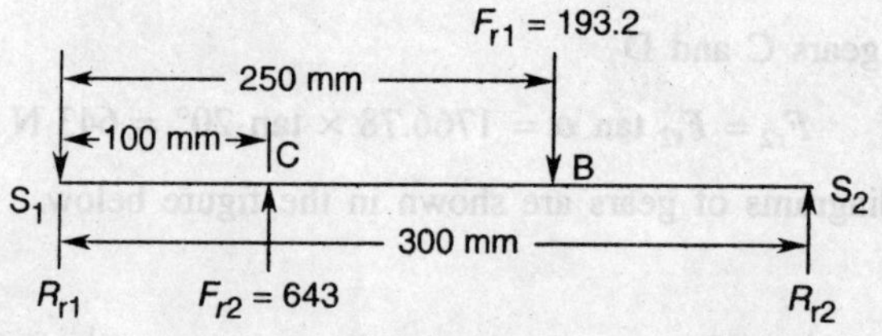

$$R_{r2} \times 350 - 193.2 \times 250 + 643 \times 100 = 0$$

or

$$R_{r2} = -45.70 \text{ N}$$

Therefore,

$$R_{r1} = -404.1 \text{ N}$$

The negative reaction means that the direction of reaction is opposite to what has been marked on the figure.

Example 12.2 A motor shaft rotating at 1440 rpm has to transmit 15 kW power to a low speed shaft rotating at 500 rpm. A 20° pressure angle involute tooth gear-pinion is used. The pinion has 25 teeth. Both gear and pinion are made of cast iron having allowable strength of 55 N/mm^2. Design a suitable gear drive.

Solution Ratio of speeds $= \dfrac{N_1}{N_2} = \dfrac{1440}{500} = 2.88$

Since the ratio of speeds is small (less than 5), we can select a spur gear pair.

Beam strength equation for spur gear is given as

$$F_{\text{beam}} = bm\sigma_d Y$$

Since the face width b and module m are the same for both pinion and gear, the beam strength will be a function of σ_d and Y.

Strength factor for pinion $= \sigma_{d1} Y_1$

Strength factor for gear $= \sigma_{d2} Y_2$

In the present case, both pinion and gear are made of the same material ($\therefore \sigma_d = \sigma_{d1} = \sigma_{d2}$) and the Lewis form factor for pinion is smaller than that for gear (the reason being that less number of teeth are on the pinion). Therefore, the strength of the pinion will be less than that of the gear. Hence the design basis shall be the pinion.

$$\text{Pitch line velocity, } v = \frac{\pi d_1 N_1}{60} = \frac{\pi m Z_1 N_1}{60 \times 1000}$$

$$= \frac{\pi \times m \times 25 \times 1440}{60 \times 1000} = 1.885m \text{ m/s}$$

Average tangential force on the gear tooth

$$F_t = \frac{P}{v} = \frac{15 \times 1000}{1.885m} = \frac{7957.5}{m} \text{ N}$$

Maximum tangential force on the gear tooth

$$F_{max} = \frac{C_S F_t}{C_v}$$

We assume that a steady load is applied and the gear drive works continuously for 24 h/day. Therefore, service factor C_S = 1.25 and C_v = 0.35 (assumed a trial value as pitch line velocity is not known initially). Thus,

$$F_{max} = \frac{1.25 \times 7957.5}{m \times 0.35} = \frac{28419.6}{m} \text{ N}$$

Lewis form factor for 20° full-depth involute teeth

$$Y_1 = \pi\left(0.154 - \frac{0.912}{Z_1}\right)$$

$$= \pi\left(0.154 - \frac{0.912}{25}\right) = 0.3692$$

Lewis beam strength, $F_{beam} = bmY\sigma_d$

where b is the face width whose value varies between 9.5m and 12.5m. We adopt b = 10m. Therefore,

$$F_{beam} = 10m \times m \times 0.3692 \times 55 = 203.06m^2 \text{ N}$$

For a statically safe design the beam strength should be greater than or equal to the maximum tangential force, i.e.

$$F_{beam} \geq F_{max}$$

or

$$203.06m^2 = \frac{28419.6}{m}$$

or

$$m = 5.2 \text{ mm, say, 6 mm}$$

$$\text{Number of teeth on gear, } Z_2 = \frac{N_1}{N_2} \times Z_1$$

$$= \frac{1440}{500} \times 25 = 72 \text{ teeth}$$

Pitch circle diameter (pcd) of pinion

$$d_1 = mZ_1 = 6 \times 25 = 150 \text{ mm}$$

$$\text{pcd of gear, } d_2 = mZ_2 = 6 \times 72 = 432 \text{ mm}$$

$$\text{Pitch line velocity, } v = 1.885m = 1.885 \times 6 = 11.31 \text{ m/s}$$

Velocity factor C_v for a carefully cut gear with pitch line velocity up to 13 m/s

$$C_v = \frac{4.58}{4.58 + 11.31} = 0.288$$

Therefore maximum tangential force

$$F_{max} = \frac{C_s\, F_t}{C_v} = \frac{1.25 \times 7957.5}{0.288 \times 6} = 5756.3 \text{ N}$$

and beam strength of tooth, $F_{beam} = bmY\sigma_d$
or

$$F_{beam} = 203.06 \times m^2 = 203.06 \times 6^2 = 7310.1 \text{ N}$$

Since $F_{beam} > F_{max}$, the tooth is safe in static load conditions.

Dynamic load

The dynamic load on the gear tooth can be found by the Buckingham equation

$$F_{dyn} = F_t + \frac{K_3 v\,(cb + F_t)}{K_3 v + \sqrt{cb + F_t}}$$

$$K_3 = \text{constant} = 20.67$$

$$c = \text{dynamic load factor} = \frac{e \times E_1 E_2}{K_1(E_1 + E_2)}$$

where

e = permissible profile error; for class II gears with 6 mm module, the maximum error e = 0.032 (refer Table 12.6)

E_1, E_2 = Young's elastic modulus
= 0.1×10^6 N/mm² (for CI)

Thus

$$c = \frac{0.032 \times 0.1 \times 10^6 \times 0.1 \times 10^6}{9 \times 2 \times 0.1 \times 10^6} \qquad (\because K_1 = 9 \text{ for } 20^\circ \text{ full-depth teeth})$$

$$= 177.78$$

Also,

$$cb + F_t = 177.78 \times 60 + \frac{7957.5}{6} = 11{,}993$$

$$\sqrt{cb + F_t} = 109.5$$

Therefore,

$$\text{Dynamic force, } F_{dyn} = 1326.25 + \frac{20.67 \times 11.3 \times 11{,}993}{20.67 \times 11.3 + 109.5} = 9491.3 \text{ N}$$

Endurance strength of the gear

$$F_{en} = bmY\sigma_{en}$$

For steady load condition, $F_{en} = 1.25F_{dyn}$

or

$$1.25 \times 9491.3 = 60 \times 6 \times 0.3692 \times \sigma_{en}$$

or

$$\sigma_{en} = 89.26 \text{ N/mm}^2$$

$$\text{Core BHN} = \frac{\sigma_{en}}{1.75} = \frac{89.26}{1.75} = 51 \text{ BHN}$$

which is reasonably smaller than the core BHN of CI. Hence the gear is safe in the dynamic load condition.

Wear load

Limiting wear load on gear–pinion

$$F_{wear} = d_1 bQK$$

$$Q = \text{ratio factor} = \frac{2Z_2}{Z_1 + Z_2} = \frac{2 \times 72}{72 + 25} = 1.4845$$

We assume that the factor of safety is 1.5, therefore, the wear load should be 1.5 times the dynamic load, i.e.

$$F_{wear} = 1.5F_{dyn}$$

or

$$1.5 \times 9491.3 = 150 \times 60 \times 1.4845 \times K$$

or

$$K = 1.0656$$

$$= \text{load-stress factor}$$

$$= \frac{\sigma_{es}^{\;2} \times \sin\alpha}{1.4}\left[\frac{1}{E_1} + \frac{1}{E_2}\right]$$

or

$$1.0656 = \frac{\sigma_{es}^{\;2} \times \sin 20^\circ}{1.4} \times \frac{2 \times 0.1 \times 10^6}{0.1 \times 10^6 \times 0.1 \times 10^6}$$

or

$$\sigma_{es} = 467 \text{ N/mm}^2$$

Therefore the recommended surface hardness for C.I. gear–pinion is 200 BHN.

Gear construction

The limit for a pinion or gear with a web is given by $d \le 23.5m + 85$ mm $= 23.5 \times 6 + 85 = 226$ mm and the limit for the solid disc gear or pinion is

$$d \le (14.8m + 60) \text{ mm} = 14.8 \times 6 + 60 = 148.8 \text{ mm}$$

Since the pinion diameter $d_1 = 150$ mm, it should be of solid disc type. However, the gear diameter $d_2 = 432$ mm is greater than the limiting diameter of the web type. Therefore the gear should be of arm type.

Thickness of the rim, $h = (2-4)m$, say, $h = 3m = 18$ mm

Rim depth, $q = h = 18$ mm

$$\text{Average torque, } T = \frac{60\,P}{2\pi N_2} = \frac{60 \times 15.0 \times 10^3}{2\pi \times 500}$$

$$= 286.47 \text{ N} \cdot \text{m}$$

Gear shaft is subjected to bending moment due to normal load,

$$F_n = \frac{F_t}{\cos\alpha} = \frac{1326.25}{\cos 20°} = 1411.36 \text{ N}$$

Thus bending moment, $M = F_n \times \text{overhang}$

$= 1411.36 \times 100$ (assume overhang = 100 mm)

$= 141{,}136 \text{ N} \cdot \text{mm}$

Gear shaft

$$\text{Equivalent torque, } T_{eq} = \sqrt{M^2 + T^2}$$

$$= \sqrt{141{,}136^2 + 286{,}470^2} = 319{,}350 \text{ N} \cdot \text{mm}$$

Let the shaft material be C30 having the allowable shear strength, $\tau = 100$ N/mm^2

$$\text{Shaft diameter, } d_S = \left(\frac{16\,T_{eq}}{\pi\tau}\right)^{1/3}$$

$$= \left[\frac{16 \times 319{,}350}{\pi \times 100}\right]^{1/3} = 25.33 \text{ mm, say, 28 mm (standard size)}$$

Hub diameter, $d_h = 1.8 d_S = 1.8 \times 28 = 50.4$ mm, say, 50 mm

Length of the hub, $l_h = b$ or $2d_S$, say, 60 mm

$$\text{Overhang} = d_S + \frac{l_h}{2} + \text{clearance}$$

$$= 28 + \frac{60}{2} + 25 = 83 \text{ mm}$$

which is less than the assumed overhang of the gear shaft.

Pinion shaft

$$\text{Torque, } T = F_t \times \frac{d_1}{2} = 1326.25 \times \frac{150}{2}$$

$$= 99{,}468.75 \text{ N} \cdot \text{mm}$$

Bending moment, $M = 141{,}136.0 \text{ N} \cdot \text{mm}$

$$\text{Equivalent torque, } T_{eq} = \sqrt{141{,}136^2 + 99{,}468.75^2}$$

$$= 172{,}665.6 \text{ N} \cdot \text{mm}$$

$$\text{Pinion shaft diameter, } d_p = \left(\frac{16T_{eq}}{\pi\tau}\right)^{1/3}$$

$$= \left[\frac{16 \times 172{,}665.6}{\pi \times 100}\right]^{1/3} = 20.64 \text{ mm, say, 22 mm}$$

$$\text{Hub diameter, } d_{ph} = 2 \times d_p = 2 \times 22 = 44 \text{ mm}$$

Since $2d_p < b$, the hub length is equal to face width b.

Arms

For gears with diameters up to 500 mm, four arms may be used. We assume that the arms section is elliptical with major axis equal to twice the minor axis.

$$\text{Section modulus, } Z = \frac{F_0(d_2 - d_h)}{2j\,\sigma_d}$$

where

F_0 = stalling load, a load on the teeth at zero pitch line velocity = $bmY_2\sigma_d$

$Y_2 = \pi(0.154 - 0.912/72) = 0.444$

Therefore,

$$Z = \frac{60 \times 6 \times 0.444 \times 55\,(432 - 50)}{2 \times 4 \times 55}$$

$$= 7632.36 \text{ mm}^3$$

$$\text{Major axis, } a_1 = \left(\frac{64Z}{\pi}\right)^{1/3}$$

$$= \left(\frac{64 \times 7632.36}{\pi}\right)^{1/3}$$

$$= 53.77 \text{ mm, say, 56 mm}$$

$$\text{Minor axis } a_2 = \frac{a_1}{2} = 28 \text{ mm}$$

Specification of the drive

Spur gear teeth 20° pressure angle, full depth

Number of teeth on pinion, $Z_1 = 25$

Number of teeth on gear, $Z_2 = 72$

module, $m = 6$ mm

Face width, $b = 60$ mm

pcd of pinion, $d_1 = 150$ mm

pcd of gear, $d_2 = 432$ mm

Addendum = m = 6 mm

Dedendum = $1.25m$ = 7.5 mm

Tooth thickness = $1.5708m$ = 9.42 mm

Working depth = $2m$ = 12 mm

Whole depth = $2.25m$ = 13.5 mm

Blank diameter of the pinion = $(Z_1 + 2)m$ = 162 mm

Blank diameter of the gear = $(Z_2 + 2)m$ = 444 mm

Pinion: solid disc type having width = 60 mm
pinion shaft diameter, d_p = 22 mm
hub diameter, d_{ph} = 44 mm
length of the hub, l_h = 60 mm

Gear: arm type with 4 number of arms
gear shaft diameter, d_S = 28 mm
hub diameter, d_h = 50 mm
length of the hub, l_h = 60 mm
elliptical section arm = 56 mm × 28 mm

Example 12.3 Design a spur gear which is required to transmit 10 kW power. The speeds of the driving motor and the driven machine are 400 rpm and 200 rpm, respectively. The approximate centre-distance may be taken as 600 mm. The teeth have 20° full-depth involute profile. Assume that the gear is made of cast iron FG200, having allowable strength of 75 N/mm² and 180 BHN core hardness.

Solution Let the pitch circle diameters of pinion and gear be d_1 and d_2 respectively. Then the centre-distance

$$C = \frac{d_1 + d_2}{2} = 600 \qquad \text{(i)}$$

and the ratio of speeds, $$\frac{N_1}{N_2} = \frac{d_2}{d_1} = 2 \qquad \text{(ii)}$$

Solving Eqs. (i) and (ii), we get

$$d_1 = 400 \text{ mm} \quad \text{and} \quad d_2 = 800 \text{ mm}$$

$$\text{Pitch line velocity, } v = \frac{\pi d_1 N_1}{60} = \frac{\pi \times 0.4 \times 400}{60} = 8.37 \text{ m/s}$$

Tangential force on gear tooth is

$$F_t = \frac{P}{v} = \frac{10 \times 10^3}{8.37} = 1194.74 \text{ N}$$

Velocity factor C_v, using Barth's formula for a carefully cut gear with a pitch line velocity up to 13 m/s

$$C_v = \frac{4.58}{4.58 + v} = \frac{4.58}{4.58 + 8.37} = 0.3536$$

Since pinion and gear are made of the same material, the pinion will be weaker because $Y_1 < Y_2$.

Lewis form factor for 20° full-depth involute profile

$$Y_1 = \pi\left(0.154 - \frac{0.912}{Z_1}\right)$$

where

$$Z_1 = \text{number of teeth on pinion} = \frac{d_1}{m}$$

or

$$Y_1 = \pi\left(0.154 - \frac{0.912 \times m}{400}\right)$$

$$= 0.4838 - 0.007163m$$

$$\text{Beam strength, } F_{\text{beam}} = bmY\sigma_d$$

We assume face width, $b = 10m$. Therefore,

$$F_{\text{beam}} = 10m^2(0.4838 - 0.007163m)75$$

$$= 362.85m^2 - 5.372m^3$$

Now maximum tangential force on gear tooth

$$F_{\max} = \frac{C_S F_t}{C_v}$$

We assume that a steady load condition for 8 to 10 h/day is the working condition. Hence the service factor $C_S = 1.0$. Therefore,

$$F_{\max} = \frac{1.0 \times 1194.74}{0.3536} = 3378.78 \text{ N}$$

For safe design, the beam strength should be greater than or equal to maximum tangential force. Therefore,

$$362.85m^2 - 5.372m^3 = 3378.78$$

By trial and error, the solution of above equation gives

$$\text{module, } m = 4 \text{ mm}$$

Lewis form factor, $Y_1 = 0.4838 - 0.007163 \times 4 = 0.455$

Number of teeth on pinion, $Z_1 = \dfrac{d_1}{m} = \dfrac{400}{4} = 100$

Number of teeth on gear, $Z_2 = \dfrac{d_2}{m} = \dfrac{800}{4} = 200$

Dynamic load

Buckingham equation to determine the dynamic load on gear tooth

$$F_{\text{dyn}} = F_t + \frac{K_3 v(cb + F_t)}{K_3 v + \sqrt{cb + F_t}}$$

$$K_3 = 20.67, \text{ a constant}$$

$$c = \text{dynamic load factor} = \frac{e \times E_1 \times E_2}{K_1(E_1 + E_2)}$$

$$e = \text{profile error}$$

For 8 m/s pitch line velocity, maximum allowable error is 0.05, which can be obtained by class I industrial or commercial cut gear having 4 mm module. Therefore,

$$c = \frac{0.05 \times 0.1 \times 10^6 \times 0.1 \times 10^6}{9.0 \times 2 \times 0.1 \times 10^6} \quad (\because K_1 = 9.0 \text{ for } 20° \text{ full-depth teeth})$$

$$= 277.78$$

Face width, $b = 10m = 40$ mm

$$cb + F_t = 277.78 \times 40 + 1194.74 = 12{,}305.94$$

$$\sqrt{cb + F_t} = 110.93$$

Therefore,

$$F_{dyn} = 1194.74 + \frac{20.67 \times 8.37 \times 12{,}305.94}{20.67 \times 8.37 + 110.93}$$

$$= 8692.94 \text{ N}$$

Endurance strength of the gear

$$F_{en} = bmY\sigma_{en}$$

Assuming steady-load condition

$$F_{en} \ge 1.25F_{dyn}$$

or

$$1.25 \times 8692.94 = 40 \times 4 \times 0.455 \times \sigma_{en}$$

or

$$\sigma_{en} = 149.26 \text{ N/mm}^2$$

$$\text{Required core BHN} = \frac{\sigma_{en}}{1.75} = \frac{149.26}{1.75} = 85.29$$

which is less than the core BHN of the gear–pinion material. Hence it is safe in dynamic load.

Wear check

Limiting wear load on the gear–pinion

$$F_{wear} = d_1 bQK$$

where

$$Q = \text{ratio factor} = \frac{2Z_2}{Z_1 + Z_2} = \frac{2 \times 200}{200 + 100} = 1.34$$

We adopt a factor of safety 1.25, therefore,

$$F_{wear} = 1.25F_{dyn}$$

or

$$400 \times 40 \times 1.34 \times K = 1.25 \times 8692.94$$

or

$$K = 0.5068$$

$$= \frac{\sigma_{es}^2 \sin \alpha}{1.4} \left(\frac{1}{E_1} + \frac{1}{E_2} \right)$$

or

$$0.5068 = \frac{\sigma_{es}^2 \sin 20^\circ}{1.4} \times \frac{2 \times 0.1 \times 10^6}{0.1 \times 10^6 \times 0.1 \times 10^6}$$

or

$$\sigma_{es} = 322.06 \text{ N/mm}^2$$

Recommended surface hardness

$$\text{BHN} = \frac{\sigma_{es} + 70}{2.75} = \frac{322.06 + 70}{2.75} = 142.56$$

We adopt surface hardness, BHN = 150 which is still less than the core BHN 180; hence the gear–pinion is safe under wear load conditions and does not require any special surface hardening.

Note: Gear construction, design of shafts, keys, and list of specifications are left as an exercise for the students.

12.7 HELICAL GEARS

When the teeth of a gear are cut in the form of a helix on the pitch cylinder, such a gear is called the helical gear. A helical gear may be considered to be composed of an infinite number of infinitesimally narrow staggered spur gears. The result is that each tooth slants across the face so as to form a cylindrical helix. In helical gears, there is progressive engagement of the tooth and gradual pick up of the load by it. This results into a smoother engagement and quiet operation even at a high pitch line velocity. These gears are used in high-speed applications having pitch line velocity up to 30 m/s. Helical gears are of two types—parallel helical gears used for parallel shafts and cross-helical gears used for non-parallel shafts. In a parallel helical gear, the slope of the helix may be either upwards or downwards. The terms right-hand and left-hand helical gear are used to differentiate between the two types. A general rule to determine whether a helical gear is right-handed or left-handed is the same as that used for the screw. In a helical gear pair, the angle of helix is the same for both pinion and gear. However, the hand of helix is opposite to one another.

12.7.1 Kinematics of Helical Gear Tooth

Figure 12.12 shows the top view of a helical gear in which lines AB and CD are the centre lines of two adjacent helical teeth taken on the pitch plane and line AD is perpendicular to the edge. The angle between the axis of the shaft and the centre line of the tooth taken on the pitch line is known as the *helix angle*, i.e. angle ADC. The distance AC measured in the plane of rotation X--X is called the *transverse circular pitch* p_{tr} and the distance AE measured in the

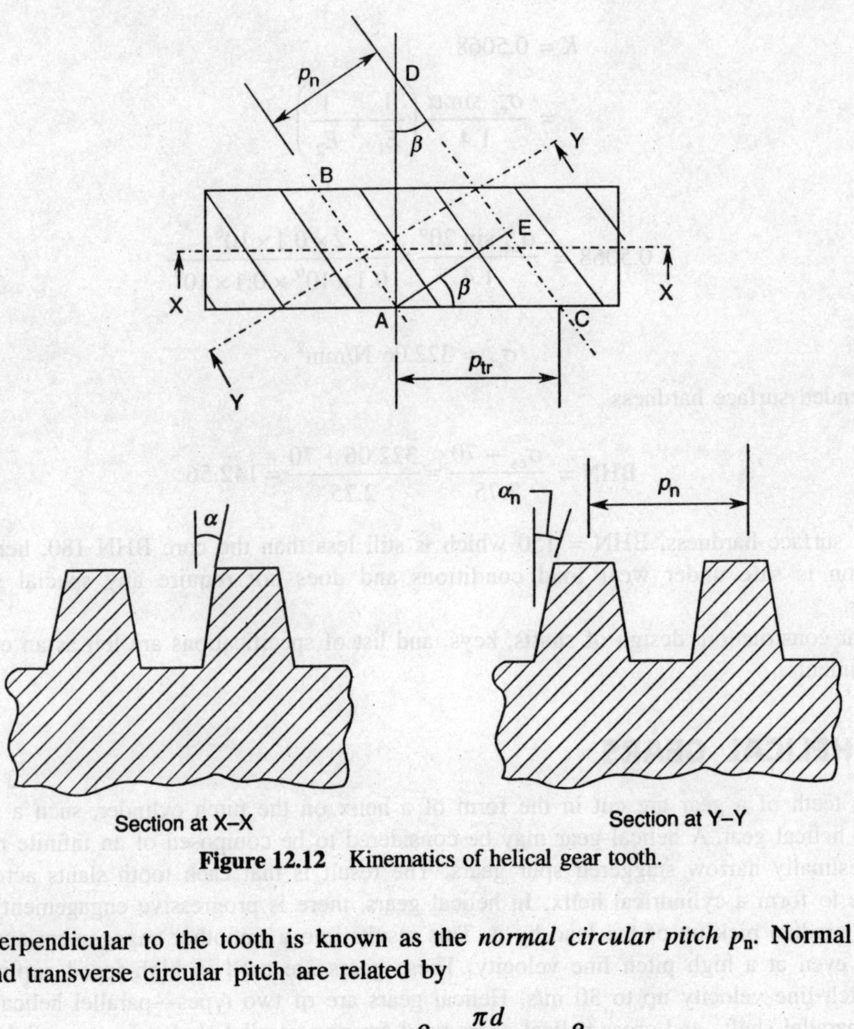

Figure 12.12 Kinematics of helical gear tooth.

plane perpendicular to the tooth is known as the *normal circular pitch* p_n. Normal circular pitch and transverse circular pitch are related by

$$p_n = p_{tr} \cos \beta = \frac{\pi d}{Z} \cos \beta \tag{12.22}$$

and normal module

$$m_n = m \cos \beta$$

The distance AD, called the axial pitch p_a, is related to transverse pitch

$$p_a = \frac{p_{tr}}{\tan \beta} \tag{12.23}$$

The centre-distance C between two helical gears having Z_1 and Z_2 as the number of teeth is computed by the following relation:

$$C = \frac{d_1 + d_2}{2} = \frac{m_n (Z_1 + Z_2)}{2 \cos \beta} \tag{12.24}$$

If the pitch cylinder of a helical gear is cut by an oblique plane X–X at an helix angle β, which is normal to the tooth, as shown in Figure 12.13, the oblique plane cuts out an arc

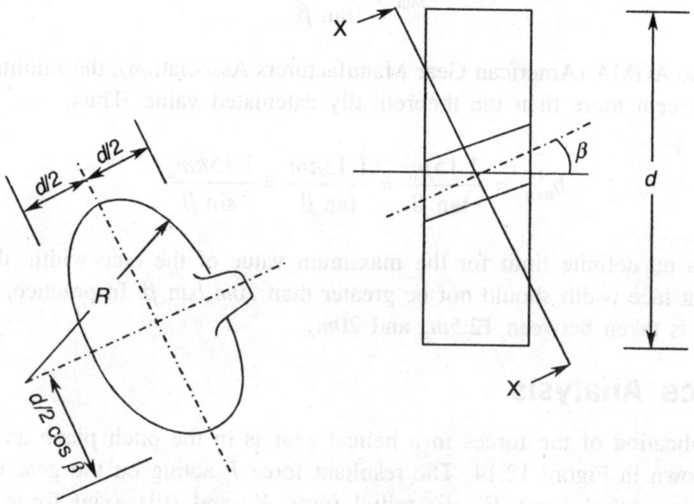

Figure 12.13 Formative pitch circle.

having a radius of curvature R. When the helix angle β is zero, the radius of curvature is equal to $d/2$, as in the case of a spur gear. If the helix angle β is increased from 0° to 90°, the radius of curvature approaches infinity. This radius is the apparent pitch circle radius of the helical gear when viewed in the direction of the tooth element. Thus it is the radius of an equivalent spur gear having a greater number of teeth. In the helical gear terminology, it is called the *formative* or *virtual number* of teeth. It is defined as the number of teeth on an equivalent spur gear which will give the same tooth profile when measured normal to the helix. The selection of a cutter for milling a helical gear is based on the formative number of teeth. The formative number of teeth Z' is related to the actual number of teeth Z by the relation

$$Z' = \frac{Z}{\cos^3 \beta} \tag{12.25}$$

12.7.2 Helical Tooth Proportions

Helical gears are not interchangeable. Therefore, there is no standard value of helix angle. As a general guideline, tooth proportions should be based on normal pressure angle of 20°. Most of the proportions used for spur gears are also applicable to helical gears. However, tooth dimensions should be calculated using the normal module m_n instead of the transverse module m.

In the helical gear, a larger helix angle for a particular face width ensures a larger contact ratio giving a larger overlap of teeth. Hence load is transferred gradually. However, the axial force F_a increases as the helix angle increases. For this reason, it is desirable to limit its maximum value to 30°. The most commonly used helix angles are 15°, 23°, and 25°. For smooth and quiet operation, the leading edge of the tooth should be advanced ahead of the trailing edge by a distance slightly greater than the circular pitch. Therefore, the minimum face width of a helical gear should be

$$b_{min} = \frac{p_{tr}}{\tan \beta}$$

According to AGMA (American Gear Manufacturers Association), the minimum face width should be 15 per cent more than the theoretically calculated value. Thus,

$$b_{min} = \frac{1.15 p_{tr}}{\tan \beta} = \frac{1.15 \pi m}{\tan \beta} = \frac{1.15 \pi m_n}{\sin \beta} \tag{12.26}$$

Although there is no definite limit for the maximum value of the face width, the AGMA has recommended that face width should not be greater than $20m_n/\sin \beta$. In practice, the face width of a helical gear is taken between $12.5m_n$ and $20m_n$.

12.7.3 Force Analysis

The point of application of the forces in a helical gear is in the pitch plane and at the centre of the face as shown in Figure 12.14. The resultant force F acting on the gear tooth has three components: (i) tangential force F_t, (ii) radial force F_r, and (iii) axial force F_a. From the geometry of Figure 12.14, the three components of resultant force F are

$$F_r = F \sin \alpha_n \tag{12.27a}$$

$$F_t = F \cos \alpha_n \cos \beta \tag{12.27b}$$

$$F_a = F \cos \alpha_n \sin \beta \tag{12.27c}$$

In a gear drive, the tangential force can be found from the available torque or power.

The direction of the axial thrust force depends upon the hand of helix and the direction of rotation of the gear. To determine the direction of the axial thrust, first select the driving gear and mark the hand of helix. Secondly, keep fingers in the direction of rotation of the gear, the thumb will then indicate the direction of the thrust force for the driving gear. The direction of the thrust for the driven gear will be opposite to that of the driving gear.

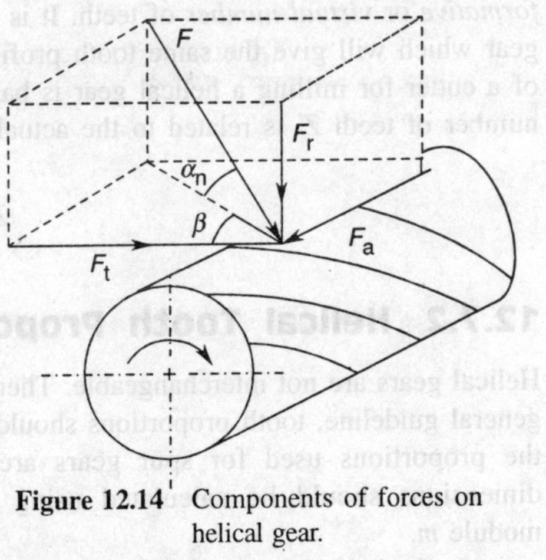

Figure 12.14 Components of forces on helical gear.

12.7.4 Beam Strength of Helical Gear Tooth

In the spur gear, a concentrated load is assumed to act upon the upper edge of the tooth. Thus the force acts with a leverage equal to height of the tooth, whereas in the helical gear the point of contact is at all times distributed over the entire working surface of the tooth. Therefore, the average lever arm of bending is about one-half the height of the tooth. This makes the helical gear tooth much stronger than the spur tooth and also decreases tooth deflection.

The design procedure of the helical gear is similar to that of the spur gear, except that the equations for beam strength, dynamic load, and wear are modified to account for the helix angle.

In order to find the beam strength equation of a helical gear, it is considered to be an equivalent spur gear having the formative number of teeth.

The beam strength of the spur gear is given as

$$F_{beam} = bmY\sigma_d$$

which can be modified for the helical gear by replacing the following parameters:

F_{beam} with $F_{beam(n)}$, i.e. beam strength normal to the tooth
m with m_n, i.e. normal module
b with $b/\cos\beta$, i.e. normal face width
Y = Lewis form factor, based upon the formative number of teeth.

Therefore,

$$F_{beam(n)} = \frac{bm_nY\sigma_d}{\cos\beta} \tag{12.28}$$

Now, the beam strength of the helical gear tooth F_{beam} is a component of $F_{beam(n)}$ in the plane of rotation, as shown in Figure 12.15, and is given by

$$F_{beam} = F_{beam(n)}\cos\beta \tag{12.29}$$

From Eqs. (12.28) and (12.29), the modified beam strength equation of the helical gear may be written as

$$F_{beam} = bm_nY\sigma_d \tag{12.30}$$

The maximum force between two meshing gear teeth is given by

$$F_{max} = \frac{C_{wear}F_t}{C_v} \tag{12.31}$$

where

C_{wear} is the wear and lubrication factor (1.0 to 1.4)

C_v is the velocity factor

Now,

$$C_v = \frac{4.58}{4.58 + v} \quad \text{for low helix angle gears with pitch line velocity up to 5 m/s}$$

$$= \frac{6.1}{6.1 + v} \quad \text{for all helical gears with pitch line velocity between 5 and 10 m/s}$$

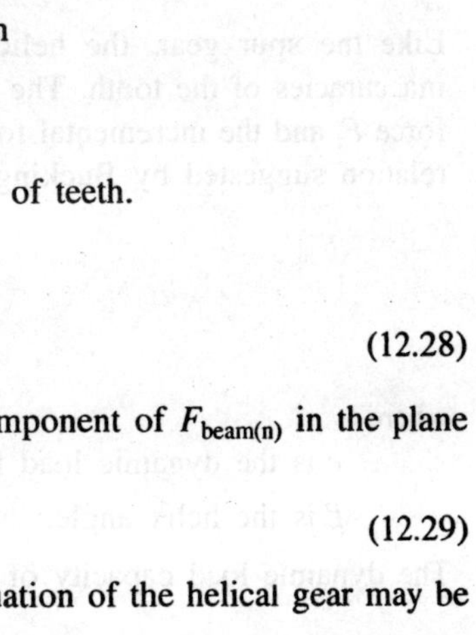

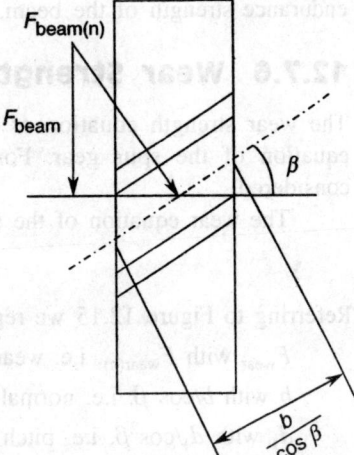

Figure 12.15 Beam strength force acting on the helical gear tooth.

$$= \frac{15.25}{15.25 + v} \quad \text{for gears with pitch line velocity between 10 and 20 m/s}$$

$$= \frac{5.55}{5.55 + \sqrt{v}} \quad \text{for precision gears with pitch line velocity greater than 20 m/s}$$

For safe design, the beam strength of the helical gear tooth must be greater than or equal to maximum force, i.e.

$$F_{beam} \geq F_{max}$$

12.7.5 Dynamic Load on Helical Gear

Like the spur gear, the helical gear is also subjected to dynamic load caused by profile inaccuracies of the tooth. The dynamic load may be determined by summing up the tangential force F_t and the incremental force F_i. The incremental force may be computed by the following relation suggested by Buckingham:

$$F_i = \frac{K_3 v\,(cb\cos^2\beta + F_t)\cos\beta}{K_3 v + \sqrt{cb\cos^2\beta + F_t}} \tag{12.32}$$

where

c is the dynamic load factor

β is the helix angle.

The dynamic load capacity of a helical gear is given as

$$F_{dyn} = F_t + F_i$$

In order to determine the degree of safety, the dynamic load may be compared with the endurance strength of the beam.

12.7.6 Wear Strength of Helical Gear

The wear strength equation of the helical gear is obtained by modifying the corresponding equation of the spur gear. For this purpose, an equivalent formative pinion and gear is considered.

The wear equation of the spur gear is given as

$$F_{wear} = d_1 bQK$$

Referring to Figure 12.15 we replace the following parameters.

F_{wear} with $F_{wear(n)}$, i.e. wear strength normal to the tooth

b with $b/\cos\beta$, i.e. normal face width

d_1 with $d_1/\cos^2\beta$, i.e. pitch circle diameter of the formative pinion.

Therefore,

$$F_{wear(n)} = \frac{d_1 bQK}{\cos^3\beta}$$

The component of the wear strength in the plane of rotation is called the *limiting wear load*. F_{wear}. Thus,

$$F_{wear} = F_{wear(n)} \cos \beta$$

or

$$F_{wear} = \frac{d_1 bQK}{\cos^2 \beta} \tag{12.33}$$

where

Q is the ratio factor, based upon the formative teeth of gear and pinion

K is the load stress factor $\left[= \frac{\sigma_{es}^2 \sin \alpha_n}{1.4} \left(\frac{1}{E_1} + \frac{1}{E_2} \right) \right]$

According to Buckingham, the limiting wear load must be greater than the dynamic load for satisfactory design.

Example 12.4 Design a pair of helical gears to transmit 30 kW power at a speed reduction ratio of 4:1. The input shaft rotates at 2000 rpm. Take helix and normal pressure angles equal to 25° and 20°, respectively. Both pinion and gear are made of steel. (The following data is given.)

Name of the part	*Permissible stress*	*BHN*
Pinion	55 MPa	340
Gear	40 MPa	300

The number of teeth on the pinion may be taken as 30.

Solution Number of teeth on gear, Z_2 = speed reduction ratio × Z_1 = 4 × 30 = 120

Lewis form factor on the basis of formative number of teeth

$$Y = \pi \left(0.154 - \frac{0.912}{Z'} \right)$$

$$Z_1' = \frac{Z_1}{\cos^3 \beta} = \frac{30}{\cos^3 25°} = 40.3, \text{ say, } 41$$

$$Z_2' = \frac{Z_2}{\cos^3 \beta} = \frac{120}{\cos^3 25°} = 161.2, \text{ say, } 162$$

$$\text{Lewis form factor, } Y_1 = \pi \left(0.154 - \frac{0.912}{41} \right) = 0.414$$

$$Y_2 = \pi \left(0.154 - \frac{0.912}{162} \right) = 0.4661$$

$$\text{Strength factor of the pinion} = \sigma_{d1} Y_1$$

$$= 55 \times 0.414 = 22.77$$

$$\text{Strength factor of the gear} = 40 \times 0.4661 = 18.64$$

Since the strength of the gear is less than that of the pinion, the gear is the weaker element.

$$\text{Tangential force on the gear, } F_t = \frac{P \times 1000}{v}$$

$$v = \text{pitch line velocity}$$

$$= \frac{\pi d_2 N_2}{60 \times 1000} = \frac{\pi \times m \times 120 \times 500}{60 \times 1000} = 3.141m \text{ m/s}$$

$$\text{Tangential force, } F_t = \frac{30 \times 1000}{3.141m} = \frac{9551}{m} \text{ N}$$

$$\text{Maximum tangential force, } F_{max} = \frac{C_{wear} F_t}{C_v}$$

C_{wear} = wear and lubrication factor

= 1.25 for scant lubrication with frequent inspection

C_v = 0.35 (assumed trial value)

$$F_{max} = \frac{1.25 \times 9551}{0.35m} = \frac{34,110.7}{m} \text{ N}$$

Since $m_n = m \cos \beta$

$$F_{max} = \frac{34,110.7}{m_n} \times \cos 25° = \frac{30,914.8}{m_n} \text{ N}$$

Lewis beam strength, $F_{beam} = b m_n Y \sigma_d$

Let us adopt face width, $b = 15m_n$. Therefore,

$$F_{beam} = 15m_n \times m_n \times 0.4661 \times 40$$
$$= 279.66m_n^2$$

For safe design, the beam strength should be greater than or equal to maximum force. Therefore,

$$\frac{30,914.8}{m_n} = 279.66\, m_n^2$$

or

$$m_n = 4.8 \text{ mm, say, 5 mm}$$

$$\text{Pitch diameter of the pinion, } d_1 = \frac{Z_1 m_n}{\cos \beta} = \frac{30 \times 5.0}{\cos 25°} = 165.5 \text{ mm}$$

$$\text{pcd of the gear, } d_2 = \frac{120 \times 5}{\cos 25°} = 662 \text{ mm}$$

$$\text{Pitch line velocity, } v = \frac{\pi \times 0.662 \times 500}{60} = 17.33 \text{ m/s}$$

$$\text{Tangential force, } F_t = \frac{9551.0}{m} = \frac{9551.0 \times \cos 25^\circ}{5} = 1731.23 \text{ N}$$

$$\text{Velocity factor, } C_v = \frac{15.25}{15.25 + 17.33} = 0.468$$

$$\text{Maximum tangential force, } F_{max} = \frac{1.25 \times 1731.23}{0.468} = 4624 \text{ N}$$

Lewis beam strength

$$F_{beam} = bm_n Y \sigma_d$$

$$= 15 \times 5 \times 5 \times 0.4661 \times 40 = 6991.5 \text{ N}$$

The beam strength is greater than the maximum tangential force (1.512 times). Hence the design is safe under static load conditions.

Dynamic load

Buckingham equation for dynamic load

$$F_{dyn} = F_t + \frac{K_3 v\,(cb \cos^2 \beta + F_t) \cos \beta}{K_3 v + \sqrt{cb \cos^2 \beta + F_t}}$$

K_3 = constant = 20.67

$$c = \text{dynamic load factor} = \frac{e}{K_1\left(\dfrac{1}{E_1} + \dfrac{1}{E_2}\right)}$$

e = profile error

A permissible profile error for 17 m/s velocity is 0.01925. This order of error can be obtained in class III gears with 5 mm module, which gives e = 0.015 mm. Thus,

$$c = \frac{0.015 \times 0.2 \times 10^6 \times 0.2 \times 10^6}{9.0 \times 2 \times 0.2 \times 10^6} = 166.67$$

$$cb \cos^2 \beta + F_t = 166.67 \times 15 \times 5 \times \cos^2 25^\circ + 1731.23 = 11{,}998.85$$

and

$$\sqrt{cb \cos^2 \beta + F_t} = 109.54$$

Hence dynamic load,

$$F_{dyn} = 1731.23 + \frac{20.67 \times 17.33 \times 11{,}998.85 \times \cos 25^\circ}{20.67 \times 17.33 + 109.54} = 10{,}059.2 \text{ N}$$

For safe design, the endurance strength of the gear should be equal to 1.25 times the dynamic load for steady load conditions, i.e.

$$F_{en} = bm_n Y \sigma_{en} \geq 1.25 F_{dyn}$$

or

$$\sigma_{en} = \frac{1.25 \times 10,059.2}{15 \times 5 \times 5 \times 0.4661} = 71.93 \text{ N/mm}^2$$

Therefore,

$$\text{Core BHN required} = \frac{\sigma_{en}}{1.75} = \frac{71.93}{1.75} = 41.1$$

which is less than the core BHN of steel.

Wear load

The limiting wear load

$$F_{wear} = \frac{d_1 bQK}{\cos^2 \beta} \geq F_{dyn}$$

$$Q = \frac{2Z_2'}{Z_2' + Z_1'} = \frac{2 \times 162}{162 + 41} = 1.596$$

$$\text{Load stress factor, } K = \frac{F_{dyn} \cos^2 \beta}{d_1 bQ}$$

$$= \frac{10,059.2 \times \cos^2 25°}{165.5 \times 75 \times 1.596} = 0.417$$

Also,

$$K = \frac{\sigma_{es}^2 \times \sin \alpha}{1.4} \times \frac{E_1 + E_2}{E_1 E_2}$$

Therefore,

$$0.417 = \frac{\sigma_{es}^2 \times \sin 20°}{1.4} \times \frac{2 \times 0.2 \times 10^6}{0.2 \times 10^6 \times 0.2 \times 10^6}$$

or

$$\sigma_{es} = 413.15 \text{ N/mm}^2$$

Required surface hardness

$$\text{BHN} = \frac{\sigma_{es} + 70}{2.75} = \frac{413.15 + 70}{2.75} = 175.7$$

which is less than the core hardness. Hence there is no need for surface hardening.

12.8 BEVEL GEARS

Bevel gears are used to transmit power between two intersecting shafts. These gears are cut on conical pitch surfaces. Two types of bevel gears—straight tooth and spiral tooth—as shown in Figure 12.16 are commonly used. The teeth on a straight tooth bevel gear are on straight lines which converge to a common point called the *apex* of the pitch cone, which is also the point

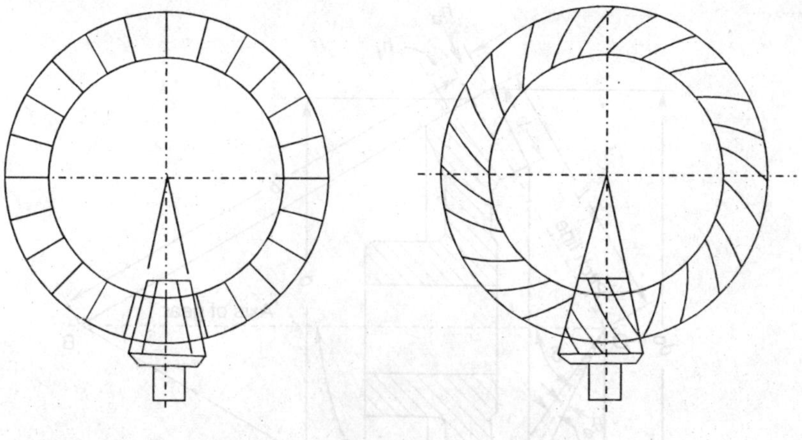

(a) Straight tooth bevel gear (b) Spiral bevel gear

Figure 12.16 Types of bevel gears.

of intersection of the gear axes. Involute profile straight bevel gears are used for relatively low-speed applications with pitch line velocity up to 10 m/s. When smooth tooth engagement, quiet operation, greater strength, and high pitch line velocity are the major requirements, spiral bevel gears with curved teeth are used.

Bevel gears are not interchangeable. Hence these gears are designed in pair. In the majority of applications, the angle between the axes of two intersecting shafts is 90°, however, the intersecting angle may be acute, or obtuse angle. These gears are manufactured either by casting, machining, or generating process. Gears manufactured by the generating process have smooth tooth profile and they can be used for transmitting power at high pitch line velocity.

12.8.1 Kinematics of Bevel Gears

The definitions and dimensions relating to bevel gear are shown in Figure 12.17. In a bevel gear if the pitch line distance L called *cone distance* is revolved about the axis of the gear, it generates an imaginary pitch cone with the apex at O. The angle formed between the pitch line and the axis is called the *pitch angle* δ. The angles θ_a and θ_d are called addendum and dedendum angles, respectively. The sum of the pitch and addendum angles ($\delta + \theta_a$) is known as the face angle. Dimensions of the bevel gear such as addendum h_a, dedendum h_f and pitch diameter d are specified at the larger end of the tooth. A line drawn perpendicular to the pitch line intersects the axis at point B and forms a cone called the back cone. The length of the back cone element is called the back cone radius r_b. The distance C is called the crown height. The B is backing distance and M is known as mounting distance.

In bevel gear, for the purpose of design, an imaginary spur gear in a plane perpendicular to the tooth at the larger end having r_b pitch circle radius is considered for finding the formative or virtual number of teeth, i.e.

$$\text{Formative number of teeth, } Z' = \frac{Z}{\cos\delta} \tag{12.34}$$

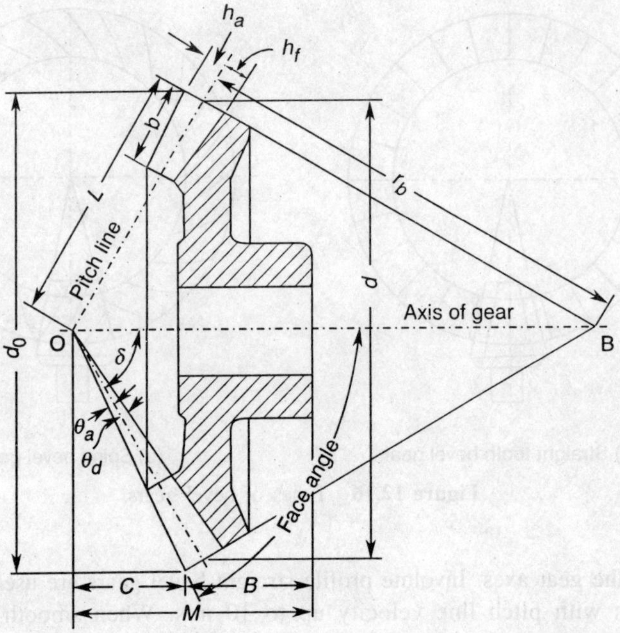

Figure 12.17 Dimensions of bevel gear.

The shaft angle for any pair of bevel gears is an angle between two intersecting axes which meet at an apex. It is equal to the sum of the pitch angles of two mating gears, i.e. $\theta = \delta_1 + \delta_2$. The relationships from gear geometry, shown in Figure 12.18, for shaft angle θ are given below:

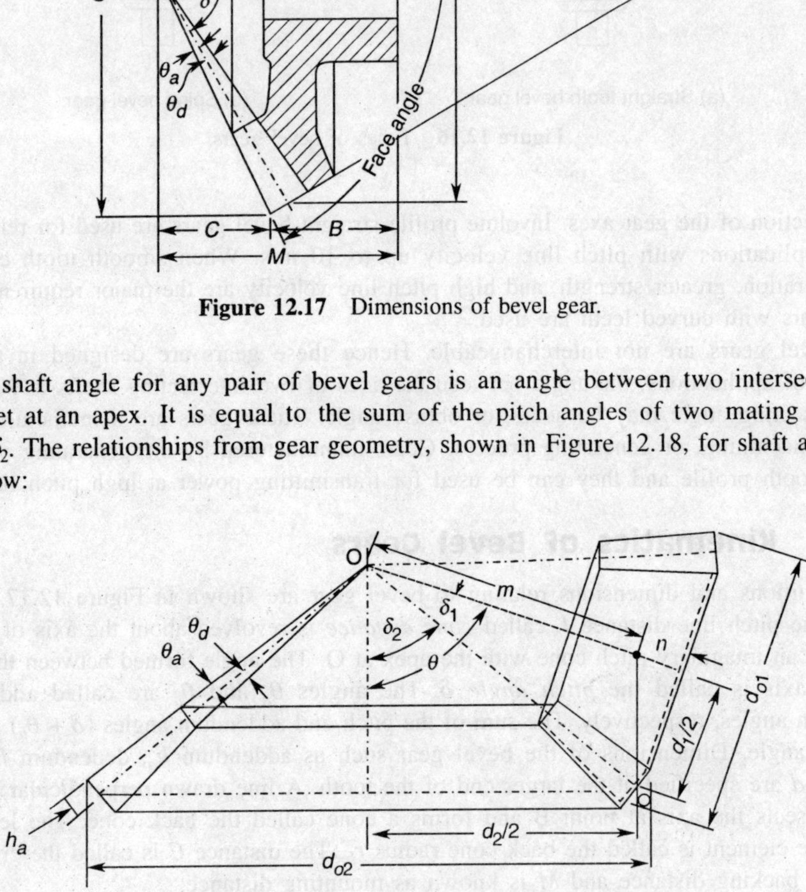

Figure 12.18 Acute angle bevel gear.

1. *Pitch angle* (δ)
   (a) For acute angle bevel gear ($0° < \theta < 90°$)

   (i) For pinion, $\tan \delta_1 = \dfrac{\sin \theta}{\dfrac{Z_2}{Z_1} + \cos \theta}$

(ii) For gear, $\tan \delta_2 = \dfrac{\sin\theta}{\dfrac{Z_1}{Z_2} + \cos\theta}$

(b) For right-angled bevel gear, $\theta = 90°$

(i) For pinion, $\tan \delta_1 = \dfrac{Z_1}{Z_2}$

(ii) For gear, $\tan \delta_2 = \dfrac{Z_2}{Z_1}$

(c) For obtuse angle bevel gear ($90° < \theta \le 180°$)

(i) For pinion, $\tan \delta_1 = \dfrac{\sin(180° - \theta)}{\dfrac{Z_2}{Z_1} - \cos(180° - \theta)}$

(ii) For gear, $\tan \delta_2 = \dfrac{\sin(180° - \theta)}{\dfrac{Z_1}{Z_2} - \cos(180° - \theta)}$

2. *Pitch diameter at the larger end*
   (i) For pinion, $d_1 = mZ_1$
   (ii) For gear, $d_2 = mZ_2$
3. *Outer diameter at the larger end*
   (i) For pinion, $d_{o1} = d_1 + 2h_a \cos\delta_1$
   (ii) For gear, $d_{o2} = d_2 + 2h_a \cos\delta_2$
4. *Length of the cone distance*

$$L = 0.5\left(d_1^2 + d_2^2\right)^{0.5}$$

12.8.2 Force Analysis

In a bevel gear, the resultant force acting between two meshing teeth is assumed to be a concentrated force acting at the mid-point along the face width of the tooth, while the actual force acts somewhere between the mid-point and the larger end of the tooth. Thus there is a small error in making this assumption. The resultant force has two components—tangential force F_t and separating force F_s as shown in Figure 12.19. The tangential force shown perpendicular to the plane of rotation is the driving force. Its magnitude can be computed by the following relation:

$$F_t = \frac{T}{r_{mid}} \tag{12.35}$$

where

T is the torque transmitted by a gear

r_{mid} is the pitch radius of the gear under consideration at the mid-point of the tooth

$$\left(= \frac{d_1}{2} - \frac{b\sin\delta_1}{2}\right) \tag{12.35a}$$

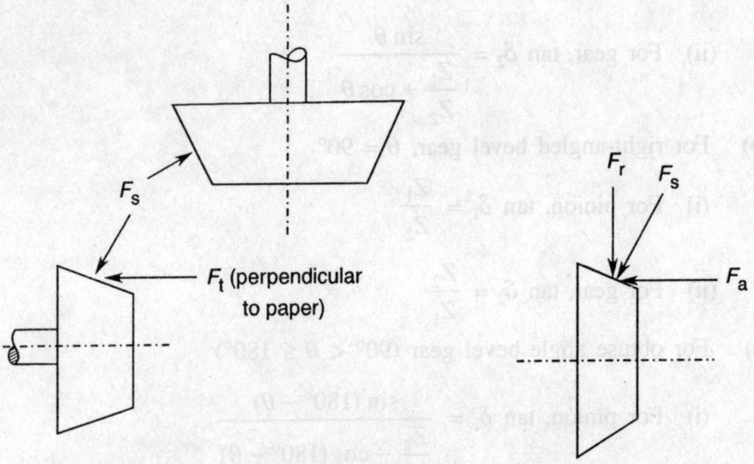

Figure 12.19 Components of forces on bevel gear.

The separating force which acts perpendicular to the pitch line is determined by the relation

$$F_s = F_t \tan \alpha \tag{12.36}$$

where α is the pressure angle.

The separating force can be further resolved into two components, namely along the axis of the gear called the axial force F_a and perpendicular to the axis of the gear called the radial force F_r. Therefore,

$$F_a = F_s \sin \delta_1 = F_t \tan \alpha \sin \delta_1 \tag{12.37a}$$

$$F_r = F_s \cos \delta_1 = F_t \tan \alpha \cos \delta_1 \tag{12.37b}$$

The forces acting on the mating gear are equal in magnitude but opposite in direction.

12.8.3 Beam Strength of Bevel Gear Tooth

As mentioned earlier, the size of the bevel gear tooth varies across the face width such that the largest tooth height is at the pitch circle and it continuously reduces towards the cone centre. Further these gears are so mounted that the portion of the gear tooth near the pitch circle, i.e. the larger end shares the maximum force compared to the portion of the tooth near the smaller end. Therefore, the beam strength of the bevel gear is computed for the elemental portion of the equivalent formative spur gear in a plane perpendicular to the tooth element and then integrated over its face width.

Consider a small tooth element of width dx at distance x from the apex over which the force oad is assumed to be uniformly distributed, as shown in Figure 12.20. According to Lewis, the beam strength of the elemental section is expressed as

$$\delta F_{\text{beam}} = m_x Y \sigma_d \, dx \tag{12.38}$$

where

m_x is the module of the section

dx is the face width of the elemental section

Y is the Lewis form factor.

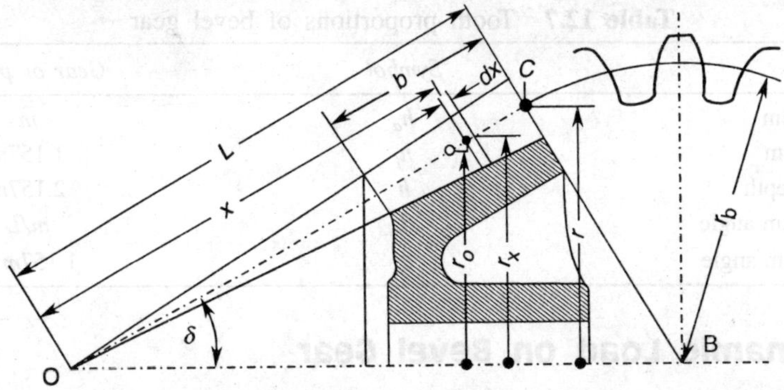

Figure 12.20 Strength determination of a bevel gear tooth.

By integrating Eq. (12.38) within limits $(L - b)$ to L and neglecting the higher order terms, the modified Lewis beam strength equation of the bevel gear tooth can be written as

$$F_{\text{beam}} = bmY\sigma_d \left(1 - \frac{b}{L}\right) \tag{12.39}$$

where

b is the face width (= $15m$ or $L/3$ whichever is smaller)

L is the cone distance

m is the module.

In the design of the bevel gear, the beam strength indicates the maximum value of the tangential force at the larger end of the gear that can be transmitted without failure. Therefore, the tangential force at the larger end should be computed instead of that at the mid-point as calculated in force analysis. The maximum tangential force is computed by the equation

$$F_{\max} = \frac{2TC_S}{d_1 C_v}$$

where

T is the torque to be transmitted

C_S is the service factor (refer to Table 12.3)

C_v is the velocity factor

$$\left(= \frac{3.05}{3.05 + v} \text{ for teeth cut by form cutter}\right)$$

$$\left(= \frac{6.1}{6.1 + v} \text{ for generated teeth}\right)$$

For safe design, the beam strength should be greater than the maximum force on the gear. Standard module m for the bevel gear can be selected from the values: 1, 1.25, 1.5, 2, 2.5, 3, 4.5, 6, 8, 10, 12, 16, 20, 25, 32, 40, and 50. The tooth proportions are listed in Table 12.7.

Table 12.7 Tooth proportions of bevel gear

Element	*Symbol*	*Gear or pinion*
Addendum	h_a	m
Dedendum	h_f	$1.157m$
Whole depth	h	$2.157m$
Addendum angle	θ_a	m/L
Dedendum angle	θ_d	$1.157m/L$

12.8.4 Dynamic Load on Bevel Gear

The dynamic load on a bevel gear tooth is caused due to profile error of the tooth, which can be computed on the lines of the spur gear except that the velocity at the largest pitch circle must be computed and tangential force should then be based on this velocity.

$$F_{\text{dyn}} = F_{\text{t}} + \frac{K_3 v\,(cb + F_{\text{t}})}{K_3 v + \sqrt{cb + F_{\text{t}}}} \tag{12.40}$$

For safe design against dynamic load, the endurance beam strength of the bevel gear should be greater than the applied dynamic load. The endurance beam strength can be computed by the equation

$$F_{\text{en}} = bmY\sigma_{\text{en}}\left(1 - \frac{b}{L}\right) \tag{12.41}$$

12.8.5 Wear Strength of Bevel Gear

The wear strength of bevel gear is the maximum force at the larger end of the tooth that can be transmitted without pitting failure. In the bevel gear, the type of contact and Hertz stresses between two meshing teeth are similar to those in the spur gear. Therefore, limiting wear load is computed on the assumption that the bevel gear is treated as an equivalent formative spur gear in a plane perpendicular to the tooth at the larger end.

The modified Buckingham equation of wear load for the bevel gear is expressed as

$$F_{\text{wear}} = \frac{d_1 bQK}{\cos \delta_1} \tag{12.42}$$

where

Q is the ratio factor $\left(= \dfrac{2Z_2'}{Z_2' + Z_1'}\right)$

K is the load-stress factor.

According to Buckingham, the limiting wear load should be greater than the dynamic load for satisfactory design.

Example 12.5 A pair of bevel gears is required to transmit 10 kW power at 500 rpm from a motor shaft to a machine shaft. The speed reduction is 3:1 and the shafts are inclined at 60°.

The pinion is to have 24 teeth with pressure angle 20° and is to be made of cast steel having strength of 75 N/mm^2. The gear is to be made of cast iron with static stress of 55 N/mm^2. The pinion is mounted midway on the shaft which is supported between two bearings having span of 200 mm. Design the gear pair and the pinion shaft.

Solution Let δ_1 and δ_2 be the pitch angles of the pinion and the gear, respectively.

$$\text{Shaft angle, } \theta = \delta_1 + \delta_2 = 60°$$

$$\text{Pitch angle of the pinion, } \tan \delta_1 = \frac{\sin \theta}{\dfrac{Z_2}{Z_1} + \cos \theta}$$

or

$$\delta_1 = \tan^{-1}\left(\frac{\sin 60°}{3 + \cos 60°}\right) = 13.9°$$

Pitch angle of the gear, $\delta_2 = 60° - 13.9° = 46.1°$

Number of teeth on the pinion, $Z_1 = 24$

Number of teeth on the gear, Z_2 = speed reduction ratio × $Z_1 = 3 \times 24 = 72$

Formative number of teeth, $Z' = Z/\cos \delta$. Therefore,

$$Z'_1 = \frac{Z_1}{\cos \delta_1} = \frac{24}{\cos 13.9°} = 24.72$$

$$Z'_2 = \frac{Z_2}{\cos \delta_2} = \frac{72}{\cos 46.1°} = 103.84$$

$$\text{Lewis form factor, } Y = \pi\left(0.154 - \frac{0.912}{Z'}\right)$$

$$\text{For pinion, } Y_1 = \pi\left(0.154 - \frac{0.912}{24.72}\right) = 0.3679$$

$$\text{For gear, } Y_2 = \pi\left(0.154 - \frac{0.912}{103.84}\right) = 0.4562$$

$$\text{Strength factor for pinion} = Y_1\sigma_{d1} = 0.3679 \times 75 = 27.59$$

$$\text{Strength factor for gear} = Y_2\sigma_{d2} = 0.4562 \times 55 = 25.09$$

Since the strength factor of the gear is less than that of the pinion, the gear is weaker.

$$\text{Tangential force, } F_t = \frac{P \times 1000}{v} = \frac{10 \times 1000}{\dfrac{\pi \times m \times 24 \times 500}{60 \times 1000}} = \frac{15{,}915.5}{m} \text{ N}$$

We assume service factor $C_S = 1.0$ and velocity factor $C_v = 0.35$ (a trial value). Therefore,

$$F_{max} = \frac{C_S\,F_t}{C_v} = \frac{1.0 \times 15{,}915.5}{0.35 \times m} = \frac{45{,}472.8}{m}\ \text{N}$$

$$\text{Lewis beam strength, } F_{beam} = bmY\sigma_d\left(\frac{L-b}{L}\right)$$

$$L = \text{slant height} = \frac{d_1}{2\sin\delta_1} = \frac{mZ_1}{2\sin 13.9^\circ} = 49.46m$$

b = face width, which should not be larger than $L/3$, say, $L/4$ (= 12.5m)

Therefore,

$$F_{beam} = 12.5m \times m \times 0.4562 \times 55 \times 0.75 = 235.23m^2$$

For safe design, the beam strength should be greater than or equal to the tangential force, i.e.

$$F_{beam} \geq F_{max}$$

or

$$235.23m^2 = \frac{45{,}472.8}{m}$$

or

$$m = 5.782\ \text{mm}$$

We adopt the standard module m = 6 mm as per IS: 5037–1969.

$$\text{pcd of pinion, } d_1 = mZ_1 = 144\ \text{mm}$$

$$\text{pcd of gear, } d_2 = mZ_2 = 432\ \text{mm}$$

$$\text{Pitch line velocity, } v = \frac{\pi\,d_1\,N_1}{60} = \frac{\pi \times 0.144 \times 500}{60} = 3.77\ \text{m/s}$$

$$\text{Velocity factor, } C_v = \frac{3.05}{3.05 + v} = 0.4472$$

$$\text{Face width, } b = 12.5m = 75\ \text{mm}$$

$$\text{Slant height, } L = 4b = 300\ \text{mm}$$

$$\text{Tangential force, } F_t = \frac{15{,}915.5}{m} = \frac{15{,}915.9}{6} = 2652.6\ \text{N}$$

$$F_{max} = \frac{1.0 \times 2652.6}{0.4472} = 5931.5\ \text{N}$$

$$\text{Beam strength, } F_{beam} = 75 \times 6 \times 0.4562 \times 55 \times 0.75 = 8468.2\ \text{N}$$

Since beam strength is approximately 1.43 times the maximum tangential force, the design is safe under static loading.

Dynamic load

Buckingham equation for dynamic load is

$$F_{dyn} = F_t + \frac{K_3 v\,(cb + F_t)}{K_3 v + \sqrt{cb + F_t}}$$

$$K_3 = 26.67, \text{ a constant}$$

$$c = \text{dynamic load factor} = \frac{e}{K_1\left(\dfrac{1}{E_1} + \dfrac{1}{E_2}\right)}$$

For the 6 mm module, pitch line velocity up to 4 m/s, class I gears have $e = 0.0652$. Therefore,

$$c = \frac{0.0652 \times 0.2 \times 10^6 \times 0.2 \times 10^6}{9.0 \times 0.2 \times 10^6 \times 2} = 724.4$$

$$cb + F_t = 724.4 \times 75 + 2652.6 = 56{,}982.6$$

$$\sqrt{cb + F_t} = 238.71$$

Thus,

$$F_{dyn} = 2652.6 + \frac{20.67 \times 3.77 \times 56{,}982.6}{20.67 \times 3.77 + 238.71} = 16{,}676.3 \text{ N}$$

We assume that for light shock load, endurance strength $F_{en} = 1.5F_d$. Therefore,

$$1.5F_{dyn} = bmY\sigma_{en}\left(\frac{L-b}{L}\right)$$

or

$$1.5 \times 16{,}676.3 = 75 \times 6 \times 0.4562 \times \sigma_{en} \times 0.75$$

or

$$\sigma_{en} = 162.5 \text{ N/mm}^2$$

Core BHN required $= \dfrac{162.5}{1.75} = 92.8$, say, 100 BHN, which is quite reasonable for the CI material.

Wear load

The limiting value of the wear load

$$F_{wear} = \frac{d_1 bQK}{\cos \delta_1}$$

$$Q = \text{ratio factor} = \frac{2Z_2'}{Z_2' + Z_1'} = \frac{2 \times 103.84}{24.72 + 103.84} = 1.615$$

For safe design, the limiting wear load should be greater than the dynamic load, i.e. $F_{wear} \geq F_{dyn}$.

$$\text{Load stress factor, } K = \frac{F_{dyn} \cos \delta_1}{b\,Q\,d_1} = \frac{16{,}676.3 \times \cos 13.9°}{75 \times 1.615 \times 144} = 0.928$$

$$\text{where, } K = \frac{\sigma_{es}^2 \sin \alpha}{1.4}\left(\frac{1}{E_1} + \frac{1}{E_2}\right)$$

Thus,

$$0.928 = \frac{\sigma_{es}^2 \sin 20°}{1.4}\left(\frac{2 \times 0.2 \times 10^6}{0.2 \times 10^6 \times 0.2 \times 10^6}\right)$$

or

$$\sigma_{es} = 616.3 \text{ N/mm}^2$$

$$\text{Required surface hardness} = \frac{\sigma_{es} + 70}{2.75} = \frac{616.3 + 70}{2.75} = 249.6 \text{ BHN, say, 250 BHN}$$

Pinion shaft design

The radial load on the pinion

$$F_r = F_{te} \tan \alpha \cos \delta_1$$

F_{te} = tooth load acting somewhere between the mid-point and the larger end.

$$= \frac{F_t \times L}{L - 0.5b} = \frac{2652.6 \times 300}{300 - 0.5 \times 75} = 3031.5 \text{ N}$$

Therefore,

$$F_r = 3031.5 \times \tan 20° \times \cos 13.9° = 1071 \text{ N}$$

$$\text{Axial thrust, } F_a = F_{te} \tan \alpha \times \sin \delta_1$$

$$= 3031.5 \times \tan 20° \times \sin 13.9° = 265 \text{ N}$$

Maximum BM due to radial force

$$M_r = \frac{F_r l}{4} = \frac{1071 \times 0.2}{4} = 53.55 \text{ N} \cdot \text{m}$$

Maximum BM due to tangential force

$$M_{te} = \frac{F_{te} l}{4} = \frac{3031.5 \times 0.2}{4} = 151.57 \text{ N} \cdot \text{m}$$

$$\text{Resultant BM, } M = \sqrt{M_r^2 + M_{te}^2}$$

$$= \sqrt{53.55^2 + 151.57^2} = 160.75 \text{ N} \cdot \text{m}$$

Average torque transmitted, $T = \frac{P}{\omega} = \frac{10 \times 1000}{\left(\frac{2\pi \times 500}{60}\right)} = 191 \text{ N} \cdot \text{m}$

$$\text{Equivalent torque, } T_{\text{eq}} = \sqrt{M^2 + T^2}$$

$$= \sqrt{160.75^2 + 191^2} = 249.6 \text{ N} \cdot \text{m}$$

We assume that the shaft is made of cold-rolled steel for which the allowable shear strength, including the effect of the keyway, $\tau = 75$ N/mm^2.

$$\text{Shaft diameter, } d_S = \left(\frac{16T_{\text{eq}}}{\pi\tau}\right)^{1/3}$$

$$= \left(\frac{16 \times 249.6 \times 1000}{\pi \times 75}\right)^{1/3} = 25.7 \text{ mm, say, } 28 \text{ mm (standard size)}$$

Axial stress induced,

$$\sigma_a = \frac{4F_a}{\pi d_S^2} = \frac{4 \times 265}{\pi \times 28^2} = 0.43 \text{ N/mm}^2$$

Actual torsional shear stress

$$\tau = \frac{16T_{\text{eq}}}{\pi d_S^3} = \frac{16 \times 249.6 \times 1000}{\pi \times 28^3} = 57.9 \text{ N/mm}^2$$

Since axial stress is very small compared to torsional shear stress, the maximum shear stress is approximately equal to torsional shear stress, which is less than the allowable strength of the shaft material. Hence the design of the shaft is satisfactory. Proportions of gear tooth may be computed from Table 12.7.

12.9 WORM GEAR SET

A worm gear set (Figure 12.21) is used to transmit power between two non-parallel, non-intersecting shafts. It consists of a worm, which is very much similar to a threaded screw, and a worm gear. It is widely used in machine tools, automobiles, material handling equipment, and as a speed reducer in cement plants for rotating the kiln. The worm gear has the following important features:

1. Reduction of high speeds in the smallest possible space.
2. Tooth engagement occurs without shock, hence operation is quieter.
3. The provision for self-locking can also be made.
4. The transmission efficiency is very low compared to spur and helical gears and generates considerable amount of heat.

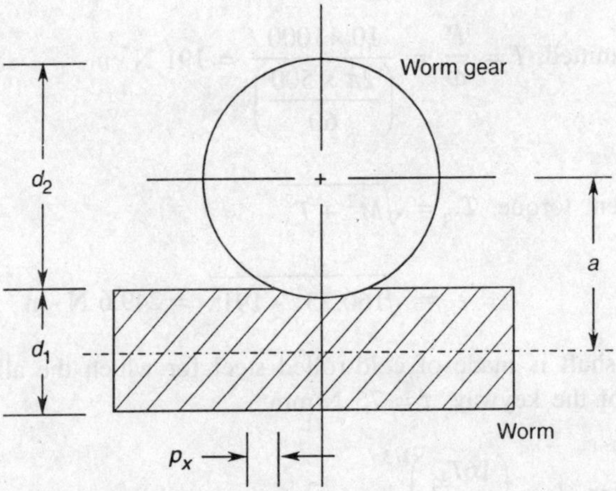

Figure 12.21 Schematic diagram of a worm gear set.

Worm gearing may be of either single or double enveloping type. In the single enveloping set, the worm gear has its width cut into a concave surface, thus, partially enclosing the worm in a mesh. In the double enveloping set, the worm tooth and the worm gear tooth are both cut concavely. Thus, both worm and worm gear are partially enclosed. The single enveloping set has fewer teeth in contact and thus has line contact. In the double enveloping set, more teeth come in contact; it therefore has area contact and can transmit a greater load.

12.9.1 Kinematics of Worm

In a worm gear set, the worm is very much similar to the power screw, therefore, the terms used for power screw are also applicable to the worm. A worm gear set is designated by four quantities as given below:

$$Z_1 / Z_2 / q / m \tag{12.43}$$

where

Z_1 is the number of starts on the worm

Z_2 is the number of teeth on the worm gear

q is the diametral quotient ($= d_1/m$)

(generally its value varies between 6 and 13)

d_1 is the pitch circle diameter of the worm

m is the module.

For higher mechanical advantage, the worm is cut with single thread. However for higher efficiency, the multistart worm is generally used. Figure 12.22 shows the nomenclature of the worm and the worm gear.

Axial pitch (p_x). It is the distance measured axially between two corresponding points on adjacent teeth. This is equal to transverse circular pitch (p_{cir}) of the mating worm gear.

$$p_x = p_{cir} = \frac{\pi d_2}{Z_2} \tag{12.44}$$

where d_2 is the pitch circle diameter of the worm gear.

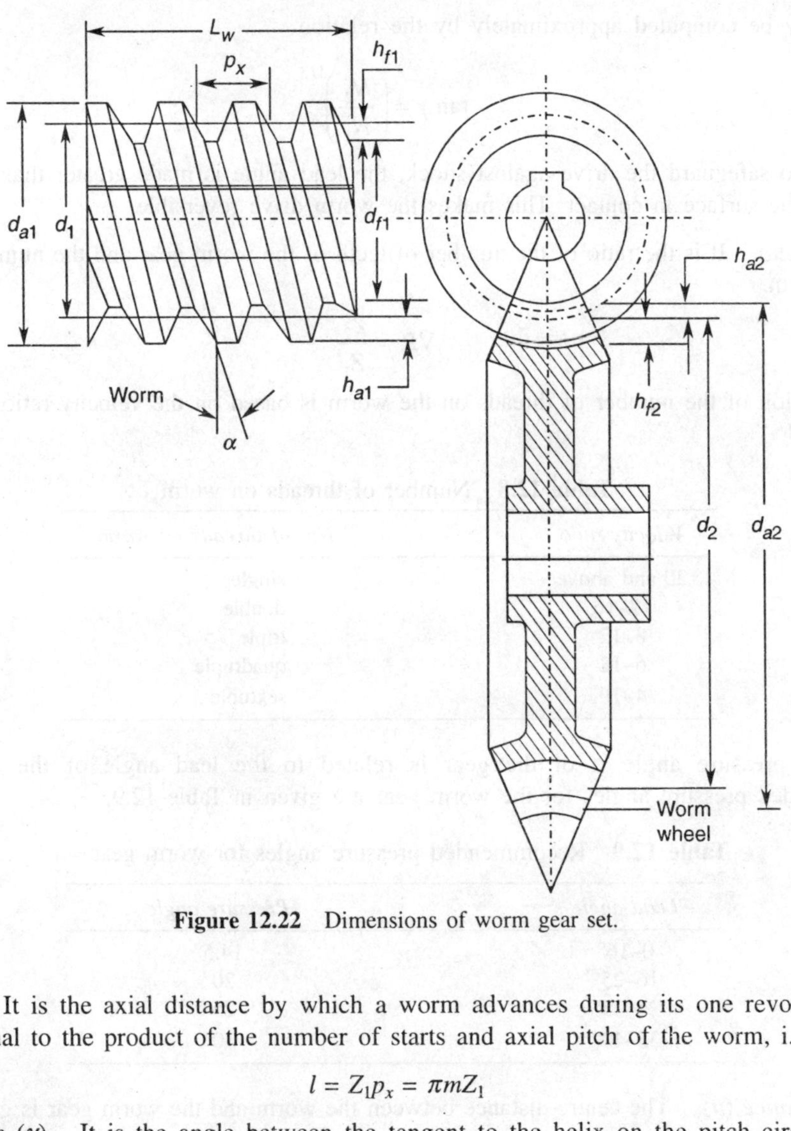

Figure 12.22 Dimensions of worm gear set.

Lead (l). It is the axial distance by which a worm advances during its one revolution. The lead is equal to the product of the number of starts and axial pitch of the worm, i.e.

$$l = Z_1 p_x = \pi m Z_1 \tag{12.45}$$

Lead angle (γ). It is the angle between the tangent to the helix on the pitch circle and the plane normal to the worm axis. It is equal to

$$\gamma = \frac{\pi}{2} - \text{helix angle}$$

For a shaft with 90° shaft angle, the lead angle is equal to the helix angle.

$$\tan \gamma = \frac{l}{\pi d_1} = \frac{Z_1}{q} \tag{12.46}$$

The lead angle of a worm may vary from 7° to 45°. Past experience has shown that lead angle less than 9° results in rapid wear, therefore, a safe value is 13°. For a compact design, the lead angle may be computed approximately by the relation

$$\tan\gamma = \left(\frac{N_2}{N_1}\right)^{1/3} \tag{12.47}$$

In order to safeguard the drive against shock, the lead angle is made greater than the friction angle of the surface in contact. This makes the worm drive reversible.

Velocity ratio. It is the ratio of the number of teeth of the worm gear and the number of starts of the worm.

$$\text{VR} = \frac{Z_2}{Z_1} \tag{12.48}$$

The selection of the number of threads on the worm is based on the velocity ratio as given in Table 12.8.

Table 12.8 Number of threads on worm

Velocity ratio	*No. of threads on worm*
20 and above	single
12–16	double
8–12	triple
6–12	quadruple
4–10	sextuple

The pressure angle α of the gear is related to the lead angle of the worm. The recommended pressure angles for the worm gear are given in Table 12.9.

Table 12.9 Recommended pressure angles for worm gear set

Lead angle	*Pressure angle*
0–16°	14.5°
16–25°	20°
25–35°	25°
35–40°	30°

Centre-distance (a). The centre-distance between the worm and the worm gear is given by the relation

$$a = 0.5(d_1 + d_2) \tag{12.49}$$

The AGMA has recommended an equation for mean pitch diameter of the worm, which is the function of the centre distance, i.e.

$$d_1 = \frac{a^{0.875}}{1.466} \tag{12.50}$$

The ends of the gear teeth are cut either parallel to the axis or radially towards the worm axis with a face angle (2δ) of 60–75°. This shape is used for worms with small lead angles.

The recommended face angle is given by

$$\tan \delta \le \frac{\tan \alpha}{\tan \gamma} \tag{12.51}$$

Parallel or straight teeth cut with a form cutter are not efficient and are used only for intermittent service and transmission of small powers. Standard proportions of worm and worm gear as recommended by AGMA for industrial use are given in Table 12.10.

Table 12.10 Standard proportions of worm gear set

Dimensions	*Single and double threads*	*Triple and quadruple threads*
Worm		
Pitch diameter, d_1 (for bored shaft)	$7.54m + 28$	$7.54m + 28$
Pitch diameter, d_1 (for integral shaft)	$7.39m + 10$	$7.39m + 10$
Face length, L_w	$14.14m + 0.063Z_1m$	$14.14m + 0.063Z_1m$
Tooth depth, h	$2.16m$	$1.96m$
Addendum, h_a	m	$0.9m$
Worm gear		
Normal pressure angle	14.5°	20°
Outside diameter, d_{o2}	$d_2 + 3.1854m$	$d_2 + 2.7982m$
Throat diameter, d_{th}	$d_2 + 2m$	$d_2 + 1.7978m$
Face width, b	$7.48m + 6.35$	$6.758m + 5.08$

12.9.2 Force Analysis of Worm Gear Set

Forces acting on the worm gear pair are shown in Figure 12.23(b). The resultant force F acting on the worm has two components:

$$\text{Normal force, } F_n = F \cos \alpha \tag{12.52a}$$

$$\text{Radial force, } F_r = F \sin \alpha \tag{12.52b}$$

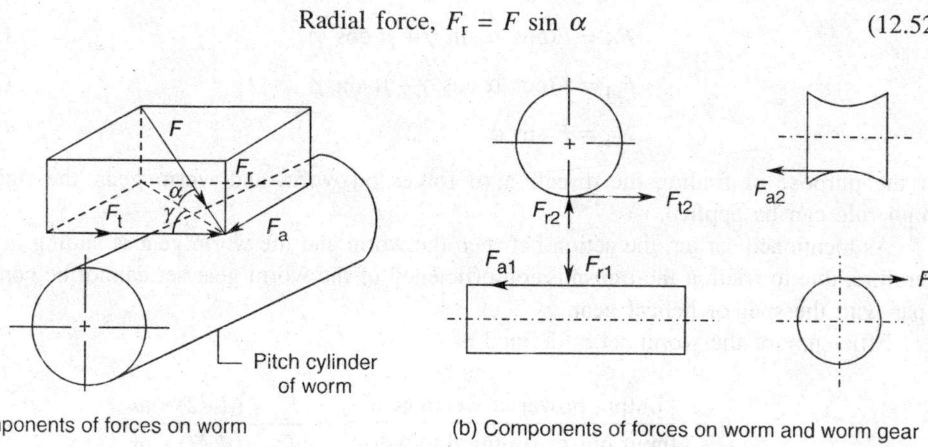

(a) Components of forces on worm (b) Components of forces on worm and worm gear

Figure 12.23 Force analysis of worm and worm gear.

The normal force can be further resolved into two components. Hence, the following three types of forces act on the worm gear as shown in Figure 12.23(a):

$$\text{Tangential force, } F_t = F \cos \alpha \sin \gamma \tag{12.53a}$$

$$\text{Axial force, } F_a = F \cos \alpha \cos \gamma \tag{12.53b}$$

$$\text{Radial force, } F_r = F \sin \alpha \tag{12.53c}$$

In the worm gear drive besides the above forces, the frictional force is quite significant due to sliding motion between the worm and the worm gear. The frictional force μF acts along the pitch helix and opposite to the direction of motion as shown in Figure 12.24. This

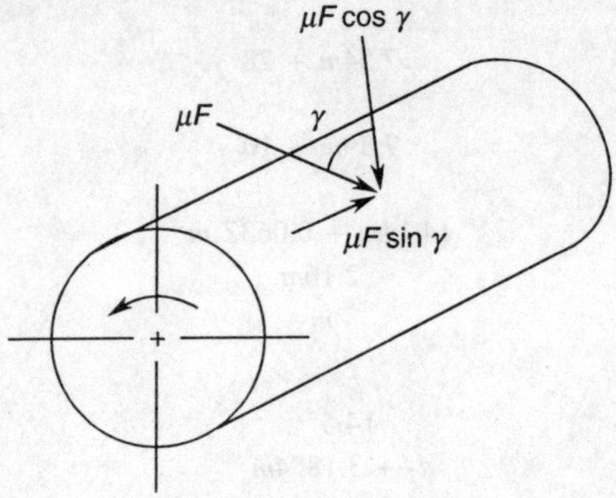

Figure 12.24 Components of frictional forces on worm.

frictional force has two components:

Along the tangential direction, $\mu F \cos \gamma$

Along the axial direction, $\mu F \sin \gamma$

Therefore, the total forces acting along the tangential and axial directions are:

$$F_{t1} = F(\cos \alpha \sin \gamma + \mu \cos \gamma) \tag{12.54a}$$

$$F_{a1} = F(\cos \alpha \cos \gamma - \mu \sin \gamma) \tag{12.54b}$$

$$F_{r1} = F \sin \alpha \tag{12.54c}$$

For the purpose of finding the direction of forces on worm and worm gear, the right-hand thumb rule can be applied.

As mentioned earlier, the action between the worm and the worm gear is sliding in nature. Therefore, due to friction the transmission efficiency of the worm gear set cannot be considered at par with the spur or helical gear.

Efficiency of the worm set is defined as

$$\eta = \frac{\text{output power at worm gear}}{\text{Input power supplied to worm}} = \frac{F_{t2} \times (d_2/2) \times \omega_2}{F_{t1} \times (d_1/2) \times \omega_1}$$

Substituting the value of the tangential force F_t, we get

$$\eta = \frac{\cos\theta - \mu\tan\gamma}{\cos\theta + \mu\cot\gamma} \tag{12.55}$$

where $\tan\theta = \tan\alpha\cos\gamma$

In the worm gear, the coefficient of friction is generally expressed in terms of rubbing or sliding velocity, which is defined as the velocity along the direction of sliding of the worm and worm gear teeth. Thus,

$$v_{\text{rubbing}} = \frac{\pi d_1 N_1}{\cos\gamma} \tag{12.56}$$

The coefficient of friction in terms of the rubbing velocity is given by the relations

$$\mu = \frac{0.0422}{(v_{\text{rubbing}})^{0.28}} \quad \text{for } 0.2 \le v_{\text{rubbing}} \le 2.8 \text{ m/s} \tag{12.57a}$$

$$\mu = 0.025 + \frac{3.281}{1000} v_{\text{rubbing}} \quad \text{for } v_{\text{rubbing}} > 2.8 \text{ m/s} \tag{12.57b}$$

12.9.3 Design of Worm Gear

The engagement tooth in the worm set is through the worm thread sliding into contact with the worm gear teeth. Thus the dynamic force on the worm gear is not so severe. The high sliding velocity between the worm and the worm gear causes excessive wear and produces heat due to friction. Therefore, besides strength considerations, the wear and heat dissipation are the major criteria for the design of the worm gear.

Beam strength

From the layout of the worm gear set, it is quite clear that the teeth of the worm gear are weaker than the threads on the worm. Therefore, the design criterion should be the worm gear. The Lewis equation for beam strength of worm gear is used to compute the load carrying capacity. The equation is

$$F_{\text{beam}} = bmY\sigma_d \tag{12.58}$$

The above equation is based upon the assumption that the entire load is taken by one tooth. The Lewis form factor Y may be computed on the basis of spur gear. However, if the number of teeth in the worm gear plus the number of teeth (threads) in 25 mm length of worm is greater than 40, the Lewis form factor Y can be determined from the equation

$$Y = 0.314 + 0.015(\alpha - 14.5°) \tag{12.59}$$

For safe design, the beam strength of the tooth computed by Eq. (12.58) should be greater than the maximum tangential force on the worm gear, which may be computed by the relation

$$F_{\max} = \frac{2T}{d_2} \times \frac{C_S}{C_v} \tag{12.60}$$

where C_v (dynamic load correction or velocity factor) $= \dfrac{6.1}{6.1 + v}$

Wear strength

The limiting wear load capacity of the worm gear is estimated by the Buckingham's equation

$$F_{wear} = d_2 bK \tag{12.61}$$

where

d_2 is the pcd of the worm gear

b is the face width

K is the load stress factor, a constant which depends upon the material used for the worm and worm gear (see Table 12.11).

Table 12.11 Load stress factor for wear load

Material		*Load stress factor, K*		
Worm	*Gear*	γ = 0–10°	γ = 10–25°	γ > 25°
Steel	Phosphor bronze	0.412	0.517	0.618
Hardened steel	Phosphor bronze	0.549	0.687	0.824
Hardened steel	Chilled bronze	0.824	1.03	1.236
Hardened steel	Antimony bronze	0.824	1.03	1.236
CI	Phosphor bronze	1.03	1.285	1.746

Thermal capacity

The efficiency of the worm gear set is quite low. Therefore, considerable amount of power is lost in friction which in turn is converted into heat due to frictional resistance. To prevent overheating of the drive, heat generated should not exceed the rate of heat transfer.

The rate of heat generated is given by

$$Q = 1000(1 - \eta) \times P \tag{12.62}$$

where

η is the efficiency of the worm gear set

P is the input power

This heat is dissipated by the housing through convection and radiation. The heat dissipation capacity of the housing depends upon its area, temperature difference conductivity of the material, and velocity of air, etc. The rate of heat dissipation can be estimated by the equation

$$Q = \frac{0.407}{10^3}(A_{gear} + A_{worm})(t_2 - t_1) \tag{12.63}$$

where

A_{worm} is the projected area of the worm $(= l_{worm} d_1)$

A_{gear} is the area of the gear $\left(= \frac{\pi}{4} d_2^2\right)$

t_2 is the gear temperature

t_1 is the room temperature

The gear temperature t_2 must not exceed 80°C, otherwise the lubricating oil may become thin and loose its properties and the worm gear may fail due to seizure.

Example 12.6 Design a worm gear speed reducer to transmit 22.5 kW at 1440 rpm. The desired speed ratio is 24:1 and efficiency 85%. Assume that the worm is made of hardened steel and gear of phosphor bronze.

Solution For efficiency of the order of 85%, a triple- or quadruple-threaded worm should be selected. We assume a quadruple start worm. Therefore, $Z_1 = 4$.

We also assume that the normal pressure angle α is 20° and the lead angle γ is 25°, which is the maximum allowed for the 20° pressure angle.

$$\text{Number of teeth on gear, } Z_2 = \text{speed ratio} \times Z_1 = 24 \times 4 = 96$$

$$\text{Speed of the worm gear, } N_2 = \frac{N_1}{24} = \frac{1440}{24} = 60 \text{ rpm}$$

$$\text{Pitch line velocity, } v = \frac{\pi d_2 N_2}{60 \times 1000} = \frac{\pi m Z_2 \times N_2}{60 \times 1000}$$

$$= \frac{\pi \times m \times 96 \times 60}{60 \times 1000} = 0.3016m \text{ m/s}$$

Tangential force

$$F_t = \frac{P \times 1000}{v} = \frac{22.5 \times 1000}{0.3016m} = \frac{74{,}602.1}{m}$$

$$\text{Maximum tangential force, } F_{max} = \frac{C_S F_t}{C_v}$$

$$\text{Let } C_S = \text{service factor} = 1.0$$

$$C_v = \text{velocity factor} = 0.5 \text{ (assumed)}$$

Therefore,

$$F_{max} = \frac{1.0 \times 74{,}602.1}{0.5 \times m} = \frac{149{,}204.2}{m} \text{ N}$$

$$\text{Lewis beam strength, } F_{beam} = bmY\sigma_d$$

$$\text{For phosphor bronze, } \sigma_d = 80 \text{ N/mm}^2$$

$$\text{Lewis form factor, } Y = \pi \times y = \pi \times 0.125 = 0.3927$$

$$\text{Face width, } b = (6.758m + 5.08) \text{ (refer Table 12.10)}$$

Therefore,

$$F_{beam} = (6.758m + 5.08)m \times 0.3927 \times 80$$

$$= (6.758m + 5.08) \times 31.416m$$

For safe design, the beam strength should be greater than the maximum tangential force. Thus,

$$(6.758m + 5.08) \times 31.416m = \frac{149{,}204.2}{m}$$

or

$$212.3m^3 + 159.6m^2 - 149{,}204.2 = 0$$

By trial and error, we get m = 8 to 9 mm, say, 9 mm

Therefore,

$$\text{Tangential force, } F_t = \frac{74{,}602.1}{9} = 8290 \text{ N}$$

$$\text{Pitch diameter, } d_2 = mZ_2 = 9 \times 96 = 864 \text{ mm}$$

$$\text{Pitch line velocity, } v = \frac{\pi d_2 N_2}{60} = \frac{\pi \times 0.864 \times 60}{60} = 2.714 \text{ m/s}$$

$$\text{Actual velocity factor, } C_v = \frac{3.05}{3.05 + 2.714} = 0.529$$

which is greater than the assumed value (0.5). Thus,

$$F_{max} = \frac{1.0 \times 8290}{0.529} = 15{,}671 \text{ N}$$

$$F_{beam} = (6.758 \times 9 + 5.08) \times 31.416 \times 9 = 18{,}633.4 \text{ N}$$

As $F_{beam} > F_{max}$, the design is safe under static load conditions.

$$\text{Pitch of the worm} = \pi m = 28.27 \text{ mm}$$

$$\text{Lead} = Z_1 \times \text{pitch} = 4 \times 28.27 = 113 \text{ mm}$$

$$\text{Face width, } b = 6.758 \times 9 + 5.08 = 65.9 \text{ mm}$$

$$\text{Diameter of worm, } d_1 = 7.54m + 28 = 95.86 \text{ mm}$$

$$\text{Face length of worm, } l_{worm} = (14.14 + 0.063 \times Z_1) \times m = 129.52 \text{ mm}$$

$$\text{Lead angle, } \gamma = \tan^{-1}\left(\frac{113}{\pi \times 95.86}\right) = 20.56°$$

$$\text{Rubbing velocity, } v_{rubbing} = \frac{\pi d_1 N_1}{60 \times 1000 \times \cos\gamma} = \frac{\pi \times 95.86 \times 1440}{60 \times 1000 \times \cos 20.56°} = 7.72 \text{ m/s}$$

$$\text{Coefficient of friction, } \mu = 0.025 + \frac{3.281}{1000} \times v_{rubbing}$$

$$= 0.025 + \frac{3.281}{1000} \times 7.72 = 0.05$$

Efficiency of the drive when worm is the driver

$$\eta = \frac{\cos\theta - \mu \tan\gamma}{\cos\theta + \mu \cot\gamma}$$

where

$$\tan\theta = \tan\alpha \cos\gamma$$

or

$$\theta = \tan^{-1}(\tan 20° \times \cos 20.56°) = 18.81°$$

Therefore,

$$\eta = \frac{\cos 18.81° - 0.05 \times \tan 20.56°}{\cos 18.81° + 0.05 \times \cot 20.56°} = 85.91\%, \text{ which is satisfactory}$$

$$\text{Heat generated} = (1 - \eta) \times \text{input power}$$

$$= (1 - 0.8591) \times 22.5 = 3.17 \text{ kW}$$

$$\text{Heat dissipated} = \frac{0.407}{10^3} (A_{gear} + A_{worm}) (t_2 - t_1)$$

$$A_{gear} = \text{area of the gear} = \frac{\pi}{4} d_2^2 = \frac{\pi}{4} \times 864^2 = 586{,}296.6 \text{ mm}^2$$

$$A_{worm} = \text{worm projected area} = l_{worm}\, d_1 = 129.5 \times 95.86 = 12{,}413.87 \text{ mm}^2$$

$$\text{Temperature rise, } \Delta T = \frac{\text{heat generated}}{\frac{0.407}{10^3} \times (A_{gear} + A_{worm})}$$

$$= \frac{3.17 \times 1000}{\frac{0.407}{10^3}(586{,}296.6 + 12{,}413.87)} = 13°\text{C}$$

which is quite low compared to the permissible temperature rise.

Wear load.

The limiting wear load

$$F_{wear} = d_{gear}\, bK$$

$$K = \text{load stress factor} = 0.687 \text{ for hardened steel and bronze}$$

or

$$F_{wear} = 864 \times 65.9 \times 0.687 = 39{,}116.1 \text{ N}$$

which is larger than F_{max}. Hence the design is safe.

12.10 COMPUTER-AIDED DESIGN OF GEAR TOOTH

The program 12.1 named "sgear.cpp" is a C++ program for the design of a spur gear pair. This program requires the following input data: (i) power, (ii) pinion speed, (iii) gear speed, (iv) number of teeth on pinion, (v) ratio of face width to module, (vi) strength of gear and pinion material and their elastic modulus, (vii) core hardness, and (viii) service factor according to operating conditions.

The static design is performed using the Lewis beam strength equation. For dynamic load consideration, the data for profile error given in Tables 12.5 and 12.6 was converted into equations using the least square curve fitting method of Microsoft Excel 2000. The indication of the quality of the curve fit is determined by the so-called r-squared value, which is defined as

$$r^2 = 1 - (\text{SSE/SST})$$

where

SSE is the sum of the squares of the errors

SST is the sum of the squares of the deviation about the mean

The fitted curve equations are given as under:

(i) Curve fitted between profile error *e* and pitch line velocity *v*

$$e = 0.1055\ e^{-0.0974v} \quad (12.64)$$
$$r^2 = 0.9986$$

(ii) Equations between profile error and module *m* for a particular class of manufacturing

(a) For gears of class I

$$e = 0.044 \ln (m) - 0.0132 \quad (12.65)$$
$$r^2 = 0.9925$$

(b) For gears of class II

$$e = 0.0219 \ln (m) - 0.0066 \quad (12.66)$$
$$r^2 = 0.9911$$

(c) For gears of class III

$$e = 0.0104 \ln (m) - 0.0017 \quad (12.67)$$
$$r^2 = 0.9961$$

The calculated dynamic load is compared with Lewis endurance load capacity. The design of the gear is also checked for the limiting wear load capacity.

A sample run of Program 12.1 is also given.

Program 12.1

```
//    sgear.cpp A PROGRAM FOR DESIGN OF SPUR GEAR
# include <iostream.h>
# include <math.h>
# include <conio.h>
//    define parameters
#define KN 20.67
#define KM    9.00
#define PI    3.1415
double modu (double module);
void main ( )
{
      int       zp, zg, gear_class;
      double    p, np, ng, v, sigmadp, sigmadg, sigmad, dp, dg, bbm, yp, yg;
      double    pinion_strength, gear_strength, module, m, lewis_factor;
      double    ft, fmax, fd, beam_strength, ep, eg, bhnc, bhns, cs, cv;
      double    emax, error, c, b, cbft, corebhn, sigmaen, ratio_fac, k, sigmaes;
      char      design_basis;
//    input of design variables
      cout << "\n    Power .......................................... P (kW) = ";
      cin >> p;
      cout << "\n    Pinion Speed .................................... NP (RPM) = ";
```

```
       cin   >> np;
       cout  << "\n    Gear Speed ........................................ NG (RPM) = ";
       cin   >> ng;
       cout  << "\n    Number of Teeth on Pinion ................................... ZP  = ";
       cin   >> zp;
       cout  << "\n    Face Width to Module Ratio ............................ BBM  = ";
       cin   >> bbm;
input: cout<< '\n    Strength of Pinion Material ........ SIGMADP (N/mm2) = ";
       cin   >> sigmadp;
       cout  << "\n    Strength of Gear Material .......... SIGMADG (N/mm2) = ";
       cin   >> sigmadg;
       cout  << "\n    Elastic Modulus of Pinion Material ........ EP (N/mm2) = ";
       cin   >> ep;
       cout  << "\n    Elastic Modulus of Gear Material .......... EG (N/mm2) = ";
       cin   >> eg;
       cout  << "\n    Core BHN of Material ...................................... BHNC  = ";
       cin   >> bhnc;
       cout  << "\n    Service Factor ................................................ Cs = ";
       cin   >> cs;
//     calculate the number of teeth on gear Lewis form factor and
//     decide design basis
       zg=int (zp*np/ng+0.5);
       yp=PI* (0.154–0.912/zp);
       yg=PI* (0.154–0.912/zg);
       pinion_strength=sigmadp*yp;
       gear_strength=sigmadg*yg;
       if (pinion_strength > gear_strength)
              {
                     design_basis='g';
              }
       else
              {
                     design_basis='p';
              }
       module=1.0;
repeat:
       if (design_basis = = 'g')
              {
                     v=PI*module*zg*ng/(60.0*1000.0);
                     lewis_factor=yg;
                     sigmad=sigmadg;
              }
       else
              {
                     v=PI*module*zp*np/(60.0*1000.0);
                     lewis_factor=yp;
                     sigmad=sigmadp;
              }
```

```
//      Compute velocity factor Cv and maximum force on gear tooth
        if (v<=8.0)
                cv=3.05/(3.05+v);
        else
                if (v<=13.0)
                        cv=4.58/(4.58+v);
        else
                if (v<=20.0)
                        cv=6.1/(6.1+v);
                else
                        cv=5.55/(5.55+sqrt(v));
        ft=p*1000.0/v;
        fmax=cs*ft/cv;
                beam_strength=bbm*module*module*lewis_factor*sigmad;
                if (beam_strength < fmax)
                {
                        m=modu (module);
                        module=m;
                        goto repeat;
                }
//      check for dynamic load on gear
        emax=0.1055*exp (-0.0974*v);
        error=0.044*log (module)-0.0132;
        if (emax > error)
        {
                gear_class=1;
        }
        else
        {
                error=0.0219*log(module)-0.0066;
                if (emax > error)
                {
                gear_class=2;
                }
                else
                {
                error=0.0104*log(module)-0.0017;
                gear_class=3;
                }
        }
        c=error*ep*eg/(KM* (ep+eg));
        b=bbm*module;
        cbft=c*b+ft;
        fd=ft + (KN*v*cbft) / (KN*v+sqrt (cbft));
        sigmaen=1.25*fd/(b*module*lewis_factor);
```

```
        corebhn=sigmaen/1.75;
        if (corebhn > bhnc)
        {
                goto input;
        }
//      check wear load
        ratio_fac=2.0*zg/(zp+zg);
        dp=zp*module;
        k=1.5*fd/(dp*b*ratio_fac);
        sigmaes=sqrt (1.4*k*ep*eg/(sin(PI*20/180)*(ep+eg)));
        bhns=(sigmaes+70)/2.75;
        print design specifications
        dg=module*zg;
        cout << "\n              DESIGN SPECIFICATIONS" <<endl;
        cout << "\n      Number of Teeth on Pinion =" << zp;
        cout << "\n      Number of Teeth on Gear =" << zg;
        cout << "\n      Module =" << module << "mm";
        cout << "\n      Face Width        =" << b << "mm";
        cout << "\n      Pitch Circle Diameter of Pinion =" << dp << "mm";
        cout << "\n      Pitch Circle Diameter of Gear =" << dg << "mm";
        cout << "\n      Addendum =" << module << "mm";
        cout << "\n      Dedendum =" << 1.25*module << "mm";
        cout << "\n      Working Depth  =" << 2.0*module << "mm";
        cout << "\n      Total Depth       =" << 2.25*module << "mm";
        cout << "\n      Surface BHN       =" <<bhns << endl;
}
//      function to find module
double modu (double mm)
{
        if (mm <3.0)
                mm=mm+0.5;
        else
                if (mm < 6.0)
                        mm=mm+1.0;
                else
                        if (mm <12)
                                mm=mm+2.0;
                        else
                                if (mm <20.0)
                                        mm=mm+4.0;
                                else
                                        goto stop;
        return(mm);
stop:
        cout << "\n Required Module is more than 20mm";
        return (mm);
}
```

Sample Run of Program 12.1

Power P(kW) =15
Pinion Speed NP(RPM) =1440
Gear Speed NG(RPM) =500
Number of Teeth on Pinion ZP =25
Face Width to Module Ratio BBM =10
Strength of Pinion Material SIGMAP(N/mm2)=55
Strength of Gear Material SIGMAG(N/mm2)=55
Elastic Modulus of Pinion Material EP(N/mm2)=0.1e06
Elastic Modulus of Gear Material EG(N/mm2)=0.1e06
Core BHN of Material BHNC=180
Service Factor Cs=1.25

DESIGN SPECIFICATIONS
Number of Teeth on Pinion =25
Number of Teeth on Gear =72
Module =6mm
Face Width =60mm
Pitch Circle Diameter of Pinion =150mm
Pitch Circle Diameter of Gear =432mm
Addendum =6mm
Dedendum =7.5mm
Working Depth =12mm
Total Depth =13.5mm
Surface BHN =330

EXERCISES

1. Discuss the various types of gear tooth failure. Explain how will you take care of the failure in your design?
2. What are the advantages and disadvantages of using involute profile vis-à-vis cycloidal profile?
3. Define contact ratio. What are its limits?
4. Define the virtual number of teeth for a helical gear.
5. What is interference in involute profile?
6. Compare the 14.5° and 20° pressure angle systems in gear drives.
7. What are the advantages of the helical gear over the spur gear?
8. What is the usual cross-section of the arms of a gear and why?
9. Discuss the suitability of CI as the gear material.
10. What is diametral quotient?
11. Why is the worm made of hardened steel?
12. Why is multistart worm more efficient than the single start one?
13. Why is dynamic load not a problem in worm gears?
14. Why is heat balance an important aspect in the design of worm gears?

15. A certain gear pair in a machine tool gear box is required to transmit 7.5 kW. The driving gear runs at 600 rpm and the speed reduction ratio is 3. Design and draw the gear pair and the shaft.
16. A reciprocating compressor is required to be driven by an electric motor running at 800 rpm through a pair of spur gears. The compressor is to be run at about 200 rpm and requires a torque of 6 kN · m with a starting overload of 25 per cent. Design the gear drive with 20° full-depth tooth. Take the same material for both pinion and gear.
17. Design completely a pair of spur gears required to drive an air compressor with the following specifications:

    Power = 5 kW
    Motor speed = 1400 rpm
    Compressor speed = 300 rpm

    Select a suitable material for pinion and gear. Specify their heat treatment requirements as well.
18. The velocity ratio of two gears is 1:4. They are arranged to transmit 30 kW at speed of 250 rpm of the pinion and are made of 20° full-depth involute teeth. The pinion is made of cast steel, while the gear is made of cast iron. Design a gear pair.
19. Design a pair of helical gears for transmitting 20 kW. The pinion rotates at 1800 rpm and the gear at 400 rpm.
20. A pair of bevel gears is to be used to transmit 14 kW from a pinion rotating at 400 rpm to a gear mounted on a shaft running at 200 rpm. The axes of the two shafts are at 90°. Design the pair of bevel gears.
21. A pair of bevel gears is to be used to transmit 11 kW from a pinion rotating at 500 rpm to a gear mounted on a shaft which intersects the pinion shaft at an angle of 120°. The speed reduction ratio is 3. The gear is mounted on a 225 mm long shaft and at a distance of 100 mm from the left-hand bearing. Design the gear drive.
22. Design and sketch a suitable drive along with the shaft and the key for an input of 11 kW at 1440 rpm and speed reduction of 27.

MULTIPLE CHOICE QUESTIONS

1. In an involute gear, the base circle must be ______.

   (a) at root circle (b) under root circle
   (c) above root circle (d) above pitch circle

2. For the spur gear, the product of the circular pitch and diametral pitch is equal to

   (a) unity (b) $1/\pi$
   (c) π (d) module

3. In which type of teeth, variation in centre-distance within limits does not affect the velocity ratio of the mating gears.

   (a) Cycloidal (b) Involute
   (c) Hypoid (d) None of the above

4. Lewis equation in gears is used to find the
   (a) tensile stress (b) fatigue stress
   (c) contact stress (d) bending stress
5. Low pressure angle gears
   (a) have stronger teeth (b) have weaker teeth
   (c) produce no effect (d) have none of the above
6. The value of the form factor depends upon
   (a) the number of teeth (b) the pressure angle
   (c) both (a) and (b) (d) the beam strength
7. Interference is inherently absent in the following type of gears.
   (a) Involute (b) Cycloidal
   (c) Stub (d) Hypocycloidal
8. In helical gears, the right-hand helix gear will mesh with
   (a) right-hand helix (b) left-hand helix
   (c) both (a) and (b) (d) zero helix
9. If both pinion and gear are made of the same material, then the load transmission capacity is decided by
   (a) the gear (b) the pinion
   (c) both (a) and (b) (d) none of the above
10. Surface hardness of the gear material is helpful in
    (a) static mode of design (b) dynamic mode of design
    (c) wear mode of design (d) all of the above
11. The initial contact in a helical gear is
    (a) a point (b) a line
    (c) a surface (d) unpredictable
12. The pressure angle recommended by BIS for gears is
    (a) 14.5° (b) 20°
    (c) 25° (d) 30°
13. Diametral quotient is defined as
    (a) axial module/reference diameter
    (b) pitch diameter/module
    (c) module/pitch diameter
    (d) pitch/pitch diameter
14. In the miter bevel gear set
    (a) the axes of gears are at 90°
    (b) the gears are of the same size
    (c) both (a) and (b)
    (d) the axes of gears are at more than 90°

Chapter 13 Shafts, Keys and Splines

13.1 INTRODUCTION

A shaft is a rotating member which transmits power. It is one of the most common and basic machine elements which is used in a variety of ways in all kinds of mechanical equipment, for example, power shaft, cam shaft, line shaft, etc. Power shafts and line shafts are designed to transmit torque and to support rotating members such as pulleys, gears, and flywheels, etc. These elements produce bending moment in addition to torque. A shaft must not only be strong enough to sustain static and dynamic stresses but also be sufficiently rigid to prevent harmful torsional and lateral deflections. Shafts are classified as follows according to their industrial applications.

Line shaft. A shaft used to transmit power to several machines is called the line shaft.

Spindle. A short shaft is called spindle, for example, head stock spindle of a lathe machine.

Stub shaft. A shaft that is integral with an engine, motor or prime mover is known as the stub shaft.

Counter shaft. A short shaft that connects a prime mover to a line shaft or a machine is known as counter shaft.

Flexible shaft. A shaft which permits transmission of motion between two points where rotational axes are at an angle with respect to each other. The power transmission capacity of such a shaft is relatively small.

13.2 SHAFT MATERIALS

Commercial power transmitting shafts up to 75 mm diameter are commonly made of cold drawn, low carbon steel with carbon content 0.1 to 0.35 per cent. A cold drawn steel shaft is somewhat stronger than a rolled steel shaft, however, it has some disadvantages in that the shaft does not remain precisely straight and the fibres near the outer surface remain under stress. These stresses are released at the time of cutting keyways with the result that the shaft gets distorted. Due to this reason, shafts of diameters more than 75 mm are generally made of hot rolled or forged steel. Proper finish is obtained through turning, grinding and other finishing operations. Machining must be deep enough to remove all of the decarburized scale caused by hot rolling. Heat treatment using carburizing, nitriding, cyaniding, flame and induction hardening methods can be used to obtain the desired mechanical properties including producing the wear resistance surface. The shaft material should conform to IS: 1570 – 1978.

For high-speed machinery where toughness, shock resistance and greater strength are needed, shafts are made of alloy steel. The most commonly used alloying materials are nickel, chromium, and vanadium. Alloy steel shafts are always heat treated to obtain the desired mechanical properties. For spline shafts subjected to severe shock conditions, oil quenched SAE 3245 steel is particularly recommended. Shafts for special purpose, those having integral flange, are generally forged. Commercial shafts are available in the following preferred sizes as per IS: 3688 – 1977. 8, 10, 12, 14, 16, 18, 20, 22, 25, 28, 30, 36, 40, 45, 50, 60, 63, 65, 70, 71, 75, 80, 85, 90, 95, 100, 110, 122, 140, 160, 180, 200, 220, 260, 280, and 300 mm.

13.3 DESIGN OF SHAFTS AGAINST STATIC LOADING

Shaft, being a rotating member, is always subjected to a cyclic or fluctuating load but as a first approximation it may be treated as subjected to the static load condition. The load to which shafts are subjected may cause simple torsion, simple bending, combined torsion and bending or combined torsion, bending and axial loading with or without column effect.

13.3.1 Simple Torsion

When a shaft is subjected to pure torsional moment T, the shaft diameter can be found from the torsional shear strength equation

$$\text{Shear strength, } \tau = \frac{16T}{\pi d^3} \quad \text{for solid shafts} \tag{13.1}$$

$$= \frac{16T}{\pi d_o^3} \frac{1}{(1-K^4)} \quad \text{for hollow shafts} \tag{13.2}$$

where

T is the design torque (N · mm)

K is the ratio of inside diameter d_i to outside diameter d_o

τ is the allowable shear strength of the shaft material, which is taken as (0.5–0.577) × tensile yield strength.

13.3.2 Simple Bending Moment

Shafts are also subjected to transverse forces due to belt, pulley, gears, flywheel, etc. These forces may cause bending. For a given bending moment M and allowable tensile strength σ_d, the shaft diameter can be determined by the theory of simple bending. Accordingly,

$$\sigma_d = \frac{32M}{\pi d^3} \quad \text{for solid shafts} \tag{13.3}$$

$$= \frac{32M}{\pi d_o^3} \frac{1}{(1-K^4)} \quad \text{for hollow shafts} \tag{13.4}$$

While calculating bending moment, it is customary to measure each moment arm to the middle of the bearing. It is also assumed that the clearance between the shaft and the bearing allows the shaft to deflect up to the middle of the bearing. This is a reasonable assumption which gives the result on the safer side.

13.3.3 Combined Torque and Bending Moment

A rotating shaft transmitting power and carrying pulleys, gears, sprocket, etc. is subjected to combined torsion and bending moment. The combined action of bending and torsion is essentially a case of biaxial loading. Therefore, the design of such a shaft must be based on either the maximum shear stress or the maximum distortion energy theory of failure, being made of ductile material.

According to maximum shear stress theory, the following relation must be satisfied.

$$\tau_{max} = \sqrt{\left(\frac{\sigma_b}{2}\right)^2 + \tau^2} \le \frac{\sigma_y}{2\text{FoS}}$$

or

$$\tau_{max} = \frac{16}{\pi d_o^3 (1 - K^4)} \sqrt{M^2 + T^2} \tag{13.5}$$

According to maximum distortion energy theory for biaxial stress state, the following relation must be satisfied.

$$\sigma_1^2 - \sigma_1 \sigma_2 + \sigma_2^2 \le \left(\frac{\sigma_{yt}}{\text{FoS}}\right)^2 \tag{13.6}$$

where σ_1 and σ_2 are maximum and minimum principal stresses calculated by the relation

$$\sigma_{1,2} = \frac{\sigma_b}{2} \pm \sqrt{\left(\frac{\sigma_b}{2}\right)^2 + \tau^2} \tag{13.7}$$

13.4 ASME CODE FOR SHAFT DESIGN

Power transmission shafts are used so often that the American Society of Mechanical Engineers (ASME) has worked out a special procedure for their design. The ASME Code for the design of shafts was based on the maximum shear stress theory of failure and employed combined shock and fatigue factors as multiplying coefficients for the applied bending moment and torque. This code was later withdrawn in 1954 and has not been replaced since then. However, this code is still extensively used by many practising engineers. This code provides a good approximation when a shaft is subjected to shock load in addition to fatigue load. According to the ASME Code, the bending moment M and torque T in Eq. (13.5) are multiplied by combined shock and fatigue factors C_m and C_t, respectively, whose values depend upon the condition of a particular application. Therefore,

$$\tau_{max} = \frac{16}{\pi d_o^3 (1 - K^4)} \sqrt{(C_m M)^2 + (C_t T)^2} \tag{13.8}$$

The recommended values of C_m and C_t are listed in Table 13.1 for various load conditions. The ASME Code further states that 25 per cent reduction in strength should be done for a higher margin of safety, when the shaft has a keyway which produces stress concentration.

Table 13.1 Shock and fatigue factors for the ASME Code

Nature of loading	*Constant, C_m*	*Constant, C_t*
A. Stationary shaft		
Gradually applied load	1.0	1.0
Suddenly applied load	1.5–2.0	1.5–2.0
B. Rotating shaft		
Steady applied load	1.5	1.0
Minor shock load	1.5–2.0	1–1.5
Heavy shock load	2.0–3.0	1.5–3.0

13.5 COMBINED AXIAL FORCE, BENDING MOMENT, AND TORQUE

In some applications such as a shaft carrying a helical or bevel gear, the shaft is subjected to direct axial force in addition to torque and bending moment. When the length of the shaft is small, direct force will produce either tensile or compressive stresses only. However, when the length of the shaft is quite large, due to direct force, it may behave like a column and hence start buckling.

If the axial force F is presented and buckling of the shaft is accounted for, then the buckling stress is given by

$$\sigma_x = -\frac{4F\alpha}{\pi d_o^2 (1 - K^2)}$$

The bending and torsional shear stress relations can be written as

$$\sigma_y = \frac{32M}{\pi d_o^3 (1 - K^4)} \quad \text{and} \quad \tau_{xy} = \frac{16T}{\pi d_o^3 (1 - K^4)}$$

For the two-dimensional biaxial stress element subjected to direct stress σ_x, bending stress σ_y, and torsional shear stress τ_{xy}, maximum shear stress is given by

$$\tau_{\max} = \frac{1}{2}\sqrt{(\sigma_x - \sigma_y)^2 + 4\tau_{xy}^2} \tag{13.9}$$

Substituting the values of σ_x, σ_y, and τ_{xy} in Eq. (13.9), we get

$$\tau_{\max} = \frac{16}{\pi d_o^3 (1 - K^4)}\sqrt{\left(\frac{\alpha d_o (1 + K^2) F}{8} + M\right)^2 + T^2} \tag{13.10}$$

where α is the ratio of maximum intensity of stress resulting from axial force to the average axial stress given by

$$= \frac{1}{1 - 0.0044\left(\dfrac{l}{k}\right)} \quad \text{when } \frac{l}{k} \le 115 \tag{13.11a}$$

$$= \frac{\sigma_{yc}}{n\pi^2 E}\left(\frac{l}{k}\right)^2 \quad \text{when } \frac{l}{k} > 115 \tag{13.11b}$$

with

l = length of the shaft
k = radius of gyration
σ_{yc} = yield stress in compression
n = constant for the type of column end conditions
= 1.0 for both ends hinged
= 2.25 for fixed ends
= 1.6 for both ends pinned, guided, and partly restrained.

13.6 SHAFT DESIGN FOR VARIABLE LOAD

A rotating shaft loaded by stationary bending moment and torsional moment (torque) is stressed by completely reversed bending stress because of shaft rotation but torsional stress remains steady. This type of fatigue loading is very common which occurs quite often and more than any other type of loading. However, in this section a generalized case of fatigue loading, where bending stress and torsional shear stress fluctuate between their maximum and minimum values, is considered as shown in Figure 13.1.

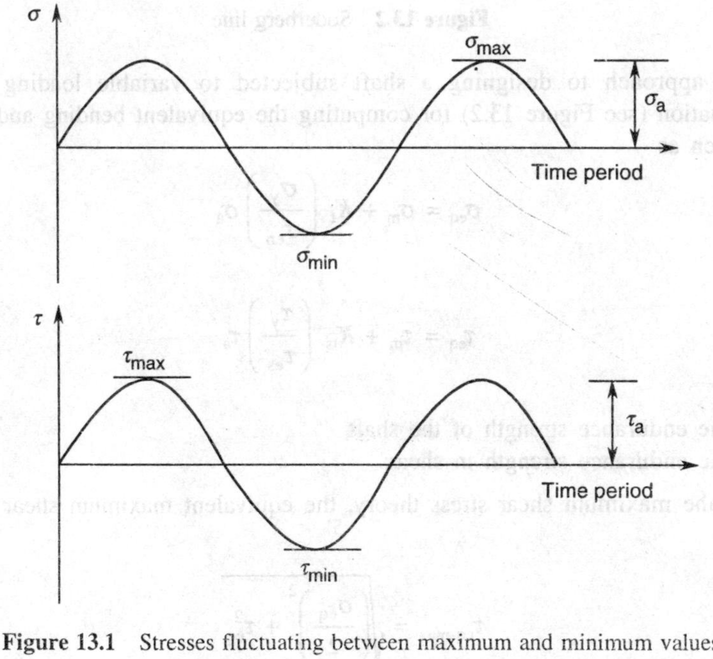

Figure 13.1 Stresses fluctuating between maximum and minimum values.

The following relations for mean stress σ_m and stress amplitude σ_a are evident from Figure 13.1.

$$\sigma_m = \frac{\sigma_{max} + \sigma_{min}}{2} \tag{13.12}$$

$$\sigma_a = \frac{\sigma_{max} - \sigma_{min}}{2} \tag{13.13}$$

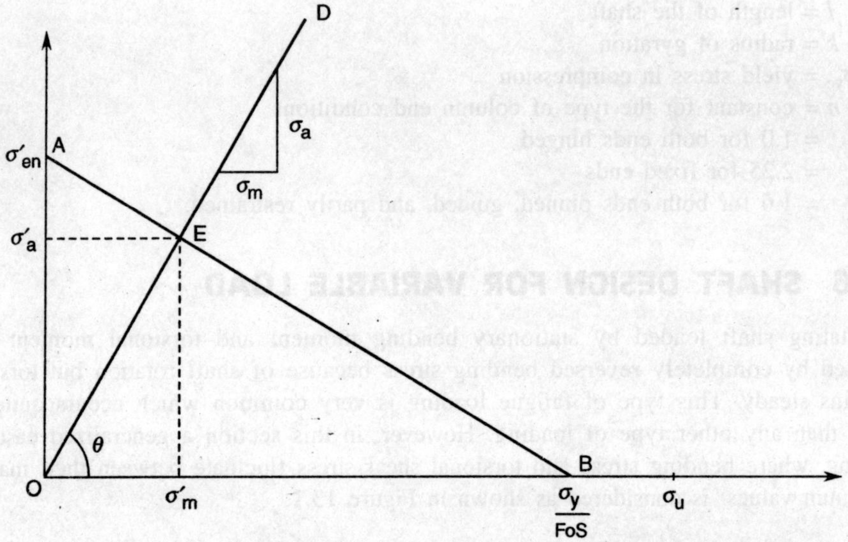

Figure 13.2 Soderberg line.

The simplest approach to designing a shaft subjected to variable loading is to use the Soderberg equation (see Figure 13.2) for computing the equivalent bending and shear stresses, which are given as

$$\sigma_{eq} = \sigma_m + K_t \left(\frac{\sigma_y}{\sigma_{en}}\right) \sigma_a \tag{13.14}$$

and

$$\tau_{eq} = \tau_m + K_{ts} \left(\frac{\tau_y}{\tau_{es}}\right) \tau_a \tag{13.15}$$

where

σ_{en} is the endurance strength of the shaft
τ_{es} is the endurance strength in shear.

According to the maximum shear stress theory, the equivalent maximum shear stress is given by

$$\tau_{eq\text{-}max} = \sqrt{\left(\frac{\sigma_{eq}}{2}\right)^2 + \tau_{eq}^2} \tag{13.16}$$

Substituting the values of σ_{eq} and τ_{eq} from Eqs. (13.14) and (13.15) into Eq. (13.16), we get

$$\tau_{\text{eq-max}} = \left[\frac{1}{4}\left(\sigma_m + \frac{K_t\sigma_y}{\sigma_{en}}\sigma_a\right)^2 + \left(\tau_m + \frac{K_{ts}\tau_y}{\tau_{en}}\tau_a\right)^2\right]^{0.5} \tag{13.17}$$

where $\tau_{es} = 0.5\sigma_{en}$ and $\tau_y = 0.5\sigma_y$.

Equation (13.17) can be transformed into moment M and torque T. For hollow shafts it can be reduced to

$$\tau_{\max} = \frac{16}{\pi d_o^3 (1-K^4)}\left[\left(M_m + \frac{K_t\sigma_y}{\sigma_{en}}M_a\right)^2 + \left(T_m + \frac{K_{ts}\sigma_y}{\sigma_{en}}T_a\right)^2\right]^{0.5} \tag{13.18}$$

Equation (13.18) does not take into account fatigue and shock factors. In some applications, e.g. punch press where heavy shock load acts, this equation may be modified to include fatigue and shock factors.

According to maximum distortion energy theory, the diameter of the shaft can be found from the following equation:

$$d_o = \left[\frac{16}{\pi\sigma_{\max}}\left(\sqrt{4M_m^2 + 3T_m^2} + \frac{K_t\sigma_y}{\sigma_{en}}\sqrt{4M_a^2 + 3T_a^2}\right)\frac{1}{1-K^4}\right]^{1/3} \tag{13.19}$$

13.7 EFFECT OF STRESS CONCENTRATION

A shaft in actual practice is frequently subjected to stress concentration caused by various reasons, such as clamping and shrinkage effect, keyways, oil holes, fillets, change in cross-section, and snap ring grooves. According to H.F. Moore, the static weakening effect of keyways is given by the formula

$$K_e = 1.0 - \frac{0.2b}{d} - \frac{1.1h}{d} \tag{13.20}$$

where

- K_e is the shaft strength factor or ratio of strength of shaft with the keyway to the same shaft without the keyway
- b is the width of the keyway
- h is the depth of the keyway
- d is the diameter of the shaft.

Alternatively, the stress–concentration factor for fatigue determined by various experiments can be used to modify the allowable strength. Typical values of stress-concentration factors for various types of stress-raisers vary from 1.2 to 2.0.

13.8 DESIGN OF SHAFT FOR RIGIDITY

In some applications, the components are required to be designed on the basis of rigidity rather than strength. Shaft is one such component. A shaft is generally designed to resist two types of deflections: (i) torsional deflection due to torque and (ii) lateral deflection due to transverse forces.

13.8.1 Shaft Design for Torsional Rigidity

The design of a shaft for torsional rigidity requires that it should be able to transmit power without causing excessive angular displacement which may contribute to vibrations and cause bearing failure. In a machine tool spindle, it may adversely affect the process capability of a machine tool.

Although no standard torsional deflection has ever been established for different shaft applications, it has become a standard practice to limit the torsional deflection to 1° in a length of 20 times the shaft diameter. In the case of cam shaft of I.C. engines, it should be less than 0.5° regardless of the shaft length. The shaft diameter can be determined from the following relation:

Angle of twist in degrees

$$\theta = \frac{584\,Tl}{Gd^4} \quad \text{for solid shafts} \tag{13.21a}$$

$$= \frac{584\,Tl}{G\left(d_o^4 - d_i^4\right)} \quad \text{for hollow shafts} \tag{13.21b}$$

where

l is the length of the shaft
G is the modulus of rigidity.

13.8.2 Shaft Design for Lateral Deflection

The design of a shaft for lateral deflection is another important criterion. The information about lateral deflection of a shaft due to various transverse forces is most important because it is used to establish the minimum permissible clearance between the pulley, gears and housing for the shaft assembly. The deflection at gear location may increase backlash between gear teeth, increase the pressure angle and reduce the length of contact. This results in uneven distribution of force which ultimately causes early failure. Deflection of shaft also results in misalignment between the axes of shaft and bearing, resulting in failure of the bearing. The minimum radial clearance for hydrodynamic bearing also depends on the amount of lateral deflection of the shaft. In practice, lateral deflection of a shaft can be controlled by the following measures:

(i) Reducing the span length
(ii) Increasing the number of supports
(iii) Selecting the cross-section in which the area moment of inertia is large such as in the case of a tubular or hollow shaft.

There are no clearly defined standards or restrictions concerning lateral deflection of shafting. In the absence of more specific information, the designer may use the following criteria:

(i) For machinery shafts, the maximum permissible deflection is generally taken as $(0.001–0.003)l$, where l is the span between two bearings.

(ii) For a shaft mounting a good quality spur gear, the deflection at the gear should not exceed $0.01m$ (module) and the slope should be limited to 0.0286°.

(iii) The permissible misalignment tolerance for deep groove ball bearing is ± 0.25°.

Various methods for the determination of lateral deflection of a shaft or a beam are extensively discussed in books on mechanics of materials. The reader is advised to refer to any standard book* on the subject for further details. However a brief review is presented in Chapter 4.

13.9 CRITICAL SPEED OF SHAFT

When shafts are manufactured and assembled, the centre of mass of a rotating shaft may not coincide with the actual centre of rotation due to practical limitations. When the rotational speed of such a shaft is increased, it starts vibrating violently. This speed of the shaft is called the *critical speed* or *natural frequency* of vibration. The critical speed of a shaft is a function of its stiffness and mass of the shaft and components being carried by it. Increasing the diameter of a shaft increases its stiffness by fourth power of diameter and its weight by only square; the critical speed is therefore increased. Whereas increasing the length of a shaft, increases its flexibility and lowers the critical speed.

In the case of a shaft of negligible mass carrying a concentrated mass, the force is proportional to the deflection of the mass from the equilibrium position and the relation for natural frequency can be expressed as

$$f_n = \frac{1}{2\pi}\sqrt{\frac{g}{\Delta}} \tag{13.22}$$

where

Δ = static deflection

$= \dfrac{mg\,a^2\,b^2}{3EI\,l}$ when the disc is mounted at a distance a from the left support and at b from the right support. The shaft is rotating in antifriction bearings and is treated as a simply supported beam.

$= \dfrac{mg\,a^3\,b^3}{3EI\,l^3}$ same as the above case with the condition that shaft is supported in journal bearings and it is assumed to be fixed ended.

When a shaft carries n discs, the critical speed of the shaft may be found, approximately, by the Dunkerley's equation

$$\frac{1}{\omega_c^2} = \frac{1}{\omega_1^2} + \frac{1}{\omega_2^2} + \cdots + \frac{1}{\omega_n^2} \tag{13.23}$$

where $\omega_1, \omega_2, \ldots, \omega_n$ are the critical speeds of the shaft when each mass is considered.

*Popov—*Engineering Mechanics of Materials*, 2nd ed., Prentice-Hall of India, New Delhi.

Example 13.1 A turbine shaft transmits 500 kW at 900 rpm. The permissible shear stress is 80 N/mm^2 while twist is limited to 0.5° in a length of 2.5 m. Calculate the diameter of shaft. Take $G = 0.8 \times 10^5$ N/mm^2. If the shaft choosen is hollow with $d_i/d_o = 0.6$, calculate the percentage saving in the material.

Solution

(A) *Solid shaft*

(i) *Strength criterion*

$$\text{Average torque transmitted by shaft, } T = \frac{60P}{2\pi N} = \frac{60 \times 500 \times 10^3}{2\pi \times 900}$$

$$= 5305.16 \text{ N} \cdot \text{m}$$

$$\text{Shaft diameter, } d = \left(\frac{16T}{\pi\tau}\right)^{1/3} = \left[\frac{16 \times 5305.16 \times 10^3}{\pi \times 80}\right]^{1/3}$$

$$= 69.6 \text{ mm, say, 70 mm}$$

(ii) *Rigidity criterion*

$$\text{Angular twist, } \theta = \frac{Tl}{GJ}$$

or

$$0.5 \times \frac{\pi}{180} = \frac{5305.16 \times 10^3 \times 2500}{0.8 \times 10^5 \times \frac{\pi}{32} d^4}$$

or

$$d = 117.9 \text{ mm, say, 120 mm}$$

Thus based upon the above two criteria, the shaft diameter is, $d = 120$ mm

(B) *Hollow shaft*

(i) *Strength criterion*

Outside diameter of the shaft

$$d_o = \left[\frac{16T}{\pi\tau} \times \frac{1}{1 - K^4}\right]^{1/3}$$

where $K = d_i/d_o = 0.6$ (given). Therefore,

$$d_o = \left[\frac{16 \times 5305.16 \times 10^3}{\pi \times 80} \times \frac{1}{1 - 0.6^4}\right]^{1/3}$$

$$= 72.93 \text{ mm, say, 75 mm}$$

(ii) *Rigidity criterion*

$$\text{Angular twist, } \theta^\circ = \frac{584\, Tl}{G\, d_o^4\,(1 - K^4)}$$

or

$$0.5 = \frac{584 \times 5305.16 \times 10^3 \times 2500}{0.8 \times 10^5 \times d_o^4\,(1 - 0.6^4)}$$

or

$$d_o = 122.12 \text{ mm, say, } 125 \text{ mm}$$

Thus, based upon the above two criteria, the outside diameter of the shaft d_o is 125 mm.

$$\text{Inside diameter, } d_i = 0.6 d_o = 75 \text{ mm}$$

$$\text{Weight of the solid shaft} = \frac{\pi}{4} d^2 l \rho$$

For constant length and density, the weight of the shaft is proportional to the area of cross-section, therefore,

$$A_{\text{solid}} = \frac{\pi}{4} d^2 = \frac{\pi}{4} 120^2 = 11{,}309.7 \text{ mm}^2$$

$$A_{\text{hollow}} = \frac{\pi}{4}\left(d_o^2 - d_i^2\right) = \frac{\pi}{4}\left(125^2 - 75^2\right) = 7853.98 \text{ mm}^2$$

$$\text{Percentage saving in material} = \frac{A_{\text{solid}} - A_{\text{hollow}}}{A_{\text{solid}}}$$

$$= \frac{11{,}309.7 - 7853.98}{11{,}309.7} \times 100 = 30.55\%$$

Example 13.2 The shaft of a rolling machine is driven by means of a motor placed horizontally. The flywheel which also acts as pulley is of 1.5 m diameter and has belt tensions 5.4 kN and 1.8 kN on tight side and slack side, respectively. The weight of the flywheel is 15 kN. Determine the shaft diameter if the maximum allowable shear strength is 50 N/mm². The overhang of the flywheel is 250 mm.

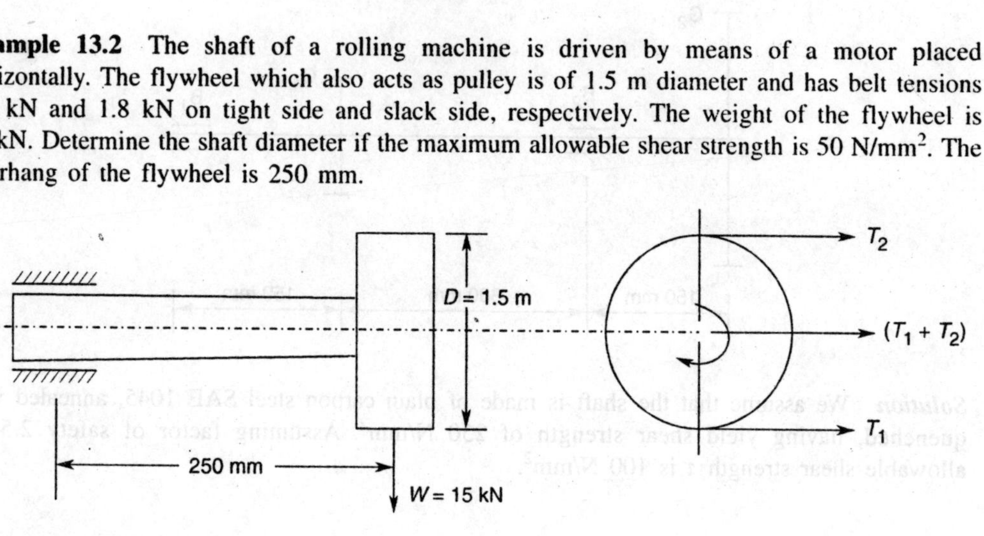

Solution Torque transmitted by shaft

$$T = (T_1 - T_2) \times \frac{D}{2} = (5.4 - 1.8) \times 750 = 2700 \text{ N} \cdot \text{m}$$

Neglecting the weight of the shaft, the total vertical load is the weight of the pulley. Considering shaft as cantilever beam with a point load of 15 kN at one end, the vertical bending moment

$$M_V = W \times l = 15 \times 250 = 3750 \text{ N} \cdot \text{m}$$

$$\text{Horizontal force} = T_1 + T_2 = 5.4 + 1.8 = 7.2 \text{ kN}$$

$$\text{Horizontal moment, } M_H = 7.2 \times 250 = 1800 \text{ N} \cdot \text{m}$$

$$\text{Resultant bending moment, } M = \sqrt{{M_H}^2 + {M_V}^2} = \sqrt{1800^2 + 3750^2} = 4159.62 \text{ N} \cdot \text{m}$$

Using the ASME Code for medium shock load condition, the values of shock and fatigue factors, $C_m = 2.0$ and $C_t = 1.5$

Thus,

$$T_{eq} = \sqrt{(C_m M)^2 + (C_t T)^2} = \sqrt{(2 \times 4159.62)^2 + (1.5 \times 2700)^2} = 9252.69 \text{ N} \cdot \text{m}$$

$$\text{The diameter of the shaft, } d = \left(\frac{16 T_{eq}}{\pi \tau}\right)^{1/3} = \left[\frac{16 \times 9252.69 \times 10^3}{\pi \times 50}\right]^{1/3} = 98 \text{ mm, say, } 100 \text{ mm}$$

Example 13.3 A shaft transmitting 25 kW at 125 rpm from gear G_1 to gear G_2 is mounted on two bearings B_1 and B_2 as shown in the figure below. Pitch circle diameters of gears G_1 and G_2 are 300 mm and 750 mm, respectively. Design the shaft using the strength criterion.

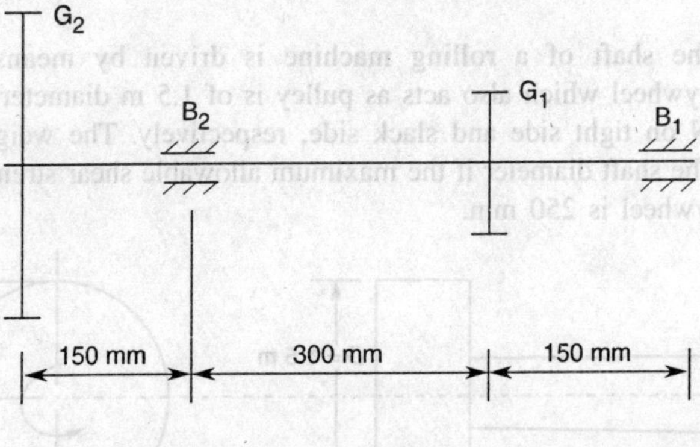

Solution We assume that the shaft is made of plain carbon steel SAE 1045, annealed water quenched, having yield shear strength of 250 N/mm². Assuming factor of safety 2.5, the allowable shear strength τ is 100 N/mm².

Load at gear G_1

$$\text{Average torque, } T = \frac{60P}{2\pi N} = \frac{60 \times 25 \times 10^3}{2\pi \times 125} = 1909.85 \text{ N} \cdot \text{m}$$

$$\text{Tangential force, } F_{t_1} = \frac{2T}{D_1} = \frac{2 \times 1909.85}{0.3} = 12{,}732.3 \text{ N}$$

Radial force, $F_{r_1} = F_{t_1} \tan \alpha$ (assuming gear with 20° pressure angle).

$$= 12{,}732.3 \tan 20° = 4634.0 \text{ N}$$

Load at gear G_2

$$\text{Tangential force, } F_{t_2} = \frac{2T}{D_2} = \frac{2 \times 1909.85}{0.75} = 5092.9 \text{ N}$$

$$\text{Radial force, } F_{r_2} = F_{t_2} \tan \alpha = 5092.9 \tan 20° = 1853.6 \text{ N}$$

Assuming that gear G_1 rotates in anticlockwise direction when viewed from the right-hand side and it receives power from top, the free body diagram of gear forces will be as shown in the figure below.

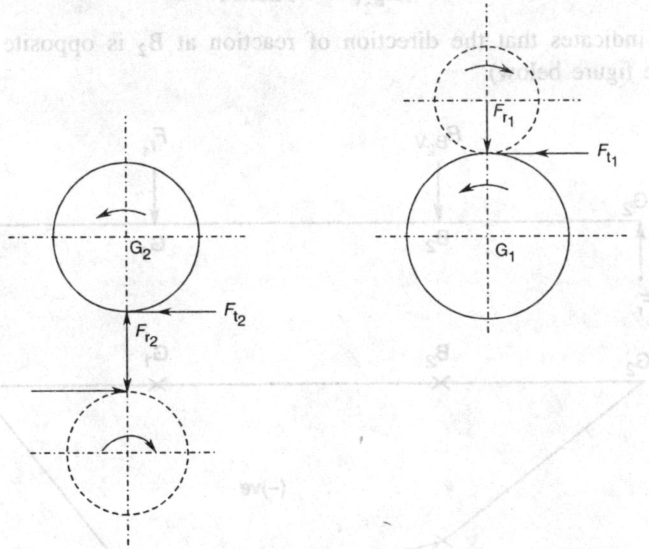

The gear forces F_{t_1}, F_{r_1} acting at G_1 and F_{t_2}, F_{r_2} acting at G_2 locations are shown in the figure below.

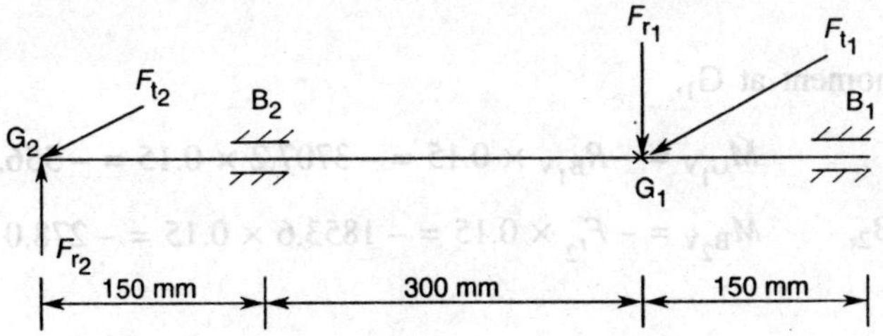

Bending moment

(i) *Vertical forces* (see figure below)

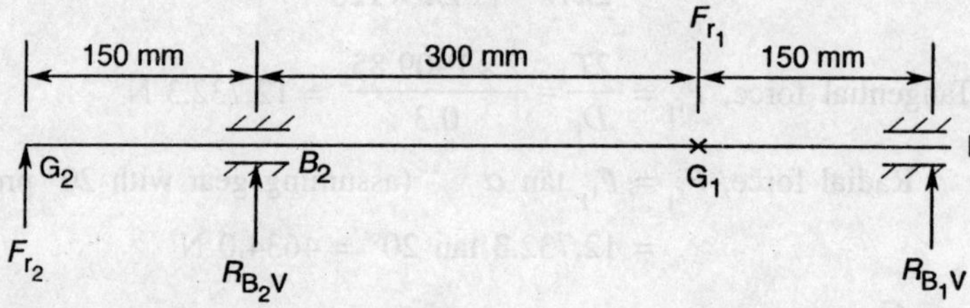

$$\text{For equilibrium, } R_{B_1V} + R_{B_2V} = F_{r_1} - F_{r_2} = 2780.4 \text{ N}$$

Taking moments about B_2

$$R_{B_1V} \times 450 - 4634.0 \times 300 - 1853.6 \times 150 = 0$$

or

$$R_{B_1V} = 3707.2 \text{ N}$$

and

$$R_{B_2V} = -\,926.8 \text{ N}$$

The (–)ve sign indicates that the direction of reaction at B_2 is opposite to what has been assumed (see the figure below)

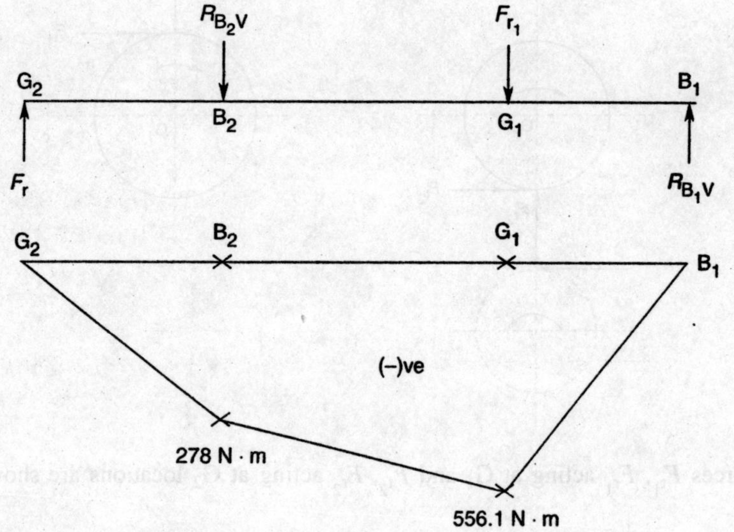

Vertical bending moment at G_1,

$$M_{G_1V} = -\,R_{B_1V} \times 0.15 = -\,3707.2 \times 0.15 = -\,556.1 \text{ N} \cdot \text{m}$$

at B_2, $$M_{B_2V} = -\,F_{r_2} \times 0.15 = -\,1853.6 \times 0.15 = -\,278.0 \text{ N} \cdot \text{m}$$

(ii) *Horizontal forces* (see the figure below)

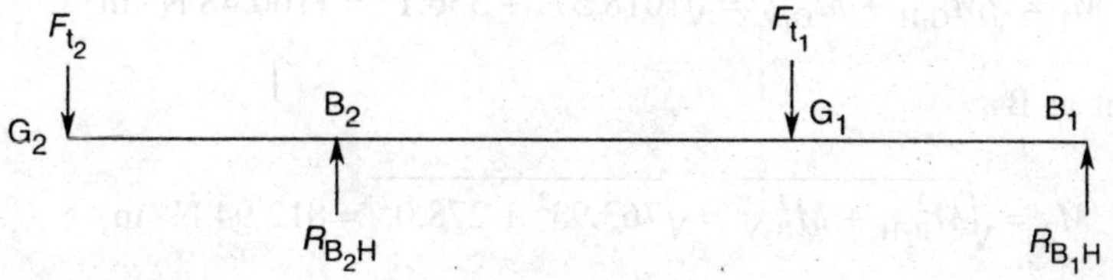

For equilibrium,

$$R_{B_1H} + R_{B_2H} = F_{t_1} + F_{t_2} = 17825.1 \text{ N}$$

Taking moment about B_2,

$$R_{B_1H} \times 450 - 12{,}732.3 \times 300 + 5092.9 \times 150 = 0$$

or

$$R_{B_1H} = 6790.5 \text{ N}$$

and

$$R_{B_2H} = -\ 11034.6 \text{ N}$$

The direction of reaction forces are same as what has been assumed (see the figure below).

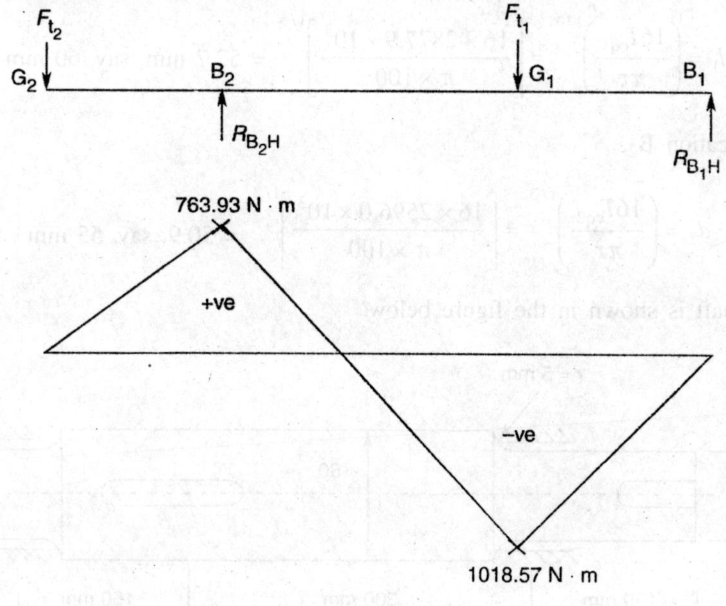

Horizontal bending moment at G_1,

$$M_{G_1H} = -\ R_{B_1H} \times 0.15 = -\ 6790.5 \times 0.15 = -\ 1018.57 \text{ N} \cdot \text{m}$$

at B_2, $$M_{B_2H} = F_{t_2} \times 0.15 = 5092.9 \times 0.15 = 763.93 \text{ N} \cdot \text{m}$$

Resultant moment at G_1,

$$M_1 = \sqrt{M_{G_1H}^2 + M_{G_1V}^2} = \sqrt{1018.57^2 + 556.1^2} = 1160.48 \text{ N} \cdot \text{m}$$

Resultant moment at B_2,

$$M_2 = \sqrt{M_{B_2H}^2 + M_{B_2V}^2} = \sqrt{763.93^2 + 278.0^2} = 812.94 \text{ N} \cdot \text{m}$$

According to the ASME Code, for light shock loads, $C_m = 1.5$ and $C_t = 1.2$, thus the equivalent torque at B_1

$$T_{eq1} = \sqrt{(C_m M_1)^2 + (C_t T)^2}$$

$$= \sqrt{(1.5 \times 1160.48)^2 + (1.2 \times 1909.85)^2} = 2877.9 \text{ N} \cdot \text{m}$$

Equivalent torque at B_2

$$T_{eq2} = \sqrt{(1.5 \times 812.94)^2 + (1.2 \times 1909.85)^2} = 2596.0 \text{ N} \cdot \text{m}$$

Required diameter of shaft at location G_1

$$d_1 = \left(\frac{16T_{eq1}}{\pi\tau}\right)^{1/3} = \left[\frac{16 \times 2877.9 \times 10^3}{\pi \times 100}\right]^{1/3} = 52.7 \text{ mm, say, 60 mm}$$

Diameter at location B_2,

$$d_2 = \left(\frac{16T_{eq2}}{\pi\tau}\right)^{1/3} = \left[\frac{16 \times 2596.0 \times 10^3}{\pi \times 100}\right]^{1/3} = 50.9 \text{, say, 55 mm}$$

The stepped shaft is shown in the figure below.

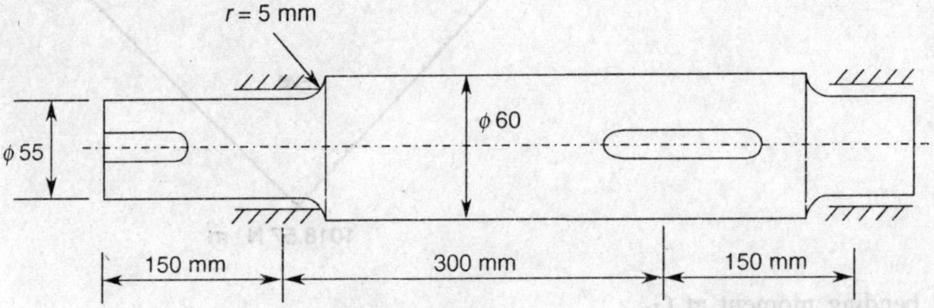

For $r/d \simeq 0.1$ and $D/d = 1.1$, the stress-concentration factor is 1.45 and the notch sensitivity factor, $q = 0.8$, the fatigue stress-concentration factor $K_f = 1 + 0.8(1.45 - 1) = 1.36$ which has already been taken into account while increasing the diameter of the shaft.

Example 13.4 A line shaft receives power through a gear and pinion. The pinion is connected to an electric motor delivering 30 kW at 1200 rpm, of which 20 kW is supplied to a milling machine through a horizontal pulley drive at P_1 and the remainder of the power is supplied to a planer through pulley P_2 by a vertical belt. The diameters of gear and pinion are 300 mm and 100 mm, respectively. The diameters of pulleys P_1 and P_2 are 750 mm and 900 mm, respectively. If the layout of the shaft is as shown in the figure below and the ratio of belt tensions in both drives is 2.0, design the shaft on the basis of strength.

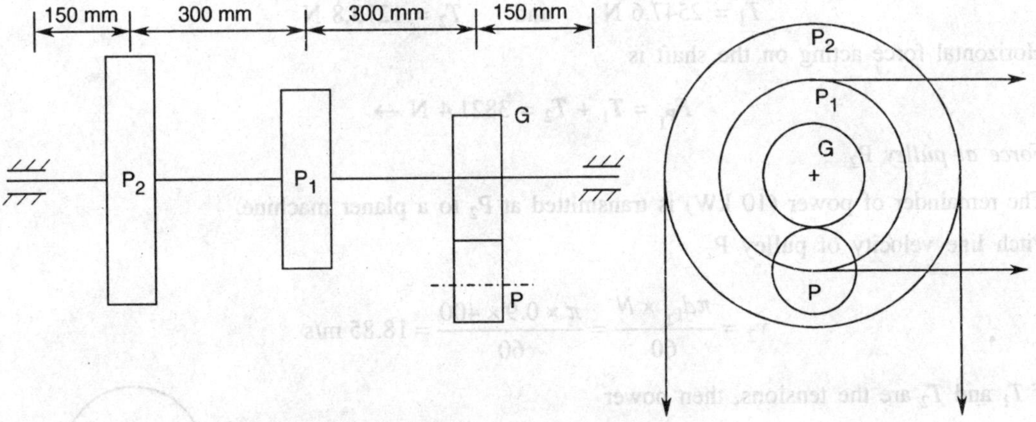

Solution We assume that the pinion is rotating in anticlockwise direction and the gear in clockwise direction so that the tangential force will oppose belt tension, and the upper belt on pulley P_1 will be under slack side tension.

Forces at gear G

The free body diagram is shown in the adjoining figure.

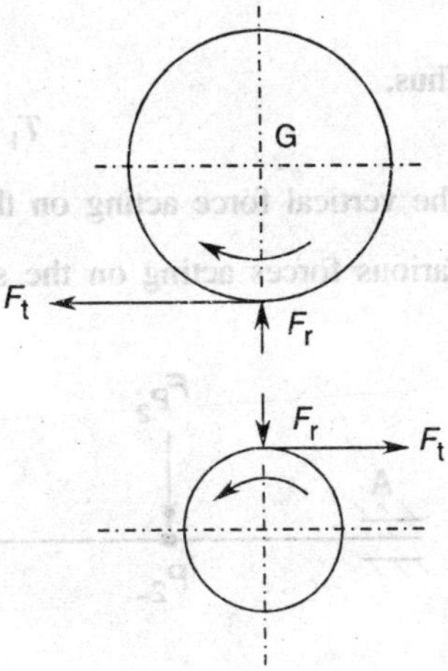

$$\text{Torque, } T = \frac{60\,P}{2\pi N} = \frac{60 \times 30 \times 10^3}{2\pi \times 1200/3} = 716.2 \text{ N} \cdot \text{m}$$

$$\text{Tangential force, } F_t = \frac{2T}{d_g} = \frac{2 \times 716.2}{0.3} = 4774.6 \text{ N} \leftarrow$$

Radial force, $F_r = F_t \tan \alpha$

(assuming gear of 20° pressure angle)

$$= 4774.6 \tan 20° = 1737.8 \text{ N} \uparrow$$

Forces at pulley P_1

At pulley P_1, 20 kW at 400 rpm is transmitted to a milling machine. The pitch line velocity of pulley P_1

$$v_1 = \frac{\pi d_{P1} \times N}{60} = \frac{\pi \times 0.75 \times 400}{60} = 15.7 \text{ m/s}$$

If T_1 and T_2 are the tight side and slack side tensions of the belt, then power

$$P_1 = \frac{(T_1 - T_2)v}{1000} \quad \text{or} \quad T_1 - T_2 = \frac{20 \times 1000}{15.7} = 1273.8 \text{ N}$$

and

$$\frac{T_1}{T_2} = 2 \text{ (given)}$$

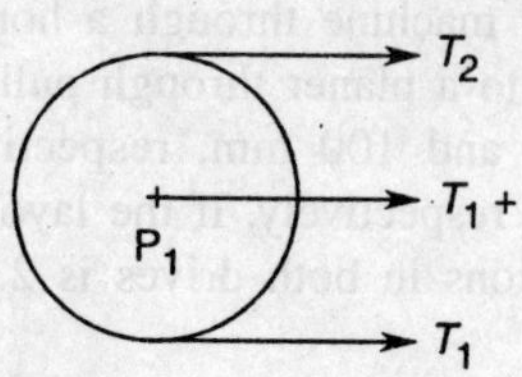

Thus,

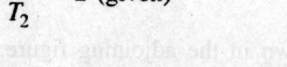

$$T_1 = 2547.6 \text{ N} \quad \text{and} \quad T_2 = 1273.8 \text{ N}$$

Horizontal force acting on the shaft is

$$F_{P_1} = T_1 + T_2 = 3821.4 \text{ N} \rightarrow$$

Force at pulley P_2

The remainder of power (10 kW) is transmitted at P_2 to a planer machine.

Pitch line velocity of pulley P_2

$$v_2 = \frac{\pi d_{P_2} \times N}{60} = \frac{\pi \times 0.9 \times 400}{60} = 18.85 \text{ m/s}$$

if T_1 and T_2 are the tensions, then power

$$P_2 = \frac{(T_1 - T_2)\, v_2}{1000} \quad \text{or} \quad T_1 - T_2 = \frac{10 \times 1000}{18.85} = 530.5 \text{ N}$$

and

$$\frac{T_1}{T_2} = 2 \text{ (given)}$$

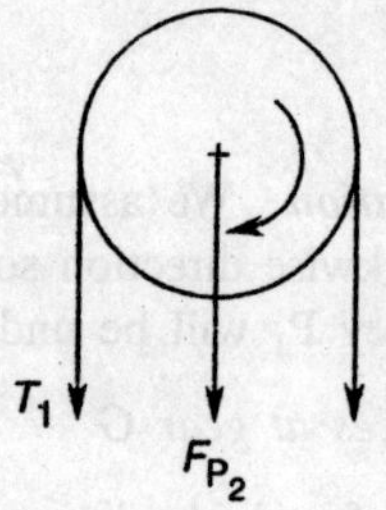

Thus,

$$T_1 = 1061.0 \text{ N} \quad \text{and} \quad T_2 = 530.5 \text{ N}$$

The vertical force acting on the shaft is $F_{p2} = T_1 + T_2 = 1591.5 \text{ N} \downarrow$

Various forces acting on the shaft are thus shown below:

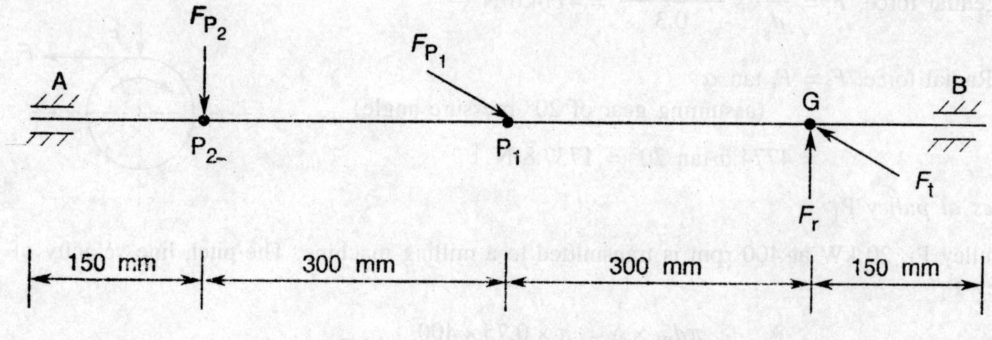

Bending moment

(i) *Horizontal bending moment*

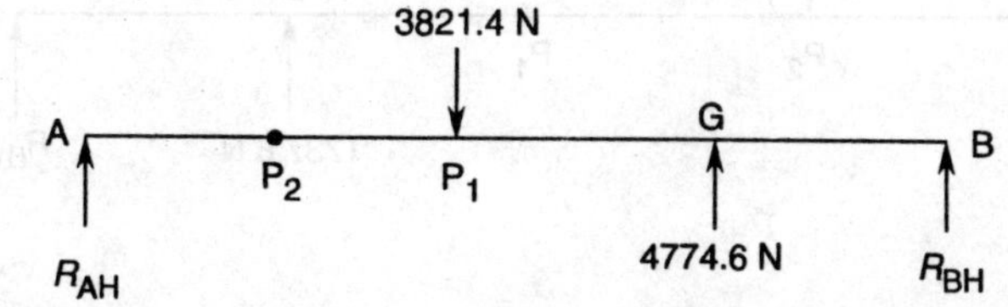

For equilibrium

$$R_{AH} + R_{BH} = + 3821.4 - 4774.6 = - 953.2 \text{ N}$$

Taking moments about A

$$R_{BH} \times 900 = - 4774.6 \times 750 + 3821.4 \times 450$$

or

$$R_{BH} = - 2068.1 \text{ N} \qquad \text{and} \qquad R_{AH} = 1114.9 \text{ N}$$

The direction of reaction at B is opposite to what has been assumed.

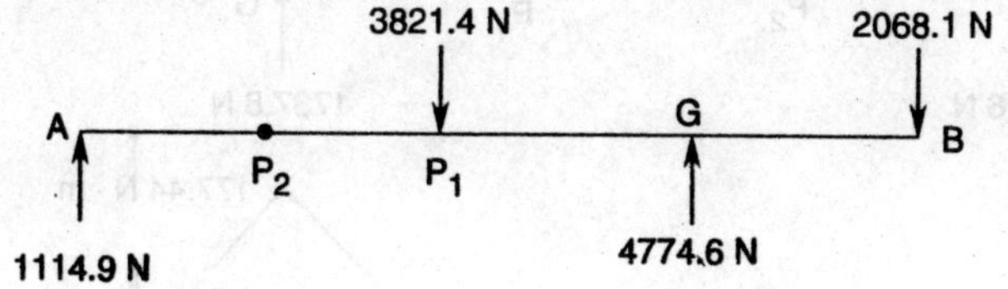

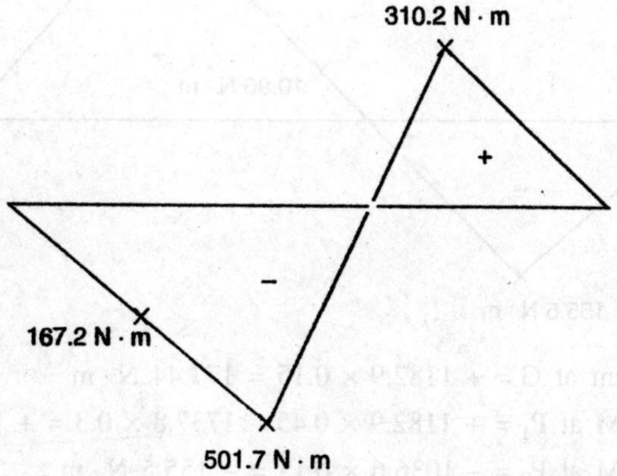

$$\text{Bending moment at G} = + 2068.1 \times 0.15 = + 310.2 \text{ N} \cdot \text{m}$$

$$\text{BM at } P_1 = - 1114.9 \times 0.45 = - 501.7 \text{ N} \cdot \text{m}$$

$$\text{BM at } P_2 = - 1114.9 \times 0.15 = - 167.2 \text{ N} \cdot \text{m}$$

(ii) *Vertical bending moment*

$$R_{AV} + R_{BV} = - 1737.8 + 1591.5 = - 146.3 \text{ N}$$

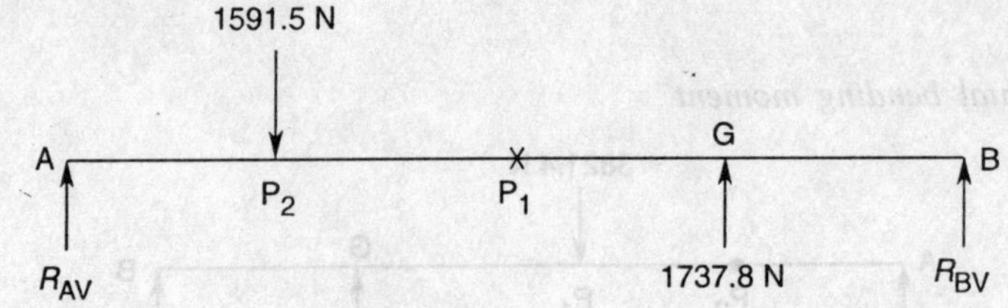

Taking moments at A,

$$R_{BV} \times 900 + 1737.8 \times 750 - 1591.5 \times 150 = 0$$

or

$$R_{BV} = -\ 1182.9 \text{ N} \qquad \text{and} \qquad R_{AV} = 1036.6 \text{ N}$$

The direction of reaction at support B is opposite to what has been assumed. So the revised force diagram is

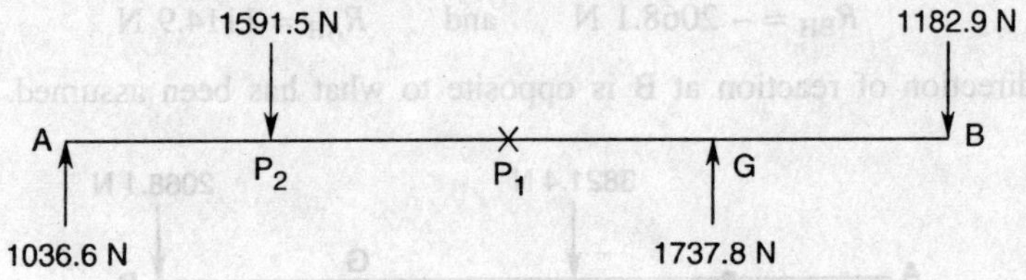

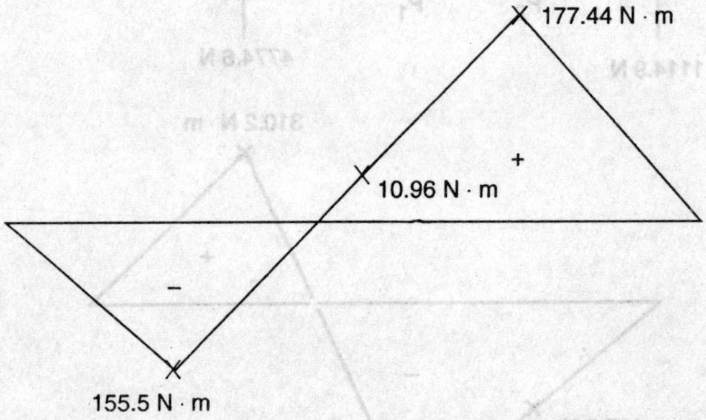

$$\text{Bending moment at G} = +\ 1182.9 \times 0.15 = 177.44 \text{ N} \cdot \text{m}$$

$$\text{BM at } P_1 = +\ 1182.9 \times 0.45 - 1737.8 \times 0.3 = +\ 10.96 \text{ N} \cdot \text{m}$$

$$\text{BM at } P_2 = -\ 1036.6 \times 0.15 = -\ 155.5 \text{ N} \cdot \text{m}$$

$$\text{Resultant BM} = \sqrt{M_V^2 + M_H^2}$$

Therefore,

$$\text{Resultant BM at G, } M_G = \sqrt{177.44^2 + 310.2^2} = 357.36 \text{ N} \cdot \text{m}$$

$$\text{Resultant BM at } P_1,\ M_{P1} = \sqrt{10.96^2 + 501.7^2} = 501.82 \text{ N} \cdot \text{m}$$

$$\text{Resultant BM at } P_2,\ M_{P2} = \sqrt{155.5^2 + 167.2^2} = 228.3 \text{ N} \cdot \text{m}$$

Torque diagram

$$\text{Torque at G} = F_t \times \frac{d_g}{2} = 4774.6 \times 0.15 = 716.2 \text{ N} \cdot \text{m}$$

$$\text{Torque at P}_1 = (T_1 - T_2)\frac{d_{P1}}{2} = (2547.6 - 1273.8)\frac{0.75}{2} = 477.67 \text{ N} \cdot \text{m}$$

$$\text{Torque at P}_2 = (1061 - 530.5)\frac{0.9}{2} = 238.72 \text{ N} \cdot \text{m}$$

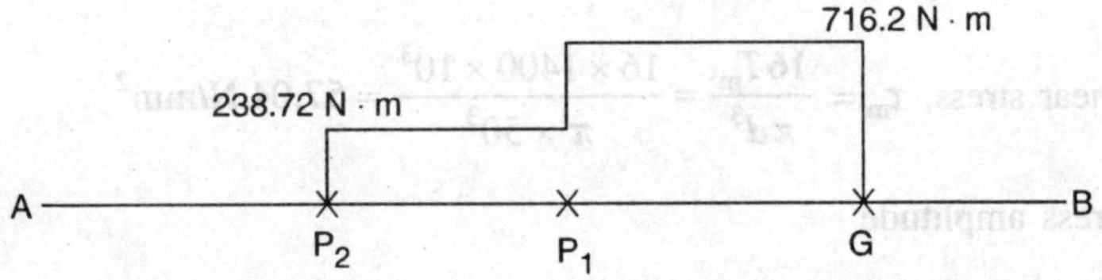

Looking into the torque diagram and the values of resultant bending moments, we can conclude that pulley P_1 on the shaft is critically loaded. According to the ASME Code with $C_m = 1.5$ and $C_t = 1.2$ for light shock loads, the equivalent torque at location P_1

$$T_{eq} = \sqrt{(1.5 \times 501.82)^2 + (1.2 \times 716.2)^2} = 1142.47 \text{ N} \cdot \text{m}$$

Assuming that the shaft is made of SAE 1045 annealed oil quenched steel having torsional yield strength 300 N/mm^2 and assuming factor of safety 4.0 which includes stress concentration due to keyway and fatigue load condition, the allowable shear strength

$$\tau = \frac{300}{4} = 75 \text{ N/mm}^2$$

The diameter of shaft, $d = \left(\frac{16 T_{eq}}{\pi \tau}\right)^{1/3}$

or

$$d = \left(\frac{16 \times 1142.47 \times 10^3}{\pi \times 75}\right) = 42.65 \text{ mm, say, 45 mm}$$

Example 13.5 A 50 mm diameter shaft is made of carbon steel having yield tensile strength of 400 N/mm^2 and ultimate strength of 600 N/mm^2. It is subjected to torque which fluctuates between 2000 N · m to 800 N · m. Using the Soderberg equation, calculate the factor of safety.

Solution The Soderberg equation for fluctuating torque is given as

$$\frac{\tau_m}{\tau_y} + \frac{\tau_a}{\tau_{en}} = \frac{1}{n}$$

where

τ_m is the mean shear stress

τ_a is the shear stress amplitude

τ_{en} is the shear endurance strength (= $0.25 \times \sigma_{ut} = 0.25 \times 600 = 150$ N/mm², where σ_{ut} = 600 N/mm² for carbon steel)

τ_y is the yield shear strength (= $0.57 \times \sigma_{yt} = 0.57 \times 400 = 228$ N/mm²)

$$\text{Mean torque, } T_m = \frac{T_{max} + T_{min}}{2} = \frac{2000 + 800}{2} = 1400 \text{ N} \cdot \text{m}$$

$$\text{Amplitude torque, } T_a = \frac{T_{max} - T_{min}}{2} = \frac{2000 - 800}{2} = 600 \text{ N} \cdot \text{m}$$

$$\text{Mean shear stress, } \tau_m = \frac{16 T_m}{\pi d^3} = \frac{16 \times 1400 \times 10^3}{\pi \times 50^3} = 57.04 \text{ N/mm}^2$$

Similarly, shear stress amplitude

$$\tau_a = \frac{16 T_a}{\pi d^3} = \frac{16 \times 600 \times 10^3}{\pi \times 50^3} = 24.44 \text{ N/mm}^2$$

Substituting in the Soderberg equation, we have

$$\frac{57.04}{228} + \frac{22.44}{150} = \frac{1}{n} \quad \text{or} \quad n = 2.5$$

Thus the factor of safety is 2.5.

Example 13.6 A pulley is keyed to a shaft midway between two bearings. The shaft is made of cold drawn steel for which the ultimate strength is 600 N/mm² and the yield strength is 450 N/mm². The bending moment at the pulley varies from –200 N · m to 400 N · m and the torque on the shaft varies from –100 N · m to 250 N · m. Design a suitable shaft for infinite life. Assume the following additional parameters:

Factor of safety	= 1.5
Load correction factor	
(a) in bending	= 1.0
(b) in torsion	= 0.6
Size factor	= 0.85
Surface factor	= 0.9
Stress correction factor	
(a) in bending	= 1.6
(b) in torsion	= 1.3

Solution Endurance strength for steel, $\sigma_{en} = 0.5\sigma_{ut} = 0.5 \times 600 = 300$ N/mm²

Endurance strength of component

$$\sigma_e = \frac{k_a \, k_b \, k_c}{k_{tf}} \times \sigma_{en}$$

where

k_a is the load correction factor
k_b is the size factor
k_c is the surface finish factor
k_{tf} is the fatigue stress concentration factor

Endurance strength of shaft in bending

$$\sigma_{eb} = \frac{1.0 \times 0.85 \times 0.9}{1.6} \times 300 = 143.43 \text{ N/mm}^2$$

Endurance strength of shaft in torsion

$$\sigma_{et} = \frac{0.6 \times 0.85 \times 0.9}{1.3} \times 300 = 106 \text{ N/mm}^2$$

$$\text{Mean bending moment, } M_m = \frac{400 + (-200)}{2} = 100 \text{ N} \cdot \text{m}$$

$$\text{Amplitude bending moment, } M_a = \frac{400 - (-200)}{2} = 300 \text{ N} \cdot \text{m}$$

$$\text{Similarly, mean torque } T_m = \frac{250 - 100}{2} = 75.0 \text{ N} \cdot \text{m}$$

$$\text{Amplitude torque } T_a = \frac{250 + 100}{2} = 175.0 \text{ N} \cdot \text{m}$$

Equivalent bending moment

$$M_{eq} = M_m + \frac{\sigma_y}{\sigma_{eb}} M_a$$

$$= 100 + \frac{450}{143.43} \times 300 = 1041.2 \text{ N} \cdot \text{m}$$

Equivalent torque

$$T_{eq} = T_m + \frac{\sigma_y}{\sigma_{et}} \times T_a$$

$$= 75 + \frac{450}{106} \times 175 = 817.9 \text{ N} \cdot \text{m}$$

Maximum equivalent torque

$$T_{eq\text{-}max} = \sqrt{(M_{eq})^2 + (T_{eq})^2}$$

$$= \sqrt{(1041.2)^2 + (817.9)^2} = 1324.0 \text{ N} \cdot \text{m}$$

$$\text{The diameter of shaft, } d = \left[\frac{16T_{\text{eq-max}}}{\pi\tau}\right]^{1/3}$$

where τ is the allowable shear strength $\left(= \dfrac{0.57 \times \sigma_{yt}}{\text{FoS}} = \dfrac{0.57 \times 450}{1.5} = 171.0 \text{ N/mm}^2\right)$

Therefore,

$$d = \left(\frac{16 \times 1324.0 \times 10^3}{\pi \times 171.0}\right)^{1/3} = 34.03 \text{ mm, say, } 36 \text{ mm}$$

13.10 COMPUTER-AIDED DESIGN OF SHAFT

As discussed in preceding sections, the shaft is a most common mechanical component which is required to support gears, pulleys, flywheel, and so forth. These members contribute to a transverse force resulting in bending moment, in addition to torque and axial force. Therefore, the most commonly used design criteria are: (i) torsional strength, (ii) bending moment, (iii) torsional rigidity, and (iv) flexural rigidity. The selection of a particular criterion depends upon the analysis of shaft for a given load condition. This means that the shaft is first required to be analyzed for all four criteria and depending upon the outcome a particular criterion is selected. This requires a lot of computation, particularly when the shaft is subjected to a large number of transverse loads. In this section, a generalized computer program called shaft, cpp is given (Program 13.1). This program is user-friendly and asks input data, namely the length of the shaft, the number of load points, power, speed, permissible deflection, and so on.

The transverse load can only be either in the vertical direction or in the horizontal direction. However, the program can be modified to supply bidirectional forces. The diameter of the shaft is computed for all the four criteria and the maximum diameter and the corresponding criterion is selected.

The use of Program 13.1 is demonstrated with a sample run.

Program 13.1

```
//	A PROGRAM TO DESIGN A SHAFT
#	include <iostream.h>
#	include <math.h>
#	include <conio.h>
#	define PI 3.1415
void main ( )
{
	int i, j, n, m, countl, count2, count;
	double lt, p, rpm, emod, smod, sigu, sigy, cm, ct, k;
	double ptwist, ymax, para, numl, den1, nm2, var2, tha, thb;
	double bmji, bmmax, torque, teq, bmeq, ft1, ft2, ft, fs, acdef;
	double d1, d2, d3, d4, d, dmax, del1mx, cs, var1, den, slopa, slopb;
	double l[10], w[10], bm[10], del1[10], b[10] [10];
```

```
//     input design parameters
       cout <<"\n DESIGN INPUT DATA" << endl;
       cout <<"\n Total length of shaft (mm): ";
       cin >>lt;
       cout <<"\n Number of loads: ";
       cin >>n;
       cout <<"\n Power (kW) : ";
       cin >>p;
       cout <<"\n Speed of shaft (rpm) : ";
       cin >>rpm;
       cout <<"\n Modulus of elasticity (N/mm2) : ";
       cin >> emod;
       cout <<"\n Shear modulus (N/mm2) : ";
       cin >> smod;
       cout <<"\n Ultimate strength (N/mm2) : ";
       cin >> sigu;
       cout <<"\n Yield strength (N/mm2) : ";
       cin >>sigy;
       cout <<"\n Shock factor in bending moment (Cm) : ";
       cin >> cm;
       cout <<"\n Shock factor in torsion (Ct) : ";
       cin >> ct;
       cout <<"\n Ratio of inner to outer dias (K) : ";
       cin >> k;
       cout <<"\n Permissible transverse deflection (mm) : ";
       cin >> ymax;
       cout <<"\n Permissible twist (degree) : ";
       cin >> ptwist;
//     input load and point of application of loads
       for (i=1; i<=n; i=i+1)
       {
          cout <<"\n Load case: " << i;
          cout <<"\n Point of application of load: ";
          cin >> l[i];
          cout <<"\n Load (newton) : ";
          cin >> w[i];
       }
//     bending moment
       for (j=1; j<=n; j=j+1)
       {
          bm[j]=0.0;
          for (i=1; i<=n; i=i+1)
          {
                  if (1[i] >= 1[j])
                          bmji=w[i]*1[j]*(lt-1[i])/lt;
                  else
                          bmji=w[i]*(lt-1[j])*1[i]/lt;
```

```
                bm[j]=bm[j]+bmji;
        }
    }
//  find maximum bending moment
    bmmax=bm[1];
    count1=1;
    for (j=1; j<=n; j=j+1)
    {
        if (bm[j] > bmmax)
        {
                bmmax=bm[j];
                count1=j;
        }
    }
//  torque calculation
    torque=9.55e+06*p/rpm;
    teq=pow ((pow(cm*bmmax, 2.0) + pow (ct*torque, 2.0)), 0.5);
    bmeq=(cm*bmmax+teq)/2.0;
//  strength calculation
    ft1=0.6*sigy;
    ft2=0.36*sigu;
    if (ft1 >= ft2)
        ft=ft1;
    else
        ft=ft2;
    ft=0.75*ft;
    fs=ft/2.0;
//  deflection calculation
    for (i=1; i<=n; i=i+1)
    {
        m=i;
        for (j=m; j<=n; j=j+1)
        {
                para=(2.0*(1[j]/lt) – pow ((1[i]/lt), 2.0) – pow ((1[j]/lt), 2.0));
                b[i][j] = (1[i]/lt)*(1 – (1[j]/lt))*para;
                b[j][i] =b[i][j];
        }
    }
    for (i=1; i<=n; i=i+1)
    {
        del1 [i]=0.0;
        for (j=1; j<=n; j=j+1)
        {
                del1[1] = del1[i] +b[i][j]*w[j];
        }
    }
```

```
//      max deflection
        del1mx=del1[1];
        count2=1;
        for (j=1; j<=n; j=j+1)
        {
           if (del1[j] > del1mx)
           {
                   del1mx=del1[j];
                   count2=j;
           }
        }
//      calculate shaft dia for different criteria
        d1=pow ((16*teq/(PI*fs)/(1-pow(k,4))), 0.3333);
        d2=pow ((32*bmeq/(PI*ft)/(1-pow(k,4))), 0.3333);
        d3=pow((584*torque*1000/(smod*ptwist)/(1-pow(k, 4))), 0.25);
        d4=pow ((del1mx*lt*3*64/(ymax*6*emod*PI*(1-pow(k, 4)))), 0.25);
        dmax=d1;
        count=1;
        if (d2 >= dmax)
        {
           dmax=d2;
           count=2;
        }
        else
        {
           if (d3 >= dmax)
           {
                   dmax=d3;
                   count=3;
           }
           else
           {
                   if (d4 > dmax)
                   {
                           dmax=d4;
                           count=4;
                   }
           }
        }
//      find standard size
        if (dmax < 25)
           d=(int((dmax-0.5)/2)+1)*2;
        else
           if (dmax <60)
                   d=(int((dmax-0.5)/5)+1)*5;
           else
                   if (dmax < 120)
```

```
                d=(int((dmax–0.5)/10)+1)*10;
          else
                d=(int((dmax–0.5)/20)+1)*20;
//    calculate freqjuency
      for (j=1; j<=n; j=j+1)
      {
         num1=w[j] *del1[j];
         den1=w[j]*pow (del1[j], 2.0);
         nm2=nm2+num1;
         den=den+den1;
      }
      var1=9810*nm2*6*emod*PI*pow(d, 4)*(1–pow (k, 4));
      cs = (60/(2*PI))*pow((var1/(den*64*pow(lt, 3))), 0.25);
//    slope calculation
      slopa=0.0;
      slopb=0.0;
      for (j=1; j<=n; j=j+1)
      {
         var2=6*emod*PI*pow(d, 4)*(1–pow (k, 4))*lt;
         tha=64*w[j]*l[j]*(lt–1[j])*(2*lt–1[j]/var2;
         tha=tha*180/PI;
         thb=64*w[j]*1[j]*(lt–1[j])*(lt+1[j])/var2;
         thb=thb*180/PI;
         slopa=slopa+tha;
         slopb=slopb+thb;
      }
//    max deflection
      acdef=64*pow(lt, 3) *del1mx/(6*emod*PI*pow(d,4)*(1–pow (k, 4)));
      cout << "\n   DESIGN REPORT" <<endl;
      cout <<"\n Design criterion = ";
      if (count = =1)
         cout << " Torsional strength" << endl;
      if (count = = 2)
         cout << " Bending strength" << endl;
      if (count = = 3)
         cout << " Torsional rigidity" << endl;
      if (count = = 4)
         cout << " Flexural rigidity" << endl;
      cout << "\n          Shaft diameter =" << d << "mm";
      cout << "\n          Maximum deflection =" << acdef << "mm";
      cout << "\n          Slope at support A =" << slopa << "deg";
      cout << "\n          Slope at support B =" << slopb << "deg";
      cout << "\n          Critical speed =" << cs << "Hz" << endl;
```

Sample run. Design a shaft to transmit 5 kW at 500 rpm. The shaft has 300 mm span and a 10 kN load acts at mid-span as shown in the figure below.

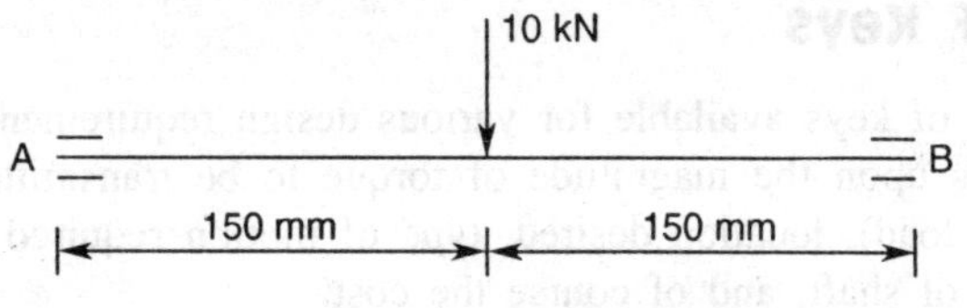

The sample input data and design report is given below:

DESIGN INPUT DATA

Total length of shaft (mm): 300
Number of loads: 1
Power (kW): 5
Speed of shaft (rpm): 500
Modulus of elasticity (N/mm2): 0.2e06
Shear modulus (N/mm2): 0.8e05
Ultimate strength (N/mm2): 600
Yield strength (N/mm2): 400
Shock factor in bending moment (Cm): 1.5
Shock factor in torsion (Ct): 1
Ratio of inner to outer dias (K): 0.6
Permissible transverse deflection (mm): 5
Permissible twist (degree): 3

Load case: 1
Point of application of load: 150
Load (newton): 10000

DESIGN REPORT

Design criterion : Torsional strength
Shaft diameter = 45 mm
Maximum deflection = 0.1605 mm
Slope at support A = 0.0919 deg.
Slope at support B = 0.0919 deg.
Critical speed = 150.14 Hz

13.11 KEYS

A key is a mechanical element which is used to connect a shaft to another mechanical element, namely pulley, gear, disc, flywheel, etc. so that there is no rotational slip between the two members. Its main function is to transmit torque between a shaft and a machine part assembled on it. In other words, a key is a piece of metal fitted in an axial direction into the mating groove, called keyway, cut in the shaft. The mating members prevent relative motions—both rotary and axial. In some cases they allow an axial motion between the shaft and the hub. Keys are used as temporary fasteners which can be easily disassembled. They are generally made of hardened and tempered steels such as C30, C35, C40, C50, and 55Mn75, etc. These materials should conform to IS: 1570–1978.

13.11.1 Types of Keys

There are numerous kinds of keys available for various design requirements. The type of key for an application depends upon the magnitude of torque to be transmitted, type of loading (steady, varying, or shock load), location desired, type of motion required to be restricted, fit required, limiting strength of shaft, and of course the cost.

Keys are often required to be replaced. The BIS has laid down the standard dimensions of different types of keys and keyways. Keys are therefore mainly classified into two categories: (i) light and medium power transmission keys which include square key, rectangular key, taper key, gib-headed key, Woodruff key, flat key, saddle key and (ii) heavy duty power transmission keys, namely round key, Barth key, Kennedy key, and feather key, and so forth.

A brief description of some of these keys is given below:

Light and medium duty keys

Square key. A sunk key having square cross-section is called the square key. The key is sunk half in the shaft and half in the hub. This type of key is used in light industrial machinery [Figure 13.3(a)].

Rectangular key. It is a modified sunk key having rectangular cross-section. It is used where higher stability is desired in connection. A rectangular key may be tapered, whose width is kept constant while the height is tapered by 1:100. The keyway in the hub has the same taper as that in the key while the keyway in the shaft has uniform depth [Figure 13.3(b)].

Gib-headed key. A gib-headed key is similar to a square or rectangular key but it has a head at one end, generally at the larger end of the taper rectangular key. The gib head is used for driving the key while assembling or disassembling [(Figure 13.3(c)].

Woodruff key. A Woodruff key is used to transmit small values of torque. The keyway in the shaft is milled in a curved shape whereas the keyway in the hub is usually straight. The main advantage of this key is that it will align itself in the keyway. It is used extensively in machine tool and automobile industry. Figure 13.3(e) shows two views of the Woodruff key.

Flat key. It is flat cross-section which does not require a keyway in the shaft. Generally, these keys are held in place by set screws. These keys are not useful for heavy duty load or sudden reversal of torque. When the bottom surface of a flat key is shaped concavely, the resulting key is called the *saddle key*. The radius of concave is approximately equal to the shaft radius. The saddle key is usually tapered because the torque transmission capacity depends upon the friction between the shaft and the hub. Figure 13.3(f) shows the saddle key.

Heavy duty keys

Round key. It is commonly known as pin key or Nordberg key. A round taper pin having a taper of about 1:200 has proved satisfactory for heavy duty torque transmission. The main advantage is that the round key produces less stress-concentration at the keyway in the shaft. The mean diameter of the pin is generally taken as one-sixth of the shaft diameter [(Figure 13.3(g)].

Barth key. The Barth key is a modification of rectangular key which has two bevelled surfaces, as shown in Figure 13.3(h). The bevelled surface ensures that the key will fit tightly.

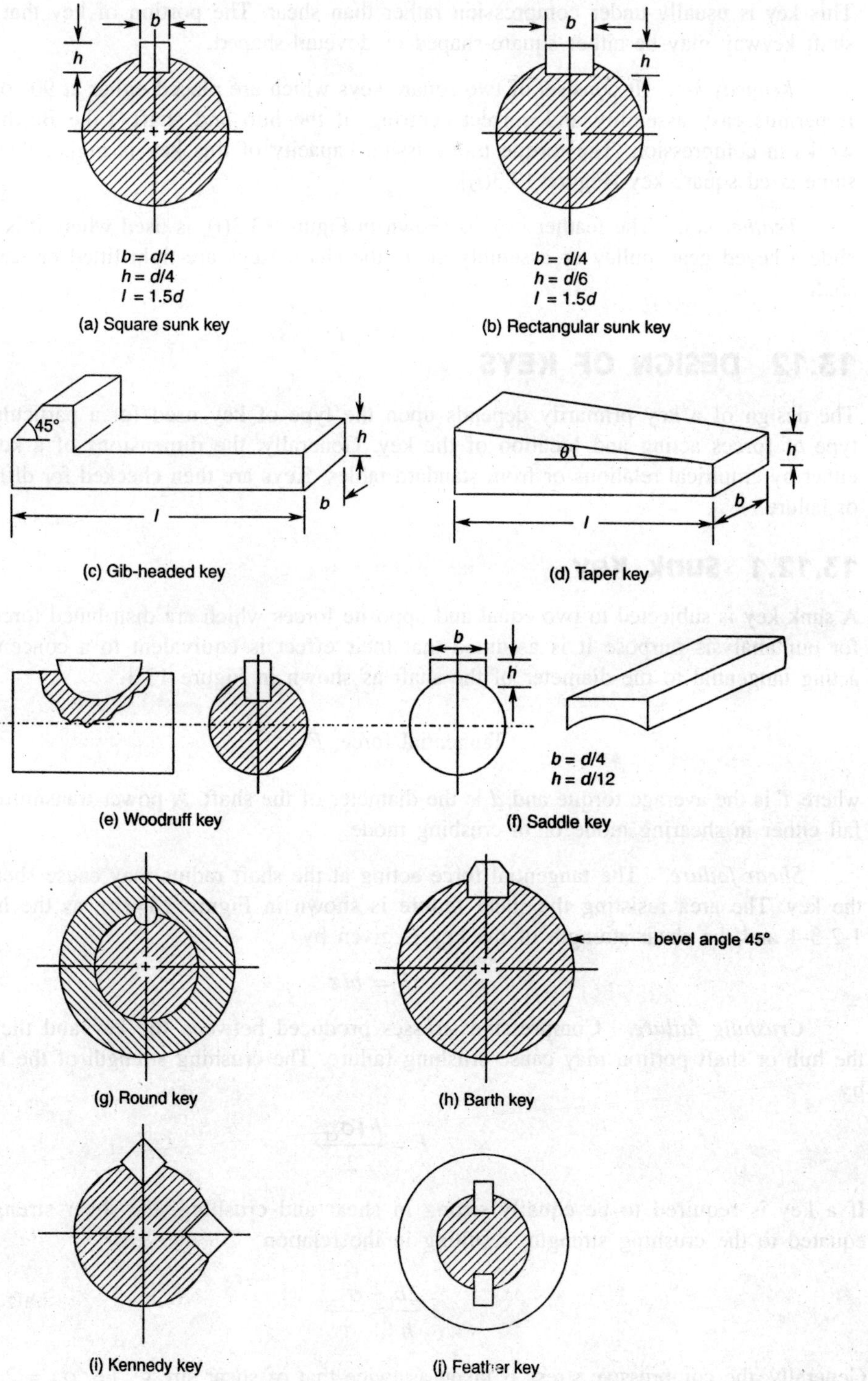

Figure 13.3 Types of keys.

This key is usually under compression rather than shear. The portion of key that fits into the shaft keyway may be either square-shaped or dovetail-shaped.

Kennedy key. It consists of two square keys which are placed either at 90° or 120° apart. It permits easy assembly and correct centring of the hub and shaft. Like Barth key, it also works in compression. The torque transmission capacity of this key is higher than that of the same sized square key [Figure 13.3(i)].

Feather key. The feather key, as shown in Figure 13.3(j), is used where it is necessary to slide a keyed gear, pulley or assembly along the shaft. Keys are tight-fitted or screwed on the shaft.

13.12 DESIGN OF KEYS

The design of a key primarily depends upon the type of key used for a particular fastening, type of forces acting and location of the key. Generally, the dimensions of a key are found either by empirical relations or from standard tables. Keys are then checked for different modes of failure.

13.12.1 Sunk Key

A sunk key is subjected to two equal and opposite forces which are distributed forces, however, for our analysis purpose it is assumed that their effect is equivalent to a concentrated force acting tangential to the diameter of the shaft as shown in Figure 13.4.

$$\text{Tangential force, } F = \frac{2T}{d} \tag{13.24}$$

where T is the average torque and d is the diameter of the shaft. A power transmitting key may fail either in shearing mode or in crushing mode.

Shear failure. The tangential force acting at the shaft radius may cause shear failure of the key. The area resisting the shear failure is shown in Figure 13.4(b), as the hatched area 1-2-3-4 and the shear strength of the key is given by

$$F = bl\tau \tag{13.25}$$

Crushing failure. Compressive stresses produced between the key and the keyway of the hub or shaft portion may cause crushing failure. The crushing strength of the key is given by

$$F = \frac{hl\sigma_{cr}}{2} \tag{13.26}$$

If a key is required to be equally strong in shear and crushing, the shear strength may be equated to the crushing strength, resulting in the relation

$$\frac{2b}{h} = \frac{\sigma_{cr}}{\tau} \tag{13.27}$$

Generally, the compressive stress is taken as twice that of shear stress , i.e. $\sigma_{cr} = 2\tau$, therefore, Eq. (13.27) results into $b = h$, i.e. a square key.

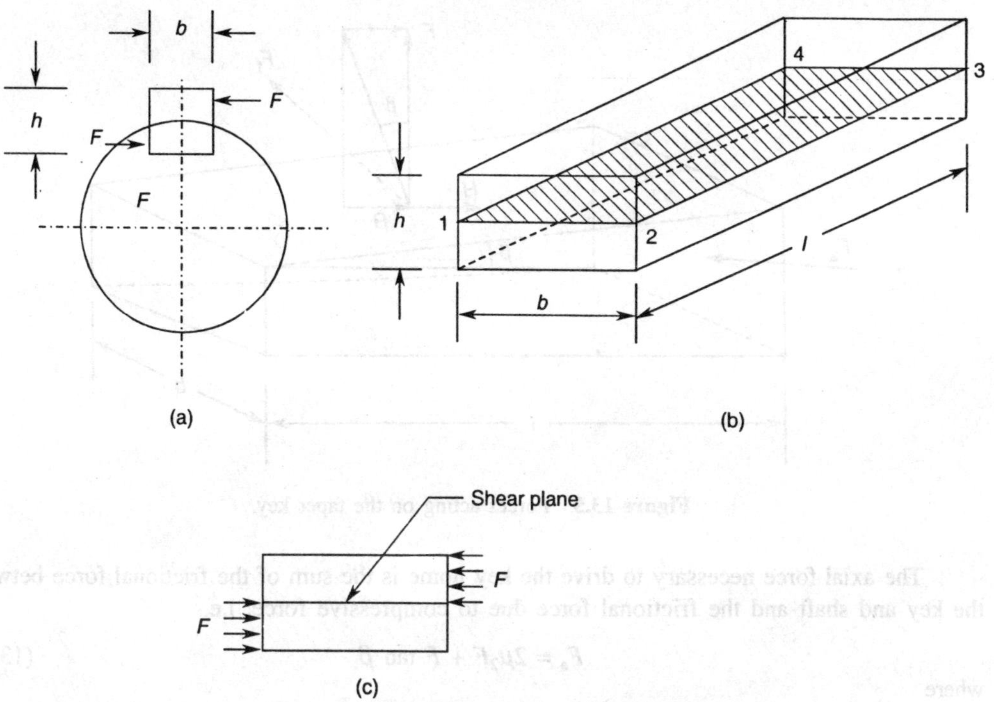

Figure 13.4 Failure of key.

Further, if a key is required to be equally strong in shear like the shaft, the shear strength of the key can be equated to shear strength of the shaft, i.e.

$$\frac{bld\tau}{2} = \frac{\pi}{16}d^3\tau$$

or the required length of the key

$$l = \frac{\pi d^2}{8b} \tag{13.28}$$

The industrial practice is to use the following proportions for the sunk key.

	Square key	*Rectangular key*
Width (b)	$d/4$	$d/4$
Height (h)	b	$d/6$
Length (l)	$1.5d$	$1.5d$

13.12.2 Taper Key

In a taper key, the torque is transmitted primarily because of the wedging action between the taper key and the hub. Forces acting on the taper key are shown in Figure 13.5. The torque transmission capacity of the taper key is

$$T = 0.5\mu_1 bld\sigma_c \tag{13.29}$$

where μ_1 is the coefficient of friction between the shaft and the hub.

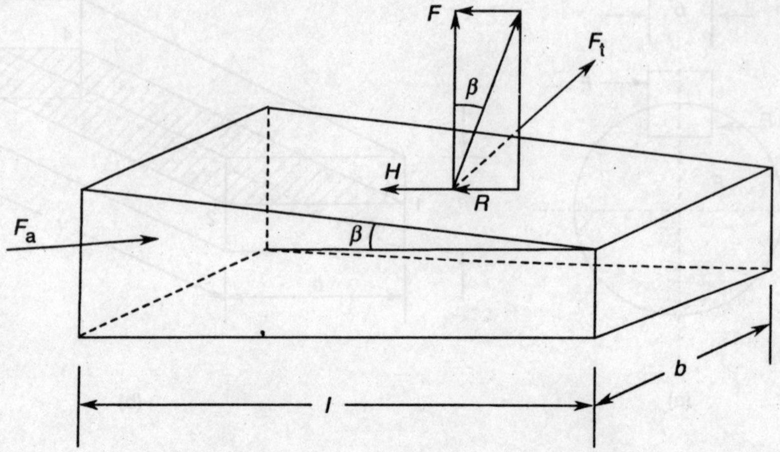

Figure 13.5 Forces acting on the taper key.

The axial force necessary to drive the key home is the sum of the frictional force between the key and shaft and the frictional force due to compressive force, i.e.

$$F_a = 2\mu_2 F + F \tan \beta \tag{13.30}$$

where

F is the compressive force acting on the key $(= bl\sigma_c)$

μ_2 is the coefficient of friction between the key and the shaft(= 0.1 for the greased lubricated key)

13.13 DESIGN OF SPLINES

Spline is a natural extension of the feather key. It consists of a number of parallel sides, i.e. integral keys milled on the surface of the shaft. That is why it is called the spline shaft. A connecting element called the spline hub (Figure 13.6) is bored to the minor diameter of the shaft and has keyways broached in it to receive the keys on the shaft. Splines are most frequently used in machine tools and automobile gear boxes to move the gear axially on the shaft. The BIS has standardized four types of splines, namely four, six, ten and sixteen splines

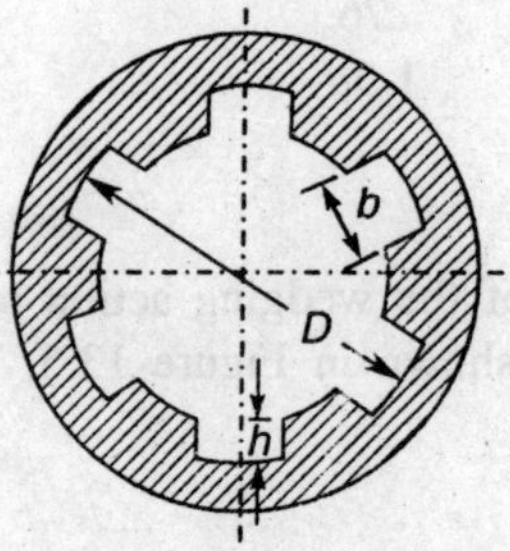

Figure 13.6 Typical spline hub.

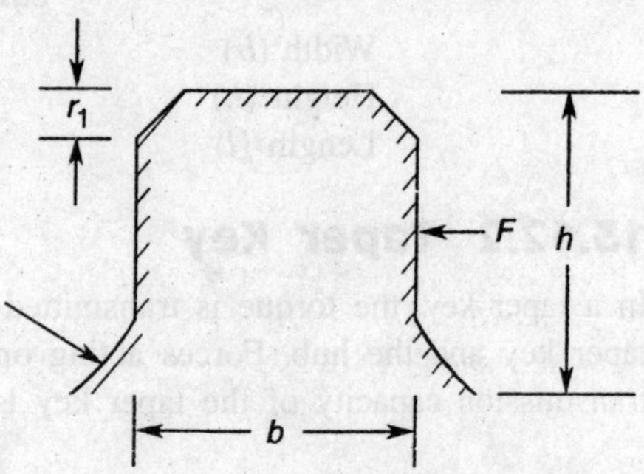

Figure 13.7 Straight spline

and all dimensions are defined in terms of major diameter of the spline shaft D. Table 13.2 shows the standard proportions of spline.

Table 13.2 Standard proportions of spline

Number of splines	*Width, b*	*Height, h*		
		Fitting A	*Fitting* B	*Fitting* C
4	0.241D	0.075D	0.125D	—
6	0.25D	0.05D	0.075D	0.1D
10	0.15D	0.045D	0.07D	0.095D
16	0.098D	0.045D	0.07D	0.095D

According to BIS, splines are specified by four characters

$$nK \times d \times D \text{ (IS: 2327 – 1963)}$$

where

n is the number of splines

K is the type of fit

= A for permanent fit, p = 21 N/mm^2

= B for a hub which is to slide when not under load, p = 14 N/mm^2

= C for a hub which is to slide when under load, p = 7 N/mm^2

d is the minor diameter of the spline shaft

D is the major diameter of the spline shaft.

For example 4A × 26 × 30 (IS: 2327 – 1963)

The torque transmission capacity of a spline shaft depends upon the bearing capacity of the spline. Considering bearing failure of spline, the torque capacity is given as

$$T = \frac{1}{2} phln(D - h) \tag{13.31}$$

where p is the bearing pressure.

Besides bearing failure, the spline may also fail in bending and shearing. The stresses induced in these modes of failure are given as

(i) bending stress, $\sigma_b = \dfrac{3Fh}{b^2 l}$ (13.32)

(ii) shear stress, $\tau = \dfrac{F}{bl}$ (13.33)

where F is the tangential force acting at the mid of spline height.

The main disadvantage of the spline connection is stress concentration, which mainly arises at the root of keys and the core diameter of shaft. Generally, a fillet of small radius (0.2–0.5 mm) is provided to overcome stress concentration.

Example 13.7 It is required to design a square key for fixing a gear on the shaft which transmits 10 kW power at 720 rpm. The shaft and the key are both made of plain carbon steel C45 and the factor of safety is 3.0.

Solution The yield tensile strength of C45 material is 360 N/mm^2. The allowable tensile strength $\sigma_t = \sigma_{yt}/\text{FoS} = 120$ N/mm^2, and maximum shear strength according to maximum distortion energy theory is given by

$$\tau = 0.577\sigma_u = 0.577 \times 120 = 69 \text{ N/mm}^2$$

$$\text{Torque transmitted, } T = \frac{60P}{2\pi N} = \frac{60 \times 10 \times 1000}{2\pi \times 720} = 132.6 \text{ N} \cdot \text{m}$$

$$\text{Shaft diameter, } d = \left(\frac{16T}{\pi\tau}\right)^{1/3} = \left[\frac{16 \times 132.6 \times 1000}{\pi \times 69}\right]^{1/3} = 21.39 \text{ mm, say, 25 mm}$$

Key dimensions. By empirical relations, the proportions of the key are:

$$\text{Width, } b = \frac{d}{4} = \frac{25}{4} = 6.25 \text{ mm, say, 6 mm}$$

$$\text{Height, } h = b = 6 \text{ mm}$$

$$\text{Length of key, } l = 1.5d = 1.5 \times 25 = 37.5 \text{ mm, say, 38 mm}$$

To check shear strength of the key,

$$\text{Torque, } T = bl\tau\frac{d}{2}$$

or

$$\text{Induced shear stress, } \tau = \frac{2T}{bld} = \frac{2 \times 132.6 \times 1000}{6 \times 38 \times 25} = 46.5 \text{ N/mm}^2$$

which is less than the allowable shear strength (69 N/mm^2).

Now considering crushing failure of the key

$$\text{Torque, } T = \frac{lh}{2}\sigma_{cr}\frac{d}{2}$$

Assuming that for ductile material, the crushing strength is equal to tensile strength

$$\text{Induced crushing stress, } \sigma_{cr} = \frac{4T}{lhd} = \frac{4 \times 132.6 \times 1000}{38 \times 6 \times 25} = 93.05 \text{ N/mm}^2$$

which is also less than the allowable strength, hence the selected proportions of the square sunk key are safe.

Design specification. Square key 6 × 6 × 38 (IS : 2048 – 1975)

Example 13.8 Design a taper key for a shaft of diameter 75 mm transmitting 45 kW at 225 rpm. The allowable compressive stress may be taken as 160 N/mm^2.

Solution We select a rectangular cross-section taper key. The proportions of the key are determined empirically.

$$\text{Width, } b = \frac{d}{4} = \frac{75}{4} = 18.75\text{, say, 20 mm}$$

$$\text{Thickness, } h = \frac{d}{6} = \frac{75}{6} = 12.5\text{ mm, say, 12 mm}$$

$$\text{Length of the taper key, } l = \frac{2T}{\mu_1 b \sigma_c d}$$

where

μ_1 is the coefficient of friction between the shaft and the hub (= 0.25 (say))

T is the average torque $\left(= \frac{60P}{2\pi N} = \frac{60 \times 45 \times 1000}{2\pi \times 225} = 1909.85 \text{ N} \cdot \text{m} \right)$

Therefore,

$$\text{length of the key, } l = \frac{2 \times 1909.85 \times 1000}{0.25 \times 20 \times 160 \times 75} = 63.66 \text{ mm}$$

Let us adopt length of the key, $l = 1.25d = 1.25 \times 75 = 93.75$ mm, say, 94 mm

Maximum compressive force on the key

$$F = bl\sigma_c = \frac{20 \times 94 \times 160}{1000} = 300.8 \text{ kN}$$

Assuming that the taper key has a standard slope of 1 : 100, the axial force required to drive the key home is

$$F_a = 2F\mu_2 + F \tan \beta$$

For a greased lubricated key, $\mu_2 = 0.1$. Therefore,

$$F_a = 2 \times 300.8 \times 0.1 + 300.8 \times \frac{1}{100} = 63.17 \text{ kN}$$

Specification of the key: Taper key 20 × 12 × 94 (IS: 2292 – 1974)

Example 13.9 Design a pair of Kennedy key for transmitting 30 kW at 360 rpm. The shaft and key are both made of C50 steel having yield strength of 390 N/mm^2 and factor of safety of 3.0.

Solution

$$\text{Allowable yield strength } \sigma_t = \frac{\sigma_{yt}}{\text{FoS}} = \frac{390}{3} = 130 \text{ N/mm}^2$$

Allowable shear strength, according to maximum shear stress theory

$$\tau = 0.5\sigma_t = 0.5 \times 130 = 65 \text{ N/mm}^2$$

$$\text{Average torque, } T = \frac{60P}{2\pi N} = \frac{60 \times 30 \times 1000}{2\pi \times 360} = 795.78 \text{ N} \cdot \text{m}$$

$$\text{Shaft diameter, } d = \left(\frac{16T}{\pi\tau}\right)^{1/3}$$

$$= \left[\frac{16 \times 795.78 \times 1000}{\pi \times 65}\right]^{1/3} = 39.65 \text{ mm, say, 40 mm}$$

Kennedy key is similar to flat square key. In this case a pair of keys is mounted between the shaft and the hub as shown in the adjoining figure.

Let us adopt the standard proportions of the key from empirical relations. Therefore, width, $b = \dfrac{d}{4} = 10$ mm, thickness, $h = b = 10$ mm. Since there are two keys, the torque transmitted by each key is one-half of the total torque.

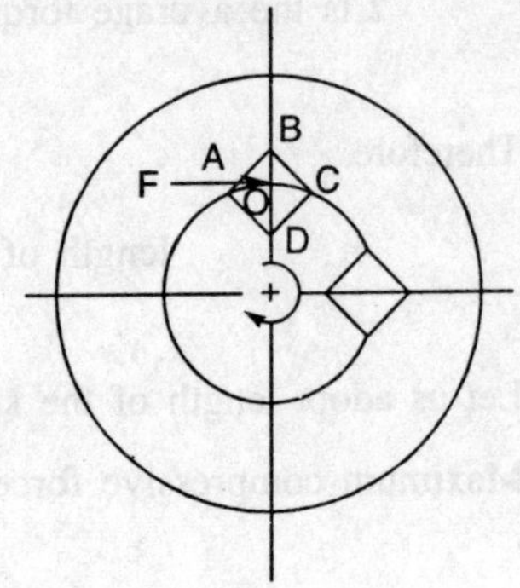

Kennedy key

Therefore, design torque

$$T_d = \frac{T}{2} = \frac{795.78}{2} = 397.9 \text{ N} \cdot \text{m}$$

$$\text{Tangential force, } F = \frac{2T_d}{d} = \frac{2 \times 397.9 \times 1000}{40} = 19{,}895 \text{ N}$$

Shear failure of the key will occur in plane AC. The area A resisting shear is

$$A = \text{AC} \times l = \sqrt{2}\, bl$$

$$\text{Induced shear stress, } \tau = \frac{F}{A} = \frac{F}{\sqrt{2}\, bl}$$

or

$$\text{Required length of the key, } l = \frac{F}{\sqrt{2}\, b\tau} = \frac{19{,}895}{\sqrt{2} \times 10 \times 65} = 21.64 \text{ mm}$$

Let us adopt the length of key at least equal to the shaft diameter, that is, $l = d = 40$ mm.

The key may fail due to crushing. The compressive stress induced is

$$\sigma_c = \frac{F}{OB \times l} = \frac{\sqrt{2}F}{bl} = \frac{\sqrt{2} \times 19,895}{10 \times 40} = 70.34 \text{ N/mm}^2$$

which is less than the allowable strength (130 N/mm^2). Hence the design of the key is safe.

Design specification: Two square keys 10 × 10 × 40 (IS: 2048 – 1975)

Example 13.10 Design a spline which is required in an automotive vehicle gear box to transmit 45 kW at 1200 rpm. The shaft is made of SAE 1045 annealed oil quenched steel.

Solution Yield shear strength of SAE 1045, $\tau_y = 240$ N/mm^2

Assuming factor of safety = 2.4 and approximate value of stress-concentration factor 1.5, the allowable shear strength

$$\tau = \frac{\tau_y}{FoS \times K_{ts}} = \frac{240}{2.4 \times 1.5} = 66.67 \text{ N/mm}^2$$

Average torque transmitted

$$T = \frac{60P}{2\pi N} = \frac{60 \times 45 \times 10^3}{2\pi \times 1200} = 358.1 \text{ N} \cdot \text{m}$$

Let us assume a shaft having 6 splines with class C fitting (where the hub can slide under the load). The permissible bearing pressure is 7 N/mm^2. The standard proportions of the spline are:

Width, $b = 0.25D$

Height, $h = 0.1D$

The torque transmission capacity of the hub is given by the following relation.

$$T = \frac{1}{2} phln(D - h); \quad n = \text{no. of splines}$$

Assuming length of hub, $l = 1.5D$

$$T = \frac{1}{2} \times 7.0 \times 0.1D \times 1.5D \times 6 \times (D - 0.1D) = 2.835D^3$$

or

$$D = \left(\frac{T}{2.835}\right)^{1/3} = \left[\frac{358.1 \times 1000}{2.835}\right]^{1/3} = 50.17 \text{ mm, say, 56 mm (standard major diameter)}$$

$$\text{Minor diameter, } d = \left(D - 2 \times \frac{D}{10}\right) = \left(56 - 2 \times \frac{56}{10}\right) = 44.8 \text{ mm}$$

$$\text{Width of key, } b = \frac{D}{4} = \frac{56}{4} = 14 \text{ mm}$$

$$\text{Height, } h = 0.1D = 5.6 \text{ mm}$$

$$\text{Length of spline, } l = 1.5D = 1.5 \times 56 = 84 \text{ mm}$$

Now checking for the shear failure of the shaft

$$T = \frac{\pi}{16} d^3 \tau$$

or

$$\text{Induced shear stress, } \tau = \frac{16T}{\pi d^3} = \frac{16 \times 358.1 \times 1000}{\pi \times 44.8^3} = 20.28 \text{ N/mm}^2$$

The value of the induced shear stress is less than the allowable shear strength, hence the shaft is safe in shear.

Shear stress in splines

$$\text{Shear force, } F = \frac{2T}{d} = \frac{2 \times 358.1 \times 1000}{44.8} = 15{,}986.6 \text{ N}$$

Induced shear stress in splines,

$$\tau = \frac{F}{bln} = \frac{15{,}986.6}{14 \times 84 \times 6} = 2.26 \text{ N/mm}^2$$

which is also within the allowable limit. Hence the design is safe.

Force required to move the gear axially

$$F_a = \frac{\mu F}{n}; \qquad \mu = 0.15$$

$$= \frac{0.15 \times 15{,}986.6}{6} = 399.6 \text{ N} \approx 400 \text{ N}$$

Design specification of the spline: 6C × 44.8 × 56 (IS: 2327 – 1963)

EXERCISES

1. What is the difference between the shaft and the axle?
2. What are the requirements of a shaft material?
3. What are the different criteria of designing a shaft?
4. How is the strength of a shaft affected by the keyway?
5. What do you mean by stiffness and rigidity of a shaft?
6. What types of stresses are induced in a key?
7. What do you understand by whirling of a shaft?

8. A shaft transmits 10 kW at 720 rpm. The overhang of the shaft is 300 mm and at the extreme right end a force, due to a pulley, of 2.0 kN magnitude is applied. Design a suitable shaft.

9. A turbine shaft transmits 20 kW at 500 rpm. If the permissible twist is limited to 0.5° in 1.5 m length, calculate the diameter of the shaft. The permissible shear strength of shaft material is 70 N/mm^2.

10. Determine the required diameter of a shaft which carries two pulleys; it is 600 mm long, is simply supported at the two ends, and the two pulleys are so located that they divide the shaft in three equal parts. Belt pull on the left pulley is 12 kN vertical, while the pull on the right pulley is 12 kN horizontal. The shaft transmits a torque of 2 kN · m between pulleys.

11. A shaft is supported on bearings A and B, 800 mm between centres. A 20° straight tooth spur gear having 600 mm pitch diameter is located at 200 mm to the right of the left-hand bearing A, and a 700 mm diameter pulley is mounted at 250 mm towards the left of bearing B. The gear is driven by a pinion with a downward tangential force while the pulley drives a horizontal belt having 180° angle of wrap. The pulley also serves as a flywheel and weighs 3000 N. The maximum belt tension is 3000 N and the tension ratio is 3. Determine the maximum bending moment. If the shaft receives 20 kW at 500 rpm, design the shaft.

12. Design a shaft to transmit power from an electric motor to a lathe head stoke through a pulley by means of a belt drive. The pulley weighs 200 N, is located at 305 mm from the centre of bearing. The diameter of the pulley is 200 mm and the maximum power transmitted is 10 kW at 120 rpm. The angle of lap of the belt is 180° and the coefficient of friction between the belt and pulley is 0.3. The allowable shear stress in the shaft may be taken as 35 N/mm^2.

13. A shaft is subjected to mean torque of 3×10^6 N · mm superimposed with variable torque of 3×10^6 N · mm. It is also subjected to bending moment of mean zero and a superimposed variable of 6×10^6 N · mm. Design the shaft to support the load for an infinite number of cycles. The shaft material is SAE 1020 steel.

14. A commercial cold-rolled C30 steel rotating shaft is to be designed to transmit variable torque ranging from 1 kN · m to 2 kN · m. Considering loading of the shaft with minor shock, specify the diameter of the shaft.

15. A shaft of 45 mm diameter is made of steel having yield strength of 400 N/mm^2. Design a parallel key if the maximum power to be transmitted is 20 kW at 800 rpm.

16. Design a key for a shaft to transmit 45 kW at 250 rpm. The allowable yield strength is 150 N/mm^2.

17. A taper key is required to be fastened to a gear on a shaft which is to transmit 15 kW at 360 rpm. Both the shaft and key are made of SAE 1020 steel. Desigh the key.

18. Design a spline shaft which is required to transmit 25 kW at 1000 rpm. The shaft is made of C30 steel.

MULTIPLE CHOICE QUESTIONS

1. A shaft directly coupled to a power source is called
   (a) line shaft (b) counter shaft
   (c) flexible shaft (d) jack shaft
2. Which of the following methods is preferred to make a shaft?
   (a) Cold rolling (b) Cold drawing
   (c) Hot rolling (d) Turning
3. The critical speed of a shaft depends on
   (a) torsional rotation (b) tranverse deflection
   (c) axial elongation (d) all of above
4. Two shafts A and B are of the same length and material. If the diameter of the shaft A is three times the diameter of shaft B, the ratio of torsional stiffness of shafts A and B is
   (a) 3 (b) 9
   (c) 27 (d) 81
5. A shaft used for distribution of power is called
   (a) line shaft (b) axle
   (c) counter shaft (d) jack shaft
6. The design of a power transmitting shaft is based upon the
   (a) maximum shear stress theory (b) maximum principle stress theory
   (c) strain energy theory (d) Rankine theory
7. A round key is used for
   (a) light duty (b) medium duty
   (c) heavy duty (d) extra heavy duty
8. The ratio of width of a rectangular key to the diameter of the shaft on which the key is fitted is
   (a) 1/2 (b) 1/4
   (c) 1/8 (d) 4
9. The Woodruff key is generally used in
   (a) a machine tool (b) an automobile
   (c) a textile machine (d) an agriculture machine
10. It is advisable to use a rectangular key of width/thickness ratio which is
    (a) less than one (b) more than one
    (c) one (d) unknown
11. The most critically stressed point in a rotating shaft with a keyway is
    (a) start of the keyway (b) middle of the keyway
    (c) end of the keyway (d) any point on the keyway

12. Which of the following keys transmits power through frictional resistance only?
    (a) Square key (b) Taper key
    (c) Kennedy key (d) Saddle key
13. Which of the following keys is usually strong in failure against shear and crushing?
    (a) Rectangular (b) Flat
    (c) Square (d) Kennedy
14. Feather keys are generally
    (a) loose in hub and shaft (b) tight in shaft and loose in hub
    (c) tight in hub and shaft (d) loose in shaft and tight in hub
15. The Kennedy key is used in applications such as
    (a) precision duty (b) light duty
    (c) medium duty (d) heavy duty
16. The average value of the weakening effect of a keyway is to reduce the strength of the shaft in shear to
    (a) 50 per cent (b) 60 per cent
    (c) 70 per cent (d) 75 per cent
17. What should be the speed of rotation of a shaft?
    (a) Below critical speed (b) Critical speed
    (c) Above critical speed (d) Any speed
18. As the length of a shaft increases, the rigidity
    (a) increases (b) decreases
    (c) remains unchanged (d) first increases and then decreases

Chapter 14 Couplings, Clutches and Brakes

14.1 COUPLINGS

A coupling is a device that connects two shafts semi-permanently, for example, an electric motor shaft connected to a driven machinery shaft. A coupling also serves some other useful functions as follows:

1. It allows easy disconnection of shafts for repair and maintenance.
2. It tolerates a small amount of misalignment between the connecting shafts.
3. It can prevent transmission of overload power.
4. Due to a flexible element like rubber that is used, it can modify the shock and vibration characteristics of the drive.

Couplings may be classified into two broad groups: (i) rigid couplings and (ii) flexible couplings.

14.2 RIGID COUPLINGS

Rigid couplings are used only in low-speed applications where good axial alignment between the connecting shafts can be achieved. Misaligned shafts, when connected with a rigid coupling, can lead to bearing or fatigue failure, or to worn flanges and broken flange bolts. When such couplings are used to connect line shafts, support bearings should be located near the couplings and checked for both static and dynamic balance. The most commonly used rigid couplings for various applications are (i) muff coupling, (ii) split muff coupling, (iii) marine or solid flange coupling, and (iv) flange coupling.

14.2.1 Muff Coupling

The muff or sleeve coupling is a hollow cylindrical piece which is fitted over both shaft ends connected by means of either keys, taper pins, or set screws, as shown in Figure 14.1. It is simple in design and easy to manufacture. It has a perfect smooth exterior which is good from the safety point of view. These couplings must be fitted with an equal depth of the key placed in the shaft and in the muff to prevent bending of the key. Materials used for the muff and the key are cast iron and plain carbon steel, respectively.

The standard dimensions of the muff coupling, shown in Figure 14.1, are given empirically in terms of the shaft diameter d.

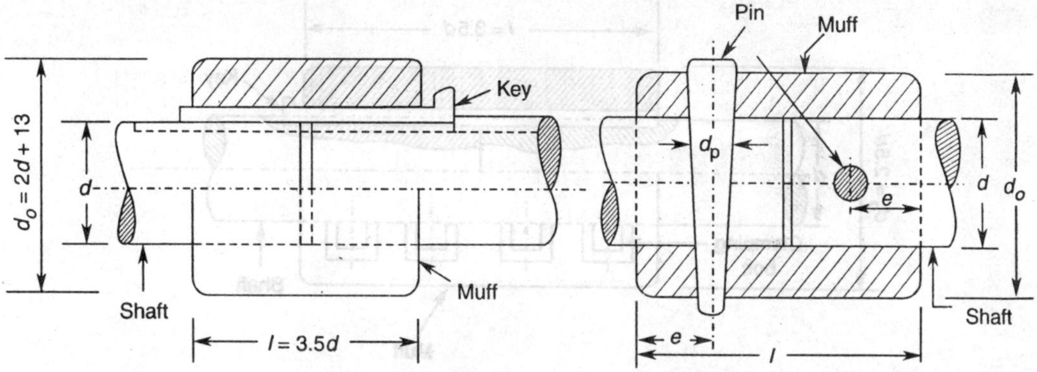

Figure 14.1 Muff or sleeve coupling.

Outer diameter of the muff, $d_o = 2d + 13$ mm
Length of the muff, $l = 3.5d$
Width of the key, $b = d/4$
Thickness of the key, $h = d/4$ or $d/6$
Mean diameter of the pin, $d_p = (0.2–0.3)d$
Taper of the pin, 1 : 20 to 1 : 30
Edge distance, $e = 0.75d$

The dimensions of the muff coupling, calculated by empirical relations, are required to be checked for *torsional failure of the muff*. We neglect the effect of the keyway in the muff while computing the torsional strength T by the equation

$$T = \frac{\pi}{16} \frac{\left(d_o^4 - d^4\right)}{d_o} \tau_{\text{muff}} \tag{14.1}$$

where τ_{muff} is the allowable shear strength of the muff material.

The key used in coupling should be checked for its capacity to withstand failure as discussed in Chapter 13.

14.2.2 Split Muff Coupling

A split muff coupling is made of two cylindrical halves. The connecting shafts are placed in such a way that their ends form a butt and a single key is fitted directly in the keyways of both the shafts. One-half portion of the muff is fixed from the bottom and the other half is placed on the shaft from the top; the both halves are then clamped by means of bolts and nuts as shown in Figure 14.2. To provide additional strength and to serve as a protection to the operator against injury from rotating bolt heads and nuts, muffs are generally ribbed.

The usual proportions of the split muff coupling are:

Outer diameter of the muff, $d_o = 2.5d$

Length of the muff, $l = 3.5d$ or higher depending upon the number of bolts used.

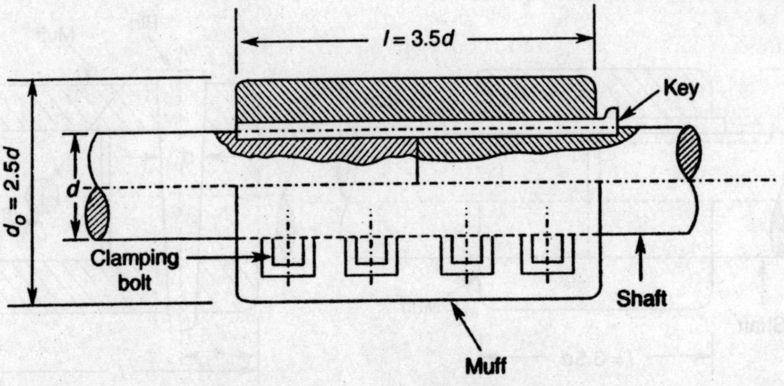

Figure 14.2 Split muff coupling.

The design procedure of the split muff coupling is similar to that of the ordinary muff coupling, except for the design of the bolt. The torque transmission capacity of the bolt is given by the relation

$$T = \frac{\pi^2}{16} \mu n d d_b^2 \sigma_t \tag{14.2}$$

where

μ is the coefficient of friction
n is the number of bolts
d is the diameter of the shaft
d_b is the diameter of the bolt
σ_t is the allowable tensile strength of the bolt material

14.2.3 Marine or Solid Flange Coupling

The marine or solid flange couplings are forged at the ends of the shafts, which are connected together by bolts as shown in Figure 14.3. The bolts are usually tapered 1 : 32 and may be with or without heads. The bolts fit perfectly in reamed holes and carry an equal load. These types of couplings are generally used for connecting marine engine shafts. The proportions of marine

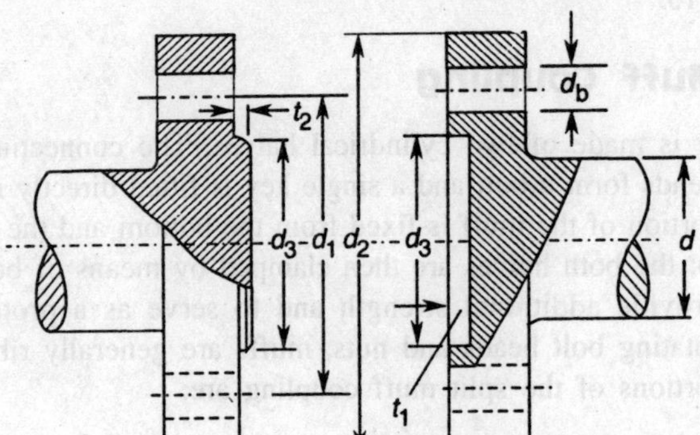

Figure 14.3 Marine flange coupling.

or solid flange couplings should conform to IS: 3653 – 1966. Generally, the following proportions in terms of the shaft diameter d are adopted:

Pitch circle diameter, $d_1 = d + 2d_b + 20$ mm

Outer diameter, $d_2 = d + 5d_b + 20$ mm

Flange thickness, $t = 0.3d + 5$ mm

Diameter of the bolt, $d_b = 0.2d + 5$ mm

Number of bolts, $n = \dfrac{d}{50} + 3$

The design procedure of marine coupling is almost similar to that of flange coupling as discussed in Section 14.2.4.

14.2.4 Flange Coupling

The flange coupling is a widely used rigid coupling which is capable of transmitting large torques. It consists of two cast iron flanges, as shown in Figure 14.4, which are keyed to the shaft ends and bolted together.

Flange couplings have the advantage of simplicity in design, low cost and are available in standard sizes conforming to IS: 6196 – 1971. Flange-connected shafts must be accurately aligned to prevent bending or any other type of failure. In order to ensure true axial alignment, a register is provided on the flange. For safety against openly exposed rotating nuts and bolt heads, a circumferential flange is provided. Such a coupling is generally called protected flange coupling [Figure 14.4(b)].

In order to design a rigid flange coupling, its various dimensions are either selected from the manufacturer's catalogue or calculated by empirical proportions as given below:

Diameter of the hub, $d_1 = (2–2.5)d$

Length of the hub, $l = 1.5d$

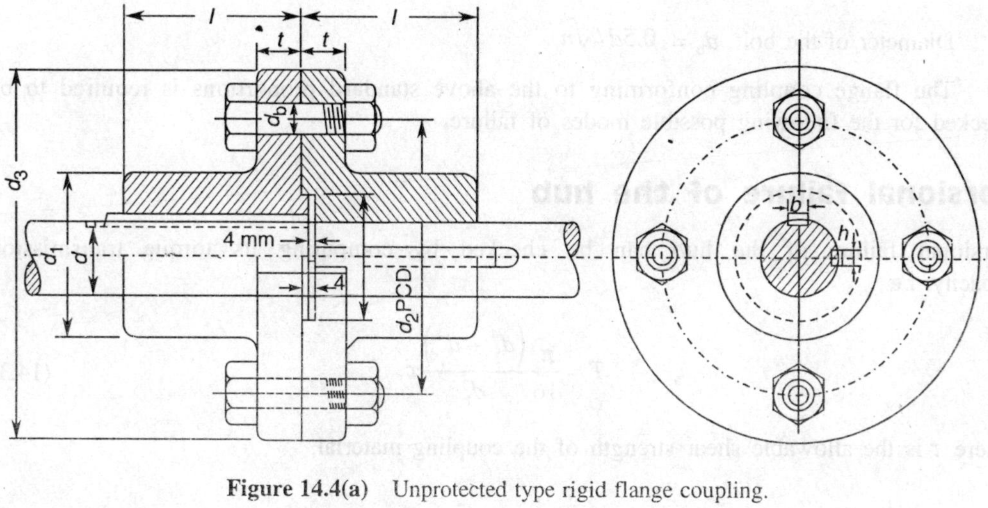

Figure 14.4(a) Unprotected type rigid flange coupling.

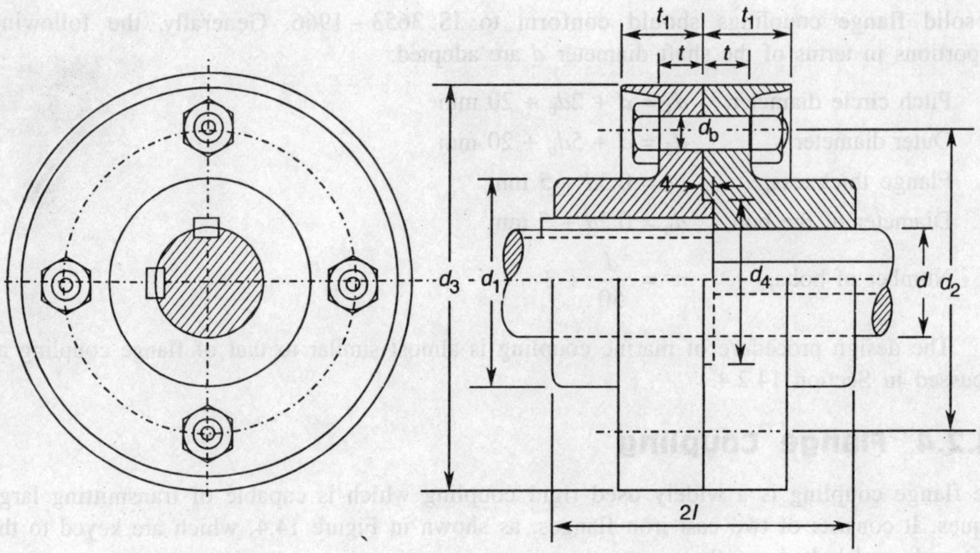

Figure 14.4(b) Protected type rigid flange coupling.

Pitch circle diameter of the bolt, d_2 = 2(radius of the hub + clearance + radius of the head of socket wrench)

$= d_1 + 1.85d_b + 18$ mm

or $d_2 \geq 3d$ (empirically)

Outer diameter of the flange, $d_3 \geq 4d$

Thickness of the flange, $t \geq 0.5d$

Rim thickness, $t_1 \geq t + 0.15d$

Number of bolts, $n = \dfrac{d}{50} + 3$

Diameter of the bolt, $d_b = 0.5d/\sqrt{n}$

The flange coupling conforming to the above standard proportions is required to be checked for the following possible modes of failure.

Torsional failure of the hub

Torsional failure of the hub can be checked by computing its torque transmission capacity, i.e.

$$T = \frac{\pi}{16}\frac{\left(d_1^4 - d^4\right)}{d_1}\tau \tag{14.3}$$

where τ is the allowable shear strength of the coupling material.

Shear failure of the flange

The torque transmission capacity based on shear failure of the flange is given by the relation

$$T = \frac{\pi d_1^2 t\tau}{2} \tag{14.4}$$

Shear failure of the bolt

The torque transmission capacity of the bolt is given by the relation

$$T = \frac{n\pi}{4} d_b^2 \, \tau_1 \, \frac{d_2}{2} \tag{14.5}$$

where

n is the number of bolts

τ_1 is the allowable shear strength of the bolt.

Crushing failure of the bolt

Bolts, while transmitting power, are subjected to compressive stress between the bolt and the flange causing crushing failure. The torque transmission capacity based upon the crushing failure of the bolt is given by the relation

$$T = n d_b t \, \sigma_{cr1} \frac{d_2}{2} \tag{14.6}$$

where σ_{cr1} is the crushing strength of the bolt.

For design of keys refer Chapter 13.

14.3 FLEXIBLE COUPLINGS

A flexible coupling is used to connect two shafts which have one or more types of misalignments (see Figure 14.5) and also to reduce the effect of shock and impact load. In practice, perfect alignment is very seldom found in power transmitting shafts due to improper assembly, workmanship or some kind of desired kinematic flexibility. When such shafts are connected by a rigid coupling, they may be subjected to continuous reversal of bending stresses which ultimately leads to fatigue failure. Therefore, it is desirable to use a flexible coupling to join such shafts.

Flexible couplings can be broadly classified into two types: (i) couplings with kinematic flexibility that employ rigid parts, e.g. Oldham coupling and (ii) couplings having a flexible or resilient member, e.g. bush type coupling.

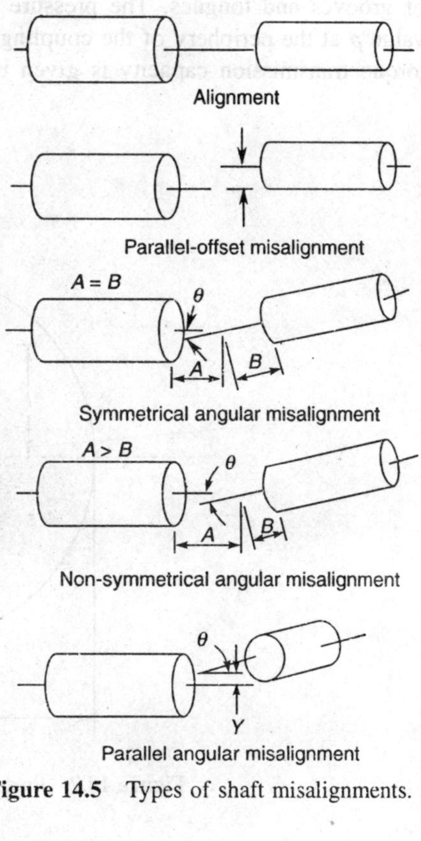

Figure 14.5 Types of shaft misalignments.

14.3.1 Oldham Coupling

In Oldham coupling, as shown in Figure 14.6, flexibility is kinematically obtained by the use of a rigid member in which constraint is absent in certain directions. In this type of coupling, the tongues of the centre piece 'b' are located at right angles to each other. These tongues are fitted in the grooves carved out in each hub. One tongue can slide up and down while the other to and fro. The combined action of these two motions can connect the shafts which are parallel but not collinear. The Oldham coupling is intended to be operated at low speeds.

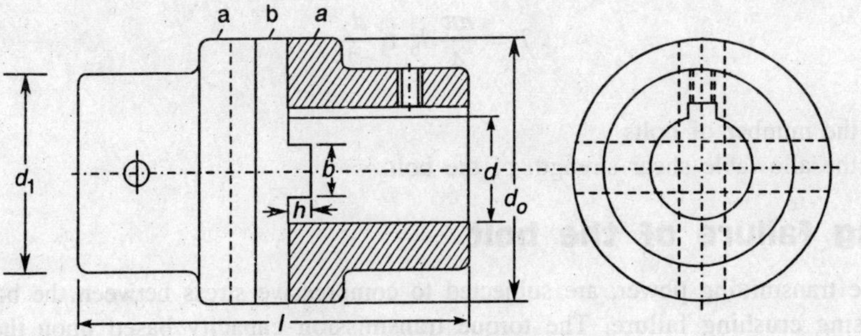

Figure 14.6 Oldham flexible coupling.

The design of the Oldham coupling is based on the allowable pressure between the faces of grooves and tongues. The pressure distribution can be assumed to vary from a maximum value p at the periphery of the coupling to zero at the centre line, as shown in Figure 14.7. The torque transmission capacity is given by the relation

$$T = \frac{1}{6} p d_o^2 h \tag{14.7}$$

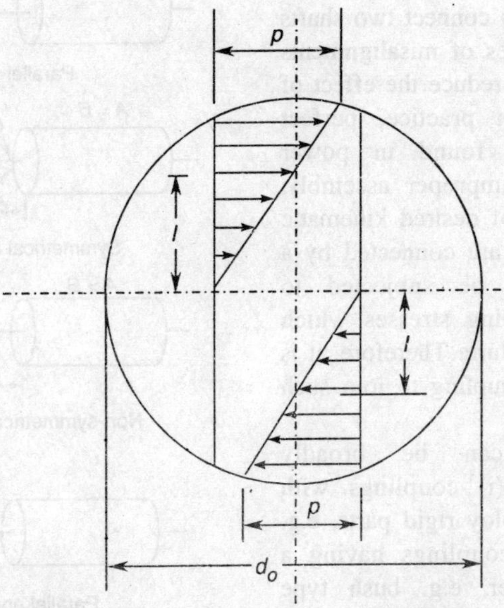

Figure 14.7 Pressure distribution in Oldham coupling.

where

d_o is the outer diameter

h is the axial dimension of the contact area (see Figure 14.6).

The other dimensions of the Oldham coupling can be found from the following empirical relations:

Hub diameter, $d_1 = 2d$

Outer diameter $d_o = (3–4)d$

Thickness of the tongue, $b = 0.45d$

Axial dimension, $h = b/2$

14.3.2 Bush Type Flexible Coupling

A bush type flexible coupling is commonly used where the driving and the driven members are mounted on a monoblock. This coupling can take up parallel misalignment of shafts up to 2.5 mm per 100 mm outside diameter of the coupling and angular misalignments up to 0.5°. The construction of the bush type flexible coupling is similar to that of a rigid flange coupling except for some minor modifications as shown in Figure 14.8 and described below:

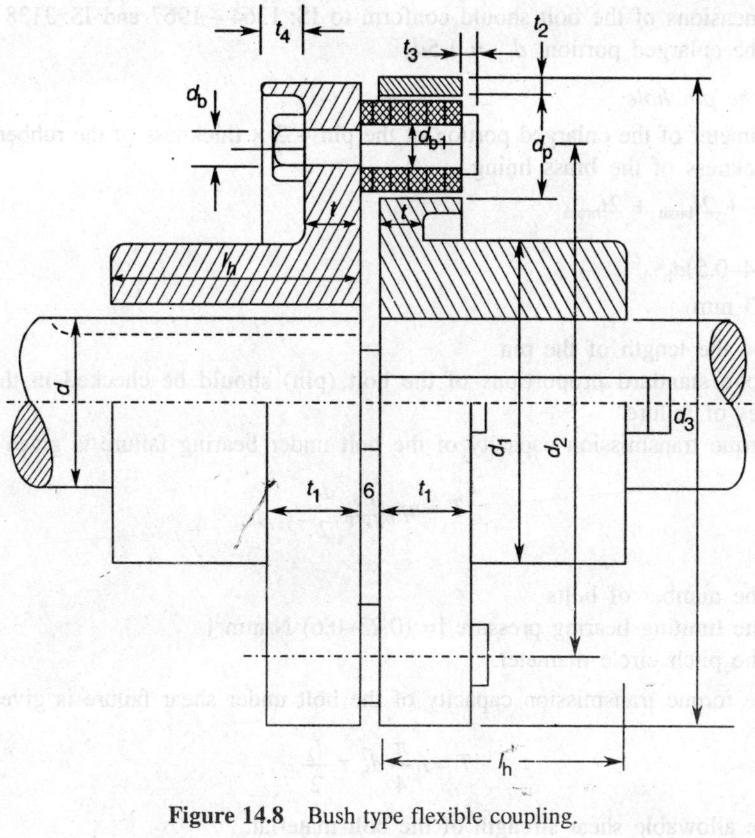

Figure 14.8 Bush type flexible coupling.

(i) The two halves of the coupling flanges are not identically similar.

(ii) The coupling bolts known as pins are specially shaped. One end of the pin is fastened rigidly to one of the flanges with the help of a nut while the other end (enlarged end) rests in the resilient rubber or leather bush. The rubber bush is provided with a brass lining to avoid excessive wear.

The flanges of flexible couplings are made of cast iron grade FG 200 of IS: 210 – 1978. The maximum allowable peripheral velocity is 30 m/s. The other dimensions of the coupling should conform to IS: 2693 – 1964.

The design procedure of the bush type flange coupling is similar to that of rigid flange coupling except for the design of the pin, which is described below:

Design of bolt or pin

The bolt, also called the pin, of a flexible coupling is enlarged at one end. It is supported by a resilient bush with brass lining. The empirical relations for the pin are:

$$\text{Number of bolts, } n = 0.04d + 3$$

$$\text{Bolt diameter, } d_b = \frac{0.423d}{\sqrt{n}} + 7.5 \text{ mm}$$

The dimensions of the bolt should conform to IS: 1364 – 1967 and IS: 3138 – 1966. The diameter of the enlarged portion, $d_{b1} = 1.5d_b$

Diameter of the pin hole

d_p = diameter of the enlarged portion of the pin + 2 × thickness of the rubber bush + 2 × thickness of the brass lining.

$= d_{b1} + 2t_{bush} + 2t_{brass}$

where

$t_{bush} = (0.4–0.5)d_b$

$t_{brass} = 2–3$ mm

Let l_p be the length of the pin.

The above standard proportions of the bolt (pin) should be checked in the following possible modes of failure.

(i) Torque transmission capacity of the bolt under bearing failure is given by

$$T = np_b d_p l_p \frac{d_2}{2} \tag{14.8}$$

where

n is the number of bolts

p_b is the limiting bearing pressure [= (0.25–0.6) N/mm^2]

d_2 is the pitch circle diameter.

(ii) The torque transmission capacity of the bolt under shear failure is given by

$$T = n\frac{\pi}{4} d_b^2 \, \tau \frac{d_2}{2} \tag{14.9}$$

where τ is the allowable shear strength of the bolt material.

(iii) The bolt is subjected to transverse force. Its one end is held fixed in the flange. Therefore, it will act as a cantilever beam as shown in Figure 14.9, subjected to a

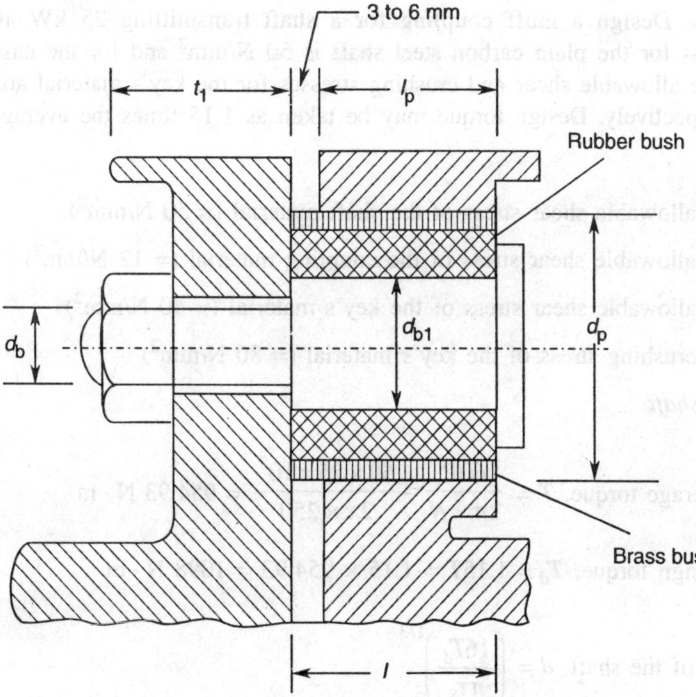

Figure 14.9 Details of pin and bush.

force acting perpendicular to its axis. The bending moment

$$M = Fl \tag{14.10}$$

where

F = transverse force = $2T / d_2$
l = $(l_p/2)$ + clearance.

(iv) Since the bolt is subjected to tensile stress due to bending and to shear stress due to transverse shear force, it is a case of biaxial loading. Therefore, the maximum principal stress and the maximum shear stress produced must be within the allowable limits. Accordingly, the following relations must be satisfied for safe design:

$$0.5\left(\sigma_b + \sqrt{\sigma_b^2 + 4\tau^2}\right) \le \frac{\sigma_{yt}}{\text{FoS}} \tag{14.11}$$

$$0.5\left(\sqrt{\sigma_b^2 + 4\tau^2}\right) \le \frac{0.5\,\sigma_{yt}}{\text{FoS}} \tag{14.12}$$

The other proportions of the flange can be computed empirically and these should be checked in different modes of failure as described in Section 14.2.4 for the flange coupling.

Example 14.1 Design a muff coupling for a shaft transmitting 25 kW at 250 rpm. The safe shear stress for the plain carbon steel shaft is 50 N/mm^2 and for the cast iron muff it is 12 N/mm^2. The allowable shear and crushing stresses for the key's material are 40 N/mm^2 and 80 N/mm^2, respectively. Design torque may be taken as 1.15 times the average torque.

Solution Let

τ_S be the allowable shear stress of the shaft material (= 50 N/mm^2)

τ_C be the allowable shear stress of the coupling material (= 12 N/mm^2)

τ_K be the allowable shear stress of the key's material (= 40 N/mm^2)

σ_{cr} be the crushing stress of the key's material (= 80 N/mm^2)

Design of the shaft

$$\text{Average torque, } T = \frac{60P}{2\pi \times N} = \frac{60 \times 25 \times 10^3}{2\pi \times 250} = 954.93 \text{ N} \cdot \text{m}$$

$$\text{Design torque, } T_d = 1.15T = 1.15 \times 954.93 = 1098 \text{ N} \cdot \text{m}$$

$$\text{Diameter of the shaft, } d = \left(\frac{16T_d}{\pi\tau_S}\right)^{1/3}$$

$$= \left[\frac{16 \times 1098 \times 1000}{\pi \times 50}\right]^{1/3} = 48.18 \text{ mm, say, 50 mm}$$

Design of the muff coupling

The proportions of the muff coupling are given by the following empirical relations:

$$\text{Outer diameter of the muff, } d_o = 2d + 13 \text{ mm}$$

$$= 2 \times 50 + 13 = 113 \text{ mm, say, 115 mm}$$

$$\text{Length of the muff, } l = 3.5d = 3.5 \times 50 = 175 \text{ mm}$$

$$\text{Width of the key, } b = \frac{d}{4} = 12.5 \text{ mm, say, 15 mm}$$

Since the given crushing stress of the key's material is twice the allowable shearing stress, a square sunk key would be suitable.

$$\text{Thickness of the key, } h = b = 15 \text{ mm}$$

$$\text{Effective length of the key } l_1 = \frac{l}{2} = \frac{175}{2} = 87.5 \text{ mm}$$

The dimensions of the muff coupling, calculated by empirical relations, are required to be checked in the following mode of failure:

Torsional failure of muff

The torque transmission capacity of the muff

$$T_d = \frac{\pi}{16} \frac{\left(d_o^4 - d^4\right)}{d_o} \tau_{muff}$$

or

$$1098 \times 10^3 = \frac{\pi}{16}\left[\frac{115^4 - 50^4}{115}\right]\tau_{muff}$$

or

$$\tau_{muff} = 3.81 \text{ N/mm}^2$$

which is less than the allowable stress τ_C of the CI muff. Hence the design is safe in torsion.

Note: The stresses induced in key may be checked as discussed in Chapter 13.

Example 14.2 Design a split muff coupling which is required to connect two shafts of 36 mm diameter transmitting 15 kW at 360 rpm. The shaft and the key are both made of plain carbon steel whose allowable shear stress is 50 N/mm^2. Four bolts of the same material are used to connect the two halves of the coupling. The allowable tensile/compressive strength of the bolt is 100 N/mm^2. The coefficient of friction between the shaft and the coupling is 0.3. The coupling is made of CI having allowable shear strength of 12 N/mm^2.

Solution

Design of the muff

The usual proportions of the split muff coupling are:

$$\text{Outer diameter of the muff, } d_o = 2.5d = 2.5 \times 36 = 90 \text{ mm}$$

$$\text{Length of the muff, } l = 3.5d = 3.5 \times 36 = 126 \text{ mm}$$

Design of the bolt

Torque transmission capacity of the bolt

$$T = \frac{60P}{2\pi \times N} = \frac{\pi^2}{16}\mu\, d_b^2\, nd\, \sigma_t$$

or

$$\frac{60 \times 15 \times 10^6}{2\pi \times 360} = 397.89 \times 10^3 = \frac{\pi^2}{16} \times 0.3 \times d_b^2 \times 4 \times 36 \times 100$$

or

$$d_b = 12.22 \text{ mm}$$

Therefore, the nominal diameter of the bolt as per IS: 4218 (Part III)—1967 is 14 mm or M14 × 1.

Note: For the design of the key, refer Chapter 13.

Example 14.3 Design and draw a cast iron protected type flange coupling to connect two shafts of 36 mm diameter transmitting 15 kW at 720 rpm. The overload capacity is 1.25 times the average torque. The bolts and keys are made of C20 steel and the flanges are made of FG 200.

Solution For C20 steel having the factor of safety 2, maximum shear stress $\tau = 50$ N/mm^2 and for cast iron $\tau = 10$ N/mm^2.

$$\text{Average torque, } T = \frac{60P}{2\pi \times N} = \frac{60 \times 15 \times 10^3}{2\pi \times 720} = 198.95 \text{ N} \cdot \text{m}$$

$$\text{Design torque, } T_d = 1.25T = 1.25 \times 198.95 = 248.7 \text{ N} \cdot \text{m}$$

Dimensions of the flange coupling

Using the empirical relations:

Hub diameter, $d_1 = 2d = 2 \times 36 = 72$ mm

Length of the hub, $l = 1.5d = 1.5 \times 36 = 54$ mm

pcd of the flange, $d_2 = 3d = 3 \times 36 = 108$ mm, say, 110 mm

Flange thickness, $t = 0.5d = 0.5 \times 36 = 18$ mm

Outer diameter of the flange, $d_3 = 4d = 4 \times 36 = 144$ mm

Key dimensions, $b \times h = 10$ mm $\times$ 10 mm

(Assuming that shear stress is half the crushing stress, a square key is used.)

$$\text{Number of bolts, } n = \frac{d}{50} + 3 = \frac{36}{50} + 3 = 3.72, \text{ say, } 4.$$

$$\text{Diameter of the bolt, } d_b = \frac{0.5d}{\sqrt{n}} = \frac{0.5 \times 36}{\sqrt{4}} = 9 \text{ mm}$$

Let us adopt M10 × 1 IS : 4218 (Part III)—1967 bolt.

The flange coupling of the above standard proportions is required to be checked for the following possible modes of failure.

(i) Torsional shear strength of the hub

$$T_d = \frac{\pi}{16} \frac{\left(d_1^4 - d^4\right)}{d_1} \tau$$

or

$$248.7 \times 10^3 = \frac{\pi}{16} \frac{72^4 - 36^4}{72} \times \tau$$

or

$$\tau = 3.62 \text{ N/mm}^2$$

(ii) Torsional shear strength based on shear failure of the flange

$$T_d = \frac{\pi d_1^2 t \tau}{2}$$

or

$$\tau = \frac{2T_d}{\pi d_1^2 t} = \frac{2 \times 248.7 \times 10^3}{\pi \times 72^2 \times 18} = 1.69 \text{ N/mm}^2$$

The induced shear stress in the above two modes of failure is less than the allowable shear strength (10 N/mm^2) of the CI flange. Hence the design is safe.

(iii) Torque capacity based on crushing strength of the bolt

$$T_d = d_b \, t \, n \, \sigma_{cr} \frac{d_2}{2}$$

or

$$\sigma_{cr} = \frac{2T_d}{n d_b \, t d_2} = \frac{2 \times 248.7 \times 10^3}{4 \times 9 \times 18 \times 110} = 6.98 \text{ N/mm}^2$$

(iv) Torque capacity based on shearing strength of the bolt

$$T_d = \frac{n\pi}{4} d_b^2 \, \tau \frac{d_2}{2}$$

or

$$\text{Induced shear stress, } \tau = \frac{8T_d}{n\pi \, d_b^2 \, d_2} = \frac{8 \times 248.7 \times 10^3}{4 \times \pi \times 9^2 \times 110} = 17.77 \text{ N/mm}^2$$

Since the induced crushing and shearing stresses in the bolt are less than the allowable limit, the selection of the bolt M10 × 1 is satisfactory.

Thickness of the protecting rim, $t_1 = t + 0.15d = 18 + 0.15 \times 36 = 23.4$ mm, say, 24 mm which also increases the rigidity of the coupling flanges.

For the sketch, refer Figure 14.4(b).

Example 14.4 Design a flexible coupling to connect two shafts which transmit 10 kW at 500 rpm.

Solution We assume that the shaft is made of plain carbon steel C40 having yield tensile strength, $\sigma_{yt} = 320$ N/mm^2

Assuming factor of safety, FoS = 2, the allowable shear strength

$$\tau = \frac{0.5\,\sigma_{yt}}{\text{FoS}} = \frac{0.5 \times 320}{2} = 80 \text{ N/mm}^2$$

$$\text{Average torque transmitted, } T = \frac{60P}{2\pi \times N}$$

$$= \frac{60 \times 10 \times 10^3}{2\pi \times 500} = 191 \text{ N} \cdot \text{m}$$

$$\text{Diameter of the shaft, } d = \left(\frac{16T}{\pi\tau}\right)^{1/3}$$

$$= \left[\frac{16 \times 191 \times 10^3}{\pi \times 80}\right]^{1/3} = 23 \text{ mm, say, } 25 \text{ mm}$$

Design of coupling

The standard proportions of flanges of flexible coupling are:

Diameter of the hub, $d_1 = 2d = 2 \times 25 = 50$ mm

Length of the hub, $l = 2d = 50$ mm

pcd of the bolt circle, $d_2 = 3d = 75$ mm

Thickness of the flange, $t_1 = 0.5d = 0.5 \times 25 = 12.5$ mm, say, 15 mm

Number of bolts, $n = 0.04d + 3 = 0.04 \times 25 + 3 = 4$, say, 6.
Diameter of the bolt at the small end

$$d_b = \frac{0.423d}{\sqrt{n}} + 7.5$$

$$= \frac{0.423 \times 25}{\sqrt{6}} + 7.5 = 11.81 \text{ mm, say, } 12 \text{ mm}$$

Diameter of the enlarged portion of the bolt

$$d_{b1} = 1.5d_b = 1.5 \times 12 = 18 \text{ mm}$$

On the enlarged portion, there will be a brass sleeve of 2 mm thickness and a rubber bush of thickness

$$t_{bush} = 0.4d_b = 0.4 \times 12 = 4.8 \text{ mm, say, } 6 \text{ mm}$$

Overall diameter of the pin hole

$$d_p = d_{b1} + 2 \times t_{bush} + 2 \times t_{brass}$$
$$= 18 + 2 \times 6 + 2 \times 2 = 34 \text{ mm}$$

Revised pcd of the pin hole

$$d_2 = d_1 + d_p + \text{some clearance to account for wrench space.}$$

$$= 50 + 34 + 16 = 100 \text{ mm}$$

$$\text{Outer diameter, } d_3 = (4\text{–}5)d = 5 \times 25 = 125 \text{ mm}$$

Check for pin

(i) Torque capacity of the pin under bearing failure:

Assume that the permissible value of bearing pressure between the rubber bush and steel bolt, $p_b = 0.5$ N/mm^2. Therefore,

$$T = np_b d_p l_p \frac{d_2}{2}$$

or

$$191 \times 10^3 = 6 \times 0.5 \times 34 \times l_p \times \frac{100}{2}$$

or

$$l_p = 37.4 \text{ mm, say, } 38 \text{ mm}$$

(ii) Torque capacity of the pin under shear failure:

$$T = n\frac{\pi}{4} d_b^2 \, \tau \frac{d_2}{2}$$

or

$$191 \times 10^3 = 6 \times \frac{\pi}{4} \times 12^2 \times \tau \times \frac{100}{2}$$

or

$$\tau = 5.63 \text{ N/mm}^2$$

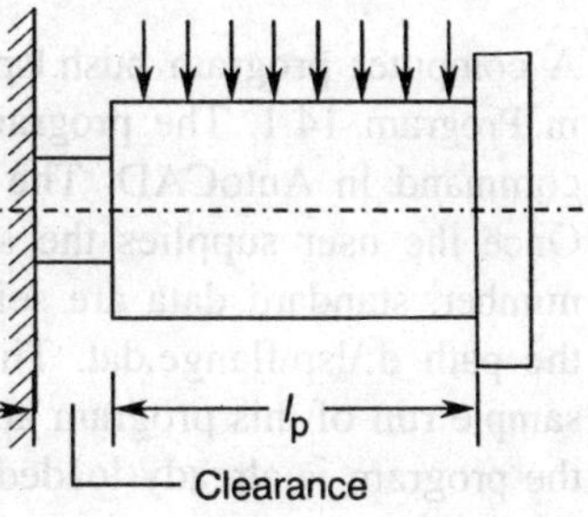

(iii) Bending stress in the bolt:

$$\text{Bending moment, } M = \frac{2T}{nd_2} \times L$$

where, L = moment arm $= \dfrac{l_p}{2}$ + clearance $= \dfrac{38}{2} + 5 = 24$ mm

Therefore,

$$\text{Moment, } M = \frac{2 \times 191 \times 24}{6 \times 100} = 15.28 \text{ N} \cdot \text{m}$$

$$\text{Bending stress, } \sigma_b = \frac{M}{Z} = \frac{32M}{\pi d_b^3}$$

$$= \frac{32 \times 15.28 \times 10^3}{\pi \times 12^3} = 90.07 \text{ N/mm}^2$$

Now considering bolt as a biaxial stress element, maximum principal stress

$$\sigma_1 = 0.5\left[\sigma_b + \sqrt{\sigma_b^2 + 4\tau^2}\right]$$

$$= 0.5\left[90.07 + \sqrt{90.07^2 + 4 \times 5.63^2}\right] = 90.42 \text{ N/mm}^2$$

which is less than the permissible yield strength.

According to maximum shear stress theory,

$$\tau_{max} \le \frac{0.5\,\sigma_{yt}}{\text{FoS}}$$

or

$$\tau_{max} = \sqrt{\left(\frac{\sigma_b}{2}\right)^2 + \tau^2}$$

$$= \sqrt{\left(\frac{90.07}{2}\right)^2 + 5.63^2} = 45.38 \text{ N/mm}^2$$

which is less than the allowable shear strength. Hence the design of the pin is safe.

Checking of flanges and key for different modes of failure is left as an exercise for the student.

14.4 COMPUTER-AIDED DESIGN

A computer program bush.lsp, for making a drawing of a bush type flexible coupling is given in Program 14.1. The program is written in AutoLISP which can be easily used as a standard command in AutoCAD. The program asks for shaft diameter and starting point coordinates. Once the user supplies the shaft diameter, a serial number is assigned and as per the serial number, standard data are selected from a data file. In this program, the data file is placed in the path d:\lsp\flange.dat. The data contained in this file are taken from IS: 2693–1964. The sample run of this program at command prompt of AutoCAD is given below: (It is assumed that the program is already loaded into memory using the load option of AutoCAD, available in the Tools menu.)

```
Command: bush ↵
Shaft diameter: 30 ↵
Start point: 100, 100 ↵
```

The sample drawing produced is shown in Figure 14.10.

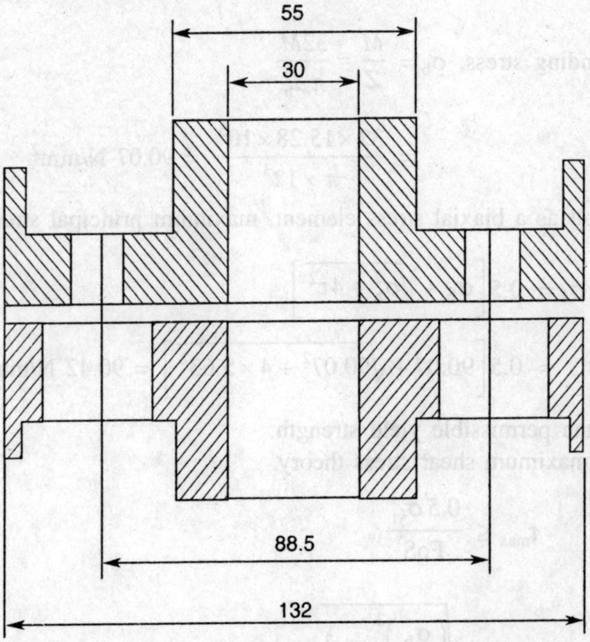

Figure 14.10 Sample output of bush.lsp program.

Program 14.1

```
;      bush type flexible coupling
;
;      function to calculate degree to radian conversion
(defun dtr (ang) (* pi(/ ang 180.0)))
;
(defun bushinp ( )
;      bush coupling input function
;
       (setq dia (getreal "\nShaft Diameter: "))
       (setq d dia)
;      define serial number for specified diameter range
          (if (and (> = d 12) (< d 16)) (setq cn 1))
          (if (and (> = d 16) (< d 22)) (setq cn 2))
          (if (and (> = d 22) (< d 30)) (setq cn 3))
          (if (and (> = d 30) (< d 45)) (setq cn 4))
          (if (and (> = d 45) (< d 56)) (setq cn 5))
          (if (and (> = d 56) (< d 75)) (setq cn 6))
          (if (and (> = d 75) (< d 85)) (setq cn 7))
          (if (and (> = d 85) (< d 110)) (setq cn 8))
          (if (and (> = d 110) (< d 130)) (setq cn 9))
          (if (and (> = d 130) (< = d 150)) (setq cn 10))
;      open flange.dat file available in d:\lisp path
       (setq file (open "D:/lisp/flange.dat" "r"))
       (setq line (read-line file))
       (setq loop t)
       (while loop
          (setq sr (atoi (substr line 1 2))); isolate first two character and convert into integer
;      convert them into integer and check whether it is equal to serial
;      number assigned to it. if yes then isolate other data
;
          (if (= cn sr)
                   (progn

                         (setq d3 (atof (substr line 3 4))) (print d3)
                         (setq d1 (atof (substr line 7 4))) (print d1)
                         (setq lh (atof (substr line 11 4))) (print lh)
                         (setq t1 (atof (substr line 15 3))) (print t1)
                         (setq t2 (atof (substr line 18 3))) (print t2)
                         (setq d2 (atof (substr line 21 4))) (print d2)
                         (setq db (atof (substr line 25 3))) (print db)
                         (setq dp (atof (substr line 28 3))) (print dp)
                         (setq c (atof (substr line 31 2))) (print c)
                         (setq loop nil)
                   ); end progn and then part
```

```
        ); end if
                        (setq line (read-line file))
                        (if (= line nil)
                            (progn
                            (promopt "\n DATA does not exist . . .")
                            (setq loop nil)
                            ); end progn
                        ); end if
    ); end while
); end bushinp function
;
(defun bushleft ( )
;
;   initialize variables
    (setq pt1 nil) (setq pt2 nil) (setq pt3 nil) (setq pt4 nil)
    (setq pt5 nil) (setq pt6 nil) (setq pt7 nil) (setq pt8 nil)
    (setq pt9 nil) (setq pt10 nil) (setq pt11 nil) (setq pt12 nil)
    (setq pt13 nil) (setq pt14 nil) (setq pt15 nil) (setq pt16 nil)
;
    (setq pt1 (getpoint "\nStart Point: "))
;
;   calculate coordinates of points
;
    (setq pt16 (polar pt1 (dtr 180) lh))
    (setq pt6 (polar pt1 (dtr 270) (/ d3 2)))
    (setq pt7 (polar pt6 (dtr 180) t1))
    (setq pt2 (polar pt1 (dtr 270) (/ dia 2)))
    (setq pt15 (polar pt2 (dtr 180) lh))
    (setq pt5 (polar pt1 (dtr 270) (/ (+ d2 db) 2.0)))
    (setq gh (- t1 t2))
    (setq pt10 (polar pt5 (dtr 180) gh))
    (setq pt3 (polar pt5 (dtr 90) db))
    (setq pt12 (polar pt3 (dtr 180) gh))
    (setq pt8 (polar pt7 (dtr 90) (/ t2 3)))
    (setq pt14 (polar pt16 (dtr 270) (/ d1 2)))
    (setq pt13 (polar pt14 (dtr 0) (- lh gh)))
    (setq pt9 (polar pt8 (dtr 0) t2))
;
;   draw left flange
;
      (command "line" pt2 pt15 " ")
      (setq ent1 (entlast))
      (command "mirror" ent1 " " pt1 pt16 "n")
      (command "line" pt3 pt12 " ")
      (setq ent2 (entlast))
      (command "mirror" ent2 " " pt1 pt16 "n")
```

```
        (command "line" pt5 pt10 " ")
        (setq ent3 (entlast))
        (command "mirror" ent3 " " pt1 pt16 "n")
        (command "pline" pt1 pt2 pt3 pt5 pt6 pt7 pt8 " ")
        (setq ent4 (entlast))
        (command "mirror" ent4 " " pt1 pt16 "n")
        (command "pline" pt16 pt15 pt14 pt13 pt9 pt8 " ")
        (setq ent5 (entlast))
        (command "mirror" ent5 " " pt1 pt16 "n")
        (command "zoom" "e")
); end bushleft function
;
(defun bushright ( )
;
;     initialize variables
      (setq qt1 nil) (setq qt2 nil) (setq qt3 nil) (setq qt4 nil)
      (setq qt5 nil) (setq qt6 nil) (setq qt7 nil) (setq qt8 nil)
      (setq qt9 nil) (setq qt10 nil) (setq qt11 nil) (setq qt12 nil)
      (setq qt13 nil) (setq qt14 nil) (setq qt15 nil) (setq qt16 nil)
;
      (setq qt1 (polar pt1 (dtr 0) c))
      (setq qt16 (polar qt1 (dtr 0) lh))
      (setq qt6 (polar qt1 (dtr 270) (/ d3 2)))
      (setq qt7 (polar qt6 (dtr 0) t1))
      (setq qt2 (polar qt1 (dtr 270) (/ dia 2)))
      (setq qt15 (polar qt2 (dtr 0) lh))
      (setq qt5 (polar qt1 (dtr 270) (/ (+ d2 dp) 2.0)))
      (setq gh (- t1 (* db 0.666)))
      (setq qt10 (polar qt5 (dtr 0) gh))
      (setq qt3 (polar qt5 (dtr 90) dp))
      (setq qt12 (polar qt3 (dtr 0) gh))
      (setq qt8 (polar qt7 (dtr 90) (/ t2 4)))
      (setq qt14 (polar qt16 (dtr 270) (/ d1 2)))
      (setq qt13 (polar qt14 (dtr 180) (- lh gh)))
      (setq qt9 (list (car qt10) (cadr qt8)))
;
;     draw right flange
;
        (command "line" qt2 qt15 " ")
        (setq ent1 (entlast))
        (command "mirror" ent1 " " qt1 qt16 "n")
        (command "line" qt3 qt12 " ")
        (setq ent2 (entlast))
        (command "mirror" ent2 " " qt1 qt16 "n")
        (command "line" qt5 qt10 " ")
        (setq ent3 (entlast))
```

```
        (command "mirror" ent3 " " qt1 qt16 "n")
        (command "pline" qt1 qt2 qt3 qt5 qt6 qt7 qt8 " ")
        (setq ent4 (entlast))
        (command "mirror" ent4 " " qt1 qt16 "n")
        (command "pline" qt16 qt15 qt14 qt13 qt12 qt10 qt9 qt8 " ")
        (setq ent5 (entlast))
        (command "mirror" ent5 " " qt1 qt16 "n")
); end bushright function
;
;     main program bush.lsp
(defun c:bush ( )
        (setvar "cmdecho" 0)
        (setq oldosmode (getvar "osmode"))
        (setvar "osmode" 0)
;       call bushinp
        (bushinp)
;       call bushleft
        (bushleft)
;       call bushright
        (bushright)
        (setvar "osmode" oldosmode)
        (setvar "cmdecho" 1)
); end main program
```

14.5 CLUTCHES

A clutch is a machine element which facilitates control over the flow of power and motion from the prime-mover to the driven machinery. Like a shaft coupling, a clutch is also used for connecting two shafts. However, it differs in construction and purpose. Whereas a coupling forms a semi-permanent connection between two shafts, a clutch provides the flexibility of readily engaging and disengaging two shafts. The shafts connected by a clutch should be strictly collinear. No misalignment is allowed. Any misalignment between shafts will seriously deteriorate the performance of a clutch, eventually resulting into failure. A clutch, in a mechanical power transmission system, can be used for any of the following functions:

1. To connect or disconnect the source power from the remaining parts of the power transmission system, at the will of the operator, as in an automobile vehicle.
2. To serve as a safety device by slipping when the torque transmitted through it exceeds a safe value.
3. To reduce or to absorb the impact and shock load and to alter the vibrational characteristics of two connecting shafts. Hydraulic clutches are most commonly used for this purpose.

Clutches may be broadly classified into two categories:

Positive contact clutches. These clutches transmit power without any slip, e.g. a jaw or claw clutch.

Friction clutches. These types of clutches transmit power through frictional force. They can be further classified into two groups:

Axial clutch. One in which the contact pressure is applied in a direction parallel to its axis of rotation. Generally, three types of axial clutches are available: (1) single disc, (2) multi-disc, and (3) cone clutch.

Radial clutch. In a radial clutch, the contact pressure is applied upon a rim in a radial direction.

For the proper selection of a clutch it is very important that the designer be familiar with the design requirements and performance characteristics of various clutches. Some of the important factors which must be taken into consideration while deciding the type of clutch to be used in a specific application are given below:

Torque. For heavy or fluctuating torques, a multidisc oil immersed friction clutch is suitable. For high torques at low speeds, a cone clutch or rim type clutch may be used. A jaw clutch may be used for small torques at low speeds.

Speed. For high rotational speeds, the multidisc type clutches are best suited as they are light, compact and dynamically balanced. For the medium range of speeds, a single plate/disc clutch is more suitable.

Space limitation. When space limitation is a major requirement, multidisc or double cone clutches are a natural choice as they are compact.

Frequency of operation. Clutches which are in frequent or continuous use should have a small travel, simple mechanism to operate and should also have a large cooling area which can dissipate heat energy. In this respect the single disc clutch is most suitable.

14.6 POSITIVE CONTACT CLUTCHES

In a positive contact type clutch, the contacting clutch surfaces are interlocked by mechanical means, namely jaws, to produce a rigid joint. Such a clutch is also called jaw or claw clutch. A jaw clutch, as shown in Figure 14.11, is quite simple and is a common type of positive contact clutch. It transmits power as long as the jaws are in contact. These clutches are simple in construction, small in size and light in weight compared to other types. They need little or no

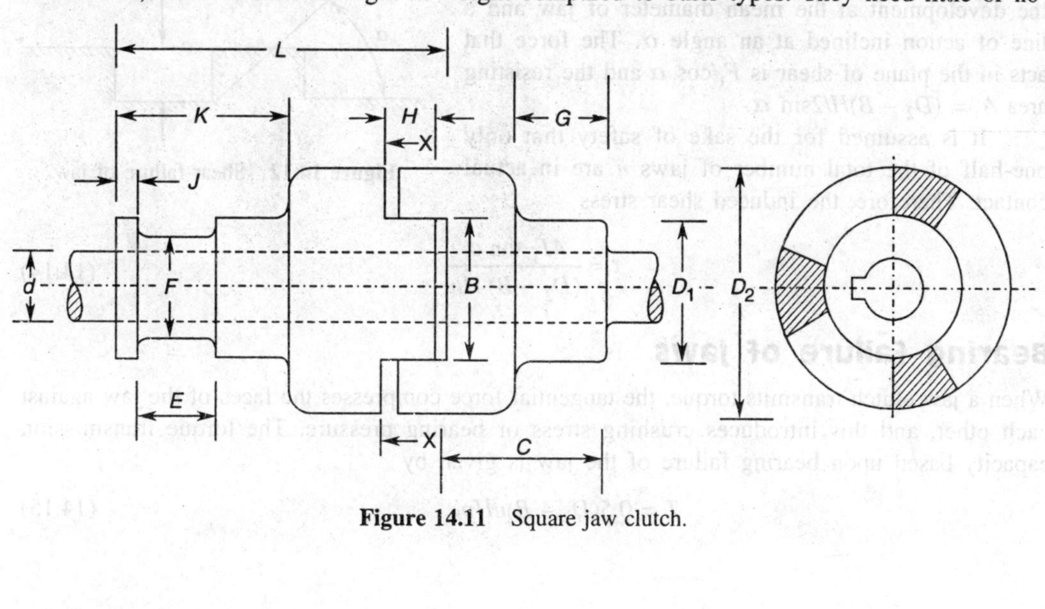

Figure 14.11 Square jaw clutch.

maintenance. For a given torque capacity, their performance is better than that of the friction clutches. However, these clutches can be used only to transmit small torques at low speeds. Further, frequent engagement or disengagement may cause wearing of jaws which may hamper performance. Positive contact clutches find use in automotive transmission, presses, punches, and household appliances like kitchen grinding machines, and so on.

The design of a jaw clutch is relatively simple. The dimensions of the clutch are determined empirically and later checked for different modes of failure. The usual proportions of a square jaw clutch are given below (for nomenclature: see Figure 14.11).

$D_2 = 2.2d + 25$ mm $H = 0.3d + 12.5$ mm
$C = 1.2d + 30$ mm $E = 0.4d + 6$ mm
$F = 1.4d + 8$ mm $J = 0.2d + 4$ mm
$G = d + 5$ mm $L = 1.7d + 58$ mm
$K = 1.2d + 20$ mm $B = 1.25d + 5$ mm
$D_1 = (1.6\text{–}2.0)d$

Some of the dimensions, calculated empirically, are required to be checked in the following possible modes of failure. Let T be the design torque and d the shaft diameter.

The tangential force F_t acting on the jaw is calculated as

$$F_t = \frac{2T}{d_{mean}} \tag{14.13}$$

where d_{mean} is the mean diameter [= $(D_2 + B)/2$].

Shear failure of jaws

The tangential force F_t which transmits torque is mainly responsible for shear failure in ductile materials. However, the test results show that a cast iron block loaded in compression fails by shearing along a plane making an angle α of about 35° with the direction of pressure. Figure 14.12 shows the development at the mean diameter of jaw and a line of action inclined at an angle α. The force that acts in the plane of shear is $F_t/\cos\alpha$ and the resisting area $A_s = (D_2 - B)H/2\sin\alpha$.

It is assumed for the sake of safety that only one-half of the total number of jaws n are in actual contact. Therefore, the induced shear stress

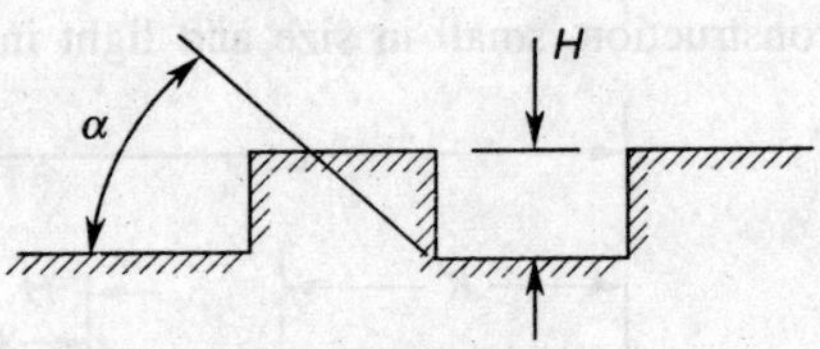

Figure 14.12 Shear failure of jaw.

$$\tau = \frac{4F_t \tan\alpha}{(D_2 - B)\,Hn} \tag{14.14}$$

Bearing failure of jaws

When a jaw clutch transmits torque, the tangential force compresses the faces of the jaw against each other, and this introduces crushing stress or bearing pressure. The torque transmission capacity based upon bearing failure of the jaw is given by

$$T = 0.5(D_2 - B)nHp_b \tag{14.15}$$

where

n is the number of jaws; usually two or three jaws are used

p_b is the allowable bearing pressure between the jaw faces

(≤ 20 N/mm^2 for small size clutches)

(≤ 40 N/mm^2 for large size clutches)

14.7 FRICTION CLUTCHES

A friction clutch transmits power under the influence of friction contact between two or more members. It generally has two or more rotating concentric surfaces, with at least one surface lined with a friction material. When these surfaces are faced against each other firmly, a tangential friction force is produced between them which transmits torque from the input shaft to the output shaft. When a friction clutch is engaged, its members tend to rotate as a single unit. However, under certain conditions, namely when the members are not in equilibrium, they may have relative motion. This phenomenon is called *slip*. In a power transmission system, the slippage of clutch is undesirable for obvious reasons.

The friction clutch offers the following advantages:

1. It can be easily engaged or disengaged at high speeds.
2. The friction element gets slipped during engagement which enables the drive to pick up and accelerate gradually without any major shock.
3. It also slips when the torque transmitted through it exceeds a safe value, thus acting as a safety device as well.

14.7.1 Design Requirements

While designing a friction clutch, the following considerations must be taken into account.

1. Selection of a suitable friction material which can transmit the desired torque.
2. Engagement and acceleration without shock and quick disengagement without any drag.
3. Provision for holding the contact surfaces together by the clutch itself.
4. Lightweight and dynamic balancing of movable parts, especially when the clutch is required to operate at high speeds.
5. Provision to compensate for backlash which may result due to wearing of the friction surfaces.
6. Easy repair and maintenance.

14.7.2 Friction Material

A material selected for friction lining on clutch surfaces must meet the following requirements:

(i) It must have a high coefficient of friction.

(ii) It should not get affected by moisture or oil.

(iii) It should have the requisite strength and wear resistance.

(iv) It must have low thermal expansion and high heat soaking capacity to prevent thermal distortion and heat spotting.

The most commonly used friction materials are wood, woven and moulded asbestos, cork, cast iron, leather, felt, phosphor bronze, and powder metals.

14.8 DISC CLUTCHES

In disc clutches, the friction surfaces are always kept normal to the axis of the connecting shaft and are brought in contact by an axial force.

Figure 14.13 shows the schematic diagrams of single-plate and multiplate clutches, consisting of flywheel, clutch plate with lining, pressure plate, thrust springs, and actuating

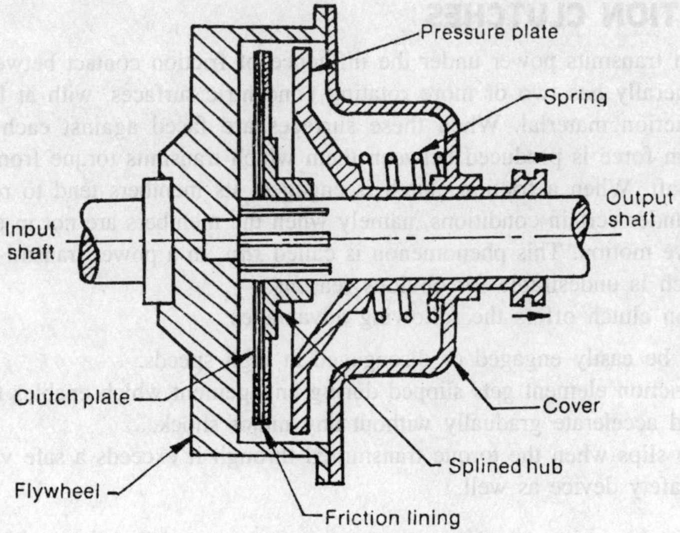

Figure 14.13(a) Single-plate friction clutch.

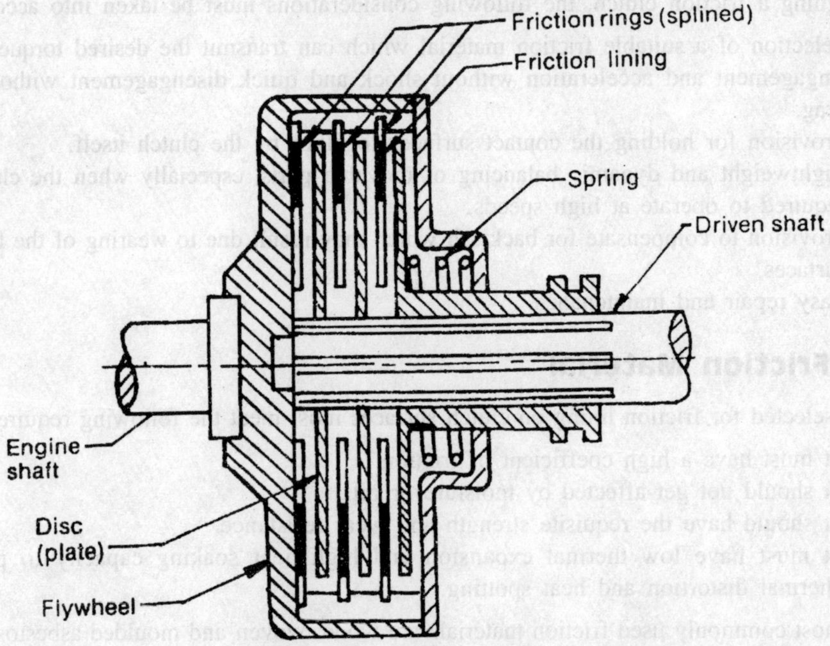

Figure 14.13(b) Multiplate friction clutch.

mechanism. In a disc clutch, on the shaft of a prime mover, a flywheel or disc is mounted. The clutch plate is generally free to slide on the splined shaft. In an ordinary situation, a clutch always remains engaged. The axial force, which is introduced through springs, is applied to the pressure plate which keeps the pressure plate, the clutch plate and the flywheel surfaces in contact to transmit power.

When the clutch pedal is pressed down, the pressure on the pressure plate is released and it moves back slightly and relieves pressure on the clutch plate. The clutch plate then slips back from the flywheel and the driven shaft is disconnected. It is known as disengagement. When pressure on the clutch pedal is released, it builds up pressure on the pressure plate which ultimately forces the pressure plate and the clutch plate to bring them in contact with the flywheel surface and the power transmission is restored.

14.8.1 Power Transmission Capacity

The power or torque transmission capacity of a friction clutch depends upon the friction force, radius at which it acts and the number of friction surfaces.

In order to compute the axial thrust and the torque transmission capacity, we consider an elemental ring of width dr on the friction surface in contact, at a radius r, as shown in Figure 14.14. If p is the intensity of pressure and $2\pi rdr$ is the area of the elemental ring, the axial force F_{ax} and the torque transmission capacity T can be computed by the relations:

$$F_{ax} = \int_{r_i}^{r_o} 2\pi prdr \tag{14.16}$$

$$T = 2\pi\mu \int_{r_i}^{r_o} pr^2 dr \tag{14.17}$$

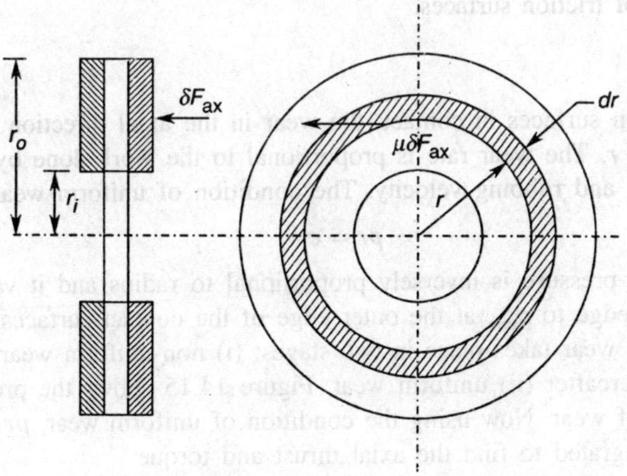

Figure 14.14 Force on clutch plate.

In clutch design, generally two cases are considered. In one case it is assumed that the intensity of the pressure on friction surfaces is constant. This assumption is valid only when the discs are relatively flexible. On the other hand, if the discs are rigid, wearing of the friction surface is approximately uniform after initial wearing-in has taken place. In practical situations, neither of these assumptions (uniform pressure or uniform wear) is correct. So the designer has to choose a hypothesis which is more close to the actual situation. Alternatively, it is better to assume a uniform wear rate because it is more conservative than assuming uniform pressure but it results into lower torque transmission capacity.

Uniform pressure

For new clutches and rigid mountings, the assumption of uniform pressure distribution is more realistic; hence pressure p may be regarded as constant. Therefore, the total axial thrust and torque can be computed by integrating Eqs. (14.16) and (14.17) as follows:

$$F_{ax} = 2\pi p \int_{r_i}^{r_o} r\,dr = \pi\left(r_o^2 - r_i^2\right)p \tag{14.18}$$

$$T = 2\pi \mu p \int_{r_i}^{r_o} r^2 dr$$

$$= \mu\, F_{ax}\, r_{mean}$$

In a multidisc clutch, the torque capacity is

$$T = \mu n F_{ax}\, r_{mean} \tag{14.19}$$

where

$$r_{mean} = \frac{2}{3}\left(\frac{r_o^3 - r_i^3}{r_o^2 - r_i^2}\right)$$

n = number of friction surfaces.

Uniform wear

In order to keep friction surfaces in contact, the wear in the axial direction must be the same for all values of radius r. The wear rate is proportional to the work done by friction which is the product of pressure and rubbing velocity. The condition of uniform wear is given as

$$pr = c \tag{14.20}$$

For uniform wear rate, pressure is inversely proportional to radius and it varies parabolically from p_{max} at the inner edge to p_{min} at the outer edge of the contact surfaces.

In actual practice, wear takes place in two stages: (i) non-uniform wear to settle down to a constant wear and thereafter (ii) uniform wear. Figure 14.15 shows the pressure distribution during the two stages of wear. Now using the condition of uniform wear, $pr = c$, Eqs. (14.16) and (14.17) can be integrated to find the axial thrust and torque.

$$F_{ax} = \int_{r_i}^{r_o} 2\pi p r\,dr$$

$$= 2\pi p r(r_o - r_i) \tag{14.21}$$

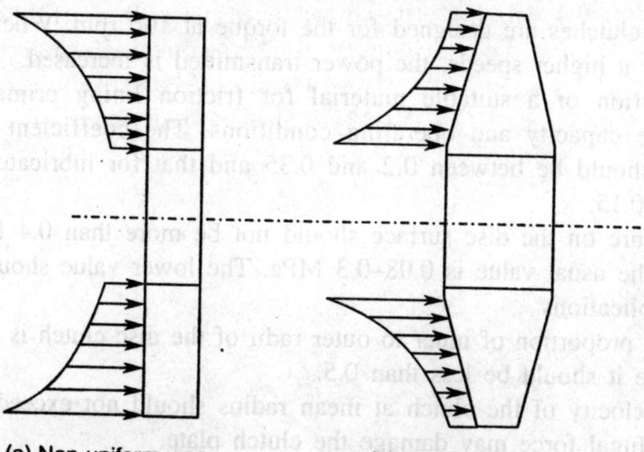

Figure 14.15 Stages of wear.

$$T = 2\pi \mu pr \int_{r_i}^{r_o} r dr$$

$$= \mu F_{ax}\, r_{mean} \quad (14.22)$$

where, r_{mean} is the mean radius $[= (r_o + r_i)/2]$.

In a multiplate clutch, the torque capacity is given as

$$T = \mu n F_{ax}\, r_{mean} \quad (14.23)$$

where

n is the number of friction surfaces
(is the 2 for a single-plate clutch with friction lining on both sides)
(is the $N_1 + N_2 - 1$ for a multi-plate clutch)

with

n_1 = number of discs on the driving shaft
n_2 = number of discs on the driven shaft

Power transmitted by the clutch

$$P = \frac{2\pi NT}{60 \times 1000} \text{ kW} \quad (14.24)$$

14.8.2 Design Considerations

While designing a disc clutch, the following points must be considered.

1. The average torque calculated from the given power needs to be modified. The design torque

$$T_d = T_{average} \times K \quad (14.25)$$

where K is the overload factor (1.25 to 1.5).

Industrial clutches are designed for the torque at 100 rpm. When these clutches are employed at higher speeds, the power transmitted is increased.

2. The selection of a suitable material for friction lining primarily depends upon the torque capacity and operating conditions. The coefficient of friction for dry clutches should be between 0.2 and 0.35 and that for lubricated clutches between 0.08 and 0.15.
3. The pressure on the disc surface should not be more than 0.4 MPa. For maximum disc life the usual value is 0.08–0.3 MPa. The lower value should be used for high speed applications.
4. The usual proportion of inner to outer radii of the disc clutch is 0.5 to 0.7, however, in no case it should be less than 0.5.
5. Surface velocity of the clutch at mean radius should not exceed 30 m/s, otherwise, the centrifugal force may damage the clutch plate.
6. A clutch plate consists of a core plate and friction lining. The thickness of the core plate depends upon its outer diameter and may vary between 1 and 3 mm. The thickness of the friction lining can be taken between 1 and 6 mm depending upon the friction material. The smaller thickness value should be used for wet or lubricated clutch plates.
7. Friction lining on the clutch plate may be grooved either radially or spirally to accommodate wear debris and to prevent vacuum formation between plates.
8. Multidisc clutches are normally of the wet type and the reasonable number of plates are 4 to 8.

14.9 CONE CLUTCHES

In a cone clutch, as shown in Figure 14.16, the contact surfaces are in the form of cones. In the engaged position, the friction surfaces of two cones A and B are in complete contact due to external spring pressure which keeps one cone pressed against the other all the time.

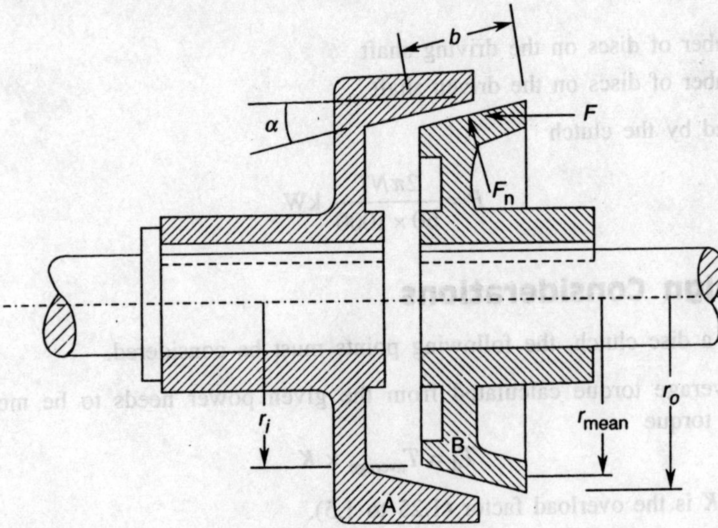

Figure 14.16 Cone clutch.

When a clutch is engaged, torque is transmitted from the driving shaft to the driven shaft through the flywheel and friction cones. For disengaging the clutch, the cone B is pulled back through an actuating lever mechanism against the spring force.

In cone clutches, the normal force on the contact surfaces is larger than that on the disc clutches, therefore, for the same axial thrust, the cone clutches can transmit more torque. However, a cone clutch exposed to dust and dirt may create some difficulty in disengagement. Further, score marks on the friction lining surface may impede sliding.

The power transmission capacity and axial thrust required can be computed by treating a cone clutch as a conical pivot as shown in Figure 14.17. In uniform wear theory, the intensity

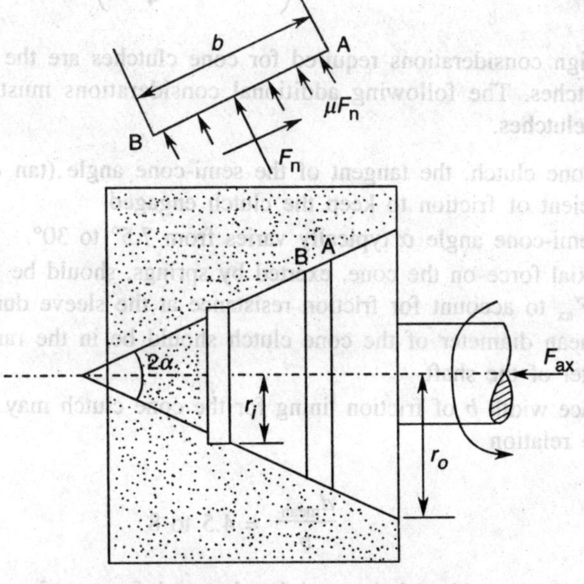

Figure 14.17 Forces on cone clutch.

of pressure at given radius r is given by

$$p = \frac{F_{ax}}{2\pi r b \sin\alpha} \tag{14.26}$$

$$\text{Torque, } T = \frac{\mu}{\sin\alpha} F_{ax} r_{mean}$$

$$= \mu' F_{ax} r_{mean} \tag{14.27}$$

where

μ' is the equivalent coefficient of friction (= $\mu/\sin\alpha$)

r_{mean} is the mean radius [= $(r_o + r_i)/2$]

The equation of axial force, i.e. $F_{ax} = F_n \sin\alpha$ is valid only under steady operation and after the clutch is engaged. However, during engagement there is an additional component of force which is mainly due to friction force as shown in Figure 14.18. This component resists the action of engagement. Therefore, the axial force required for engaging the clutch increases. Thus,

$$F'_{ax} = F_n \sin \alpha + \mu F_n \cos \alpha$$

However, the various experimental results have indicated that the $\mu F_n \cos \alpha$ term is only 25 per cent effective. Therefore, the actual axial force required to keep the clutch engaged is

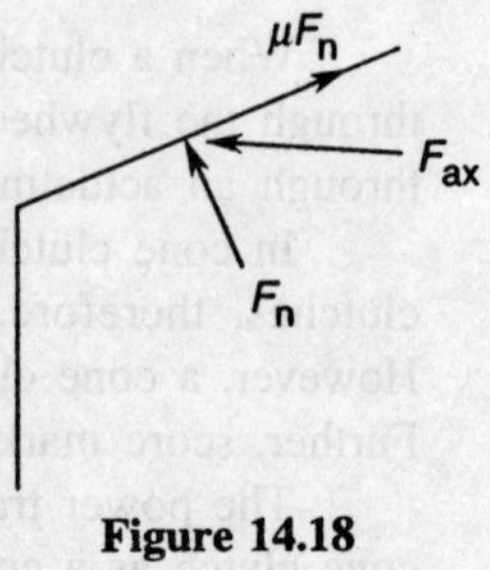

Figure 14.18

$$F'_{ax} = F_n \left(\sin \alpha + \frac{\mu \cos \alpha}{4} \right) \tag{14.28}$$

The general design considerations required for cone clutches are the same as those discussed for the disc clutches. The following additional considerations must be taken care of while designing cone clutches.

1. In a cone clutch, the tangent of the semi-cone angle ($\tan \alpha$) must be less than the coefficient of friction to keep the clutch engaged.
2. The semi-cone angle α typically varies from 7.5° to 30°.
3. The axial force on the cone, exerted by springs, should be 15 to 20 per cent greater than F'_{ax} to account for friction resistance at the sleeve during engagement.
4. The mean diameter of the cone clutch should be in the range of 5 to 10 times the diameter of the shaft.
5. The face width b of friction lining for the cone clutch may be computed empirically by the relation

$$\frac{d_{mean}}{b} = 4.5 \text{ to } 8 \tag{14.29}$$

6. The limiting value of the peripheral speed for metal to metal contact clutch is 5–15 m/s, whereas for leather or asbestos faced lining, it is 15–25 m/s.

Example 14.5 Design a suitable clutch for the speed gear box of a lathe machine to transmit 15 kW at 1000 rpm. Due to space limitation, the outer diameter is limited to 150 mm.

Solution We select the axial friction clutch with woven asbestos friction lining having $\mu = 0.2$. The maximum operating temperature is 260°C.

Allowable bearing pressure, $p = (0.3\text{–}0.7)\ \text{N/mm}^2$

$$\text{Average torque, } T = \frac{60P}{2\pi \times N} = \frac{60 \times 15 \times 10^3}{2\pi \times 1000} = 143.24\ \text{N} \cdot \text{m}$$

Assuming 25% overload capacity, design torque

$$T_d = 1.25 \times 143.24 = 179\ \text{N} \cdot \text{m}$$

The torque transmission capacity using uniform wear rate

$$T = n\mu F_{ax} r_{mean}$$

where

$$F_{ax} = 2\pi p_{max} r_i(r_o - r_i)$$

Let

$$p_{max} = \text{maximum pressure at the inner radius} = 0.35 \text{ N/mm}^2$$

$$r_i = 0.6r_o = 45 \text{ mm}$$

$$n = \text{number of friction surfaces}$$

Therefore,

$$179 \times 10^3 = n \times 0.2 \times 2\pi \times 0.35 \times 45(75 - 45) \times \frac{(75 + 45)}{2}$$

or

$$n = 5.02$$

So, let us select a multiplate clutch having the following numbers of discs:

Driving shaft, $n_1 = 4$

Driven shaft, $n_2 = 3$

Therefore, the total number of friction surfaces, $n = n_1 + n_2 - 1$

$$= 4 + 3 - 1 = 6$$

which is larger than the required number of surfaces.

$$\text{Axial force, } F_{ax} = \frac{T_d}{\mu n r_{mean}} = \frac{179 \times 10^3}{0.2 \times 6 \times 60} = 2486.1 \text{ N}$$

$$\text{Operating pressure, } p = \frac{F_{ax}}{2\pi r_i(r_o - r_i)}$$

$$= \frac{2486.1}{2\pi \times 45(75 - 45)} = 0.293 \text{ N/mm}^2$$

which is reasonable.

Note: The design of shaft, spline, and springs is left as an exercise for the student.

Example 14.6 Design a cone clutch to transmit 10 kW at maximum speed of 1000 rpm. The outer cone is of cast iron and forms the part of the I.C. engine flywheel. The overall dimension restricts the mean diameter of the cone to 300 mm. The semi-cone angle is 15°. The inner cone is positioned by means of a centrally placed helical spring.

Solution Average torque, $T = \frac{60P}{2\pi N} = \frac{60 \times 10 \times 10^3}{2\pi \times 1000} = 95.5 \text{ N} \cdot \text{m}$

Let the design torque be 25% more than the average torque. Hence,

$$T_d = 1.25 \times T = 1.25 \times 95.5 = 119.375 \approx 120 \text{ N} \cdot \text{m}$$

Let us assume that the shaft is made of plain carbon steel SAE1020 having allowable shear strength, $\tau = 45$ N/mm^2.

$$\text{Shaft diameter, } d_S = \left(\frac{16T_d}{\pi\tau}\right)^{1/3}$$

$$= \left[\frac{16 \times 120 \times 10^3}{\pi \times 45}\right]^{1/3}$$

$$= 23.8 \text{ mm, say, 36 mm to account for rigidity.}$$

The ratio $d_{mean}/d_S = 8.34$, which is within limit.

Torque capacity of the cone clutch

$$T = \frac{\mu F_{ax}\, r_{mean}}{\sin\alpha}$$

where

$$F_{ax} = 2\pi r_{mean} \times b \times p \times \sin\alpha$$

Let the friction lining be cork against CI, having coefficient of friction $\mu = 0.2$.

$$\text{Bearing pressure, } p = 0.09 \text{ N/mm}^2$$

Therefore,

$$120 \times 10^3 = 0.2 \times 2\pi \times 0.09 \times b \times 150^2$$

or

$$b = 47.15 \text{ mm, say, 50 mm}$$

The ratio d_{mean}/b should be within 4.5 and 8.

Here, $\dfrac{d_{mean}}{b} = \dfrac{300}{50} = 6$, satisfies the condition.

$$\text{Outer diameter of the cone } d_o = d_{mean} + b \sin\alpha$$

$$= 300 + 50 \sin 15° = 312.94 \text{ mm, say, 313 mm}$$

$$\text{Inner diameter of the cone, } d_i = d_{mean} - b \sin\alpha$$

$$= 300 - 50 \sin 15° = 287 \text{ mm}$$

$$\text{Axial force, } F_{ax} = F_n(\sin\alpha + 0.25\mu \cos\alpha)$$

$$F_n = 2\pi\, r_{mean}\, bp$$

$$= 2\pi \times 150 \times 50 \times 0.09 = 4241.15 \text{ N}$$

Therefore,

$$F_{ax} = 4241.15 (\sin 15° + 0.25 \times 0.2 \cos 15°) = 1302.52 \text{ N}$$

Note: The design of the spline and the spring is left as an exercise for the student.

14.10 BRAKES

A brake is a mechanical device which is used to apply external frictional resistance to a moving body to stop or retard it by absorbing its kinetic energy or potential energy or both in the form of heat which is ultimately dissipated by conduction or convection. In other words, the function of a brake is to regulate the speed of a machine by transforming, through friction, the kinetic energy of the moving body into heat and then dissipating it.

The brakes are usually classified according to the means by which kinetic energy is transformed into heat energy. Thus there are three basic types of brakes:

Mechanical brakes

In this type of brake, the physical contact is used for energy transformation. The main types of mechanical brakes are: (i) block brake, (ii) band brake, (iii) band and block brake, and (iv) internal expanding shoe.

Hydraulic brakes

This type of brake utilizes the fluid friction created by whirling of the fluid instead of the mechanical surface friction. Fluid pump and agitator are most commonly used for this purpose.

Electric brakes

When a moving member is transmitting high torque at high speed, mechanical brakes cause thermal distortion and failure of the braking system. In such cases electric brakes are used. In these brakes, the kinetic energy is absorbed by driving one or more generators; the output of generators is either converted into useful work or dissipated through resistive heating.

In this chapter, we shall discuss in detail mechanical brakes which are useful for low to moderate speeds.

14.10.1 Block Brakes

A block brake consists of a block or shoe, which is pressed against a rotating drum by means of a lever, as shown in Figure 14.19. The frictional force acting between the block and the drum causes the retardation of the brake drum. In block brakes, the block is either rigidly fixed or pivoted to the lever. In certain applications, if only one block is used, a side thrust on the bearing of the shaft supporting the drum will act, which can be prevented by using two blocks—one on each side of the drum as shown in Figure 14.20. This also doubles the braking torque.

In block brakes, the angle of contact is usually small. When it is less than 45°, the distribution of pressure between the block and the drum may be assumed to be uniform.

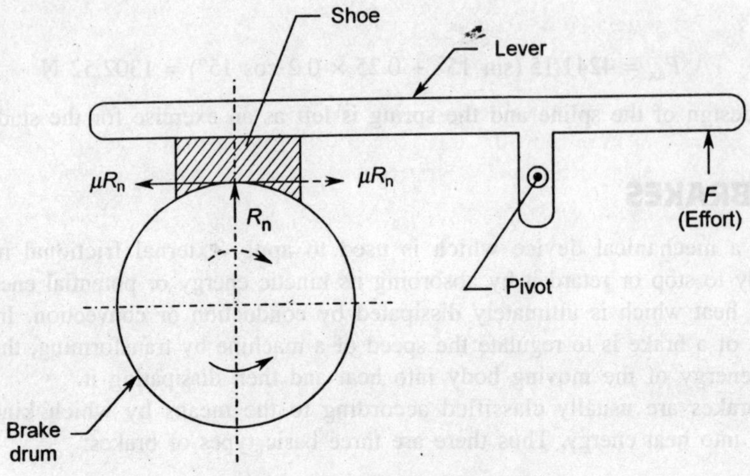

Figure 14.19 Single shoe brake.

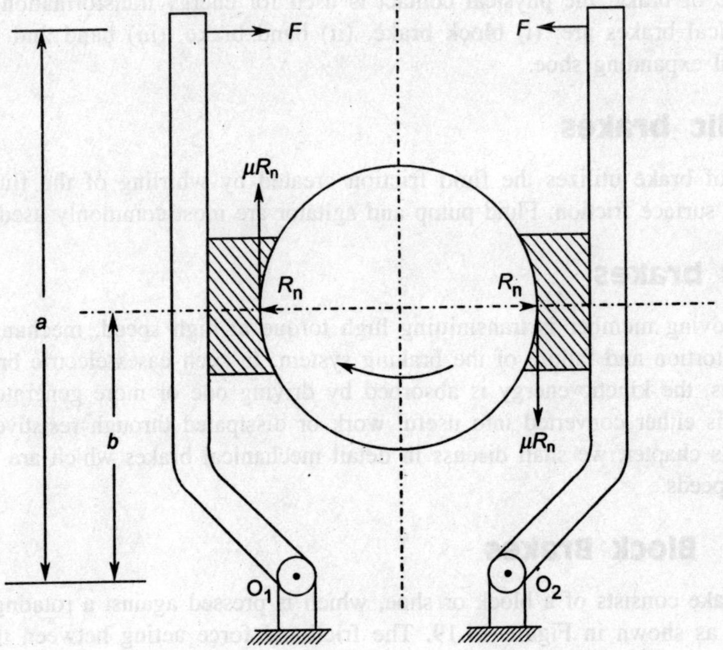

Figure 14.20 Double shoe brake.

Suppose

r is the radius of the brake drum

μ is the coefficient of friction between the block and the drum

R_n is the normal reaction on the block
F is the applied force at the lever end
F' is the frictional force $(= \mu R_n)$

Assuming that the normal force R_n and the friction force F' act at the contact point between the block and the drum, the direction of the friction force on the drum is such as to oppose its motion, while on the block the friction force acts in the direction of rotation.

$$\text{Braking torque} = \text{friction force} \times \text{radius} = \mu R_n \times r \tag{14.30}$$

Analysis of block brake is presented below under three cases:

Case I. *When the fulcrum point passes through the line of action of forces on the shoe.*

Let us assume that the brake drum rotates in the clockwise direction. For the brake lever to remain in equilibrium, the sum of moments of forces about the fulcrum point should be zero (Figure 14.21). Thus,

$$F \times a = R_n \times b$$

or

$$R_n = F\left(\frac{a}{b}\right) \tag{14.31}$$

$$\text{Braking torque, } T = \mu R_n \times r = \mu Fr\left(\frac{a}{b}\right) \tag{14.32}$$

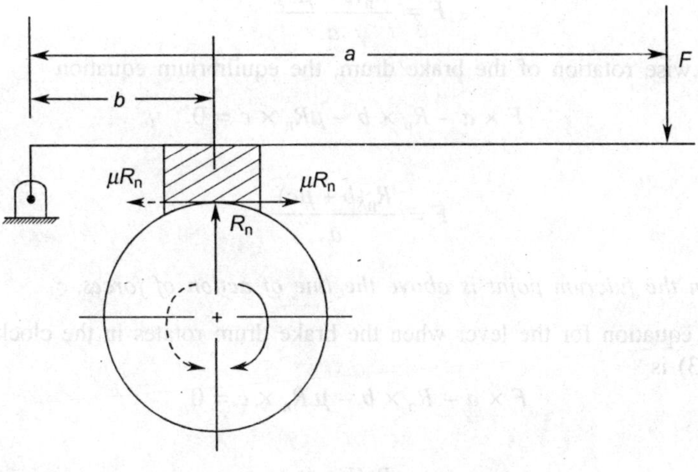

Figure 14.21 Block brake—Case I.

From Eq. (14.32), it is evidently clear that the brake can be applied in either direction and still the braking action can be obtained by the same lever force.

Case II. *When the fulcrum point is below the line of action of forces.*

The equilibrium equation of the lever for clockwise rotation of the brake drum can be found out by summing the moments (see Figure 14.22). That is,

$$F \times a - R_n \times b + \mu R_n \times c = 0$$

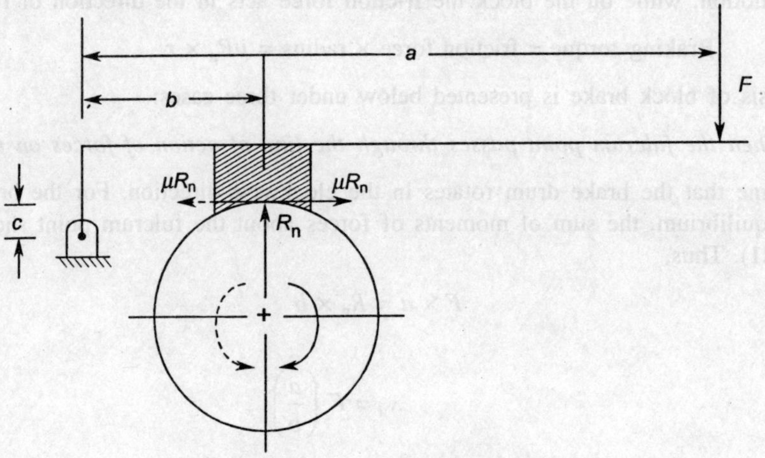

Figure 14.22 Block brake—Case II.

or

$$F = \frac{R_n(b - \mu c)}{a} \tag{14.33}$$

For counterclockwise rotation of the brake drum, the equilibrium equation

$$F \times a - R_n \times b - \mu R_n \times c = 0$$

or

$$F = \frac{R_n(b + \mu c)}{a} \tag{14.34}$$

Case III. *When the fulcrum point is above the line of action of forces.*

The equilibrium equation for the lever when the brake drum rotates in the clockwise direction (see Figure 14.23) is

$$F \times a - R_n \times b - \mu R_n \times c = 0$$

or

$$F = \frac{R_n(b + \mu c)}{a} \tag{14.35}$$

For counterclockwise rotation of the brake drum, the equilibrium equation

$$F \times a - R_n \times b + \mu R_n \times c = 0$$

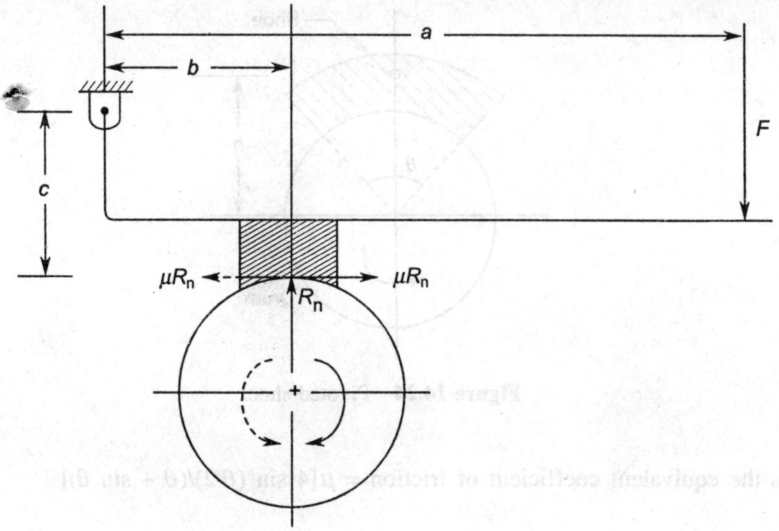

Figure 14.23 Block brake—Case III.

or

$$F = \frac{R_n(b - \mu c)}{a} \tag{14.36}$$

In Eqs. (14.33) and (14.36) when $b = \mu c$, the effort required on the lever is zero, which implies that the force needed to apply the brake is virtually zero or in other words once a contact is made between the block and the drum, the brake is applied itself. Such a brake is known as the *self-locking* brake.

It is further observed from equilibrium equations that the moments of the force μR_n about the fulcrum is in the same direction as that of the applied effort F. Therefore, the moment due to μR_n aids in applying the brake. Such a brake is known as the *self-energized* brake.

In the above analysis, it is assumed that the normal reaction and the frictional forces act at the midpoint of the block. However, this is true only when the angle of contact is small, i.e. $\theta < 45°$. When the angle of contact is more than 45°, the normal pressure distribution is not uniform and is less at the ends than at the centre. In that case the effective radius of the brake drum h, as shown in Figure 14.24, is given by

$$h = r\left[\frac{4\sin(\theta/2)}{\theta + \sin\theta}\right] \tag{14.37}$$

Therefore, frictional torque

$$T = \mu R_n h = \mu R_n r \left[\frac{4\sin(\theta/2)}{\theta + \sin\theta}\right]$$

or

$$T = \mu' R_n r \tag{14.38}$$

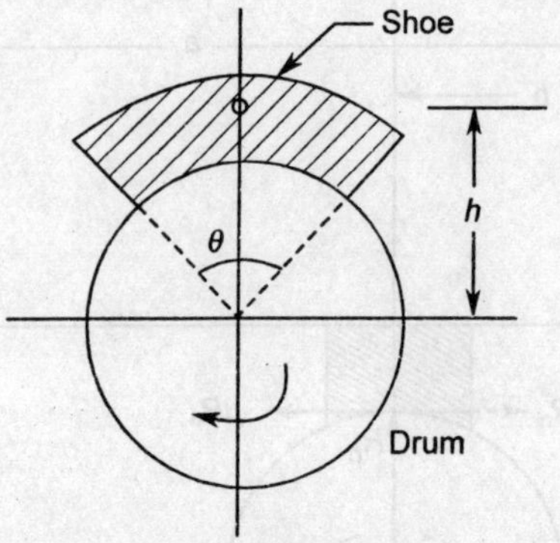

Figure 14.24 Pivoted shoe.

where μ' is the equivalent coefficient of friction $= \mu[4 \sin (\theta/2)/(\theta + \sin \theta)]$ (14.39)

14.10.2 Band Brakes

A band brake consists of a rope, belt, or a flexible steel band, lined with a friction material, which is pressed against the external surface of a cylindrical drum when the brake is applied. When a braking force is applied at the free end of the lever, as shown in Figure 14.25, tensions are produced in the band just like in the belt drive. The ratio of the tight side and the slack side tensions in the band is given by

$$\frac{T_1}{T_2} = e^{\mu\theta} \tag{14.40}$$

where

μ is the coefficient of friction between the band and the brake drum

θ is the angle of contact.

The braking torque on the drum is given by

$$T = (T_1 - T_2)r \tag{14.41}$$

where r is the effective radius of the drum.

In the band brake, the effectiveness of applied force F depends upon (i) the direction of rotation of the drum, (ii) the ratio of lengths a and b, and (iii) the direction of applied force F. To apply the brake to a rotating drum, the band has to be tightened on the drum; this is possible if

(a) the applied force acts in the downward direction when $a > b$, and
(b) the applied force acts in the upward direction when $a < b$.

Band brakes may be classified into two categories: (a) simple band brake and (b) differential band brake.

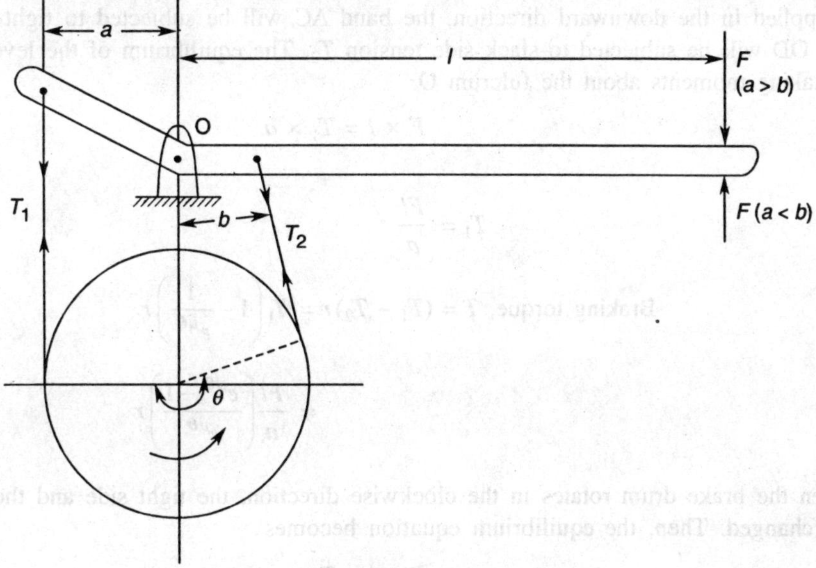

Figure 14.25 Band brake.

Simple band brake

A simple band brake is one in which one end of the band is attached to the fulcrum and other end to the brake lever. There are various arrangements of the simple band brake. One typical arrangement is shown in Figure 14.26.

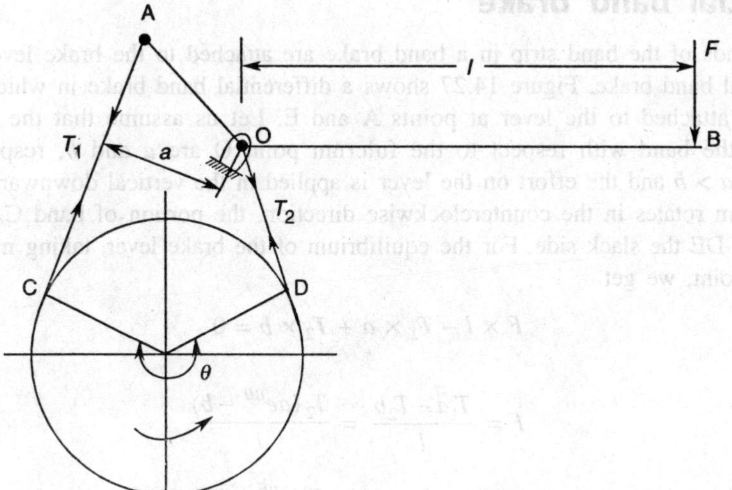

Figure 14.26 Simple band brake.

When the drum rotates in the counterclockwise direction and the effort on the brake lever is applied in the downward direction, the band AC will be subjected to tight side tension T_1 and OD will be subjected to slack side tension T_2. The equilibrium of the lever can be found by taking moments about the fulcrum O.

$$F \times l = T_1 \times a$$

or

$$T_1 = \frac{Fl}{a} \tag{14.42}$$

$$\text{Braking torque, } T = (T_1 - T_2)r = T_1\left(1 - \frac{1}{e^{\mu\theta}}\right)r$$

$$= \frac{Fl}{a}\left(\frac{e^{\mu\theta} - 1}{e^{\mu\theta}}\right)r \tag{14.43}$$

When the brake drum rotates in the clockwise direction, the tight side and the slack side are interchanged. Then, the equilibrium equation becomes

$$F \times l = T_2 \times a$$

and

$$\text{Braking torque, } T = \frac{Fl}{a}(e^{\mu\theta} - 1)\,r \tag{14.44}$$

A comparison of Eqs. (14.43) and (14.44) reveals that when the brake drum rotates in the clockwise direction and the brake lever pulls the slack side of the band, the braking torque is $e^{\mu\theta}$ times greater than when the drum rotates in the counterclockwise direction.

Differential band brake

When both ends of the band strip in a band brake are attached to the brake lever, it is called the differential band brake. Figure 14.27 shows a differential band brake in which the ends of the band are attached to the lever at points A and E. Let us assume that the perpendicular distances of the band with respect to the fulcrum point O are a and b, respectively. It is assumed that $a > b$ and the effort on the lever is applied in the vertical downward direction. If the brake drum rotates in the counterclockwise direction, the portion of band CA become the tight side and DE the slack side. For the equilibrium of the brake lever, taking moments about the fulcrum point, we get

$$F \times l - T_1 \times a + T_2 \times b = 0 \tag{14.45}$$

or

$$F = \frac{T_1 a - T_2 b}{l} = \frac{T_2(ae^{\mu\theta} - b)}{l} \tag{14.46}$$

$$\text{Braking torque, } T = (T_1 - T_2)r = \frac{Fl(e^{\mu\theta} - 1)\,r}{ae^{\mu\theta} - b}$$

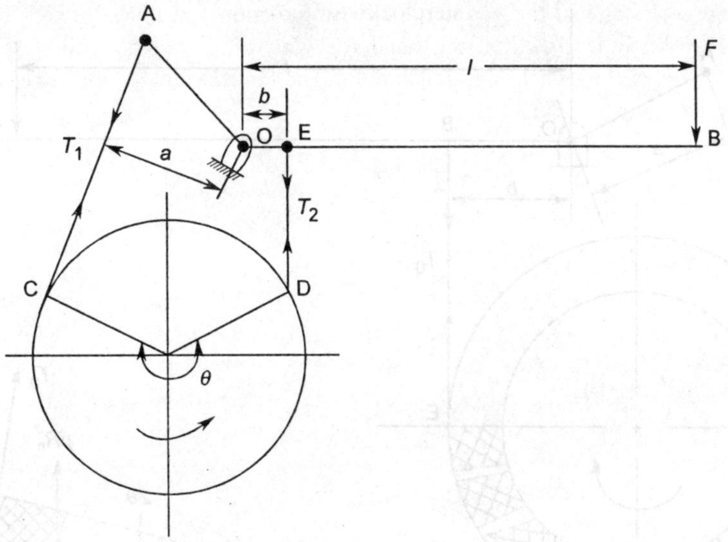

Figure 14.27 Differential band brake.

From Eq. (14.46) it is found that when $b > ae^{\mu\theta}$ or $T_2 b > T_1 a$, the brake is called *self-locking* because the effort required to brake becomes negative. It means some effort is required to disengage it.

When the brake drum rotates in the clockwise direction, the tight and slack side tensions are interchanged. The portion of band CA slacks while DE gets tightened. The equilibrium equation of the lever is found by taking moments about the fulcrum, i.e.

$$Fl - T_2 \times a + T_1 \times b = 0 \tag{14.47}$$

or

$$F = \frac{T_2 a - T_1 b}{l} = \frac{T_2(a - be^{\mu\theta})}{l}$$

$$\text{Braking torque, } T = (T_1 - T_2) r = \frac{Fl\,(e^{\mu\theta} - 1)\, r}{a - be^{\mu\theta}} \tag{14.48}$$

Comparison of Eqs. (14.46) and (14.48) also reveals that the denominator of Eq. (14.48) is smaller than that of Eq. (14.46). It means that when the brake drum rotates in the clockwise direction, it is more effective.

14.10.3 Band and Block Brake

A band and block brake consists of a number of brake shoes secured inside a flexible steel band as shown in Figure 14.28. When the brake is applied, the shoes are pressed against the drum. The two sides of the band DA and BE become tight and slack, respectively, as usual. Block shoes have higher coefficient of friction, thus they increase the effectiveness of the brake.

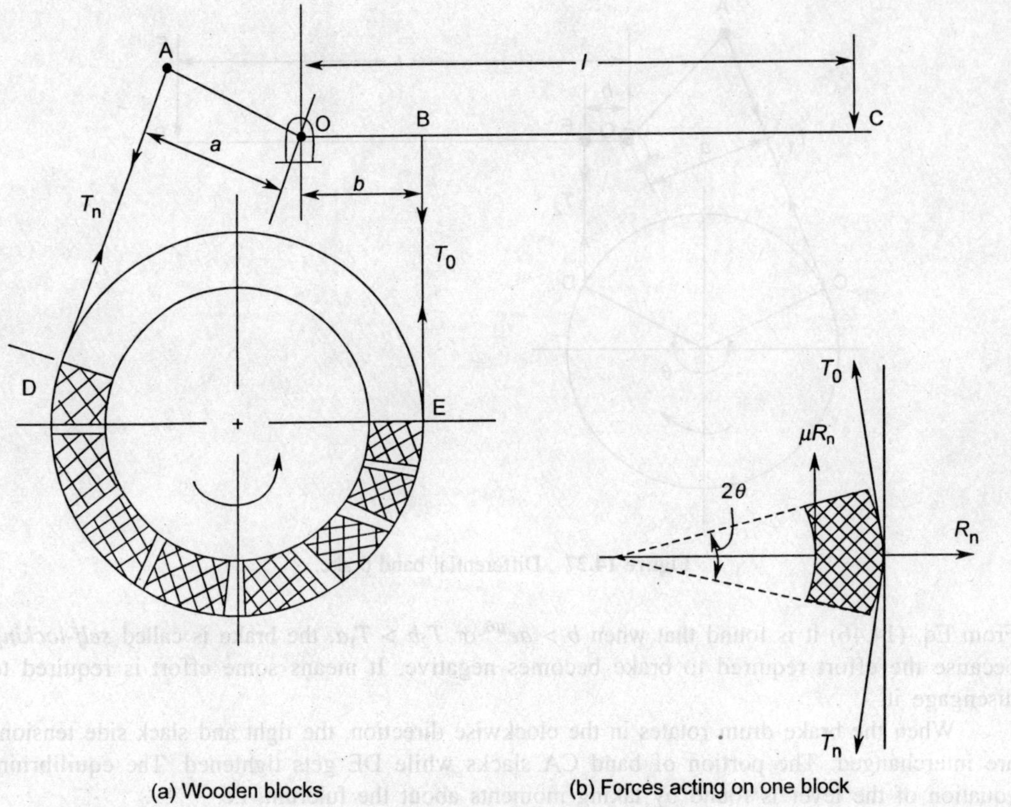

Figure 14.28 Band and block brake.

The ratio of tensions in the band and block brake is given by

$$\frac{T_n}{T_0} = \left(\frac{1 + \mu \tan \theta}{1 - \mu \tan \theta}\right)^n \tag{14.49}$$

where

2θ is the angle subtended by each block

μ is the coefficient of friction

n is the number of blocks.

14.10.4 Internal Expanding Shoe Brake

Figure 14.29 shows a typical internal shoe brake currently used in automobiles. It consists of two semi-circular shoes which are lined with friction material of high coefficient of friction and good wearing properties. The shoes are normally held in the inactive position by light springs *S*. To apply the brakes, either the cam is rotated or fluid pressure is applied through a cylinder-plunger assembly. This forces the shoes against the inner surface of the brake drum which is integral to wheel casing.

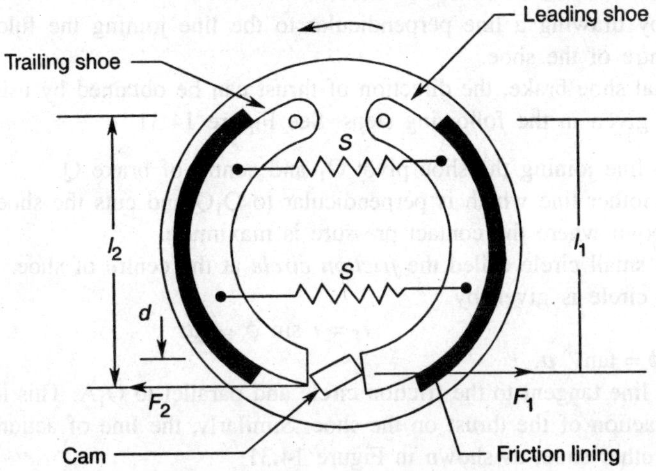

Figure 14.29 Internal expanding shoe brake.

Assuming that each shoe is rigid compared to the friction surface, the friction lining on the shoe when compressed obeys the Hooke's law. The compressive pressure p at any point A, as shown in Figure 14.30, on the contact surface will be proportional to its distance l from the pivot.

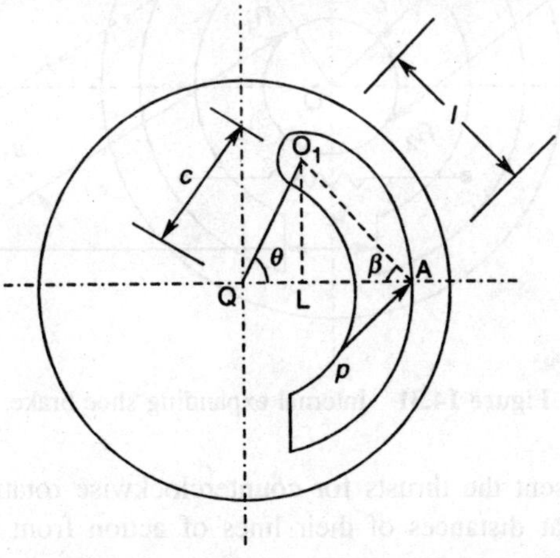

Figure 14.30 Forces acting on the internal expanding shoe brake.

Considering the leading edge shoe, $p \propto l$ or $p = k_1 l$, where k_1 is a constant. The direction of pressure p is perpendicular to O_1A. The normal pressure

$$p_n = k_1 l \cos (90° - \beta) = k_1 l \sin \beta$$

$$= k_1 c \sin \theta \qquad (\because \; O_1L = l \sin \beta = c \sin \theta)$$

$$= k_2 \sin \theta$$

where k_2 is another constant. The normal pressure p_n is maximum when $\theta = 90°$. It means that the location of the point where the contact pressure is maximum, can be determined graphically just by drawing a line perpendicular to the line joining the fulcrum point of the shoe with the centre of the shoe.

In an internal shoe brake, the direction of thrust can be obtained by using the concept of friction circle, as given in the following steps. See Figure 14.31.

1. Draw a line joining the shoe pivot O_1 and centre of brake Q.
2. Draw another line which is perpendicular to O_1Q and cuts the shoe at point A. This is the point where the contact pressure is maximum.
3. Draw a small circle called the *friction circle* at the centre of shoe. The radius of the friction circle is given by

$$r_f = r \sin \phi \simeq \mu r$$

where $\phi = \tan^{-1} \mu$.

4. Draw a line tangent to the friction circle and parallel to O_1A. This line represents the line of action of the thrust on the shoe. Similarly, the line of action can be obtained for the other shoe, as shown in Figure 14.31.

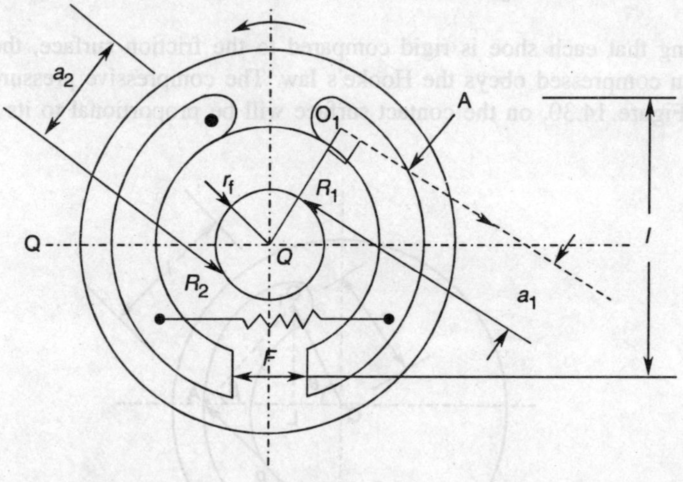

Figure 14.31 Internal expanding shoe brake.

Let R_1 and R_2 represent the thrusts for counterclockwise rotation of the drum and let a_1 and a_2 be the perpendicular distances of their lines of action from the respective shoe pivot.

$$\text{Braking torque, } T = (R_1 + R_2)r_f = \mu r(R_1 + R_2) \tag{14.50}$$

Assuming that the forces applied on the shoe are equal in magnitude ($F_1 = F_2 = F$) and act in one line (i.e. $l_1 = l_2$), the equilibrium equations of the shoe are

$$F \times l = R_2 \times a_2$$

and

$$F \times l = R_1 \times a_1$$

Therefore,

$$\text{Braking torque, } T = \mu r \left(\frac{Fl}{a_1} + \frac{Fl}{a_2}\right) = \mu F l r \left(\frac{1}{a_1} + \frac{1}{a_2}\right) \tag{14.51}$$

14.11 THERMAL ASPECT OF BRAKE DESIGN

The energy that a brake is required to absorb for a general-purpose braking system is given by

$$E = (\Delta KE)_{trans} + (\Delta KE)_{rot} + (\Delta PE) \tag{14.52}$$

where

$(\Delta KE)_{trans}$ = change in kinetic energy of translating masses

$= \frac{1}{2}m\left(v_1^2 - v_2^2\right)$, with v_1 and v_2 as linear velocities of the mass m before and after braking

$(\Delta KE)_{rot}$ = change in kinetic energy of rotating masses

$= \sum \frac{1}{2}I\left(\omega_1^2 - \omega_2^2\right)$, with ω_1 and ω_2 as angular velocities of rotating parts

I = mass moment of inertia = mk^2

$(\Delta PE) = \Sigma mgh$, with h as the height being lowered or raised.

The rate of heat generation is given by

$$H_{gen} = \frac{E}{t} \times q \tag{14.53}$$

where

q is the load factor $\left(= \dfrac{\text{actual braking time}}{\text{total time of operating cycle}}\right)$

t is the time of operation of cycle.

This heat generated is required to be dissipated continuously so that the brakes do not overheat. Therefore, while designing a brake, the braking area, radiating surface and the air circulation should be so proportioned that overheating of the brake is prevented.

Since cooling laws for braking systems have not been well established, it is difficult to compute the actual time required to cool the brake. It requires a lot of experimental tests. However, the rate of heat dissipation at a given temperature can be computed approximately by the relation

$$H_{diss} = KA\Delta T \tag{14.54}$$

where

K is the heat transfer coefficient. It depends on the temperature difference and the rate of circulation of the air. Typical values vary between 30 and 45 W/m^2 °C

A is the surface area which transfers heat

ΔT is the difference in temperature between the heat dissipating surfaces and the surrounding air.

Thus for satisfactory performance of a brake, the heat dissipated must be greater than or equal to heat generated.

14.12 OTHER DESIGN CONSIDERATIONS

While designing a brake, besides aspects such as kinematic, torque capacity and thermal, the following considerations should also be taken into account.

1. The efficiency of a brake also depends upon the rate of wear of friction lining which is equal to the work done or is proportional to the product of pressure and velocity. For satisfactory life of a brake, the value of the product *pv* must be limited to lie within the range 1–3.
2. The width of the shoe is generally selected empirically. It is defined in terms of the brake drum diameter as

$$0.25D \le b \le 0.5D$$

3. The most commonly used material for brake drums is martensitic cast iron having nickel content of 5–6 per cent and Brinell hardness of 400–450. The materials for band and brake lever are generally plain carbon steels 30C8 and 40C8 with tensile strength of 400–500 N/mm^2.
4. In a band brake, the thickness of the band is taken as 0.005 time the diameter of the brake drum.
5. The width of the band is determined from considerations of strength using a high factor of safety.
6. The brake band ends should be fastened to the lugs either by welding or by riveting.

Example 14.7 A double shoe brake, as shown in the following figure, is capable of absorbing a torque of 1400 N · m. The diameter of the brake drum is 350 mm and the angle of contact for each shoe is 100°. If the coefficient of friction between the brake drum and the lining is 0.4, find: (a) the force necessary to set the brake, and (b) the width of the brake shoe, if the bearing pressure on the lining material is not to exceed 0.3 N/mm^2.

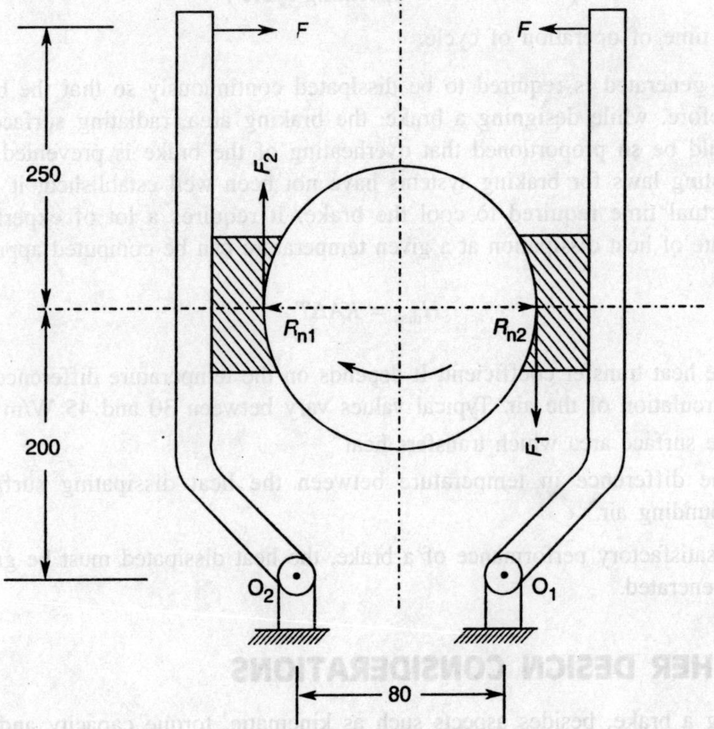

Solution Equivalent coefficient of friction

$$\mu' = \frac{4\mu \sin(\theta/2)}{\theta + \sin\theta} = \frac{4 \times 0.4 \times \sin 50^\circ}{1.75 + \sin 100^\circ} = 0.45$$

(a) Suppose F = force applied to set the brake

R_{n1}, R_{n2} = normal reactions

F_1, F_2 = frictional forces

Taking moments about the fulcrum O_1,

$$F \times 450 = R_{n1} \times 200 + F_1(175 - 40)$$

$$\mu' R_{n1} = F_1$$

or

$$F_1 = 0.776F$$

Similarly, taking moments about O_2, we get

$$F \times 450 = R_{n2} \times 200 - F_2(175 - 40)$$

$$\mu' R_{n2} = F_2$$

or

$$F_2 = 1.454F$$

Torque capacity of the brake

$$T = (F_1 + F_2) \times r$$

or

$$1400 \times 10^3 = (0.776 + 1.454)F \times 175$$

or

$$F = 3587.4 \text{ N}$$

(b) Let

b = width of the brake shoe

p_b = 0.3 N/mm^2 (given)

Projected area, $A = b(2r \sin \theta)$

$= b \times 2 \times 175 \times \sin 50^\circ = 268.1b$ mm^2

Maximum normal force, $R_{n2} = \dfrac{F_2}{\mu'} = \dfrac{1.454 \times 3587.4}{0.45} = 11{,}591.3$ N

Resisting area, $A = \dfrac{R_{n2}}{p_b}$

or

$$268.1b = \frac{11{,}591.3}{0.3}$$

or

$$b = 144.1 \text{ mm, say, } 145 \text{ mm}$$

Example 14.8 A band brake is to be designed for a winch to lift a load of 20 kN through a 15 m height by a rope wire wound on a barrel of 450 mm diameter. The hoisting cycle is 3 minutes, out of which the actual braking time is 60 seconds. The angle of contact between the band and the brake drum is 210°. The brake drum may be keyed to the same shaft. Give the complete design.

Solution As the brake drum should be larger than the barrel, let the brake drum diameter d be 650 mm.

$$\text{External torque, } T = \text{load} \times \text{barrel radius}$$

$$= 20 \times \frac{450}{2} = 4500 \text{ N} \cdot \text{m}$$

Let

T_1 = tension in the tight side of the band

T_2 = tension in the slack side of the band

$$\text{Braking torque, } T = (T_1 - T_2)r$$

Let r be the radius of the brake drum which should be greater than the radius of the barrel. We assume this to be 0.375 m. Therefore,

$$(T_1 - T_2) \times 0.375 = 4500 \qquad \text{(i)}$$

Now, the ratio of tensions

$$\frac{T_1}{T_2} = e^{\mu\theta} = e^{0.25 \times \frac{\pi}{180} \times 210°} = 2.5 \qquad \text{(ii)}$$

Solving Eqs. (i) and (ii),

$$T_1 = 20{,}000 \text{ N} \qquad T_2 = 8000 \text{ N}$$

The layout of the band brake is shown in the figure below.

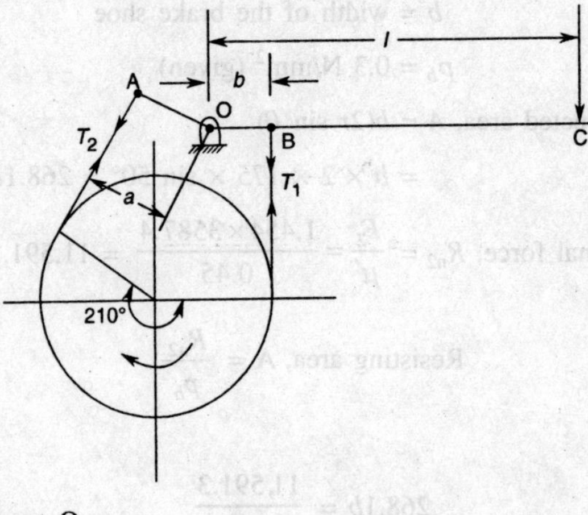

Taking moments about O

$$F \times l + T_1 b - T_2 a = 0$$

or

$$F = \frac{T_2 a - T_1 b}{l} = \frac{T_2\,(a - e^{\mu\theta}\; b)}{l}$$

To avoid force F to be zero or of negative value,

$$a \geq be^{\mu\theta}$$

Let us assume, $a = 1.5be^{\mu\theta} = 1.5 \times 2.5b \approx 3.8b$

Let us assume fulcrum distance, $b = 50$ mm. Therefore, $a = 3.8 \times 50 = 190$ mm.

Let effort $F = 600$ N, assuming that one person can apply it manually

$$\text{Moment arm, } l = \frac{T_2\,(a - be^{\mu\theta})}{F} = \frac{8000\,(190 - 2.5 \times 50)}{600 \times 1000} = 0.867 \text{ m, say, } 0.9 \text{ m}$$

$$\text{Revised effort, } F = \frac{8000\,(190 - 2.5 \times 50)}{0.9 \times 1000} = 577.7 \text{ N}$$

Band design

The width of the band can be determined by the maximum allowable pressure between the band and the drum. The limiting value of pressure for steel band with friction lining and CI drum is (0.5–0.7) N/mm^2.

$$\text{Bearing pressure, } p_b = \frac{T_1}{b \times r}. \text{ Let } p_b = 0.6 \text{ N/mm}^2$$

Therefore,

$$\text{Band width, } b = \frac{20{,}000}{0.6 \times 375} = 88.8 \text{ mm, say, } 120 \text{ mm (to account for greater heat dissipation).}$$

The thickness of the band can be determined by the strength consideration, $\sigma_t = \dfrac{T_1}{bt}$

where σ_t is the allowable strength, assumed 60 N/mm^2.

Therefore,

$$\text{Band thickness, } t = \frac{T_1}{b\sigma_t} = \frac{20{,}000}{120 \times 60} = 2.77 \text{ mm, say, } 3 \text{ mm}$$

Heat transfer

$$\text{Heat generated, } H_{gen} = \frac{E}{t} \times q$$

$$E = \text{change in potential energy} = mgh = 20 \times 15 = 300 \text{ kN} \cdot \text{m}$$

$$q = \text{heat load factor} = \frac{\text{actual braking time}}{\text{total cycle time}} = \frac{60}{3 \times 60} = 0.33$$

Therefore,

$$H_{gen} = \frac{300 \times 10^3 \times 0.33}{60} = 1650 \text{ J/s}$$

$$\text{Heat dissipated, } H_{diss} = KA\Delta T$$

where

ΔT is the temperature difference [= $(t_2 - t_1)$]

t_2 is the operating temperature, say, 105°C

t_1 is the atmospheric temperature, say, 30°C

Let B' = band width + 50 mm = 120 + 50 = 170 mm

$$A = \text{heat dissipating area} = 2(\pi DB') - \pi DB' \times \frac{\theta}{360°}$$

$$= 2\pi \times 0.75 \times 0.17 - \pi \times 0.75 \times 0.17 \times \frac{210}{360} = 0.5674 \text{ m}^2$$

K = heat transfer coefficient = 41 W/m °C

Therefore,

$$H_{diss} = KA\Delta T = 41 \times 0.5674 \times 75 = 1744.7 \text{ J/s}$$

Since $H_{gen} < H_{diss}$, the brake is operating within safe thermal limits. Hence the design is safe for the given conditions.

Note: The design of the lever is left as an exercise for the student (refer Chapter 10).

EXERCISES

1. What is function of couplings and clutches? How does a coupling differ from a clutch?
2. What is the use of register in a flange coupling?
3. What type of flexibility does an Oldham coupling provide?
4. Discuss the various types of misalignments, which normally occur between two shafts.
5. Discuss the suitability of friction clutch for high torque transmission.
6. What are the design requirements for a friction clutch?
7. What do you understand by uniform pressure theory and uniform wear theory in the design of a friction clutch? Which theory is most suitable?
8. Why does a cone clutch transmit more power than a plate clutch?
9. What do you mean by a self-energizing brake and a self-locking brake?
10. What are the important factors in brake design? Explain them.
11. Design a split muff coupling to transmit 10 kW at 1000 rpm. The muff is made of FG200 CI and the shaft and the key are made of C20 steel. The coefficient of friction between the shaft and coupling is 0.25.
12. Design a protected type flange coupling to connect two shafts in order to transmit 110 kW at 250 rpm. The following permissible stresses may be used:

    Shear stress for shaft, bolt and key = 50 MN/m²
    Crushing stress = 150 MN/m²
    Shear stress for flange = 10 MN/m²
13. Two shafts of the same diameter running at 100 rpm transmit 150 kW through a marine type coupling. Design the coupling including bolts.

14. Design a flexible coupling to connect two shafts which transmit 10 kW at 700 rpm. The materials for various parts are:

Part name	*Material*
Shaft and key	C40
Bolts	35Mn2
Flange	FG250

15. Design a bush type flexible coupling to transmit 50 kW at 300 rpm to a compressor. Select your own materials and factor of safety.
16. Design a friction clutch for a car to transmit 32 kW at 4000 rpm. The maximum outer space is 200 mm. Assume all other data suitably.
17. Determine the main dimensions of a single plate clutch, with both sides effective, to transmit 20 kW at 2000 rpm. The ratio of the outer diameter to the inner diameter is 1.4.
18. Design a suitable clutch for a speed gear box of a milling machine to transmit 7.5 kW at 1440 rpm.
19. A cone clutch is to transmit 30 kW at 900 rpm. Determine the main dimensions of the clutch and the thrust spring. The mean radius is 160 mm. The semi-cone angle is 10°.
20. A leather-faced cone clutch is used to transmit 40 kW at 800 rpm. The larger end diameter is 300 mm. The semi-cone angle is 15° and the coefficient of friction μ is 0.2. Determine
    (a) the dimensions of the clutch,
    (b) the axial thrust, and
    (c) the dimensions of the single spring used to apply axial thrust.
21. Design a double shoe brake to absorb 5 kW at 700 rpm. Choose all other relevant data.
22. Design a band brake to absorb 20 kW at 1000 rpm. The brake drum diameter is 500 mm. The actual braking time is 40 per cent of the cycle period which is 4 minutes. During the braking period, the peripheral velocity reduces to one-fourth. The mass of the brake drum is 100 kg.
23. Determine the maximum braking torque for a band and block brake having 12 blocks, each of which subtends an angle of 16° at the centre. The brake is applied to a rotating drum of diameter 600 mm. The blocks are 75 mm thick. The two ends of the band are attached to pins on the opposite sides of the brake fulcrum at distances 40 mm and 150 mm from the fulcrum. A force of 250 N is applied at a distance of 900 mm from the fulcrum.

MULTIPLE CHOICE QUESTIONS

1. Which of the following couplings is a flexible coupling?

(a) Muff	(b) Split muff
(c) Flange	(d) Bush type

2. The pin type flexible coupling can accommodate
   (a) parallel misalignment up to 0.5 mm
   (b) axial displacement up to 2 mm
   (c) angular misalignment up to 1°
   (d) all of the above
3. Which of the following coupling provides kinematic flexibility?
   (a) Pin type bush (b) Oldham
   (c) Split muff (d) Flange
4. In the flange coupling, the two flanges are coupled by means of bolts fitted in
   (a) cast holes (b) drilled holes
   (c) threaded holes (d) reamed holes
5. Self-locking is not possible in the case of
   (a) simple band brake (b) differential band brake
   (c) simple block brake (d) internal expanding shoe brake
6. Short shoe brakes have angle of contact less than
   (a) 10° (b) 20°
   (c) 60° (d) 45°
7. In block brakes, the normal reaction and frictional force act at the midpoint of the block, this is true only if the angle is
   (a) less than 45° (b) less than 60°
   (c) greater than 45° (d) greater than 60°
8. The type of clutch used on motors with low starting torque is
   (a) single plate clutch (b) multiple plate clutch
   (c) cone clutch (d) centrifugal clutch
9. In the case of a single plate clutch, six springs are used for
   (a) avoiding spring failure
   (b) distributing axial load uniformly
   (c) reducing the size of the spring
   (d) none of the above
10. The number of effective surfaces with five steel and four brass plates in a multiplate clutch is
    (a) 5 (d) 4
    (c) 9 (d) 8
11. The ratio of the inner to the outer radii of the plate-type clutch varies between
    (a) 0.1–0.25 (b) 1–3
    (c) 0.5–0.7 (d) 5–10

Chapter **Bearings**

15

15.1 INTRODUCTION

Rotating shafts are required to be supported at suitable places. The mechanical device which can take up the load and support the shaft is called the bearing. The bearing is so named because the surface of support is subjected to a bearing load. In bearings, the relative motion between the two mating surfaces causes friction and generates heat. Any substance placed between the two surfaces which reduces friction, wear and takes away heat, is known as lubricant.

The classification of bearings depends upon the type of their construction, the type of the relative motion between two parts, and the nature of load. A general classification of bearings is given below:

Based upon the direction of load. On the basis of this criterion, bearings may be broadly classified into two categories.

Radial bearings. Radial bearings carry external load perpendicular to the axis of rotation of the shaft. Journal bearings and some types of ball and roller bearings belong to this category.

Thrust bearings. In these bearings, the load acts along the axis of rotation, e.g. hydrostatic bearings and antifriction thrust bearings.

Based upon the nature of contact. Bearings are also classified according to the nature of relative motion between members.

Sliding contact bearings. Bearings with sliding friction between the members are called sliding bearings, e.g. journal bearings and hydrostatic bearings.

Rolling contact bearings. Bearings with rolling contact friction between the members are known as rolling contact bearings, e.g. ball, roller, and taper roller bearings. These bearings have either point contact or line contact.

15.2 SELECTION OF BEARINGS

The selection of a bearing for a particular application is based upon several characteristics relating to mechanical, environmental, and economic requirements.

15.2.1 Mechanical Requirements

The following considerations are of paramount importance in the selection of type of bearing:

Load. The rolling element bearings are superior to slider bearings because they can carry high unidirectional and cyclic loads whereas the load-carrying capacity of slider bearings depends upon the speed of the journal.

Speed. Both sliding and rolling element bearings have a practical limit to the peripheral velocity, fixed by different criteria. In sliding bearings, the speed limitation is on account of the rise in temperature of lubricating oil caused by high speed shearing action. In rolling element bearings, the limiting speed is a function of the product of shaft diameter (mm) and speed (DN) in rpm. The limiting value of the DN factor is one million.

Misalignment. A slider bearing can tolerate misalignment better than what the rolling element bearing can because the latter has rigid structure and close tolerance.

Frictional loss. Rolling element bearings have low starting friction but journal bearings attain lower coefficient of friction only when sufficient pressure is built up.

Failure. A properly designed journal bearing can work even after failure of supply of lubricating oil because of boundary lubrication. In ball bearings, the failure of the lubricant film can cause very serious damage to moving parts.

Lubrication. Rolling element bearings require only a small amount of lubricant. The ball-bearings operating at DN value less than 0.3 million are pre-lubricated and can run for years. Journal bearings, on the other hand, require a large amount of lubricant to maintain the oil film.

Damping capacity. Slider bearings which carry a thick lubrication oil film have a higher damping capacity compared to the rolling element bearings which virtually have no damping capacity of their own.

Space. Journal bearings occupy less space in radial direction whereas rolling element bearings require less space in axial direction.

15.2.2 Environmental Conditions

The major environmental conditions under which a bearing is required to operate satisfactorily are temperature and corrosion. The maximum operating temperature is an important parameter. The rolling element bearings can operate at 200°C for an extended period of time whereas in slider bearings the lubricating oil loses its viscosity at high temperatures and thereby affects the load-carrying capacity. Both types of bearings are also subjected to corrosion due to chemical reaction with humidity and due to other corrosive atmospheric conditions.

15.2.3 Cost Economics

The overall cost economics of a bearing for a particular application is a function of its initial cost, life, maintenance, and cost of replacement. A properly designed slider bearing operating under uniform load has virtually unlimited life whereas the life of rolling element bearing under any given load and speed conditions is limited. The cost of a slider bearing is low compared to that of the rolling element bearing.

15.3 JOURNAL BEARINGS

A journal bearing is a sliding contact bearing which gives lateral support to the rotating shaft. Journal is that part of the shaft which runs in a sleeve or bushing. Usually, the sleeve or bushing is at rest and the journal rotates. In a journal bearing, the diameter of the journal is kept less than the diameter of the bearing to allow the flow of lubricant between the surfaces.

A journal bearing may be the full bearing in which the bearing surface is over full 360°. This bearing is capable of supporting radial force in any direction. A bearing which covers less than 360° is called partial bearing. It can support only the unidirectional load. If in the partial bearing, the diameters of the journal and bearing are equal, it is called fitted bearing. Figure 15.1 shows three different types of bearings.

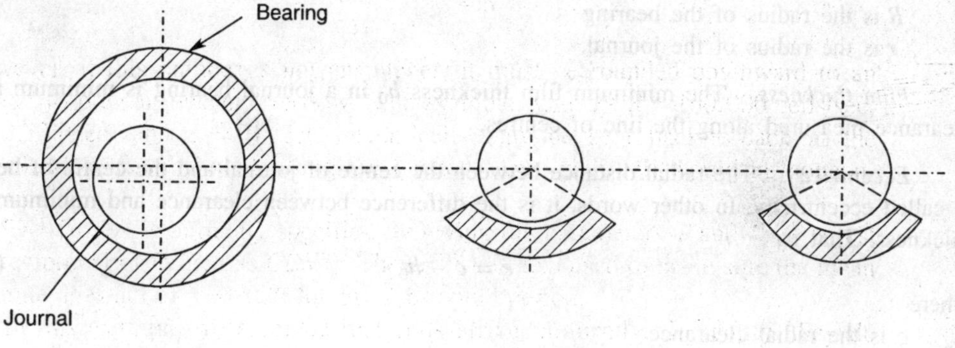

Figure 15.1 Types of journal bearings.

15.3.1 Journal Bearing Terminology

The most common terms and definitions used in connection with journal bearings, as shown in Figure 15.2, are given below:

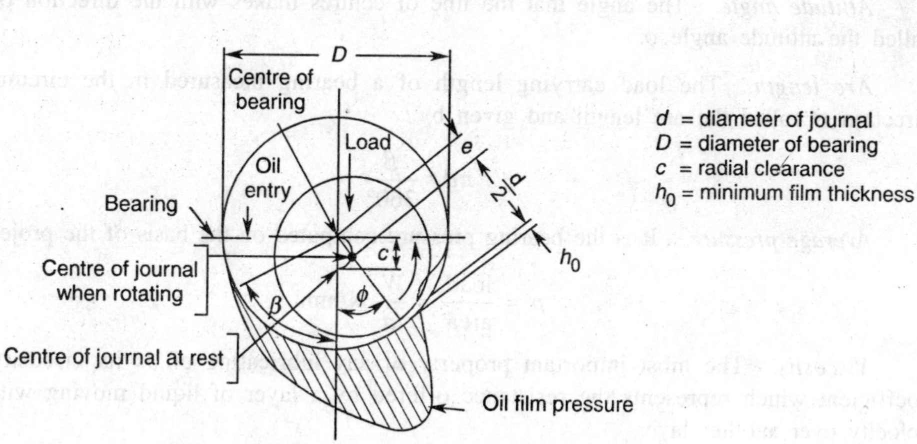

Figure 15.2 Journal bearing terminology.

Journal. It is that portion of the shaft which is supported by the bearing.

Bearing. The circular surface supporting the journal is called the bearing. It is generally stationary but may rotate in some applications.

Line of centre. It is the line joining the centres of the journal and the bearing. This line is usually taken as the reference line.

Clearance. It is the radial distance measured between the bearing and the journal. The clearance measured along the line of centres gives maximum clearance on one side and minimum on the other side. The radial clearance c is given by

$$c = R - r \tag{15.1}$$

where

R is the radius of the bearing
r is the radius of the journal.

Film thickness. The minimum film thickness h_0 in a journal bearing is minimum radial clearance measured along the line of centres.

Eccentricity. The radial distance between the centre of journal and the centre of bearing is called eccentricity. In other words, it is the difference between clearance and minimum film thickness. That is,

$$e = c - h_0 \tag{15.2}$$

where

c is the radial clearance
h_0 is the minimum film thickness.

Eccentricity ratio. It is the ratio of eccentricity to radial clearance. The eccentricity ratio is given by

$$\varepsilon = \frac{e}{c} = \frac{c - h_0}{c} = 1 - \frac{h_0}{c} \tag{15.3}$$

It is also known as attitude of bearings.

Attitude angle. The angle that the line of centres makes with the direction of load is called the attitude angle ϕ.

Arc length. The load carrying length of a bearing measured in the circumferential direction is called the arc length and given by

$$\pi d \times \frac{\beta}{360°} \tag{15.4}$$

Average pressure. It is the bearing pressure computed on the basis of the projected area

$$p = \frac{\text{load}}{\text{area}} = \frac{W}{ld} \text{ N/mm}^2 \tag{15.5}$$

Viscosity. The most important property of any lubricating oil is its viscosity. It is a coefficient which represents the resistance offered by a layer of liquid moving with certain velocity over another layer.

In practice, viscosity is measured by the Saybolt universal viscometer. It gives the time t in seconds required for a certain volume of the oil to flow under a certain head through a tube of standard diameter and length. The measured time is converted into kinematic viscosity Z_k, which is a ratio of the absolute viscosity and density. Thus,

$$Z_k = \left(0.22t - \frac{180}{t}\right) \times 10^{-6} \text{ m/s} \tag{15.6}$$

To convert kinematic viscosity to dynamic or absolute viscosity, the former is multiplied by the density.

The viscosity of lubricating oil depends upon the oil grade and operating temperature. Figure 15.3 shows that viscosity of lubricants drops off significantly with the rise in temperature.

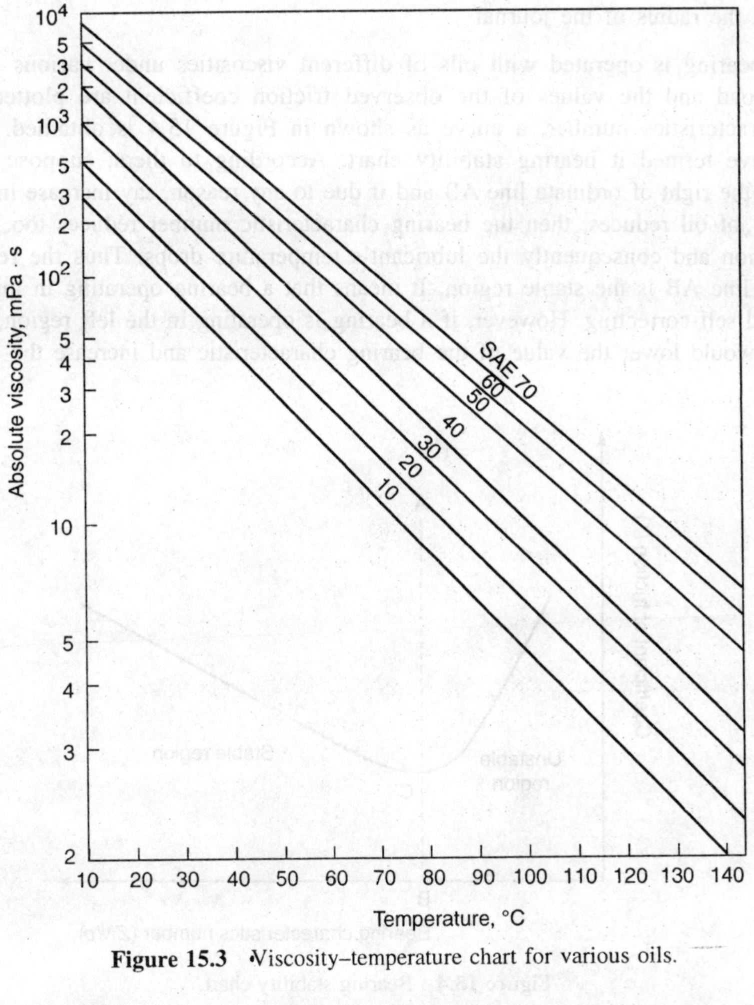

Figure 15.3 Viscosity–temperature chart for various oils.

Friction coefficient. The coefficient of friction in the design of a bearing depends upon many factors, namely materials of bearing and journal, surface finish, oil viscosity, operating temperature and speed, dimensions of journal and bearing including the relative radial clearance. Petroff has developed the following theoretical relation for determining the coefficient of friction.

$$f = \frac{\pi^2}{0.5 \times 10^6}\left(\frac{r}{c}\right)\left(\frac{ZN}{p}\right) \tag{15.7}$$

where

ZN/p is the bearing characteristics number
Z is the absolute viscosity (N · s/m^2)
N is the speed, in rpm
p is the bearing pressure (N/mm^2)
c is the radial clearance
r is the radius of the journal

If a given bearing is operated with oils of different viscosities under various conditions of speed and load and the values of the observed friction coefficient are plotted against the bearing characteristics number, a curve as shown in Figure 15.4 is obtained. Mackee and Mackee* have termed it bearing stability chart. According to them, suppose a bearing is operating to the right of ordinate line AB and if due to any reason, say increase in temperature, the viscosity of oil reduces, then the bearing characteristic number reduces too. This reduces heat generation and consequently the lubricant's temperature drops. Thus the region right to the ordinate linc AB is the stable region. It means that a bearing operating in this region will be stable and self-correcting. However, if a bearing is operating in the left region, the decrease in viscosity would lower the value of the bearing characteristic and increase the coefficient of

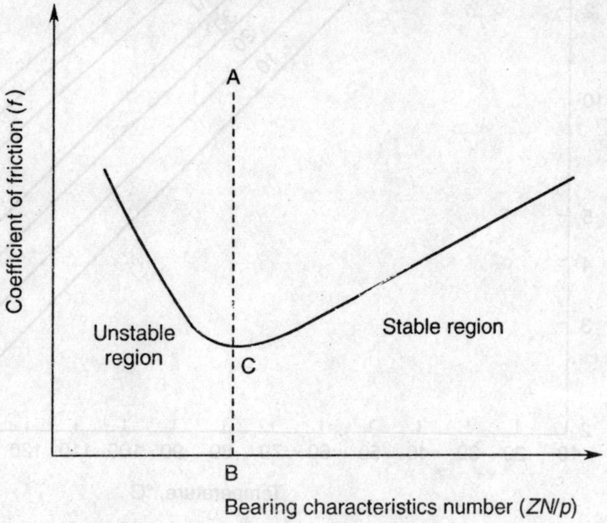

Figure 15.4 Bearing stability chart.

* Mackee S.A. and T.R. Mackee, "Journal bearing friction in the region of thin film," *SAE Journal*, Vol. **13**, pp. 371–377, 1937.

friction. A temperature rise would ensue and the viscosity be further reduced, ultimately resulting into metal-to-metal contact. Therefore, the region to the left of line AB represents the unstable region. The point C on the curve indicates the beginning of the metal-to-metal contact or boundary lubrication.

15.4 HYDRODYNAMIC THEORY OF LUBRICATION

The hydrodynamic theory of lubrication was developed first time by Tower in 1883. According to him, when a partial bearing of angle 157°, 100 mm in diameter and 150 mm in length, with bath types of lubrication was set in motion, the pressure inside the bearing increased and almost all oil flowed out. The pressure distribution, as obtained in Tower's experiment, is shown in Figure 15.5. According to Tower, the factor which directly affects the load carrying capacity of the bearing is the flow of a lubricant in a converging circular channel. Figure 15.6(a) shows a journal in the rest position subjected to static load. When this journal starts rotating under load without the presence of oil, friction causes the journal to climb up the side of the bearing [Figure 15.6(b)]. Under these conditions equilibrium will be obtained when the friction force is balanced by the tangential component of the bearing load. In this case there will be metal-to-metal rubbing and wearing of parts will take place. If sufficient lubricating oil is filled in the bearing, the rotating action of the journal causes the lubricant to be pumped into a wedge-shaped space and this forces the journal over to the other side, as shown in Figure 15.6(c). This action builds up sufficient pressure to lift and support the journal. Hence there is no metal-to-metal rubbing and only a small friction value due to shearing action of the lubricating oil will occur. This phenomenon is termed *hydrodynamic theory of lubrication*.

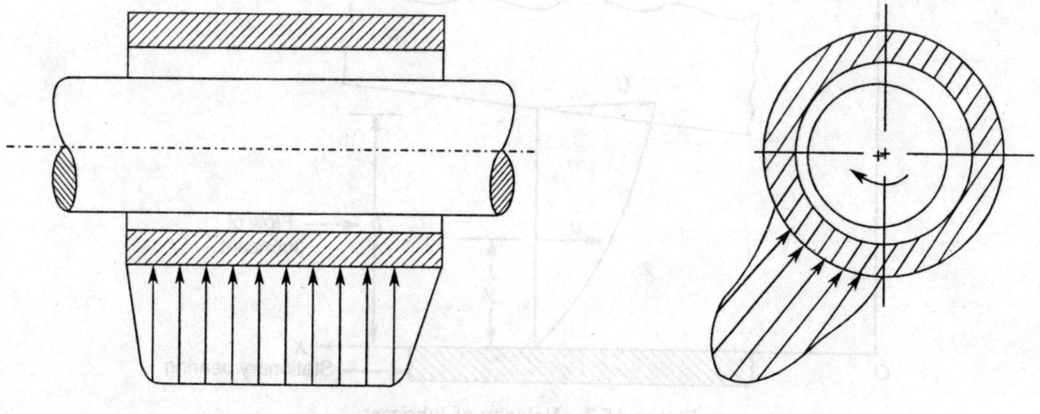

Figure 15.5 Oil film pressure distribution.

The results obtained by Tower were theoretically investigated by Osborne Reynold. He realized that the fluid film in the bearing is so thin compared to the bearing radius that the effect of curvatures can be neglected. This enabled him to replace the curved partial bearing with the flat bearing.

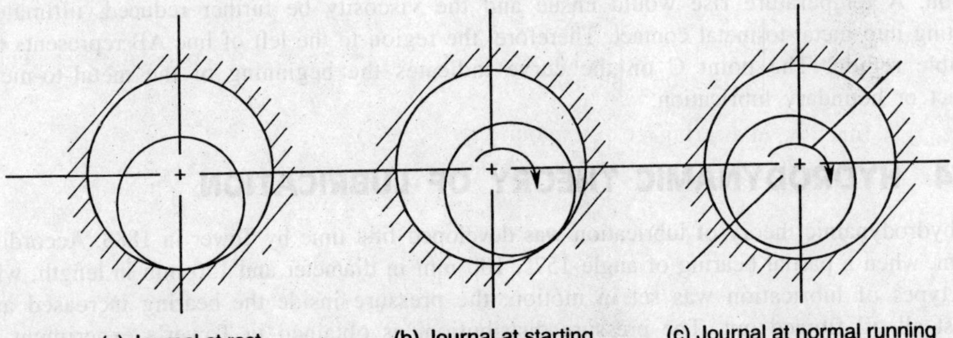

Figure 15.6 Journal at various positions and formation of oil film.

Reynold analyzed the flat bearing, as shown in Figure 15.7. He developed an equation for the flow of lubricant, called the Reynold equation and expressed it as

$$\frac{d}{dx}\left(\frac{h^3}{Z}\cdot\frac{dp}{dx}\right) = -6U\frac{dh}{dx} \tag{15.8}$$

where

p is the pressure distribution

U is the peripheral velocity

dp/dx is the pressure gradient.

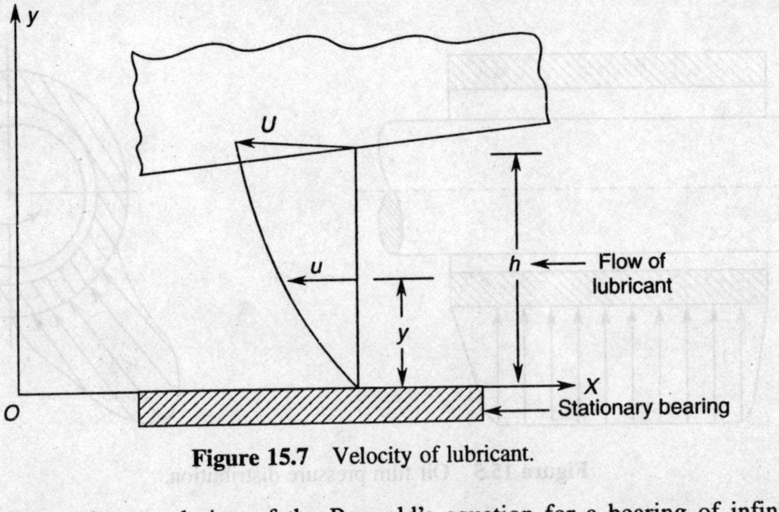

Figure 15.7 Velocity of lubricant.

Later, an approximate solution of the Reynold's equation for a bearing of infinite length was attempted by Sommerfeld using numerical and electrical analogy methods. The important solution determined by him is a relation between the friction coefficient and a non-dimensional number given as

$$\frac{r}{c}f = \phi\left[\left(\frac{ZN}{p}\right)\times\left(\frac{r}{c}\right)^2\right] \tag{15.9}$$

where ϕ represents the functional relationship between the coefficient of friction and a constant. The quantity $\left[\frac{ZN}{p} \times \left(\frac{r}{c}\right)^2\right]$ is known as Sommerfeld number S. This non-dimensional number is a function of bearing design parameters over which the designer has control. The bearing arc is the only parameter which is not included in the Sommerfeld number.

15.5 BEARING DESIGN FACTORS

While designing a journal bearing, there are a large number of parameters which a designer has to choose. They can be divided into two groups. The first group has all those parameters which are either given or are under the control of the designer. These parameters are load, speed viscosity, and bearing dimensions.

The designer usually has no control over the load and speed of the journal because these two parameters are application-dependent. However, the designer can choose the remaining parameters over which he has direct control. Table 15.1 gives general guidelines for selection of the bearing parameters.

Table 15.1 Design data for bearings

Machine	*Bearing type*	*Maximum pressure,* p_{max} (MN/m^2)	*Viscosity, Z* ($N \cdot s/m^2$)	*Minimum value of ZN/p*	*c/r*	*l/d*
Gas and oil engines (four-stroke)	Main	5.5–12.0		2.9	0.001	0.6–2.0
	Crank pin	10–24	0.002–0.0065	1.45	< 0.001	0.6–1.5
	Wrist pin	16–30		0.76	< 0.001	1.5–2.0
Gas and oil engines (two-stroke)	Main	3.4–5.5		3.63	0.001	0.6–2.0
	Crank	6.9–10.3	0.02–0.065	1.73	< 0.001	0.6–1.5
	Wrist	8.2–12.4		1.45	< 0.001	1.5–2.0
Low speed steam turbines	Main	2.8	0.06	2.9	< 0.001	1–2
Steam turbines	Main	0.7–1.9	0.002–0.016	14.5	0.001	1.0–2.0
Machine tools	Main	2.1	0.04	5.8	0.001	1–4.0
Rolling mills	Main	20.6	0.05	1.45	0.0015	1.1–1.5
Reciprocating pumps and compressors	Main	1.8		4.36	0.001	1.0–2.2
	Crank	4.1	0.03–0.08	2.90	< 0.001	0.9–1.7
	Wrist	6.9		1.45	< 0.001	1.5–2.0
Transmission shafts	Light	0.2		145.0	0.001	2–3
	Self-aligning	1.0	0.025–0.06	4.36	0.001	2.5–4.0
	Heavy	1.0		4.36	0.001	2–3
Generators, motors, centrifugal pumps	Rotor	0.7–1.4	0.025	29.01	0.0013	1–2

The parameters in the second group are dependent parameters. These are coefficient of friction, minimum film thickness, oil flow rate, and temperature rise in the bearing. These parameters are used as an index for measuring the performance of the designed bearing. Thus they put some limitations on the variables of the first group to enable satisfactory performance of the bearing to be achieved.

Raimondi and Boyd have obtained comprehensive computer solutions of the Reynold's equation to evaluate the performance of the bearing. These solutions have been plotted in the form of various curves between the non-dimensional function parameters and the Sommerfeld number. A brief description of a few important curves is presented below.

15.5.1 Minimum Film Thickness

The minimum oil film thickness, according to Raimondi and Boyd, can be obtained from the chart plotted for full journal bearing as shown in Figure 15.8. From this chart, for the known value of the Sommerfeld number, a non-dimensional film thickness parameter h_0/c is obtained and thereafter the minimum film thickness h_0 can be calculated for the known value of the radial clearance c. A journal bearing is usually designed for two conditions—minimum power loss due to friction and maximum load-carrying capacity. Figure 15.8 shows these conditions by two separate curves drawn in dotted lines. The zone between the boundaries defined by these two conditions may therefore be considered as the recommended operating zone.

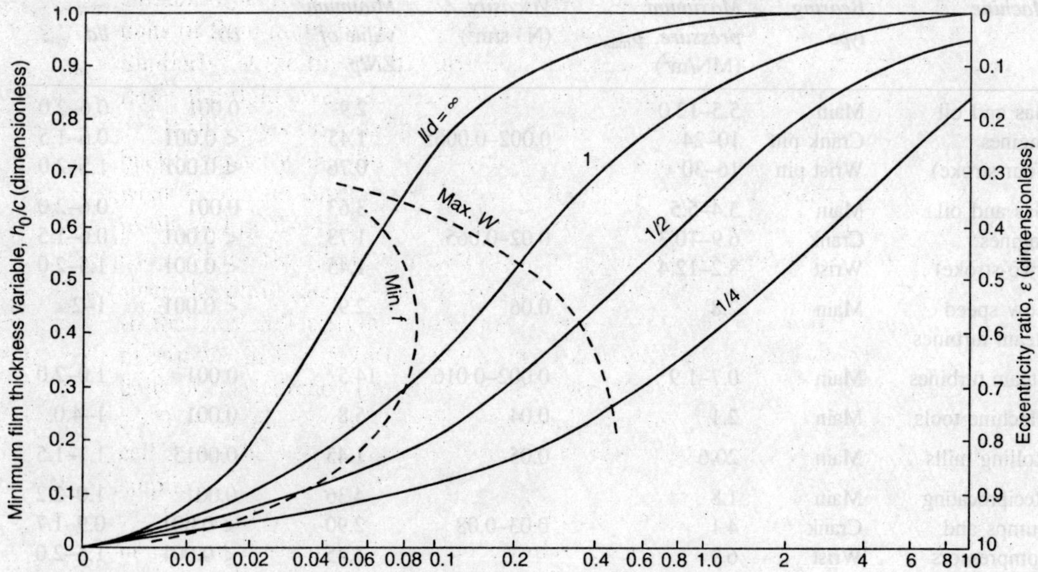

Figure 15.8 Minimum film thickness and eccentricity ratio.

Figure 15.9 gives the minimum film thickness for bearings having arc angles between 60° and 360°.

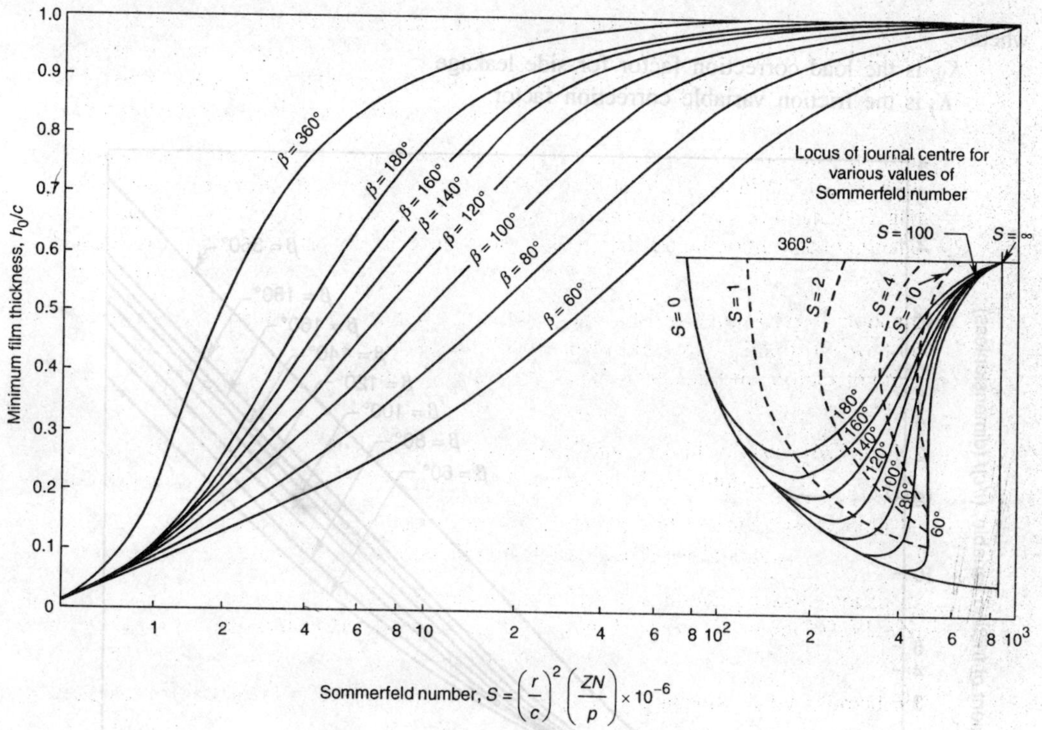

Figure 15.9 Minimum film thickness for different values of Sommerfeld number.

15.5.2 Friction Variable

In a friction variable chart, shown in Figure 15.10, the dimensionless friction variable $\left(\frac{r}{c}\right) f$ is plotted against the Sommerfeld number. The coefficient of friction in a bearing is one of the important parameters which decides the energy lost in the bearing. Although the initial assessment of the coefficient of friction is generally done by the Petroff equation, its actual value is estimated from the friction variable.

$$\text{Coefficient of friction, } f = \text{FV} \times \frac{c}{r} \tag{15.10}$$

where FV is the friction variable measured from Figure 15.10.

If a journal bearing is a short bearing and there is leakage of lubricating oil, the coefficient of friction in such a circumstance can be determined by modifying the Sommerfeld number and the friction variable. The modified Sommerfeld number is given by

$$S = \left(\frac{r}{c}\right)^2 \times \left(\frac{ZN}{p}\right) \times K_W \tag{15.11}$$

Coefficient of friction

$$f = \text{FV} \times \frac{c}{r} \times \frac{K_f}{K_W} \tag{15.12}$$

where

K_W is the load correction factor for side leakage
K_f is the friction variable correction factor.

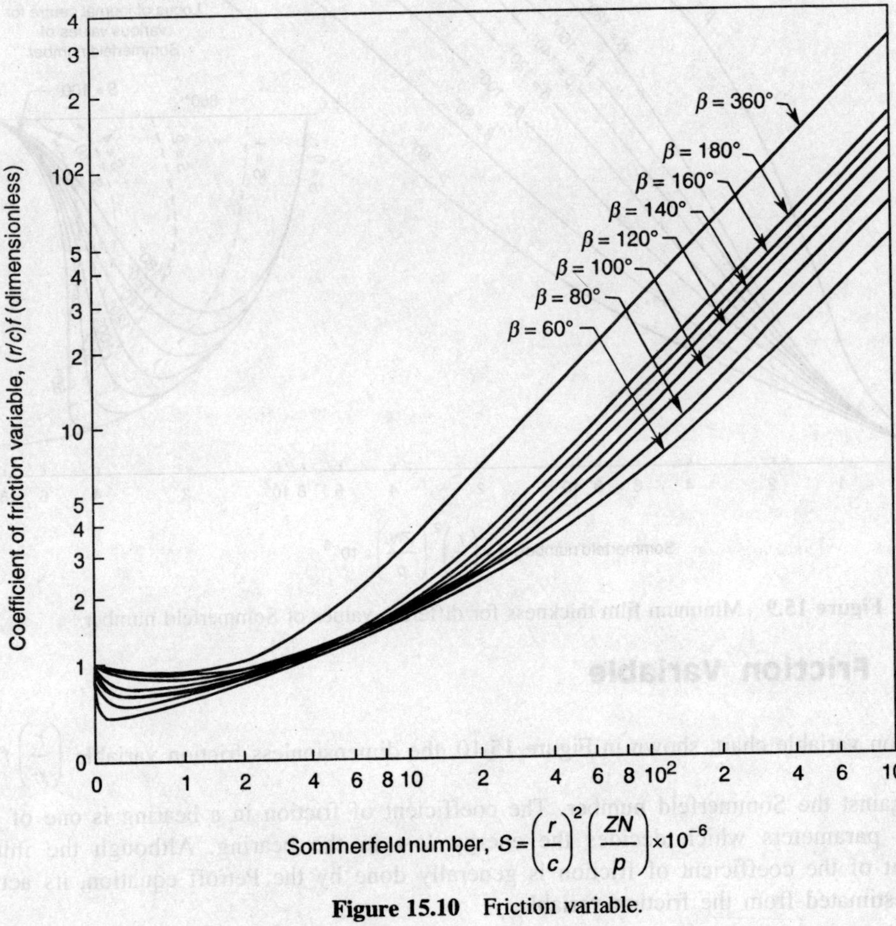

Figure 15.10 Friction variable.

15.5.3 Flow Variable

The pressure developed in the bearing and its load carrying capacity depends upon the volume of the lubricating oil fed into the bearing. Raimondi and Boyd have solved the Reynold's equation for determining the quantity of lubricating oil required to be pumped into the converging space of the rotating journal. This solution is plotted as the dimensionless parameter called the *flow variable*, as shown in Figure 15.11. This chart is based on the following assumptions:

(i) The lubricant is supplied at atmospheric pressure.
(ii) There is no lubricating oil groove.
(iii) There is no side leakage of oil.

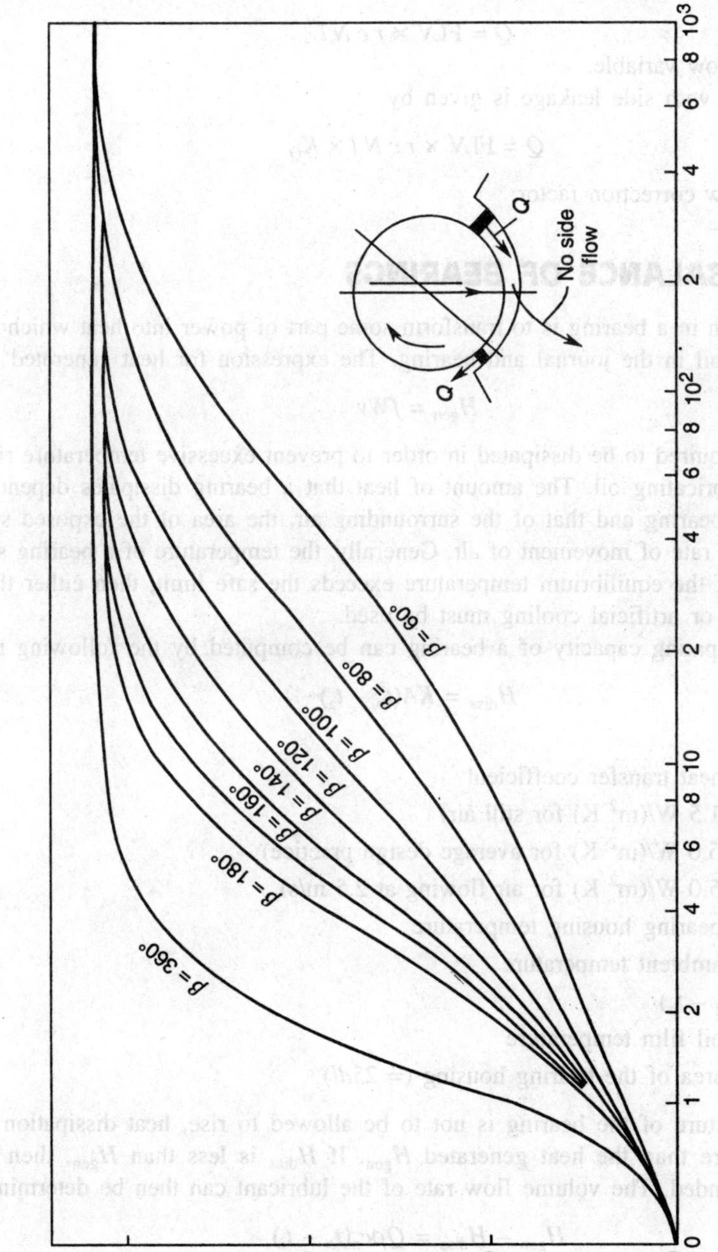

Figure 15.11 Flow variable.

The flow rate without side leakage is given by

$$Q = \text{FLV} \times r\,c\,N\,l \tag{15.13}$$

where FLV is the flow variable.

The flow rate with side leakage is given by

$$Q = \text{FLV} \times r\,c\,N\,l \times K_Q \tag{15.14}$$

where K_Q is the flow correction factor.

15.6 HEAT BALANCE OF BEARINGS

The effect of friction in a bearing is to transform some part of power into heat which increases the temperature of oil in the journal and bearing. The expression for heat generated is

$$H_{\text{gen}} = fWv \tag{15.15}$$

The heat is required to be dissipated in order to prevent excessive temperature rise which may oxidize the lubricating oil. The amount of heat that a bearing dissipates depends on the temperature of the bearing and that of the surrounding air, the area of the exposed surface of the bearing, and the rate of movement of air. Generally, the temperature of a bearing should be kept below 75°C. If the equilibrium temperature exceeds the safe limit, then either the design should be modified or artificial cooling must be used.

The heat dissipating capacity of a bearing can be computed by the following relation.

$$H_{\text{diss}} = KA(t_{\text{b}} - t_{\text{a}}) \tag{15.16}$$

where

K is the heat transfer coefficient

(= 11.5 W/(m^2 K) for still air)

(= 15.0 W/(m^2 K) for average design practice)

(= 35.0 W/(m^2 K) for air flowing at 2.5 m/s)

t_{b} is the bearing housing temperature

t_{a} is the ambient temperature

$(t_{\text{b}} - t_{\text{a}}) = 0.5(t_{\text{f}} - t_{\text{a}})$

t_{f} is the oil film temperature

A is the area of the bearing housing (= $25dl$)

If the temperature of the bearing is not to be allowed to rise, heat dissipation capacity H_{diss} should be more than the heat generated H_{gen}. If H_{diss} is less than H_{gen}, then artificial cooling is recommended. The volume flow rate of the lubricant can then be determined from

$$H_{\text{gen}} - H_{\text{diss}} = Q\rho c_p(t_o - t_i) \tag{15.17}$$

where

Q is the volume flow rate (m^3/s)

ρ is the density (kg/m^3)(= 850–930 kg/m^3)

c_p is the specific heat at constant pressure (J/(kg °C))(= 1850–1960 J/(kg °C))

t_o is the outlet temperature

t_i is the inlet temperature.

15.7 DESIGN PROCEDURE

The design of a journal bearing involves a large number of parameters. We first select reasonable values of various parameters and then apply the available equations to establish their validity. Raimondi and Boyd have suggested the following procedure to design a bearing using various design charts:

1. Select a suitable value of the *l/d* ratio. Determine the length of the bearing and the bearing pressure comparable to the maximum value given in Table 15.1.
2. Select a suitable bearing material from Table 15.2 based upon the bearing pressure selected.
3. Select the clearance ratio *c/r*.
4. Select a suitable lubricating oil and its viscosity at operating temperature, which is limited to 75°C because most oils get oxidized above this temperature.
5. Determine the bearing characteristic number *ZN/p*. It should be greater than the minimum value.
6. Determine the Sommerfeld number and thereafter the minimum film thickness from Figure 15.9. The minimum film thickness should be compatible with surface finish and it should be more than one-fourth of the radial clearance.
7. Calculate the coefficient of friction for the pre-determined value of the Sommerfeld number from Figure 15.10.
8. Calculate the quantity of the lubricating oil required from Figure 15.11.
9. If a bearing is a short bearing or the side leakage condition is given, then the Sommerfeld number should be revised. The friction and flow variables should also be modified using the respective correction factors.
10. Determine the heat generated and the heat dissipation capacity to establish the thermal equilibrium.

15.8 MECHANICAL ASPECTS OF BEARING DESIGN

In addition to determining the basic design parameters of a journal bearing, namely diameter, length, radial clearance, minimum film thickness, proper oil viscosity and heat transfer rate, the following additional information is needed, too, to specify a bearing.

(i) Type of the bearing material
(ii) Type of the bearing housing
(iii) Bearing shell
(iv) Strength and rigidity of the bearing cap and bolts
(v) Provision of lubrication

15.8.1 Bearing Material

A bearing, however carefully and properly designed, may fail either due to deflection, temperature distortion, surface roughness, contamination of oil or due to faulty selection of the bearing material. Therefore, a proper selection of the bearing material is very important. A good bearing material possesses the following properties:

1. Good compatibility between the journal and the bearing surface.
2. Good conformability, which is obtained through those materials which have low elastic modulus.
3. High compressive and fatigue strength.
4. High heat conductivity, high score, wear and corrosion resistance and low coefficient of friction.

The properties of some of the bearing materials are listed in Table 15.2.

Table 15.2 Properties of some bearing materials

Material	*Characteristics*	*Applications*
Babbitt	Poor corrosion and fatigue resistant, good thermal conductivity, expensive	I.C. engine main and crank bearings. Allowable pressure range 7–15 N/mm^2.
Plastic bronze	Resistant to seizure, good conformability and embedability, low coeffcient of friction	Automobiles, locomotive engines, rolling mills. Allowable pressure range 5–10 N/mm^2.
Phosphor bronze	High tensile and fatigue strength, low coefficient of friction	Heavy duty machinery such as turbines, rolling mills, machine tools, pumps, compressors. Allowable pressure range 15–60 N/mm^2.
Gun metal	Low cost, lower strength, reasonable good thermal conductivity	Light duty machinery such as fans, small size compressors, transmission shafts. Allowable pressure range 1.0–5 N/mm^2.
Brass	Lower cost, high strength, low coefficient of friction	Moderate load conditions with allowable pressure range 10–30 N/mm^2.
Aluminium	Reasonable corrosion and fatigue resistant, high thermal expansion, low embedability	Suitable for moderate load conditions with allowable pressure range 15–30 N/mm^2.
Cast Iron	High wear resistance, good lubrication, low embedability, low cost.	Cam shafts, light transmission shafts. Maximum allowable pressure is 3.5 N/mm^2.

15.8.2 Bearing Housings

A bearing housing is a mechanical structure which carries a bush and supports the shaft. It is mounted on the frame or structure of the machine. Generally for up to 50 mm journal diameters, solid housing is used. However, a split-type housing with split bushes is preferred for ease of assembly and replacement. When a housing is of the split-type, the two pieces called the housing and the cap are joined by nut-bolt fasteners. The most common type of split-housing for journal bearings is the plummer block. It is available in two series—light and medium. The dimensions of plummer block should conform to IS : 4773 – 1979.

15.8.3 Bearing Shell

The bearing housings are generally lined with bearing shell or bushes for easy replacement of the worn-out bearings. The bearing shell of a journal bearing is constructed in two ways: (i) solid bearing shell (or solid bush) and (ii) lined bearing shell (or lined bush). For a small-sized journal which operates at very low speed, generally a solid bearing shell is used. The solid bearing shells are made either by casting or by machining from a bar. A lined bearing shell consists of a steel backing with a thin lining of bearing materials like white metal, moulded plastic or powder metals, etc. It is usually split into two halves. The thicknesses of the bearing shell and lining are found empirically (see Figure 15.12) as follows:

(i) For small to medium sized journals having diameters up to 75 mm, the thickness of brass or bronze shell is, $h = 0.08d + 2.5$ mm

(ii) For a medium-sized bearing, steel backing $h_1 = 0.15d$ and lining thickness $h_2 = 0.01d + 3.0$ mm

(iii) For aluminium shells, $h = 0.044d + 0.5$ mm

The average thickness of the liner should be at least 2.5 to 3.5 mm for journals up to 80.0 mm diameter.

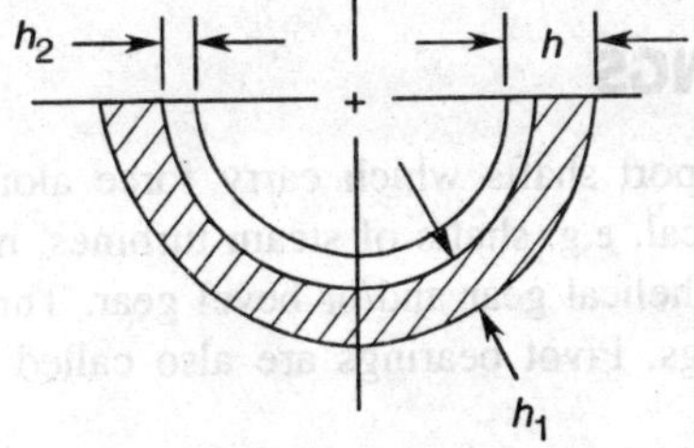

Figure 15.12 Split lined bush.

15.8.4 Bearing Cap and Bolts

The cap of a journal bearing is generally not subjected to heavy loads. However, in certain cases, namely the main bearing of automobiles and big end of the connecting rods, etc. the load acts upon the cap. In such cases, the cap should be checked for strength and lateral deflection by assuming it to act as a simply supported beam loaded at the centre and supported at the bolt centres, as shown in Figure 15.13.

$$\text{Thickness of the cap, } t = \sqrt{\frac{3WC}{2L\sigma_b}} \tag{15.18}$$

where

C is the distance between the bolt centres
L is the length of the bearing
t is the thickness of the cap
σ_b is the allowable bending strength of the cap material
W is the force acting upon the cap

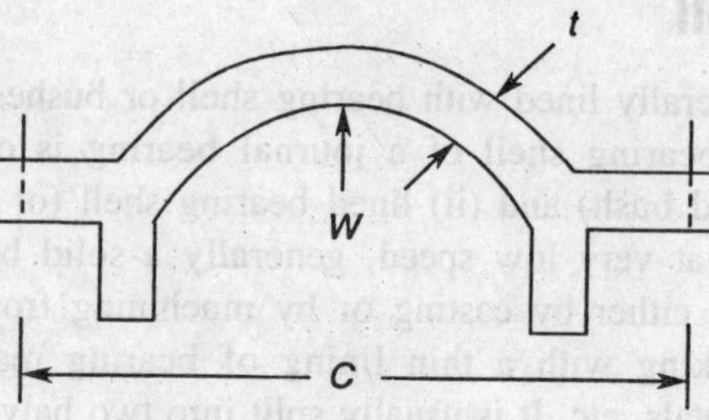

Figure 15.13 Bearing cap.

The deflection of the cap is given by

$$\delta = \frac{WC^3}{48EI} \tag{15.19}$$

and this deflection should be limited to 0.0125 – 0.025 mm. The bolts or studs which hold down the cap may be assumed to be subjected to tension force. The load on each bolt is taken as $1.33W/i$, where i is the number of bolts. The constant 1.33 takes into account the uneven load distribution on the bolt due to friction of the journal.

15.9 THRUST BEARINGS

Thrust bearings are used to support shafts which carry force along their axes. The shaft axis may be either horizontal or vertical, e.g. shafts of steam turbines, motors, pumps, machine tools and all those shafts which carry helical gear and/or bevel gear. Thrust bearings are classified as pivot bearings and collar bearings. Pivot bearings are also called step bearings.

15.9.1 Pivot Bearings

A simple form of thrust bearing is the plain pivot bearing, as shown in Figure 15.14, which is mostly used for vertical shafts. In these bearings, maximum wear takes place at the outer radius where the rubbing velocity is maximum. As a result, the pressure at the centre becomes excessively high for a constant wear rate ($pv = c$) which ultimately leads to overheating and failure of lubrication. Therefore, in order to eliminate this problem a hole of some diameter d_i is carved out at the centre of the thrust disc.

The total axial force on the shaft, for a flat pivot bearing, as shown in Figure 15.14, is given by

$$F = \frac{\pi}{4}(d_o^2 - d_i^2)p \tag{15.20}$$

The friction torque in the flat pivot bearing is given by

$$T = \mu F r_{\text{mean}} \tag{15.21}$$

where

μ is the coefficient of friction ($= 84\ p^{-0.67}\ v^{0.5}$)
F is the axial force on the bearing
p is the bearing pressure
v is the rubbing velocity

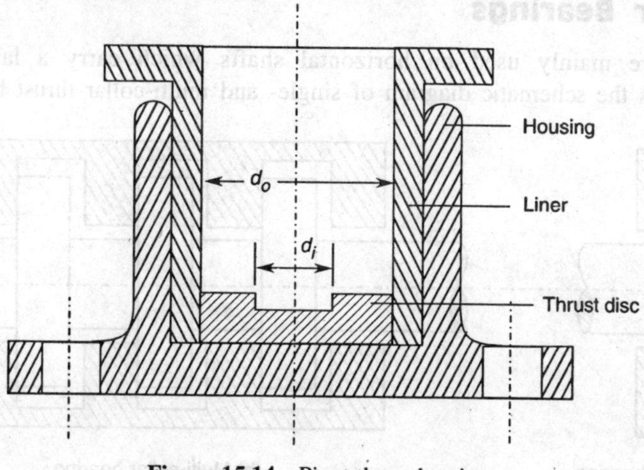

Figure 15.14 Pivot thrust bearing.

r_{mean} is the mean radius of the flat pivot bearing

$$= \frac{2}{3}\left(\frac{r_o^3 - r_i^3}{r_o^2 - r_i^2}\right) \text{ for the case when it is assumed that pressure distribution is uniform}$$

$$= \frac{r_o + r_i}{2} \text{ for uniform wear rate}$$

$$\text{Power lost} = \frac{2\pi NT}{60 \times 1000} \text{ kW} \quad (15.22)$$

In case when the end of the bearing is a frustum of cone, i.e. a conical pivot, as shown in Figure 15.15, the frictional torque is given by the relation

$$T = \frac{\mu F r_{mean}}{\sin \alpha} \quad (15.23)$$

where 2α is the cone angle of the conical pivot.

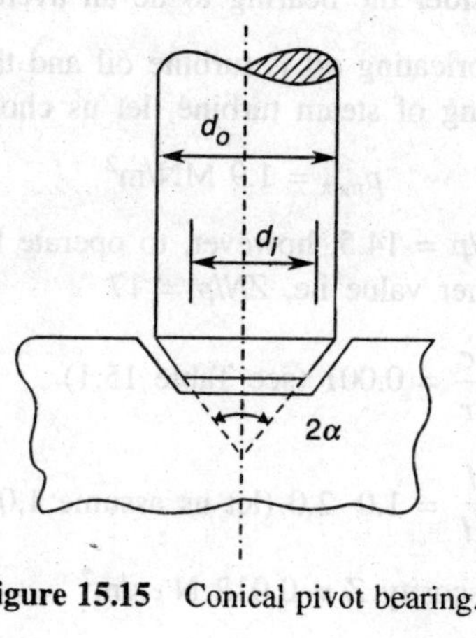

Figure 15.15 Conical pivot bearing.

15.9.2 Collar Bearings

Collar bearings are mainly used on horizontal shafts which carry a large axial force. Figure 15.16 shows the schematic diagram of single- and multi-collar thrust bearings.

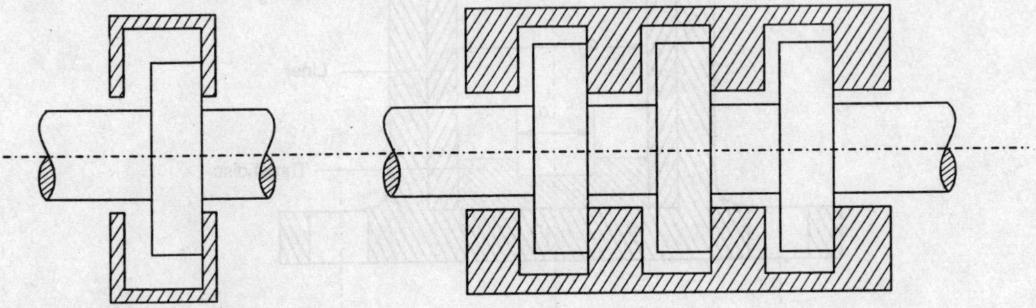

(a) Single-collar bearing (b) Multi-collar bearing

Figure 15.16 Collar type thrust bearing.

In these bearings, the collar should be placed near the point of application of the axial force so that the effect of column action upon the shaft can be avoided. In multi-collar thrust bearings, the allowable bearing pressure between the collar and the bearing surface is slightly less than that for pivot bearings because the force is not likely to be evenly distributed among all collars.

The ratio of the collar diameter to shaft diameter (d_o/d_i) is usually kept between 1.3–1.7. The width of the collar is usually taken as $d_i/6$ to $d_i/4$.

The allowable bearing pressure for pivot and collar bearings should be maintained in such a way that the wear rate (the product of pressure and velocity (pv)) does not exceed 57.0×10^4 for the shafts operating at speeds between 0.25 and 1 m/s. That is, for low-speed shafts, the bearing pressure may be as high as 14 MN/m^2, for intermittent service 10 MN/m^2, and for speeds over 1 m/s the bearing pressure should not exceed 0.69 MN/m^2. In multi-collar bearings, the bearing pressure should be halved.

Example 15.1 Design a journal bearing for a 10 MW, 1000 rpm steam turbine which is supported by two bearings. Consider the bearing to be an average industrial bearing.

Solution Let us assume that lubricating oil is turbine oil and the operating temperature 60°C. From Table 15.1, the main bearing of steam turbine, let us choose the following data:

$$p_{max} = 1.9 \text{ MN/m}^2$$

The minimum value of ZN/p = 14.5, however, to operate bearing under the stable region we should choose a slightly higher value i.e. ZN/p = 17

$$\frac{c}{r} = 0.001 \text{ (see Table 15.1)}$$

$$\frac{l}{d} = 1.0\text{–}2.0 \text{ (let us assume 1.0) (see Table 15.1)}$$

For turbine oil at 60°C, viscosity $Z = 0.012 \text{ N} \cdot \text{s/m}^2$

Therefore,

$$\frac{ZN}{p} = \frac{0.012 \times 1000}{p} = 17$$

or

$$p = 0.7059 \text{ N/mm}^2$$

$$\text{Sommerfeld number, } S = \frac{ZN}{p} \times \left(\frac{r}{c}\right)^2 \times 10^{-6} = 17 \times (1000)^2 \times 10^{-6} = 17$$

$$\text{Torque transmitted, } T = \frac{60 \times 10 \times 10^6}{2\pi \times 1000} = 95{,}493 \text{ N} \cdot \text{m}$$

Assuming that the rotor of the steam turbine is made of alloy steel 40Ni2 Crl Mo28 having tensile yield strength of about 1275 N/mm^2 and assuming factor of safety FoS 1.5, the permissible shear strength

$$\tau = \frac{0.577\sigma_{yt}}{\text{FoS}} = \frac{0.577 \times 1275}{1.5} = 490.45 \text{ N/mm}^2$$

$$\text{Diameter of rotor, } d = \left(\frac{16T}{\pi\tau}\right)^{1/3} = \left[\frac{16 \times 95{,}493 \times 10^3}{\pi \times 490.45}\right]^{1/3}$$

$$= 99.70 \text{ mm, say, } 100 \text{ mm}$$

$$\text{From Figure 15.10, for } S = 17 \text{ and } \beta = 360°, \quad \frac{r}{c} f = 6.25 \quad \text{or} \quad f = 0.00625$$

$$\text{Linear velocity, } v = \frac{\pi dN}{60} = \frac{\pi \times 0.1 \times 1000}{60} = 5.236 \text{ m/s}$$

$$\text{Heat generated, } H_{gen} = fWv$$

$$\text{Load, } W = pdl = 0.7059 \times 100 \times 100 = 7059 \text{ N}$$

$$H_{gen} = 0.00625 \times 7059 \times 5.236 = 231 \text{ W}$$

Heat dissipated by the body of journal

$$H_{diss} = KA(t_b - t_a)$$

$$t_b - t_a = 0.5(t_f - t_a)$$

Let ambient temperature, $t_a = 25°\text{C}$

Temperature of the oil film, $t_f = 60°\text{C}$

Heat transfer coefficient, $K = 15 \text{ W/(m}^2 \text{ K)}$

Therefore,

$$H_{diss} = 15 \times 20 \times 0.1 \times 0.1 \times 0.5 \times (60 - 25) = 52.5 \text{ W}$$

Since heat generated is greater than heat dissipated, artificial cooling by oil is needed to carry away the excess heat. Thus,

$$\text{Excess heat} = H_{\text{gen}} - H_{\text{diss}} = 231 - 52.5 = 178.5 \text{ W}$$

$$178.5 = Q\rho c_p(t_o - t_i)$$

$$= Q \times 890 \times 1950 \times 35$$

or

$$Q = 2.938 \times 10^{-6} \text{ m}^3\text{/s} = 0.176 \text{ litres/min}$$

From Figure 15.9, for $S = 17$ and $\beta = 360°$, $\dfrac{h_o}{c} = 0.94$ or $h_0 = 0.94 \times c = 0.94 \times 0.05 = 0.047$ mm ($\because c = 0.001 \times r = 0.05$ mm)

Bearing cap. Let us assume that the distance between the bolt centres $C = 3l = 3 \times 100 = 300$ mm. Assuming that the cap is made of grey CI of FG250 with tensile strength 250 N/mm² and FoS = 4, we have

$$\sigma_b = \frac{250}{4} = 62.5 \text{ N/mm}^2$$

$$\text{Therefore, thickness of the cap, } t = \sqrt{\frac{3WC}{2l\sigma_b}} = \sqrt{\frac{3 \times 7059 \times 300}{2 \times 100 \times 62.5}}$$

$$= 22.54 \text{ mm, say, } t = 25 \text{ mm}$$

Bolt. Load on each bolt, $F = \dfrac{4}{3} \times \dfrac{W}{i}$

$$= \frac{4}{3} \times \frac{7059}{2} = 4706 \text{ N}$$

Assuming bolt material 45C8 having allowable strength 80 N/mm²,

$$\text{Resisting area} = \frac{4706}{80} = 58.8 \text{ mm}^2$$

As per IS : 4218 – 1967, we select the bolt M10 × 1

Example 15.2 Design a journal bearing to carry a radial load of 3000 N. The journal having 50 mm diameter rotates at 1500 rpm. The viscosity of oil at the operating temperature is 25 cP.

Solution Assuming that the journal bearing is designed for electric motor or generator, from Table 15.1,

$$p_{\max} = 1.4 \text{ MN/m}^2; \left(\frac{ZN}{p}\right)_{\min} = 29.01; \frac{c}{r} = 0.0013; \text{ and } \frac{l}{d} = 1.0$$

$$\text{Bearing pressure, } p = \frac{W}{ld} = \frac{3000}{50 \times 50} = 1.2 \text{ N/mm}^2$$

$$\frac{ZN}{p} = \frac{0.025 \times 1500}{1.2} = 31.25 > 29.01 \quad (\because \ Z = 25 \times 10^{-9} \text{ N} \cdot \text{s/mm}^2 = 25 \times 10^{-3} \text{ N} \cdot \text{s/m}^2)$$

So the bearing will operate in the stable region.

$$\text{Sommerfeld number, } S = \frac{ZN}{p} \times \left(\frac{r}{c}\right)^2 \times 10^{-6}$$

$$= 31.25 \times \left(\frac{1}{0.0013}\right)^2 \times 10^{-6} = 18.49$$

From Figure 15.10, for $S = 18.49$ and $\beta = 360°$, we get friction variable, FV = 6. Therefore,

coefficient of friction, $f = FV \times \dfrac{c}{r} = 6.0 \times 0.0013 = 0.0078$

$$\text{Linear velocity, } v = \frac{\pi \times 0.05 \times 1500}{60} = 3.927 \text{ m/s}$$

$$\text{Heat generated, } H_{gen} = fWv$$

$$= 0.0078 \times 3000 \times 3.927 = 91.89 \text{ W}$$

From Figure 15.11, for $S = 18.49$ and $\beta = 360°$, we get flow variable, FLV = 3.1

$$\text{Oil flow rate, } Q = \text{FLV} \times rcNl$$

$$= 3.1 \times 25 \times 0.0013 \times 25 \times 1500 \times 50/60$$

$$= 3148.4 \text{ mm}^3\text{/s}$$

From Figure 15.9, for $S = 18.49$ and $\beta = 360°$, we get $\dfrac{h_0}{c} = 0.94$

$$\text{Minimum film thickness, } h_0 = 0.94 \times 0.0325 \quad (\because c = 0.0013 \times 25 = 0.0325 \text{ mm})$$

$$= 0.0305 \text{ mm} = 30.5 \text{ µm}$$

Assuming that the entire heat is carried away by lubricating oil, the temperature rise

$$\Delta T = \frac{H_{gen}}{Q\rho c_p} = \frac{91.89}{3148.4 \times 10^{-9} \times 900 \times 1700} = 19°\text{C}$$

If the ambient temperature is 30°C, the oil film temperature will be 49°C, which is within the permissible limit.

Example 15.3 Design a self-contained journal bearing for the crank shaft of a four-stroke petrol engine to carry a radial load of 10 kN. The journal diameter is 50 mm and it rotates at 1000 rpm. SAE 30 oil may be used.

Solution From Table 15.1, for four-stroke gas and oil engines the maximum pressure, p_{max} = 5.5–12.0 MN/mm². Let us assume p_{max} = 5.5 N/mm².

$$\left(\frac{ZN}{p}\right)_{\min} = 2.9; \quad \frac{c}{r} = 0.001; \quad \frac{l}{d} = 0.6\text{–}2, \text{ say, } 1.5 \quad \text{or} \quad l = 75 \text{ mm}$$

$$\text{Bearing pressure, } p = \frac{W}{ld} = \frac{10 \times 1000}{50 \times 75} = 2.66 \text{ N/mm}^2$$

We assume that the operating temperature is 70°C for SAE 30 oil. At 70°C, $Z = 0.022 \text{ N} \cdot \text{s/m}^2$. Thus,

$$\frac{ZN}{p} = \frac{0.022 \times 1000}{2.66} = 8.2 > \left(\frac{ZN}{p}\right)_{\min}$$

$$\text{Sommerfeld number, } S = \frac{ZN}{p} \times \left(\frac{r}{c}\right)^2 \times 10^{-6}$$

$$= 8.2 \times (1000)^2 \times 10^{-6} = 8.2$$

From Figure 15.10, for $S = 8.2$ and $\beta = 360°$, we get friction variable FV = 2.8

$$\text{Coefficient of friction, } f = \text{FV} \times \frac{c}{r} = 2.8 \times \frac{1}{1000} = 0.0028$$

$$\text{Linear velocity, } v = \frac{\pi dN}{60} = \frac{\pi \times 0.05 \times 1000}{60} = 2.618 \text{ m/s}$$

$$\text{Heat generated, } H_{\text{gen}} = fWv = 0.0028 \times 10{,}000 \times 2.618 = 73.30 \text{ W}$$

$$\text{Heat dissipated, } H_{\text{diss}} = KA(t_b - t_a)$$

$$t_b - t_a = 0.5(t_f - t_a)$$

$$\text{Let } t_a = 25°\text{C} \quad \text{and} \quad t_f = 70°\text{C}$$

$$\text{Heat transfer coefficient, } K = 20 \text{ W/(m}^2\text{K)}$$

$$\text{Area of bearing housing, } A = 25dl$$

$$H_{\text{diss}} = 20 \times 25 \times 0.05 \times 0.075 \times 0.5(70 - 25) = 42.2 \text{ W}$$

Since $H_{\text{diss}} < H_{\text{gen}}$, let us revise the operating temperature. Let us assume, $t_f = 80°\text{C}$

Viscosity of SAE 30 oil at 80°C, $Z = 0.018 \text{ N} \cdot \text{s/m}^2$. Thus,

$$\frac{ZN}{p} = \frac{0.018 \times 1000}{2.66} = 6.77 > \left(\frac{ZN}{p}\right)_{\min}$$

$$\text{Sommerfeld number, } S = \frac{ZN}{p} \times \left(\frac{r}{c}\right)^2 \times 10^{-6} = 6.77$$

From Figure 15.10, for $S = 6.77$ and $\beta = 360°$, we get, friction variable FV = 2.2

$$\text{Coefficient of friction, } f = \text{FV} \times \frac{c}{r} = 2.2 \times 0.001 = 0.0022$$

$$\text{Heat generated, } H_{gen} = 0.0022 \times 10{,}000 \times 2.618 = 57.59 \text{ W}$$

$$\text{Heat dissipated, } H_{diss} = 20 \times 25 \times 0.05 \times 0.075 \times 27.5$$

$$= 51.56 \text{ W} < H_{gen}$$

So it is further required to raise the operating temperature. Let $t_f = 90°C$. Now, the viscosity of oil at 90°C, $Z = 0.014 \text{ N} \cdot \text{s/m}^2$. Thus,

$$\frac{ZN}{p} = \frac{0.014 \times 1000}{2.66} = 5.26 > \left(\frac{ZN}{p}\right)_{min}$$

$$\text{Sommerfeld number, } S = 5.26$$

From Figure 15.10, for $S = 5.26$ and $\beta = 360°$, we get friction variable FV = 1.8

$$\text{Coefficient of friction, } f = \text{FV} \times \frac{c}{r} = 1.8 \times 0.001 = 0.0018$$

$$H_{gen} = 0.0018 \times 10{,}000 \times 2.618 = 47.124 \text{ W}$$

$$\text{Heat dissipated, } H_{diss} = 20 \times 25 \times 0.05 \times 0.075 \times 32.5$$

$$= 60.93 \text{ W} > H_{gen}$$

In order to obtain equilibrium temperature, we plot two curves between H_{gen}, H_{diss} and the operating temperature.

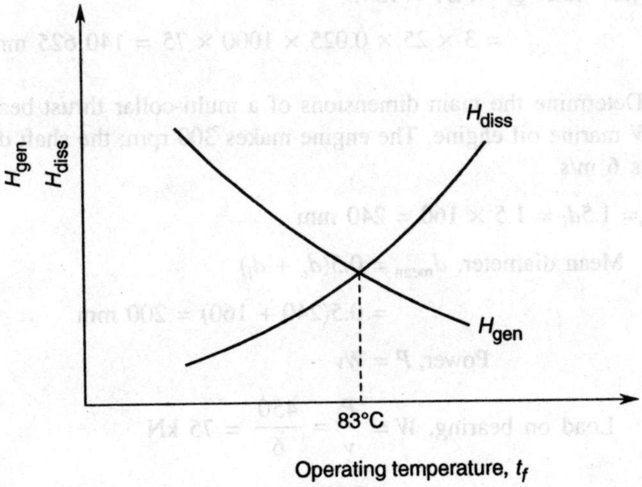

Equilibrium temperature is obtained by the point of intersection of these curves.

$$t_f = 83°C$$

At 83°C,

$$\text{oil viscosity, } Z = 0.016 \text{ N} \cdot \text{s/m}^2$$

$$\frac{ZN}{p} = \frac{0.016 \times 1000}{2.66} = 6$$

$$\text{Sommerfeld number, } S = \frac{ZN}{p} \times \left(\frac{r}{c}\right)^2 \times 10^{-6} = 6$$

From Figure 15.10, for $S = 6$ and $\beta = 360°$, we get friction variable FV = 2

$$\text{Coefficient of friction, } f = \text{FV} \times \frac{c}{r} = 0.002$$

$$\text{Heat generated, } H_{\text{gen}} = 0.002 \times 10{,}000 \times 2.618 = 52.36 \text{ W}$$

$$t_b - t_a = 0.5(83 - 25) = 29$$

$$H_{\text{diss}} = 20 \times 25 \times 0.05 \times 0.075 \times 29$$

$$= 54.3 \text{ W} \approx H_{\text{gen}}.$$

Hence the bearing is self-contained.

From Figure 15.9 for $S = 6$ and $\beta = 360°$, we get $\frac{h_0}{c} = 0.815$

$$\text{Minimum film thickness, } h_0 = 0.815c = 0.815 \times 0.001 \times 25 = 0.0203 \text{ mm}$$

From Figure 15.11 for $S = 6$ and $\beta = 360°$, we get flow variable FLV = 3. Therefore,

$$\text{Oil flow rate, } Q = \text{FLV} \times rcNl$$

$$= 3 \times 25 \times 0.025 \times 1000 \times 75 = 140{,}625 \text{ mm}^3\text{/min}$$

Example 15.4 Determine the main dimensions of a multi-collar thrust bearing for a propeller shaft of a 450 kW marine oil engine. The engine makes 300 rpm; the shaft diameter is 160 mm. The boat speed is 6 m/s.

Solution Let $d_o = 1.5d_i = 1.5 \times 160 = 240$ mm

$$\text{Mean diameter, } d_{\text{mean}} = 0.5(d_o + d_i)$$

$$= 0.5(240 + 160) = 200 \text{ mm}$$

$$\text{Power, } P = Wv$$

$$\text{Load on bearing, } W = \frac{P}{v} = \frac{450}{6} = 75 \text{ kN}$$

$$\text{Average rubbing velocity, } v_r = \frac{\pi d_{\text{mean}} N}{60} = \frac{\pi \times 0.2 \times 300}{60} = 3.141 \text{ m/s}$$

$$\text{For constant wear, } pv_r < 57 \times 10^4$$

Therefore, bearing pressure, $p = \dfrac{57 \times 10^4}{3.141} = 181{,}436.6 \text{ N/m}^2 \approx 0.1814 \text{ N/mm}^2$

$$\text{Load, } W = \frac{\pi}{4}(d_o^2 - d_i^2) \times p \times i$$

or

$$\text{Number of collars, } i = \frac{4W}{\pi(d_o^2 - d_i^2)p} = \frac{4 \times 75 \times 1000}{\pi(240^2 - 160^2)0.1814}$$

$$= 16.4, \text{ say, } 17$$

$$\text{Actual pressure, } p = \frac{4 \times 75 \times 1000}{\pi(240^2 - 160^2) \times 17} = 0.1755 \text{ N/mm}^2$$

Coefficient of friction

$$\mu = 84 p^{-0.67} v^{0.5}$$

$$= 84 \times (0.1755 \times 10^6)^{-0.67} \times (3.141)^{0.5}$$

$$= 0.0456$$

$$\text{Frictional torque, } T = \mu W r_{\text{mean}}$$

$$= 0.0456 \times 75{,}000 \times \frac{120 + 80}{2 \times 1000}$$

$$= 342 \text{ N} \cdot \text{m}$$

$$\text{Power loss} = \frac{2\pi NT}{60} = \frac{2\pi \times 300 \times 342}{60 \times 1000} = 10.744 \text{ kW}$$

15.10 ROLLING ELEMENT BEARINGS

In a rolling element bearing, the contact between the bearing elements is rolling instead of sliding as found in journal bearings. The shaft is supported on rollers or balls. A bearing of this type has very small friction, the reason being that the friction due to rolling of surfaces over each other is considerably less than the sliding friction. These bearings are sometimes called *antifriction* bearings which, of course, is a misnomer because some amount of friction is always present. Figure 15.17 shows the comparison of friction for three types of bearings. Rolling element bearings have low starting friction which is not much greater than the usual running friction. The coefficient of friction for rolling element bearings varies from 0.001 to 0.0045 depending upon the type of bearing.

A well manufactured rolling element bearing in properly designed applications offers the following advantages over the sliding bearing.

(i) Starting friction is low.
(ii) Load-carrying capacity is approximately constant.
(iii) Can carry overload for a limited period of time.

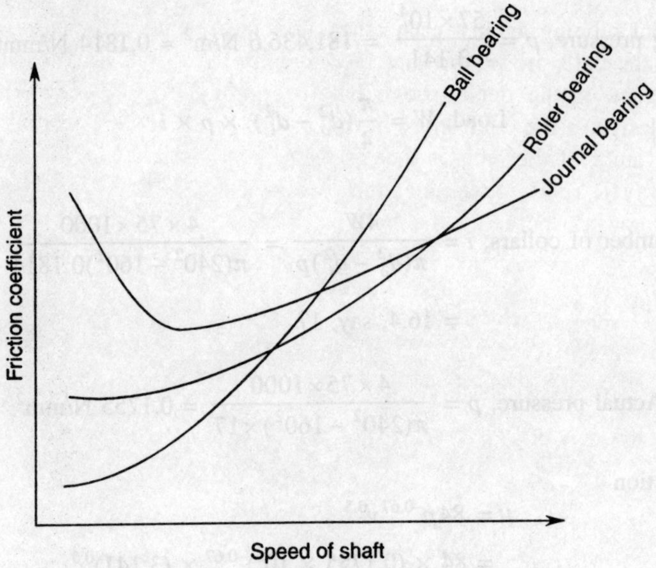

Figure 15.17 Variation of friction with speed.

(iv) Can maintain accurate shaft alignment.
(v) Requires less axial space.
(vi) Lubrication is simple and a pre-greased bearing can run for a long period.

Rolling element bearings have the following disadvantages:

(i) High initial cost.
(ii) Require large diametral space compared to journal bearings.
(iii) High running friction.
(iv) Resistance to shock load is poor.

15.11 TYPES OF BEARINGS

Rolling element bearings may be classified as ball or roller depending upon whether the primary rolling elements are balls or rollers. Sometimes needle bearings are reported as the third type of rolling element bearing; however it belongs to the category of roller bearing.

Depending upon the load to be carried, the rolling element bearings are classified as (i) radial, (ii) angular contact, and (iii) thrust bearings.

15.11.1 Ball Bearings

The ball bearing, which is mainly used to take up radial and thrust load, consists of four parts: (i) outer race, (ii) inner race, (iii) balls, and (iv) retaining cage or separator. In the ball bearing, the outer race and the inner race are held concentric with each other by inserting spherical balls circumferentially at equal intervals of space. These balls are held in their positions by the retaining cage, which also serves the purpose of always keeping the balls separated and thereby preventing them from rubbing against each other. In some special cases these bearings do not have an inner race; the balls are placed in a race directly cut in the shaft. Figure 15.18 shows the schematic diagram of the deep-groove ball bearing. One of the important aspects to successful bearing design is the conformity of ball radius to the raceway radius (Figure 15.19). If the radius of curvature of the balls is increased, the unit surface stress induced between the balls and the raceway is reduced, thereby the load-carrying capacity is increased. However, increasing the ball radius increases the bearing friction. Therefore, a proper selection of conformity of ball and raceway is an important design consideration.

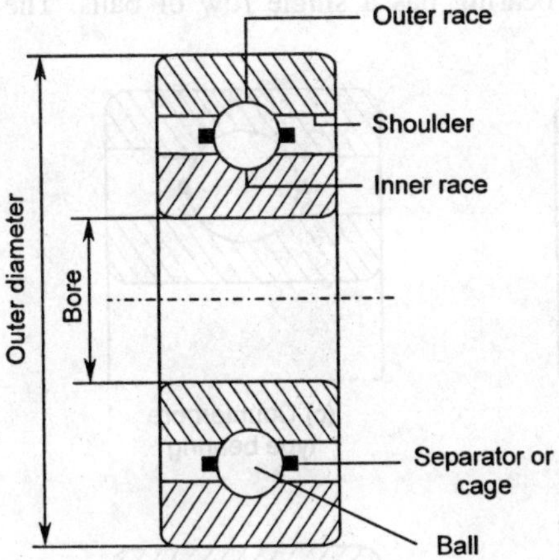

Figure 15.18 Nomenclature of ball bearing.

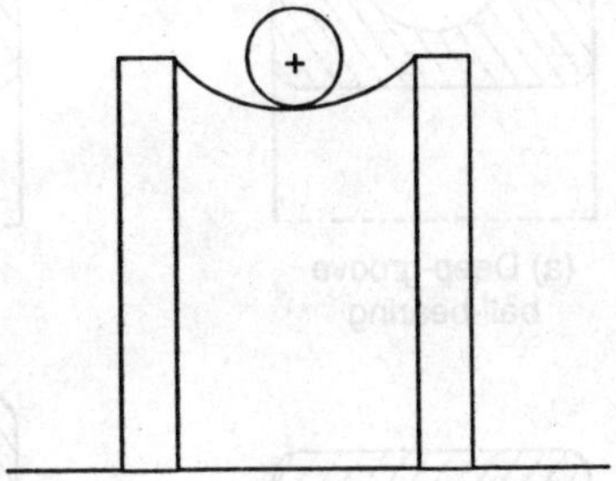

Figure 15.19 Conformity of ball bearing.

Ball bearings are classified into three types: (i) radial ball bearings, (ii) angular-contact bearings, and (iii) thrust ball bearings.

Radial ball bearings. Radial ball bearings have been primarily designed to take up radial force; they can also take up axial thrust to a certain extent. Various types of ball bearings are shown in Figure 15.20. A brief description of these bearings is given below:

Deep-groove bearing. It has deep continuous raceway all over the circumference. Construction of the deep-groove ball bearing permits the bearing to support high radial load along with thrust load [Figure 15.20(a)].

Filling notch type bearing. In order to increase the radial load-carrying capacity of the deep-groove ball bearing, its construction is slightly modified and a filling notch or loading groove, as shown in Figure 15.20(b) is created. Through this filling notch, additional balls are inserted which increase its radial load capacity. The thrust load-carrying capacity of these bearings is reduced.

Counterbore type bearing. This bearing is similar to deep-groove bearing except that the outer ring has one shoulder, which makes it separable. On account of this feature, this bearing permits the inner and outer races to be mounted separately. This bearing can support one direction thrust in addition to the radial load [Figure 15.20(c)].

Self-aligning bearing. The self-aligning bearing is used in those applications where a misalignment between the axes of shaft is likely to exist. These bearings are available in two types: (i) self-aligning internal and (ii) self-aligning external bearings as shown in Figures 15.20 (d) and (e).

Double-row ball bearing. A double-row deep-groove bearing can be used to carry high radial and thrust loads. The line of contact for these types of bearings converges outside the bearing envelope which increases the rigidity of the bearing. Figures 15.20(f) and (g) show double-row ball bearings.

Angular-contact ball bearing. The angular-contact bearing is mainly used to take up high axial thrust. This type of bearing is available in two types: (i) one-directional and (ii) two-directional bearings. A one-directional angular-contact bearing has a single row of balls. The

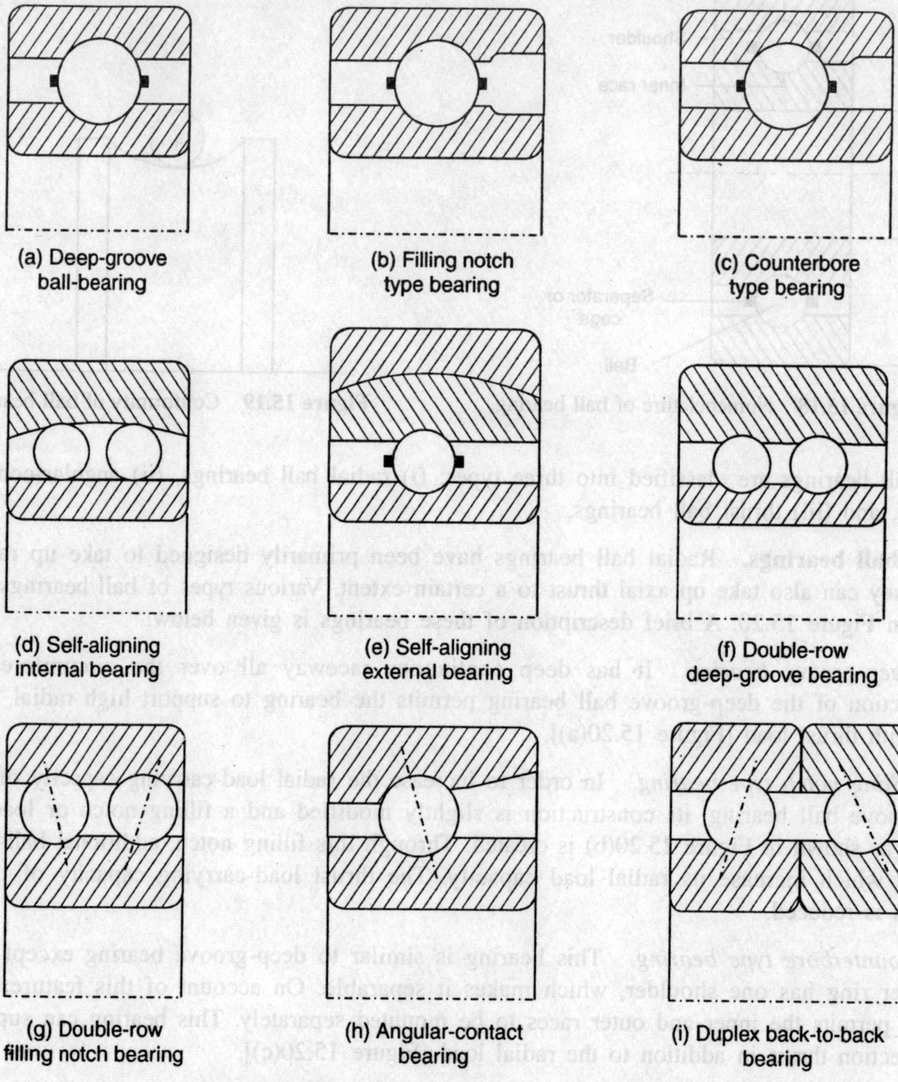

(a) Deep-groove ball-bearing
(b) Filling notch type bearing
(c) Counterbore type bearing
(d) Self-aligning internal bearing
(e) Self-aligning external bearing
(f) Double-row deep-groove bearing
(g) Double-row filling notch bearing
(h) Angular-contact bearing
(i) Duplex back-to-back bearing

Figure 15.20 Different types of ball bearings.

centre of contact between the ball and the race makes an angle which is called the contact angle. The outer race of these bearings has one heavy raceway shoulder, while the other raceway shoulder is removed by counterbore. This design permits the bearing to carry higher radial and one-directional thrust loads than what the deep groove bearing can carry [Figure 15.20(h)].

In certain cases the where-rigid bearing supports are required to support both axial and radial deflection, the angular contact bearing in duplexing form may be used as shown in Figure 15.20(i).

The relative load-carrying capacities of various types of ball bearings are listed in Table 15.3. The load-carrying capacity is defined in terms of the radial load capacity of deep-groove bearing.

Table 15.3 Relative load-carrying capacity of bearings

Type	*Radial capacity*	*Thrust capacity*	*Direction*
Single-row deep groove	F_{rad}	$0.7F_{rad}$	2*d*
Filling notch	$(1.2–1.4)F_{rad}$	$0.2F_{rad}$	
Angular contact	$(1.0–1.15)F_{rad}$	$(1.5–2.3)F_{rad}$	1*d*
Double-row deep-groove	$1.5F_{rad}$	$1.5F_{rad}$	2*d*
Self-aligning	$0.7F_{rad}$	$0.2F_{rad}$	1*d*

Thrust ball bearings. Thrust ball bearings are designed to carry pure thrust load. These bearings are available in three types: (i) one-directional flat race, (ii) one-directional grooved race, and (iii) two-directional grooved race. The one-directional flat race bearing consists of two ungrooved flat washers, balls, and ball separator. The friction in this type of bearing is very small; its speed is limited due to centrifugal force. In the one-directional grooved race bearing, the races are grooved which provide passage to the balls. It can be used at moderate speeds and can carry higher thrust loads. The coefficient of friction for this type of bearing is larger than that for flat race ball bearings.

A two-directional grooved race bearing consists of two separators and a middle-grooved race which rotates with balls. These bearings can withstand high axial thrust in both directions. Figure 15.21 shows the schematic diagrams of thrust bearings.

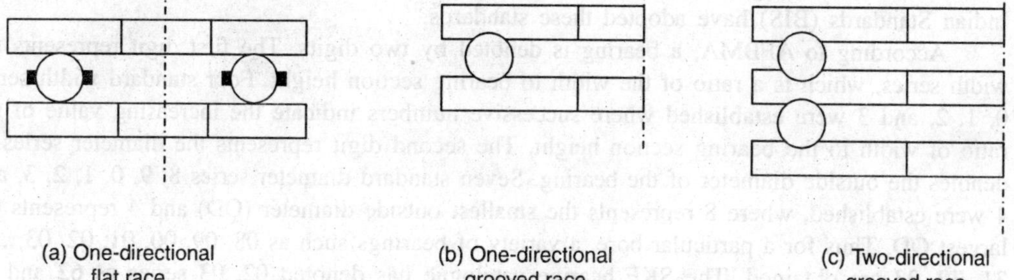

Figure 15.21 Different types of thrust bearings.

15.11.2 Roller Bearings

A roller bearing consists of an outer race, an inner race, and a set of rollers with or without a roller cage. In some cases one or both races may be absent. In that case the shaft and/or a bored hole serves as the recess. Roller bearings serve the same purpose as ball bearings but can carry much higher loads and support large shaft diameters, because of their line contact instead of point contact. If the axes of the rollers are parallel to the rotational axis of the supported shaft, the bearing usually is a cylindrical roller bearing. The rollers of cylindrical roller bearing have length to diameter ratio from 1:1 to 3:1. The outside diameter of rollers is often crowned to increase the load-carrying capacity. If the axes of the rollers are inclined to and intersect the rotational axis at a common point, the bearing is called the taper roller bearing. These bearings are designed to carry high radial and thrust loads at moderate to high speeds. If the length to diameter ratio of cylindrical rollers is large, say 6 or more, this type of bearing is called needle roller bearing; it can carry higher radial loads. Figure 15.22 shows the line diagram of various types of roller bearings.

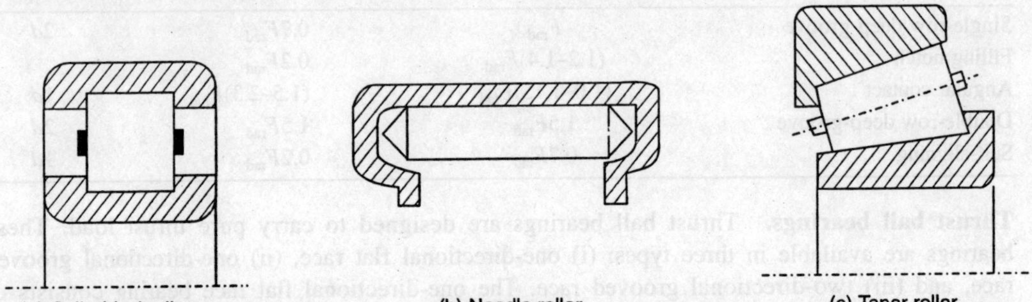

Figure 15.22 Different types of roller bearings.

15.12 STANDARD DIMENSIONS

Standardization of mechanical elements and devices is always desirable to reduce the cost of design, and the cost of production and maintenance. Being the most commonly used machine elements, a need was felt to standardize bearings so that they can be easily replaced or interchanged. Antifriction Bearings Manufacturers' Association (AFBMA) of the USA has done a lot of work in this regard. Later on, International Standards Organization (ISO) and Bureau of Indian Standards (BIS) have adopted these standards.

According to AFBMA, a bearing is denoted by two digits. The first digit represents the width series, which is a ratio of the width to bearing section height. Four standard width series 0, 1, 2, and 3 were established where successive numbers indicate the increasing value of the ratio of width to the bearing section height. The second digit represents the diameter series. It denotes the outside diameter of the bearing. Seven standard diameter series 8, 9, 0, 1, 2, 3, and 4 were established, where 8 represents the smallest outside diameter (OD) and 4 represents the largest OD. Thus for a particular bore, a variety of bearings such as 08, 09, 00, 01, 02, 03, . . ., 31, 32, 33 are obtained. The SKF bearing catalogue has denoted 02, 03 series as 62 and 63 series, respectively. Figure 15.23 shows the relative proportions of boundary dimensions of different types of bearings. Besides the width and diameter series, the bearing nomenclature also consists of a two or three digit number which represents the bore of the inner race and two letters of the alphabet which represent the type of the bearing.

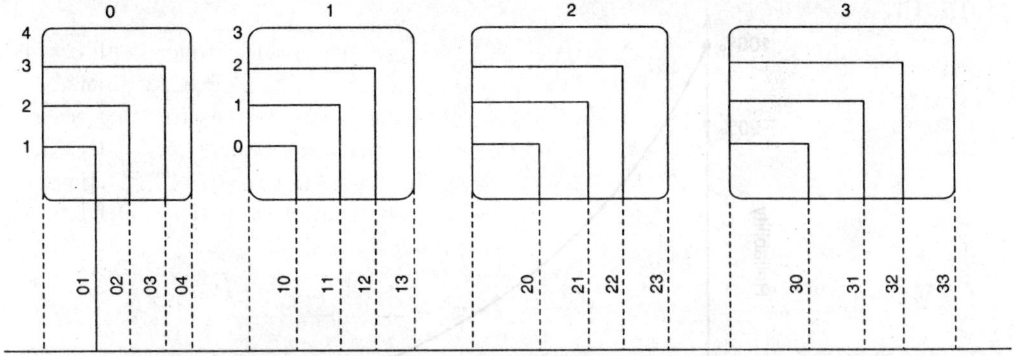

Figure 15.23 Proportions of boundary dimensions of different bearings.

To illustrate the AFBMA procedure, suppose a bearing is designated as 20BC03. The number 20 represents 20 mm bore size. The letters BC represent a single-row deep-groove ball bearing and 03 represents the width and diameter series. The details of bearing designation procedure are given in IS : 5669 – 1970.

In the SKF way of bearing designation, the last two digits represent the code for bore. The digits 00, 01, 02, and 03 represent 10, 12, 15, and 17 mm bore, respectively. The digits starting from 04 represent one-fifth of the bore in millimetres. To illustrate the SKF procedure, suppose a bearing is designated as SKF 6304. The first two digits 63 represent a deep-groove ball-bearing having 03 width–diameter series. The next two digits 04 represent 20 mm (4 × 5) inner race bore, which designation is similar to 20BC03 in accordance with AFBMA and ISO.

15.13 LOAD AND LIFE RATING OF BEARINGS

The type and size of a bearing to be used for a particular application is selected on the basis of its load-carrying capacity and other requirements such as life and reliability. An installed bearing usually fails by fatigue failure caused by contact stresses.

15.13.1 Rated Life

The life of a radial ball bearing is the number of revolutions or the number of hours at given constant speed that a bearing runs before the first evidence of fatigue develops in the material of either the raceway or the ball. However it is evident from both laboratory tests and practical experience that seemingly identical bearings operating under similar conditions have different lives. Therefore the life of bearings is expressed as statistical life. The rated life of a group of identical bearings is defined as the number of revolutions or hours at some constant speed that 90 per cent of a group of identical bearings will complete or exceed before the first evidence of fatigue failure occurs. The rated life is represented as L_{10} life. In other words, the reliability of the bearing is 90 per cent.

In certain applications, where there is a greater risk to human life or to equipment, it becomes necessary to select a bearing having a reliability greater than 90 per cent. Based upon various experimental results it is found that statistically the relation between the bearing life and reliability follows the Wiebull distribution curve, as shown in Figure 15.24. Since bearings

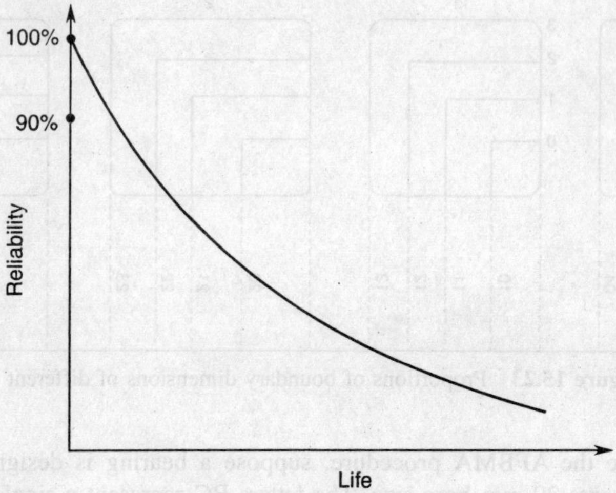

Figure 15.24 Life–reliability curve.

are available at 90 per cent reliability, the expected life of the bearing at any given reliability, other than 90 per cent, can be found by the relation

$$\frac{L_{rel}}{L_{90}} = \left[\frac{\log_e\left(\dfrac{1}{R_{rel}}\right)}{\log_e\left(\dfrac{1}{R_{90}}\right)}\right]^{\frac{1}{b}} \tag{15.24}$$

where

L_{rel} is the bearing life at the required reliability
R_{rel} is the required reliability
L_{90} is the bearing life at 90 per cent (R_{90}) reliability
b is the constant (= 1.17).

Using Eq. (15.24), the expected life at the required reliability is transformed into life at 90% reliability and then such a bearing is selected from the manufacturer's catalogue.

15.13.2 Load Rating

The load-carrying capacity of a rolling element bearing is called its load rating. In bearing terminology, there are two types of load ratings: (i) basic static load rating and (ii) dynamic load rating.

Static load rating. It is defined as that load which causes at one contact area between the rolling element and the raceway a permanent deformation of 0.0001 time the diameter of the rolling element. The basic static load rating C_0 is used in calculation when bearings are to rotate at very slow speed or are to be stationary under load for extended periods of time. The static load rating can be computed either from the formulae given in IS : 2823 (Parts I–IV) —1966 or from the manufacturer's catalogue.

Dynamic load rating. The dynamic load rating is defined as the constant stationary radial load which a group of apparently identical bearings with stationary outer race can carry for minimum rated life of 1 million revolutions of the inner race. The rated life in this definition is based upon the assumption that 90 per cent of the bearings will complete or exceed 1 million revolutions before the first evidence of fatigue failure develops. The dynamic load rating is usually provided by the bearing manufacturers.

15.14 EQUIVALENT BEARING LOAD

In many practical applications, bearings have to carry both radial and axial loads. In addition, they are sometimes required to operate with a rotating outer race and a stationary inner race. It is, therefore, necessary to convert the two loads and the rotating outer race condition into an hypothetical equivalent load which satisfies the above conditions. An equivalent load is that stationary radial load which if applied to the bearing with the rotating inner race and stationary outer race, would give the same life as a bearing operating under actual conditions.

The equivalent dynamic load P, which is applicable to ball and roller bearings, is given by

$$P = VXF_{rad} + YF_{ax} \tag{15.25}$$

where

F_{rad} is the radial load
F_{ax} is the axial thrust
V is the rotation factor
(= 1.0 for inner race rotating)
(= 1.2 for outer race rotating)
(= 1.0 for self-aligning bearings)
X is the radial load factor
Y is the thrust load factor

The values of factors X, Y, and V can be obtained from Table 15.4.

15.15 LOAD–LIFE RELATION

Extensive laboratory testing and subsequent statistical analysis have shown that the relationship between the dynamic load C, equivalent load P and the rated bearing life L_{10} can be expressed by the equation

$$L_{10} = \left(\frac{C}{P}\right)^b \tag{15.26}$$

where b = a constant
(= 3 for ball bearings)
(= 10/3 for roller bearings)

The relationship between life in millions of revolutions and life in working hours at a constant rpm is given as

$$L_{10} = \frac{60NL_h}{10^6} \tag{15.27}$$

where

N is the speed in rpm
L_h is the bearing life in hours.

Table 15.4 Values of factors X, Y and V

Bearing type	$\frac{iF_{ax}}{C_0}$	$\frac{F_{ax}}{C_0}$	V: If in relation to the load, the inner ring is: Rotating	V: Stationary	X: For single-row bearing* when $\frac{F_{ax}}{VF_{rad}} > e$	X: For double-row bearings when $\frac{F_{ax}}{VF_{rad}} \le e$	X: For double-row bearings when $\frac{F_{ax}}{VF_{rad}} > e$	Y: For single-row bearing* when $\frac{F_{ax}}{VF_{rad}} > e$	Y: For double-row bearing when $\frac{F_{ax}}{VF_{rad}} \le e$	Y: For double-row bearing when $\frac{F_{ax}}{VF_{rad}} > e$	e
(1)	(2)	(3)	(4)	(5)	(6)	(7)	(8)	(9)	(10)	(11)	(12)
Non-filling slot assembly, radial contact groove ball bearings	—	0.014						2.30		2.30	0.19
		0.028						1.99		1.99	0.22
		0.056						1.71		1.71	0.26
	—	0.084						1.55		1.55	0.28
		0.110	1	1.2	0.56	1	0.56	1.45	0	1.45	0.30
		0.170						1.31		1.31	0.34
		0.280						1.15		1.15	0.38
		0.420						1.04		1.04	0.42
		0.560						1.00		1.00	0.44
Angular-contact groove ball bearings	0.014								2.78	3.74	0.23
	0.028								2.40	3.23	0.26
	0.056								2.07	2.78	0.30
	0.085								1.87	2.52	0.34
	0.110		1	1.2	For this type, use X, Y and e values applicable to single row, non-filling slot assembly, radial contact, groove ball-bearings	1	0.78	For this type, use X, Y and e values applicable to single row, non-filling slot assembly, radial contact, groove ball bearings	1.75	2.36	0.36
	0.170								1.58	2.13	0.40
	0.280								1.39	1.87	0.45
	0.420								1.26	1.69	0.50
	0.560								1.21	1.63	0.52

* For single row bearings, when $\frac{F_{ax}}{VF_{rad}} \le e$, use $X = 1$ and $Y = 0$

i = Number of rows

15.16 SELECTION OF BEARINGS

A general guideline for the selection of a bearing from the manufacturer's catalogue is given as under :

1. Compute the radial and axial thrust forces acting on the shaft.
2. Compute the shaft diameter.
3. Select a suitable bearing from a manufacturer's catalogue. This selection depends upon the shaft diameter and magnitude of the radial and axial forces.
4. Decide whether the inner race is rotating or stationary. Accordingly, select the value of the rotation factor V.
5. Determine the value of the radial load factor X and the axial load factor Y. Refer Table 15.4. The values of these factors depend upon two ratios (F_{ax}/F_{rad}) and (F_{ax}/C_0), where C_0 is the static load capacity.
6. Compute the equivalent dynamic load factor from the following equation

$$P = VXF_{rad} + YF_{ax}$$

7. Decide the expected life of the bearing. Convert the expected life in hours into millions of revolutions.
8. Calculate the dynamic load capacity from the load–life equation

$$L_{10} = \left(\frac{C}{P}\right)^b$$

9. Check whether the selected bearing has the required dynamic load capacity. If yes, the selected bearing is suitable for this purpose. Otherwise, select another bearing from the next series and go back to step 3 and continue.

The above procedure of bearing selection is illustrated in various example problems.

15.17 BEARINGS FOR CYCLIC LOADS AND SPEEDS

In certain applications, bearings are subjected to cyclic loads at different speeds. For example, a bearing supporting a shaft is subjected to radial force F_{rad1}, thrust load F_{ax1} at speed N_1 for r_1 fraction of the cyclic period. Later on, the radial and axial forces change to F_{rad2} and F_{ax2}, the speed also changes to N_2 for r_2 fraction of the cycle period. Similarly, the radial and axial forces and operating speed change for subsequent portions of the cycle period in such a way that the sum of $r_1, r_2, \ldots, r_n$ is 1.0 at the end of the cycle. In such a circumstance, therefore, there is a need to define a relationship that will relate to the dynamic load rating C and rated life L_{10} for a varying equivalent load.

The load–life equation for ball bearings is given as

$$L_{10} = \left(\frac{C}{P}\right)^3$$

or

$$C^3 = L_{10} \times P^3 = \frac{60\,NL_h}{10^6}\,P^3 \qquad (15.28)$$

For the case of cyclic load and speed variation, if during a fraction of cycle r_i the speed N_i and the equivalent load P_i are constant, we can express Eq. (15.28) as a summation of the effect of N_i and load P_i for each fraction of the cycle. Thus, we obtain the dynamic load rating as

$$C^3 = \frac{60L_h}{10^6}\sum_{i=1}^{m} r_i N_i \,(K_S P_i)^3 \tag{15.29}$$

where

r_i is the fraction of the ith cycle period
N_i is the speed during the ith cycle
P_i is the equivalent load during the ith cycle
m is the total number of cycles
K_S is the service factor (1.0–2.5)

15.18 TAPER ROLLER BEARINGS

Tapered roller bearings are mainly designed to withstand high radial and thrust loads. These bearings can operate at moderate to high speeds (850–9000 rpm). Figure 15.25 shows the schematic diagram of the single-row tapered roller bearing capable of resisting thrust in one direction only. It consists of a tapered inner race called *cone*, the outer race called *cup*, and a *cage*. The cup is separable from the remaining assembly of the bearing. In this type of bearing it is possible to make adjustments for radial clearance.

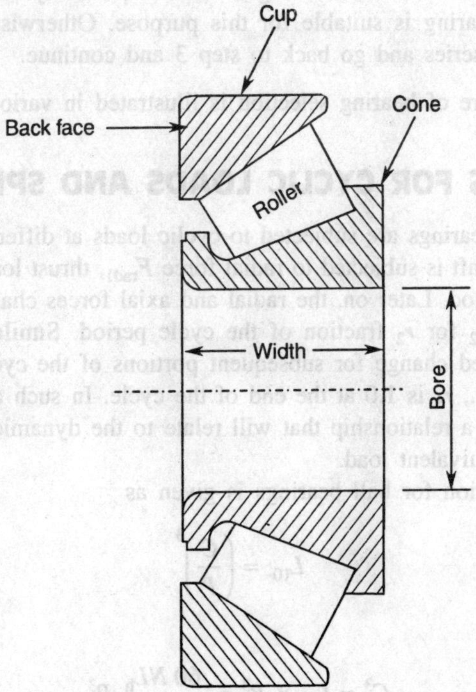

Figure 15.25 Nomenclature of taper roller bearing.

In a taper roller bearing, even if the external force is radial force and acts normal to the surface, it induces a thrust reaction within the bearing. To avoid separation of the cup, this thrust reaction must be balanced by an equal and opposite force. Generally, two taper roller bearings with back-to-back face are mounted on the shaft.

The axial thrust force F_{ax} due to pure radial load F_{rad} is approximately given by

$$F_{ax} = \frac{0.5F_{rad}}{Y} \tag{15.30}$$

where Y is the thrust load factor, which is the ratio of the radial force rating to thrust force rating of the bearing whose value may be assumed 1.5.

The equivalent dynamic load for tapered roller bearings may be computed by the following formulae:

(1) Single tapered roller bearing

$$P = F_{rad} \quad \text{when} \quad \frac{F_{ax}}{F_{rad}} \le e \tag{15.31a}$$

$$P = 0.4F_{rad} + YF_{ax} \quad \text{when} \quad \frac{F_{ax}}{F_{rad}} > e \tag{15.31b}$$

(2) Paired tapered roller bearing

$$P = F_{rad} + YF_{ax} \quad \text{when} \quad \frac{F_{ax}}{F_{rad}} \le e \tag{15.32a}$$

$$P = 0.67F_{rad} + YF_{ax} \quad \text{when} \quad \frac{F_{ax}}{F_{rad}} > e \tag{15.32b}$$

The values of the thrust factor Y and constant e are given by manufacturers of bearings. The load–life relationship for tapered roller bearings is given by

$$L_{10} = \left(\frac{C}{P}\right)^{\frac{10}{3}} \tag{15.33}$$

15.19 LUBRICATION OF BEARINGS

The choice of lubricant is primarily determined by the operating temperature and speed of the bearing. Under normal operating conditions, grease is the most widely used lubricant. Grease contains oil plus the thickener, generally in the form of metallic soap. While selecting a suitable grease it is necessary to consider the consistency, operating temperature range, and rust inhibiting properties. Consistency is classified according to the National Lubricating Grease Institute (NLGI) scale. Metallic soap based greases of consistency 1, 2, or 3 may be used for rolling bearings. In lubricating grease, three types of thickeners, namely (i) calcium, (ii) sodium, and (iii) lithium are the most commonly used ones.

When the temperature rise due to bearing operation is high, then lubricating grease is replaced by oil. Solvent refined mineral oils are most commonly used for lubrication of rolling bearings. At temperatures more than 125°C, the use of synthetic oil, namely polyglycol type, is recommended. Additives to improve certain operating properties are generally required only when the operating conditions are exceptional.

Example 15.5 A shaft rotating at 1440 rpm is supported by two bearings. The forces acting on each bearing are 6000 N radial load and 3500 N axial thrust. If the shaft diameter is 40 mm and the expected life of the bearing is 500 h, select a suitable bearing.

Solution Since $F_{ax}/F_{rad} < 0.7$, so a single-row deep-groove ball bearing may be suitable.

Let us select the SKF 6208 bearing whose static load rating, C_0 = 16,600 N and dynamic load rating, C = 30,700 N. Hence

$$\frac{F_{ax}}{C_0} = \frac{3500}{16,600} = 0.2108$$

and the corresponding value of e from Table 15.4 is 0.3545.

Since $F_{ax}/F_{rad} = 0.5834 > e$, from Table 15.4, the radial load factor, $X = 0.56$ and the thrust load factor, $Y = 1.252$ (by interpolation). Let us assume that the inner race is rotating, therefore, the rotation factor $V = 1$.

$$\text{Equivalent load, } P = XVF_{rad} + YF_{ax}$$

$$= 0.56 \times 1 \times 6000 + 1.252 \times 3500 = 7742 \text{ N}$$

$$\text{Life in million of revolutions, } L_{10} = \frac{60NL_h}{10^6}$$

$$= \frac{60 \times 1440 \times 500}{10^6} = 43.2$$

The required dynamic load capacity

$$C = (L_{10})^{1/3} \cdot P$$

$$= (43.2)^{1/3} \times 7742 = 27,165.3 \text{ N}$$

Since the dynamic load rating of the bearing SKF 6208 is larger than the required dynamic load capacity, the selected bearing is suitable.

Example 15.6 Select a suitable bearing for the load conditions and various data given in Example 15.5, if the required reliability of the bearing is to be 99 per cent.

Solution Let us select the SKF 6308 bearing for which C = 41,000 N and C_0 = 22,400 N

For $\dfrac{F_{ax}}{C_0} = \dfrac{3500}{22,400} = 0.1562$, $e = 0.3308$ (refer Table 15.4).

Since $F_{ax}/F_{rad} = 0.5834 > e$, from Table 15.4, the radial load factor $X = 0.56$ and the axial thrust factor, $Y = 1.35$ (by interpolation) and the rotation factor $V = 1$, for inner race rotating.

Equivalent load, $P = XVF_{rad} + YF_{ax} = 0.56 \times 1 \times 6000 + 1.35 \times 3500 = 8085$ N

Expected life at 99 per cent reliability

$$L_{99} = \frac{60 \times 1440 \times 500}{10^6} = 43.2 \text{ million revolutions}$$

Expected life at 90 per cent reliability (L_{90}) is obtained from

$$\frac{L_{99}}{L_{90}} = \left[\frac{\log_e\left(\frac{1}{0.99}\right)}{\log_e\left(\frac{1}{0.9}\right)}\right]^{\frac{1}{1.17}} = 0.1342$$

or

$$L_{90} = \frac{L_{99}}{0.1342} = \frac{43.2}{0.1342} = 321.9 \text{ million revolutions}$$

Required dynamic load rating

$$C = (321.9)^{1/3} \times 8085 = 55{,}410 \text{ N}$$

which is greater than the rated dynamic load of bearing SKF 6308. Hence this bearing is not suitable. Let us try another bearing, i.e. SKF 6408, for which $C = 63{,}700$ N and $C_0 = 36{,}500$ N.

$$\text{For } \frac{F_{ax}}{C_0} = \frac{3500}{36{,}500} = 0.0959, \quad e = 0.289$$

Since $\frac{F_{ax}}{F_{rad}} > e$, therefore, $X = 0.56$, $Y = 1.5$, and $V = 1.0$.

$$\text{Equivalent load, } P = XVF_{rad} + YF_{ax}$$

$$= 0.56 \times 1 \times 6000 + 1.5 \times 3500 = 8610 \text{ N}$$

Required dynamic load capacity

$$C = (321.9)^{1/3} \times 8610 = 59{,}007.9 \text{ N}$$

which is smaller than the rated dynamic load capacity of the bearing. Hence the SKF 6408 bearing is suitable.

Example 15.7 The spindle of a wood-working machine runs at 1000 rpm. It is mounted on two single-row ball bearings, one of which is required to carry a radial load of 2250 N and thrust load of 1900 N. The machine runs 8 h per day. Assuming a life of four years and spindle diameter equal to 30 mm, select a suitable bearing.

Solution Here $F_{ax}/F_{rad} = 1900/2250 = 0.8445$. Since $F_{ax}/F_{rad} > 0.7$, we select an angular-contact bearing. From the SKF catalogue we select the bearing 7306 BE for which

$$C = 34{,}500 \text{ N} \quad \text{and} \quad C_0 = 19{,}000 \text{ N}$$

For $\dfrac{iF_{ax}}{C_0} = \dfrac{1 \times 1900}{19,000} = 0.1$, from Table 15.4, $e = 0.292$.

Since $F_{ax}/F_{rad} > e$, therefore, we get

$$X = 0.56 \text{ and } Y = 1.5, \text{ and } V = 1.0 \text{ for inner race rotating}$$

$$\text{Equivalent load, } P = XVF_{rad} + YF_{ax} = 0.56 \times 1 \times 2250 + 1.5 \times 1900 = 4110 \text{ N}$$

The life of the bearing in hours, assuming 300 working days/year

$$L_h = 4 \times 300 \times 8 = 9600 \text{ h}$$

Life in million of revolutions

$$L_{10} = \frac{60 N L_h}{10^6} = \frac{60 \times 1000 \times 9600}{10^6}$$

$$= 576 \text{ million revolutions}$$

$$\text{The required dynamic load, } C = (576)^{1/3} \times 4110$$

$$= 34,196.5 \text{ N}$$

Since the required dynamic load is approximately equal to the dynamic load capacity, the angular-contact bearing SKF 7306 BE is suitable.

Example 15.8 A 75 mm diameter machine shaft is to be supported at the ends. It operates continuously for 8 h/day, 300 days/year for 10 years. The load and speed cycle for one of the bearings is given below:

S. no.	*Fraction of cycle*	*Radial load* (N)	*Thrust load* (N)	*Speed* (rpm)	*Load condition*
1	0.25	3500	2000	1000	Steady
2	0.25	2500	2000	1500	Steady
3	0.5	4000	2000	800	Light shock

Select a suitable bearing.

Solution Service factor $K_S = 1.0$ for steady load

$= 1.5$ for light shock load

$F_{ax}/F_{rad} = 0.57$, 0.8 and 0.5 for all three fractions of cycle. Considering the values of F_{ax}/F_{rad}, let us select a deep-groove ball bearing, SKF 6315, for which

$$C = 114,000 \text{ N} \quad \text{and} \quad C_0 = 72,000 \text{ N}$$

For $\dfrac{F_{ax}}{C_0} = \dfrac{2000}{72,000} = 0.0277$, from Table 15.4, $e = 0.219$.

Since $F_{ax}/F_{rad} > e$, $X = 0.56$, $Y = 1.99$ (by interpolation) and $V = 1$ for inner race rotating

$$\text{Equivalent load, } P = XVF_{rad} + YF_{ax}$$

Therefore,

$$P(0.25) = 0.56 \times 1 \times 3500 + 1.99 \times 2000 = 5940 \text{ N}$$

$$P(0.25) = 0.56 \times 1 \times 2500 + 1.99 \times 2000 = 5380 \text{ N}$$

$$P(0.5) = 0.56 \times 1 \times 4000 + 1.99 \times 2000 = 6220 \text{ N}$$

$$\text{Life of bearing in hours, } L_h = 8 \times 300 \times 10 = 24{,}000 \text{ h}$$

The required dynamic load capacity

$$C^3 = \frac{60 \times L_h}{10^6} \sum_{i=1}^{3} N_i r_i (K_S P_i)^3$$

$$= \frac{60 \times 24{,}000}{10^6} \times [0.25 \times 1000 \times (1.0 \times 5940)^3 + 0.25 \times 1500 \times (1.0 \times 5380)^3 + 0.5 \times 800 \times (1.5 \times 6220)^3]$$

or

$$C = 85{,}606 \text{ N}$$

which is less than the dynamic load rating of SKF 6315 bearing. Hence the selection is OK.

Example 15.9 A bearing, supporting a power transmitting shaft, is subjected to 3000 N radial load and 4500 N axial thrust. The shaft rotates at 400 rpm and the expected life of the bearing is 10,000 h. Select a suitable bearing, if the diameter of the shaft is 40 mm.

Solution Here $F_{ax}/F_{rad} = 4500/3000 = 1.5$. Therefore a tapered roller bearing will be suitable for the given load condition. Let us select a single-row tapered roller bearing SKF 30208 from the catalogue, having

$$C = 58{,}300 \text{ N} \quad \text{and} \quad C_0 = 73{,}600 \text{ N}$$

From the manufacturer's catalogue, $Y = 1.6$ and $e = 0.37$

Since $\dfrac{F_{ax}}{F_{rad}} > e$

$$\text{Equivalent load, } P = 0.4F_{rad} + YF_{ax}$$

$$= 0.4 \times 3000 + 1.6 \times 4500 = 8400 \text{ N}$$

$$\text{Life of bearing, } L_{10} = \frac{60 \times 400 \times 10{,}000}{10^6} = 240 \text{ million revolutions}$$

The required dynamic load capacity

$$C = (L_{10})^{0.3} \times P$$

$$= (240)^{0.3} \times 8400 = 43{,}485.3 \text{ N}$$

which is less than the rated dynamic load capacity of the bearing. Hence the selection of the SKF 30208 bearing is satisfactory.

EXERCISES

1. What do you mean by hydrodynamic lubrication?
2. What are the effects of clearance on the performance of a bearing?
3. List the important considerations for the selection of a bearing.
4. Define the following terms:

   (i) Journal, (ii) Eccentricity ratio, (iii) Clearance, (iv) Minimum film thickness, (v) Attitude angle, (vi) Bearing characteristic number, (vii) Sommerfeld number
5. Explain the stability chart of the journal bearing.
6. Describe the AFBMA method of specifying antifriction bearings.
7. Why does a filling-notch bearing support a higher radial load than a deep-groove ball bearing?
8. Explain the following terms:

   (i) Rated life, (ii) Basic load rating, (iii) Dynamic load
9. Why are ball bearings preferred to journal bearings for a shaft mounted on a gear box?
10. Why do we prefer taper roller bearings to cylindrical roller bearings?
11. Select the suitable dimensions for a journal bearing which is subjected to 3.5 kN radial load. The diameter and length of the bearing are 75 mm each. The journal rotates at 400 rpm.
12. A sleeve bearing is 80 mm in diameter and 60 mm in length. The journal speed is 600 rpm. The oil supply is SAE 30 at the inlet temperature of 40°C. The radial load on the bearing is 2.5 kN. Design a suitable journal bearing.
13. Determine the bearing characteristic number, coefficient of friction, heat generated and dissipated for a journal bearing with the following data:

    Load = 3500 N; Speed = 1200 rpm; Journal diameter = 50 mm; l/d = 1.0; c/r = 0.001; Oil viscosity = 128 cP at the operating temperature 55°C.
14. A journal bearing is to be used for a centrifugal pump. The diameter of the journal is 100 mm and the load on it is 30 kN. The journal speed is 900 rpm. Complete the design. Assume that the ambient temperature is 30°C.
15. Select a ball bearing to be mounted on a 35 mm diameter shaft. The radial and axial loads are 2.5 kN and 1.5 kN, respectively. The speed of the shaft is 200 rpm and the required life is 10,000 rpm.
16. Select a suitable ball bearing for the data given in Exercise 15, if the reliability of the bearing is 95 per cent.
17. A shaft is mounted on two bearings 400 mm apart and carries at its middle a gear of 200 mm pitch diameter. The gear causes a 10 kN radial and a 2.5 kN thrust load on the shaft when rotating at 500 rpm. Assume 20° pressure angle and 25° helix angle for the helical gear. The allowable stress is 40 MPa in shear. Determine the suitable shaft diameter and select the appropriate bearing.

18. Select a single-row ball bearing with the operation cycle given below, which will have a life of 10,000 h. The shaft diameter is 50 mm.

Fraction of cycle	*Type of load*	*Radial load* (kN)	*Thrust load* (kN)	*Speed* (rpm)
1/10	Heavy shock	7	4	400
1/10	Light shock	5	3	500
1/5	Moderate shock	4	5	600
3/5	Steady	3	7	600

19. Select a suitable bearing with inner race rotating and having a 10-second work cycle as under:

For 3 seconds	*For 7 seconds*
$F_{rad} = 40$ kN	$F_{rad} = 20$ kN
$F_{ax} = 20$ kN	$F_{ax} = 0$
Speed = 900 rpm	Speed = 1200 rpm
Light shock	Steady load

The average expected life is 5000 h. The shaft diameter is 75 mm.

MULTIPLE CHOICE QUESTIONS

1. In order to realize the advantage of fluid friction, it is necessary to have:
   (a) parallel oil film (b) converging oil film
   (c) diverging oil film (d) any type of oil film
2. A journal rotating in the anticlockwise direction at slow speed inside a fluid bearing will be
   (a) at the bottom-most position of the bearing
   (b) towards the left side of the bearing
   (c) towards the right side of the bearing
   (d) at the centre of the bearing
3. The bearing characteristic number is defined as
   (a) ZN/p (b) p/ZN (c) Zp/N (d) pN/Z
4. The type of contact between the ball and the raceway in a ball bearing is
   (a) a point (b) a line (c) a surface (d) none of the above
5. A journal bearing is operating in the left region of the stability chart. A decrease in viscosity will cause
   (a) an increase in the bearing characteristic number
   (b) a decrease in the coefficient of friction
   (c) an increase in the coefficient of friction
   (d) no change

6. Which of the following materials is used as bearing liner?

   (a) Cast iron (b) Babbitt (c) Brass (d) All of the above

7. An oil hole in a journal bearing should be provided where the pressure is

   (a) maximum (b) minimum (c) average (d) any other value

8. The antifriction bearings are

   (a) journal bearings (b) gas lubricated bearings
   (c) ball and roller bearings (d) none of the above

9. Which of the following bearings has a low starting friction?

   (a) Ball bearing (b) Roller bearing
   (c) Journal bearing (d) Taper roller bearing

10. Which of the following bearings can take up large thrust loads?

    (a) Deep-groove ball bearing (b) Filling-notch ball bearing
    (c) Self-aligning bearing (d) Angular-contact bearing

11. According to AFBMA, the two-digit bearing designation represents:

    (a) width–diameter series (b) diameter–width series
    (c) width–bore series (d) one-fifth of the bore diameter

12. In ball and roller bearings, the life vs. reliability curve follows the

    (a) normal distribution (b) straight line
    (c) Wiebull distribution (d) none of the above

Chapter 16

Pressure Vessels

16.1 INTRODUCTION

The term pressure vessels refers to those vessels of different shapes and constructions which are used to store or supply liquids, vapours, or gases and are subjected to internal or external pressure more than 0.7 atmospheric gauge. Pressure vessels are extensively used in thermal power plants, nuclear power plants, process and chemical industries, and for the supply of water, steam, gas, and air in various industries. Pressure vessels are fabricated from steel plates welded together by the fusion welding process. These vessels are classified according to the following criteria:

Statutory regulations. According to statutory regulations, the pressure vessels are classified into the following three categories:

(a) *Class I.* These vessels are used to store poisonous and toxic gases and liquids and are designed to operate below −20°C temperature.

(b) *Class II.* Boilers and high pressure tanks pertain to the Class II category. These vessels are designed as per provisions of the Indian Boiler Regulation Act, 1961.

(c) *Class III.* These vessels are used for relatively light duties where the operating temperature is less than 250°C and the maximum pressure is limited to 1.75 MPa. These vessels are designed to conform to IS : 2825 – 1969.

Geometric shape. Based upon the geometric shape, pressure vessels may be classified into (a) cylindrical, (b) conical, and (c) spherical.

End construction. According to the type of end construction, the vessels may be classified into:

(a) Open-ended vessels, namely cylinders, cylinder liners, and so on.
(b) Closed-ended vessels with various types of ends, namely. flat end, hemispherical, semi-ellipsoidal, or dish end.

Thickness of vessel. A vessel with a rauo of inner diameter to wall thickness greater than 20 is treated as thin vessel for all practical purposes. For example, boiler drum, hydraulic accumulator, air tank, and so on. Pressure vessels having diameter to thickness ratio less than 20 are called thick vessels. for example, cylinder liners, gun barrels, and so forth.

16.2 THIN PRESSURE VESSELS

In this section, we shall discuss stress analysis of two commonly used thin pressure vessels, namely (i) cylindrical vessels and (ii) spherical vessels.

16.2.1 Cylindrical Vessels

The stress analysis of a thin-wall cylinder subjected to internal pressure is based upon the assumption that the induced tensile stresses are uniformly distributed over the cross-section.

A closed-ended thin cylinder subjected to internal pressure experiences the following stresses:

1. Tensile stress called *hoop stress*, σ_t, acting tangential to the circumference.
2. Tensile stress called *longitudinal stress*, σ_L, acting along the axis.

Consider a seamless cylindrical vessel of inner diameter D_i, thickness t, and length l. The vessel is subjected to internal pressure p_i. Figure 16.1(a) shows the section of the cylindrical vessel. Consider two elementary strips on the vessel, each at an angle θ, on either side of the vertical axis and subtending an angle $d\theta$.

The normal force on each element of length l is $prd\theta l$ and the resultant vertical force on the element is $2prl \cos \theta \, d\theta$. The net effect of the radial pressure on the upper- or lower-half portion of the shell, known as the bursting force, can be found as follows:

$$F = 2p_i r_i l$$

$$= p_i \times \text{projected area } (D_i l) \tag{16.1}$$

For the equilibrium of forces acting on half-portion of the cylinder as shown in Figure 16.1(b), the hoop stress is given as

$$\sigma_t = \frac{p_i D_i}{2t} \tag{16.2}$$

If the efficiency of the joint is less than 100 per cent and some provision for corrosion is to be made, then the thickness of the vessel is computed by the following relation:

$$t = \frac{p_i D_i}{2\sigma_t \eta} + \text{corrosion allowance} \tag{16.3}$$

According to IS : 2825 – 1969, a code for unfired pressure vessels, the thickness of a thin vessel should be determined from the following equation:

$$t = \frac{p_i D_i}{2\sigma J - p_i} \tag{16.4}$$

where J is the weld joint efficiency factor (see Table 16.1).

As per IS : 2825 – 1969, a minimum corrosion allowance of 1.5 mm should be provided unless a protective lining is employed.

Considering the equilibrium of forces in the longitudinal direction [Figure 16.1(c)], we have

$$\frac{\pi}{4} D_i^2 p_i = \pi D_i t \times \sigma_L$$

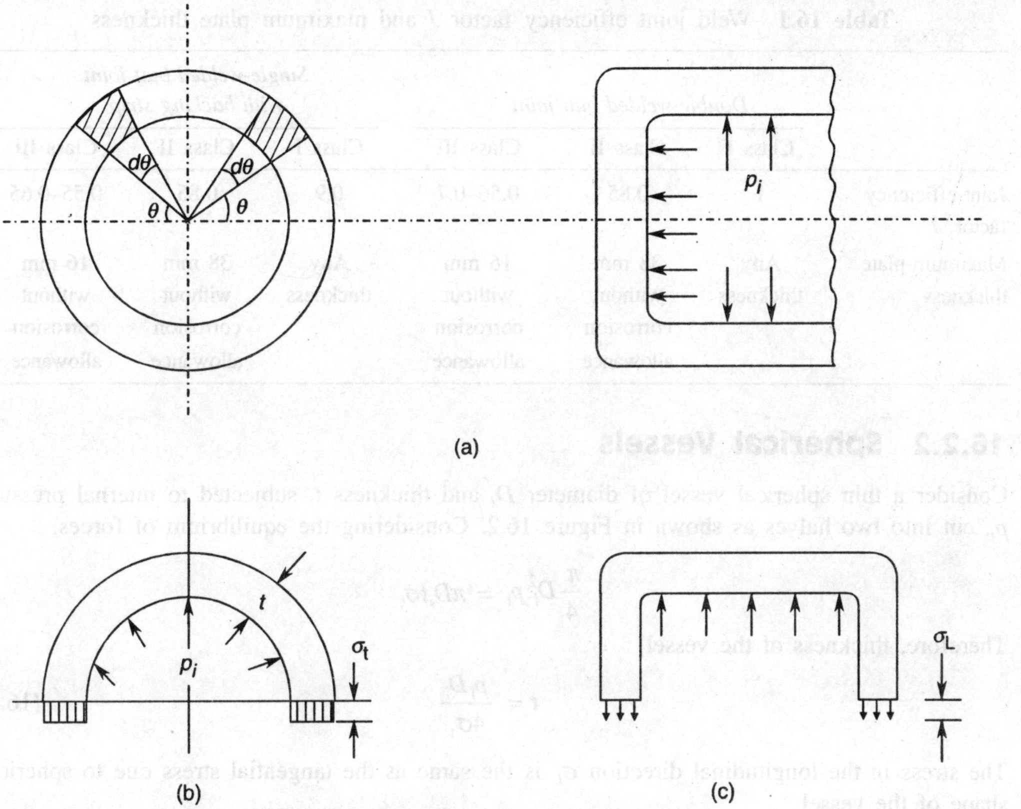

Figure 16.1 Stresses in thin cylinder.

or

$$\sigma_L = \frac{p_i D_i}{4t} \tag{16.5}$$

or

$$t = \frac{p_i D_i}{4\sigma_L} \tag{16.6}$$

Thus the intensity of the longitudinal stress is half that of the hoop stress. However, both stresses act normal to each other and they are also normal to the internal pressure. It means that at any point we have three principal stresses—σ_t, σ_L, and p_i—although the magnitude of the pressure p_i is very small compared to other two stresses. The maximum shear stress is equal to $\sigma_t/2$, if the internal pressure p_i is neglected.

The change in the volume of a cylindrical vessel can be computed by knowing the volumetric strain, which is given by

$$\varepsilon_{vol} = \frac{dV}{V} = \frac{\delta l}{l} + \frac{2\delta d}{d} = \varepsilon_l + 2\varepsilon_\theta \tag{16.7}$$

where $\varepsilon_l = \dfrac{\sigma_L}{E} - \dfrac{\mu\sigma_t}{E} + \dfrac{\mu p_i}{E}$.

Table 16.1 Weld joint efficiency factor *J* and maximum plate thickness

	Double-welded butt joint			*Single-welded butt joint with backing strip*		
	Class I	Class II	Class III	Class I	Class II	Class III
Joint efficiency factor, *J*	1	0.85	0.56–0.7	0.9	0.85	0.55–0.65
Maximum plate thickness	Any thickness	38 mm without corrosion allowance	16 mm without corrosion allowance	Any thickness	38 mm without corrosion allowance	16 mm without corrosion allowance

16.2.2 Spherical Vessels

Consider a thin spherical vessel of diameter D_i and thickness t, subjected to internal pressure p_i, cut into two halves as shown in Figure 16.2. Considering the equilibrium of forces,

$$\frac{\pi}{4} D_i^2 p_i = \pi D_i t \sigma_t$$

Therefore, thickness of the vessel,

$$t = \frac{p_i D_i}{4\sigma_t} \tag{16.8}$$

The stress in the longitudinal direction σ_L is the same as the tangential stress due to spherical shape of the vessel.

As per IS: 2825–1969, the thickness of the spherical shell is given by the following equation

$$t = \frac{p_i D_i}{4\sigma J - p_i} \tag{16.9}$$

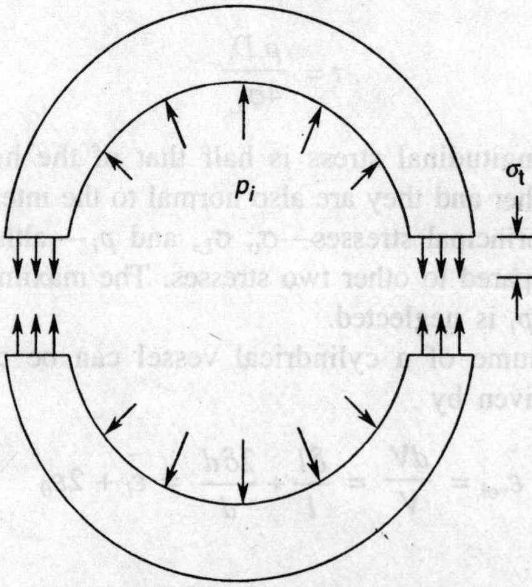

Figure 16.2 Stresses in spherical vessel.

16.3 THICK CYLINDERS

While analyzing the thin cylinder we had assumed that the tangential or hoop stress σ_t is constant throughout the thickness of the vessel and the radial stress ($\sigma_{rad} = p_i$) is negligible compared to hoop stress. However in a thick cylinder, these assumptions do not hold good. The problem of the thick cylinder is more complex in nature and is solved making the following assumptions:

(i) The material is homogeneous and isotropic.

(ii) The plane section of the cylinder perpendicular to the longitudinal axis remains plane under the pressure. It means that the longitudinal strain is constant at all the points on the cylinder and is independent of the radius of cylinder.

Figure 16.3 shows a thick cylinder subjected to internal and external pressures. Consider a small annular ring of internal radius *r* and thickness *dr*. Let the radial stresses due to pressure on the internal and external surfaces be σ_{rad} and ($\sigma_{rad} + d\sigma_{rad}$), respectively. Let σ_t be the tangential stress. Considering the equilibrium of the vertical forces on a unit length ring, the equilibrium equation after neglecting the small terms is given as

$$\sigma_t + \sigma_{rad} + r\frac{d}{dr}(\sigma_{rad}) = 0 \tag{16.10}$$

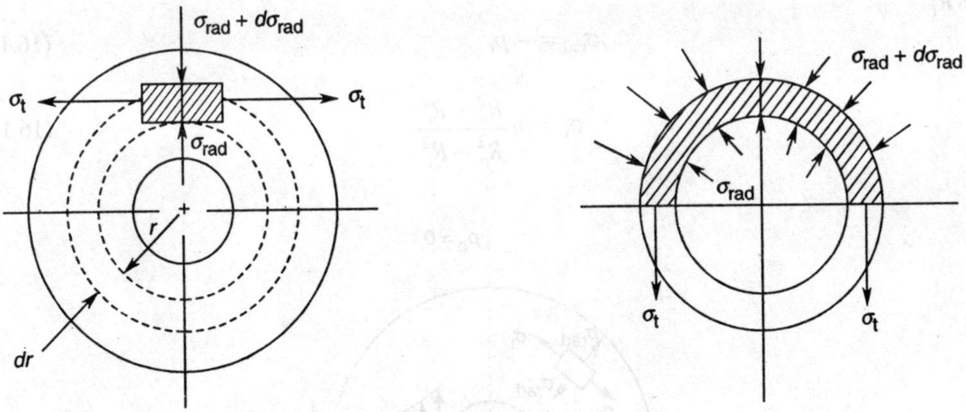

Figure 16.3 Stresses in thick cylinder.

Another relation based on the assumption that the longitudinal strain is constant and independent of radius *r* is written as

$$\sigma_t - \sigma_{rad} = 2\text{A, a constant} \tag{16.11}$$

Substituting the value of σ_t in Eq. (16.10), we get

$$\frac{d\sigma_{rad}}{\sigma_{rad} + \text{A}} = -\frac{2dr}{r} \tag{16.12}$$

Integrating Eq. (16.12), the following relations for radial and tangential stresses, called the Lame's equations, are obtained:

$$\text{Radial stress, } \sigma_{rad} = \frac{B}{r^2} - A \tag{16.13}$$

$$\text{Tangential stress, } \sigma_t = \frac{B}{r^2} + A \tag{16.14}$$

where A and B are constants to be evaluated from the known values of pressure.

Vessel subjected to internal pressure (p_i)

Let R_i and R_o be the internal and external radii of the cylinder. If the cylinder is subjected to internal pressure p_i, then for the following conditions $r = R_i$, $\sigma_{rad} = -p_i$, and $r = R_o$, $\sigma_{rad} = 0$ the value of the constants A and B are computed as

$$A = p_i \frac{R_i^2}{R_o^2 - R_i^2} \quad \text{and} \quad B = p_i \frac{R_i^2 R_o^2}{R_o^2 - R_i^2}$$

The distribution of stresses is shown in Figure 16.4. The stresses at the inner surface, i.e. at $r = R_i$

$$\sigma_{rad} = -p_i \tag{16.15}$$

$$\sigma_t = p_i \frac{R_o^2 + R_i^2}{R_o^2 - R_i^2} \tag{16.16}$$

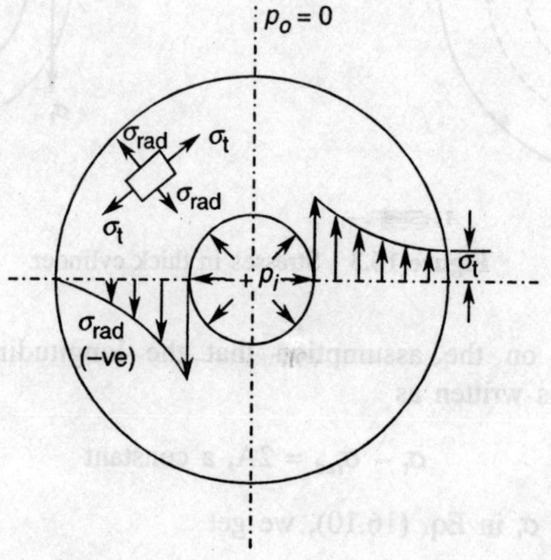

Figure 16.4 Stress distribution in a thick cylinder subjected to internal pressure.

The stresses at the outer surface, i.e. at $r = R_o$

$$\sigma_{\text{rad}} = 0 \tag{16.17}$$

$$\sigma_t = \frac{2p_i R_i^2}{R_o^2 - R_i^2} \tag{16.18}$$

The principal stress in the longitudinal direction is assumed to be uniform over the cylinder wall thickness. Considering equilibrium of forces in the axial direction

$$\pi R_i^2 p_i = \pi\left(R_o^2 - R_i^2\right) \times \sigma_L$$

or

$$\text{Longitudinal stress, } \sigma_L = \frac{p_i R_i^2}{R_o^2 - R_i^2} \tag{16.19}$$

Vessel subjected to external pressure (p_o)

If a cylinder is subjected to external pressure p_o only, then the radial stresses at the following conditions are:

When $r = R_o$, $\sigma_{\text{rad}} = -p_o$
and when $r = R_i$, $\sigma_{\text{rad}} = 0$

Substituting these conditions in Eqs. (16.13) and (16.14) and solving for constants A and B, we get

$$A = -\frac{p_o R_o^2}{R_o^2 - R_i^2} \quad \text{and} \quad B = -p_o \frac{R_o^2 R_i^2}{R_o^2 - R_i^2}$$

The distribution of stresses is shown in Figure 16.5 and the magnitude of stresses are given by:

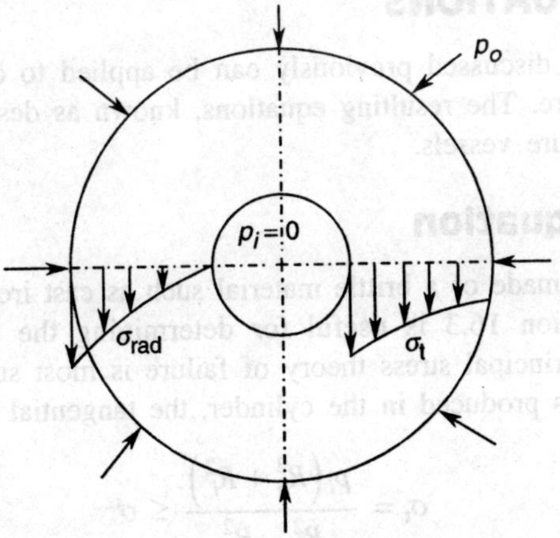

Figure 16.5 Stress distribution in a thick cylinder subjected to external pressure.

(a) At the inner surface, i.e. $r = R_i$

$$\sigma_{rad} = 0 \tag{16.20}$$

$$\sigma_t = \frac{-2p_o R_o^2}{R_o^2 - R_i^2} \tag{16.21}$$

(b) At the outer surface, i.e. $r = R_o$

$$\sigma_{rad} = -p_o \tag{16.22}$$

$$\sigma_t = \frac{-p_o R_o^2}{R_o^2 - R_i^2}\left(\frac{R_i^2}{R_o^2} + 1\right) \tag{16.23}$$

Vessel subjected to both internal and external pressures

In vessels subjected to both internal and external pressures, the radial stresses at the boundary conditions are

At $r = R_i$, $\sigma_{rad} = -p_i$
and at $r = R_o$, $\sigma_{rad} = -p_o$

Substituting these conditions in Eqs. (16.13) and (16.14) and solving for constants A and B, we get

$$A = \frac{p_i R_i^2 - p_o R_o^2}{R_o^2 - R_i^2} \quad \text{and} \quad B = (p_i - p_o)\frac{R_i^2 R_o^2}{R_o^2 - R_i^2}$$

The maximum hoop stress at radius R_i is given by

$$\sigma_t = \frac{p_i\left(R_o^2 + R_i^2\right) - 2p_o R_o^2}{R_o^2 - R_i^2} \tag{16.24}$$

16.4 DESIGN EQUATIONS

The basic stress analysis discussed previously can be applied to different applications under different theories of failure. The resulting equations, known as design equations, are used for designing the thick pressure vessels.

16.4.1 Lame's Equation

When a thick cylinder is made of a brittle material such as cast iron or cast steel, the Lame's equation derived in Section 16.3 is useful for determining the wall thickness. For brittle materials the maximum principal stress theory of failure is most suitable. Accordingly, out of the three principal stresses produced in the cylinder, the tangential stress is maximum

$$\sigma_t = \frac{p_i\left(R_o^2 + R_i^2\right)}{R_o^2 - R_i^2} \le \sigma$$

where σ is the allowable tensile strength of the material.

Or in terms of the wall thickness of the cylinder

$$t = R_i\left[\left(\frac{\sigma + p_i}{\sigma - p_i}\right)^{0.5} - 1\right] \tag{16.25}$$

16.4.2 Clavarino's Equation

When a closed-ended cylindrical vessel is made of ductile material, namely plain carbon steel, alloy steel, etc. its design is based upon the maximum strain theory of failure. Accordingly, the maximum strain will occur when the maximum principal stress is applied in the tangential direction.

Thus, the stress induced in the vessel is written as

$$\sigma = p_i \frac{(1+\mu)R_o^2 + (1-2\mu)R_i^2}{R_o^2 - R_i^2} \tag{16.26}$$

or in terms of the cylinder wall thickness

$$t = R_i\left[\left(\frac{\sigma + (1-2\mu)p_i}{\sigma - (1+\mu)p_i}\right)^{0.5} - 1\right] \tag{16.27}$$

16.4.3 Birnie's Equation

When a cylinder is made of ductile material and its ends are open, the longitudinal stress σ_L in the cylinder will be zero. Using the maximum strain theory of failure

$$\sigma = \sigma_t - \mu\sigma_{rad}$$

Substituting the value of principal stresses, we get

$$\sigma = p_i \frac{(1+\mu)R_o^2 + (1-\mu)R_i^2}{R_o^2 - R_i^2}$$

Therefore, the Birnie's equation for thickness of the cylinder is

$$t = R_i\left[\left(\frac{\sigma + (1-\mu)p_i}{\sigma - (1+\mu)p_i}\right)^{0.5} - 1\right] \tag{16.28}$$

16.5 COMPOUND CYLINDER

In many applications, the vessel is subjected to a very high internal pressure which produces a high amount of tangential stress (or hoop stress) at the inner surface of the cylinder. Sometimes these stresses are more than the allowable strength of the vessel material. For example, if a closed-ended vessel is subjected to internal pressure p_i which is approximately equal to the allowable strength ($p_i \approx \sigma$), then in such a case, according to Eqs. (16.27) and (16.28), the denominator will be either zero or have a negative sign, which indicates that no thickness for a single thick cylinder will ever be suitable. Hence in such cases the compound cylinder is used.

A compound cylinder has two cylinders, fitted with interference fit as shown in Figure 16.6. The inner diameter of the outer cylinder D_i (sometimes called the jacket) is slightly smaller than the outer diameter of the inner cylinder D_h. Therefore, while assembling these cylinders, the outer cylinder is first heated to a suitable temperature so that it expands

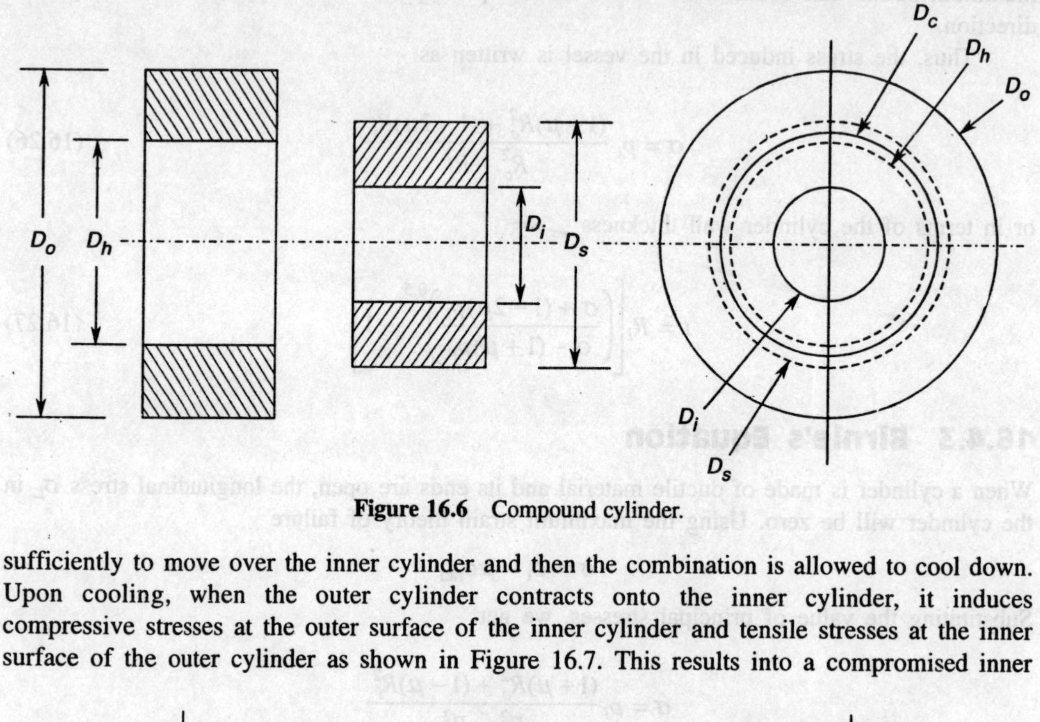

Figure 16.6 Compound cylinder.

sufficiently to move over the inner cylinder and then the combination is allowed to cool down. Upon cooling, when the outer cylinder contracts onto the inner cylinder, it induces compressive stresses at the outer surface of the inner cylinder and tensile stresses at the inner surface of the outer cylinder as shown in Figure 16.7. This results into a compromised inner

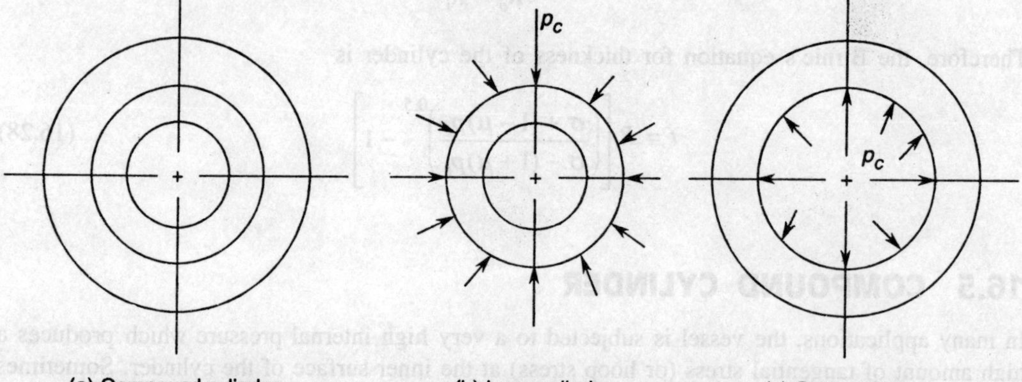

Figure 16.7 Shrink stresses in a compound cylinder.

diameter of outer cylinder which is equal to the outer diameter of the inner cylinder, called the junction diameter D_c.

In a compound cylinder open at both ends, subjected to shrinkage pressure p_c between two cylinders, the tangential stress at the inner surface of the inner cylinder according to Birnie's equation is

$$\sigma_{ti} = -\frac{2p_c D_c^2}{D_c^2 - D_i^2} \quad (16.29)$$

The tangential stress at the outer surface of the inner cylinder

$$\sigma_{toi} = -p_c\left(\frac{D_c^2 + D_i^2}{D_c^2 - D_i^2} - \mu\right) \quad (16.30)$$

Tangential stress at the inner surface of the outer cylinder

$$\sigma_{tio} = -\sigma_{toi} \quad (16.31)$$

and tangential stress at the outer surface of the outer cylinder

$$\sigma_{to} = \frac{2p_c D_c^2}{D_o^2 - D_c^2} \quad (16.32)$$

The distribution of stresses on the cylinder's wall thickness is shown in Figure 16.8.

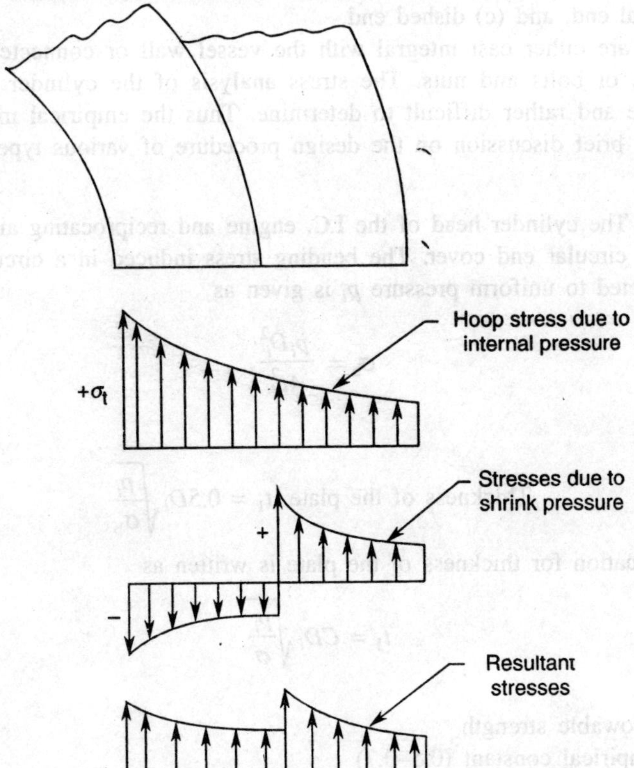

Figure 16.8 Stress distribution in a compound cylinder.

The total shrinkage allowance δ between the two cylinders is the sum of increase in the inner diameter of the outer cylinder δ_o and decrease in the outer diameter of the inner cylinder δ_i.

$$\text{Shrinkage allowance, } \delta = \frac{p_c D_c}{E}\left[\frac{2D_c^2\left(D_o^2 - D_i^2\right)}{\left(D_o^2 - D_c^2\right)\left(D_c^2 - D_i^2\right)}\right] \tag{16.33}$$

The shrinkage pressure p_c can be determined for a given amount of interference δ or vice versa. The temperature by which the outer cylinder must be heated for ease of assembly is given by the relation

$$T_2 - T_1 \geq \frac{\delta}{\alpha D_c} \tag{16.34}$$

where α is thermal expansion coefficient.

16.6 END COVERS

The ends of cylindrical pressure vessels such as the cylinder of an I.C. engine, boiler, air tank, etc. may be either of the two types: (i) flat plate and (ii) formed end. The formed ends are most commonly used for unfired cylindrical pressure vessels. They are: (a) hemispherical end, (b) semi-ellipsoidal end, and (c) dished end.

These ends are either cast integral with the vessel wall or connected to the vessel by riveting, welding, or bolts and nuts. The stress analysis of the cylinder end cover is very complex in nature and rather difficult to determine. Thus the empirical method of design is generally used. A brief discussion on the design procedure of various types of end covers is presented below.

Flat end cover. The cylinder head of the I.C. engine and reciprocating air compressor is an example of a flat circular end cover. The bending stress induced in a circular flat plate with fixed edges subjected to uniform pressure p_i is given as

$$\sigma_b = \frac{p_i D_i^2}{4t_1^2}$$

or

$$\text{Thickness of the plate, } t_1 = 0.5 D_i \sqrt{\frac{p_i}{\sigma_b}} \tag{16.35}$$

In general, the equation for thickness of the plate is written as

$$t_1 = C D_i \sqrt{\frac{p_i}{\sigma}} \tag{16.36}$$

where

σ is the allowable strength
C is the empirical constant (0.5–0.7)
D_i is the inside diameter of the vessel.

Hemispherical end. The hemispherical end, as shown in Figure 16.9, of the cylindrical vessel has the minimum plate thickness, minimum weight and consequently lower material cost. However, the cost of forming the hemispherical end is very high due to difficulties in the forming process. The thickness of the hemispherical end is computed by Eq. (16.9).

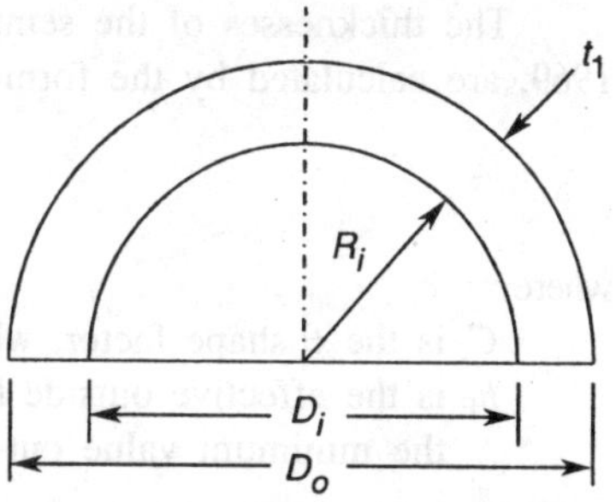

Figure 16.9 Hemispherical end.

Semi-ellipsoidal end. In these types of ends, the ratio of the major axis to the minor axis is about 2:1 and the thickness of the end shell is about two times the corresponding hemispherical end. Due to its shallow dished shape, the cost of forming is less than that of the hemispherical end (Figure 16.10).

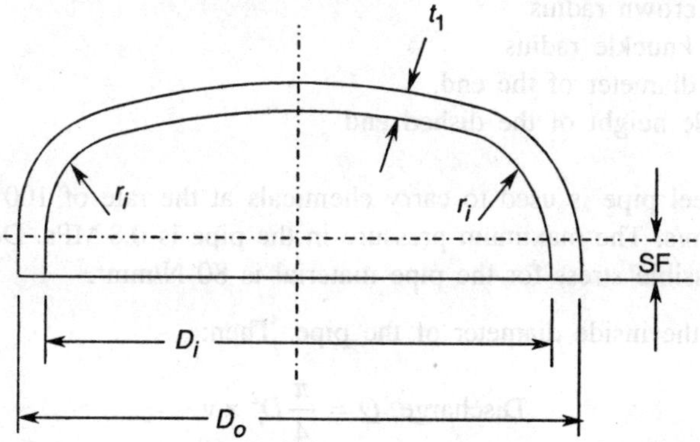

Figure 16.10 Semi-ellipsoidal end.

Dished end. This type of end is extensively used for many types of cylindrical vessels. These ends are shaped using the crown radius R_C and the knuckle radius r_i. The knuckle radius is used at the corners that join the crown of the dished end with the straight portion of the end (SF). The dished ends require less forming than that by the semi-ellipsoidal ends; however stress concentration at the joining of the crown and the knuckle is very high which may lead to failure (Figure 16.11).

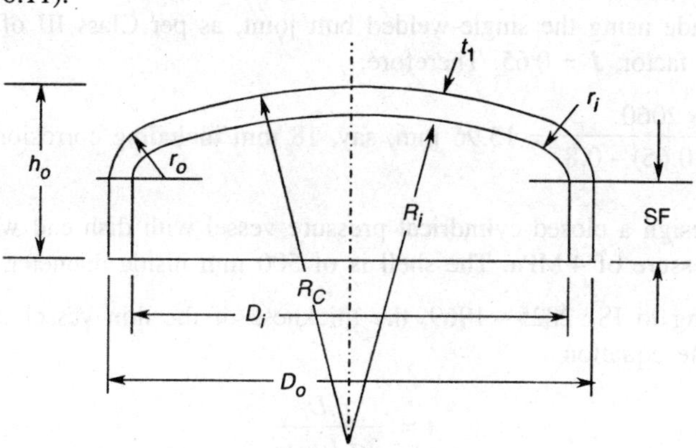

Figure 16.11 Dished end.

The thicknesses of the semi-ellipsoidal end and the dished end, according to IS : 2825 – 1969, are calculated by the formula

$$t_1 = \frac{p_i D_o C_i}{2\sigma J} \tag{16.37}$$

where

C_i is the θ shape factor, which can be obtained from Figure 16.12

h_E is the effective outside height of the end (a parameter used in Figure 16.12) and has the minimum value out of the following:

$$\left(h_o, \frac{D_o^2}{4R_C^2}, \sqrt{\frac{D_o r_o}{2}}\right)$$

with

R_C is the outer crown radius

r_o is the outer knuckle radius

D_o is the outer diameter of the end.

h_o is the outside height of the dished end.

Example 16.1 A steel pipe is used to carry chemicals at the rate of 100 m^3/min and with a flow velocity of 0.5 m/s. The maximum pressure in the pipe is 0.8 MPa. Determine the size of the pipe if the permissible stress for the pipe material is 80 N/mm^2.

Solution Let D_i be the inside diameter of the pipe. Then:

$$\text{Discharge, } Q = \frac{\pi}{4} D_i^2 \times v$$

or

$$D_i = \sqrt{\frac{4Q}{\pi v}} = \sqrt{\frac{4 \times 1.667}{\pi \times 0.5}} \quad (\because 100 \text{ m}^3/\text{min} = 1.667 \text{ m}^3/\text{s})$$

$$= 2.06 \text{ m}$$

$$\text{Thickness of the pipe, } t = \frac{p_i D_i}{2\sigma J - p_i}$$

Let the pipe be made using the single-welded butt joint, as per Class III of Table 16.1. Then, the joint efficiency factor, J = 0.65. Therefore,

$$t = \frac{0.8 \times 2060}{(2 \times 80 \times 0.65) - 0.8} = 15.96 \text{ mm, say, 18 mm including corrosion allowance.}$$

Example 16.2 Design a closed cylindrical pressure vessel with dish end which is required to contain air at a pressure of 4 MPa. The shell is of 500 mm inside diameter.

Solution According to IS : 2825 – 1969, the thickness of the thin vessel of Class III can be determined from the equation

$$t = \frac{p_i D_i}{2\sigma J - p_i}$$

where

J is the joint efficiency factor (= 0.7 for Class III vessel with double-welded butt joint)
σ is the allowable strength (= 100 N/mm^2 for plain carbon steel)

Therefore,

$$t = \frac{4 \times 500}{(2 \times 100 \times 0.7) - 4.0} = 14.7 \text{ mm, say, 16 mm}$$

which is within the limit of Class III vessel.

End cover. The type of end is the dished end.

Outside diameter, $D_o = D_i + 2t = 500 + 2 \times 16 = 532$ mm

The thickness of the dish end as shown in the figure below is determined by the following equation:

$$t_1 = \frac{p_i D_o C_i}{2\sigma J}$$

where C_i the shape factor is obtained from Figure 16.12.

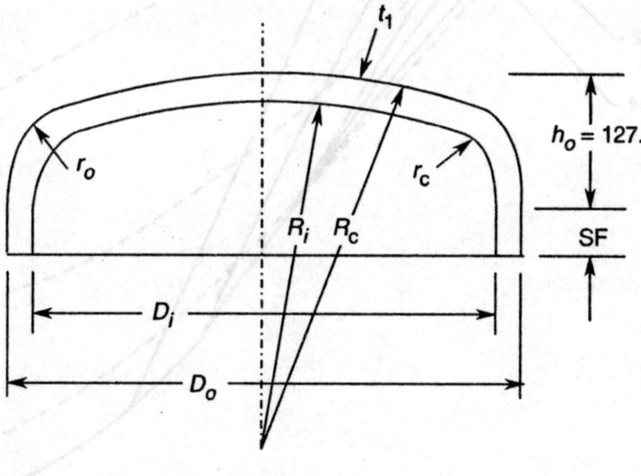

Let us adopt the following dimensions of the dished end. Empirically, the inner radius of the dished end

$$R_i = 0.9D_i = 0.9 \times 500 = 450 \text{ mm}$$

The outer radius of the dished end

$$R_C = R_i + \text{thickness of the dished end}$$

Let us assume the thickness of the dished end to be 30 mm (the actual value will be determined later). Thus,

$$R_C = R_i + 30 = 480 \text{ mm}$$

Inner radius of the knuckle, $r_i = 0.1D_o = 53.2$ mm

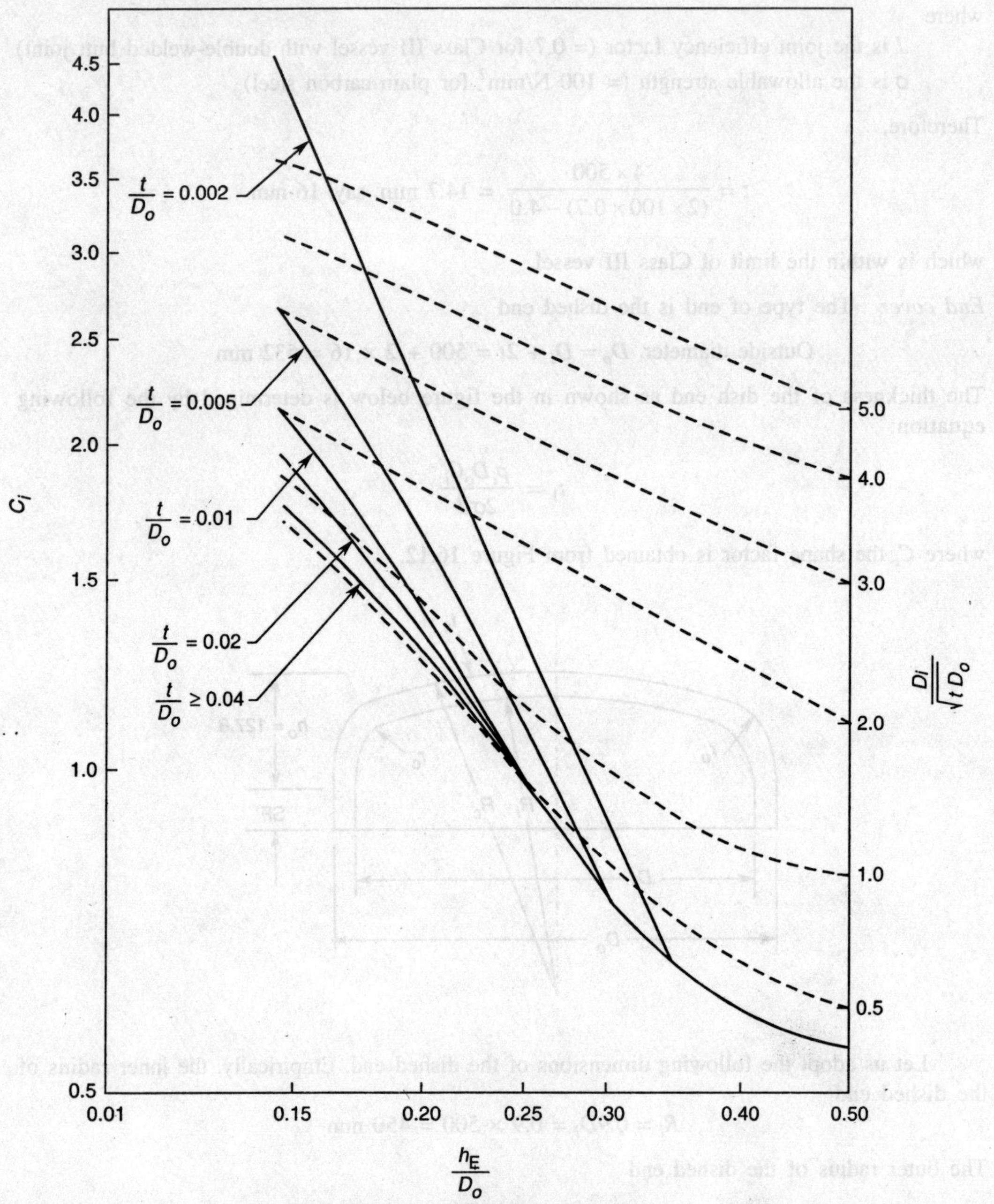

Figure 16.12 Shape factor for dished end.

Outer radius of the knuckle, $r_o = r_i$ + thickness of the end cover

$$= 53.2 + 30 = 83.2 \text{ mm}$$

Outside height of the dished end

$$h_o = R_C - \sqrt{\left(R_C - \frac{D_o}{2}\right)\left(R_C + \frac{D_o}{2} - 2r_o\right)}$$

$$= 480 - \sqrt{\left(480 - \frac{532}{2}\right)\left(480 + \frac{532}{2} - 2 \times 83.2\right)}$$

$$= 127.8 \text{ mm}$$

Effective outside height of the dished end

$$h_E = \min\left[\frac{D_o^2}{4R_C}, \sqrt{\frac{D_o r_o}{2}}, h_o\right]$$

$$= \min\left[\frac{532^2}{4 \times 480}, \sqrt{\frac{532 \times 83.2}{2}}, 127.8\right]$$

$$= \min[147.4, 148.76, 127.8] = 127.8 \text{ mm}$$

Factor, $\dfrac{h_E}{D_o} = \dfrac{127.8}{532} = 0.24$. From Figure 16.12 for $h_E/D_o = 0.24$ and $t/D_o = 0.03$, the value of the shape factor, $C_i = 1.075$. Therefore,

$$\text{Thickness of the dish end, } t_1 = \frac{p_i D_o C_i}{2\sigma J}$$

$$= \frac{4.0 \times 532 \times 1.075}{2 \times 100 \times 0.7}$$

= 16.34 mm, say, 18 mm including corrosion allowance.

Example 16.3 The cylinder of a portable hydraulic riveter is 250 mm in diameter. The pressure of the fluid is 12 MPa gauge. Determine the suitable thickness of the cylinder wall assuming that the cylinder is made of 20Mn6 steel having allowable strength of 150 N/mm^2.

Solution Assuming that the cylinder is thick and open at one end, the thickness of the cylinder can be determined by the Birnie's equation:

$$t = R_i\left[\left(\frac{\sigma + (1-\mu)p_i}{\sigma - (1+\mu)p_i}\right)^{0.5} - 1\right]$$

$$= \frac{250}{2}\left[\left(\frac{150+(1-0.3)\times 12}{150-(1+0.3)\times 12}\right)^{0.5} - 1\right]$$

= 10.7 mm, say, 12.5 mm including corrosion allowance.

Example 16.4 A cylindrical vessel having 300 mm inner diameter is subjected to internal pressure of 150 N/mm². The cylinder is made of 20Mn6 steel having allowable strength of 160 N/mm². Design the pressure vessel. Take Poissons ratio, μ = 0.3.

Solution Assuming that the vessel is closed-ended, the wall thickness may be determined using the Clavarino's equation:

$$t = \frac{D_i}{2}\left[\left(\frac{\sigma+(1-2\mu)p_i}{\sigma-(1+\mu)p_i}\right)^{0.5} - 1\right]$$

In the above equation, $\sigma < (1 + \mu)p_i$, therefore, no thickness of the cylinder will prevent failure. The only solution in this case is that of a compound cylinder.

For a compound cylinder, as shown in Figure 16.6, the design equation is

$$p_i = \tau\left[1-\left(\frac{D_i}{D_c}\right)^2 + 1-\left(\frac{D_c}{D_o}\right)^2\right] \tag{i}$$

where, $\tau = 0.577 \times \sigma_{yt} = 0.577 \times 160 = 92.32$ N/mm²

For maximum pressure, the interface diameter of the compound cylinder

$$D_c = \sqrt{D_i D_o} \tag{ii}$$

Substituting this value of D_c in Eq. (i)

$$p_i = 2\tau\left(1-\frac{D_i}{D_o}\right)$$

or

$$D_o = \frac{D_i}{1-p_i/2\tau} = \frac{300}{(1-150/2\times 92.32)}$$

= 1599 mm, say, 1600 mm

Interface diameter, $D_c = \sqrt{D_i D_o} = \sqrt{300\times 1600} = 692.82$ mm

The tangential stress at the inner surface of the inner cylinder due to pressure 150 N/mm² is given by

$$\sigma_t = (1-\mu)\frac{p_i D_i^2}{D_o^2 - D_i^2} + (1+\mu)\frac{p_i D_o^2}{D_o^2 - D_i^2}$$

$$= (1-0.3)\left(\frac{150\times 300^2}{1600^2 - 300^2}\right) + (1+0.3)\left(\frac{150\times 1600^2}{1600^2 - 300^2}\right) = 205.92 \text{ N/mm}^2$$

To limit the design stress to 0.75 × σ_{yt}, i.e. 120 N/mm^2, the compressive stress due to shrinkage pressure should be, 205.92 – 120 = 85.92 N/mm^2

Tangential stress due to shrinkage

$$\sigma_{ti} = \frac{-2p_c D_c^2}{D_c^2 - D_i^2}$$

where p_c = shrinkage pressure

$$= \frac{\sigma_{ti}\left(D_c^2 - D_i^2\right)}{2D_c^2}$$

$$= 85.92 \times \frac{692.82^2 - 300^2}{2 \times 692.82^2} = 34.9 \text{ N/mm}^2$$

Shrinkage allowance

$$\delta = \frac{p_c D_c}{E}\left[\frac{2D_c^2\left(D_o^2 - D_i^2\right)}{\left(D_o^2 - D_c^2\right)\left(D_c^2 - D_i^2\right)}\right]$$

$$= \frac{34.9 \times 692.82}{2.1 \times 10^5}\left[\frac{2 \times 692.82^2\,(1600^2 - 300^2)}{(1600^2 - 692.82^2)(692.82^2 - 300^2)}\right]$$

$$= 0.3365 \text{ mm}$$

Thus, we have the design specification as:

Inside diameter of the inner cylinder, D_i = 300 mm

Outside diameter of the inner cylinder, $D_s = D_c$ = 692.82 mm

Inside diameter of the outer cylinder, D_h = 692.82 – 0.3365 = 692.435 mm

Outside diameter of the outer cylinder, D_o = 1600 mm

The temperature to which the outer cylinder needs to be heated is

$$T_2 \geq \frac{\delta}{\alpha D_c} + T_1$$

$$= \frac{0.3365}{10 \times 10^{-6} \times 692.82} + T_1$$

or

$$T_2 - T_1 = 48.44°\text{C}$$

If the ambient temperature is 30°C, the cylinder is required to be heated to 80°C. Then it is pressed onto the inner cylinder and allowed to cool.

16.7 PIPE JOINTS

Pipes are used for the transportation of fluids, for example, a liquid or a gas in process industries, and for water supply and sewage disposal, and so forth. Generally pipes are made of cast iron, steel, wrought iron, copper, brass, plastic, or concrete depending upon the pressure and temperature of the carrying fluids. Cast iron pipes are best suited for underground water supply, sewage, and gas main line. These pipes can be used for pressures up to 0.7 MPa and temperatures not exceeding 250°C. Steel pipes are used to carry water, oil, steam, and fuel oils. For corrosive mediums, stainless steel pipes are mainly used. Vinyl and polyethylene plastic pipes are widely used in chemical industries because of their corrosion resistance property.

The size of a pipe is dependent upon the flow rate and the losses allowed in the system. It is specified by its inside diameter and wall thickness. If Q is the quantity of the fluid to be conveyed by a pipe of internal diameter d_i and v is the velocity of the fluid, then

$$Q = \frac{\pi}{4} d_i^2 v \quad \text{or} \quad d_i = 1.13\sqrt{\frac{Q}{v}} \tag{16.38}$$

The thickness t of the pipe is generally determined on the basis of the thin cylinder pressure vessel theory in which the stresses across the section are assumed to be constant. Accordingly:

$$t = \frac{pd_i}{2\sigma\eta} + C \tag{16.39}$$

where C= corrosion allowance (= 1 mm for $t \le 6$ mm and $1.8t$ for $t > 6$ mm).

The thickness calculated by Eq. (16.39) in certain cases may not be sufficient from the rigidity point of view, therefore, it should be suitably increased.

When the variation of stress across the thickness of the pipe is taken into account, the pipes may be treated as thick cylinders and the Lame's equation may be used to compute the pipe thickness.

Pipes are available in small lengths, therefore, several pipes are needed to be joined to suit the different requirements. A variety of joints and fittings are available to connect two pipes. The most commonly used methods of joining pipes are:

(i) Socket and spigot joints
(ii) Welding
(iii) Screwed connections
(iv) Expansion joints
(v) Flange joints

The flange joints are classified into three categories—circular, square and oval—depending upon the shape of the flange. According to IBR – 1961, all dimensions of flanges are standardized and classified into five classes of pressure lying within the overall range from 0.35 MPa to 2.5 MPa.

16.8 OVAL FLANGED JOINT

The oval flanged joint, as shown in Figure 16.13, is most widely used in pipes carrying fluids at high pressure. The joint flanges are of socket and spigot type so that pipes can be easily fitted coaxially. To make a joint leakproof, a packing of trapezoidal section is used and the

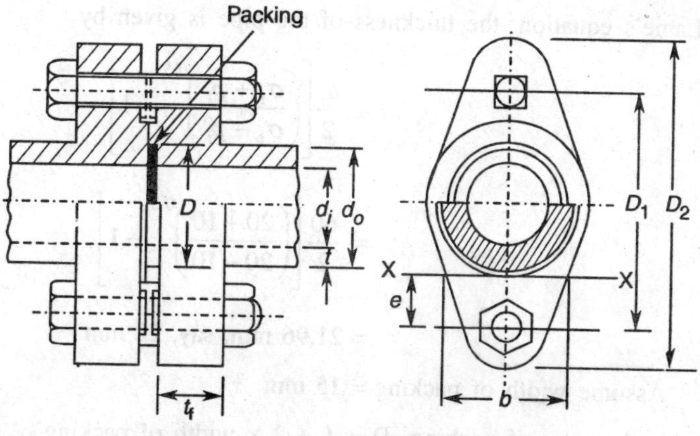

Figure 16.13 Oval type flange joint.

packing material is compressed to the same pressure as that of the fluid. Thus the total separating force is the sum of (i) the force owing to fluid pressure and (ii) the pressure owing to compression of the packing.

Suppose

d_i is the inside diameter of the pipe
d_o is the outside diameter of the pipe
D is the outside diameter of the packing

Total force, $F = F_1 + F_2$

$$= \frac{\pi}{4} d_i^2 p_i + \frac{\pi}{4}\left(D^2 - d_i^2\right) p_i = \frac{\pi}{4} D^2 p_i \quad (16.40)$$

Generally, an oval flange is fastened by two bolts, therefore, the force on each bolt is

$$F = 2 \times A_{\text{bolt}} \times \sigma_{\text{t, bolt}} \quad (16.41)$$

The thickness of the flange is obtained by considering bending of the flange due to force acting on the bolt.

$$\text{Bending moment, } M = Fe \quad (16.42)$$

where e is the distance between the centre of the bolt and outside diameter of the pipe, as shown in Figure 16.13.

$$\text{The section modulus, } Z = \frac{1}{6} b t_f^2$$

where

b is the width of the flange at section X–X
t_f is the thickness of the flange.

Example 16.5 Design an oval flanged joint for a pipe of 60 mm bore. It is subjected to fluid pressure of 10 MPa. The allowable strength of the pipe and flange is 20 N/mm^2 and that for the bolt is 60 N/mm^2.

Solution Using Lame's equation, the thickness of the pipe is given by

$$t = \frac{d_i}{2}\left[\left(\frac{\sigma_t + p_i}{\sigma_t - p_i}\right)^{0.5} - 1\right]$$

$$= \frac{60}{2}\left[\left(\frac{20 + 10}{20 - 10}\right)^{0.5} - 1\right]$$

$$= 21.96 \text{ mm, say, } 22 \text{ mm}$$

Assume width of packing = 15 mm

Outside diameter of packing, $D = d_i + 2 \times$ width of packing

$$= 60 + 2 \times 15 = 90 \text{ mm}$$

Outside diameter of the pipe, $d_o = d_i + 2t$

$$= 60 + 2 \times 22 = 104 \text{ mm}$$

$$\text{Separating force, } F = \frac{\pi}{4} D^2 \times p_i$$

$$= \frac{\pi}{4} \times 90^2 \times 10 = 63{,}617.25 \text{ N}$$

$$\text{Force on each bolt} = \frac{F}{2} = 31{,}808.6 \text{ N}$$

$$\text{Tensile strength of the bolt} = A_{\text{bolt}} \times \sigma_{\text{t, bolt}} = 31{,}808.6 \text{ N}$$

or

$$A_{\text{bolt}} = 530.14 \text{ mm}^2$$

As per IS : 4218 (Part III) – 1967, the suitable size of the bolt is M30 × 3.

Outside diameter of the flange (empirically) (see adjoining figure).

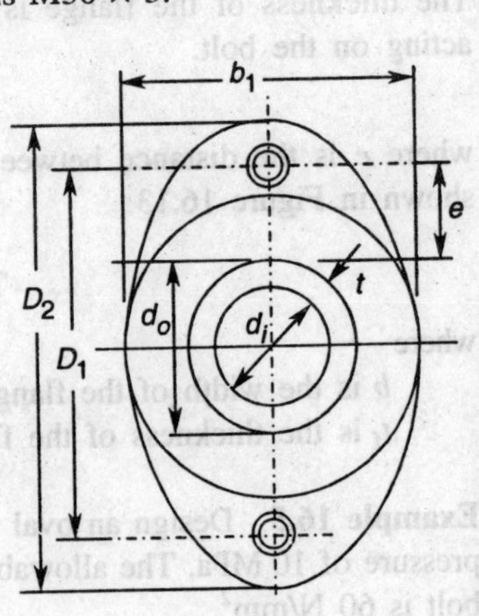

$$D_2 = d_i + 2t + 4.6d_{\text{bolt}}$$

$$= 60 + 2 \times 22 + 4.6 \times 30 = 242 \text{ mm}$$

Pitch circle diameter, $D_1 = D_2 - (3t + 20)$

$$= 242 - (3 \times 22 + 20) = 156 \text{ mm}$$

Let us select elliptical profile of the flange as shown in the adjoining figure.

$$\text{Semi-major axis, } a_1 = \frac{D_2}{2} = 121 \text{ mm}$$

Semi-minor axis, $b_1 = \dfrac{D_1 - d_{\text{bolt}}}{2} = \dfrac{156 - 30}{2} = 63$ mm

Using equation of ellipse, $\dfrac{x^2}{a_1^2} + \dfrac{y^2}{b_1^2} = 1$

For $x = \dfrac{d_o}{2} = 52$ mm, we get

$$\frac{52^2}{121^2} + \frac{y^2}{63^2} = 1 \quad \text{or} \quad y = 56 \text{ mm}$$

Width of flange, $b = 2y = 2 \times 56 = 112$ mm

Eccentricity, $e = 0.5(D_1 - d_o) = 0.5(156 - 104) = 26$ mm

Bending moment $M = Fe$

$= 31{,}808.6 \times 26 = 827{,}023.6$ N · mm

Considering bending failure, the section modulus

$$Z = \frac{M}{\sigma_b} = \frac{827{,}023.6}{20} = 41{,}351.2 \text{ mm}^3$$

or

$$\frac{1}{6} b t_f^2 = 41{,}351.2$$

or

$$t_f = \left(\frac{6 \times 41{,}351.2}{112} \right)^{0.5}$$

$= 47$ mm, say, 50 mm

EXERCISES

1. What is a thick cylinder pressure vessel?
2. How are pressure vessels classified?
3. What are the different methods of prestressing a thick cylinder?
4. Compare stress distribution in a thin vessel with that in a thick vessel.
5. Sketch the different types of ends used for pressure vessels and list their applications.
6. Show the distribution of stresses in a compound cylinder.
7. A mild steel cylinder needs to be built up by shrinking one cylinder on the outside of another. The inner, outer, and mating surface diameters are 150, 250, and 200 mm respectively. The shrinkage is 0.08 mm. Find the safe internal pressure to which the cylinder can be subjected to if the allowable stress is 200 N/mm^2.
8. A cylindrical vessel has an inner diameter of 1 m. It is subjected to an internal pressure of 12 MPa. If the allowable stress is 50 MPa, find the required thickness for this vessel.

9. Show by a diagram how stress varies over the wall of a compound steel cylinder subjected to 100 N/mm^2 internal pressure. The compound cylinder is built up by shrinking one cylinder on the outside of another. The inner, outer and mating diameters are 120, 160, and 220 mm respectively. Shrinkage = 0.1 mm.

10. A shrink-fit assembly is to be formed by shrinking one steel tube over the other. The maximum internal pressure may be 400 bar. The internal and external diameters are 100 mm and 200 mm, respectively. The diameter at the junction is 140 mm and maximum stress is to be limited to 60 MN/m^2. Determine the resulting stress distribution.

11. A thin cylindrical vessel of inside diameter 300 mm is subjected to 400 kPa. Determine the wall thickness if the allowable stress is 150 N/mm^2. Also determine the thickness of the hemispherical end cover.

12. A cylindrical vessel of 500 mm internal diameter is subjected to an internal pressure of 8 MPa. If the allowable stress is 50 N/mm^2, determine the thickness of the cylinder. If this cylinder is covered by dish ends, design the dish end.

13. A cylinder head of steam engine is 250 mm and is subjected to 1.75 MPa. Design the cylinder and the end cover.

14. A cylindrical steel vessel having an internal radius of 200 mm is required to resist an internal pressure of 140 N/mm^2. The tensile and compressive strength of the material is 160 N/mm^2. Design the vessel.

MULTIPLE CHOICE QUESTIONS

1. Pressure vessels used to store toxic substances belong to the category of
   (a) Class I (b) Class II
   (c) Class III (d) Class IV
2. The minimum corrosion allowance as per IS: 2825–1969 is
   (a) 0.5 mm (b) 1.0 mm
   (c) 1.5 mm (d) 2.0 mm
3. The wall thickness of a cylindrical pressure vessel is given by
   (a) $pD/2t\eta$ (b) $pt/2D\eta$
   (c) $pD/2\eta$ (d) $pD/2\eta$
4. The maximum wall thickness for a Class III vessel is
   (a) 8 mm (b) 16 mm
   (c) 24 mm (d) 38 mm
5. The maximum diameter opening which does not require compensation is
   (a) 100 mm (b) 150 mm
   (c) 200 mm (d) 250 mm
6. The most economical way of prestressing is
   (a) compounding (b) auto frettage
   (c) shrink fitting (d) jacketing

Chapter 17

Design of I.C. Engine Components

17.1 INTRODUCTION

The I.C. engine is one of the most widely used prime movers which operates on fossil fuel. These engines are classified into four-stroke and two-stroke engines. Further, on the basis of the operating cycle they are also called petrol engines (based on the Otto cycle) and diesel engines (based on the diesel cycle or dual cycle). An I.C. engine has the following major components: cylinder block, cylinder liner, cylinder head, piston, piston rings, piston pin, connecting rod, crank shaft, main bearings, cam shaft, and valve operating mechanism, etc. However, in this chapter a few principal parts only, namely cylinder, piston, connecting rod, and crank shaft are covered. It is assumed that the reader has the background knowledge of elementary thermodynamics and dynamics of reciprocating mechanisms.

17.2 CYLINDER

The cylinder of an I.C. engine acts as the structural member and retains the working fluid in a closed space with movable piston wall. It is tested to high explosive pressure, which is approximately 3–8 times the maximum compression pressure, and high temperature ranging between 1800 K and 2400 K. Thus the cylinder of an I.C. engine should be able to withstand high working pressure and should be able to transfer heat efficiently without thermal distortion taking place. For small engines operating at low speed, the cylinder block is cast as one piece. However for large engines a separate cylinder liner is used. It facilitates easy repair or replacement of the liner in the event of wear and tear of the cylinder.

Cylinder liners are generally made of closed grained pearlitic cast iron, nickel CI, nickel–chrome CI, cast steel, and forged alloy steel. The inner surface of the liner is heat treated to obtain a hard surface. Sometimes it is chrome plated to obtain a hard and smooth surface.

Generally two types of cylinder liners are used in I.C. engines: dry liners and wet liners (Figure 17.1). A cylinder liner which does not come in direct contact with the cooling medium is called the dry liner, whereas a wet liner comes in contact with the cooling medium. This type of liner is supported at two positions and a water jacket is formed between the liner and the cylinder block in which cooling water is circulated.

The basic dimensions of the cylinder liner are determined on the basis of strength and rigidity to prevent ovalization of the liner during assembly and operation. The dimensions should conform to IS : 6750 – 1972. A cylinder liner should be designed and/or checked in the following possible modes of failure.

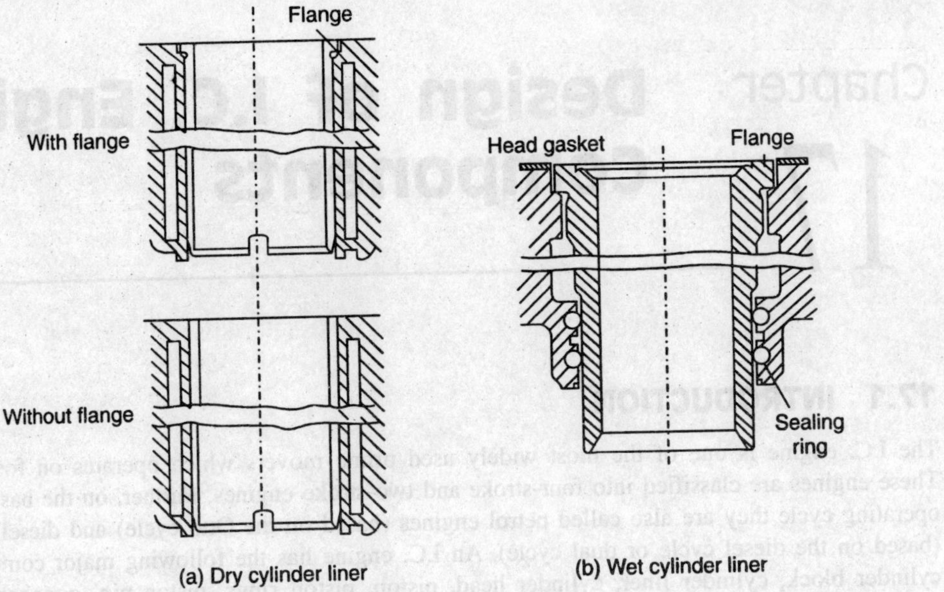

Figure 17.1 Cylinder liners.

(1) A cylinder liner is designed by treating it as either thick cylinder or thin cylinder depending upon the bore to thickness ratio.

(a) *Thick cylinder.* The thickness of the liner for a thick cylinder is computed by the Birnie's equation:

$$t = 0.5D\left[\left(\frac{\sigma + (1-\mu)p_{\max}}{\sigma - (1+\mu)p_{\max}}\right)^{0.5} - 1\right] \tag{17.1}$$

where

σ is the permissible strength (= 50–60 N/mm^2 for CI and 70–100 N/mm^2 for steel)

D is the cylinder bore

$p_{\max}$ is the maximum pressure (N/mm^2).

(b) *Thin cylinder.* For a thin cylinder, the thickness of the liner is computed by the relation

$$t = \frac{p_{\max} D}{2\sigma_\theta} + C \tag{17.2}$$

where

σ_θ is the permissible hoop stress

C is the reboring allowance (1.5–15.0 mm).

(2) A cylinder liner should be checked for thermal stress caused by high temperature difference between the outer and inner surfaces of the liner.

$$\sigma_{\text{th}} = \frac{E\alpha\Delta T}{2(1-\mu)} \tag{17.3}$$

where

E is the elastic modulus (N/mm^2)

α is the thermal expansion coefficient, (10–12) × 10^{-6} mm/°C

μ is the Poisson's ratio

ΔT is the temperature difference (= 100–150°C for high heat zone, i.e. top portion of the liner.)

The total stress in the cylinder liner ($\sigma_\theta + \sigma_{th}$) should be less than 100–130 N/mm^2 for cast iron and less than 160–200 N/mm^2 for alloy steel liner.

(3) In a cylinder liner, longitudinal stress is produced in addition to hoop stress, though marginal, which causes extension of the cylinder, i.e.

$$\sigma_L = \frac{p_{max} D}{4t} \tag{17.4}$$

(4) The side thrust caused by obliquity of the connecting rod on the cylinder liner induces bending stresses. Considering that the liner is supported at two points as shown in Figure 17.2 and maximum side thrust R_{max} acts at a distance a from the TDC position of the piston, the bending moment

$$M = R_{max} \times \frac{ab}{a+b} \tag{17.5}$$

where

a is the distance between the piston pin axis and the TDC position

b is the distance between the piston pin axis and the BDC position.

$$\text{Bending stress, } \sigma_b = \frac{M}{Z}$$

where

$$Z = \frac{\pi}{32}\left(\frac{D_o^4 - D^4}{D_o}\right)$$

with D_o (= $D + 2t$) as the outside diameter of the liner.

Total tensile stress due to longitudinal and bending stress

$$\sigma = \sigma_L + \sigma_b \le 60 \text{ N/mm}^2 \text{ for CI}$$
$$\le 100 \text{ N/mm}^2 \text{ for steel}$$

Other dimensions of the cylinder block can be determined empirically as follows:

(a) Thickness of the cylinder block wall

$$t_1 = 0.045D + 2 \text{ mm} \tag{17.6}$$

(b) Thickness of the cylinder flange

$$t_2 = (1.2\text{–}1.4)t_1 \tag{17.7}$$

(c) Thickness of the jacket wall

$$t_3 = 0.032D + 1.5 \text{ mm} \tag{17.8}$$

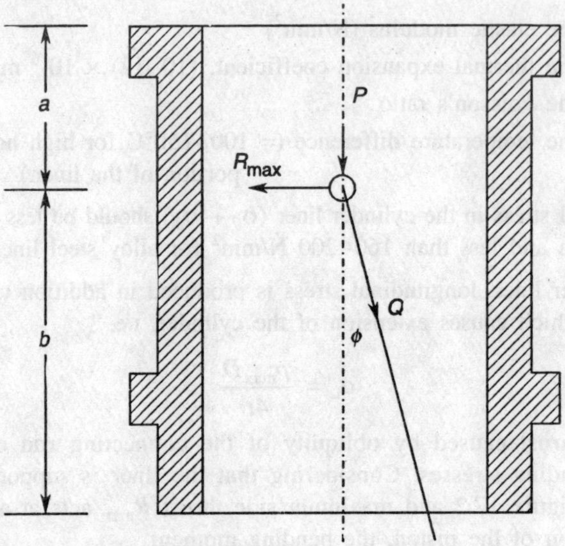

Figure 17.2 Forces acting on cylinder liner.

(d) Water space between the outer cylinder wall and the inner jacket wall

$$t_4 = 0.08D + 6.5 \text{ mm} \tag{17.9}$$

Cylinders are attached to the crank case by means of the flange, studs, and nuts. The diameter of the studs may be obtained by the relation

$$d_{bolt} = D\left(\frac{p_{max}}{N\sigma_t}\right)^{0.5} \tag{17.10}$$

where

N is the number of studs $\left[= (0.25-0.5)\dfrac{D_p}{10} + 4\right]$

D_p is the pitch circle diameter (mm)

σ_t is the allowable tensile strength of the bolt material.

17.2.1 Cylinder Head

The construction of the cylinder head is very complicated due to the presence of inlet and exhaust valves, spark plug, fuel injector, and the shape of the combustion chamber. For an approximate analysis, the cylinder head may be assumed as a flat circular plate held rigidly at the circumference by a suitable number of studs. The thickness of this plate may be computed by the relation

$$t_h = D\left(\frac{Cp_{max}}{\sigma}\right)^{0.5} \tag{17.11}$$

where C = constant = 0.162.

Example 17.1 Design a cylinder for a four-stroke water-cooled diesel engine developing 4 kW at 1500 rpm. Assume that the indicated mean effective pressure at the full load condition is 700 kN/m^2.

Solution Assuming the mechanical efficiency, $\eta_{mech} = 0.8$, indicated power, IP = $\dfrac{BP}{\eta_{mech}} = \dfrac{4}{0.8} = 5$ kW

From thermodynamics, we know IP = $p_{mi}\,LAN$

Assuming length of stroke, $L = 1.1D$, we have

$$5 = 700 \times 1.1D \times \frac{\pi}{4} D^2 \times \left(\frac{1500}{2 \times 60}\right)$$

or

$$D = 0.087 \text{ m or } 87 \text{ mm}$$

$$L = 1.1 \times 87 = 95.7 \text{ mm, say, } 96 \text{ mm}$$

Assuming that the maximum explosion pressure

$$p_{max} = 8 \times p_{mi} = 8 \times 700 = 5600 \text{ kN/m}^2 = 5.6 \text{ N/mm}^2$$

For cast iron cylinder, allowable strength σ = 60 N/mm^2 and Poisson's ratio, μ = 0.21

The thickness of the cylinder is given by

$$t = 0.5D\left[\left(\frac{\sigma + (1-\mu)p_{max}}{\sigma - (1+\mu)p_{max}}\right)^{0.5} - 1\right]$$

$$= 0.5 \times 87\left[\left(\frac{60 + (1-0.21)5.6}{60 - (1+0.21)5.6}\right)^{0.5} - 1\right]$$

$$= 4.35 \text{ mm, say, } 6 \text{ mm}$$

which includes the reboring allowance.

(i) The hoop stress produced in the cylinder

$$\sigma_\theta = \frac{p_{max} D}{2t} = \frac{5.6 \times 87}{2 \times 6} = 40.6 \text{ N/mm}^2$$

(ii) Thermal stress, $\sigma_{th} = \dfrac{E\alpha\Delta T}{2(1-\mu)}$

Let $\alpha = 11 \times 10^{-6}$ mm/°C and ΔT = 120°C. Therefore,

$$\sigma_{th} = \frac{1 \times 10^5 \times 11 \times 10^{-6} \times 120}{2(1-0.21)} = 83.5 \text{ N/mm}^2$$

Total stress, $\sigma_\theta + \sigma_{th} = 40.6 + 83.5 = 124.1$ N/mm^2 which is less than the tensile strength of cast iron (130 N/mm^2).

(iii) Longitudinal tensile stress

$$\sigma_L = \frac{p_{max} \cdot D}{4t} = \frac{5.6 \times 87}{4 \times 6} = 20.3 \text{ N/mm}^2$$

(iv) The side thrust in the cylinder

$$R = Q \sin \phi = P \tan \phi$$

assuming that for small obliquity of the connecting rod the maximum side thrust is 10% of the gas force, we have

$$R_{max} = 0.1 \times \frac{\pi}{4} \times 87^2 \times 5.6 = 3329 \text{ N}$$

Length of the cylinder = 1.25 × stroke length = 1.25 × 96 = 120 mm

Let the position of the piston pin from TDC be

$$a = 50 \text{ mm and } b = 70 \text{ mm (refer Figure 17.2)}$$

Therefore, the bending moment is

$$M = R_{max} \times \frac{ab}{a+b}$$

$$= 3329 \times \frac{50 \times 70}{120} = 97{,}095.8 \text{ N} \cdot \text{mm}$$

Section modulus

$$Z = \frac{\pi}{32}\left(\frac{D_o^4 - D^4}{D_o}\right)$$

where $D_o = D + 2t = 87 + 2 \times 6 = 99$ mm. Therefore,

$$Z = \frac{\pi}{32}\left(\frac{99^4 - 87^4}{99}\right) = 38{,}446.6 \text{ mm}^3$$

$$\text{Bending stress, } \sigma_b = \frac{M}{Z} = \frac{97{,}095.8}{38{,}446.6} = 2.5 \text{ N/mm}^2$$

Total tensile stress = $\sigma_b + \sigma_L = 22.8$ N/mm^2, which is less than the allowable strength. Hence the design is satisfactory.

Other dimensions

(i) Thickness of the cylinder block wall

$$t_1 = 0.045D + 2 = 0.045 \times 87 + 2 \approx 6 \text{ mm}$$

(ii) Thickness of the cylinder flange

$$t_2 = 1.3t_1 = 1.3 \times 6 \approx 8 \text{ mm}$$

(iii) Water space, $t_4 = 0.08D + 6.5 \text{ mm} = 0.08 \times 87 + 6.5 \approx 14 \text{ mm}$

Assuming that the cylinder is fitted to the crank case with studs, the pcd of studs, $D_p = 1.6D$, say, 140 mm

$$\text{Number of studs, } N = (0.25\text{–}0.5)\frac{D_p}{10} + 4$$

$$= 0.35 \times \frac{140}{10} + 4 = 8.9, \text{ say, } 10$$

$$\text{Core diameter of the stud, } d_{bolt} = D\left(\frac{p_{max}}{N\sigma_t}\right)^{0.5}$$

Let us assume that the allowable strength of studs, $\sigma_t = 80 \text{ N/mm}^2$. Then,

$$d_{bolt} = 87\left(\frac{5.6}{10 \times 80}\right)^{0.5} = 7.2 \text{ mm, say, of nominal size 10 mm}$$

$$\text{Thickness of the cylinder head, } t_h = D\left(\frac{Cp_{max}}{\sigma}\right)^{0.5}$$

$$= 87\left(\frac{0.162 \times 5.6}{60}\right)^{0.5} = 10.7 \text{ mm, say, 12.5 mm}$$

17.3 PISTON

The piston is a reciprocating element of the I.C. engine which is subjected to the most severe stress conditions. It receives an impulse from expanding gases and transmits force to a crank shaft via the connecting rod. The piston is also subjected to inertial and heat loads. The pistons of I.C. engines are either of trunk (full skirt) type or of short skirt type which are open at one end. A piston consists of (i) a crown or head which carries the gas force and major portion of the heat load, (ii) a skirt to contain the side thrust, (iii) piston rings to create proper compression, and (iv) a piston pin to connect to the connecting rod which finally transfers force to the crank shaft (Figure 17.3).

A well-designed piston has the following characteristics:

1. It should form a closed vessel of variable capacity to prevent blow-by; i.e. escape of flue gases from the combustion chamber past the piston and into the crank case.
2. The weight should be small to reduce the inertia force due to reciprocating parts.
3. It should be scuff-resistant; scuffing is a type of wear in which erosion of metal from piston takes place due to localized spot welding.
4. The construction should be rigid enough to withstand thermal and mechanical distortions.

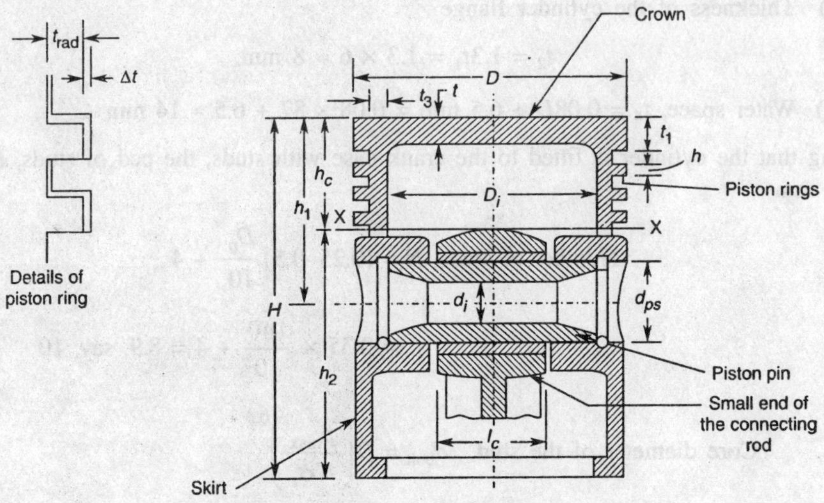

Figure 17.3 Piston of I.C. engine.

5. The design should be such that heat flow occurs from the head of the piston to the skirt. The head of the piston should be hot enough to prevent loss of indicated thermal efficiency; at the same time its temperature should not be very high. Too hot a piston head may cause pre-ignition of charge specially in petrol engines.
6. The temperature distribution in the skirt portion of the piston should be controlled. A high temperature may cause thermal deformation and result into seizure of the piston. However, a very low temperature may cause a large clearance between the cylinder liner and the piston. At TDC position of the piston, a sudden shift in the direction of the side thrust may cause piston blow to the cylinder wall. This action is usually called the piston 'slap'. To take care of this phenomenon, the top diameter of the piston is kept slightly smaller than the skirt diameter. A typical temperature distribution curve around the piston is shown in Figure 17.4.

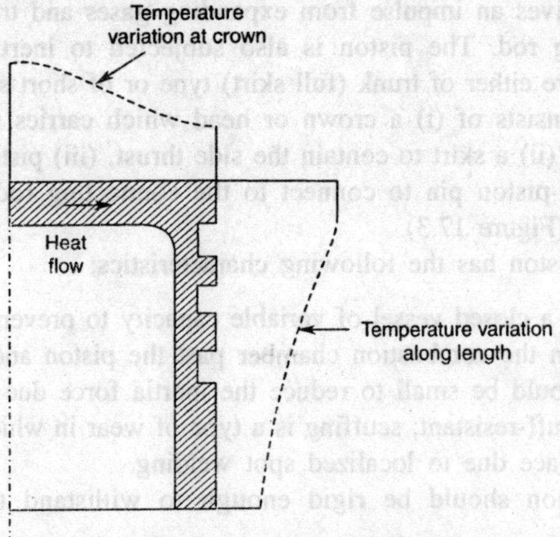

Figure 17.4 Temperature distribution around piston.

17.3.1 Piston Material

The piston material should be light weight, have low expansion coefficient, high conductivity, low coefficient of friction, and good wear resistant properties.

Cast iron is the most popular material used for the construction of pistons for low-speed engines. Mostly, the closed grain pearlitic cast iron is used. Tin coating of cast iron pistons reduces cylinder wear and scuffing. However, the main limitation of cast iron is the high weight density which increases the mass of the reciprocating parts and thereby the inertia force. Modern high-speed engines use aluminium alloys containing silicon, copper, magnesium and nickel. However, the strength of the aluminium pistons is low compared to those made of cast iron.

17.3.2 Design of Piston

The detailed design of various elements of the piston is discussed below:

Piston crown

The simplest piston crown is a flat plate. In some cases it may be of cup type. The selection of the piston crown primarily depends upon the requirement of volume for the combustion chamber and the valve arrangement required. For engines having length/bore ratio up to 1.5, a cup is provided at the top of the piston head, as shown in Figure 17.5.

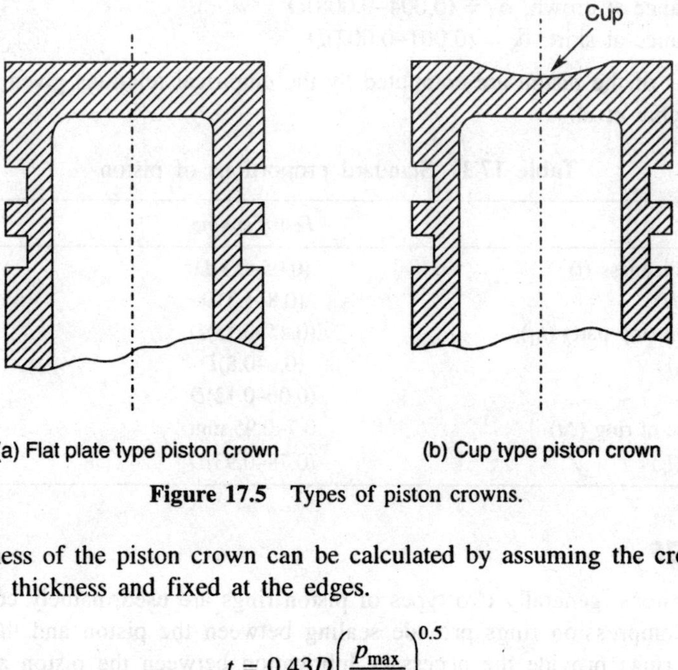

Figure 17.5 Types of piston crowns.

The thickness of the piston crown can be calculated by assuming the crown to be a flat plate of uniform thickness and fixed at the edges.

$$t = 0.43D\left(\frac{p_{max}}{\sigma_t}\right)^{0.5} \qquad (17.12)$$

where

p_{max} is the maximum explosion pressure

D is the piston diameter

σ_t is the allowable tensile strength (40–50 N/mm^2 for CI and 20–50 N/mm^2 for aluminium)

Besides the strength requirement, the piston crown has to carry the heat that is generated during combustion of the fuel. Assuming a flat crown, the required thickness of the crown to dissipate heat can be calculated by the equation

$$t = \frac{D^2 q}{1600K(T_C - T_e)} \tag{17.13}$$

where

q is the heat flow from gases (J/s-m^2) which depends upon the piston material, mean effective pressure and stroke to bore ratio.

K is the thermal conductivity (460 J/s m^2°C/mm length for CI and 1600 J/s m^2°C/mm length for aluminium)

$T_C - T_e$ is the temperature difference (222°C for CI and 111°C for aluminium).

The piston at section X–X as shown in Figure 17.3 is weakened by oil holes. Therefore, it should be checked for both compressive load due to gas force and tensile force due to inertia force. Further, to prevent seizure of the piston or piston slap during operation, the diameter of the crown and skirt should have some diametral clearance between the cylinder wall and the piston. The usual values of clearances are:

(i) Clearance at crown, $\delta_C = (0.004–0.008)D$
(ii) Clearance at skirt, $\delta_S = (0.001–0.002)D$

Other dimensions of the piston are computed by the empirical relations given in Table 17.1 as reported by Kolchin et al.

Table 17.1 Standard proportions of piston

Dimension	*Petrol engine*	*Diesel engine*
Piston crown thickness (t)	$(0.05–0.1)D$	$(0.12–0.2)D$
Piston height (H)	$(0.8–1.3)D$	$(1.0–1.7)D$
Height of piston (top part) (h_1)	$(0.45–0.75)D$	$(0.6–1.0)D$
Skirt length (h_2)	$(0.6–0.8)D$	$(0.7–1.1)D$
Top land (t_1)	$(0.06–0.12)D$	$(0.11–0.2)D$
Radial clearance of ring (Δt)	0.7–0.95 mm	0.7–0.95 mm
Length of pin (l_p)	$(0.78–0.93)D$	$(0.8–0.93)D$

Piston rings

In I.C. engine pistons, generally two types of piston rings are used, namely compression rings and oil rings. Compression rings provide sealing between the piston and the cylinder wall, whereas the oil rings provide the necessary lubrication between the piston and the cylinder. These rings also help to transfer heat from piston to cylinder liner.

The compression rings are usually made of rectangular cross-section and their diameter is made slightly bigger than that of the cylinder bore. Different types of piston rings are shown in Figure 17.6. A portion of the piston ring is cut off so that it can go into the cylinder against

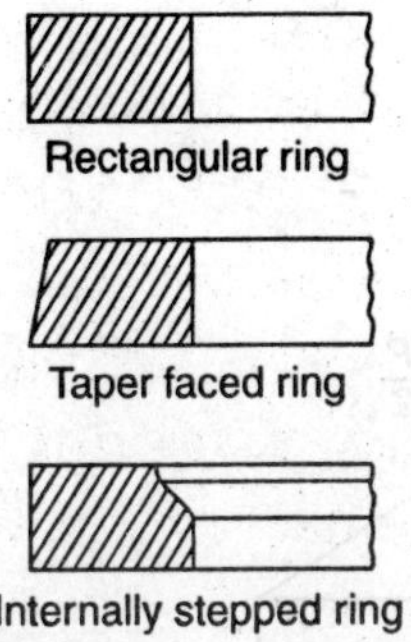

Figure 17.6 Different types of piston rings.

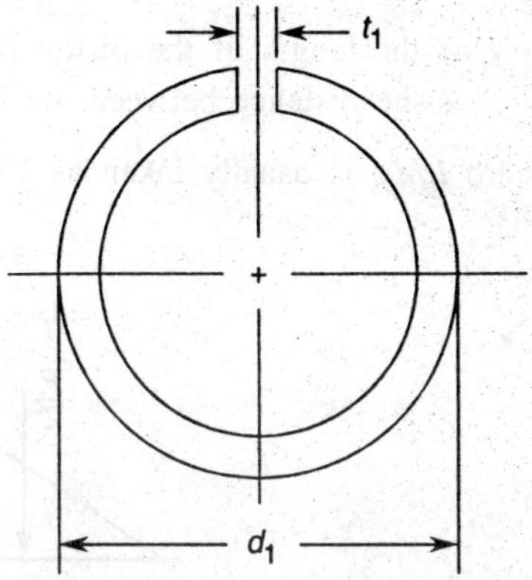

Figure 17.7 Piston ring.

the cylinder wall (Figure 17.7). Piston rings are usually made of chrome-plated grey cast iron or alloy CI to prevent their wear. They should also conform to IS : 5791 – 1977.

The radial thickness of cast iron piston ring may be computed by the equation

$$t_{\text{rad}} = D\left(\frac{3p_{\text{rad}}}{\sigma}\right)^{0.5} \tag{17.14}$$

where

σ is the allowable strength of cast iron [(80–100) N/mm^2]

p_{rad} is the radial pressure on the ring [(0.025–0.035) N/mm^2].

Piston pin

The design of the piston pin is based upon the following criteria:

Bearing failure. The piston pin is subjected to maximum force, being the sum of the gas force and the inertia force. This force is resisted by the bearing pressure between the piston pin and the support length depending upon the nature of the connection between the piston, the piston pin, and the small end of the connecting rod.

$$\text{Bearing pressure, } p_{\text{br}} = \frac{P}{d_{ps} l_s} \tag{17.15}$$

where

P is the maximum gas force

l_s is the length of small end of the connecting rod

d_{ps} is the outer diameter of the piston pin.

The permissible value of the bearing pressure is (8–16) N/mm^2 for petrol and diesel engines.

If the piston pin is employed as floating on the bosses, the bearing pressure exerted is calculated as

$$p_{\text{br}} = \frac{P}{d_{ps}(l_{\text{p}} - b)} \tag{17.16}$$

where

l_p is the length of the piston pin

b is the distance between the boss ends as shown in Figure 17.8.

The ratio l_p/d_{ps} is usually taken as 1.5–2.5.

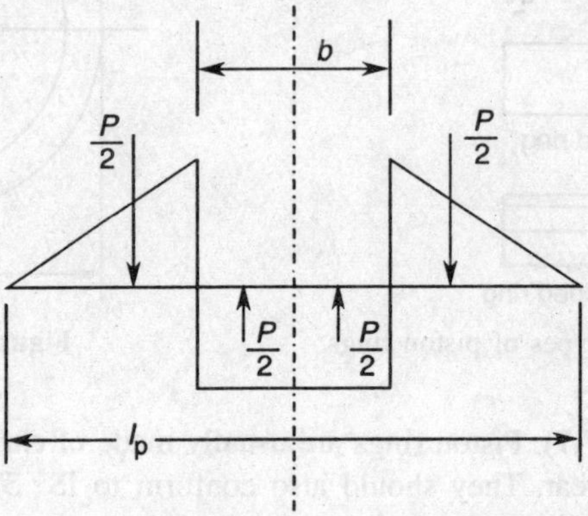

Figure 17.8 Pressure distribution on piston pin.

Bending failure. The pressure distribution of piston pin is as shown in Figure 17.8, assuming that pressure distribution is replaced by the equivalent point load.

The maximum bending moment

$$M = \frac{P}{2} \times \frac{b}{4} - \frac{P}{2}\left(\frac{b}{2} + \frac{l_p - b}{6}\right) \tag{17.17}$$

$$\text{Bending stress } \sigma_b = \frac{M}{Z}$$

where

$$Z \text{ is the section modulus} = \frac{\pi}{32} \frac{\left(d_{ps}^4 - d_i^4\right)}{d_{ps}}$$

d_i is the hollow diameter of the piston pin.

Shear failure of pin. The piston pin may fail in double shear near the bosses. The induced shear stress is

$$\tau = \frac{2P}{\pi\left(d_{ps}^2 - d_i^2\right)} \tag{17.18}$$

The induced shear stress should not exceed (60–70) N/mm^2 for alloy steel.

Ovalization of pin*. Due to nonlinear distribution of forces applied on the piston pin, it is sequeezed into elliptical shape as shown in Figure 17.9. This deformation is called *ovalization.*

* For details, refer to: Kolchin V. and Demidov V., *Design of Automotive Engines*, MIR Publishers, Moscow.

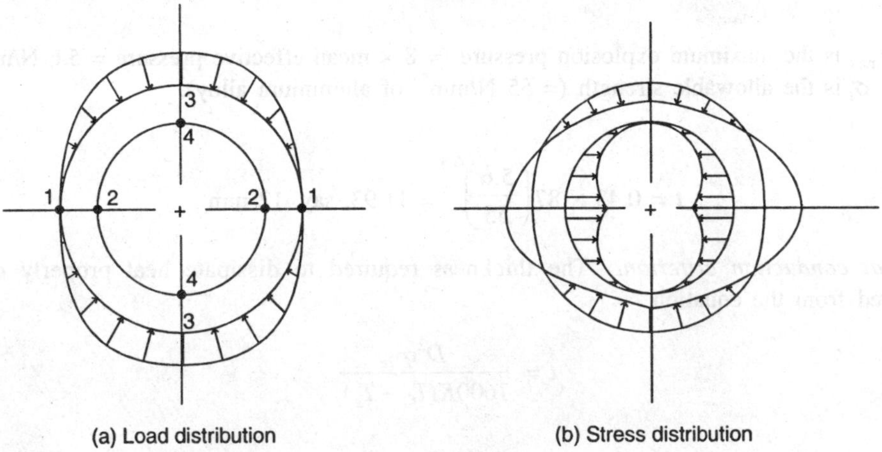

Figure 17.9 Load and stress distribution on piston pin.

The maximum ovalization, i.e. change in diameter takes place in the horizontal plane. The magnitude of ovalization can be determined by the equation

$$\delta d_{\mathrm{p}} = \frac{1.35P}{El_{\mathrm{p}}}\left(\frac{1+\alpha}{1-\alpha}\right)^3 [0.1 - (\alpha - 0.4)^3] \tag{17.19}$$

where α is the ratio of the inner diameter to the outer diameter i.e. d_i/d_{ps}.

The maximum value of ovalization lies between 0.02 and 0.05 mm.

Example 17.2 Design a piston for a four-stroke diesel engine developing power at 1500 rpm. Other related data are the following:

(i) Piston diameter = 87 mm
(ii) Length of the stroke = 96 mm
(iii) Mean effective pressure = 0.7 N/mm^2
(iv) bsfc = 0.26 kg/kWh
(v) L/r ratio = 4
(vi) Heat conducted through the piston crown = 10% of heat generated during combustion.
(vii) Calorific value of the fuel (CV) = 42 MJ/kg

Assume that the piston is made of aluminium alloy.

Solution **Piston crown.** The thickness of the piston crown may be computed by the following two criteria:

(a) *Strength criterion.* The thickness of the piston crown based upon the strength criterion can be calculated by assuming the crown to be a flat plate of uniform thickness and fixed at the edges

$$t = 0.43D\left(\frac{p_{\max}}{\sigma_{\mathrm{t}}}\right)^{0.5}$$

where

p_{max} is the maximum explosion pressure (= 8 × mean effective pressure = 5.6 N/mm²)

σ_t is the allowable strength (= 55 N/mm², of aluminium alloy).

Thus,

$$t = 0.43 \times 87\left(\frac{5.6}{55}\right)^{0.5} = 11.93, \text{ say, } 12 \text{ mm}$$

(b) *Heat conduction criterion.* The thickness required to dissipate heat properly can be computed from the equation

$$t = \frac{D^2 q}{1600K(T_C - T_e)}$$

where

q is the heat flow from gases (J/s m²)

K is the thermal conductivity (= 1600 J/s m² °C/mm for aluminium alloy)

$T_C - T_e$ is the temperature difference (=111°C for aluminium piston)

We know that

$$\text{Indicated power, IP} = \frac{p_{mi} LAN}{60 \times 2}$$

$$= 0.7 \times 10^3 \times \frac{96}{1000} \times \frac{\pi}{4} \times 0.087^2 \times \frac{1500}{60 \times 2}$$

$$= 4.99, \text{ say, } 5 \text{ kW}$$

Assuming mechanical efficiency of engine as 80%

$$\text{Brake power, BP} = \text{IP} \times \eta_{mech} = 5 \times 0.8 = 4 \text{ kW}$$

$$\text{Fuel consumption} = \text{bsfc} \times \text{BP} = 0.26 \times 4 = 1.04 \text{ kg/h}$$

$$\text{Heat supplied} = \text{fuel consumption} \times \text{CV}$$

$$= \frac{1.04 \times 42 \times 10^6}{3600} = 12{,}133.3 \text{ J/s}$$

$$\text{Heat conducted through crown} = 10\% \text{ of heat supplied}$$

$$= 0.1 \times 12{,}133.3 = 1213.33 \text{ J/s}$$

$$\text{Heat flow rate, } q = \frac{\text{heat conducted}}{\text{cross-section area}}$$

$$= \frac{4 \times 1213.33}{\pi \times 0.087^2} = 204{,}103.55 \text{ J/s m}^2$$

$$\text{Thickness of the crown, } t = \frac{87^2 \times 204{,}103.55}{1600 \times 1600 \times 111} = 5.43 \text{ mm, say, } 12 \text{ mm}$$

Other dimensions of the piston are calculated as follows:

(i) Assuming that the piston ring is made of chrome-plated CI, the radial thickness of the piston ring

$$t_{rad} = D\left(\frac{3p_{rad}}{\sigma}\right)^{0.5}$$

where

p_{rad} is the radial pressure on the ring (= 0.025 N/mm^2 for four-stroke diesel engine)
σ is the allowable strength of cast iron (= 85 N/mm^2).

Therefore,

$$t_{rad} = 87\left(\frac{3\times 0.025}{85}\right)^{0.5} = 2.58 \text{ mm, say, 4 mm}$$

(ii) Width of the ring, $h = (0.7\text{–}1.0)t_{rad} \approx 0.8t_{rad} = 0.8 \times 4 = 3.2$ mm

(iii) Number of rings, $i = \dfrac{D}{10h} = \dfrac{87}{10\times 3.2} = 2.7$

Let us adopt 3 compression rings and one oil ring. The shape of the piston ring is shown in Figure 17.7.

(iv) Distance between the first ring groove and top surface, i.e. top land, t_1 = ((0.08–0.2)D $\approx 0.08D = 0.08 \times 87 = 6.96$ mm, say, 7 mm

(v) Thickness of the piston crown wall
$t_3 = (0.05 - 0.1)D \approx 0.1D = 8.7$ mm

(vi) Piston inner diameter
$D_i = D - 2(t_3 + t_{rad} + \Delta t)$
where Δt = radial clearance of ring = 0.7–1.1 mm $\approx$ 0.8 mm
Thus,

$$D_i = 87 - 2(8.7 + 4 + 0.8) = 60 \text{ mm}$$

(vii) Thickness of the skirt wall
$t_2 = (2\text{–}5) \text{ mm} \approx 4$ mm

Design of the piston pin. Piston pin is generally made of alloy steel, say 37Mn2, having allowable tensile strength of 120 N/mm^2. The pin may be designed by considering the following modes of failure.

(a) *Bearing failure.* Induced bearing pressure, $p_{br} = \dfrac{P}{d_{ps} \times l_s}$

$$P = \text{gas force} = \frac{\pi}{4} \times 87^2 \times 5.6 = 33.29 \text{ kN}$$

Assuming l_s/d_{ps} = 1.5 and permissible bearing pressure p_{br} = 20 N/mm^2

$$d_{ps} \times 1.5d_{ps} = \frac{P}{p_{br}} = \frac{33.29\times 10^3}{20} \text{ or } d_{ps} = 33.3 \text{ mm}$$

Let us adopt outer diameter of pin, d_{ps} = 35 mm, and inside diameter, $d_i = 0.6 \times d_{ps}$ = 21 mm

(b) *Failure due to bending.* By empirical relation, l_p = (0.80–0.93)D

Let us choose, $l_p = 0.9D = 0.9 \times 87 = 78.3$ mm, say, 80 mm.

b = length of small end of the connecting rod + clearance

$= 1.5d_{ps} + 4$ mm

$= 1.5 \times 35 + 4 = 56.5$ mm, say, 56 mm

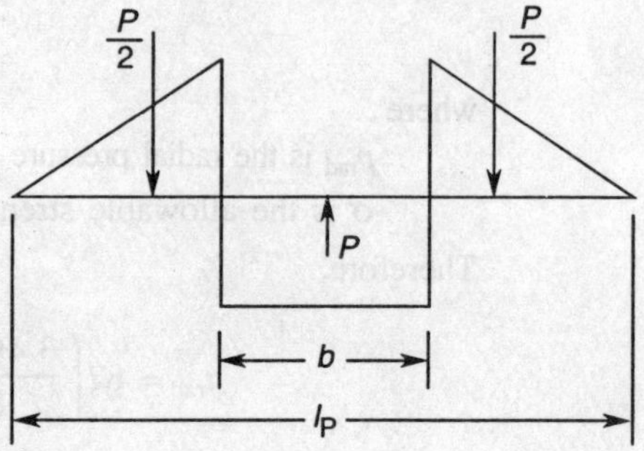

Bending moment

$$M = \frac{P \times b}{4} - \frac{P}{2}\left(\frac{b}{2} + \frac{l_p - b}{6}\right)$$

$$= \frac{33{,}290 \times 56}{4} - \frac{33{,}290}{2}\left(\frac{56}{2} + \frac{80 - 56}{6}\right)$$

$= 66{,}580$ N · mm

$$\text{Section modulus, } Z = \frac{\pi}{32}\left(\frac{35^4 - 21^4}{35}\right) = 3663.7 \text{ mm}^2$$

$$\text{Bending stress, } \sigma_b = \frac{M}{Z} = \frac{66{,}580}{3663.7} = 18.17 \text{ N/mm}^2$$

which is less than the allowable strength.

(c) *Shear failure.* The piston pin may fail in double shear. The induced shear stress is

$$\tau = \frac{2P}{\pi\left(d_{ps}^2 - d_i^2\right)} = \frac{2 \times 33{,}290}{\pi(35^2 - 21^2)}$$

$= 27.0$ N/mm^2, which is reasonable.

(d) *Ovalization.* The maximum change in diameter of the pin is

$$\delta d_p = \frac{1.35P}{El_p}\left(\frac{1+\alpha}{1-\alpha}\right)^3 [0.1 - (\alpha - 0.4)^3]$$

$$= \frac{1.35 \times 33{,}290}{2 \times 10^5 \times 80}\left(\frac{1+0.6}{1-0.6}\right)^3 [0.1 - (0.6 - 0.4)^3]$$

$= 0.0165$ mm which is less than the permissible value of 0.02 mm; hence it is safe.

17.4 CONNECTING ROD

The connecting rod is an intermediate link between the piston and the crank shaft of an I.C. engine. The basic purpose of it is to transmit motion and force from piston to the crank pin. It also carries the lubricating oil from the crank pin end to the piston pin end and provides lubrication to the piston cylinder assembly. The connecting rod converts the reciprocating motion of the piston to oscillatory motion of itself which is finally converted to rotary motion of the crank shaft. The main parts of the connecting rod, as shown in Figure 17.10, are: (i) the small end which connects the connecting rod to the piston through piston pin, (ii) the shank, usually of I-section, and (iii) the big end which is usually split to surround the crank pin.

The length of the connecting rod is usually kept 3 to 4.5 times the crank radius. The shorter length of the connecting rod increases obliquity and thereby the side thrust on the cylinder, whereas the longer length increases the height of the engine.

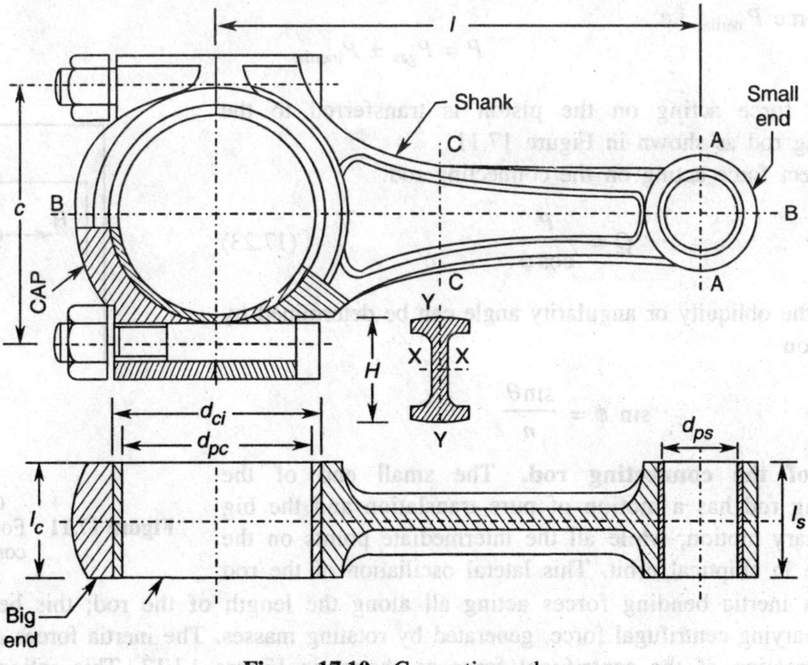

Figure 17.10 Connecting rod.

17.4.1 Force Analysis

The stresses set up in the connecting rod are due to the following forces acting on it.

Gas force. The direct force due to gas pressure acting on the piston is transferred to the connecting rod. This force can be calculated with the help of the indicator diagram. Though the gas force varies in magnitude with the crank rotation angle, the maximum explosion pressure is about 8 to 10 times the indicated mean effective pressure. Thus, the gas force on the piston is

$$P_{gas} = p_{max} \times \text{area of cross-section} \tag{17.20}$$

Inertia force due to reciprocating parts. The inertia force due to reciprocating mass is computed by the relation

$$P_{\text{inertia}} = -\, m\omega^2 r\left(\cos\theta + \frac{\cos 2\theta}{n}\right) \qquad (17.21)$$

where

m is the mass of the piston + 0.33 × mass of the connecting rod to account for the small end portion of the connecting rod which is assumed to be reciprocating
ω is the angular speed of the crank shaft
θ is the crank angle
n is the ratio of length of the connecting rod to crank radius (l/r)

and the (–)ve sign represents the fact that inertia opposes acceleration.

The net force acting on the piston will be the algebraic sum of the gas force P_{gas} and the inertia force $P_{inertia}$, i.e.

$$P = P_{\text{gas}} \pm P_{\text{inertia}} \qquad (17.22)$$

This net force acting on the piston is transferred to the connecting rod as shown in Figure 17.11.

Direct force acting on the connecting rod

$$Q = \frac{P}{\cos\phi} \qquad (17.23)$$

where ϕ the obliquity or angularity angle can be determined by the relation

$$\sin\phi = \frac{\sin\theta}{n}$$

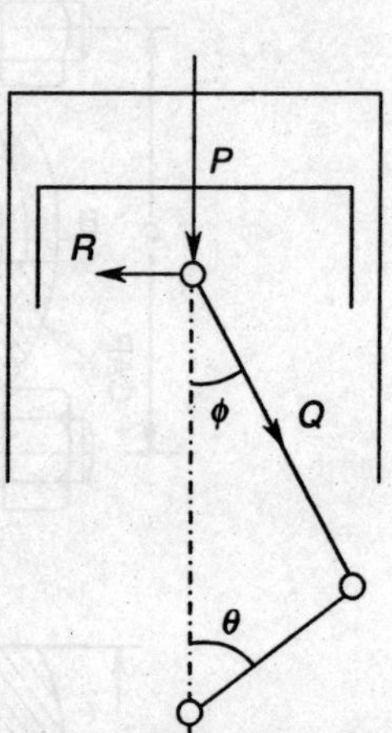

Figure 17.11 Force on the connecting rod.

Inertia of the connecting rod. The small end of the connecting rod has a motion of pure translation and the big end a rotary motion, while all the intermediate points on the rod move in elliptical orbit. This lateral oscillation of the rod results in inertia bending forces acting all along the length of the rod; this being due to linearly varying centrifugal force, generated by rotating masses. The inertia forces act opposite to the direction of the centrifugal force as shown in Figure 17.12. This action is termed *whipping* and the stress induced in the rod is called *whipping stress*.

Let ρ be the mass density. Then,

$$\text{Inertia force, } F_{\text{inertia}} = \frac{\rho A\omega^2 rl}{2} \qquad (17.24)$$

The maximum bending moment, which acts at distance $l/\sqrt{3}$ from the piston pin end, is

$$M_{\max} = \frac{2F_{\text{inertia}}\, l}{9\sqrt{3}} \qquad (17.25)$$

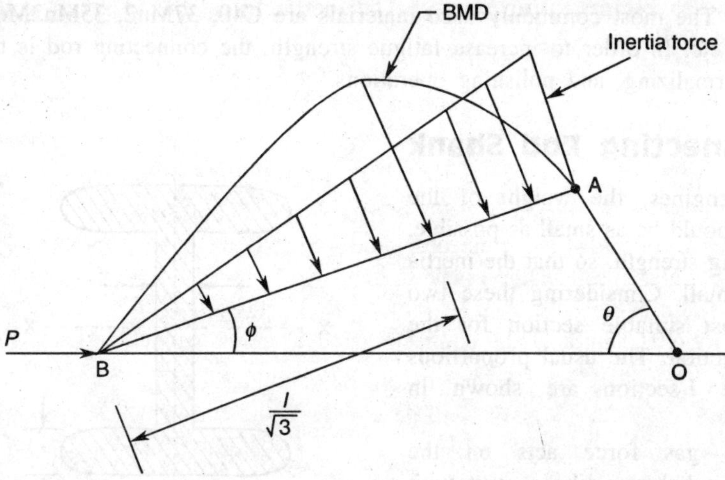

Figure 17.12 Inertia force and bending moment on the connecting rod.

The value of the crank angle θ at which the bending moment is maximum is given by

$$\theta = 90° - \frac{3500}{(n + 7.82)^2} \tag{17.26}$$

The graphical representations of distribution of stresses on the shank are shown in Figure 17.13. Therefore, the shank of the connecting rod must be designed for maximum resultant stress.

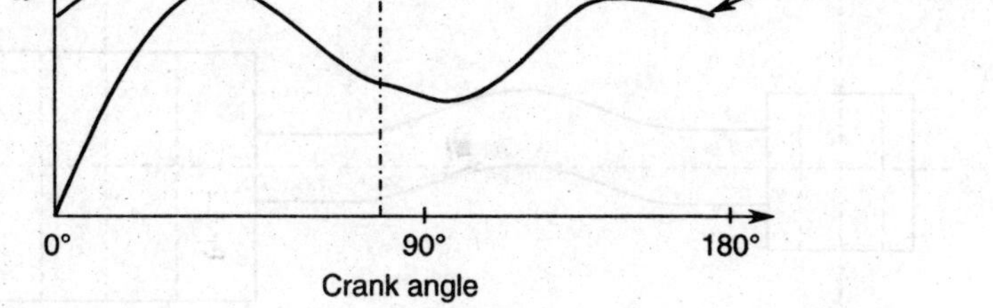

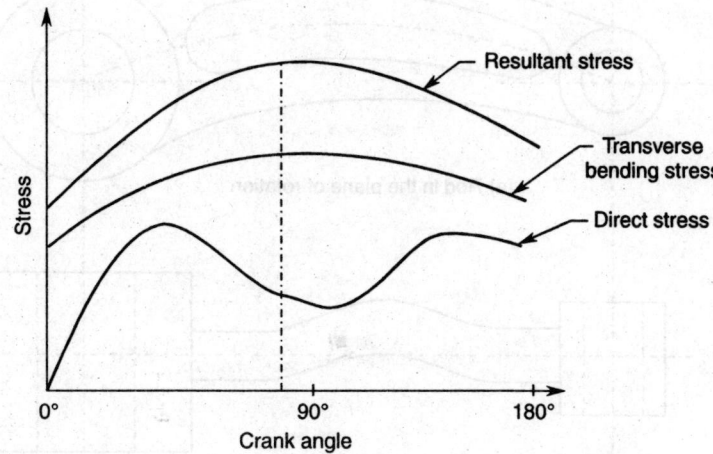

Figure 17.13 Stresses on the connecting rod at various crank positions.

The connecting rod is subjected to fluctuating load condition and has to operate for millions of cycles. Thus manganese steel, chrome steel and other alloy steels are most commonly used. The most commonly used materials are C40, 37Mn2, 35Mn2Mo28, 40Cr1, 40Ni1Cr1Mo15, etc. In order to increase fatigue strength, the connecting rod is treated with shot blasting, normalizing, and polishing operations.

17.4.2 Connecting Rod Shank

In high-speed engines, the weight of the connecting rod should be as small as possible, without sacrificing strength, so that the inertia forces remain small. Considering these two aspects, the most suitable section for the shank is the I-section. The usual proportions chosen for the I-section are shown in Figure 17.14.

When the gas force acts on the connecting rod, it behaves like a strut with both ends hinged in the plane of rotation and both ends fixed in the plane perpendicular to the plane of rotation as shown in Figure 17.15. Therefore, a connecting rod can be equally strong if it satisfies the condition:

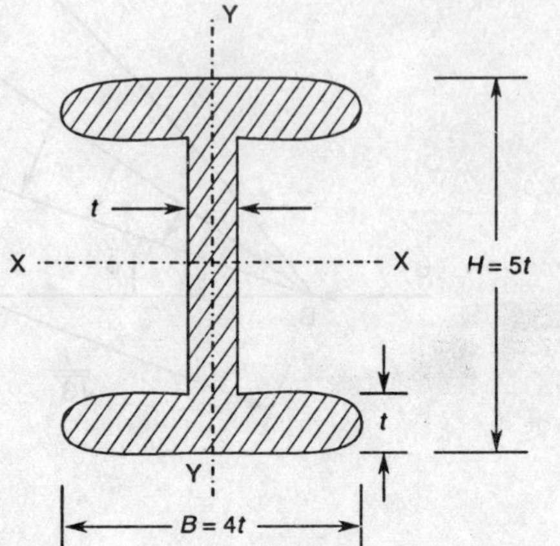

Figure 17.14 Section of the connecting rod shank.

$$I_{XX} = 4I_{YY} \tag{17.27}$$

where I_{XX}, I_{YY} is the moment of inertia about the X-X and Y-Y axis, respectively.

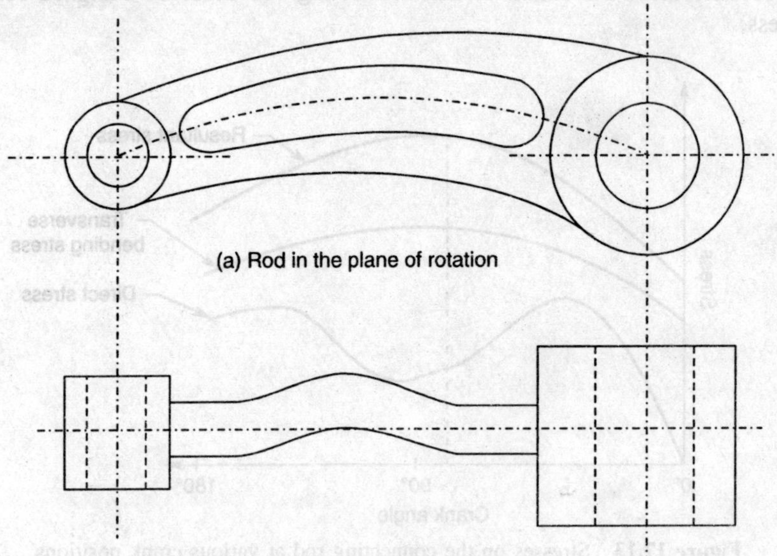

Figure 17.15 Crippling of connecting rod.

The crippling stress induced in the rod can be computed by the Rankine formula given below:

$$\sigma_{cr} = \frac{P}{A}\left[1 + a\left(\frac{l}{k}\right)^2\right] \tag{17.28}$$

where a is the Rankine constant (= 1/6250 for both ends hinged and 1.95/25,000 for both ends fixed).

17.4.3 Small End

The dimensions of the small end can be found empirically by the following relations:

(i) Inner diameter of the small end, $d_{si} = (1.1–1.25)d_{ps}$
(ii) Outer diameter of the small end, $d_{so} = (1.25–1.65)d_{ps}$
(iii) Length of the small end, $l_s = (0.3–0.45)D$

where

d_{ps} is the outer diameter of the piston pin

D is the cylinder bore.

Figure 17.16 shows the small end of the connecting rod. It is designed/checked for the following considerations:

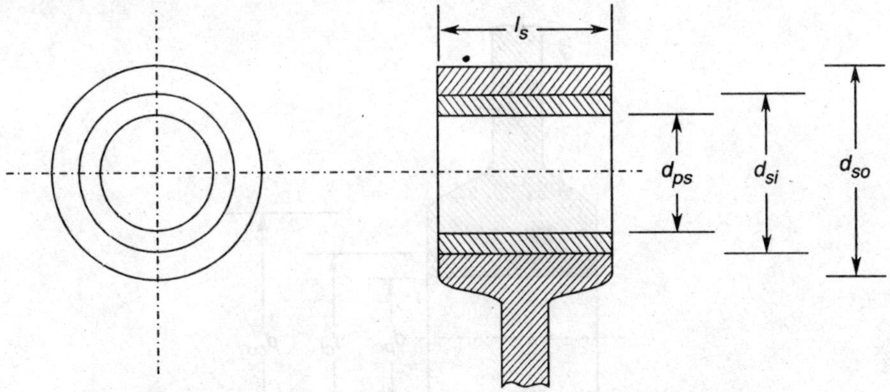

Figure 17.16 Small end of the connecting rod.

(i) *Bearing failure of pin.* The limiting bearing pressure is calculated by the equation

$$p_{br} = \frac{P}{l_s d_{ps}} \le p_{allowable} \tag{17.29}$$

(ii) *Bending of pin.* The bending of the piston pin due to gas force should be checked as follows:

$$\sigma_b = \frac{M}{Z} \le \sigma_{allowable} \tag{17.30}$$

where M and Z are bending moment and section modulus respectively.

(iii) The upper part of the small end is subjected to tensile stress due to inertia of the reciprocating masses. This inertia force is maximum when the crank rotation angle is 0°.

$$\sigma_t = \frac{P_{inertia}}{(d_{so} - d_{si})l_s} \leq \sigma_{allowable} \tag{17.31}$$

The bronze bush is generally press fitted inside the small end. When the engine operates, due to high temperature, the bronze bush expands more than the steel ring end. This difference in thermal expansion produces stresses, which are neglected.

17.4.4 Big End

The dimensions of the big end can be found empirically by the following relations:

(i) Crank pin diameter, $d_{pc} = (0.55\text{–}0.75)D$
(ii) Length of the big end, $l_c = (0.45\text{–}1.0)d_{pc}$
(iii) Bush thickness, $t_{bush} = (0.03\text{–}0.1)d_{pc}$
(iv) Distance between the centre of bolts, $c = (1.3\text{–}1.75)d_{pc}$

The big end of the connecting rod, as shown in Figure 17.17, is checked for bearing and bending considerations. In the absence of detailed design of the crank shaft, the following check may be used.

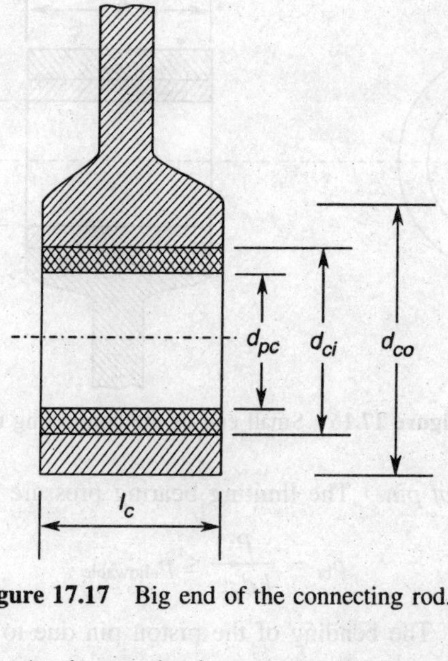

Figure 17.17 Big end of the connecting rod.

Bearing consideration. At the crank pin end, the bearing pressure varies between 7.5 and 15.0 N/mm^2 depending upon the bearing material and the method of lubrication.

$$\text{Bearing pressure, } p_{br} = \frac{P}{l_c d_{pc}} \tag{17.32}$$

17.4.5 Bolts for the Big End

The big end of connecting rod is usually made in two halves, which are fastened by two bolts. The maximum force on the bolt and on the big end will be the inertia force at TDC of the suction stroke.

Usually, two bolts are used to fasten the big end cap. The initial tightening force F_{initial} must be 2 to 3 times the inertia force per bolt. Thus, the total force on the bolt

$$P_{\text{bolt}} = F_{\text{initial}} + K\frac{P_{\text{inertia}}}{N} \tag{17.33}$$

where

F_{initial} is the initial tightening force
K is the gasket factor
N is the number of bolts.

17.4.6 Cap of the Big End

The cap of the big end is designed as a beam supported at the bolt centre and may be assumed to be loaded with concentrated load.

The thickness of the cap is computed by the relation

$$t_C = \left(\frac{P_{\text{inertia}}c}{l_c\sigma_y}\right)^{0.5} \tag{17.34}$$

where c is the distance between the bolt centres (Figure 17.18).

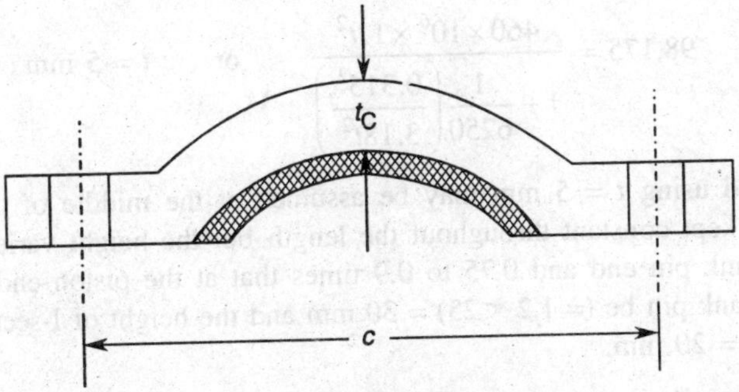

Figure 17.18 Cap of the big end.

Example 17.3 Design a connecting rod for a petrol engine from the following data:

(i) Diameter of piston = 100 mm
(ii) Weight of reciprocating part = 1.8 kg
(iii) Length of connecting rod = 315 mm
(iv) Stroke = 140 mm
(v) Speed = 1500–2500 rpm
(vi) Compression ratio = 4:1
(vii) Maximum explosion pressure = 2.5 MPa

Solution As discussed in Section 17.4.2, the I-section is the most suitable for the connecting rod, with the usual proportions as shown the adjoining figure.

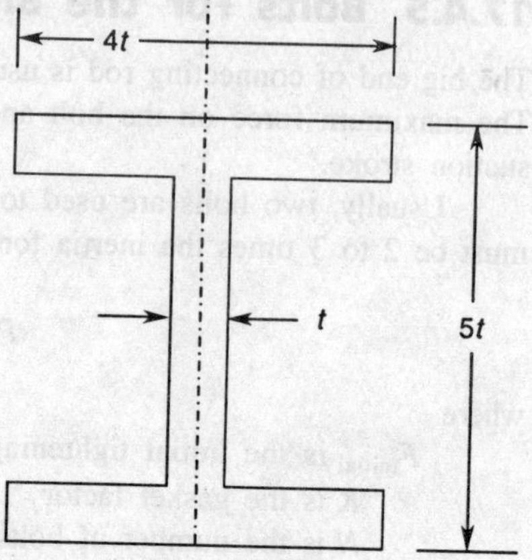

Width of flange, $B = 4t$
Height of I-section, $H = 5t$
Web thickness = t

The maximum gas force

$$P_{gas} = \frac{\pi}{4} D^2 \times p_{max} = \frac{\pi}{4} \times 100^2 \times 2.5 = 19{,}635 \text{ N}$$

Since a connecting rod is subjected to severe load conditions including fatigue load, a high factor of safety (5–6) is used while treating it as static strut. Thus, in the Rankine formula

$$F_{cr} = \frac{\sigma_{cr} A}{1 + a\left(\dfrac{l}{k}\right)^2}$$

taking F_{cr} = crippling load = P_{gas} × FoS = 19,635 × 5 = 98,175 N, and assuming that the material for the connecting rod is 37Mn2 with yield strength σ_{cr} of 460 N/mm², Rankine constant a = 1/6250 (for both ends hinged), $k_{XX}^2 = I_{XX}/A = 3.18t^2$, and A = area of I-section = $11t^2$, we have

$$98{,}175 = \frac{460 \times 10^6 \times 11t^2}{1 + \dfrac{1}{6250}\left(\dfrac{0.315^2}{3.18t^2}\right)} \qquad \text{or} \qquad t = 5 \text{ mm}$$

The section obtained using t = 5 mm may be assumed at the middle of the rod length. The width of the rod is kept constant throughout the length but the height varies from 1.1 to 1.25 times that at the crank pin end and 0.75 to 0.9 times that at the piston end. Let the height of I-section near the crank pin be (= 1.2 × 25) = 30 mm and the height of I-section near the piston pin be (= 0.8 × 25) = 20 mm.

Small end. The dimension of the small end can be found empirically as follows:

(i) Inner diameter of the small end, d_{si} = (1.1–1.25) d_{ps}
(ii) Outer diameter of the small end, d_{so} = (1.25–1.65) d_{ps}
(iii) Length of the small end, l_s = (0.3–0.45)D

where d_{ps} is the outer diameter of the piston pin.

Considering bearing failure of the pin

$$p_{br} = \frac{P_{gas}}{l_s d_{ps}}$$

let the allowable bearing pressure, p_{br} be 15 N/mm^2 and l_s/d_{ps} be 2. Then,

$$d_{ps} = \left(\frac{P_{gas}}{2p_{br}}\right)^{0.5} = \left(\frac{19,635}{2 \times 15}\right)^{0.5} = 25.58, \text{ say, } 26 \text{ mm}$$

Therefore, length of the small end $l_s = 2d_{ps} = 52$ mm. Other dimensions of the small end are:

(i) Inner diameter of the small end, $d_{si} = 1.15d_{ps} = 1.15 \times 26 = 29.9$ mm, say, 30 mm

(ii) Outer diameter of the small end, $d_{so} = 1.4d_{ps} = 1.4 \times 26 = 36.4$ mm, say, 37 mm

A babit or bronze metal bush of $(d_{si} - d_{ps})/2$, i.e. 2 mm is inserted.

Big end. Considering bearing failure of crank and assuming empirical relations:

$$\text{Diameter of the crank pin, } d_{pc} = (0.55\text{–}0.75)D \approx 0.6D = 0.6 \times 100 = 60 \text{ mm}$$

$$\text{Length of crank pin, } l_c = 1.0d_{pc} = 60 \text{ mm}$$

$$\text{Bearing pressure, } p_{br} = \frac{P_{gas}}{l_c d_{pc}} = \frac{19,635}{60 \times 60} = 5.45 \text{ N/mm}^2$$

which is reasonable for the crank pin bearing.

$$\text{Thickness of bush, } t_{bush} = (0.03\text{–}0.1)d_{pc} = 0.05d_{pc} = 0.05 \times 60 = 3 \text{ mm}$$

Bolts for the big end. The maximum force on the bolt at the big end cap will be the inertia force at TDC of the suction side. That is,

$$P_{inertia} = m\omega^2 r\left(\cos\theta + \frac{\cos 2\theta}{n}\right)$$

where

$$n = l/r = \frac{315}{70} = 4.5; \theta = 0°$$

$$\omega = \frac{2\pi \times \text{rpm}}{60} = \frac{2\pi \times 2500}{60} = 261.79 \text{ rad/s}$$

$$m = 1.8 \text{ kg}$$

Therefore,

$$P_{inertia} = 1.8 \times (261.79)^2 \times \frac{70}{1000}\left(1 + \frac{1}{4.5}\right) = 10,554.2 \text{ N}$$

Let us assume number of bolts, $N = 2$

Material of bolts is 35Ni1Cr60, having tensile strength 600 N/mm^2. Assume FoS = 5

$$\text{Allowable strength, } \sigma = \frac{600}{5} = 120 \text{ N/mm}^2$$

$$\text{Initial tightening force, } F_{initial} = 2.5 \times \frac{P_{inertia}}{N}$$

$$= \frac{2.5 \times 10{,}554.2}{2} = 13{,}192.75 \text{ N}$$

$$\text{Total force, } P_{\text{bolt}} = F_{\text{initial}} + \frac{KP_{\text{inertia}}}{2}$$

where K = gasket factor = 0.2 for hard gasket. Thus,

$$P_{\text{bolt}} = 13{,}192.75 + 0.2 \times \frac{10{,}554.2}{2} = 14{,}248.17 \text{ N}$$

$$\text{Resisting area, } A = \frac{P_{\text{bolt}}}{\sigma_t} = \frac{14{,}248.17}{120} = 118.73 \text{ mm}^2$$

As per IS : 4218 (Part III)—1967, M14 × 1.5 mm size bolts are suitable.

Cap of the big end. It is designed as a beam supported at the bolt centre. The cap may be assumed to be loaded at the centre by a concentrated load.

$$\text{Bending moment, } M = \frac{P_{\text{inertia}} \times c}{6}$$

c = distance between the centres of bolts

= diameter of the crank pin + 2 × thickness of the bush liner + diameter of the bolt + clearance

Thickness of bush liner = thickness of shell + thickness of bearing metal

= $0.05D$ + 2 mm = 7 mm

Therefore,

$$c = 60 + 2 \times 7 + 14 + 7 = 95 \text{ mm}$$

Empirically, the distance between the centres of bolts is kept as, $c = (1.3\text{–}1.75)d_{pc}$. Thus, c = 95 mm is within this range.

$$\text{Bending moment, } M = \frac{P_{\text{inertia}} \times c}{6} = \frac{10{,}554.2 \times 95}{6} = 167{,}108.17 \text{ N} \cdot \text{mm}$$

Assuming allowable strength of cap made of 37Mn2 as 100 N/mm^2

$$\text{Section modulus, } Z = \frac{1}{6} l_c t_C^2 = \frac{1}{6} \times 60\, t_C^2 = 10 t_C^2$$

$$\text{Bending stress, } \sigma_b = \frac{M}{Z} \le \sigma_{\text{allowable}}$$

or

$$\frac{167{,}108.17}{10 t_C^2} = 100 \quad \text{or} \quad t_C = 12.9 \text{ mm, say, 13 mm.}$$

Transverse inertia bending stress

$$\text{Bending moment, } M = \frac{2F_{\text{inertia}}\,l}{9\sqrt{3}}$$

$$F_{\text{inertia}} = \text{inertia force} = \frac{\rho A\omega^2 rl}{2}$$

$$= \frac{7800 \times 11 \times \left(\frac{5}{1000}\right)^2 \times 261.79^2 \times 0.07 \times 0.315}{2} = 1621 \text{ N}$$

$$\text{Therefore, bending moment, } M = \frac{2 \times 1621 \times 315}{9\sqrt{3}} = 65{,}511.9 \text{ N}\cdot\text{mm}$$

$$\text{Bending stress, } \sigma_b = \frac{My}{I_{XX}} = \frac{M \times 5t/2}{\frac{419}{12}t^4} = \frac{65{,}511.9 \times 5 \times 5/2}{\frac{419}{12} \times 5^4} = 37.5 \text{ N/mm}^2$$

which is within the permissible limit. Hence the design is safe.

17.5 CRANK SHAFT

The crank shaft is a principal member of an I.C. engine mechanism which is used to convert the reciprocating motion of the piston into rotary motion through the connecting rod. A crank shaft is composed of the following parts: (i) crank pin, (ii) crank web, and (iii) shaft. The crank pin is used to connect the connecting rod, the shaft rotates in main bearings which are mounted on the engine structure. The crank web connects the crank pin and the shaft.

Crank shafts are usually classified according to two criteria: (i) based on the location of the crank pin—centre crank and overhang crank and (ii) the number of crank pins used in the assembly—single-throw and multi-throw cranks. A crank shaft used in a multi-cylinder engine having more than one crank pin is called the multi-throw crank shaft. Figure 17.19 shows the schematic diagram of the various types of crank shafts.

A crank shaft should be strong enough to resist fluctuating and shock forces and rigid enough to keep the deflection and distortion within permissible limits. Generally, crank shafts are forged from medium carbon steel, namely C40, C50, 37Mn6, 35Mn6Mo4 for low-speed engines. For high-speed automotive engines, chrome–nickel, chrome–vanadium and chrome–molybdenum alloys steels are widely used.

17.5.1 Force Analysis

A crank shaft is subjected to both bending moment and torque which it receives through the connecting rod. Further, manufacturing features of crank shaft cause various discontinuties which make stress analysis very complex and cumbersome. In the present analysis, these complexities have been ignored and accounted for in the form of factor of safety.

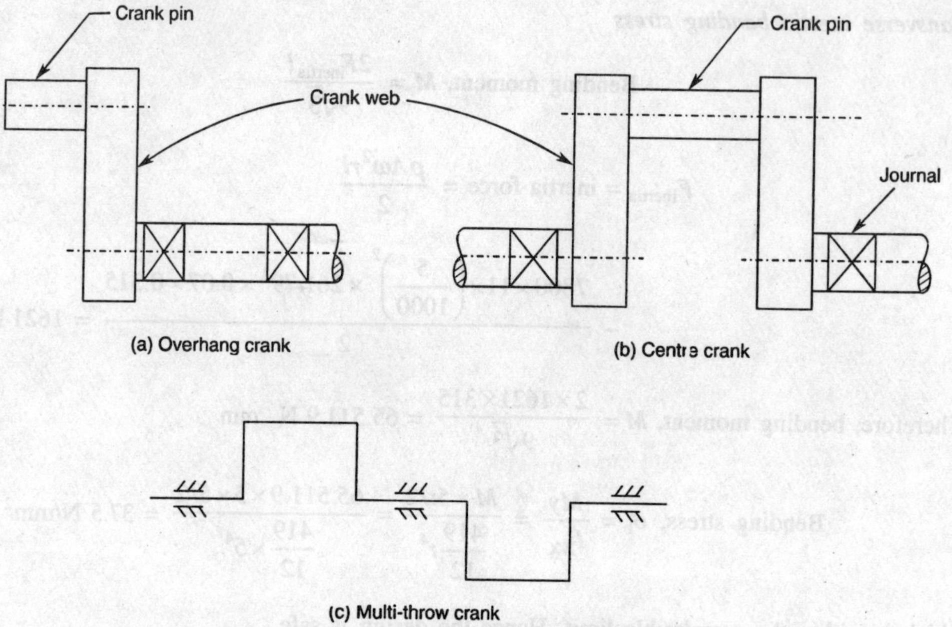

Figure 17.19 Different types of crank shafts.

In order to carryout analysis of stress due to external force, it is essential to consider the crank positions at TDC where it is subjected to maximum bending moment but the torque at that position is zero.

At other positions, both bending moment and torque act on the crank pin. Figure 17.20 shows the centre crank shaft which supports the flywheel on the right-hand side. It is also assumed that the flywheel acts as a pulley to transmit power

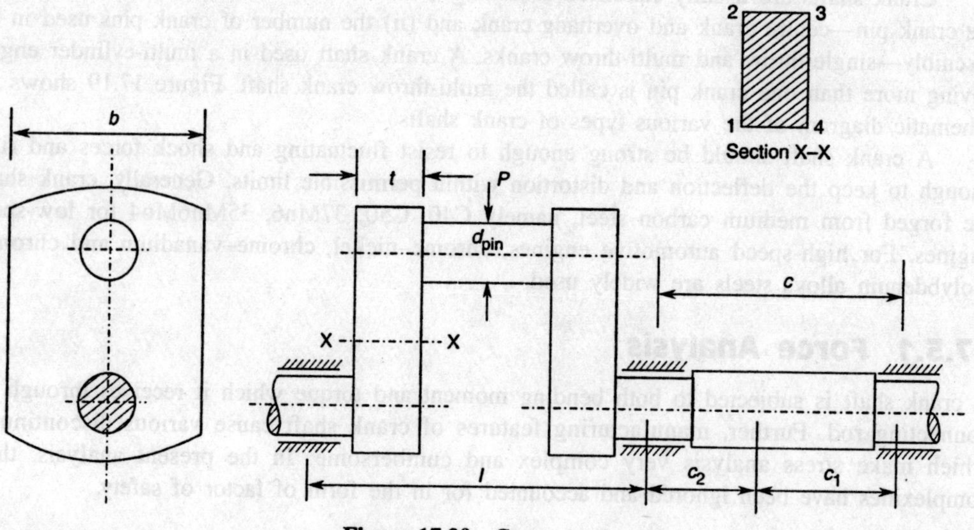

Figure 17.20 Centre crank.

Crank at TDC position

Consider a centre crank supported at three main bearings numbered 1, 2, and 3 (Figure 17.21). The crank shaft at the TDC position is subjected to gas force P, the weight of flywheel W and the sum of the tight side and slack side tensions F. Assuming that the gas force is concentrated at the midpoint of the crank pin, the reactions of force P on the bearings 1 and 2 are

$$R_{P1} = R_{P2} = \frac{P}{2} \tag{17.35}$$

The reactions at bearings 2 and 3 due to flywheel weight are

$$R_{W2} = \frac{Wc_1}{c} \quad \text{and} \quad R_{W3} = \frac{Wc_2}{c} \tag{17.36}$$

where c_1, c_2, and c are the distances as shown in Figure 17.21.

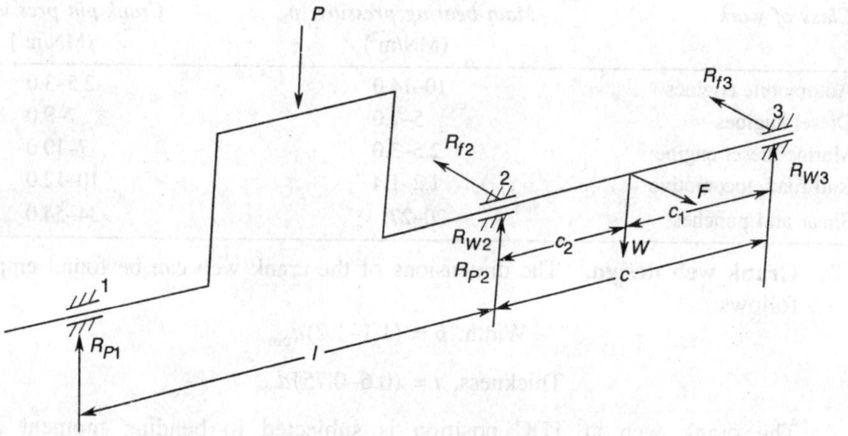

Figure 17.21 Crank shaft subjected to forces at TDC position.

Horizontal reactions at bearings 2 and 3 due to belt tension are

$$R_{f2} = \frac{Fc_1}{c} \quad \text{and} \quad R_{f3} = \frac{Fc_2}{c} \tag{17.37}$$

The resultant reaction forces at bearings 2 and 3 are

$$R_2 = \sqrt{R_{f2}^2 + \left(R_{P2} + R_{W2}\right)^2} \quad \text{and} \quad R_3 = \sqrt{R_{f3}^2 + R_{W3}^2} \tag{17.38}$$

1. **Crank pin design.** The crank pin at TDC is subjected to bending moment, bearing pressure and shear force.
   (a) *Bending stress.* Maximum bending moment on the crank pin is, $M = R_{p1} \times l/2$, and bending stress, $\sigma_b = M/Z$, where $Z = (\pi/32)\, d_{pin}^3$.
   (b) *Bearing pressure.* The crank pin is required to be checked for bearing pressure. In order to ensure proper lubrication of the rubbing surfaces, the unit bearing pressure should not exceed the permissible value (refer Table 17.2).

$$p_{br} = \frac{\text{load}}{\text{projected area}} = \frac{P}{l_{pin} \times d_{pin}} \tag{17.39}$$

For modern automobile engines, the ratio of the crank pin length to diameter (l_{pin}/d_{pin}) is generally kept between 0.8 and 1.1.

(c) *Transverse shear stress.* The maximum gas force may cause transverse shear failure of the pin, therefore, it should be checked in this mode as well. Considering double shear failure of pin

$$\tau = \frac{P}{\frac{\pi}{4} d_{pin}^2} \tag{17.40}$$

Table 17.2 Allowable bearing pressure

Class of work	*Main bearing pressure, p_{br}* (MN/m²)	*Crank pin pressure, p_{br}* (MN/m²)
Automobile engines	10–14.0	2.5–3.0
Diesel engines	5–7.0	7–9.0
Marine diesel engine	2.5–3.0	7–10.0
Rail road, locomotive	1.2–1.4	10–12.0
Shear and punches	20–27	34–54.0

2. **Crank web design.** The dimensions of the crank web can be found empirically as follows:

$$\text{Width, } b = (1.1\text{–}1.2)d_{pin} \tag{17.41}$$

$$\text{Thickness, } t = (0.6\text{–}0.75)d_{pin} \tag{17.42}$$

The crank web at TDC position is subjected to bending moment and direct compressive stresses.

(a) *Bending stress.* The bending moment on the web

$$M_1 = \frac{P}{2}\left(\frac{l}{2} - \frac{l_{pin}}{2} - \frac{t}{2}\right) \tag{17.43}$$

$$\text{Bending stress, } \sigma_{b1} = \frac{M_1}{Z_1}, \text{ where } Z_1 = \frac{1}{6}bt^2$$

The bending stress will be tensile on the face 1–2 and compressive on the face 3–4, as shown in Figure 17.20.

(b) Direct *compressive stress* due to gas force

$$\sigma_c = \frac{P}{bt} \tag{17.44}$$

Thus the maximum compressive stress at the face 3–4 should be less than the allowable limit, i.e.

$$\sigma_{c,\max} = (\sigma_{b1} + \sigma_c) \le \sigma_{allowable} \tag{17.45}$$

3. **Shaft diameter under the flywheel.** The shaft under the flywheel is subjected to maximum bending moment at the flywheel location, i.e.

$$M_{\text{shaft}} = R_3 c_1 = \frac{\pi}{32} d_{\text{shaft}}^3 \sigma_b \tag{17.46}$$

where

d_{shaft} is the shaft diameter
σ_b is the allowable strength.

Crank at an angle of maximum torque

Figure 17.22 shows the position of the crank when it transmits maximum torque. The crank angle for this position usually lies between 25° and 35° from TDC for petrol engines and

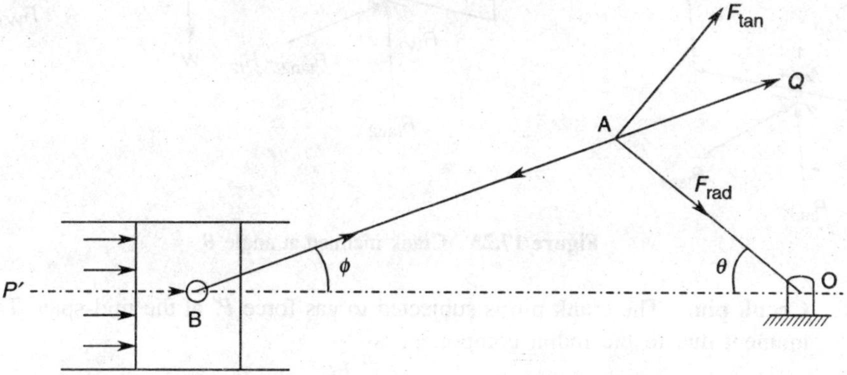

Figure 17.22 Crank at maximum torque position.

between 30° and 40° for diesel engines. At this crank angle, the gas pressure is not maximum. Let the gas force acting at that instant be P'. The force transmitted to the connecting rod is then

$$Q = \frac{P'}{\cos \phi} \tag{17.47}$$

where

$$\phi = \text{obliquity angle} = \sin^{-1}\left(\frac{\sin \theta}{n}\right)$$

$$n = \frac{l_{\text{con}}}{r}$$

The tangential force

$$F_{\tan} = Q \sin (\theta + \phi) \tag{17.48}$$

and the radial force along the crank

$$F_{\text{rad}} = Q \cos (\theta + \phi) \tag{17.49}$$

The tangential force $F_{\tan}$ will have two reactions $R_{\tan 1}$ and $R_{\tan 2}$ at bearings 1 and 2, respectively. Similarly the radial force F_{rad} will have two reactions R_{rad1} and R_{rad2}. Reactions at bearing 2 and 3 due to flywheel weight and belt tension remain the same. The free body diagram of crank shaft with various forces is shown in Figure 17.23. The crank shaft in this situation is subjected to combined bending moment and torque.

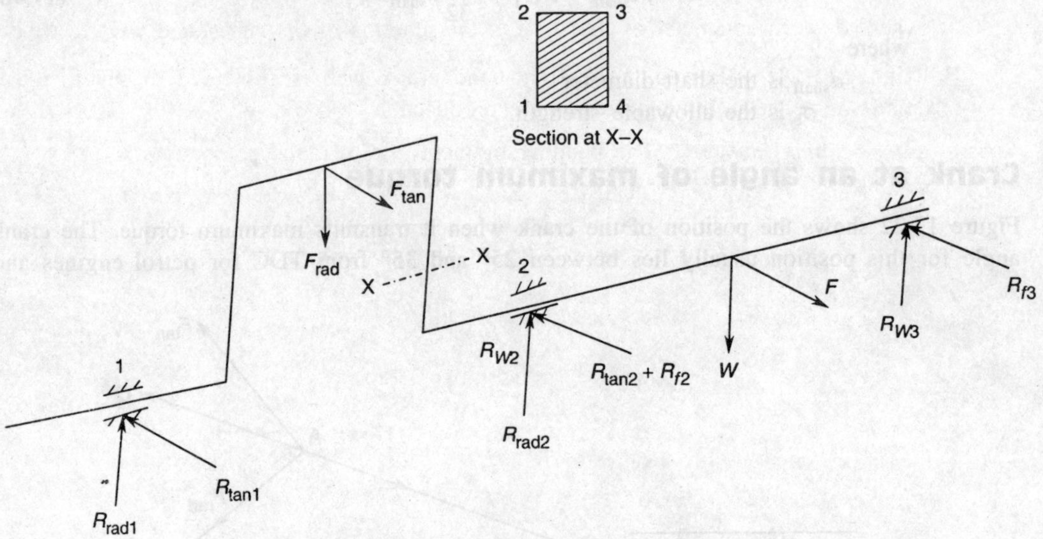

Figure 17.23 Crank inclined at angle θ.

1. **Crank pin.** The crank pin is subjected to gas force P' at the mid-span. The bending moment due to the radial component is

$$M = R_{\text{rad1}} \times \frac{l}{2}$$

and

$$\text{torque, } T = R_{\text{tan1}} \times r$$

$$\text{Equivalent torque, } T_{\text{eq}} = \sqrt{M^2 + T^2} = \frac{\pi}{16} d_{\text{pin}}^3 \times \tau \tag{17.50}$$

where τ is the allowable shear strength.

2. **Shaft under the flywheel.** The bending moment due to flywheel weight and belt tension is

$$M_{\text{shaft}} = R_3 c_1 \tag{17.51}$$

$$\text{and torque } T = F_{\tan} \times r \tag{17.52}$$

$$\text{Equivalent torque} = \sqrt{M_{\text{shaft}}^2 + T^2} \le \frac{\pi}{16} d_{\text{shaft}}^3 \times \tau \tag{17.53}$$

3. **Shaft diameter at the junction of right-hand crank web.** Bending moment

$$M = R_1 \left(\frac{l}{2} + \frac{l_{\text{pin}}}{2} + \frac{t}{2} \right) - Q \left(\frac{l_{\text{pin}}}{2} + \frac{t}{2} \right) \tag{17.54}$$

where R_1 is the resultant reaction at bearing 1.

$$\text{Torque, } T = F_{\text{tan}} \times r$$

$$\text{Equivalent torque} = \sqrt{M^2 + T^2} \le \frac{\pi}{16} d_2^3 \tau \tag{17.55}$$

where d_2 is the diameter of the shaft at the junction of the right-hand web.

4. **Right-hand crank web.** The right-hand crank web is subjected to severe load conditions. The following stresses are produced:

(a) *Bending moment.* The bending moment due to radial component is

$$M_{\text{rad}} = R_{\text{rad2}} \left(\frac{l}{2} - \frac{l_{\text{pin}}}{2} - \frac{t}{2} \right) \tag{17.56}$$

$$\text{Bending stress, } \sigma_{\text{b1}} = \frac{M_{\text{rad}}}{Z}$$

where $Z = \dfrac{bt^2}{6}$.

Face 1–2 of the right-hand web is subjected to compressive stress, whereas face 3–4 is subjected to tensile stress.

(b) *Bending moment due to F_{tan}.* The bending moment due to tangential force F_{tan} is

$$M_{\text{tan}} = \frac{F_{\text{tan}}}{2} \left(r - \frac{d_2}{2} \right) \tag{17.57}$$

where

d_2 is the shaft diameter at bearing number 2

r is the radius of crank.

$$\text{Bending stress, } \sigma_{\text{b2}} = \frac{M_{\text{tan}}}{Z}$$

where $Z = \dfrac{tb^2}{6}$.

Here σ_{b2} is the compressive stress on face 1–4 and tensile stress on face 2–3.

(c) *Radial force F_{rad}* also produces direct compressive stress, i.e.

$$\sigma_c = \frac{F_{\text{rad}}}{2bt} \tag{17.58}$$

Thus maximum compressive stress

$$\sigma'_c = \sigma_{\text{b1}} + \sigma_{\text{b2}} + \sigma_c \le \sigma_{\text{allowable}}$$

(d) Crank web is subjected to *twisting moment* due to tangential force

$$T = \frac{F_{\text{tan}}}{2} \left(\frac{l_{\text{pin}}}{2} + \frac{t}{2} \right) \tag{17.59}$$

$$\text{Torsional shear stress, } \tau = \frac{T}{Z_{\text{polar}}} \tag{17.60}$$

where Z_{polar} is the polar section modulus $\left(=\dfrac{bt^2}{4.5}\right)$

Since web is subjected to biaxial stresses, thus the maximum principal stress should be within the allowable limit.

$$\sigma_1 = 0.5\left(\sigma_c' + \sqrt{\sigma_c'^2 + 4\tau^2}\right) \le \sigma_{\text{allowable}} \tag{17.61}$$

The left-hand crank is less severely stressed than the right-hand crank web. Hence there is no need to check the stresses induced in the web. The dimensions of the left-hand web may be taken the same as those of the right-hand web.

Example 17.4 Design a single-throw centre crank for a single cylinder I.C. engine having the following data:

(i) Cylinder diameter = 120 mm
(ii) Stroke length = 160 mm
(iii) Power = 9 kW at 300 rpm
(iv) Explosion pressure = 2.5 N/mm^2
(v) Maximum torque = 25° crank angle
(vi) Explosion pressure at θ = 25° = 2.1 N/mm^2
(vii) Distance between main bearings of the crank = 360 mm
(viii) Flywheel weight = 3500 N
(ix) Flywheel is located at mid-span between bearing numbers 2 and 3 which are 400 mm apart.
(x) Belt tension = 2500 N horizontal

Solution The layout of the proposed crank shaft is shown in the figure below:

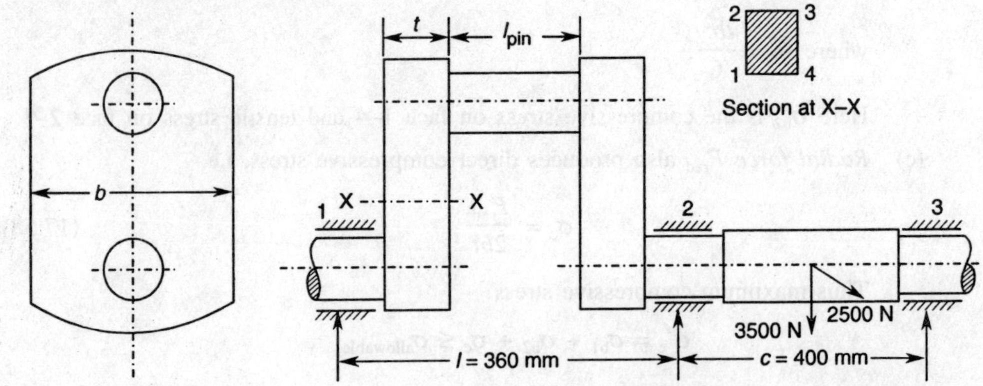

Crank at the dead centre. In this position of the crank shaft, as shown in the figure above, the thrust on the connecting rod will be equal to the maximum gas force and the same is transferred to the crank pin.

$$\text{Gas force, } P = \frac{\pi}{4} D^2 \times p_{max}$$

where

D is the bore of cylinder (= 120 mm)

p_{max} is the maximum explosion pressure (= 2.5 N/mm^2)

Therefore,

$$P = \frac{\pi}{4} 120^2 \times 2.5 = 28.27 \text{ kN}$$

Reaction at bearing numbers 1 and 2

$$R_{P1} = R_{P2} = \frac{P}{2} = 14.14 \text{ kN}$$

Reaction at bearings 2 and 3 due to flywheel weight

$$R_{W2} = R_{W3} = \frac{Wc_1}{c} = 3500 \times \frac{200}{400} = 1750 \text{ N}$$

Horizontal reaction at bearings 2 and 3 due to belt tension

$$R_{f2} = R_{f3} = \frac{2500}{2} = 1250 \text{ N}$$

The various forces acting on the crank shaft are shown in the figure below.

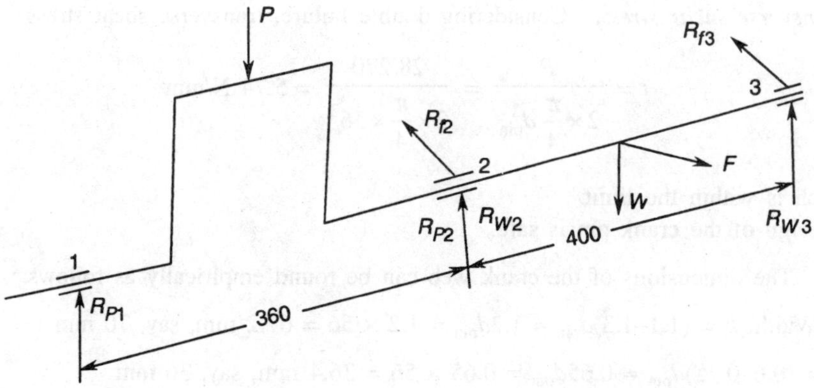

Resultant reactions at bearings 2 and 3 are

$$R_2 = \sqrt{R_{f2}^2 + \left(R_{P2} + R_{W2}\right)^2}$$

$$= \sqrt{1250^2 + (14{,}140 + 1750)^2} = 15.94 \text{ kN}$$

$$R_3 = \sqrt{R_{f3}^2 + R_{W3}^2}$$

$$= \sqrt{1250^2 + 1750^2} = 2150.6 \text{ N}$$

Crank pin. The crank pin of the crank shaft is subjected to maximum bending moment, bearing pressure, etc. when it is at the TDC position.

(a) *Bending moment.* The bending moment in crank pin is

$$M = R_{P1} \times \frac{l}{2} = 14.14 \times \frac{360}{2} = 2545.2 \text{ N} \cdot \text{m}$$

Assuming that the crank shaft is made of C40 steel having allowable strength, $\sigma_t = \frac{\sigma_{yt}}{\text{FoS}} = \frac{600}{4} = 150 \text{ N/mm}^2$

$$\text{Bending stress, } \sigma_b = \frac{M}{Z} = \frac{2545.2 \times 10^3}{\frac{\pi}{32} d_{pin}^3} = 150$$

or

$$d_{pin} = 55.7 \text{ mm, say, } 56 \text{ mm}$$

(b) *Bearing pressure.* Induced bearing pressure for $l_{pin}/d_{pin} = 1$ is

$$p_{br} = \frac{P}{l_{pin} d_{pin}} = \frac{28,270}{56 \times 56} = 9 \text{ N/mm}^2$$

which is within the acceptable limit of 7 to 15 N/mm².

(c) *Transverse shear stress.* Considering double failure, transverse shear stress

$$\tau = \frac{P}{2 \times \frac{\pi}{4} d_{pin}^2} = \frac{28,270}{2 \times \frac{\pi}{4} \times 56^2} = 5.74 \text{ N/mm}^2$$

which is within the limit.

Hence the design of the crank pin is safe.

Crank web. The dimensions of the crank web can be found empirically as follows:

$$\text{Width, } b = (1.1\text{–}1.3)d_{pin} \approx 1.2d_{pin} = 1.2 \times 56 = 67.2 \text{ mm, say, } 70 \text{ mm}$$

$$\text{Thickness } t = (0.6\text{–}0.75)d_{pin} \approx 0.65d_{pin} = 0.65 \times 56 = 36.4 \text{ mm, say, } 36 \text{ mm}$$

Stresses induced in the crank web are the following:

(a) *Bending stress.* Bending moment

$$M_1 = \frac{P}{2}\left(\frac{l}{2} - \frac{l_{pin}}{2} - \frac{t}{2}\right)$$

$$= \frac{28,270}{2 \times 10^3}\left(\frac{360}{2} - \frac{56}{2} - \frac{36}{2}\right) = 1894.1 \text{ N} \cdot \text{m}$$

$$\text{Bending stress, } \sigma_{b1} = \frac{M_1}{Z_1} = \frac{M_1}{\frac{bt^2}{6}} = \frac{1894.1 \times 10^3}{\frac{1}{6} \times 70 \times 36^2} = 125.27 \text{ N/mm}^2$$

(b) *Direct compressive stress*

$$\sigma_c = \frac{P}{bt} = \frac{28,270}{70 \times 36} = 11.2 \text{ N/mm}^2$$

Thus the maximum compressive stress at the face 3–4 is

$$\sigma_{max} = \sigma_{b1} + \sigma_c = 125.27 + 11.2 = 136.47 \text{ N/mm}^2$$

which is within the allowable limit. Hence the design is safe.

Shaft diameter under the flywheel. The portion of the shaft, i.e. that between bearing numbers 2 and 3 is subjected to maximum bending moment at the location of the flywheel.

$$M_{shaft} = R_3 \times c_1 = 2150.6 \times 0.2 = 430.12 \text{ N} \cdot \text{m}$$

$$\text{Bending stress, } \sigma_b = \frac{M_{shaft}}{Z_{shaft}} = \frac{430.12 \times 10^3}{\frac{\pi}{32} d_{shaft}^3} = 150$$

or

$$d_{shaft} = 30.8 \text{ mm}$$

Empirically the minimum diameter of the shaft should be at least equal to the crank pin diameter. Therefore,

$$d_{shaft} = d_{pin} = 56 \text{ mm}$$

Crank at an angle 25° where maximum torque occurs. The position of the crank when it transmits its maximum torque is shown in the figure below. The crank angle θ = 25°, the maximum explosion pressure at this position is p = 2.1 N/mm². Thus, the gas force at that instant

$$P' = \frac{\pi}{4} \times 120^2 \times 2.1 = 23,750.4 \text{ N}$$

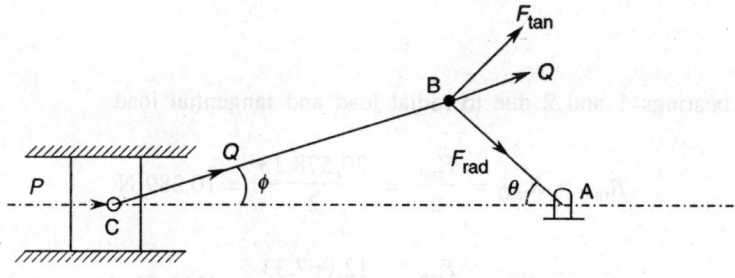

Force transmitted to the connecting rod

$$Q = \frac{P'}{\cos\phi}$$

where ϕ the obliquity angle is given by

$$\sin\phi = \frac{\sin\theta}{n}$$

If $n = \frac{l_{con}}{r} = 4.5$ (let us assume), then $\phi = 5.388°$. Therefore,

$$Q = \frac{P'}{\cos\phi} = \frac{23{,}750.4}{\cos 5.388°} = 23{,}855.4 \text{ N}$$

$$\text{Tangential force, } F_{tan} = Q \sin(\theta + \phi)$$

$$= 23{,}855.4 \sin(25° + 5.388°) = 12{,}067.33 \text{ N}$$

$$\text{Radial force, } F_{rad} = Q \cos(\theta + \phi)$$

$$= 23{,}855.4 \cos(25° + 5.388°) = 20{,}578.13 \text{ N}$$

The distribution of forces on the crank shaft is shown in the figure below:

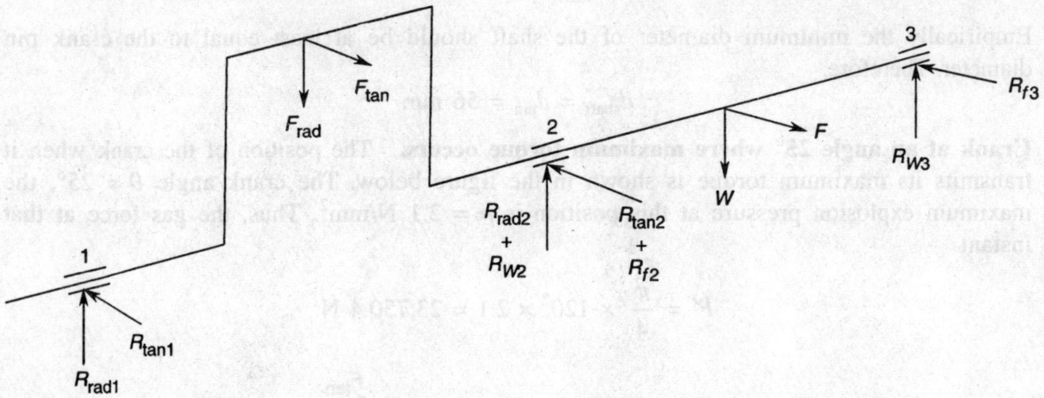

Reactions at bearings 1 and 2 due to radial load and tangential load

$$R_{rad1} = R_{rad2} = \frac{F_{rad}}{2} = \frac{20{,}578.13}{2} = 10{,}289 \text{ N}$$

$$R_{tan1} = R_{tan2} = \frac{F_{tan}}{2} = \frac{12{,}067.33}{2} = 6033.66 \text{ N}$$

Crank pin. Checking stresses in the crank pin

$$\text{Bending moment, } M = R_{\text{rad1}} \times \frac{l}{2} = 10{,}289 \times \frac{360}{2000} = 1852.02 \text{ N} \cdot \text{m}$$

$$\text{Twisting moment, } T = R_{\text{tan1}} \times r = 6033.66 \times \frac{160}{2 \times 1000} = 482.69 \text{ N} \cdot \text{m}$$

$$\text{Equivalent torque, } T_{\text{eq}} = \sqrt{M^2 + T^2} = \sqrt{1852.02^2 + 482.69^2} = 1913.89 \text{ N} \cdot \text{m}$$

$$\text{Induced shear stress, } \tau = \frac{16T_{\text{eq}}}{\pi d_{\text{pin}}^3} = \frac{16 \times 1913.89 \times 10^3}{\pi \times 56^3} = 55.5 \text{ N/mm}^2$$

which is within the allowable limit.

Shaft diameter under the flywheel. Bending moment

$$M = R_3 \times \frac{c}{2} = 2150.6 \times 0.2 = 430.12 \text{ N} \cdot \text{m}$$

Torque transmitted by shaft

$$T = F_{\text{tan}} \times r$$

$$= 12{,}067.33 \times \frac{0.16}{2} = 965.4 \text{ N} \cdot \text{m}$$

$$\text{Equivalent torque, } T_{\text{eq}} = \sqrt{430.12^2 + 965.4^2} = 1056.88 \text{ N} \cdot \text{m}$$

$$\text{Shaft diameter, } d_{\text{shaft}} = \left(\frac{16T_{\text{eq}}}{\pi\tau}\right)^{1/3} = \left(\frac{16 \times 1056.88 \times 10^3}{\pi \times 75}\right)^{1/3} = 41.55 \text{ mm}$$

Let us adopt d_{shaft} at least equal to the crank pin diameter. That is, $d_{\text{shaft}} = 56$ mm.

Shaft diameter at the junction of the right-hand web. Bending moment

$$M = R_1\left(\frac{l}{2} + \frac{l_{\text{pin}}}{2} + \frac{t}{2}\right) - Q\left(\frac{l_{\text{pin}}}{2} + \frac{t}{2}\right)$$

$$R_1 = \sqrt{R_{\text{rad1}}^2 + R_{\text{tan1}}^2} = \sqrt{10{,}280^2 + 6033.66^2} = 11{,}927.6 \text{ N}$$

Therefore,

$$M = 11{,}927.6\left(\frac{0.36}{2} + \frac{0.056}{2} + \frac{0.036}{2}\right) - 23{,}855.4\left(\frac{0.056}{2} + \frac{0.036}{2}\right)$$

$$= 1598.29 \text{ N} \cdot \text{m}$$

$$\text{Torque, } T = F_{\text{tan}} \times r = 12{,}067.33 \times \frac{0.16}{2} = 965.4 \text{ N} \cdot \text{m}$$

$$\text{Equivalent torque, } T_{eq} = \sqrt{(1598.29)^2 + (965.4)^2} = 1867.2 \text{ N} \cdot \text{m}$$

$$\text{Shaft diameter, } d_2 = \left(\frac{16T_{eq}}{\pi\tau}\right)^{1/3} = \left(\frac{16 \times 1867.2 \times 10^3}{\pi \times 75}\right)^{1/3} = 50.23 \text{ mm}$$

Let us adopt shaft diameter d_2 at least equal to 56 mm. The shaft diameter at the flywheel should be increased to 60 mm and the sharp change should be filleted by 2 mm radius. This will cause a minor rise in stresses due to stress concentration which may be neglected.

Right-hand crank web. It is subjected to severe load conditions and the following stresses are induced.

(a) *Bending moment.* The right-hand web is subjected to bending moment due to radial component

$$M_{rad} = R_{rad2}\left(\frac{l}{2} - \frac{l_{pin}}{2} - \frac{t}{2}\right)$$

$$= 10,289\left(\frac{0.36}{2} - \frac{0.056}{2} - \frac{0.036}{2}\right) = 1378.73 \text{ N} \cdot \text{m}$$

$$\text{Bending stress, } \sigma_{b1} = \frac{M}{Z} = \frac{6 \times 1378.73 \times 10^3}{70 \times 36^2} = 91.18 \text{ N/mm}^2$$

This bending stress is compressive on face 1–2 and tensile on face 3–4.

(b) *Bending due to tangential force*

$$M_{tan} = \frac{F_{tan}}{2}\left(r - \frac{d_2}{2}\right)$$

$$\approx \frac{F_{tan}}{2} \times r = \frac{12,067.33}{2} \times \frac{0.16}{2} = 482.7 \text{ N} \cdot \text{m}$$

$$\text{Bending stress, } \sigma_{b2} = \frac{M_{tan}}{Z} = \frac{6 \times 482.7 \times 10^3}{36 \times 70^2} = 16.4 \text{ N/mm}^2$$

(c) *Direct compressive stress*

$$\sigma_c = \frac{F_{rad}}{2bt} = \frac{20,578.13}{2 \times 70 \times 36} = 4.08 \text{ N/mm}^2$$

The maximum compressive stress

$$\sigma'_c = \sigma_{b1} + \sigma_{b2} + \sigma_c$$

$$= 91.18 + 16.4 + 4.08 = 111.66 \text{ N/mm}^2$$

which is less than the allowable limit. Hence the design of the crank web is safe.

(d) The crank web is also subjected to twisting moment due to tangential force

$$T = \frac{F_{\tan}}{2}\left(\frac{l_{pin}}{2} + \frac{t}{2}\right)$$

$$= \frac{12{,}067.33}{2}\left(\frac{0.056}{2} + \frac{0.036}{2}\right) = 277.5 \text{ N} \cdot \text{m}$$

$$\text{Torsional shear stress, } \tau = \frac{T}{Z_{polar}} = \frac{277.5 \times 10^3 \times 4.5}{70 \times 36^2} = 13.76 \text{ N/mm}^2$$

Considering biaxial stress condition, the maximum principal stress

$$\sigma_{max} = 0.5\left(\sigma'_c + \sqrt{\sigma_c'^2 + 4\tau^2}\right)$$

$$= 0.5\left(113.92 + \sqrt{113.92^2 + 4 \times 13.76^2}\right)$$

$$= 115.55 \text{ N/mm}^2$$

which is less than the allowable strength. Hence the design is safe. The specifications of crank shaft are thus the following:

(i) Crank pin dia, d_{pin} = 56 mm
(ii) Crank pin length, l_{pin} = 56 mm
(iii) Width of the web, b = 70 mm
(iv) Thickness of the web, t = 36 mm
(v) Shaft dia at bearings, d_{shaft} = 56 mm
(vi) Shaft dia at flywheel = 60 mm

Program 17.1

```
//    crank.cpp a program to design centre crank shaft of ic engine
#     include <iostream.h>
#     include <math.h>
#     define PI 3.1415
void main ( )
{     double bore, 1s, pmax, 1, theta, pmax_th, w, c, tension, sigmau;
      double pgas, rp1, rp2, rw2, rw3, rf2, rf3, r2, r3, bm, sigma_a, dpin;
      double 1pin, 1byd, pbr, tau, factor, b, t, bm1, sigmab, sigmac, sigmax;
      double ms, dshaft, phi, qgas, ftan, frad, c_radius, rrad1, rrad2, rtan1, rtan2;
      double torque_c, teq, torque_s, dshaft, r1, torque, d2, mr, sigmab1;
      double sigmab2, mt, tweb, sigma_p;
```

```
//    design input data
      cout <<"\n DESIGN INPUTS" << endl;
      cout <<"\n Cylinder Bore (mm)=";
      cin >> bore;
      cout <<"\n Stroke Length (mm)=";
      cin >>1s;
      cout <<"\n Explosion Pressure (N/mm2)=";
      cin >> pmax;
      cout <<"\n Distance between Crank Main Bearings=";
      cin >>1;
      cout <<"\n Crank Angle (for maximum torque)=";
      cin >> theta;
      cout << "\n Explosion Pressure (at maximum torque)=";
      cin >> pmax_th;
      cout <<"\n Flywheel Weight (N)=";
      cin >> w;
      cout <<"\n Distance Between Bearings=";
      cin >>c;
      cout <<"\n Belt Tension (N)=";
      cin >>tension;
      cout <<"\n Tensile Strength of Crank Shaft Material=";
      cin >> sigmau;
//    crank at TDC position
      pgas=(PI/4)*bore*bore*pmax;
      rp1=pgas/2;
      rp2=rp1;
//    reaction at bearing no. 2 and 3 due to flywheel weight
      rw2=w/2;
      rw3=rw2;
         rf2=tension/2;
         rf3=rf2;
//    reaction at bearing 2 & 3
      r2=pow((rf2*rf2+pow((rp2+rw2),2)),0.5);
      r3=pow((rf3*rf3+rw3*rw3),0.5);
//    crank pin design
      bm=rp1*1/2;
      sigma_a=sigmau/4; //assume fos=4;
      dpin=pow((32*bm/(PI*sigma_a)),0.3333);
      dpin=(int((dpin-0.5)/2)+1)*2;
//    bearing considerations
      1byd=1.0;
repeat:
      1pin=1byd*dpin
      pbr=pgas/(1pin*dpin);
      if (pbr > 15)
```

```
        {
           1byd=1byd+0.05;
           if (1byd<=1.3)
                   goto repeat;
           else
           {
                   dpin=dpin+1;
                   goto repeat;
           }
        }
//      check transverse shear stress
        tau=2*pgas/(PI*dpin*dpin);
        while (tau >= sigma_a/2)
        {
           dpin=dpin+1;
           tau=2*pgas/(PI*dpin*dpin);
        }
        1pin=1byd*dpin;
//      crank web design
        factor=1.1;
loop1:
        b=factor*dpin
        t=b/1.8;
//      check stresses in web
//      1. Bending Stress
        bm1=(pgas/2)*(1/2-lpin/2-t/2);
        sigmab=bm1/(b*t*t/6);
//      2. Direct compression
        sigmac=pgas/(b*t);
        sigmax=sigmab+sigmac;
        if (sigmax >=sigma_a)
        {
           factor=factor+0.05;
           if (factor <=1.3)
                   goto loop1;
           else
           {
                   dpin=dpin+1;
                   goto loop1;
           }
        }
        b=(int((b-0.5)/2)+1)*2;
        t=(int((t-0.5)/2)+1)*2;
//      shaft dia under the flywheel dsf
        ms=r3*c/2;
        dshaft=pow((32*ms/(PI*sigma_a)),0.3333);
```

```
        if (dshaft < dpin)
            dshaft=dpin;
        else
            dshaft=(int((dshaft-0.5)/2)+1)*2;
// crank at maximum torque position
        pgas=(PI/4)*bore*bore*pmax_th;
        theta=theta*PI/180;
        phi=asin(sin(theta)/4.5);//assume 1/r ratio=4.5
        qgas=pgas/cos(phi);
        ft=qgas*sin(theta+phi);
        fr=qgas*cos(theta+phi);
//      calculate reaction at main bearings
        c_radius=1s/2;
        rrad1=fr/2;
        rrad=rrad1;
        rtan=ft/2;
        rtan=rtan1;
//      check stress in crank pin
        bm=rrad1*1/2;
        torque_c=rtan1*c_radius;
        teq=pow((bm*bm+torque_c*torque_c), 0.5);
        tau=16*teq/(PI*pow(dpin,3.0));
        while(tau>=sigma_a/2)
        {
            dpin=dpin+1;
            tau=16*teq/(PI*pow(dpin,3.0));
        }
        lpin=lbyd*dpin
//      shaft under flywheel
        bm=r3*c2;
        torque_s=ft*c_radius;
        teq=pow((bm*bm+torque_s*torque_s),0.5);
        tau=sigma_a/2;
        dshaft=pow((16*teq/(PI*tau)),0.3333);
        if (dshaft <=dpin)
            dshaft=dpin;
        else
            dshaft=(int((dshaft-0.5)/2)+1)*2;
//      shaft dia at junction of right hand bearing
        r1=pow((rrad1*rrad1+rtan1*rtan1),0.5);
        bm=r1*(1/2+lpin/2+t/2)-qgas*(lpin/2+t/2);
        torque=ft*c_radius;
        teq=pow((bm*bm+torque*torque),0.5);
        d2=pow((16*teq/(PI*tau)),0.3333);
        if (d2 < dpin)
            d2=dpin;
```

```
        else
            d2=(int((d2-0.5)/2)+1)*2;
//      stress in right web
//      1. bm due to radial component
loop2;
        lpin=lbyd*dpin;
        mr=rrad2*(1/2-lpin/2-t/2);
        sigmab1=mr/(b*t*t/6);
//      2. bm due to tangential component
        mt=ft/2*(c_radius-d2/2);
        sigmab2=mt/(t*b*b/6);
//      3. direct compressive stress
        sigmac=fr/(b*t);
        sigmax=sigmab1+sigmab2+sigmac;
        if (sigmax > sigma_a)
        {
            factor=factor+0.05;
            if (factor <=1.3)
                    goto loop2;
            else
            {
                    dpin=dpin+1;
                    goto loop2;
            }
        }
//      torsional stress in web
        tweb=ft/2*(lpin/2+t/2);
        tau=tweb*4.5/(b*t*t);
        sigma_p=0.5*(sigmax+pow((sigmax*sigmax+4*tau*tau),0.5));
        if (sigma_p > sigma_a)
        {
            factor=factor+0.05;
            if (factor <=1.3)
                    goto loop2;
            else
            {
                    dpin=dpin+1;
                    goto loop2;
            }
        }
//      print design report
        cout <<"\n DESIGN REPORT" <<endl;
        cout <<"\n Crank Pin Diameter=" << dpin << "mm";
        cout <<"\n Length of Crank Pin=" <<lpin << "mm";
        cout <<"\n Width of Web="<<b<<"mm";
        cout <<"\n Thickness of Web=" <<t<< "mm";
```

```
    cout <<"\n Shaft Diameter at Main Bearing="<<d2<<"mm";
    cout <<"\n Shaft diameter at Flywheel="<<d2+4<<"mm"<<endl;
}
```

Sample run of Program 17.1

```
DESIGN INPUTS

Cylinder Bore (mm)                            =120
Stroke Length (mm)                            =160
Explosion Pressure (N/mm2)                    =2.5
Distance between Crank Main Bearings          =360
Crank Angle (for maximum torque)              =25
Explosion Pressure (at maximum torque)        =2.1
Flywheel Weight (N)                           =3500
Distance Between Bearings                     =400
Belt Tension                                  =2500
Tensile Strength of Crank Shaft Material      =600

DESIGN REPORT

Crank Pin Diameter                     =56mm
Length of Crank Pin                    =56mm
Width of Web                           =68mm
Thickness of Web                       =38mm
Shaft Diameter at Main Bearing         =56mm
Shaft Diameter at shaft Flywheel       =60mm
```

EXERCISES

1. What are the main functions of the cylinder liner?
2. What do you mean by blow-by operation?
3. What is piston slap and how can it be controlled?
4. What is the effect of piston crown thickness and diameter on heat flow?
5. What are the basic functions of piston rings?
6. How is the wear of the piston rings prevented?
7. Why is a hollow piston pin preferred to a solid one?
8. Which type of the cross-section do you prefer for the main body of the connecting rod and why?
9. Why is one end of the connecting rod bigger than the other end?
10. Enumerate the design considerations for crank pin.
11. What types of crank shafts are commonly used?

12. Design a cylinder for a four-stroke petrol engine developing 15 kW at 2000 rpm. Assume indicated mean effective pressure to be 1.2 N/mm^2.

13. Design a cylinder for an 1100 cc six-cylinder car engine with the following data:

    (i) Power 40 kW at 4400 rpm
    (ii) Mean effective pressure = 1 N/mm^2

14. Design and draw a trunk type of piston for a single cylinder four-stroke diesel engine running at 1000 rpm. Other data available are:

    (a) Maximum explosion pressure = 3.5 MN/m^2
    (b) Mean effective pressure = 0.65 MN/m^2
    (c) Diameter of piston = 150 mm
    (d) Stroke length = 200 mm
    (e) Connecting rod length = 450 mm
    (f) bsfc = 0.27 kg/kWh

15. The following projected data refer to a four-cylinder petrol engine of a car to be designed:

    Diameter of the piston = 68 mm
    Stroke length = 75 mm
    Maximum pressure = 2.5 MN/m^2
    Connecting rod length = 175 mm
    Brake power = 32 kW at 5000 rpm
    bsfc = 0.33 kg/kWh

    Design a suitable piston.

16. Design a connecting rod for a petrol engine from the following data:

    Diameter of the piston = 120 mm
    Weight of the reciprocating part = 2.0 kg
    Length of the connecting rod = 300 mm
    Stroke length = 140 mm
    Speed = 2000 rpm
    Maximum explosion pressure = 2.25 N/mm^2

17. Design a suitable connecting rod for a car with the following data:

    Piston diameter = 68 mm
    Stroke length = 80 mm
    Length of the connecting rod = 160 mm
    Maximum explosion pressure = 3.5 N/mm^2
    Weight of the reciprocating part = 2.5 kg
    Speed = 4000 rpm
    Compression ratio = 8:1

18. Design an overhang crank shaft with two main bearings for an I.C. engine with the following data:

    Cylinder bore = 250 mm
    Stroke length = 300 mm
    Flywheel weight = 27 kN

Maximum pressure = 2.5 N/mm^2
Maximum torque at crank rotation 30°, the pressure at that instant = 1.7 N/mm^2

19. Design a single-throw, double-view crank shaft made of forged steel for a single cylinder, vertical I.C. engine having cylinder diameter of 120 mm and stroke length of 160 mm. The engine develops 10 kW at 300 rpm. The explosion pressure is 2.5 N/mm^2 gauge. The maximum torque is developed when the crank shaft turns through 25° from the TDC position during the expansion stroke. The burnt gas pressure at that moment is 2.0 N/mm^2. The crank shaft main bearings are 320 mm apart.

MULTIPLE CHOICE QUESTIONS

1. Which engine has a high thermal efficiency?
   (a) Petrol engine (b) Diesel engine
   (c) Steam engine (d) None of the above
2. Which of the following is not a part of the I.C. engine?
   (a) Cylinder (b) Piston
   (c) Piston rod (d) Crank shaft
3. Low temperature at the end of combustion in the diesel engine is due to
   (a) high thermal efficiency (b) high cooling rate
   (c) high air factor (d) low air factor
4. Erosion of metal from the piston due to local welding is called
   (a) abrasive wear (b) pitting
   (c) scuffing (d) piston slap
5. To avoid the phenomenon of piston slap, the top diameter of the piston should be
   (a) larger (b) smaller
   (c) equal everywhere (d) unpredictable
6. For smooth and hard surface cylinder liner, it should be
   (a) annealed (b) hardened
   (c) chrome-plated (d) normalized
7. The design of piston crown is based on
   (a) strength and rigidity (b) rigidity and heat transfer
   (c) heat transfer (d) strength and heat transfer
8. Usually, the section of the connecting rod shank is
   (a) rectangular (b) circular
   (c) I-section (d) T-section
9. For equal strength of the connecting rod section, which of the following criterion should be satisfied?
   (a) $I_{XX} = 4I_{YY}$ (b) $I_{YY} = 4I_{XX}$
   (c) $I_{XX} = 0.25I_{YY}$ (d) $I_{XX} = I_{YY}$

10. A small length of the connecting rod results into
    (a) small obliquity and small side thrust
    (b) small obliquity and large side thrust
    (c) large obliquity and small side thrust
    (d) large obliquity and large side thrust
11. Whipping stress is called
    (a) crippling stress
    (b) stress due to transverse bending
    (c) stress due to reciprocating mass inertia
    (d) none of the above
12. In the plane of motion, the ends of the connecting rod are treated as
    (a) both fixed (b) one fixed other hinged
    (c) one fixed other free (d) both hinged
13. Maximum load on bolts of the connecting rod cap
    (a) occurs at TDC of the suction stroke
    (b) occurs at TDC of the expansion stroke
    (c) occurs at BDC of the compression stroke
    (d) is unpredictable.

Chapter 18

Flywheel and Rotating Disc

18.1 INTRODUCTION

A flywheel is a heavy rotating mass which is placed between the power source and the driven machine to act as a reservoir of energy. It stores up energy when the demand for energy is less than the availability and delivers energy when there is a lean period. In other words whenever there is excess or deficit of energy, the speed of the driven machine will fluctuate. This fluctuation of speed can be controlled by the flywheel within a certain range.

Depending upon the source of power and the type of driven machinery, there are three distinct situations where a flywheel is necessitated.

(i) When the availability of energy is at a fluctuating rate, but the requirement of it for the driven machinery is at uniform rate, a flywheel is needed to store surplus energy (shown as hatched area in Figure 18.1) to be delivered during the lean period. Figure 18.1 shows typical energy requirement and availability curves for an I.C. engine driven water pump or rotary compressor.

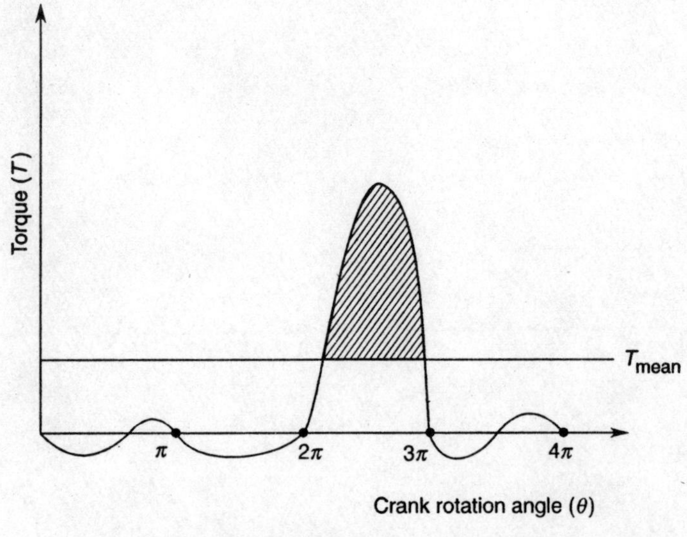

Figure 18.1 The T–θ diagram of I.C. engine.

(ii) In other applications, namely electric motor driven punching, shearing and riveting machines, rolling mills etc., though the energy is available at a uniform rate, the demand for it is variable. Figure 18.2 shows a typical energy requirement and availability curve for an electric motor driven rolling mill which shows that for a small fraction of the cycle period there is huge requirement of energy. Thus again a flywheel is needed. The variation of angular speed during the cycle period is shown in Figure 18.2(b).

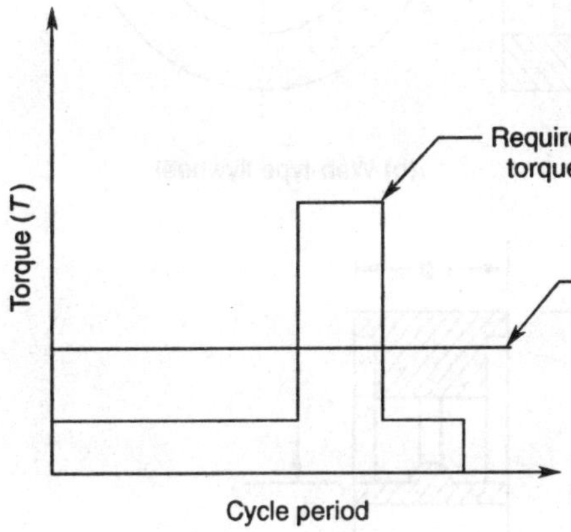

Figure 18.2(a) Torque diagram.

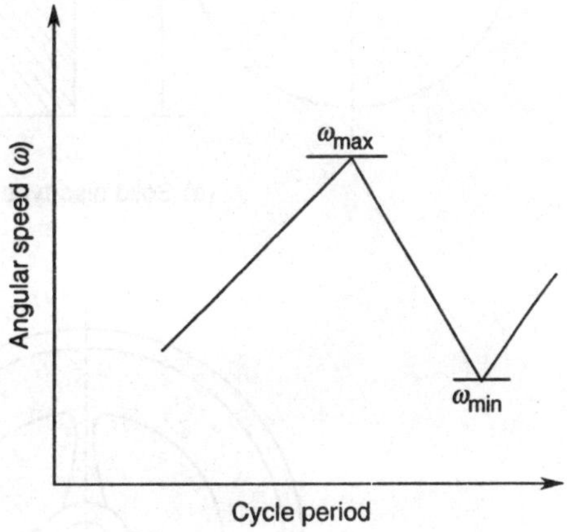

Figure 18.2(b) Angular speed variation.

(iii) In the third situation, both the requirement and availability of energy represent a variable rate, e.g. I.C. engine driven reciprocating air compressor or pump.

Flywheels are made in three different types:

Disc-type flywheel. This type of flywheel is of solid disc type, as shown in Figure 18.3(a). A typical example is the flywheel used in an automotive vehicle where one of the faces of the flywheel acts as a friction surface for the clutch.

Web-type flywheel. It consists of a heavy rim connected to the hub by a disc-shaped plate called web, as shown in Figure 18.3(b). This type of flywheel is mostly used with small-power vertical I.C. engines.

Arm-type flywheel. It is the most common type of flywheel used in cases where a large size flywheel is needed. It consists of a heavy rim which is connected to the hub by a large number of radial arms. An arm type flywheel, up to 2 m diameter, is usually casted in a single piece.

18.2 FLYWHEEL EFFECT

A flywheel stores up energy in the form of kinetic energy of a rotating mass, which is given by

$$E = \frac{1}{2} I\omega^2 \tag{18.1}$$

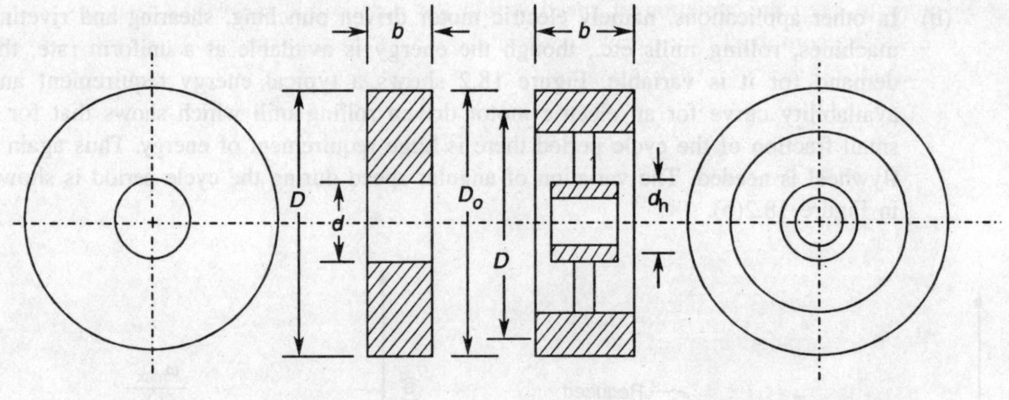

(a) Solid disc-type flywheel (b) Web-type flywheel

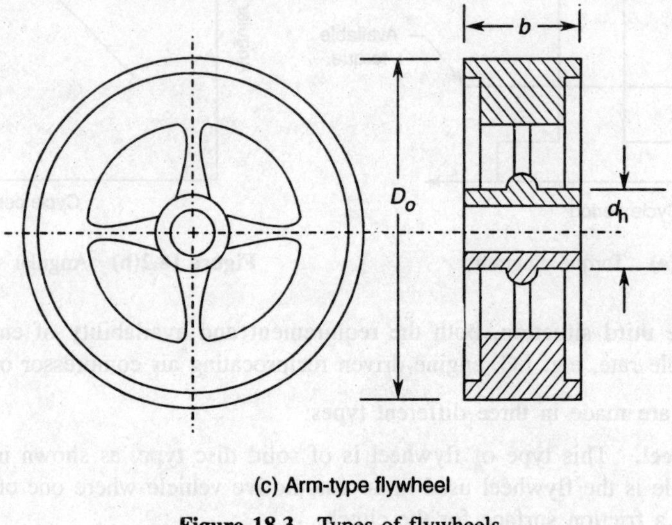

(c) Arm-type flywheel

Figure 18.3 Types of flywheels.

where

I is the mass moment of inertia of the flywheel (= mk^2)
m is the mass of the rotating body
k is the radius of gyration of the rotating body
ω is the angular speed.

The mass moment of inertia, required for a flywheel, is termed *flywheel effect.*

When the input power to an application is more than that required during a part of a cycle, the flywheel is accelerated and its angular velocity increases to ω_{max}. During the lean or power deficit period, the flywheel is retarded and its angular velocity decreases to ω_{min}, as shown in Figure 18.2(b).

Thus, the change in kinetic energy of the flywheel is given as

$$\Delta KE = \frac{1}{2} I(\omega^2_{max} - \omega^2_{min}) \tag{18.2}$$

In order to keep the variation of speed within the permissible range, the fluctuation of energy (ΔE) of the combined driver/driven system should be equal to change in kinetic energy. Thus fluctuation of energy

$$\Delta E = \frac{1}{2} I(\omega^2_{max} - \omega^2_{min})$$

or

$$\Delta E = I\omega^2 C_S \tag{18.3}$$

where

C_S is the coefficient of fluctuation of speed [= $(\omega_{max} - \omega_{min})/\omega$]

ω is the mean angular speed [= $(\omega_{max} + \omega_{min})/2$]

The value of coefficient of fluctuation of speed depends on the permissible variation between the highest and the lowest speeds during the operating cycle of the driven machine and on the method of connecting it to the driving power source. Table 18.1 gives typical values of coefficient of fluctuation of speed.

Table 18.1 Coefficient of fluctuation of speed (C_S)

Driven machinery	C_S
Punch press	0.06–0.2
Pumps	0.03–0.05
DC generator	0.007
AC generator	0.003
Automobile idling	0.2
Automobile normal running	0.1

Substituting $I = mk^2$ in Eq. (18.3), we get

$$\text{Mass of flywheel, } m = \frac{\Delta E}{k^2\omega^2 C_S} \tag{18.4}$$

or

$$\text{Weight of the flywheel, } w = \frac{g\Delta E}{k^2\omega^2 C_S} \tag{18.5}$$

The value of radius of gyration k depends upon the type of flywheel used in a particular application:

$k = R$, mean radius of the flywheel for the case where the rim thickness is small compared to diameter of the flywheel, e.g. arm and web type flywheels

$= \frac{R}{\sqrt{2}}$, for solid disc type flywheels.

In order to determine the mass/weight of the flywheel, it is essential to know the maximum value of fluctuation of energy ΔE, to be absorbed by the flywheel. This value can be computed either graphically or analytically from known characteristics of the power source and driven machinery. However in some applications, namely I.C. engine the fluctuation of energy can be determined by the use of coefficient of fluctuation of energy C_E, which is defined as the ratio of change of energy ΔE to mean energy E per revolution or per cycle, i.e.

$$C_E = \frac{E_{max} - E_{min}}{E} = \frac{\Delta E}{E} \tag{18.6}$$

The values of coefficient of fluctuation of energy C_E for various types of I.C. engines are listed in Table 18.2.

Table 18.2 Coefficient of fluctuation of energy (C_E)

Type of engine	C_E	
	Four-stroke cycle	*Two-stroke cycle*
Single-cylinder engine	2.35–2.4	0.95–1.00
Two-cylinder engine with crank 180° apart	1.5–1.6	0.2–0.25
Four-cylinder engine with crank 180° or 90° apart	0.15–0.2	0.075–0.16

18.3 DISC-TYPE FLYWHEEL

The simplest type of flywheel is a solid circular disc as shown in Figure 18.3(a). This type of flywheel is widely used in automotive vehicles because its flat face can act as friction surface for the clutch.

The mass of the flywheel is given by

$$m = \pi R^2 b\rho \tag{18.7}$$

where

b is the width of the disc

ρ is the mass density.

When a circular disc flywheel rotates at high speed, two types of stresses are set up. They are radial stress σ_{rad} and tangential stress σ_θ. The numerical equations for these stresses at any radius r are given below:

$$\sigma_{rad} = \frac{3+\mu}{8}\rho\omega^2 (R^2 - r^2) \tag{18.8}$$

$$\sigma_\theta = \frac{\rho\omega^2}{8}[(3+\mu)R^2 - (1+3\mu)r^2] \tag{18.9}$$

18.4 RIM-TYPE FLYWHEEL

A rim-type flywheel usually consists of a rim and hub which are connected either by a circular disc called *web* or a certain number of spokes called *arms*. Accordingly, a rim-type flywheel is classified either as web type or as arm type. In most cases, an arm-type flywheel is used where the required diameter of the flywheel is more than 600 mm.

While designing a rim-type flywheel, it is difficult to determine its exact mass moment of inertia due to its complex geometric shape. Therefore, the analysis of such a flywheel is done on the basis of the assumption that the effect of arms or web, hub and the shaft is to contribute only 10 per cent of the total moment of inertia and the remaining 90 per cent is contributed by the flywheel rim. Therefore,

$$\text{moment of inertia of rim, } I_{\text{rim}} = 0.9I = m_{\text{rim}}k^2 \tag{18.10}$$

where m_{rim} is the mass of the rim.

18.4.1 Stresses in Rim

In the design of flywheel it is required to decide the mean radius of the flywheel rim, which primarily depends upon two factors: (1) availability of space and (ii) the limiting value of peripheral velocity. For a grey cast iron flywheel, the limiting velocity is usually 25 m/s. If the peripheral velocity of the flywheel exceeds this limit, there is a possibility of its bursting due to centrifugal force. Therefore, it becomes necessary to compute stresses developed in the flywheel rim due to centrifugal force.

Besides the centrifugal stress, a flywheel is also subjected to other types of stresses, namely bending stress produced due to the restraining effect of arms, shrinkage stress due to solidification of casting and stresses due to variations of load and speed. Thus, the determination of actual stresses in a flywheel is very complicated and only an approximation can be made to compute stresses in the two extreme cases by neglecting stresses due to shrinkage, speed and load variation. That is: (i) rim unstrained by the arms and (ii) rim restrained by the arms.

Rim unstrained by arms

The tensile or hoop stress in the rim of a flywheel due to centrifugal force, assuming that the rim is unstrained by arms, is determined by considering it as a thin ring rotating about an axis through its centre of gravity and perpendicular to its central plane. It is assumed that the thickness of the rim is small compared with the radius of the rim and the hoop stress is nearly uniform.

When a rim of a flywheel having mean radius R rotates at angular velocity ω rad/s, at any point, say at radius r, the rim will have a radial inward acceleration equal to $\omega^2 r$ or v^2/r. This inward radial acceleration will give rise to centripetal force which is resisted by an equal and opposite force called the centrifugal force produced due to inertia of the flywheel rim. The centrifugal force acts like an internal pressure in a thin cylinder trying to burst it into two halves.

Consider a small element ABCD of the rim subtending an angle $d\theta$ at the centre as shown in Figure 18.4.

The vertical component of force dF, which tries to burst the rim along section X–X is

$$dF_V = bh\rho v^2 \sin\theta \cdot d\theta$$

$$\text{Total vertical force, } F = \int_0^{\pi} bh\rho v^2 \sin\theta \cdot d\theta = 2bh\rho v^2$$

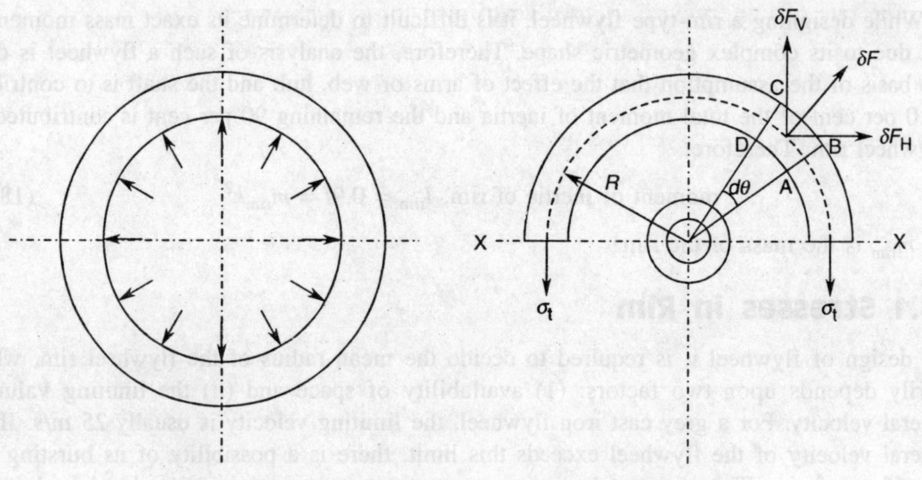

Figure 18.4 Centrifugal force acting on unstrained rim.

Therefore, hoop stress σ_t induced at section X–X due to resisting area *2bh* is

$$\sigma_t = \rho v^2 \tag{18.11}$$

The hoop stress is independent of cross-section of the rim and wholly depends upon the velocity of rim, which should not increase beyond the limiting value

$$v_{\text{limiting}} = \left(\frac{\sigma_t}{\rho}\right)^{0.5} \tag{18.12}$$

Rim Restrained by arms

In the rim of a flywheel, stresses may also be produced due to restrain of the rim between a pair of arms which behaves like a fixed beam subjected to a uniformly distributed centrifugal force, as shown in Figure 18.5.

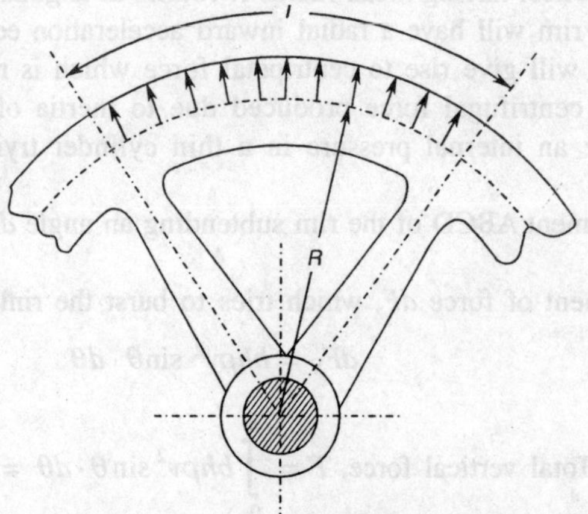

Figure 18.5 Centrifugal force in a restrained rim.

The rate of uniformly distributed centrifugal force is given by

$$w = \frac{bh\rho v^2}{R} \text{ N/m} \tag{18.13}$$

and the length of rim between two arms

$$l = 2\pi R/i$$

where i is the number of arms.

The bending stress due to bending moment for the fixed beam subjected to a uniformly distributed load ($wl^2/12$) is computed by the relation

$$\sigma_b = \rho v^2 \left(\frac{2\pi^2 R}{i^2 h}\right) \tag{18.14}$$

However, Prof. G. Lanza has reported that the actual stress lies in between the value of two extreme cases. That is: (i) unstrained rim and (ii) restrained rim. The resultant stress in the rim at the junction with the arm is approximately given by

$$\sigma_{res} = 0.75\sigma_t + 0.25\sigma_b \tag{18.15}$$

For safe design of the flywheel rim, the resultant stress should be less than or equal to allowable strength of the flywheel. Generally, for cast iron flywheels $\sigma_{res} \le 40$ MN/m^2.

18.4.2 Stresses in Arms

Stresses induced in an arm of the flywheel due to three main types of forces, besides shrinkage stresses due to unequal rate of cooling of the flywheel, are discussed below:

Bending stress due to torque

When a flywheel is accelerated from its rest position or when power supply to the flywheel is cut off, the arms have to carry the full torque. In such a situation, each arm may be treated as a cantilever beam which is fixed at the hub end and carries a concentrated tangential force F_t at the rim end. The bending stress in the cantilever arm due to tangential force is given as under:

$$\sigma_{t1} = \frac{T(D - d_h)}{iDZ} \tag{18.16}$$

where

d_h is the hub diameter

Z is the section modulus ($= (\pi/32)a_1^2 b_1$) for elliptical sections having major axis a_1 and minor axis b_1 (see Figure 18.6).

The allowable limit for bending stress is

$$\sigma_{t1} \le 7.0 \text{ to } 14.0 \text{ MN/m}^2$$

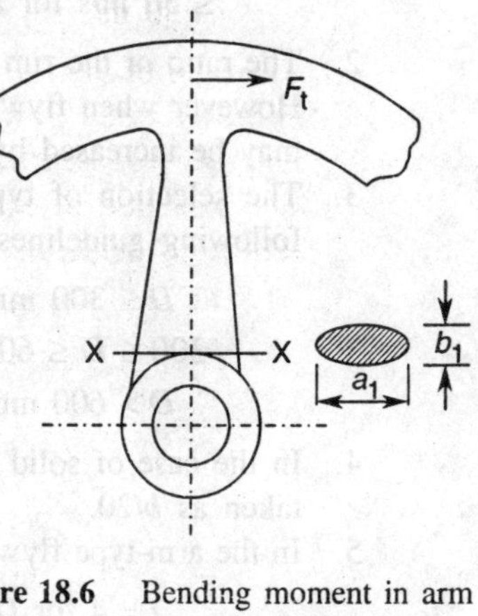

Figure 18.6 Bending moment in arm due to tangential force.

Stress due to belt tension

In some applications, the flywheel is used as a pulley of a belt drive to transmit power. In such a situation, the arms of the flywheel get bent not only due to speed variation but also due to net belt tension. If the tight side and slack side tensions in the belt are T_1 and T_2, respectively, and arms are not rigid enough, it is assumed that only half the total number of arms carry the load. The bending stress produced in the arm due to belt action is

$$\sigma_{t2} = \frac{(T_1 - T_2)(D - d_h)}{iZ} \tag{18.17}$$

Stress due to centrifugal force

Arms are subjected to direct tensile stress due to centrifugal force acting upon the rim when the flywheel is running at its maximum velocity. It is given as

$$\sigma_{t3} = \rho v^2 \tag{18.18}$$

Thus, the maximum tensile stress in the arm at the hub end is

$$\sigma_{max} = \sigma_{t1} + \sigma_{t2} + \sigma_{t3} \tag{18.19}$$

For safe design, the maximum stress in the arm should be less than the allowable strength of the flywheel material. For cast iron

$$\sigma_{max} \le 20 \text{ MN/m}^2$$

18.5 DESIGN CONSIDERATIONS

While designing a flywheel, the following points should be considered.

1. The maximum peripheral velocity of the flywheel is governed by the allowable strength of the flywheel material. Generally, the following rim velocities are considered to be safe.

   $v \le 25$ m/s for ordinary grey CI flywheel

   ≤ 35 m/s for good quality casting free from defects, viz. blow holes etc.

   ≤ 50 m/s for automotive vehicle engines.

2. The ratio of the rim width to thickness of a flywheel should range between 0.65–2.0. However when flywheel is used as a pulley to transmit power, the width of flywheel may be increased by 25 to 50 mm.
3. The selection of type of flywheel primarily depends upon its mean diameter. The following guidelines may be used.

   $D \le 300$ mm, a solid disc type flywheel

   $300 \le D \le 600$ mm, a web type flywheel

   $D > 600$ mm, an arm type flywheel

4. In the case of solid web type of flywheels, the web thickness may be approximately taken as $b/20$.
5. In the arm-type flywheel, the number of arms depends upon diameter of the flywheel.

   $i = 4$ for $D < 0.75$ m

   $i = 6$ to 8 for $0.75 \text{ m} < D \le 2.0$ m

   $i = 8$ to 12 for $D > 2.0$ m

6. Usually, the cross-section of the arm is elliptical with major axis twice the minor. The major axis is kept in the plane of rotation.

Example 18.1 Design a flywheel for a single-cylinder, four-stroke vertical cylinder diesel engine developing 4 kW at 1500 rpm. Assume coefficient of speed fluctuation, $C_S = 0.01$.

Solution From Table 18.2, for a single-cylinder, four-stroke engine, the coefficient of fluctuation of energy, $C_E = 2.35$.

$$\text{Fluctuation of energy, } \Delta E = \frac{C_E \times 1000P}{N/60}$$

$$= \frac{2.35 \times 1000 \times 4.0}{1500/60} = 376 \text{ N} \cdot \text{M}$$

$$\text{For flywheel, } \Delta E = I\omega^2 C_S = mv^2 C_S$$

For grey cast iron flywheel, the limiting speed $v = 25$ m/s

$$\text{Mean rim diameter, } D = \frac{60v}{\pi N} = \frac{60 \times 25}{\pi \times 1500} = 0.3183 \text{ m say, 320 mm}$$

$$\text{Mass of flywheel, } m = \frac{\Delta E}{v^2 C_S} = \frac{376}{25^2 \times 0.01} = 60.16 \text{ kg}$$

Assume that mass of the rim is 90% of the total mass and the remaining 10% is contributed by web and hub. Thus,

$$m_{rim} = 0.9m = 0.9 \times 60.16 = 54.144 \text{ kg} = \pi Dbh\rho$$

or

$$\text{Rim cross-section, } bh = \frac{m_{rim}}{\pi D\rho} = \frac{54.144}{\pi \times 0.32 \times 7100} = 7.586 \times 10^{-3} \text{ m}^2 \quad (\text{with } \rho = 7100 \text{ kg/m}^3)$$

Let width to thickness ratio, $b/h = 1.5$. Thus,

$$1.5h \times h = 7.586 \times 10^{-3} \quad \text{or} \quad h = 0.071 \text{ m, say, 70 mm}$$

Hence rim width, $b = 1.5 \times 70 = 105$ mm, say, 110 mm.
Since the mean rim diameter is 320 mm, we select the web-type flywheel.
Stress in the rim, assuming that the rim is unstrained

$$\sigma_t = \rho v^2 = \frac{7100 \times 25^2}{10^6} = 4.43 \text{ MN/m}^2$$

which is reasonable and less than the allowable strength; hence the design of the rim is safe.

$$\text{Web thickness, } t = b/20 = 110/20 = 5.5 \text{ mm, say, 6 mm}$$

*Shaft diameter** – Let us assume that shaft is overhang by some amount and acts as a cantilever beam

* Actual shaft diameter depends upon the design of the crank shaft. For detailed design, refer Chapter 17.

$$\text{Overhang} = \frac{\text{length of bearing}}{2} + \frac{\text{width}}{2} + \text{clearance}$$

Since the length of bearing is not known, we assume that the overhang length l = 200 mm.

$$\text{Bending moment, } M = \text{Weight of flywheel} \times \text{Overhang}$$
$$= 60.16 \times 9.81 \times 0.2 = 118.03 \text{ N} \cdot \text{m}$$

$$\text{Mean torque, } T = \frac{60P}{2\pi N} = \frac{60 \times 4.0 \times 10^3}{2\pi \times 1500} = 25.46 \text{ N} \cdot \text{m}$$

We assume that shaft is subjected to heavy shock load for which shock and fatigue factors C_m and C_t are 2.0 and 1.5, respectively. Therefore,

$$\text{Equivalent torque, } T_{\text{eq}} = \sqrt{(C_m M)^2 + (C_t T)^2}$$

$$= \sqrt{(2 \times 118.03)^2 + (1.5 \times 25.46)^2} = 239.13 \text{ N} \cdot \text{m}.$$

Assume that the crank shaft is made of medium carbon steel having torsional yield strength, τ_y = 250 N/mm². We assume factor of safety, FoS = 5. Therefore,

$$\tau_{\text{d}} = \frac{\tau_y}{\text{FoS}} = \frac{250}{5} = 50 \text{ N/mm}^2$$

$$\text{Shaft diameter, } d = \left(\frac{16T_{\text{eq}}}{\pi\tau_{\text{d}}}\right)^{1/3} = \left(\frac{16 \times 239.13 \times 10^3}{\pi \times 50}\right)^{1/3} = 28.98 \text{ mm, say, 30 mm.}$$

Hub diameter, $d_{\text{h}} = 2d$ = 60 mm

Length of the hub, $l = 2.5d$ = 75 mm

Example 18.2 A punching machine makes 30 holes/min in a steel plate of 18 mm thickness. The diameter of hole is 25 mm and shear strength of plate is 240 N/mm². If the actual punching operation takes place during one-tenth of a revolution of crank, design a suitable flywheel which is required to rotate at 9 times the speed of crank shaft. The permissible coefficient of fluctuation of speed is 0.1 and space limitation requires that the diameter of the flywheel should not exceed 1000 mm.

Solution Mean speed of flywheel, N = 9 × Number of strokes/min = 9 × 30 = 270 rpm
Maximum shear force required to punch a hole

$$= \text{shear strength} \times \text{resisting area}$$

$$= \tau \times \pi dt$$

$$= \frac{240 \times \pi \times 25 \times 18}{1000} = 339.3 \text{ kN}$$

Energy required to punch one hole

$$= \text{average force} \times \text{distance travelled by punch}$$
$$= 0.5 \times 339.3 \times 18 = 3053.7 \text{ N} \cdot \text{m}$$

Since mechanical efficiency of the system is less than 100%, we assume efficiency, $\eta = 85\%$. Therefore,

$$\text{Total energy required, } E = \frac{3053.7}{0.85} = 3592.6 \text{ N} \cdot \text{m}$$

As given, the actual punching operation lasts for one-tenth of the cycle period. Therefore, during the remaining period, i.e. 9/10th of the cycle period, the energy is stored by the flywheel. Thus, fluctuation of energy

$$\Delta E = \frac{9}{10} E = \frac{9}{10} \times 3592.6 = 3233.3 \text{ N} \cdot \text{m}$$

Maximum space available is 1000 mm, therefore, the mean rim diameter should be less than 1000 mm. We adopt mean rim diameter, D = 800 mm. Therefore,

$$\text{Rim velocity, } v = \frac{\pi \times 0.8 \times 270}{60} = 11.3 \text{ m/s}$$

which is less than the maximum permissible velocity for grey cast iron. For the flywheel, fluctuation of energy

$$\Delta E = I\omega^2 C_S = mv^2 C_S$$

or

$$\text{Mass of flywheel, } m = \frac{\Delta E}{v^2 C_S} = \frac{3233.3}{11.3^2 \times 0.1} = 253.2 \text{ kg}$$

We assume that the mass of the rim is 90% of the total mass.

$$m_{rim} = 0.9m = 0.9 \times 253.2 = 227.88 \text{ kg}$$

Also,

$$m_{rim} = \pi D b h \rho \qquad \text{where } \rho = 7100 \text{ kg/m}^3 \text{ for CI}$$

$$\text{Therefore, rim cross-section, } bh = \frac{227.88}{\pi \times 0.8 \times 7100} = 0.01277 \text{ m}^2$$

We assume that the width to thickness ratio of the flywheel, b/h = 1.5. Thus,

$$\text{Thickness of the rim} = \sqrt{0.01277/1.5} = 0.0922 \text{ m}$$

We adopt rim thickness, h = 90 mm

rim width, $b = 1.5 \times 90 = 135$ mm

We adopt rim width, b = 150 mm

Outside diameter of flywheel, $D_o = D + h = 0.89$ m which is less than the maximum space of 1 m.

Shaft diameter

$$\text{Overhang length, } l = \frac{\text{length of bearing}}{2} + \text{clearance} + \frac{\text{width}}{2}$$

Since length of the bearing is not known, we assume an overhang of 200 mm.

$$\text{Bending moment, } M = \text{Weight of flywheel} \times \text{Overhang}$$

$$= 253.2 \times 9.81 \times 0.2 = 496.78 \text{ N} \cdot \text{m}$$

$$\text{Energy required/min} = \text{Energy required per hole} \times \text{Number of holes per min}$$

$$= 3592.6 \times 30 = 107{,}778 \text{ N} \cdot \text{m}$$

$$\text{Average torque} = \frac{\text{Energy required/min}}{2\pi N}$$

$$= \frac{107{,}778}{2\pi \times 270} = 63.53 \text{ N} \cdot \text{m}$$

Assuming suddenly applied load condition for which shock and fatigue factors are, $C_m = 1.5$ and $C_t = 2.0$.

$$\text{Equivalent torque, } T_{\text{eq}} = \sqrt{(C_m M)^2 + (C_t T)^2}$$

$$= \sqrt{(1.5 \times 496.78)^2 + (2 \times 63.53)^2} = 755.92 \text{ N} \cdot \text{m}$$

We assume that the shaft is made of medium carbon steel SAE1045 annealed water quenched, for which $\tau_y = 360$ N/mm^2. Assume factor of safety = 4. Therefore,

$$\tau_{\text{d}} = \frac{\tau_y}{\text{FoS}} = \frac{360}{4} = 90 \text{ N/mm}^2$$

$$\text{Shaft diameter, } d_s = \left(\frac{16T_{\text{eq}}}{\pi\tau}\right)^{1/3} = \left(\frac{16 \times 755.92 \times 10^3}{\pi \times 90}\right)^{1/3} = 34.97 \text{ mm}$$

Let us adopt shaft size, $d_s = 40$ mm

Hub diameter $d_{\text{h}} = 2d_s = 80$ mm

Length of the hub, $l_{\text{h}} = 2.5d_s = 100$ mm

$$\text{Clearance between the flywheel and bearing} = \text{overhang} - \frac{\text{length}}{2} - \frac{\text{width}}{2}$$

Let length of bearing = $2d_s$ = 80.0 mm.

Thus, clearance = $200 - \dfrac{80}{2} - \dfrac{150}{2} = 85$ mm which is reasonable.

We choose arm-type flywheel having 4 number of arms, i.e. $i = 4$

Stresses in rim

(a) Stress in unstrained rim

$$\sigma_t = \rho v^2 = \frac{7100 \times 11.3^2}{10^6} = 0.9066 \text{ MN/m}^2$$

(b) Stress in restrained rim

$$\sigma_b = \rho v^2 \left(\frac{2\pi^2 R}{i^2 h} \right) = \frac{7100 \times 11.3^2}{10^6} \times \frac{2\pi^2 \times 0.4}{4^2 \times 0.09} = 4.97 \text{ MN/m}^2$$

Total stress in the rim

$$\sigma_{res} = 0.75\sigma_t + 0.25\sigma_b$$

$$= 0.75 \times 0.9066 + 0.25 \times 4.97 = 1.922 \text{ MN/m}^2$$

which is less than the allowable strength for CI. Hence the design of the rim is safe.

Stress in arms

(i) Bending stress in the arm due to suddenly applied load
Assuming arm as cantilever beam having elliptical cross-section

$$\sigma_{t1} = \frac{T(D - d_h)}{iDZ}$$

Let σ_{t1} = allowable strength = 10 N/mm²

$$\text{Section modulus, } Z = \frac{T(D - d_h)}{iD\sigma_{t1}} = \frac{63.53(0.8 - 0.08)}{4 \times 0.8 \times 10} \times 1000 = 1429.4 \text{ mm}^3$$

We assume that major axis of the ellipse is twice the minor axis, i.e. $a_1 = 2b_1$. Therefore,

$$Z = \frac{\pi}{32} a_1^2 b_1 = \frac{\pi}{32} \times 4b_1^3$$

or

$$b_1 = \left(\frac{32Z}{\pi \times 4} \right)^{1/3} = \left[\frac{32 \times 1429.4}{\pi \times 4} \right]^{1/3} = 15.38 \text{ mm}$$

We adopt minor axis, $b_1 = 16$ mm
major axis, $a_1 = 32$ mm

(ii) Direct stress due to contrifugal force

$$\sigma_{t2} = \rho v^2 = 0.9066 \text{ N/mm}^2$$

Total stress = $\sigma_{t1} + \sigma_{t2}$ = 10.9066 N/mm²

which is less than the allowable strength 20 N/mm². Hence design of the arms are safe.

Example 18.3* The areas of the turning moment diagram for one revolution of a multi-cylinder engine with reference to mean torque below and above the line (in mm^2) are: –32, 408, –267, 333, –310, 226, –374, 260, –244 mm^2. The scale for abscissa and ordinate are 1 mm = 2.4° and 1 mm = 650 N · m, respectively. The mean speed is 300 rpm with percentage speed fluctuation of ±1.5%. Design a flywheel.

Solution The torque–crank rotation angle for one revolution with reference to mean torque, below and above the line is shown below.

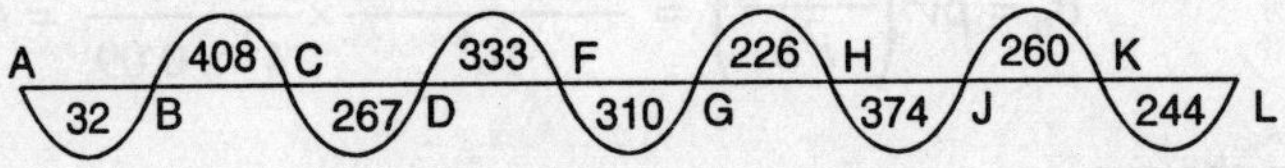

Let the energy of the flywheel at point A be E units. The energy level at different points can be computed as below:

Energy at A = E

$B = E - 32$

$C = E - 32 + 408 = E + 376$

$D = E + 376 - 267 = E + 109$

$F = E + 109 + 333 = E + 442$

$G = E + 442 - 310 = E + 132$

$H = E + 132 + 226 = E + 358$

$J = E + 358 - 374 = E - 16$

$K = E - 16 + 260 = E + 244$

$L = E + 244 - 244 = E$

Energy levels at various points indicate that maximum and minimum energy levels are at point F and B, respectively.

$$\text{Fluctuation of energy, } \Delta E = E_{max} - E_{min}$$

$$= (E + 442) - (E - 32)$$

$$= 474 \text{ mm}^2$$

or

$$\Delta E = 474 \times \text{scale of abscissa} \times \text{scale of ordinate}$$

$$= 474 \times 650 \times \frac{2.4}{180} \times \pi$$

$$= 12{,}905.6 \text{ N} \cdot \text{m}$$

$$\text{Coefficient of fluctuation of speed, } C_S = \frac{2 \times 1.5}{100} = 0.03$$

* Examples 18.3 to 18.5 are solved for finding the flywheel size and rim cross-section. Detailed design has been left as an exercise. The reader may consult Examples 18.1 and 18.2.

We assume that the flywheel is made of grey CI for which the limiting speed is, $v = 25$ m/s. Therefore,

$$\Delta E = I\omega^2 C_S = mv^2 C_S$$

or

$$m = \frac{\Delta E}{v^2 C_S} = \frac{12{,}905.6}{25^2 \times 0.03} = 688.3 \text{ kg}$$

$$\text{Mean diameter of flywheel, } D = \frac{60v}{\pi N} = \frac{60 \times 25}{\pi \times 300} = 1.591 \text{ m, say, 1.6 m}$$

Assuming that 90% mass is contributed by rim, mass of rim

$$m_{\text{rim}} = 0.9m = 0.9 \times 688.3 = 619.47 \text{ kg}$$

Also,

Mass of rim $m_{\text{rim}} = \pi Dbh\rho$. Where $\rho = 7100$ kg/m^3. Therefore, rim cross-section,

$$bh = \frac{m_{\text{rim}}}{\pi D\rho}$$

$$= \frac{619.47}{\pi \times 1.6 \times 7100} = 0.01735 \text{ m}^2$$

We assume that the ratio of width to thickness of rim, $b/h = 2$. Therefore,

$$h = \sqrt{\frac{0.01735}{2}} = 0.0931 \text{ m}$$

Let us adopt rim thickness, $h = 95$ mm
and width of rim, $b = 190$ mm.

Example 18.4 A machine that requires a torque $T_{MC} = 5000 + 600 \sin \theta$ N · m for its drive, is coupled to a three-cylinder engine that develops an effective torque $T_E = 5000 + 1500 \sin 3\theta$ at crank shaft rotating at 300 rpm. Design a suitable flywheel if it allows 1 per cent speed fluctuation.

Solution Work done by engine per revolution

$$W = \int_0^{2\pi} (5000 + 1500 \sin 3\theta)d\theta$$

$$= 5000\theta + \frac{1500 \cos 3\theta}{3}\Bigg|_0^{2\pi} = 10{,}000\pi = 31{,}415.9 \text{ N} \cdot \text{m}$$

$$\text{Mean torque, } T_m = \frac{\text{Torque/revolution}}{2\pi} = \frac{10000\pi}{2\pi} = 5000 \text{ N} \cdot \text{m}$$

The turning moment diagrams of machine and engine are both shown in the figure below.

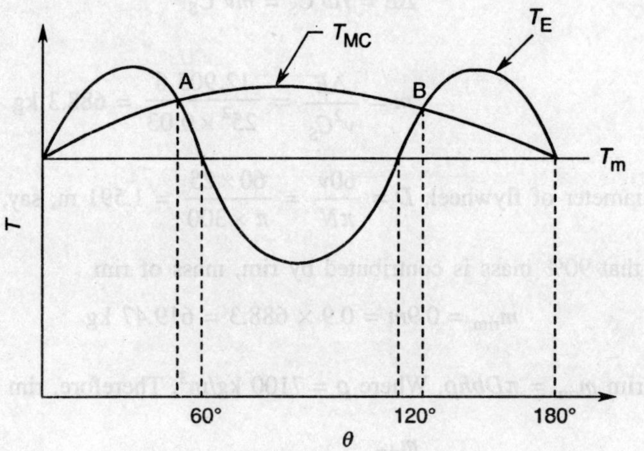

It is found from the above figure that there are two points A and B where the torque developed by the engine is equal to the torque required by the machine. Therefore, at any of these points

$$5000 + 1500 \sin 3\theta = 5000 + 600 \sin \theta$$

or

$$2.5 \sin 3\theta = \sin \theta$$

Substituting $\sin 3\theta = 3 \sin \theta - 4 \sin^3\theta$, we get,

$$2.5(3 \sin \theta - 4 \sin^3\theta) = \sin \theta$$

or

$$2.5(3 - 4 \sin^2\theta) = 1 \quad \text{or} \quad \sin \theta = 0.806$$

or

$$\theta_A = 53.7° \quad \text{and} \quad \theta_B = 126.3°$$

Thus, maximum fluctuation of energy

$$\Delta E = \int_{53.7}^{126.3} (T_E - T_{MC})d\theta$$

$$= \int_{53.7}^{126.3} (1500 \sin 3\theta - 600 \sin \theta)d\theta = -1656.4 \text{ N} \cdot \text{m}$$

The (–)ve sign indicates that there is deficit of torque and that much of torque has to be supplied by the flywheel. That is, $\Delta E = 1656.4$ N · m

For grey CI flywheel, the limiting speed $v = 25$ m/s. Let us adopt $v = 20$ m/s.

$$\text{Mean diameter of wheel, } D = \frac{60v}{\pi N}$$

or

$$D = \frac{60 \times 20}{\pi \times 300} = 1.27 \text{ m, say, } 1.3 \text{ m}$$

$$\text{Actual velocity } v = \frac{\pi \times 1.3 \times 300}{60} = 20.42 \text{ m/s}$$

$$\text{Mass of flywheel, } m = \frac{\Delta E}{v^2 C_S} = \frac{1656.4}{20.42^2 \times 0.01} = 397.24 \text{ kg}$$

Assuming that mass of the rim, $m_{\text{rim}} = 0.9m$, we have

$$m_{\text{rim}} = 0.9 \times 397.24 = 357.5 \text{ kg}$$

Also

$$\text{mass of rim, } m_{\text{rim}} = \pi D b h \rho$$

or

$$\text{Rim section, } bh = \frac{m_{\text{rim}}}{\pi D \rho} = \frac{357.5}{\pi \times 1.3 \times 7100}$$

$$bh = 0.01233 \text{ m}^2$$

Let the width to thickness ratio of the rim, $b/h = 1.5$. Therefore,

$$h = \sqrt{\frac{0.01233}{1.5}} = 0.09066 \text{ m}$$

Let us adopt rim thickness, $h = 90$ mm.
Therefore, rim width, $b = 1.5 \times 90 = 135$ mm
Let us adopt rim width, $b = 140$ mm

Example 18.5 A single-cylinder four-stroke gas engine develops 30 kW at 300 rpm. The work done by the gases during the expansion stroke is three times the work done on the gases during the compression stroke. The work during suction and exhaust strokes may be neglected. If total fluctuation of speed is ±2% of mean speed, design a suitable flywheel.

Solution The T–θ diagram of an I.C. engine is shown in the following figure. It is given that the work during suction and exhaust strokes is negligible. Assuming that compression and expansion strokes are approximately represented by triangles, the approximate T–θ diagram of the I.C. engine is also shown in the following figure.

Let mechanical efficiency be 85%.

$$\text{Indicated power, IP} = \frac{bp}{\eta_{\text{mech}}} = \frac{30}{0.85} = 35.3 \text{ kW}$$

$$= W/\text{cycle} \times \text{number of cycle/s}$$

$$\text{Number of cycles/s} = \frac{N}{2 \times 60} = \frac{300}{2 \times 60} = 2.5$$

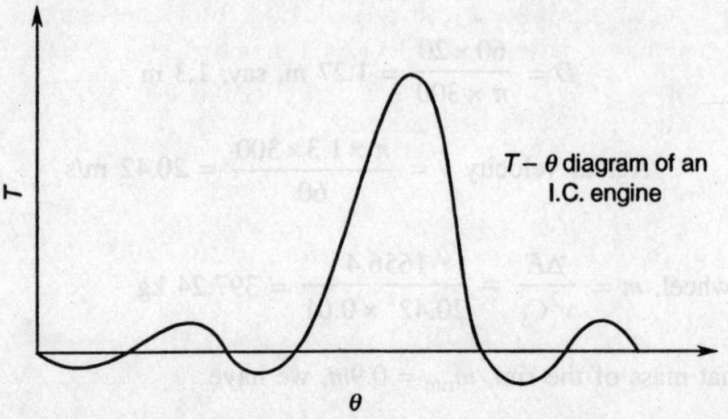

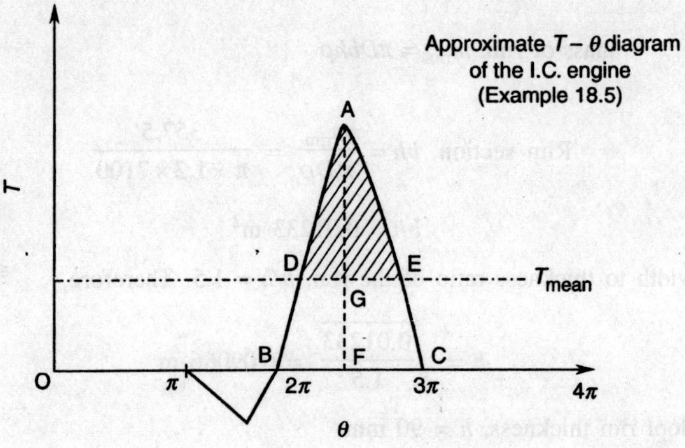

Therefore

$$W\text{/cycle} = \frac{35.3 \times 1000}{2.5} = 14{,}120 \text{ N} \cdot \text{m}$$

Work done during expansion = 3 × work consumed during compression

or

$$W_{exp} = 3W_{com}$$

$$\text{Net work done} = W_{exp} - W_{com} = \frac{2}{3} W_{exp} = 14{,}120 \text{ N} \cdot \text{m}$$

$$W_{exp} = 21{,}180 \text{ N} \cdot \text{m}$$

$$\text{area of triangle ABC} = \frac{1}{2} \text{BC} \times \text{AF} = 21{,}180$$

or

$$T_{max} = \text{AF} = \frac{21{,}180 \times 2}{\pi} = 13{,}483.6 \text{ N} \cdot \text{m}$$

$$\text{Mean torque, } T_{mean} = \frac{60P}{2\pi N} = \frac{60 \times 30 \times 10^3}{2\,\pi \times 300 \times 0.85} = 1123.45 \text{ N} \cdot \text{m}$$

Excess torque, $T_{\text{excess}} = \text{AG} = \text{AF} - \text{FG}$

$$= 13{,}483.6 - 1123.45 = 12{,}360.15 \text{ N} \cdot \text{m}$$

Now from similar triangles ABC and ADE

$$\frac{\text{DE}}{\text{BC}} = \frac{\text{AG}}{\text{AF}}$$

or

$$\text{DE} = \text{BC} \times \frac{\text{AG}}{\text{AF}} = \pi \times \frac{12{,}360.15}{13{,}483.6} = 2.88$$

Fluctuation of energy

$$\Delta E = \text{area of triangle ADE}$$

$$= 0.5 \times \text{DE} \times \text{AG}$$

$$= 0.5 \times 2.88 \times 12{,}360.15 = 17{,}798.6 \text{ N.m}$$

From space consideration, let us assume that mean rim diameter, $D = 1.5$ m. Therefore,

$$\text{Rim velocity, } v = \frac{\pi DN}{60} = \frac{\pi \times 1.5 \times 300}{60} = 23.56 \text{ m/s}$$

which is within the permissible value of 25 m/s.

For flywheel, energy fluctuation

$$\Delta E = I\omega^2 C_S = mv^2 C_S$$

or

$$\text{Mass of flywheel, } m = \frac{\Delta E}{v^2 C_S} = \frac{17{,}798.6}{23.56^2 \times 0.04} = 801.6 \text{ kg}$$

Assuming that mass of rim, $m_{\text{rim}} = 0.9m = 0.9 \times 801.6 = 721.4$ kg

Also, rim mass $m_{\text{rim}} = \pi Dbh\rho$

Therefore,

$$\text{Rim cross-section, } bh = \frac{m_{\text{rim}}}{\pi D\rho} = \frac{721.4}{\pi \times 1.5 \times 7100} = 0.02156 \text{ m}^2 \text{ (with } \rho = 7100 \text{ kg/m}^3\text{)}$$

Let the rim width to thickness ratio be $b/h = 1.5$

$$\text{Rim thickness, } h = \sqrt{\frac{0.02156}{1.5}} = 0.1198 \text{ m}$$

Let us adopt rim thickness, $h = 125$ mm

and rim width, $b = 190$ mm

18.6 COMPUTER-AIDED DESIGN OF FLYWHEEL

The application of computer-aided design of flywheel is demonstrated in this section. The listing of C++ source code is given in Program 18.1. The program requires the following inputs:

1. Fluctuation of energy, in N · m
2. Mean speed of the flywheel
3. Coefficient of speed fluctuation
4. Space limitation

If the space limitation is not given, the user should supply zero. When the space limitation is given, then 85 per cent of space is taken as the mean diameter subject to the condition that maximum peripheral speed is not more than 25 m/s for CI flywheels. The program selects the type of flywheel to be designed, i.e. (i) solid disc type, (ii) web type, or (iii) arms type. A sample run of the program is given along with listing.

Program 18.1

```
//    flywheel.cpp A program to design flywheel
#     include<iostream.h>
#     include <math.h>
#     define PI 3.1415
void main ()
{
      int code,n:
      double deltae,rpm,cs,space,d,v,mass,density,b,h,mass_r;
      double bh,bbyh,mu,omega,sigmax,sigmal,sigma2,sigma_r;
//    design input parameter
      cout <<"\n DESIGN  INPUTS" << end1;
      cout <<"\n Fluctuation of Energy(Nm)=";
      cin >>deltae;
      cout <<"\n Mean Speed(rpm)=";
      cin >>rpm;
      cout <<"\n Coefficient of Speed Fluctuation=";
      cin >>cs;
      cout <<"\n If space limitation is not given type zero(0)";
      cout <<"\n Space Limitation=";
      cin >>space;
//    calculate mean dia
      if (space > 0)
             d=0.85*space;
      else
      {
             v=20.0;
             d=v*60/(PI*rpm);//assume v=20m/s for CI flywheel
             d=d*1000;
             d=(int((d-0.5)/5)+1)*5;
      }
      v=PI*d*rpm/60000;
      if (v > 25)
```

```
	{
		d=d-5;
		v=PI*d*rpm/60000;
	}
//	calculate mass
repeat:
	mass=deltae/(v*v*cs);
	density=7100.0;
//	classify solid disc/web type flywheel
	if (d <= 300)
		code=1;
	else
		code=2;
//	calculate flywheel size
	switch (code)
	{
	case 1:
		b=mass/((PI/4)*(d/1000)*(d/1000)*density);
		b=b*1000;
		b=(int((b-0.5)/5)+1)*5;
		break;
	case 2:
		mass_r=0.9*mass;
		bh=mass_r/(PI*(d/1000)*density);
		if (space ==0)
		{
			h=pow((bh/1.5), 0.5);//assume b/h ratio=1.5
			h=h*1000;
			h=(int((h-0.5)/2)+1)*2;
			b=1.5*h;
			b=(int((b-0.5)/5)+1)*5;
		}
		else
		{
			bbyh=1.7;
loop1:
			h=pow((bh/bbyh),0.5);
			h=1000*h;
			h=int(h+0.5);
			if ((d+h) < space)
			{
				bbyh=bbyh-0.05;
				if (bbyh >=1.2)
					goto loop1;
			}

			b=bbyh*h;
```

```
                b=(int((b-0.5)/5)+1)*5;
            }
            break;
        }
//      calculate stresses
        if (d > 600)
            code=3;
        switch (code)
        {
        case 1:
            mu=0.211; // poisson's ratio for CI
            omega=2*PI*rpm/60;
            sigmax=(3+mu)/8*density*omega*omega*d*d/4;
            break;
        case 2:
            sigmax=density*v*v/1.0e06;
            break;
        case 3:
            if (d <=750)
                n=4;
            else
                if (d <=2000)
                    n=6;
                else
                    n=8;
//          stress in unstrained rim
            sigmal=density*v*v/1.0e06;
//          stress in restrained rim
            sigma2=sigmal*(2*PI*PI*(d/2000))/(n*n*(h/1000));
            sigma_r=0.75*sigmal+0.25*sigma2;
            if (sigma_r > 40)
            {
                v=v-0.5;
                d=v*60000/(PI*rpm);
                goto repeat;
            }
            break;
        }
//      print design report
        cout <<"\n DESIGN REPORT" << end1;
        if (code == 1)
            cout <<"\n Flywheel Type= Solid Disc";
        else
            if (code ==2)
                cout <<"\n Flywheel Type= Web Type";
            else
```

```
            cout <<"\n Flywheel Type= Arms Type";
//
    if (code ==1)
    {
            cout <<"\n Diameter of Flywheel= " << d << "mm";
            cout <<"\n Width of Flywheel= " <<b << "mm" << endl;
    }
    else
    {
            cout <<"\n Mean Flywheel Diameter= " << d <<"mm";
            cout <<"\n Outside Diameter = " << d+h << "mm";
            cout <<"\n Rim Width= " << b << "mm";
            cout <<"\n Rim Thickness=" << h << "mm";

    }
    if (code ==2)
    {
    cout <<"\n Web Thickness= " << (int((b/20 – 0.5)/2)+1)*2 << "mm" <<endl;
    }
    if (code == 3)
    {
            cout <<"\n Number of Arms= " << n <<endl;
    }
}
```

Sample Run of Program 18.1

DESIGN INPUTS
Fluctuation of Energy (Nm) = 376
Mean Speed (rpm) = 1500
Coefficient of Speed Fluctuation = 0.01
If space limitation is not given type zero (0)
Space Limitation = 0
DESIGN REPORT
Flywheel Type = Web Type
Mean Flywheel Diameter = 315 mm
Outside Diameter = 387 mm
Rim Width = 110 mm
Rim Thickness = 72 mm
Web Thickness = 6 mm

18.7 ROTATING DISC

In mechanical machines there are several elements, namely circular rings, wheel rims, circular discs, cylinders and cutters, etc. which rotate at high speeds to perform the given tasks. In such elements, besides functional stresses additional stresses are induced due to high speed of rotation. In this section these rotating discs are analyzed for stresses caused by high speed rotation.

Consider a circular disc rotating about its axis. It is assumed that the disc is of variable thickness, too small compared to disc diameter, and the stress variation along the thickness is negligible. It is further assumed that at the free flat surface, there can be no shear either normal or perpendicular to these faces. Thus the principal stress directions may be assumed as the directions of radial and tangential stresses.

We now consider the equilibrium of a small element ABCD cut out from the disc by the radial sections OD and OC and by two cylindrical surfaces AB and CD normal to the disc. The direct radial stresses on the sides AB and CD of the element are σ_{rad} and $(\sigma_{rad} + d\sigma_{rad})$, respectively. On the faces AD and BC tangential or hoop stress σ_θ act as shown in Figure 18.7.

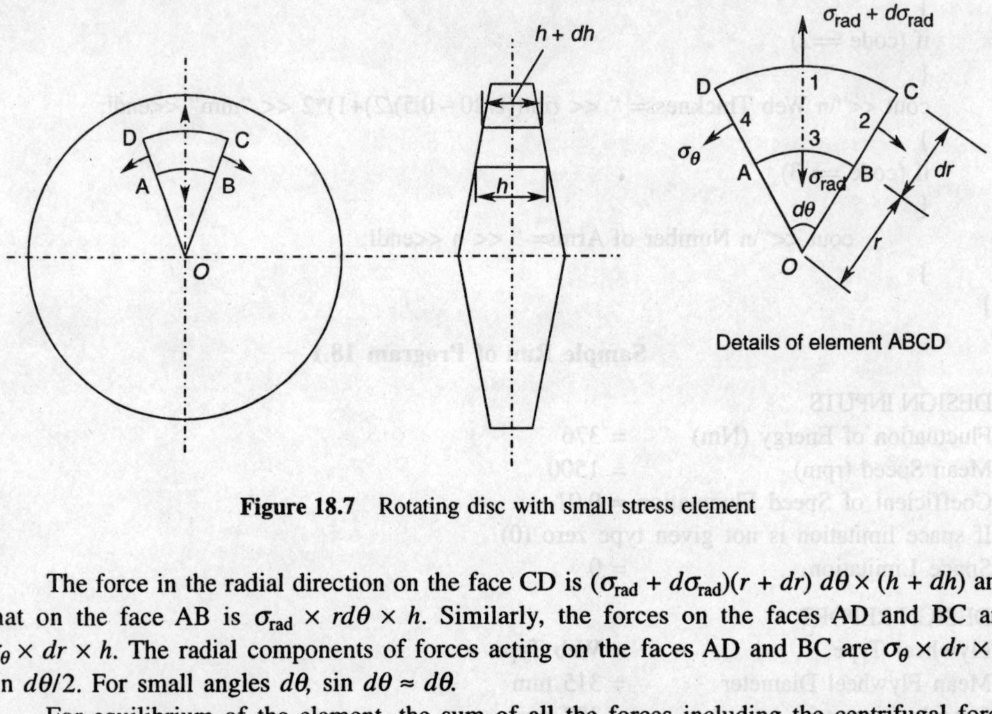

Figure 18.7 Rotating disc with small stress element

The force in the radial direction on the face CD is $(\sigma_{rad} + d\sigma_{rad})(r + dr)\, d\theta \times (h + dh)$ and that on the face AB is $\sigma_{rad} \times r d\theta \times h$. Similarly, the forces on the faces AD and BC are $\sigma_\theta \times dr \times h$. The radial components of forces acting on the faces AD and BC are $\sigma_\theta \times dr \times h$ sin $d\theta/2$. For small angles $d\theta$, sin $d\theta \approx d\theta$.

For equilibrium of the element, the sum of all the forces including the centrifugal force $\rho\omega^2 r^2 h\theta dr$ must be zero. Thus:

$$(\sigma_{rad} + d\sigma_{rad})(r + dr)d\theta\,(h + dh) - \sigma_{rad} \times rhd\theta - 2\sigma_\theta \times dr\frac{d\theta}{2} \times h + \rho\omega^2 r^2 h\, d\theta dr = 0 \quad (18.20)$$

After simplifying and neglecting small terms and dividing it by $dr \cdot d\theta$, we get the following differential equation which governs stress distribution

$$\frac{d}{dr}(h r \sigma_{rad}) - h\sigma_\theta + \rho\omega^2 r^2 h = 0 \quad (18.21)$$

18.8 DISC OF CONSTANT THICKNESS

For a disc of constant thickness, as shown in Figure 18.8, dh/dr = 0 and the governing Eq. (18.21) is reduced to:

$$\frac{d}{dr}(r\sigma_{\text{rad}}) - \sigma_\theta + \rho\omega^2 r^2 = 0 \tag{18.22}$$

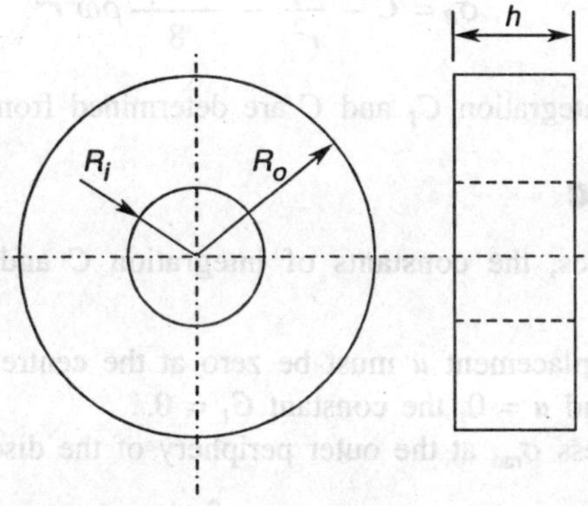

Figure 18.8 Disc of constant thickness.

We assume that if u is a displacement, the radial and tangential strains are du/dr and u/r, respectively. The radial and tangential stresses can be written as

$$\sigma_{\text{rad}} = \frac{E}{1-\mu^2}(\varepsilon_{\text{rad}} + \mu\varepsilon_\theta)$$

$$= \frac{E}{1-\mu^2}\left(\frac{du}{dr} + \mu\frac{u}{r}\right) \tag{18.23}$$

$$\sigma_\theta = \frac{E}{1-\mu^2}(\varepsilon_\theta + \mu\varepsilon_{\text{rad}})$$

$$= \frac{E}{1-\mu^2}\left(\frac{u}{r} + \mu\frac{du}{dr}\right) \tag{18.24}$$

Substituting the values of σ_{rad} and σ_θ in Eq. (18.22), we get the following governing equation:

$$r^2\frac{d^2u}{dr^2} + r\frac{du}{dr} - u = -\frac{(1-\mu^2)}{E}\rho\omega^2 r^3 \tag{18.25}$$

The general solution of Eq. (18.25) is

$$u = \frac{1}{E}\left[(1-\mu)C\cdot r - (1+\mu)C_1\frac{1}{r} - \frac{(1-\mu^2)}{8}\rho\omega^2 r^3\right] \tag{18.26}$$

where C and C_1 are constants of integration.

The corresponding stresses are found from Eqs. (18.23) and (18.24), i.e.

$$\sigma_{rad} = C + \frac{C_1}{r^2} - \frac{3+\mu}{8}\rho\omega^2 r^2 \tag{18.27}$$

and

$$\sigma_\theta = C - \frac{C_1}{r^2} - \frac{1+3\mu}{8}\rho\omega^2 r^2 \tag{18.28}$$

The constants of integration C_1 and C are determined from boundary conditions.

18.8.1 Solid Disc

In the case of solid discs, the constants of integration C and C_1 may be found from the following conditions:

(i) The radial displacement u must be zero at the centre of the disc, i.e. $r = 0$. Now, when $r = 0$ and $u = 0$, the constant $C_1 = 0$.

(ii) The radial stress σ_{rad} at the outer periphery of the disc must be zero. That is, when $r = R$, and $\sigma_{rad} = 0$, the constant $C = \frac{3+\mu}{8}\rho\omega^2 R^2$.

Substituting the value of constants C and C_1, we get

$$\sigma_{rad} = \frac{3+\mu}{8}\rho\omega^2(R^2 - r^2) \tag{18.29}$$

and

$$\sigma_\theta = \frac{\rho\omega^2}{8}[(3+\mu)R^2 - (1+3\mu)r^2] \tag{18.30}$$

These stresses are maximum at the centre of the disc where

$$\sigma_{rad} = \sigma_\theta = \frac{3+\mu}{8}\rho\omega^2 R^2 \tag{18.31}$$

18.8.2 Disc with a Circular Hole

In the case of a disc with a circular hole of radius R_i at the centre and outer radius R_o, the constants of integration in Eq. (18.27) are found from the boundary conditions: at $r = R_i$, $\sigma_{rad} = 0$ and at $r = R_o$, $\sigma_{rad} = 0$.

The relations for radial and tangential stresses at any radius r are given as under:

$$\sigma_{rad} = \frac{3+\mu}{8}\rho\omega^2\left(R_o^2 + R_i^2 - \frac{R_o^2 R_i^2}{r^2} - r^2\right) \tag{18.32}$$

$$\sigma_\theta = \frac{3+\mu}{8}\rho\omega^2\left(R_o^2 + R_i^2 + \frac{R_o^2 R_i^2}{r^2} + \frac{1+3\mu}{3+\mu}r^2\right) \tag{18.33}$$

The maximum radial stress at $r = \sqrt{R_o R_i}$ is

$$\sigma_{rad(max)} = \frac{3+\mu}{8}\rho\omega^2(R_o - R_i)^2 \qquad (18.34)$$

and the maximum tangential stress at the inner boundry, i.e. at $r = R_i$ is

$$\sigma_{\theta(max)} = \frac{3+\mu}{4}\rho\omega^2\left(R_o^2 + \frac{1-\mu}{3+\mu}R_i^2\right) \qquad (18.35)$$

The distribution of radial and tangential stresses on the disc are shown in Figure 18.9.

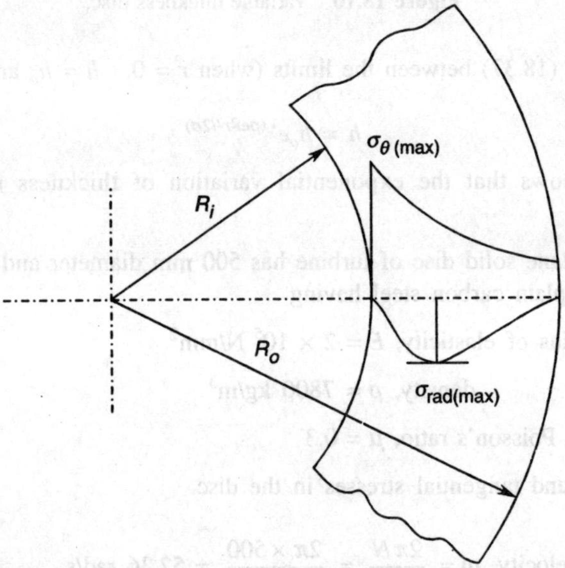

Figure 18.9 Distribution of tangential and radial stresses in a disc of uniform thickness with a centre hole.

18.9 DISC OF VARIABLE THICKNESS

A disc of variable thickness is also called *disc of uniform strength* (Figure 18.10). It is a disc which has equal radial and tangential stresses at all the points.

The governing equation for stress distribution in a disc of variable thickness is [refer Eq. (18.21)]

$$\frac{d}{dr}(hr\sigma_{rad}) - h\sigma_\theta + \rho\omega^2 r^2 h = 0 \qquad (18.36)$$

Assuming $\sigma_{rad} = \sigma_\theta = \sigma$ (a constant), Eq. (18.36) is transformed to

$$\frac{dh}{h} = \frac{\rho\omega^2 r}{\sigma}dr \qquad (18.37)$$

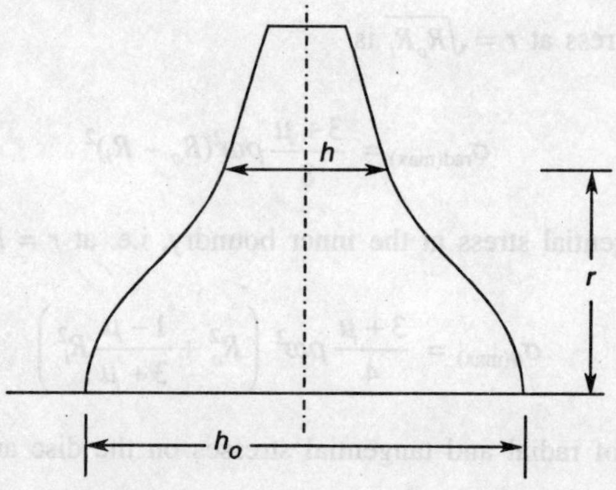

Figure 18.10 Variable thickness disc.

Integrating Eq. (18.37) between the limits (when $r = 0$, $h = h_o$ and when $r = r$, $h = h$), we get

$$h = h_o e^{-(\rho\omega^2 r^2/2\sigma)} \tag{18.38}$$

Equation (18.38) shows that the exponential variation of thickness results into a disc of uniform strength.

Example 18.6 A plane solid disc of turbine has 500 mm diameter and rotates at 500 rpm. If the disc is made of plain carbon steel having

modulus of elasticity, $E = 2 \times 10^5$ N/mm²

density, $\rho = 7800$ kg/m³

Poisson's ratio, $\mu = 0.3$

Calculate the radial and tangential stresses in the disc.

Solution Angular velocity, $\omega = \dfrac{2\pi N}{60} = \dfrac{2\pi \times 500}{60} = 52.36$ rad/s

Considering the solid disc of turbine as a rotating disc of uniform thickness without hole, the radial and tangential stresses at radius r are calculated using equations

$$\text{Radial stress, } \sigma_{rad} = \frac{3+\mu}{8}\rho\omega^2(R^2 - r^2)$$

$$= \frac{3+0.3}{8} \times \frac{7800 \times 52.36^2}{10^6}(0.25^2 - r^2)$$

or

$$\sigma_{rad} = (0.5513 - 8.82r^2) \text{ MN/m}^2 \tag{i}$$

Tangential stress at radius r

$$\sigma_\theta = \frac{\rho\omega^2}{8}[(3+\mu)R^2 - (1+3\mu)r^2]$$

$$= \frac{7800 \times 52.36^2}{8 \times 10^6} [(3 + 0.3) \times 0.25^2 - (1 + 3 \times 0.3)r^2]$$

or

$$\sigma_\theta = (0.5513 - 5.0787r^2) \text{ MN/m}^2 \qquad \text{(ii)}$$

Equations (i) and (ii) represent equations of radial and tangential stress distribution along radius r.

These stresses are maximum at the centre of the disc, where $r = 0$. Thus,

$$\sigma_{\text{rad(max)}} = \sigma_{\theta\text{(max)}} = \frac{3+\mu}{8} \times \frac{\rho\omega^2 R^2}{10^6}$$

$$= \frac{3+0.3}{8} \times \frac{7800 \times 52.36^2 \times 0.25^2}{10^6}$$

$$= 0.5526 \text{ MN/m}^2$$

Example 18.7 A thin disc is to be used as a rotating cutter. It is of uniform thickness except at the periphery where it is sharpened. The outer diameter of the disc may be taken as 250 mm. The disc is to be mounted on a 50-mm diameter shaft. Ignoring clamping force, calculate the safe speed for the disc if the maximum stress is not to exceed 200 MN/m^2.

Solution Assume cutter disc as a rotating disc of uniform thickness with centre hole. General distributions of radial and tangential stresses are shown in the adjoining figure. We find from it that the tangential stress is maximum at the inner surface and at that point the radial stress is zero.

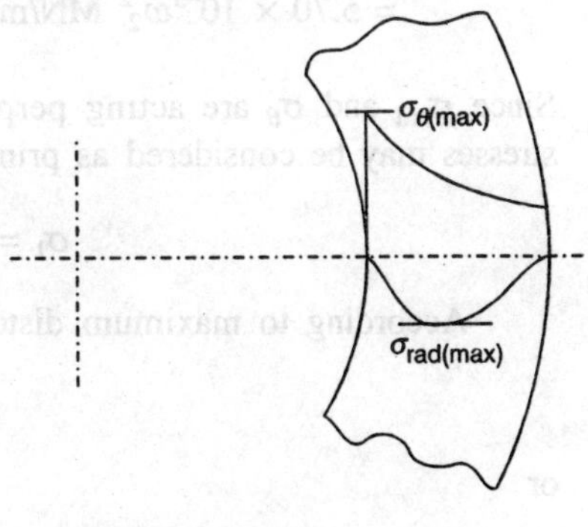

Further, radial stress is maximum at $r = \sqrt{R_i R_o}$ and at that radius, the tangential stress is not maximum. Thus we consider these two cases separately.

(i) Stress at the inner surface, i.e. when $r = R_i$

$$\sigma_{\theta\text{ (max)}} = \frac{3+\mu}{4} \rho\omega_1^2 \left(R_o^2 + \frac{1-\mu}{3+\mu} R_i^2 \right)$$

$$= \frac{3+0.3}{4} \times \frac{7800\omega_1^2}{10^6} \left(0.125^2 + \frac{1-0.3}{3+0.3} \times 0.025^2 \right)$$

$$= 1.014 \times 10^{-4}\ \omega_1^2 \text{ MN/m}^2$$

For safe design, $\sigma_{\theta\text{(max)}}$ should be less than or equal to the allowable strength, i.e.

$$1.014 \times 10^{-4}\ \omega_1^2 = 200 \qquad \text{or} \qquad \omega_1 = 1404.4 \text{ rad/s}$$

(ii) Stress at $r = \sqrt{R_i R_o}$

Radial stress is maximum at $r = \sqrt{R_i R_o}$ or $r = \sqrt{0.125 \times 0.025} = 0.05509$ m

$$\sigma_{\text{rad (max)}} = \frac{3+\mu}{8} \rho \omega_2^2 (R_o - R_i)^2$$

$$= \frac{3+0.3}{8} \times \frac{7800\, \omega_2^2}{10^6} (0.125 - 0.025)^2$$

$$= 3.2175 \times 10^{-5} \omega_2^2 \text{ MN/m}^2$$

At radius $r = 0.05509$ m, tangential stress

$$\sigma_\theta = \frac{3+\mu}{8} \rho \omega_2^2 \left(R_o^2 + R_i^2 + \frac{R_o^2 R_i^2}{r^2} - \frac{1+3\mu}{3+\mu} r^2 \right)$$

$$= \frac{3+0.3}{8} \times \frac{7800\, \omega_2^2}{10^6} \left(0.125^2 + 0.025^2 + \frac{0.125^2 \times 0.025^2}{0.05509^2} - \frac{1+3\times 0.3}{3+0.3} \times .05509^2 \right)$$

$$= 5.70 \times 10^{-5} \omega_2^2 \text{ MN/m}^2$$

Since σ_{rad} and σ_θ are acting perpendicular to each other and there is no shear stress, these stresses may be considered as principal stresses.

$$\sigma_1 = \sigma_{\text{rad(max)}} \quad \text{and} \quad \sigma_2 = \sigma_\theta$$

According to maximum distortion energy theory

$$\sigma_1^2 - \sigma_1 \sigma_2 + \sigma_2^2 \le \sigma_y^2$$

or

$$(3.2175 \times 10^{-5} \omega_2^2)^2 - (3.2175 \times 10^{-5} \omega_2^2 \times 5.70 \times 10^{-5} \omega_2^2) + (5.70 \times 10^{-5} \omega_2^2)^2 = 200^2$$

or

$$\omega_2 = 2010.19 \text{ rad/s}$$

Among these two cases, the maximum allowable angular speed is lower of the two values, i.e. $\omega_1 = 1404.4$ rad/s.

Therefore,

$$N = \frac{\omega \times 60}{2\pi} = \frac{1404.4 \times 60}{2\pi} = 13{,}414.8 \text{ rpm}$$

Therefore, the maximum speed at which the cutter can operate is 13,414.8 rpm.

EXERCISES

1. Define the following terms:
   (a) Flywheel effect (b) Coefficient of fluctuation of speed
   (c) Coefficient of fluctuation of energy
2. What type of stresses are produced in a disc flywheel?
3. List the factors on which the value of coefficient of speed fluctuation depends.
4. Describe the type of stresses that are produced in
   (a) flywheel rim (b) flywheel arms
5. While computing stress (bending) in an arm of a fly-wheel, the arm is treated as a cantilever beam. Justify the assumption.
6. How is the limiting speed of the flywheel determined?
7. What do you mean by a disc of uniform strength? Draw a sketch to show the variation of its thickness with radius.
8. Show the variation of hoop and radial stresses in a disc of constant thickness with centre hole.
9. A four-stroke, four-cylinder I.C. engine, which develops 38 kW at 750 rpm, is used to drive an AC generator. Design and sketch the flywheel. Assume it to be supported on a shaft of 60 mm diameter.
10. Design a flywheel for a press using the following data:

    Work done at the crank shaft per revolution = 10 kJ
    Duration of punching operation = 30% of revolution of the crank shaft
    Mechanical efficiency = 80%
    Coefficient of steadiness = 5
    Speed of crank shaft = 40 rpm

    The crank shaft is geared to a driving shaft and the speed reduction ratio is 6:1. The flywheel is mounted on the driving shaft.
    Maximum available space = 1 m
11. A single-cylinder double-acting steam engine delivers 185 kW at 100 rpm. The maximum fluctuation of energy per revolution is 15%. The speed variation is ±1%. The mean diameter of the flywheel is 2 m. Design and draw a suitable flywheel.
12. A bicycle driven single-acting, single-cylinder air compressor is being considered for spray painting applications for the places where there is no power. The data on capability of human beings indicates that man can provide energy in the following order:

    1000 W for 5 s
    150 W for next 5 min
    100 W for next 30 min
    75 W for next 10 min
    750 W for next 5 s, and none for the next 14 min 50 s. The cycle then repeats. Assuming 90% efficiency, calculate the average power output. A speed variation of 8% takes place either way from the mean speed of 1000 rpm. Design and draw a suitable flywheel.

13. A De-Leval steam turbine rotor is 150 mm in diameter and of 6 mm thickness at top. If the maximum speed of the turbine is 20,000 rpm and the maximum allowable stress in the rotor material is 150 N/mm^2, determine the variation of rotor thickness for uniform strength.

14. A circular cutter of 750 mm diameter, 5 mm thick, is mounted on a 75-mm shaft. If the maximum allowable strength of the cutter material is 200 N/mm^2, determine the operating speed. Also show the variation of stresses on the cutter.

MULTIPLE CHOICE QUESTIONS

1. The main function of a flywheel is to control
   (a) mean speed
   (b) fluctuation of speed
   (c) fluctuation of energy
   (d) none of the above

2. A solid-disc type flywheel is used in
   (a) a punching press
   (b) an I.C. engine coupled with generator
   (c) an automotive vehicle
   (d) any application

3. A flywheel stores up energy in the form of
   (a) kinetic energy
   (b) potential energy
   (c) electrical energy
   (d) none of the above

4. The hoop stress in the rim of flywheel is
   (a) ρv (b) ρv^2
   (c) $\rho v^2/r$ (d) ρ/v^2

5. If σ_b is the bending stress in rim and σ_t the hoop stress, the resultant stress in the rim is
   (a) $0.25\sigma_t + 0.75\sigma_b$
   (b) $\sigma_t + \sigma_b$
   (c) $0.75\sigma_t + 0.25\sigma_b$
   (d) $0.5(\sigma_t + \sigma_b)$

6. Maximum peripheral velocity of flywheel for ordinary grey cast iron is
   (a) 10 m/s (b) 25 m/s
   (c) 40 m/s (d) 50 m/s

7. The significant stress in a disc
   (a) is hoop stress (b) is radial stress
   (c) depends on the material (d) is unpredictable

8. The relation between σ_t and σ_θ stresses at the centre of a solid disc is

   (a) $\sigma_{rad} = \sigma_\theta$ (b) $\sigma_{rad} > \sigma_\theta$

   (c) $\sigma_\theta > \sigma_{rad}$ (d) $\sigma_\theta = \sigma_{rad} + 25$

9. For a disc of constant thickness with a centre hole, the hoop stress is maximum at

   (a) $r = R_i$ (b) $r = R_o$

   (c) $r = \sqrt{R_i R_o}$ (d) none of the above

10. For a disc of constant thickness with a centre hole, the radial stress is maximum at

   (a) $r = R_i$ (b) $r = R_o$

   (c) $r = \sqrt{R_i R_o}$ (d) none of the above

Chapter 19 Design Optimization

19.1 INTRODUCTION

In design methodology, for a given objective, there can be an infinite number of possible design solutions. Most of them are usually obtained to satisfy functional requirements. Such designs are termed *adequate designs*.

In any design problem there are some desirable and/or undesirable parameters; the degree of significance for each depends upon the application. An explicit design in which it is possible to minimize the most significant undesirable parameter or to maximize the most significant desirable parameters without ignoring any functional requirements, is called *optimal design*.

19.2 PROBLEM FORMULATION

Optimization is the act of obtaining the best result under given circumstances; thus it is impossible to apply a single formulation procedure for all engineering design problems. A general outline of the steps used in the optimal design formulation process is shown in Figure 19.1.

The formulation of the optimum design process involves various considerations, namely design variables, objective function, constraints and variable bounds. While trying to achieve an optimal design solution, one consideration may influence the other; therefore the optimization is an iterative procedure, though in Figure 19.1 a straight line relationship is shown.

A brief discussion of optimal design considerations is given below:

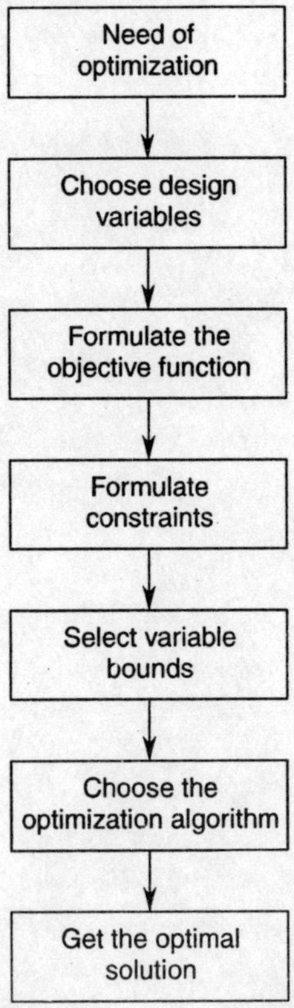

Figure 19.1 Optimization procedure.

Design variables

A design problem usually involves many design parameters. Some of them are highly sensitive to the working of design. These parameters are called *variables*. There are no rigid guidelines to choose apriori the design variables which may be important to the problem; one variable may be more important with respect to one objective function while it may not be significant with respect to another objective function. However, the speed and efficiency of optimization algorithms depend upon the number of design variables. As a thumb rule, it is suitable to choose as few a design variables as possible. The outcome of optimization may be reviewed and if necessary, the number of variables may then be increased.

Objective function

The second task of formulation procedure is to find the objective function in terms of design variables. The most common objectives are minimization of cost of manufacture, weight of component, or maximization of profit, life, production, productivity, etc. Thus the objective function can be of two types: (i) maximization and (ii) minimization. The optimization algorithms are therefore usually written either for maximization or for minimization problems. However the duality principle can be used to solve either type of problem without making structural changes in algorithms. According to the duality principle if a point x corresponds to minimum value of function $f(x)$, the same point also corresponds to maximum value of negative of the function $-f(x)$, as shown in Figure 19.2.

In real-world optimization problems, there could be more than one objective function which a designer would like to optimize, for example minimization of cost of a product while maximizing the quality of the product.

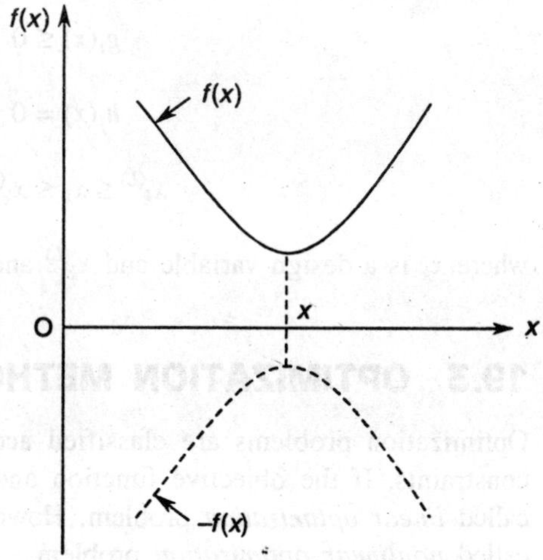

Figure. 19.2 Maxima and minima of a function.

Constraints

The constraints represent some functional relationships among design variables and other design parameters satisfying physical, functional and resource limitations. For example, a shaft subjected to transverse bending load is allowed to deflect up to a certain limit, say 2 mm, the constraint for the optimization problem is then written as

$$\delta_{max} \leq 2$$

The nature and number of constraints to be included in the problem depend upon the design problem. Two types of constraints, inequality and equality, are used. The inequality type constraints may be of less than equal to or greater than equal to type. On the other hand, the equality type constraints are of equal to type. The equality constraints help in reducing the number of design variables.

Variable bounds

The behaviour of the objective function may not be unimodal (one having unique optimal solution); hence there may be more than one solution in search space. However, design variables cannot attain any value due to functional or resource constraints. The designer has to limit the range of search space. This range of search space is called *variable bound*. Generally, the variable bounds are incorporated as additional constraints. For example, if the space available to place a shaft is between 25 mm and 75 mm, then the design variable shaft diameter d has 25 mm as the lower bound value and 75 mm as the upper bound value, represented as constraints:

$$d \geq 25; \; d \leq 75$$

Once the above four tasks are completed, the optimization problem can be mathematically written as

$$\text{Maximize or Minimize } f(x) \tag{19.1}$$

subject to constraints

$$g_i(x) \leq 0 \qquad i = 1, 2, \ldots, m \tag{19.2}$$

$$h_j(x) = 0 \qquad j = 1, 2, \ldots, n \tag{19.3}$$

$$x_k^{(L)} \leq x_k \leq x_k^{(U)} \qquad k = 1, 2, \ldots, p \tag{19.4}$$

where x_k is a design variable and $x_k^{(L)}$ and $x_k^{(U)}$ are lower and upper bounds of the variable.

19.3 OPTIMIZATION METHODOLOGY

Optimization problems are classified according to the nature of the objective function and constraints. If the objective function and the constraint functions are linear, the problem is called *linear optimization* problem. However, if these functions are nonlinear, the problem is called *nonlinear optimization* problem.

While solving an optimization problem, one has to select the criterion of optimality. Generally, one of the following three criteria is used.

Local optimum point

A point or a solution x' is said to be locally optimal minimum if no point in the neighbourhood has a function value smaller than $f(x')$.

Global optimum point

A point or a solution x'' is said to be global optimal if no point in the entire search space has a function value smaller than $f(x'')$.

Inflection point

A point or a solution x''' is said to be an inflection point if the function value increases locally as x''' increases and decreases locally as x''' reduces or vice versa.

The problem of optimization can be solved by the following three methods:

(i) Classical method
(ii) Graphical method
(iii) Numerical methods

A brief presentation of these methods is now given.

19.3.1 Classical Method

The classical method of optimization is useful in finding the optimum of continuous and differentiable functions. This method is analytical and makes use of differential calculus in locating the optimum point.

In the classical differential calculus method, assuming that the first- and second-order derivatives of the objective function $f(x)$ exist, it can be shown that conditions for a point to be minimum are:

$$\frac{df(x)}{dx} = 0 \tag{19.5}$$

and

$$\frac{d^2 f(x)}{dx^2} > 0 \tag{19.6}$$

The first condition alone suggests that the point is either a minimum, a maximum or an inflection point and the second condition suggests that the point is a minimum. The general conditions of optimality are given below.

Suppose at point x, the first derivative is zero and the first nonzero higher order derivative is denoted by n; then

(i) if n is odd, x is an inflection point;
(ii) If n is even, x is a local optimum point.

If the derivative is positive, x is a local minimum and if it is negative, x is a local maximum point.

The use of the classical method is illustrated through the following worked example.

Example 19.1 A steel bar is subjected to 10 kN axial force. The allowable strength of the material is 80 N/mm². If the cost of the material is Rs 20/kg and the cost of machining per surface is Rs 10, determine the section size for minimum cost of production. The length of the bar is 750 mm. Take ρ = 7800 kg/m³.

Solution Cost of production (C) = Cost of material (C_m) + Cost of machining (C_s). Therefore,

$$C = C_m \times bhl\rho + 2l(b + h)C_s \tag{i}$$

The selection of width b and height h cannot be arbitrary because the bar has to sustain axial force F without failure; therefore

$$\sigma_d = \frac{F}{bh} \tag{ii}$$

Eliminating h from Eq. (i) using Eq. (ii),

$$C = C_m l\rho b \times \frac{F}{\sigma_d b} + 2lC_s\left(b + \frac{F}{b\sigma_d}\right)$$

$$= k + 2lC_s\left(b + \frac{F}{b\sigma_d}\right)$$

where k = a constant = $\dfrac{C_m l\rho F}{\sigma_d}$

The objective function of the design problem is to minimize $f = k + 2lC_s(b + F/b\sigma_d)$.

The above objective function is a single variable function where except b, all others are constants.

Using the classical method of differential calculus

$$\frac{\partial F}{\partial b} = 0 + 2lC_s(1 - F/b^2\sigma_d) = 0 \qquad \text{(iii)}$$

and

$$\frac{\partial^2 F}{\partial b^2} = \frac{4lC_sF}{b^3\sigma_d} > 0$$ i.e. positive and satisfies the condition of f as minimum.

From Eq. (iii),

$$b = \sqrt{\frac{F}{\sigma_d}} = \sqrt{\frac{10{,}000}{80}} = 11.18 \text{ mm}$$

and

$$h = \frac{F}{b\sigma_d} = \sqrt{\frac{F}{\sigma_d}}$$

Therefore, the optimum section is a square section having $b = h = \sqrt{F/\sigma_d}$, i.e. $b = h = 11.18$ mm.

19.3.2 Graphical Method

A linear programming (LP) problem with only two variables can be solved by the graphical method. This method, apart from the solution, gives a physical picture of certain geometrical characteristics of the LP problem. Linear programming problems with more than two variables are solved by the simplex method. In this section, the use of the graphical method is illustrated through the following example.

Example 19.2 Optimize the design of a plastic tray which is capable of holding liquid of volume V such that the tray has fixed depth h and thickness t. The limiting width and length of the tray are b_1 and l_1, respectively.

Solution Let b = width of the tray

l = length of the tray

h = depth of the tray

Volume = bhl

t = thickness of sheet

Neglecting other costs, we assume that the tray must be of minimum weight so as to optimize for the cost of material (C). Therefore,

$$C = C_m\rho(bl + 2bh + 2lh)t \qquad \text{(i)}$$

where

C_m = cost of material/kg

ρ = density of material

Constraints

Volume, $bhl \geq v_1$ (ii)

Width, $b \leq b_1$ (iii)

Length, $l \leq l_1$ (iv)

Since Eqs. (i) and (ii) are nonlinear, they may be linearized by taking logarithms. The formulated problem is as given under:

Minimize $f = \ln [C_m \rho(bl + 2bh + 2lh)t]$ (v)

subject to

$\ln (b) + \ln (h) + \ln (l) \leq \ln (v_1)$ (vi)

$\ln (b) \leq \ln (b_1)$ (vii)

$\ln (l) \leq \ln (l_1)$ (viii)

Equations (vi) to (viii) can be plotted graphically as shown in the figure below.

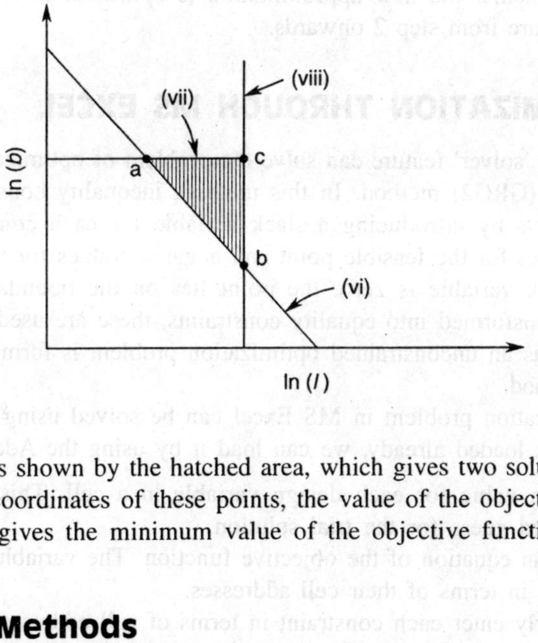

The optimal feasible region is shown by the hatched area, which gives two solutions marked as point a, b. By knowing the coordinates of these points, the value of the objective function can be found. The point which gives the minimum value of the objective function is called the optimal solution.

19.3.3 Numerical Methods

In design optimization we often come across situations where objective functions and/or constraints are either not continuous or cannot be differentiated easily. In such a situation we have to opt for numerical methods for the solution of these problems. A large number of optimization techniques, based on numerical methods, suitable for finding the optimal solution are available in literature. Some of the most commonly used techniques are the following:

1. Search methods
   (a) bounding phase (b) Fibonacci
   (c) golden section (d) Newton Raphson

2. Direct search methods
   (a) Random search (b) Univariate method
   (c) Hooke & Jeeves pattern search method
3. Descent method–steepest descent method
4. Penalty function method.

The detailed discussion of these methods is beyond the scope of this book. The reader is advised to use a good reference* on the subject for further details.

The basic philosophy of most of the numerical methods is to produce a sequence of improved approximations which eventually leads to the optimum value according to the following scheme:

1. Start with an initial trial solution.
2. Find a suitable direction which points towards the general direction of minimum.
3. Find an appropriate step length along that direction.
4. Obtain the new approximation
5. Test whether the new approximation is optimum; if yes then stop, else repeat the procedure from step 2 onwards.

19.4 OPTIMIZATION THROUGH MS EXCEL

Microsoft Excel's 'solver' feature can solve the problem of optimization. It uses the generalized reduced gradient (GRG2) method. In this method, inequality constraints are transformed into equality constraints by introducing a slack variable for each constraint. The slack variables take positive values for the feasible point and negative values for the non-feasible point. If the value of the slack variable is zero, the point lies on the boundary. Once all the inequality constraints are transformed into equality constraints, these are used for elimination of some of the variables. Thus an unconstrained optimization problem is formed, which is solved through the gradient method.

The optimization problem in MS Excel can be solved using the following procedure. If the 'solver' is not loaded already, we can load it by using the Add-in Tools menu.

1. Enter a value for each design variable in a cell. This initial value should be an educated guess for the trial solution.
2. Enter an equation of the objective function. The variables of the equation should be written in terms of their cell addresses.
3. Similarly enter each constraint in terms of cell addresses.
4. Now select the "solver" option from the "Tools" menu. It will display a dialog box as shown in Figure 19.3. Enter the following information:
   (i) In the "Set Target Cell" location enter the address of the cell containing the objective function.
   (ii) Just below the "Target cell", the options for maximization and minimization are available in radio buttons. Select any one option depending upon the requirement.

**Optimization for Engineering Design: Algorithms and Examples*, Kalyanmoy Deb, Prentice-Hall of India, New Delhi.

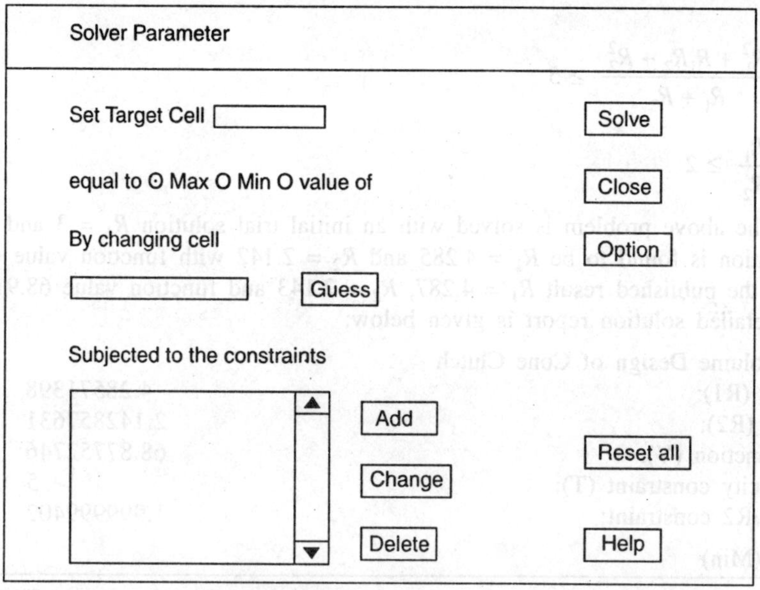

Figure 19.3 Dialog box of "Solver".

(iii) Enter the range of cell addresses containing the design variables in the area labelled "By changing cell".

(iv) Now enter the address containing each constraint; the type of constraint and the value of right-hand side. To enter these, click the "Add" button. The buttons "change" and "delete" can be used to modify or delete the constraints added through the "Add" button.

(v) If the problem is that of linear optimization, click the 'Option' button and select "Assume linear model" and select "OK" button.

(vi) When all the required information is added, select the "Solve" button. This will start the actual solution procedure.

5. When solution is over, it will display another dialog box titled "Solver Results". In this dialogue box, select the radio button of 'Keep solver solution' and 'answer report'. Press the 'OK' button. The answer report will then be generated on a separate worksheet.

Application of MS Excel to mechanical design problems is demonstrated through worked examples. In Example 19.3, a cone clutch is optimized for minimum volume. The detailed formulation is discussed in a paper titled "Application of complementary geometric programming to mechanical design problems" by SS Rao published in *International Journal of Mechanical Engineering Education*, Vol. **13**, No. 1, 1985.

Example 19.3 Determine the optimum value of inner radius R_2 and outer radius R_1 of a cone clutch for minimum volume. The following information is given.

Objective function, minimize $f(R_1, R_2) = (R_1^3 - R_2^3)$

subject to

(i) $\dfrac{R_1^2 + R_1R_2 + R_2^2}{R_1 + R_2} \geq 5$

(ii) $\dfrac{R_1}{R_2} \geq 2$

Solution The above problem is solved with an initial trial solution $R_1 = 3$ and $R_2 = 3$. The optimal solution is found to be $R_1 = 4.285$ and $R_2 = 2.142$ with function value 68.87, which agrees with the published result $R_1 = 4.287$, $R_2 = 2.143$ and function value 68.91.

The detailed solution report is given below:

Minimum Volume Design of Cone Clutch

Outer radius (R1):	4.28571398
Inner radius (R2):	2.142857631
Objective function (V):	68.87752746
Torque capacity constraint (T):	5
Ratio of R1/R2 constraint:	1.999999402

Target Cell (Min)

Cell	Name	Original value	Final value
B6	Objective function (V):	0	68.87752746

Adjustable Cells

Cell	Name	Original value	Final value
B3	Outer radius (R1):	3	4.28571398
B4	Inner radius (R2):	3	2.142857631

Constraints

Cell	Name	Cell value	Formula	Status	Slack
B8	Torque capacity constraint (T):	5	B8>=5	Binding	0
B9	Ratio of R1/R2 constraint	1.999999402	SB$9>=2	Binding	0
B3	Outer radius (R1):	4.28571398	B3>=0	Not binding	4.28571398
B4	Inner radius (R2):	2.142857631	B4>=0	Not binding	2.142857631

Example 19.4 Design a closed coil helical spring for minimum weight. The spring should have 75 N/mm stiffness and maximum deflection 60 mm. The maximum available space is 85 mm.

Solution Let k = stiffness = 75 N/mm

y = deflection = 60 mm

F = force = $k \times y = 75 \times 60 = 4500$ N

We assume that the material for the spring is patented chrome–vanadium steel having design shear stress

$$\tau_d = \frac{860}{d^{0.167}}$$

where d is the diameter of the spring wire.

Let i be the no. of active turns; the ends of the spring are square and ground. Weight of the spring wire = $\frac{\pi}{4} d^2 \times \pi D(i + 2)\rho$.

Density ρ being constant, we optimize for minimum volume.

$$\text{Volume, } V = \frac{\pi^2 d^2 D(i+2)}{4}$$

Let C = spring index = $\frac{D}{d}$. Then

$$V = \frac{\pi^2 d^3 C(i+2)}{4}$$

(i) According to the strength criterion

$$\frac{8KFD}{\pi d^3} \le \tau_d$$

or

$$\tau_d - \frac{8KFD}{\pi d^3} \ge 0$$

or

$$\frac{860}{d^{0.167}} - \frac{8 \times 4500}{\pi} \times \frac{K \cdot C}{d^2} \ge 0$$

where K = Wahl's factor = $\frac{C - 0.25}{C - 1} + \frac{0.615}{C}$

Therefore,

$$\frac{860}{d^{0.167}} - 11{,}459.1 \times \left[\frac{C - 0.25}{C - 1} + \frac{0.615}{C}\right] \times \frac{C}{d^2} \ge 0$$

(ii) According to rigidity or stiffness criterion

$$k = \frac{Gd^4}{8D^3 i}$$

or

$$k - \frac{0.8 \times 10^5 d}{8C^3 i} = 0$$

or

$$75 - \frac{10{,}000d}{C^3 i} = 0$$

The problem of optimization is now formulated as given below:

Minimize $f(d, C, i) = \pi^2 d^3 C(i + 2)/4$

subject to

(i) strength requirement

$$\frac{860}{d^{0.167}} - 11{,}459.1\left[\frac{C-0.25}{C-1} + \frac{0.615}{C}\right] \times \frac{C}{d^2} \geq 0$$

(ii) stiffness constraint

$$75 - \frac{10{,}000d}{C^3 i} = 0$$

(iii) space constraint

$$Cd + d \leq 85$$

(iv) buckling constraint

$$\frac{(i+2)d + 69}{Cd} \leq 5.2$$

(v) variable bound $C > 4$

(vi) $C \leq 10$

(vii) i = integer

The solution is initiated with the trial value

$$d = 2, \quad C = 4, \text{ and } i = 5$$

The optimal solution obtained by Excel is given below:

Wire diameter, d = 11.64 mm
Spring index, C = 5.2
No. of turns, i = 11
Value of $f(d, C, i)$ = 263,376.3 mm^3

Example 19.5 Design a solid disc-type flywheel for an engine which can absorb maximum energy with the following considerations:

(i) maximum permissible weight = 750 N
(ii) maximum permissible diameter = 1.0 m
(iii) maximum rotational speed = 7500 rpm
(iv) maximum allowable stress = 200 MN/m^2

Solution The energy absorbed, $E = \frac{1}{2} I\omega^2$

where I is the mass moment of inertia. Therefore,

$$E = \frac{mk^2\omega^2}{2}$$

with k = radius of gyration = $\frac{d}{2\sqrt{2}}$ for solid disc. Thus,

$$E = \frac{1}{2}\rho \times \frac{\pi}{4} d^2 \times b \times \left(\frac{d}{2\sqrt{2}}\right)^2 \times \left(\frac{2\pi N}{60}\right)^2$$

With ρ equal to 7800 kg/m^3, we get

$$E = 4.2d^4N^2b$$

$$\text{Weight} = \frac{\pi}{4}d^2 \times b \times 7800 = 6126.1d^2b \text{ N}$$

$$\text{Stress induced, } \sigma = \frac{3+\mu}{8}\rho\omega^2\left(\frac{d}{2}\right)^2$$

$$= \frac{3+0.3}{8} \times 7800 \times \left(\frac{2\pi N}{60}\right)^2 \times \frac{d^2}{4}$$

$$= 8.821 N^2d^2 \text{ N/m}^2$$

The optimization problem is formulated as given below:

Maximize $f(d, N, b) = 4.2d^4N^2b$

subject to

(i) $6126.1d^2b \le 750$
(ii) $8.821N^2d^2 \le 200 \times 10^6$
(iii) $d \le 1.0$
(iv) $N \le 7500$

The initial trial solution chosen is

$d = 0.16$ m
$N = 7000$ rpm
$b = 0.025$ m

and the optimal solution obtained through Excel is given below:

Flywheel diameter, $d = 0.680$ m
Flywheel width, $b = 0.264$ m
Speed, $N = 7000$ rpm
Objective function, $f(d, N, b) = 3330.97$

Example 19.6 Design a gear–pinion pair for minimum weight which can transmit 5 kW power at 1000 rpm. The velocity ratio is 3.

Solution Let d_1, d_2 = pitch diameters of pinion and gear, respectively
z_1, z_2 = number of teeth on pinion and gear, respectively
m = module.

$$\text{Approximate weight} = \rho \times \frac{\pi}{4}(d_1^2 + d_2^2)\,b$$

where

b is the face width

ρ is the density.

Density being a constant, we can minimize volume V.

$$V = \frac{\pi m^2 b(z_1^2 + z_2^2)}{4} = 0.7854m^2b\left(z_1^2 + z_2^2\right)$$

Considering static strength only

$$\text{Average torque} = \frac{60P \times 10^3}{2\pi N} = \frac{60 \times 5 \times 10^3}{2\pi \times 1000} = 47.75 \text{ N} \cdot \text{m}$$

$$\text{Tangential force, } F_t = \frac{2T}{d_1} = \frac{2T}{mz_1} = \frac{95{,}500}{mz_1}$$

Let service factor $C_S = 1.0$

Velocity factor $C_V = 0.3$

$$\text{The maximum force, } F_{max} = \frac{C_S F_t}{C_V} = \frac{1 \times 95{,}500}{0.3 \times mz_1} = \frac{318{,}333.3}{z_1}$$

$$\text{Beam strength, } F_{beam} = mb\sigma_d Y$$

Let σ_d = design stress = 75 N/mm²

Y = Lewis form factor

$= \pi(0.154 - 0.912/z_1)$ for 20° full-depth teeth

Therefore,

$$F_{beam} = mb \times 75 \times \pi(0.154 - 0.912/z_1)$$

$$= 235.62mb(0.154 - 0.912/z_1)$$

For safe design, $F_{beam} \geq F_{max}$ or $F_{beam} - F_{max} \geq 0$

or

$$235.62mb(0.154 - 0.912/z_1) - \frac{318{,}333.3}{z_1} \geq 0$$

The problem of optimization is formulated as given below:

$$\text{Minimize } f(m, b, z_1, z_2) = 0.7854\, m^2 b\left(z_1^2 + z_2^2\right)$$

subject to

(i) $235.62mb(0.154 - 0.912/z_1) - \dfrac{318{,}333.3}{z_1} \geq 0.$

(ii) $b/m \geq 9.5$

(iii) $b/m \leq 12.5$

(iv) $z_1 > 18$

(v) $z_2/z_1 = 3$

(vi) z_1 = integer

(vii) z_2 = integer

The initial solution chosen is $z_1 = 18$, $z_2 = 30$, $b = 20$ mm, $m = 2$

The optimal solution obtained is:

Number of teeth on pinion, $z_1 = 24$

Number of teeth on gear, $z_2 = 72$

Face width, $b = 77.88$ mm

Module, $m = 6.23$ mm

Objective function, $f(m, b, z_1, z_2) = 136{,}79{,}159.3$ mm³

Chapter 20 Introduction to Finite Element Method

20.1 INTRODUCTION

The advent of high-speed digital computer has enabled engineers to employ various numerical discretization techniques for approximate solution of the complex problems. The Finite Element Method (FEM) is one such technique. Although it was originally developed as a tool for static analysis of structures, its applications nowadays range from linear deformation and stress analysis to nonlinear and dynamic analysis, heat transfer, fluid mechanics, rock mechanics, magnetic flux, and various other areas of engineering.

The basic concept of finite element method is that a body or a structure, under study, is divided into smaller elements of finite length and width called finite elements. These elements are assumed to be interconnected at joints called nodes or nodal points. The properties of these elements are formulated by assuming a simple function which depicts the variation of field variables within the element in terms of nodal parameters. Such assumed functions are called *field variable functions.* A variational principle such as Rayleigh–Ritz or Galerkin principle is employed to obtain a set of equilibrium equations for an element. The equilibrium equations for the entire body are then obtained by combining the equations of individual elements in such a way that continuity is preserved at the interconnecting nodes. These equations are modified for the given boundary conditions and then solved to obtain the unknown field variables.

20.2 PROCEDURE OF FEM

The general procedure to be adopted to solve a problem by the finite element method is outlined below:

(i) Discretization of domain
(ii) Selection of an interpolation model
(iii) Formulation of characteristics matrix
(iv) Assembly of the characteristic matrices
(v) Application of boundary conditions
(vi) Solution of equations.

A brief description of these steps is given in the following subsections:

20.2.1 Discretization of Domain

The first step in the FEM is to discretize the solution region or body. The continuum or physical body, under study, is subdivided into an equivalent system of finite elements. This is equivalent to replacing the physical body having infinite degrees of freedom by discretized elements having a finite number of degrees of freedom. The process of discretization is essentially an exercise of engineering judgement.

The choice of the type of elements through which the body is discretized is dictated by the geometry of the body and the number of independent coordinates necessary to describe them. Figure 20.1 shows some of the most commonly used elements.

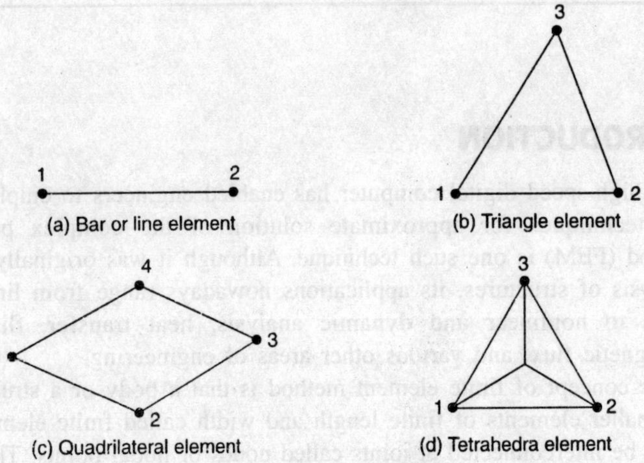

Figure 20.1 Types of elements.

The size and the number of elements influence the accuracy and convergence of the solution, and therefore need to be chosen with care. If the size of the element is small, the outcome solution is expected to be more accurate but the cost of computation increases rapidly. Therefore, sometimes, we have to use elements of different sizes in the same body to strike a balance between accuracy and computation cost. In general, whenever a steep gradient of the field variable is expected, we must choose a finer mesh in that region. However by increasing the number of elements, the accuracy of the solution increases only up to a limit.

The location of nodes is another important aspect of discretization. If a body has abrupt changes in geometry, material properties and external conditions like load, temperature, etc. nodes have to be introduced at these discontinuities as shown in Figure 20.2.

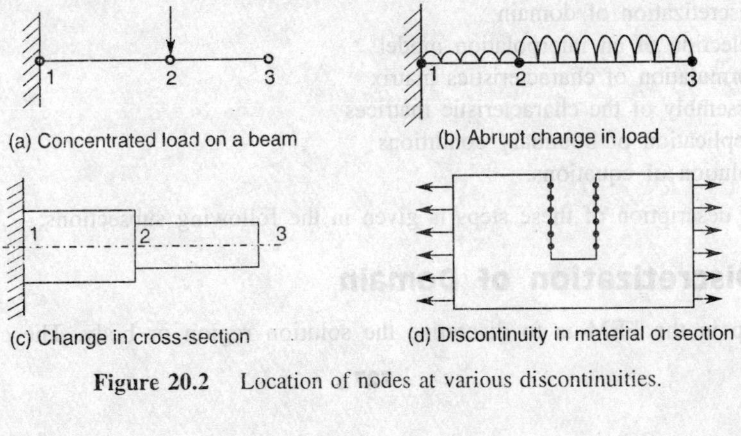

Figure 20.2 Location of nodes at various discontinuities.

20.2.2 Selection of an Interpolation Model

In finite element analysis the solution functions used to represent the behaviour of field variables in each element are called *interpolation functions.* Generally, polynominal functions are most widely used as they are easy to formulate, computerize, differentiate, and integrate. The accuracy of the solution can be improved by increasing the order of the polynomial.

Various polynomials used for one-dimensional and two-dimensional problems are given below (see also Figure 20.3).

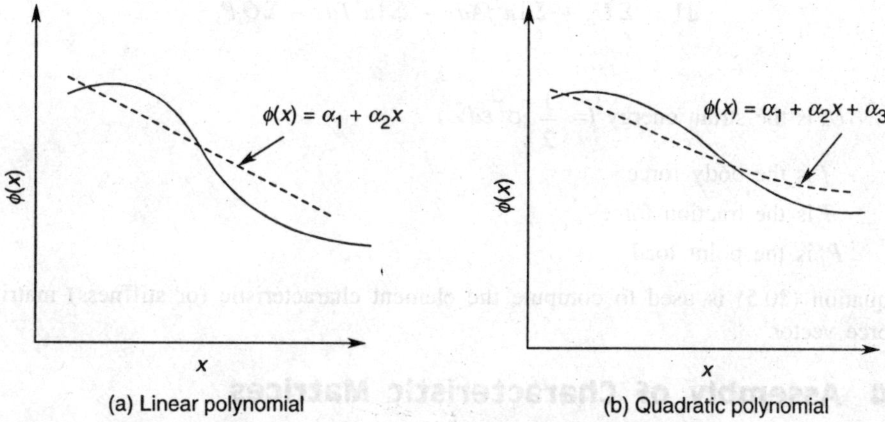

Figure 20.3 Different polynomial approximations.

1. *Linear polynomial*
   (a) For one-dimensional problems

$$\phi(x) = \alpha_1 + \alpha_2 x \tag{20.1}$$

   (b) For two-dimensional problems

$$\phi(x, y) = \alpha_1 + \alpha_2 x + \alpha_3 y \tag{20.2}$$

2. *Quadratic polynomial*
   (a) For one-dimensional problems

$$\phi(x) = \alpha_1 + \alpha_2 x + \alpha_3 x^2 \tag{20.3}$$

   (b) For two-dimensional problems

$$\phi(x,y) = \alpha_1 + \alpha_2 x + \alpha_3 y + \alpha_4 x^2 + \alpha_5 y^2 + \alpha_6 xy \tag{20.4}$$

20.2.3 Formulation of Characteristics Matrix

The element characteristics matrix of an element can be derived from the material and geometrical properties of the element obtained by minimum potential energy principle called the Rayleigh–Ritz method. According to the principle of minimum potential energy, "For a conservative system, of all the kinematically admissible displacement fields, those corresponding to equilibrium extremize the total potential energy. If the extreme condition is minimum, the equilibrium state is stable". Kinematically admissible displacements are those which satisfy the single-valued nature of displacement and boundary conditions.

In a structural problem, the characteristic matrix is called the *stiffness* matrix of the element. The forces acting upon the element are called force vectors which are converted into an equivalent nodal force.

The total potential energy of an elastic body is defined as the sum of the strain energy and the work potential

$$\Pi = U + WP$$

For a discretized system, the potential energy becomes

$$\Pi = \sum_e U_e - \sum_e \int u^T fA dx - \sum_e \int u^T T dx - \sum Q_i P_i \quad (20.5)$$

where

U_e is the strain energy $(= \frac{1}{2} \int \sigma^T \varepsilon dv)$

f is the body force

T is the traction force

P_i is the point load.

Equation (20.5) is used to compute the element characteristic (or stiffness) matrix and nodal force vector.

20.2.4 Assembly of Characteristic Matrices

The individual element characteristic matrices and vectors calculated in the local coordinate system are transformed to a common coordinate system called the global coordinate system. Later, the global characteristic matrices and vectors of all the elements are assembled for the entire body. The most common assembly technique used is called the *direct stiffness method*. In general, the basis of assembly is that the displacement at a node for all connecting elements must be the same.

The overall assembled characteristics matrix and the nodal vector can be represented as

$$[K]\{Q\} = \{F\} \quad (20.6)$$

where

$[K]$ is the assembled characteristics matrix

$\{Q\}$ is the nodal function vector

$\{F\}$ is the nodal characteristics vector.

The total potential energy in a continuum is given as

$$\Pi = \frac{1}{2} Q^T K Q - Q^T F \quad (20.7)$$

20.2.5 Application of Boundary Conditions

A problem in finite element method is considered as incomplete unless certain boundary conditions are prescribed. The kinematic constraints which ensure the equilibrium of the system are called the *boundary constraints*. In this section, we will discuss two approaches for dealing with the specified displacement boundary conditions. These approaches are described below.

Elimination approach

In the elimination approach the equilibrium equations are obtained by minimizing the potential energy function with respect to the nodal function vector Q, subjected to specified boundary conditions.

Let us assume that in a structural system the boundary condition at node point 1, i.e. Q_1 equals a_1. For an N degree of freedom system, we have the global stiffness matrix as

$$[K] = \begin{bmatrix} k_{11} & k_{12} & \cdots & k_{1n} \\ k_{21} & k_{22} & \cdots & k_{2n} \\ \vdots & \vdots & \cdots & \vdots \\ k_{n1} & k_{n2} & \cdots & k_{nn} \end{bmatrix}$$

where the displacement vector $Q = [Q_1\ Q_2\ Q_3 \ldots Q_n]^T$ and the force vector $F = [F_1\ F_2\ F_3 \ldots F_n]^T$.

To minimize the potential energy, we perform the following operation

$$\frac{d\Pi}{dQ_i} = 0 \qquad i = 2, 3, \ldots, n \tag{20.8}$$

which gives the finite element equations in the form of matrix as

$$\begin{bmatrix} k_{22} & k_{23} & \cdots & k_{2n} \\ k_{32} & k_{33} & \cdots & k_{3n} \\ \vdots & \vdots & \cdots & \vdots \\ k_{n2} & \cdots & \cdots & k_{nn} \end{bmatrix} \begin{Bmatrix} Q_2 \\ Q_3 \\ \vdots \\ Q_n \end{Bmatrix} = \begin{Bmatrix} F_2 - k_{21}a_1 \\ F_3 - k_{31}a_1 \\ \vdots \\ F_n - k_{n1}a_1 \end{Bmatrix}$$

or

$$[K]\{Q\} = \{F\} \tag{20.9}$$

We observe from Eq. (20.9) that rows and columns of the stiffness matrix corresponding to the specified displacement are eliminated to get the equilibrium stiffness matrix.

Penalty approach

In the penalty approach, a spring of a large stiffness C is used to model the boundary condition, whose one end is assumed to be displaced by a specified amount.

We assume that the displaced amount is a_1 in the direction of the global displacement vector Q_1. Therefore, in the potential energy relation, an additional term of strain energy due to this spring is added. The potential energy relation is

$$\Pi = \frac{1}{2}Q^T K Q + \frac{1}{2}C(Q_1 - a_1)^2 - Q^T F \tag{20.10}$$

The minimization of Π can be carried out by setting $\frac{d\Pi}{dQ_i} = 0$. The resulting finite element equations in matrix form are

$$\begin{bmatrix} k_{11}+C & k_{12} & \cdots & k_{1n} \\ k_{21} & k_{22} & \cdots & k_{2n} \\ \vdots & \vdots & \cdots & \vdots \\ k_{n1} & k_{n2} & \cdots & k_{nn} \end{bmatrix} \begin{Bmatrix} Q_1 \\ Q_2 \\ \vdots \\ Q_n \end{Bmatrix} = \begin{Bmatrix} F_1 + Ca_1 \\ F_2 \\ \vdots \\ F_n \end{Bmatrix} \tag{20.11}$$

The finite element equilibrium equations obtained by the application of boundary conditions are solved by the Gauss–elimination method to determine the nodal functional variables.

20.3 ONE-DIMENSIONAL PROBLEM

Consider a one-dimensional bar element, as shown in Figure 20.4, which is subjected to various forces. In one-dimensional problems, the location of the node is defined by a single coordinate and each node has one degree of freedom. The finite element discretized model is shown in Figure 20.4(b). Each discretized element has two nodes which are numbered by two coordinate systems called the local coordinate and the global coordinate system. The local coordinate system is a coordinate system of individual elements, whereas the global coordinate system is a generalized coordinate system which is common for all elements. Elements in local and global coordinates are represented in Figure 20.5.

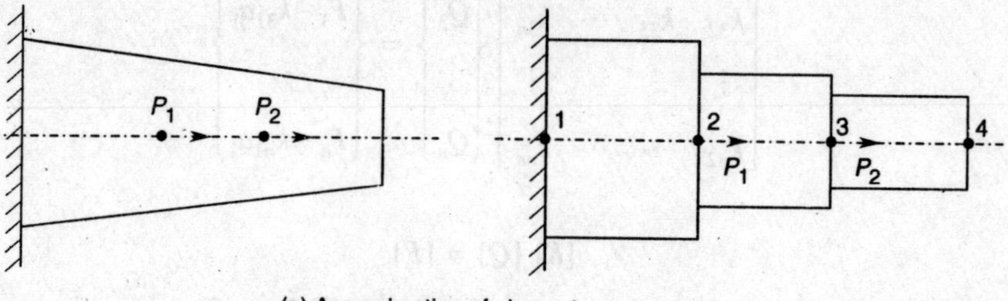

(a) Approximation of element

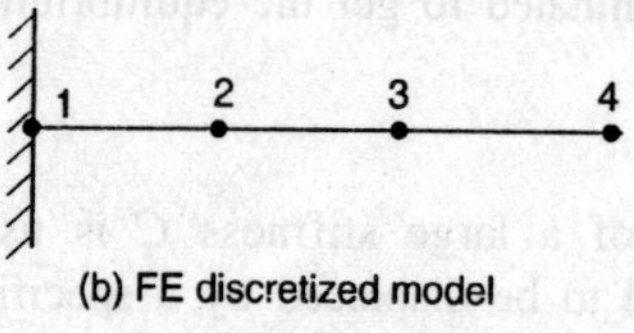

(b) FE discretized model

Figure 20.4 FE modelling of bar subjected to axial force.

In the local coordinate system the displacements of each element at the local node numbers are given by q_1 and q_2, whereas in the global coordinate system the displacement and force vector are given as

$$Q = \begin{bmatrix} Q_1 & Q_2 & Q_3 & Q_4 \end{bmatrix}^T \tag{20.12}$$

and

$$F = \begin{bmatrix} F_1 & F_2 & F_3 & F_4 \end{bmatrix}^T \tag{20.13}$$

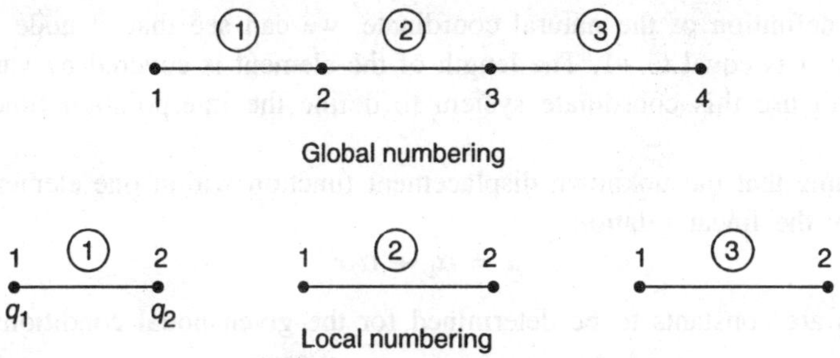

Figure 20.5 Elements in local and global coordinates

20.3.1 Shape Function

Consider a typical finite element as shown in Figure 20.6 and numbered as 1 and 2. The locations of the nodes are defined by coordinates x_1 and x_2.

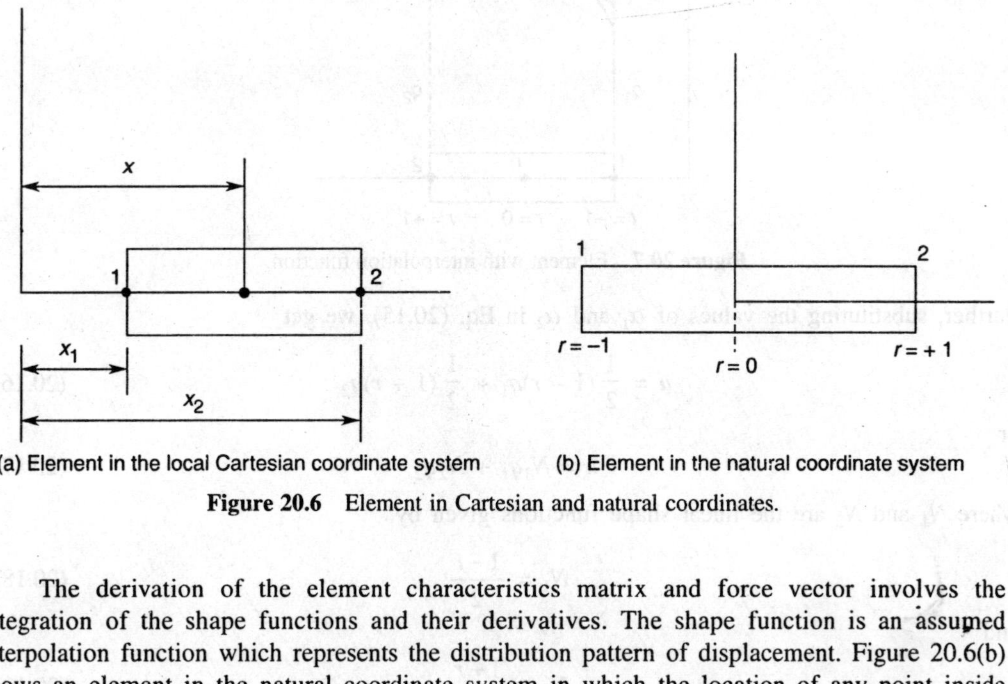

Figure 20.6 Element in Cartesian and natural coordinates.

The derivation of the element characteristics matrix and force vector involves the integration of the shape functions and their derivatives. The shape function is an assumed interpolation function which represents the distribution pattern of displacement. Figure 20.6(b) shows an element in the natural coordinate system in which the location of any point inside the element can be defined by a non-dimensional number whose magnitude lies between 0 and 1 or between −1 and +1. Usually, the natural coordinates are chosen such that some of the natural coordinates will have unity magnitude at the primary nodes of the element.

The natural coordinate denoted by r for the above one-dimensional element is given as

$$r = \frac{2(x - x_1)}{x_2 - x_1} - 1 \tag{20.14}$$

From the above definition of the natural coordinate, we can see that at node 1, r is equal to -1, and at node 2, r is equal to $+1$. The length of the element is covered by varying r between -1 to $+1$. We will use this coordinate system to define the interpolation function called the *shape function.*

Let us assume that the unknown displacement function within one element (Figure 20.7) is interpolated by the linear relation

$$u = \alpha_1 + \alpha_2 x \tag{20.15}$$

where α_1 and α_2 are constants to be determined for the given nodal conditions, which are:

$$\text{When } r = -1 \quad u = q_1; \quad \text{when } r = +1 \quad u = q_2$$

Substituting these conditions in Eq. (20.15), we get the value of α_1 and α_2.

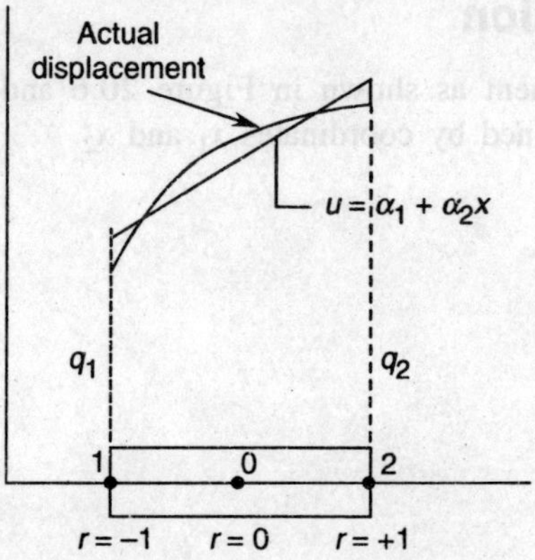

Figure 20.7 Element with interpolation function.

Further, substituting the values of α_1 and α_2 in Eq. (20.15), we get

$$u = \frac{1}{2}(1 - r)q_1 + \frac{1}{2}(1 + r)q_2 \tag{20.16}$$

or

$$u = N_1 q_1 + N_2 q_2 \tag{20.17}$$

where N_1 and N_2 are the linear shape functions given by

$$N_1 = \frac{1 - r}{2} \tag{20.18}$$

and

$$N_2 = \frac{1 + r}{2} \tag{20.19}$$

In matrix notations, Eq. (20.17) can be written as

$$\{u\} = [N]\{q\}$$

where

$[N]$ is the shape function matrix ($= [N_1 \quad N_2]$)

$\{q\}$ is the displacement vector ($= [q_1 \quad q_2]^T$).

In finite element, the element whose geometrical shape and field variables are described by the same interpolation function or shape function are known as *isoparametric elements.* These elements are most widely used in two- and three-dimensional elasticity problems. Therefore, the coordinate x of any point P within the element can be expressed as a linear combination of nodal coordinates of nodes and linear shape functions as

$$x = N_1 x_1 + N_2 x_2 \tag{20.20}$$

20.3.2 Strain–Displacement Matrix

In a one-dimensional element, the strain induced in the element can be expressed in terms of displacement u by the following relation

$$\varepsilon = \frac{du}{dx} \tag{20.21}$$

This relation can be written in terms of displacement at local node by the use of natural coordinates and shape functions. Thus, using the chain rule, we obtain

$$\varepsilon = \frac{du}{dr} \cdot \frac{dr}{dx} \tag{20.22}$$

Differentiating (20.14) and (20.16), we get the relation for strain as

$$\varepsilon = \frac{1}{x_2 - x_1}(-q_1 + q_2) \tag{20.23}$$

or

$$\varepsilon = [B]\{q\} \tag{20.24}$$

where $[B]$ is the strain–displacement matrix $\left(= \frac{1}{x_2 - x_1}[-1 \quad 1] \right)$. (20.25)

The stress–strain relation from the Hooke's Law is

$$\sigma = E[B]\{q\} \tag{20.26}$$

20.3.3 Element Characteristics Matrix

The element stiffness matrix in local coordinates can be obtained from the elements' strain energy relation

$$U_e = \frac{1}{2}\int \sigma^T \varepsilon A dx \tag{20.27}$$

Substituting the values of stress σ and strain ε, we get

$$U_e = \frac{1}{2} q^T \int [B]^T E \, [B] \, A \, dx \, q \tag{20.28}$$

The above equation can be written in the form

$$U_e = \frac{1}{2} q^T K_e q \tag{20.29}$$

where

K_e is the element stiffness matrix $\left(= \int [B]^T E\, [B]\, A\, dx \right)$ (20.30)

E is the elastic modulus; however in 2-D and 3-D problems E is replaced by the property matrix $[D]$.

In the present finite element model, the cross-sectional area of the element is constant. Further, the transformation of coordinate x to r from Eq. (20.14) is given by

$$dx = \frac{x_2 - x_1}{2} dr = \frac{l}{2} dr$$

Substituting the value of $[B]$ and dx in Eq. (20.30), we get

$$K_e = \frac{AEl}{2} [B]^T [B] \int_{-1}^{+1} dr$$

or

$$K_e = \frac{AE}{l} \begin{bmatrix} 1 & -1 \\ -1 & 1 \end{bmatrix} \tag{20.31}$$

The force vector of the element can be obtained from the force terms appearing in the total potential energy relation.

(a) *Body force*. The body force terms of the element can be obtained from

$$\int u^T \rho A dx = \rho A \int (N_1 q_1 + N_2 q_2) dx \tag{20.32}$$

where ρ is the force per unit volume or density.

Equation (20.32) can be rewritten as

$$\int u^T \rho A dx = q^T f^e$$

where f^e is the nodal body force vector $= \dfrac{A\rho l}{2} \begin{bmatrix} 1 \\ 1 \end{bmatrix}$ (20.33)

(b) *Traction force*. Similarly, the element traction force vector can be obtained from the traction force term

$$\int u^T T dx = q^T T^e \tag{20.34}$$

where T^e is the nodal traction force vector $\left(= \frac{Tl}{2} \begin{bmatrix} 1 \\ 1 \end{bmatrix} \right)$ (20.35)

with T as the fraction force per unit length.

Once the element characteristics matrices and vectors are calculated for all elements, they can be assembled according to the available information on nodal connectivity. The assembled global stiffness and global force vector for the three-element discretized model as shown in Figure 20.4(b) is given below:

$$[K] = \begin{bmatrix} \frac{A_1E}{l_1} & \frac{-A_1E}{l_1} & 0 & 0 \\ \frac{-A_1E}{l_1} & \left(\frac{A_1}{l_1} + \frac{A_2}{l_2}\right)E & -\frac{A_2E}{l_2} & 0 \\ 0 & \frac{-A_2E}{l_2} & \left(\frac{A_2}{l_2} + \frac{A_3}{l_3}\right)E & \frac{-A_3E}{l_3} \\ 0 & 0 & \frac{-A_3E}{l_3} & \frac{A_3E}{l_3} \end{bmatrix} \quad (20.36)$$

The assembled force vector is

$$\{F\} = \begin{bmatrix} \frac{A_1l_1\rho}{2} + \frac{l_1T_1}{2} \\ \left(\frac{A_1l_1\rho}{2} + \frac{l_1T_1}{2}\right) + \left(\frac{A_2l_2\rho}{2} + \frac{l_2T_2}{2}\right) \\ \left(\frac{A_2l_2\rho}{2} + \frac{l_2T_2}{2}\right) + \left(\frac{A_3l_3\rho}{2} + \frac{l_3T_3}{2}\right) \\ \left(\frac{A_3l_3\rho}{2} + \frac{l_3T_3}{2}\right) \end{bmatrix} + \begin{bmatrix} 0 \\ P_1 \\ P_2 \\ 0 \end{bmatrix} \quad (20.37)$$

The finite element equations for a continuum problem are written as

$$[K]\{Q\} = \{F\} \quad (20.38)$$

In the above equations, appropriate boundary conditions are incorporated and then they are solved for the unknown field variable, i.e. displacement.

Example 20.1 The figure below shows a steel bar element subjected to axial force at two locations. The bar is rigidly fixed at one end and free to displace by 3.5 mm at the other end. Determine the nodal displacement.

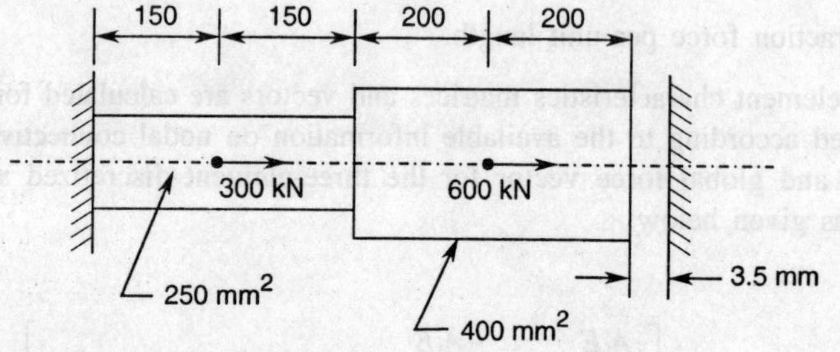

Solution The finite element model of the bar is shown below:

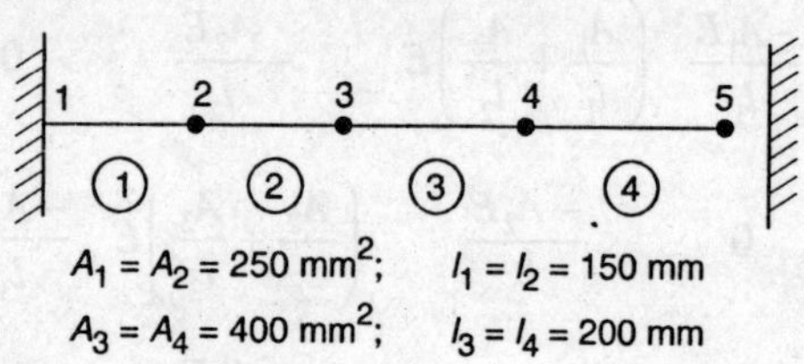

The element stiffness matrices are:

$$K^1 = \frac{EA_1}{l_1}\begin{bmatrix} 1 & -1 \\ -1 & 1 \end{bmatrix} = \frac{2 \times 10^5 \times 250}{150}\begin{bmatrix} 1 & -1 \\ -1 & 1 \end{bmatrix}$$

$$K^2 = \frac{2 \times 10^5 \times 250}{150}\begin{bmatrix} 1 & -1 \\ -1 & 1 \end{bmatrix}$$

$$K^3 = \frac{2 \times 10^5 \times 400}{200}\begin{bmatrix} 1 & -1 \\ -1 & 1 \end{bmatrix}$$

and

$$K^4 = \frac{2 \times 10^5 \times 400}{200}\begin{bmatrix} 1 & -1 \\ -1 & 1 \end{bmatrix}$$

The assembled stiffness matrix is

$$[K] = 2 \times 10^5 \begin{bmatrix} 1.667 & -1.667 & 0 & 0 & 0 \\ -1.667 & 3.334 & -1.667 & 0 & 0 \\ 0 & -1.667 & 3.667 & -2 & 0 \\ 0 & 0 & -2 & 4 & -2 \\ 0 & 0 & 0 & -2 & 2 \end{bmatrix}$$

Neglecting body force and traction force, the force vector due to point load is

$$\{F\} = 10^5 \times \begin{bmatrix} 0 & 3 & 0 & 6 & 0 \end{bmatrix}^T$$

Boundary conditions at node 1 $\quad Q_1 = 0$
at node 5 $\quad Q_5 = 3.5$ mm

By using the penalty approach, a large number $C = 8 \times 10^9$ is added to the first and fifth diagonal terms of the stiffness matrix and also to force vector as per Eq. (20.11). Thus, we get equations of finite element model as

$$2 \times 10^5 \begin{bmatrix} 80001.667 & -1.667 & 0 & 0 & 0 \\ 0 & 3.334 & -1.667 & 0 & 0 \\ 0 & -1.667 & 3.667 & -2 & 0 \\ 0 & 0 & -2 & 4 & -2 \\ 0 & 0 & 0 & -2 & 80002 \end{bmatrix} \begin{Bmatrix} Q_1 \\ Q_2 \\ Q_3 \\ Q_4 \\ Q_5 \end{Bmatrix} = \begin{Bmatrix} 0 \\ 3 \times 10^5 \\ 0 \\ 6 \times 10^5 \\ 2.8 \times 10^{10} \end{Bmatrix}$$

Solving the above equations by the Gauss-elimination approach we get the global displacement vector

$$Q = \begin{bmatrix} 2.02 \times 10^{-4} & 2.018 & 3.136 & 4.068 & 3.5 \end{bmatrix}^T$$

20.4 TWO-DIMENSIONAL PROBLEM—CST ELEMENT

There are many problems in which the geometry of the region and field variables such as displacement, forces, temperature, flow velocity or pressure, etc. are defined by two independent coordinates. Such problems can be solved by 2-D finite element formulations.

In the 2-D finite element formulation, the region can be filled by triangular or quadrilateral elements. Triangular elements are relatively simple in their formulation and can be easily programmed. They can be used along with other element shapes to discretize regions of steep strain gradient and irregular or curved boundaries. However when the curved bounded region is modelled by constant strain triangular CST elements, some region may remain unfilled. This unfilled region contributes to some part of approximation, as shown in Figure 20.8.

In solid mechanics, each node or corner of a triangular element is permitted to be displaced in two directions, i.e. along x and y axes. Thus each node has two degrees of freedom (dof) and each triangular element has six dof in the local coordinates. The displacement vectors in local and global coordinates are given by

$$\{q\} = \begin{bmatrix} q_1 & q_2 & q_3 & q_4 & q_5 & q_6 \end{bmatrix}^T \quad (20.39a)$$

$$[Q] = \begin{bmatrix} Q_1 & Q_2 & \cdots & Q_n \end{bmatrix}^T \quad (20.39b)$$

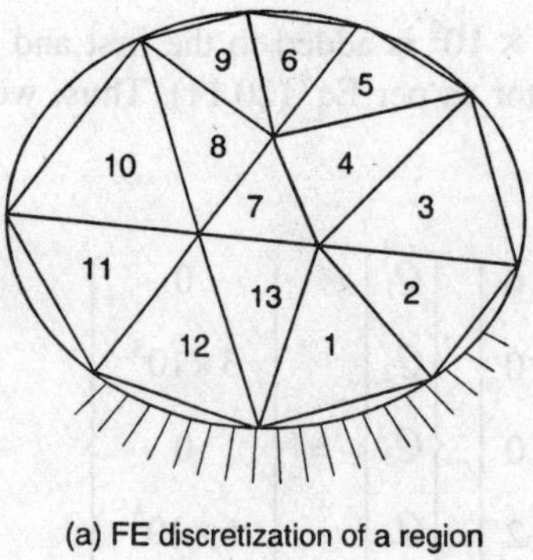

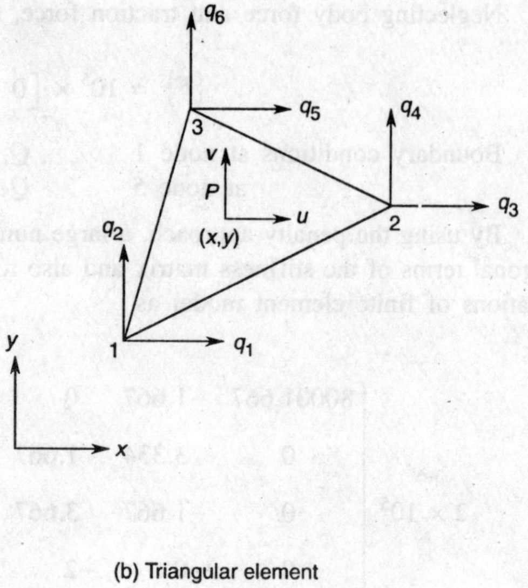

Figure 20.8 Finite element discretization of a region by triangular elements.

20.4.1 Shape Function

Consider a typical triangular element, as shown in Figure 20.9(a) in local coordinates, whose nodes are numbered 1, 2, and 3, respectively. The location of these nodes is defined by coordinates (x_1, y_1), (x_2, y_2), and (x_3, y_3), respectively. The parent element is mapped into an isosceles right-angle triangle in natural coordinates and in turn is transformed into an isoparametric triangular element shown in Figure 20.9.

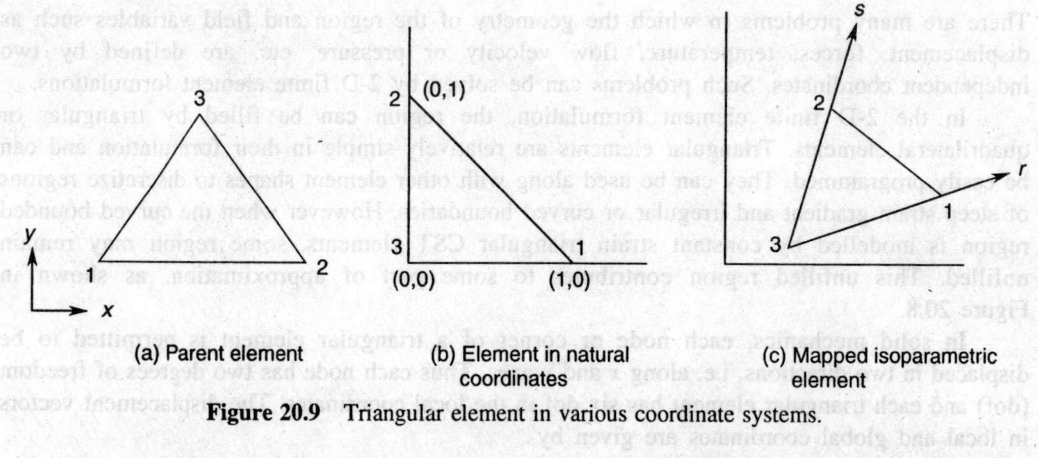

Figure 20.9 Triangular element in various coordinate systems.

The two independent natural coordinates r and s are taken for transformation to Cartesian coordinates as

$$x = \sum_{i=1}^{3} N_i x_i \quad \text{and} \quad y = \sum_{i=1}^{3} N_i y_i \tag{20.40}$$

where N_i are the shape functions which are defined by linear interpolation in natural coordinates, just similar to the one-dimensional problem.

The independent shape functions are

$$N_1 = r; \qquad N_2 = s; \qquad N_3 = 1 - r - s$$

where r and s are the natural coordinates, defined as

$$r = \frac{2(x - x_1)}{x_2 - x_1} - 1 \tag{20.41}$$

$$s = \frac{2(y - y_1)}{y_2 - y_1} - 1 \tag{20.42}$$

The field variable inside the element namely the displacement can be written using the shape function and nodal values of the unknown displacements as

$$[U] = [N]\{q\} \tag{20.43}$$

where

$[N]$ is the shape function matrix given by

$$[N] = \begin{bmatrix} N_1 & 0 & N_2 & 0 & N_3 & 0 \\ 0 & N_1 & 0 & N_2 & 0 & N_3 \end{bmatrix}$$

$\{q\}$ is the displacement vector in local coordinates given by

$$\{q\} = \begin{bmatrix} q_1 & q_2 & q_3 & q_4 & q_5 & q_6 \end{bmatrix}^T$$

The field variable displacement Eq. (20.43) can be rewritten in terms of natural coordinates as

$$u = (q_1 - q_5)r + (q_3 - q_5)s + q_5 \tag{20.44}$$

$$v = (q_2 - q_6)r + (q_4 - q_6)s + q_6 \tag{20.45}$$

In the isoparametric triangular element the coordinates of any point lying inside the triangle can be represented in terms of the shape function and nodal coordinates. Thus,

$$x = N_1x_1 + N_2x_2 + N_3x_3 \tag{20.46}$$

$$y = N_1y_1 + N_2y_2 + N_3y_3 \tag{20.47}$$

Substituting the values of the shape function, we get

$$x = (x_1 - x_3)r + (x_2 - x_3)s + x_3 \tag{20.48}$$

$$y = (y_1 - y_3)r + (y_2 - y_3)s + y_3 \tag{20.49}$$

The above equations can be written in short by using the notations

$$x_{ij} = x_i - x_j \qquad \text{and} \qquad y_{ij} = y_i - y_j$$

or

$$x = x_{13}r + x_{23}s + x_3 \tag{20.50}$$

$$y = y_{13}r + y_{23}s + y_3 \tag{20.51}$$

20.4.2 Strain–Displacement Relation

For a linear elastic element, the strain–displacement relation for the 2-D problem is given by

$$\varepsilon = [\varepsilon_x \quad \varepsilon_y \quad \gamma_{xy}]$$

$$= \left[\frac{\partial u}{\partial x} \quad \frac{\partial v}{\partial y} \quad \left(\frac{\partial u}{\partial y} + \frac{\partial v}{\partial x}\right)\right]$$

The strain function can be obtained by first differentiating displacement with respect to natural coordinates and then by transforming into Cartesian coordinates. The relation between strain and displacement is written as

$$\varepsilon = \left[B\right]\{q\} \tag{20.52}$$

where $\left[B\right]$ is the strain–displacement matrix given by

$$[B] = \frac{1}{|J|}\begin{bmatrix} y_{23} & 0 & y_{31} & 0 & y_{12} & 0 \\ 0 & x_{32} & 0 & x_{13} & 0 & x_{21} \\ x_{32} & y_{23} & x_{13} & y_{31} & x_{21} & y_{12} \end{bmatrix} \tag{20.53}$$

with $|J|$ is the determinant of Jacobian matrix.*

20.4.3 Element Characteristics Matrix

The element characteristics matrix and vectors, namely the stiffness matrix and the nodal force vectors can be obtained by potential energy expression of the system, given by Eq. (20.5). Element stiffness matrix

$$K_e = \int \left[B\right]^T \left[D\right] \left[B\right] t dA \tag{20.54}$$

For plane stress and plane strain problems, the element stiffness matrix can be determined by taking appropriate material property matrix $\left[D\right]$.

* Jacobian matrix is defined as

$$[J] = \begin{bmatrix} x_{13} & y_{13} \\ x_{23} & y_{23} \end{bmatrix}$$

For plane stress condition

$$[D] = \frac{E}{1-\mu^2}\begin{bmatrix} 1 & \mu & 0 \\ \mu & 1 & 0 \\ 0 & 0 & \frac{1-\mu}{2} \end{bmatrix} \tag{20.55}$$

and for plane strain condition

$$[D] = \frac{E}{(1+\mu)(1-2\mu)}\begin{bmatrix} 1-\mu & \mu & 0 \\ \mu & 1-\mu & 0 \\ 0 & 0 & 0.5-\mu \end{bmatrix} \tag{20.56}$$

The element body force and traction force vectors can be obtained by the terms of work potential:

(i) Body force vector

$$f^e = \frac{At}{3}\left[f_x \;\; f_y \;\; f_x \;\; f_y \;\; f_x \;\; f_y\right]^T \tag{20.57}$$

(ii) Traction force vector

$$T^e = \frac{tl}{2}\left[T_x \;\; T_y \;\; T_x \;\; T_y\right]^T \tag{20.58}$$

The stress value in the element, which is constant in magnitude is calculated from the following equation

$$\sigma = [D][B]\{q\} \tag{20.59}$$

where $\{q\}$ is the element nodal displacements extracted from global displacement vector Q and transformed into local coordinates.

Example 20.2 A two-dimensional plate is loaded by a 10 kN force as shown in the figure below. Determine the displacements at nodes 1 and 2 using the plane stress conditions. The plate may be discretized by two triangular elements for hand calculations. The body force and traction force may be neglected. The thickness of the plate is 15 mm and the elastic modulus $E = 2 \times 10^5$ N/mm^2 and $\mu = 0.3$.

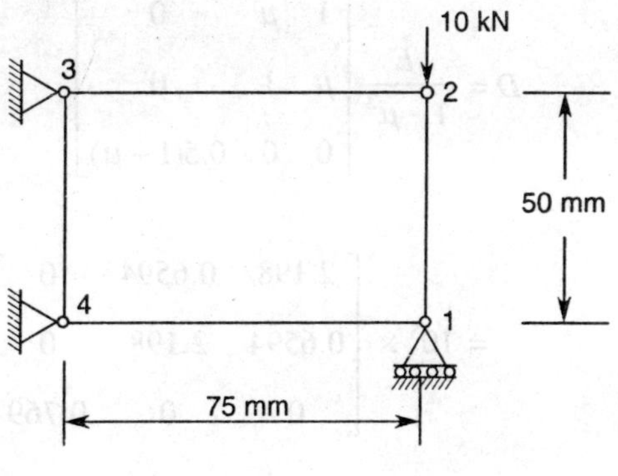

Solution The plate is modelled by two 2-D constant strain triangular elements. Their node numbers are shown in the figure below as the numbers written within circles.

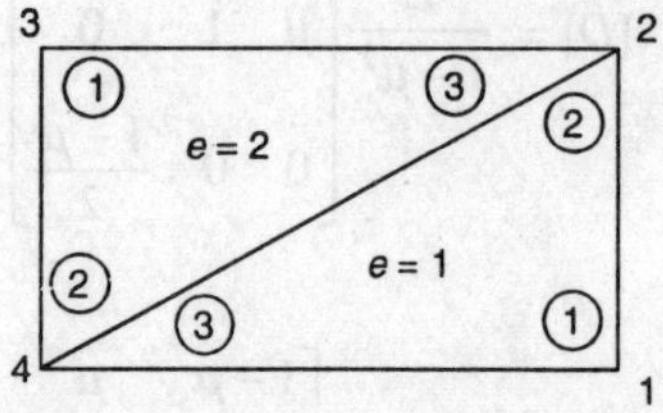

The strain–displacement matrix of element 1

$$[B^1] = \frac{1}{|J|}\begin{bmatrix} y_{23} & 0 & y_{31} & 0 & y_{12} & 0 \\ 0 & x_{32} & 0 & x_{13} & 0 & x_{21} \\ x_{32} & y_{23} & x_{13} & y_{31} & x_{21} & y_{12} \end{bmatrix}$$

where $|J| = x_{13}y_{23} - x_{23}y_{13} = 75 \times 50 - 75 \times 0 = 3750$. Therefore,

$$[B^1] = \frac{1}{3750}\begin{bmatrix} 50 & 0 & 0 & 0 & -50 & 0 \\ 0 & -75 & 0 & 75 & 0 & 0 \\ -75 & 50 & 75 & 0 & 0 & -50 \end{bmatrix}$$

$$= \begin{bmatrix} 0.0134 & 0 & 0 & 0 & -0.0134 & 0 \\ 0 & -0.02 & 0 & 0.02 & 0 & 0 \\ -0.02 & 0.0134 & 0.02 & 0 & 0 & -0.02 \end{bmatrix}$$

The properties matrix $[D]$ for plane stress condition is given by

$$D = \frac{E}{1-\mu^2}\begin{bmatrix} 1 & \mu & 0 \\ \mu & 1 & 0 \\ 0 & 0 & 0.5(1-\mu) \end{bmatrix}$$

$$= 10^5 \times \begin{bmatrix} 2.198 & 0.6594 & 0 \\ 0.6594 & 2.198 & 0 \\ 0 & 0 & 0.769 \end{bmatrix}$$

On performing matrix multiplication, $[D][B^1]$, we get

$$[D][B^1] = \begin{bmatrix} 2945.3 & -1318.8 & 0 & 1318.8 & -2945.3 & 0 \\ 883.6 & -4396.0 & 0 & 4396 & -883.6 & 0 \\ -1538 & -1030.5 & 1538 & 0 & 0 & -1538 \end{bmatrix}$$

Similarly, the $\left[D\right]\left[B^2\right]$ for the second element can be written as

$$\left[D\right]\left[B^2\right] = \begin{bmatrix} -2945.3 & 1318.8 & 0 & -1318.8 & 2945.3 & 0 \\ -883.6 & 4396 & 0 & -4396 & 883.6 & 0 \\ 1538 & -1030.5 & -1538 & 0 & 0 & 1538 \end{bmatrix}$$

Now computing the element stiffness matrix

$$K_e = tA\left[B\right]^T\left[D\right]\left[B\right]$$

where A is the area of the triangle (= 0.5 × 75 × 50 = 1875 mm^2)

$$\begin{array}{l} \text{dof} \\ \rightarrow \end{array} \quad \begin{array}{cccccc} 1 & 2 & 3 & 4 & 5 & 6 \end{array}$$

$$\left[K^1\right] = 10^7 \times \begin{bmatrix} 0.1965 & -0.107 & -0.0865 & 0.0494 & -1.099 & 0.0577 \\ & 0.2857 & 0.0577 & -0.2472 & 0.0494 & -0.0384 \\ & & 0.0865 & 0 & 0 & -0.0577 \\ & & & 0.2472 & -0.0494 & 0 \\ & & & & 0.1099 & 0 \\ \text{symmetric} & & & & & 0.0384 \end{bmatrix}$$

Similarly, the stiffness matrix of element 2

$$\begin{array}{cccccc} 5 & 6 & 7 & 8 & 3 & 4 \end{array}$$

$$[K^2] = 10^7 \times \begin{bmatrix} 0.1965 & -0.1071 & -0.0865 & 0.0494 & -0.1099 & 0.0577 \\ & 0.2857 & 0.0577 & -0.2472 & 0.0494 & -0.0384 \\ & & 0.0865 & 0 & 0 & -0.0577 \\ & & & 0.2472 & -0.0494 & 0 \\ & & & & 0.1099 & 0 \\ \text{symmetric} & & & & & 0.0384 \end{bmatrix}$$

In the above matrices, the global dof is shown at the top. The given boundary conditions are

$$Q_2 = Q_5 = Q_6 = Q_7 = Q_8 = 0$$

After performing assembly of matrices $[K^1]$ and $[K^2]$, apply the boundary conditions using the elimination approach and considering point load, $F_4 = -10000$ N. The finite element equations in matrix form are

$$[K]\{Q\} = \{F\}$$

or

$$10^7 \times \begin{bmatrix} 0.1965 & -0.0865 & 0.0494 \\ -0.0865 & 0.1965 & 0 \\ 0.0494 & 0 & 0.2856 \end{bmatrix} \begin{bmatrix} Q_1 \\ Q_3 \\ Q_4 \end{bmatrix} = \begin{bmatrix} 0 \\ 0 \\ -10000 \end{bmatrix}$$

Solving the above equations for Q_1, Q_3, and Q_4 by the Gauss-elimination approach, we get

$$Q_1 = 0.156 \times 10^{-3} \text{ mm}$$

$$Q_3 = 0.094 \times 10^{-4} \text{ mm}$$

$$Q_4 = -0.37 \times 10^{-2} \text{ mm}$$

20.5 BEAM ELEMENT

Beam is a slender member which is used for supporting transverse bending load, namely shaft supported in bearings, members used in buildings, bridges, or machine structures. Rigidly connected members of complex structure which are subjected to axial force in addition to bending moment are called plane frames or two-dimensional beam elements. These members have two displacements and one rotation, i.e. three degrees of freedom (dof) per node.

Consider a typical 2-D beam element lying in the x–y plane with its longitudinal axis parallel to the x-axis as shown in Figure 20.10. The x–y axis system of an element is called the *local coordinate* system. However, the axis system X–Y which is common to all elements of a structure is called the *global coordinate* system.

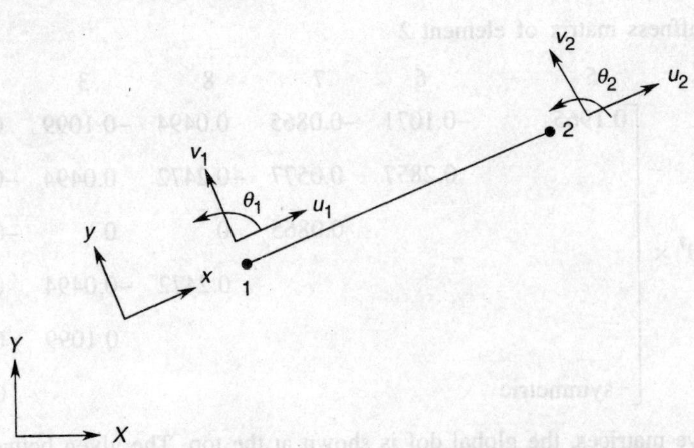

Figure 20.10 2-D beam element with three dof per node.

The degrees of freedom at the nodes are

$$\{q\} = \begin{bmatrix} u_1 & v_1 & \theta_1 & u_2 & v_2 & \theta_2 \end{bmatrix}^T \tag{20.60}$$

where u_1, v_1 and u_2, v_2 are the displacements along the x and y axes and θ_1 and θ_2 are the rotations about the z-axis at node 1 and node 2, respectively. It may be noted that the stiffness coefficient due to axial displacement u_1 and u_2 are the same as derived in Section 20.3 for the one-dimensional bar element. The stiffness coefficients due to displacement v_1, θ_1 and v_2, θ_2 need to be derived.

In a beam member, the rotation, θ, about the z-axis is defined as the displacement along the y-axis per unit length; hence it is a function of the displacement v, which can be represented as

$$\theta = \frac{dv}{dx} \tag{20.61}$$

Therefore, it is required to express the variation of v along the length of the member. Typical variations of v and θ are shown in Figure 20.11.

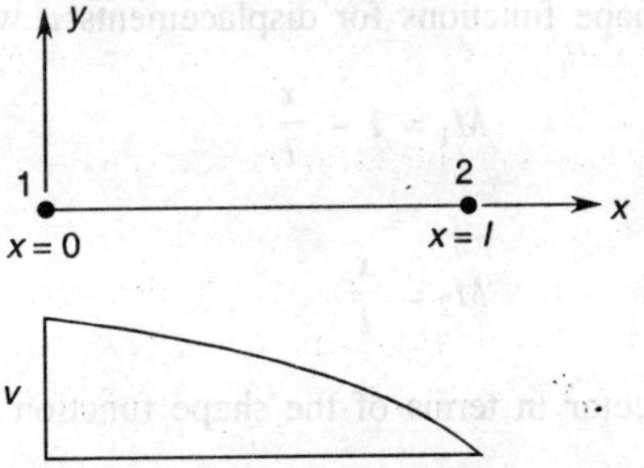

(a) Variation of displacement v

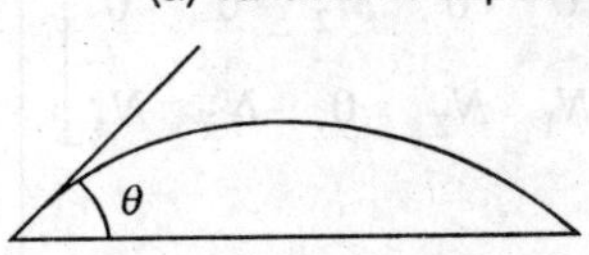

(b) Variation of slope θ

Figure 20.11 Displacement and slope variations.

The field variable v can be interpolated by the cubic polynomial

$$v = \alpha_1 + \alpha_2 x + \alpha_3 x^2 + \alpha_4 x^3 \tag{20.62}$$

and

$$\theta = \frac{dv}{dx} = \alpha_2 + 2\alpha_3 x + 3\alpha_4 x^2 \tag{20.63}$$

For known boundary conditions:

at $x = 0$ $\quad v = v_1 \quad$ and $\theta = \theta_1$

at $x = l$ $\quad v = v_2 \quad$ and $\theta = \theta_2$

Substituting these boundary conditions in Eqs. (20.62) and (20.63), we can solve for the unknown coefficients, and the displacements can be expressed as

$$v = N_1 v_1 + N_2 \theta_1 + N_3 v_2 + N_4 \theta_2 \tag{20.64}$$

where N_1, N_2, N_3, and N_4 are the shape functions as defined below:

$$\left.\begin{aligned} N_1 &= 1 - \frac{3x^2}{l^2} + \frac{2x^3}{l^3} \\ N_2 &= x - \frac{2x^2}{l} + \frac{x^3}{l^2} \\ N_3 &= \frac{3x^2}{l^2} - \frac{2x^3}{l^3} \\ N_4 &= \frac{x^3}{l^2} - \frac{x^2}{l} \end{aligned}\right\} \tag{20.65}$$

Let M_1 and M_2 be the shape functions for displacements u which are defined as

$$M_1 = 1 - \frac{x}{l}$$

$$M_2 = \frac{x}{l} \tag{20.66}$$

Thus the displacement vector in terms of the shape function can be written as

$$\begin{bmatrix} u \\ v \end{bmatrix} = \begin{bmatrix} M_1 & 0 & 0 & M_2 & 0 & 0 \\ 0 & N_1 & N_2 & 0 & N_3 & N_4 \end{bmatrix} \begin{bmatrix} u_1 \\ v_1 \\ \theta_1 \\ u_2 \\ v_2 \\ \theta_2 \end{bmatrix}$$

or in matrix notations

$$[U] = [N]\{q\} \tag{20.67}$$

where $[N]$ is the shape function matrix.

The strain–displacement relations for the beam element are:

(i) Axial strain, $$\frac{du}{dx} = M_1' u_1 + M_2' u_2 \tag{20.68}$$

(ii) Curvature strain, $\frac{d^2v}{dx^2} = N_1''v_1 + N_2''\theta_1 + N_3''v_2 + N_4''\theta_2$ (20.69)

where

$$M' = \frac{dM}{dx} \quad \text{and} \quad N'' = \frac{d^2N}{dx^2}$$

In matrix form

$$\varepsilon = \begin{bmatrix} \dfrac{du}{dx} \\ \dfrac{d^2v}{dx^2} \end{bmatrix} = \begin{bmatrix} M_1' & 0 & 0 & M_2' & 0 & 0 \\ 0 & N_1'' & N_2'' & 0 & N_3'' & N_4'' \end{bmatrix} \begin{bmatrix} u_1 \\ v_1 \\ \theta_1 \\ u_2 \\ v_2 \\ \theta_2 \end{bmatrix}$$

or

$$\varepsilon = [B]\{q\} \tag{20.70}$$

where

$[B]$ is the strain–displacement matrix
$\{q\}$ is the displacement vector.

The element stiffness matrix can be computed by the relation

$$K^e = \int_0^l [B]^T [D][B]\, dx \tag{20.71}$$

where $[D]$ is the property matrix, defined as

$$[D] = \begin{bmatrix} EA & 0 \\ 0 & EI \end{bmatrix} \tag{20.72}$$

The computed stiffness matrix of the element in the local coordinate system is given as

$$\left[K^e\right] = \begin{bmatrix} \dfrac{AE}{l} & 0 & 0 & \dfrac{-AE}{l} & 0 & 0 \\ & \dfrac{12EI}{l^3} & \dfrac{6EI}{l^2} & 0 & \dfrac{-12EI}{l^3} & \dfrac{6EI}{l^2} \\ & & \dfrac{4EI}{l} & 0 & \dfrac{-6EI}{l^2} & \dfrac{2EI}{l} \\ & & & \dfrac{AE}{l} & 0 & 0 \\ & & & & \dfrac{12EI}{l^3} & \dfrac{-6EI}{l^2} \\ \text{symmetric} & & & & & \dfrac{4EI}{l} \end{bmatrix} \tag{20.73}$$

As mentioned earlier, the local coordinate system is defined for a particular element whereas a global coordinate system refers to the entire assemblage. It is usually not possible to adopt local displacement direction that coincides with the global coordinates. Therefore, in such cases before we construct the finite element equation for the assemblage, we must transform the element stiffness and force vectors into a common frame of coordinates called the global coordinate system.

The stiffness matrix and force vector can be transformed to the global coordinate system by the transformation matrix

$$[K]_g = [T]^T [K^e][T] \tag{20.74}$$

and

$$[F]_g = [T]^T [F^e] \tag{20.75}$$

where

$[K]_g$ is the global stiffness matrix

$[F]_g$ is the global force vector

$[T]$ is the transformation matrix, which is defined as

$$[T] = \begin{bmatrix} \cos\alpha & \sin\alpha & 0 & 0 & 0 & 0 \\ -\sin\alpha & \cos\alpha & 0 & 0 & 0 & 0 \\ 0 & 0 & 1 & 0 & 0 & 0 \\ 0 & 0 & 0 & \cos\alpha & \sin\alpha & 0 \\ 0 & 0 & 0 & -\sin\alpha & \cos\alpha & 0 \\ 0 & 0 & 0 & 0 & 0 & 1 \end{bmatrix} \tag{20.76}$$

where α is the angle of inclination between the local axis and the global axis as shown in Figure 20.12.

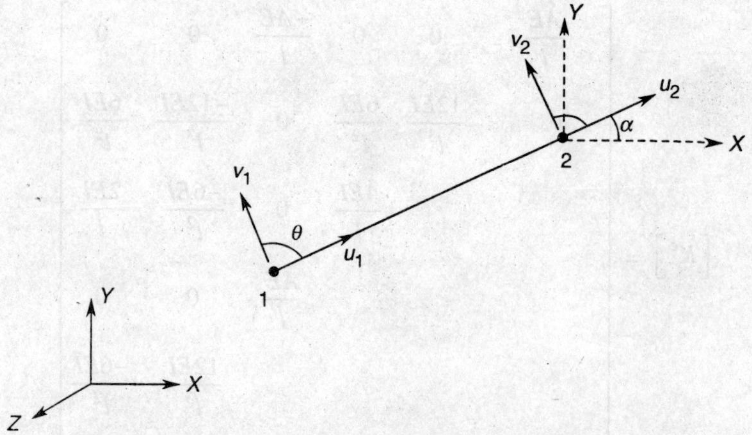

Figure 20.12 Coordinate transformation for a beam element.

Finally, the global stiffness matrices and force vectors of all the elements are assembled according to nodal connectivity to form equations of equilibrium. These equations, after applying suitable boundary conditions, are solved for displacement.

If there is distributed load on a member, as shown in Figure 20.13, the nodal forces due to the distributed load intensity w N/unit length are calculated by the following relation:

$$f = [T]^T \{f'\} \tag{20.77}$$

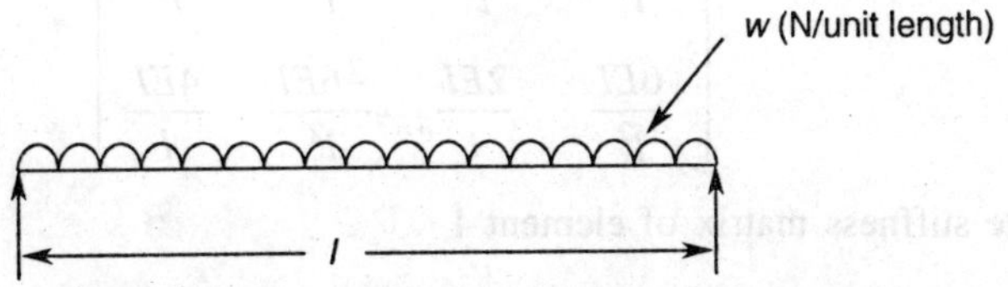

Figure 20.13 Distributed load on a beam element.

where f' is the force vector in local coordinates given by

$$f' = -\left[0 \quad \frac{wl}{2} \quad \frac{wl^2}{12} \quad 0 \quad \frac{wl}{2} \quad \frac{-wl^2}{12}\right]^T$$

Example 20.3 Determine the displacement and rotation at various nodal points of a beam as shown in the following figure. The spring stiffness is $24EI/l^3$ and $EI = 400$ kN · m^2.

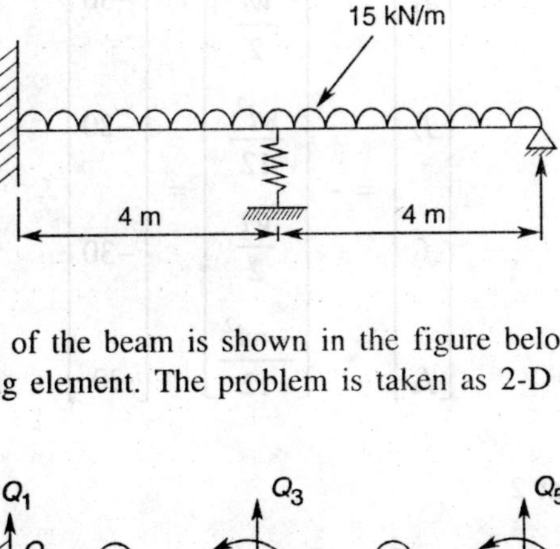

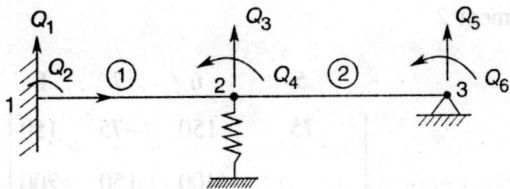

Solution The finite model of the beam is shown in the figure below where it is divided into two elements and one spring element. The problem is taken as 2-D beam element without the

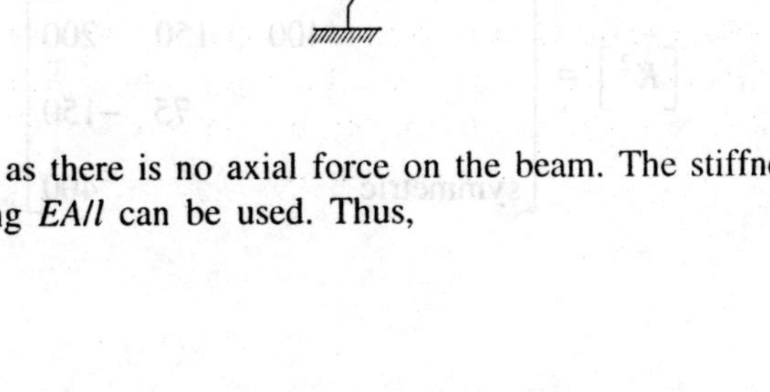

terms for axial stiffness as there is no axial force on the beam. The stiffness matrix of the 2-D beam element, neglecting EA/l can be used. Thus,

$$[K'] = \begin{bmatrix} \frac{12EI}{l^3} & \frac{6EI}{l^2} & \frac{-12EI}{l^3} & \frac{6EI}{l^2} \\ \frac{6EI}{l^2} & \frac{4EI}{l} & \frac{-6EI}{l^2} & \frac{2EI}{l} \\ \frac{-12EI}{l^3} & \frac{-6EI}{l^2} & \frac{12EI}{l^3} & \frac{-6EI}{l^2} \\ \frac{6EI}{l^2} & \frac{2EI}{l} & \frac{-6EI}{l^2} & \frac{4EI}{l} \end{bmatrix}$$

Therefore, the stiffness matrix of element 1

$$\begin{matrix} \text{global dof} & 1 & 2 & 3 & 4 \end{matrix}$$

$$[K^1] = \begin{bmatrix} 75 & 150 & -75 & 150 \\ & 400 & -150 & 200 \\ & & 75 & -150 \\ \text{symmetric} & & & 400 \end{bmatrix}$$

Force vector for element 1

$$\begin{bmatrix} f_1 \\ f_2 \\ f_3 \\ f_4 \end{bmatrix} = - \begin{bmatrix} \frac{wl}{2} \\ \frac{wl^2}{12} \\ \frac{wl}{2} \\ \frac{-wl^2}{12} \end{bmatrix} = \begin{bmatrix} -30 \\ -20 \\ -30 \\ 20 \end{bmatrix}$$

Stiffness matrix of element 2

$$\begin{matrix} & 5 & 6 & 7 & 8 \end{matrix}$$

$$[K^2] = \begin{bmatrix} 75 & 150 & -75 & 150 \\ & 400 & -150 & 200 \\ & & 75 & -150 \\ \text{symmetric} & & & 400 \end{bmatrix}$$

Force vector for element 2

$$\begin{bmatrix} f_5 \\ f_6 \\ f_7 \\ f_8 \end{bmatrix} = \begin{bmatrix} -30 \\ -20 \\ -30 \\ 20 \end{bmatrix}$$

Assembled stiffness matrix

$$[K] = \left[\begin{array}{cc|cc|cc} 75 & 150 & -75 & 150 & & \\ & 400 & -150 & 200 & & \\ \hline & & 75+75 & -150+150 & -75 & 150 \\ & & & 400+400 & -150 & 200 \\ \hline & & & & 75 & -150 \\ & & & & & 400 \end{array}\right]$$

or

$$\begin{array}{lcccccc} \text{global dof} & 1 & 2 & 3 & 4 & 5 & 6 \end{array}$$

$$[K] = \begin{bmatrix} 75 & 150 & -75 & 150 & 0 & 0 \\ & 400 & -150 & 200 & 0 & 0 \\ & & 150 & 0 & -75 & 150 \\ & & & 800 & -150 & 200 \\ & & & & 75 & -150 \\ \text{symmetric} & & & & & 400 \end{bmatrix}$$

$$\text{Spring stiffness} = \frac{24EI}{l^3} = \frac{24 \times 400}{4^3} = 150 \text{ kN/m}$$

Add the spring stiffness to location (3, 3) in the assembled matrix. Therefore,

$$[K] = \begin{bmatrix} 75 & 150 & -75 & 150 & 0 & 0 \\ & 400 & -150 & 200 & 0 & 0 \\ & & 300 & 0 & -75 & 150 \\ & & & 800 & -150 & 200 \\ & & & & 75 & -150 \\ \text{symmetric} & & & & & 400 \end{bmatrix}$$

Assembled force vector

$$f = \begin{bmatrix} -30 \\ -20 \\ -30 \\ 20 \end{bmatrix} + \begin{bmatrix} -30 \\ -20 \\ -30 \\ 20 \end{bmatrix} = \begin{bmatrix} -30 \\ -20 \\ -60 \\ 0 \\ -30 \\ 20 \end{bmatrix}$$

Therefore the finite element equations without boundary conditions are

$$\begin{bmatrix} 75 & 150 & -75 & 150 & 0 & 0 \\ & 400 & -150 & 200 & 0 & 0 \\ & & 300 & 0 & 75 & 150 \\ & & & 800 & -150 & 200 \\ & & & & 75 & 150 \\ \text{symmetric} & & & & & 400 \end{bmatrix} \begin{bmatrix} Q_1 \\ Q_2 \\ Q_3 \\ Q_4 \\ Q_5 \\ Q_6 \end{bmatrix} = \begin{bmatrix} -30 \\ -20 \\ -60 \\ 0 \\ -30 \\ 20 \end{bmatrix}$$

The boundary conditions are

$$Q_1 = Q_2 = Q_5 = 0$$

Applying the boundary conditions using the elimination approach, we get

$$\begin{bmatrix} 300 & 0 & 150 \\ 0 & 800 & 200 \\ 150 & 200 & 400 \end{bmatrix} \begin{bmatrix} Q_3 \\ Q_4 \\ Q_6 \end{bmatrix} = \begin{bmatrix} -60 \\ 0 \\ 20 \end{bmatrix}$$

Solving these equations using the Gauss-elimination method, we get

$$\left.\begin{aligned} Q_3 &= -0.29 \text{ m} \\ Q_4 &= -0.0454 \text{ rad} \\ Q_6 &= 0.1818 \text{ rad} \end{aligned}\right]$$

EXERCISES

1. What do you understand by discretization of domain?
2. What is the effect of mesh size on the accuracy of solution?
3. What is interpolation model? How does an interpolation function affect the outcome solution?

4. Define the local and global coordinate systems.
5. What is the importance of boundary conditions in a finite element problem?
6. What do you mean by shape function?
7. What do you mean by isoparametric element?
8. Determine the nodal displacement and element stresses in a bar as shown in the figure below:

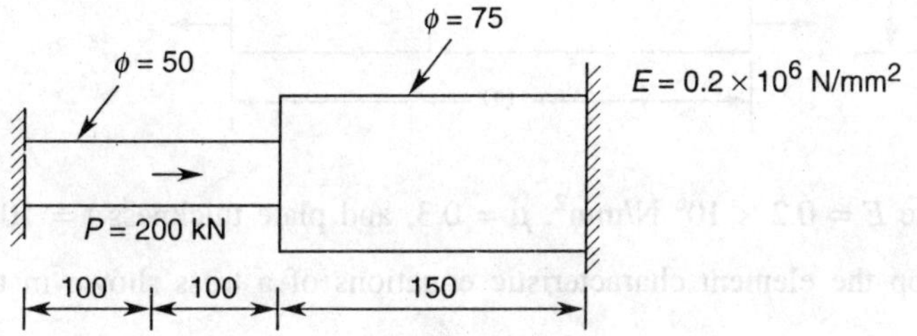

9. Derive the element characteristics matrix for a one-dimensional line element using the quadratic interpolation function.
10. Determine the nodal displacement in a bar element made of steel, thickness t = 10 mm, and as shown in the figure below.

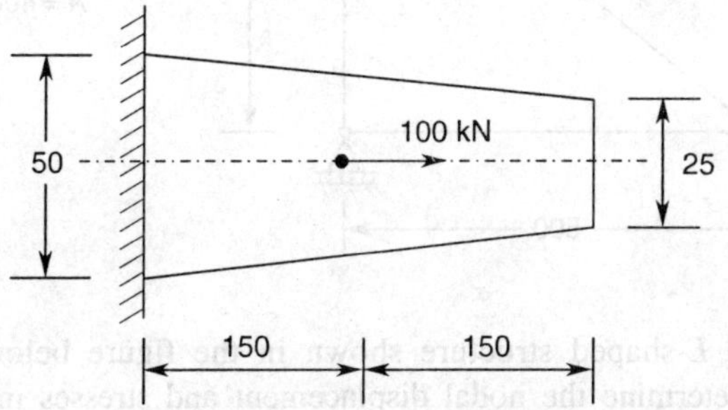

11. Considering the plane stress problem, determine the nodal displacement and stresses in a plate shown in the figure below. Model the plate with (i) two triangular elements and (ii) three triangular elements. Compare the results.

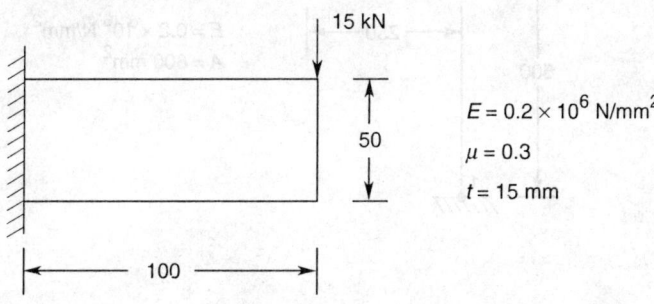

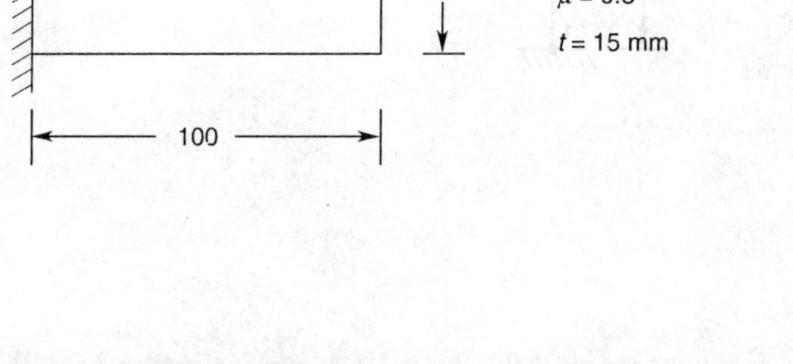

12. Determine the stresses in the plate as shown in the figure below. Model the plate by triangular elements.

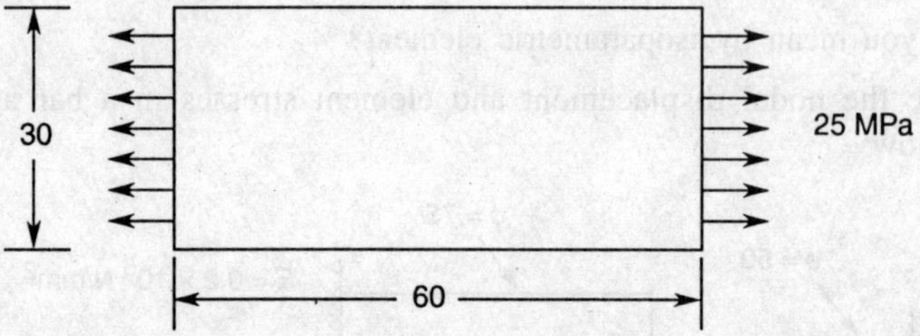

Take $E = 0.2 \times 10^6$ N/mm^2, $\mu = 0.3$, and plate thickness $t = 10$ mm.

13. Develop the element characteristic equations of a truss shown in the figure below.

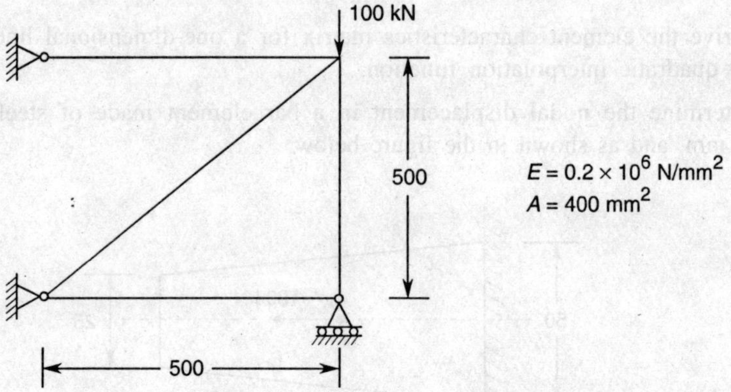

14. An inverted L-shaped structure shown in the figure below is made of 2-D beam elements. Determine the nodal displacement and stresses in elements

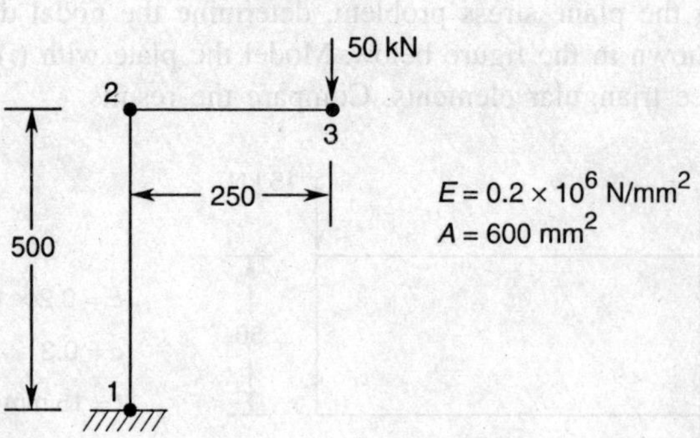

Bibliography

Abbott, W., *Machine Drawing and Design,* Blackie & Sons.

Asimow, A., *Introduction to Design,* Prentice-Hall, Englewood Cliffs, New Jersey.

AutoCad 2000 Users' Manual, Autodesk, USA.

AutoLisp Reference Manual, Autodesk, USA.

Bahl, R.C. and V.K. Goel, *Mechanical Machine Design,* Standard Publishers & Distributors, New Delhi, 1981.

Bevan, T., *Theory of Machines,* Longmans, 1967.

Bhandari, V.B., *Machine Design,* Tata McGraw-Hill, New Delhi.

Black, P.H. and O.E. Adams, *Machine Design,* McGraw-Hill Book Publishing, New York, 1968.

Boyer, H.E. et al., *Metals Handbook,* American Society of Metals, Ohio.

Chandrupatla, T.R. and A.D. Belegundu, *Introduction to Finite Elements in Engineering,* 3rd ed., Prentice-Hall of India, New Delhi, 2002.

Crouse, W.H., *Automotive Engine Design,* McGraw-Hill Book Publishing, New York, 1970.

Deb, Kalyanmoy, *Optimization for Engineering Design: Algorithms and Examples,* Prentice-Hall of India, New Delhi, 1998.

Design Data Handbook, PSG College of Technology, Coimbatore, 1995.

Deutchman, A.D. et al., *Machine Design: Theory and Practice,* Macmillan Publishing, New York, 1975.

Dixon, J.R., *Design Engineering,* McGraw-Hill Book Publishing, New York.

Dobrovolsky, V. et al., *Machine Design,* MIR Publishers, Moscow, 1968.

Faires, V.M., *Design of Machine Elements,* Macmillan Publishing, New York, 1965.

Hopkins, R.B., *Design Analysis of Shafts and Axels,* McGraw-Hill Book Publishing, New York, 1970.

Houghton, P.S., *Ball and Roller Bearings,* Allied Science Publisher, London, 1976.

Juvinall, R.C., *Fundamentals of Machine Component Design,* John Wiley & Sons, New York, 1983.

Karwa, R., *Machine Design,* Laxmi Publications, New Delhi, 1999.

Kents' *Mechanical Engineers Handbook—Design & Production*, John Wiley & Sons, New York.

Khurmi, R.S. and J.K. Gupta, *Machine Design*, Eurasia Publishing House, New Delhi.

Kolchin, A. and V. Demidov, *Design of Automotive Engines,* MIR Publishers, Moscow, 1984.

Krishnamurti, C.S., *The Finite Element Methods,* Tata McGraw-Hill, New Delhi.

Kulkarni, S.G., *Machine Design,* Tata McGraw-Hill, New Delhi.

Lakhtin, Yu, *Engineering Physical Metallurgy and Heat Treatment,* MIR Publishers, Moscow, 1975.

Kutz, Mayer, *Mechanical Engineers' Handbook,* Wiley Inter-science Publication, New York.

Litchy, L.C., *Internal Combustion Engines,* McGraw-Hill Book Publishing, Tokyo.

Mahadevan, K. and K.B. Reddy, *Design Data Handbook,* CBS Publishers & Distributors, New Delhi, 1989.

Maleev, V.L. and J.B. Hartman, *Machine Design*, CBS Publishers & Distributors, New Delhi, 1983.

Marks' *Standard Handbook for Mechanical Engineers,* McGraw-Hill Book Publishing, New York.

Mubeen, A., *Advanced Machine Design,* Khanna Publishers, New Delhi.

Oberg, E. et al., *Machinery's Handbook,* Industrial Press, New York, 1982.

Orlov, P., *Fundamentals of Machine Design*, Vols. I to V, MIR Publishers, Moscow, 1975.

Peterson, R.E., *Stress Concentration Factor*, John Wiley & Sons, New York, 1974.

Phelan, R.M., *Fundamentals of Mechanical Design,* John Wiley & Sons, London.

Popov, E.P., *Engineering Mechanics of Solids*, 2nd ed., Prentice-Hall of India, New Delhi, 2002.

Pujara, K. et al., *Machine Design,* Dhanpat Rai & Sons, New Delhi, 1984.

Ramamurti, V., *Computer Aided Design,* Tata McGraw-Hill, New Delhi.

Rao, S.S., *Optimization: Theory and Practice,* Wiley Eastern, New Delhi, 1984.

Sadhu Singh, *Machine Design,* Khanna Publishers, New Delhi, 1997.

Sharma, P.C. and D.K. Agrawal, *Machine Design*, Kataria & Sons, New Delhi.

Shigley, J.E., *Machine Design,* McGraw-Hill Book Publishing, New York, 1986.

———, *Machine Design Handbook*, McGraw-Hill Book Publishing, New York.

SKF Bearings Catalogue 2000E, SKF Bearings India Ltd., Mumbai.

Spott, M.F., *Mechanical Design Analysis,* Prentice Hall, Englewood Cliffs, New Jersey, 1964.

Vallance, A. and V.L. Doughtie, *Design of Machine Elements*, McGraw-Hill Book Publishing, New York.

Wahl, A.M., *Mechanical Springs,* McGraw-Hill Book Publishing, New York.

Zienkiewicz, O.C., *The Finite Element Methods*, Tata McGraw-Hill, New Delhi.

Answers to Multiple Choice Questions

Chapter 1

1. (b)	2. (c)	3. (d)	4. (a)
5. (a)	6. (c)	7. (c)	8. (a)
9. (c)			

Chapter 3

1. (b)	2. (a)	3. (d)	4. (d)
5. (a)	6. (b)	7. (b)	8. (c)
9. (c)	10. (a)	11. (a)	12. (b)
13. (c)	14. (a)	15. (b)	16. (a)

Chapter 4

1. (b)	2. (c)	3. (b)	4. (d)
5. (b)	6. (c)	7. (b)	8. (c)
9. (c)			

Chapter 5

1. (d)	2. (b)	3. (b)	4. (a)
5. (d)	6. (d)	7. (c)	8. (d)
9. (a)	10. (a)		

Chapter 6

1. (a)	2. (b)	3. (c)	4. (a)
5. (b)	6. (c)	7. (b)	8. (a)
9. (a)	10. (b)	11. (c)	12. (d)

Chapter 7

1. (b)	2. (c)	3. (a)	4. (d)
5. (c)	6. (b)	7. (d)	8. (a)
9. (a)	10. (c)		

Chapter 8

1. (a)	2. (b)	3. (c)	4. (c)
5. (d)	6. (b)	7. (a)	8. (c)
9. (b)	10. (a)		

Chapter 9

1. (a)	2. (c)	3. (a)	4. (b)
5. (b)	6. (d)	7. (d)	8. (b)
9. (d)	10. (b)	11. (b)	

Chapter 11

1. (a)	2. (a)	3. (b)	4. (c)
5. (b)	6. (c)	7. (b)	8. (c)

Chapter 12

1. (a)	2. (c)	3. (b)	4. (d)
5. (b)	6. (c)	7. (b)	8. (b)
9. (b)	10. (c)	11. (a)	12. (b)
13. (b)	14. (c)		

Chapter 13

1. (d)	2. (c)	3. (b)	4. (d)
5. (a)	6. (a)	7. (c)	8. (b)
9. (b)	10. (c)	11. (c)	12. (d)
13. (c)	14. (b)	15. (d)	16. (d)
17. (c)	18. (b)		

Chapter 14

1. (d)	2. (d)	3. (b)	4. (d)
5. (a)	6. (b)	7. (c)	8. (a)
9. (b)	10. (d)	11. (c)	12. (c)

Chapter 15

1. (b)	2. (c)	3. (a)	4. (a)
5. (c)	6. (d)	7. (b)	8. (c)
9. (a)	10. (d)	11. (a)	12. (c)

Chapter 16

1. (a)	2. (c)	3. (a)	4. (b)
5. (c)	6. (b)		

Chapter 17

1. (b)	2. (c)	3. (c)	4. (c)
5. (b)	6. (c)	7. (d)	8. (c)
9. (a)	10. (d)	11. (b)	12. (d)
13. (a)			

Chapter 18

1. (b)	2. (c)	3. (a)	4. (b)
5. (c)	6. (b)	7. (a)	8. (a)
9. (a)	10. (c)		

Index